LIGHT SCATTERING SPECTRA OF SOLIDS

FINANCIAL SPONSORS

United States Army Research Office (Durham)
New York State Science and Technology Foundation
New York University
United States Office of Naval Research

CO-SPONSOR

International Union of Pure and Applied Physics

LIGHT SCATTERING SPECTRA OF SOLIDS

Proceedings of the International Conference
on Light Scattering Spectra of Solids held at:
New York University, New York
September 3, 4, 5, 6, 1968

Edited by **GEORGE B. WRIGHT**
Lincoln Laboratory,
Massachussets Institute of Technology

SPRINGER-VERLAG NEW YORK INC.

1969

Library of Congress Catalog Card Number 70-79823
Printed in the United States of America
Title Number 1594

CONFERENCE ORGANIZATION AND SITE

CONFERENCE SECRETARY

Professor Joseph L. Birman, New York University

PLANNING COMMITTEE

Professor E. Burstein, Chairman, University of Pennsylvania
Professor Joseph L. Birman, New York University
Dr. C. Boghosian, United States Army Research Office (Durham)
Professor H. Z. Cummins, The Johns Hopkins University
Professor A. A. Maradudin, University of California (Irvine)
Professor A. L. McWhorter, Massachusetts Institute of Technology, Lincoln Laboratory
Professor P. S. Pershan, Harvard University
Professor S. P. S. Porto, University of Southern California
Dr. L. Rimai, Ford Motor Company Research Laboratory
Dr. J. M. Worlock, Bell Telephone Laboratories (Holmdel)

INTERNATIONAL ADVISORY COMMITTEE

Professor J. Brandmüller, University of Munich, Germany
Professor R. S. Krishnan, Indian Institute of Science, India
Professor R. Loudon, University of Essex, England
Professor J. P. Mathieu, University of Paris, France
Professor T. Moriya, University of Tokyo, Japan
Dr. A. I. Stekhanov, Ioffe Physical Technical Institute, U. S. S. R.
Professor H. Welsh, University of Toronto, Canada

INTERNATIONAL AFFAIRS COMMITTEE

Professor A. A. Maradudin, Chairman, University of California (Irvine)
Professor Joseph L. Birman, New York University
Professor S. P. S. Porto, University of Southern California

PROGRAM COMMITTEE

Dr. J. M. Worlock, Chairman, Bell Telephone Laboratories (Holmdel)
Professor G. B. Benedek, Massachusetts Institute of Technology
Professor Joseph L. Birman, New York University
Professor H. Z. Cummins, The Johns Hopkins University
Professor A. A. Maradudin, University of California (Irvine)
Dr. A. Mooradian, Massachusetts Institute of Technology, Lincoln Laboratory
Dr. J. H. Parker, Jr., Westinghouse Research Laboratories
Dr. P. A. Wolff, Bell Telephone Laboratories
Dr. G. B. Wright, Lincoln Laboratory

FINANCE COMMITTEE

Professor Joseph L. Birman, Chairman, New York University
Dr. C. Boghosian, United States Army Research Office (Durham)
Dr. J.M. Worlock, Bell Telephone Laboratories (Holmdel)

PUBLICATION COMMITTEE

Dr. G.B. Wright, Editor, Lincoln Laboratory
Professor G.B. Benedek, Massachusetts Institute of Technology
Professor A.L. McWhorter, Massachusetts Institute of Technology, Lincoln Laboratory
Professor P.S. Pershan, Harvard University

LOCAL COMMITTEE AT NEW YORK UNIVERSITY

Professor Joseph L. Birman
Mrs. H.D. Chazel
Professor H.G. Hartmann
Professor L. Yarmus

CONFERENCE SITE

Courant Institute of Mathematical Sciences
Warren Weaver Hall
New York University

FOREWORD

The International Conference on Light Scattering Spectra of Solids was held at New York University on September 3, 4, 5, 6, 1968. The Conference received financial support from the U. S. Army Research Office (Durham), The New York State Science and Technology Foundation, the U. S. Office of Naval Research, and The Graduate School of Arts and Sciences of New York University. Co-sponsoring the Conference was the International Union of Pure and Applied Physics.

The initial conception for the Light Scattering Conference arose from informal discussions held by Professor Eli Burstein, Professor Marvin Silver (representing the U. S. Army Research Office) and Professor Joseph Birman, late in 1966. In early discussions a format was put forth for a meeting to be held the following year, reviewing the state of the art, and emphasizing novel developments which had occurred since the 1965 International Colloquium on Scattering Spectra of Crystals held in Paris (proceedings published in Le Journal de Physique, Volume 26, November 1965).

Further consideration during the Spring of 1967 suggested that, owing to the rapid expansion of theoretical and experimental work in this field, the conference should be expanded to a full International Conference on Light Scattering by Solids. This change won enthusiastic support among the active workers whose opinion was solicited. The U. S. Army Research Office (Durham) generously agreed to give financial support to the projected Conference, based on the initial plans provided by Professor Burstein, who carried most of the burden of the planning during this early stage.

In December of 1967 the Planning Committee met under the Chairmanship of Professor Burstein and decided to accept the offer of New York University to be host of the Conference. At this time the membership of all working committees was also agreed upon. A policy decision was taken that the Conference would concentrate on the basic physics of light scattering by solids, rather than on "quantum optics" or "non-linear" optics - fields already covered by other conferences.

Implementation of the general philosophy guiding the Conference devolved on the Conference Secretary after December 1967. He worked with the Finance Committee in allocating our resources, and with the Program Committee Chairman, Dr. John Worlock. An attempt was made to have all active workers in the field of light scattering by solids present at the Conference: this included graduate students and post-doctoral persons entering the field.

Response to the call for papers was most gratifying and was reflected in the high quality of the presentations. Approximately 225 scientists participated in the Conference and the consensus was that it was indeed productive. All sessions were plenary and there was a great deal of active discussion after the papers, as well as in the halls. All papers which were given at the Conference appear in these Proceedings, as well as three (A-7, B-4, G-1) by authors who were not present. No record of discussions was kept.

The Conference was fortunate in having a Distinguished Guest with us: Professor Leon Brillouin. Professor Brillouin kindly prepared his remarks on the early history of Brillouin Scattering for the Proceedings and they are given in these Proceedings. Unfortunately, the rigors of travel and other pressing professional claims prevented Professor Sir. C. V. Raman from attending, but he also prepared remarks which are included in these Proceedings.

On behalf of the Conference I thank the Courant Institute of Mathematical Sciences of New York University for generously permitting us to use their Auditorium for our meeting, and for a Chamber Music Concert by the Beaux-Arts String Quartet on Wednesday Evening, September 4, 1968. The Local Committee, and particularly Mrs. Helen Chazel, worked with great energy and effect, to make the participants comfortable. The success of the Light Scattering Conference held at New York University, Washington Square Center in Greenwich Village, New York City, was gratifying to all those who put time and effort into the Conference.

It is our hope that these Proceedings will be a continued and productive reminder of the Conference in time to come.

Joseph L. Birman
Conference Secretary
Physics Department
New York University

CONTENTS

Chairman: E. Burstein

MAGNONS AND OTHER ELECTRONIC EXCITATIONS

Chairman: T. Moriya

Chairman: G. B. Wright

FREE CARRIERS

PHONONS; RESONANCE SCATTERING; METALS; MORPHIC EFFECTS

MIXED CRYSTALS AND POINT DEFECTS

Chairman: K. K. Rebane

Chairman: B. Szigeti

BRILLOUIN SCATTERING

Chairman: H. Z. Cummins

INTRODUCTORY REMARKS

Sir C. V. Raman, Nobel Laureate
Director of the Raman Research Institute
Hebbal Post, Bangalore. 6.

SCATTERING OF LIGHT IN CRYSTALS

An essay by me entitled the "Molecular Diffraction of Light" was published by the Calcutta University in February 1922. It covered a wide field of observation and theory. The concluding chapter on "Molecular Diffraction and the Quantum Theory of Light" envisaged the possibility that the corpuscular concept of the nature of light put forward by Einstein might come into evidence in the phenomena of the scattering of light.

Chapter VI of the essay dealt with the effects observed when a beam of light traverses a transparent crystal. The thermal agitation in the crystal should, according to the concepts of wave-optics, result in local fluctuations of optical density. Hence, there should be an observable light-scattering of which the intensity could be estimated following a procedure indicated in the essay. Effects of the nature and of the order of magnitude indicated by the theory were actually observed with the crystals studied, viz., quartz, ice, and rock-salt, and the reality of such scattering was thereby established.

These early observations were made using sunlight for the illuminating beam. In Chapter VIII of the essay, the suggestion was put forward that monochromatic light should be used and that spectral analysis should be applied for investigations on the scattering of light in condensed media. But the feebleness of the phenomena under study and the obvious need for intense light-sources acted as a deterrant and nothing was done to give effect to these and pursue the matter further in the direction indicated.

Later, I became aware of the remarkably brilliant monochromatic illumination which could be obtained by the aid of the commercially available mercury arcs sealed in quartz tubes. Towards the end of February 1928, I took the decision to make use of such lamps for all further studies in the field of light-scattering. The success which attended this forward step was immediate and highly gratifying. Experience in working with sunlight indicated the techniques necessary for the observation of extremely weak phenomena, viz., the rigorous exclusion of stray light and the conditioning of the observer's vision by a prolonged stay in darkness. On setting up the apparatus and making these preparations, I found that the light of the mercury arc diffused by various materials when examined through a direct vision spectroscope showed the presence, besides the lines of mercury, also of other lines the positions of which varied with the substances under study. Amongst the numerous materials thus examined was a large block of clear ice. This showed sharp displaced lines in the spectrum of the scattered light in approximately same positions as the rather diffuse bands observed with pure water. Within a few days of the discovery, photographic spectra were successfully recorded in which the additional lines showed up very clearly.

It was realized by me at once that the phenomenon which thus stood revealed was a vindication of the corpuscular concept of light and that it should be interpreted as a consequence of an exchange between the energy of the radiation field and the vibrational energy of the molecules in the substance traversed by the light.

In my lecture on "A New Radiation" delivered at Bangalore on the 16th of March 1928 and which was published and distributed on the 31st of March, the phenomenon was explained on that basis. The universality of the phenomenon was also emphasized and all its essential features were set out. The concluding paragraph of the published lecture made it clear that a vast new field of research had been opened up by the discovery.

Léon Brillouin
Member of the National Academy of Science of the United States

BIRTH AND GROWTH OF THE BRILLOUIN SCATTERING

On May 11, 1914, the "Comptes-rendus de l'Académie des Sciences" of Paris published a note on light scattering by an homogeneous transparent body (vol. 158, pp. 1331-1334). This was the very first note I published in my scientific career, although it was not the first paper I actually wrote. In 1913 I worked with Sommerfeld on group velocity, but the paper was only printed somewhat later in the Annalen der Physik. Later on, I committed more than 200 papers and 35 books, and the list is not yet closed!

This first note marked the birth of "Brillouin Scattering", and I may recall that my interest in scattering problems went back to a few years earlier, when I spent a summer on the Monte Rosa, the second highest peak in the Alps (15,000 feet), observing the blue color of the sky. This kept me busy every day in the morning and left the afternoons free for climbing all the summits in the neighborhood. The blue sky is essentially due to Rayleigh scattering from gas molecules, but the actual intensity is often twice as large as computed from Rayleigh's formulas, on account of the high reflective power from the ground. My experimental measurements were never published since much more accurate experiments had been performed at the same time in American Observatories.

Rayleigh computed the light scattered from a single molecule and simply added the intensities, assuming molecules to be distributed at random. Einstein recalculated the effect (Ann. der Physik, vol. 33, 1910, p. 1275), starting from light scattered by density fluctuations, and showing that for an ideal gas, the density fluctuations could be obtained from a random distribution of molecules; his theory thus checked with Rayleigh, but what could be said about some denser medium, a liquid or a solid for instance? This problem was on my mind, and I thought I could solve it with the help of Debye's theory of specific heat for solids (Ann. der Physik, vol. 39, 1912, p. 789). Debye analyzed thermal agitation in elastic vibrations of hyperfrequencies, propagating throughout the body, a vibration of frequency ν_o obtaining an average energy according to Planck's formula:

$$E_{\nu o} = \frac{h\nu_o}{\exp \frac{h\nu_o}{kT} - 1} \qquad (1)$$

starting from this model; assuming density fluctuations to be proportional to the energy of longitudinal waves (and postponing for a later study the role of transverse waves), I could compute the density fluctuations in a liquid or a solid, and use this expression in Einstein's scattering theory. Einstein assumed equipartition of energy, hence E_{ν_o} equal to kT. My whole computation was rather clumsy, but I was able to perform it (with some reasonable approximations) and to obtain the light scattering coefficient. Here I was stuck, and very much surprised at the result: Einstein's formula contained a factor kT, showing a scattering intensity proportional to Brownian agitation. I expected something rather similar, involving a certain average taken on Planck's energies (1), and reducing to kT for high temperatures ($T > \theta_{Debye}$). This, however, was not the case. My scattering formula contained only one expression (1), thus suggesting that only one Debye vibration was responsible for the scattering of light (at given light frequency ν and scattering angle ω) with hypersonic vibration ν_o

$$\nu_o = \frac{u}{\lambda} [2(1 - \cos \omega)]^{1/2} \qquad \begin{array}{l} u \text{, sound velocity} \\ \lambda \text{, optical wavelength} \end{array} \tag{2}$$

I simply could not understand what this strange relation might mean, and made a few attempts at possible explanations that looked promising. Furthermore, my theory gave reasonable orders of magnitude for the extreme cases of X-rays. But... two months later we were in the first world war. I became a radio engineer in the French signal corps, very busy building and trying all sorts of radio equipment for Army, Navy and Air Forces (including a pilotless automatic airplane) and completely forgetting my scattering problems.

In 1919, I was finally demobilized and rushed to my old papers, that I found tossed together in a corner, I knew I had a great deal of material ready for my Ph. D. thesis, but I was horrified when I tried to read my own old notes. The computations were there, with no explanations - and they did not make sense to me anymore. I had to rebuild everything from scratch. In the problem of scattering I decided to start it backwards: my previous computation of 1914, suggested an interaction between light ν and a single hypersonic wave ν_o; hence I started from that simple problem, leaving for later discussing the superposition of many hypersonic waves. All this discussion was published later, as a part of my Ph. D. Thesis, in the Annales de Physique (vol. 17, pp. 88-122, 1921). Ultrasonic or hypersonic plane waves propagating through a liquid or a solid create a succession of planes of higher and lower density, moving along with sound velocity u. These planes of different density do reflect light waves of frequency ν in a selective way. If the ultrasonic waves were at rest, this would be the usual problem of X-rays reflection on crystal planes according to Bragg. Since the ultrasonic waves have a frequency ν_o and propagate with velocity u, we can predict that the scattered beam will obtain a different frequency $\nu \pm \nu_o$, that may also be considered as a Doppler effect on the moving mirror planes of velocity u. This very simple mechanism did completely explain the strange result of formula (2): when light frequency ν and scattering angle ω are given, there is only one set of ultrasonic or hypersonic waves coming into play, according to Bragg's rule, and its frequency ν_o corresponds to condition (2).

The Theory of "Brillouin scattering" was thus established and the original nucleus of the whole story was contained in my original note of 1914.

The change in frequency was first doubted or even denied: it had always been taken for granted that scattering did not involve any frequency change: "when there is a frequency change it is a problem of fluorescence, not scattering!" said my professors! Remember that Raman scattering was yet unknown, and appeared only in 1928!! The frequency change in Brillouin scattering was so small that it could be observed only with the very best Perot-Fabry equipment.

With the invention of "lasers", the situation was completely changed. From a very difficult laboratory experiment, the Brillouin scattering became a regular industrial problem. The very important role played by Kastler in France and by Townes and Benedek in the U. S. A. is known to all physicists, and was recognized by Nobel prizes.

The whole story was told by Schawlow in the September, 1968 issue of the Scientific American (see p. 122 passim).

After the last war, I was teaching a course on applied mathematics for veterans, at Harvard. They had a very hard time getting back to work after so many years in the army. I told them of my own experience in 1919 and I managed to have a large number of assistants to help the students along. I could easily understand their problem!

LIGHT SCATTERING SPECTRA OF SOLIDS

A-1: LIGHT SCATTERING WITH LASER SOURCES

S. P. S. Porto
Department of Physics and Electrical Engineering,
University of Southern California, Los Angeles, Calif.

When light passes through matter weak random scattered radiation appears. In the early 16th Century, Leonardo da Vinci prophetically suggested scattering by particles of air as the explanation for the blueness of the sky. This idea was pursued by many scientists including Newton and Tyndall who tried with only limited success to identify the particles responsible for the scattering. Lord Rayleigh[1] following a suggestion by Maxwell, finally proved that the air molecules themselves were responsible for the blueness of the sky. After three centuries of thought, a correct and unambiguous explanation to all known properties of scattering--frequency dependence, critical opalescence, index of refraction, etc.--became a reality. Then came the anti-climax that usually follows the insertion of the last jagged piece in a jig-saw puzzle: most physicists turned their attention elsewhere.

A few didn't. In 1922 Brillouin[2] predicted that if monochromatic radiation was allowed to scatter from an optical medium, side bands would appear. He went on to theorize that the bands would result from a Doppler shift due to the generation of a sound wave produced by the light wave as it encountered molecules in its path. The frequency shift would be a function of the angle of observation and of the sound velocity in the medium. In 1923 Smekal[3] considered, in the Bohr theory approximation, the scattering of light by a system having two quantized energy levels and predicted the effect to be discovered in 1928 by Raman[4]. Working independently in Russia, Mandelstam and Landsberg[5] discovered the same phenomenon in quartz: appearance of lines in the spectrum in addition to those from the source. (The Russians, believing that their work actually antedated Raman's, identify the effect as "combinational scattering"). In 1930 Gross[6] confirmed the theory of Brillouin, demonstrating that the Doppler-shifted frequencies appeared as predicted for both liquids and solids.

The implication of line shifting was at once evident and a flurry of new enthusiasm broke out. By 1934, more than 500 papers related to the Raman effect had been published and by the early forties this number had increased to a few thousand.

In 1934, Placzek[7] wrote a lengthy and excellent review paper on the Raman effect. Just as Lord Rayleigh's explanation of the elastic scattering 35 years earlier had resulted in a virtual halt to further work in this area, Placzek's paper seemed to mark the end of an era in frequency-shifted spectra. Most fundamental Raman problems seemed now to be so well understood that research ground to a halt and pertinent papers in the Physical Review virtually disappeared. The Raman effect developed instead as a tool for the structural physical chemist. Only a handful of devoted physicists, notably groups under Krishnan, Raman and Baghavantam in India, Welsh in Canada, Mathieu in France and Stekhanov in Russia, remained to keep the flame of fundamental research barely flickering.

Light scattering research was adrenalized once again in this decade with the invention of laser sources. Not only was the physical chemist handed a new, more powerful, and cleaner source but, equally important, the physicist was furnished with means for testing rigorously the theories of Placzek, Rayleigh and Brillouin. Phenomena such as directional effects in scattering processes and inelastic scattering from very small cross sections which had previously defied measurement could now be studied easily with laser sources. The great level of current enthusiasm in such investigations is evident from the impressive number of laboratories and scientists engaged in related light scattering experiments.

In this paper we shall attempt to develop from very simple arguments the origin of the most common scattering mechanisms. Without pretending to be thorough--this would require a large book--we shall "island hop" from subject to subject in the hope of attracting more visitors to these research shores.

If an electric field E is applied to a medium having polarizability α, a polarization (most commonly, but not necessarily, a dipole) P will be induced obeying the relation:

$$\underline{P} = \alpha \underline{E} \text{ or } \begin{bmatrix} P_x \\ P_y \\ P_z \end{bmatrix} = \begin{bmatrix} \alpha_{xx} & \alpha_{xy} & \alpha_{xz} \\ \alpha_{yx} & \alpha_{yy} & \alpha_{yz} \\ \alpha_{zx} & \alpha_{zy} & \alpha_{zz} \end{bmatrix} \begin{bmatrix} E_x \\ E_y \\ E_z \end{bmatrix} \qquad (1)$$

where α is a tensor. If the field is applied in a certain direction i, the polarization can be induced in a different direction j. In the study of light scattering, E is the field associated with an electromagnetic radiation and it can, without loss of generalization, be expressed as $E=E_0 \cos \omega_L t$ where ω_L is the source, or laser, frequency.

Since the atomic or molecular dimension is of the order of angstroms and the laser frequency is of the order of thousands of angstroms the usual approximation that the electric field is slowly varying across the molecular dimensions (Born approximation) is valid but on top of this is the fact that the field is constant over many molecular dimensions. What we actually sample when we apply light to matter is an average polarization and an average polarizability of all those molecules:

$$< \underline{P} > = < \alpha > \underline{E} \qquad (2)$$

Each molecule has a polarizability tensor $\alpha^{(i)}$ associated with it so if we neglect interaction between molecules:

$$< \underline{P} > = < N\alpha^i > \underline{E} = V < \rho > < \alpha^i > \underline{E} \qquad (3)$$

i.e., the average polarizability of the medium can be approximated by averaging the product of the volume V, the density of the sample and the polarizability of each molecule α^i.

We know also from electromagnetic theory that a permanent polarization such as a dipole will not interact with radiation but an oscillating dipole will emit or absorb light so that in order to observe a scattering phenomenon we have to look for oscillating polarization, an electric dipole for instance.

RAYLEIGH AND BRILLOUIN SCATTERING

In Eq. (3) let us concentrate on density fluctuations and assume that the polarizability tensor associated with each molecule is a diagonal (trace) tensor i.e., only α_{xx}, α_{yy} and $\alpha_{zz} \neq 0$. We can write those fluctuations in density as

$$\delta\rho = \left(\frac{\partial\rho}{\partial P}\right)_S \delta P + \left(\frac{\partial\rho}{\partial S}\right)_P \delta_S \qquad (4)$$

i.e., the density will fluctuate with pressure or with entropy S. The pressure fluctuations of the density are those which propagate through the material while the entropy of thermal fluctuations will not propagate. Propagating pressure fluctuations will scatter the incoming photon at a displaced frequency and give rise to the Brillouin scattering while entropy fluctuations will give rise to the Rayleigh scattering. Imagine that a sound, or acoustical wave, characterized by a frequency ω_B and sound velocity V_B, travels through a medium. When this sound wave scatters a photon both energy and momentum must be conserved:

$$E_L = E_S \pm E_B \text{ or } \omega_L = \omega_S \pm \omega_B \quad \text{and} \quad \underline{k}_L = \underline{k}_S + \underline{k}_B \qquad (5)$$

The conservation of momentum diagram for all scattering processes is shown in Fig. (1). If we can assume small dispersion, i.e., $n(\omega_L) \cong n(\omega_s)$ from Fig. (1) we obtain

$$|\underline{k}_B| \simeq 2\, |\underline{k}_L| \sin\frac{\theta}{2} \quad \text{or} \quad |k_B| = \frac{\omega_B}{V_B} = 2\,\frac{\omega_L n}{C} \sin\frac{\theta}{2} \quad \text{or}$$

$$\omega_B = 2\,\frac{V_B n}{C}\,\omega_L \sin\frac{\theta}{2} \qquad (6)$$

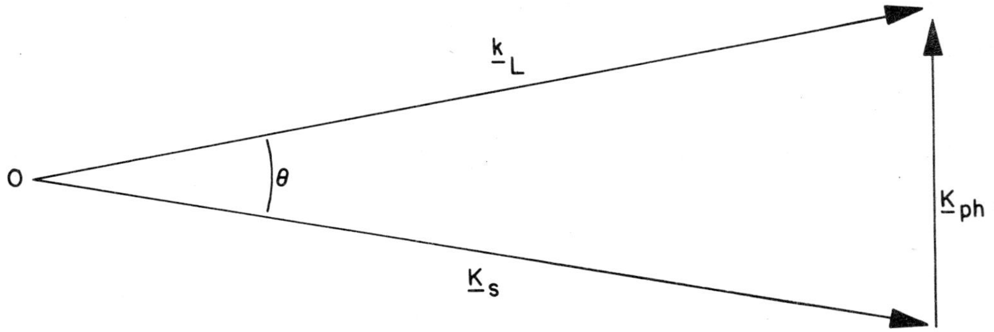

Fig. 1. Conservation of momentum diagram in any scattering experiment, K_L, K_S and K_{ph} are respectively the wave vectors of the laser light, the scattered light and of the phonon or of any scattering quasi-particle.

So the Brillouin frequency shift obeys a Bragg or grating diffraction law and its measurement at a given angle and excitation frequency provides a measure of the sound velocity in the medium. Since the sound velocities in condensed matter are of the order of a thousand meters/sec, ν_B is of the order of 3-10 kMHz or about .1 - .3 cm^{-1} for 90° scattering with a visible laser excitation source. Experimentally, Brillouin and Rayleigh spectra are studied by observing the scattering with a Fabry-Perot interferometer[8], a high resolution spectrograph, [9] or by photobeating electronic techniques[10]. Since the normal line width of an argon laser is of the order of .15 cm^{-1} and that of a He-Ne is of the order of .05 cm^{-1} much care has to be exercised to mode select the laser so that its line-width is less than that of the Brillouin line (~ 700 MHz).

From the way in which we arrived at the Brillouin and Rayleigh scattering as fluctuations in density, we can see that both Brillouin and Rayleigh lines are completely polarized. By relating the fluctuations in pressure and entropy to known thermodynamic quantities C_P and C_V one obtains the well-known Landau-Placzek relation between the intensities of the Brillouin and Rayleigh scatterings:

$$\frac{I_B}{I_R} = \frac{C_V}{C_V - C_P} \qquad (7)$$

For instance for water, where $C_V \cong C_P$, most of the intensity observed in the "Rayleigh" line, in low resolution instruments, corresponds to the two Brillouin components. Fig. (2) shows a high[9] resolution spectrum of the Rayleigh and Brillouin spectra of water.

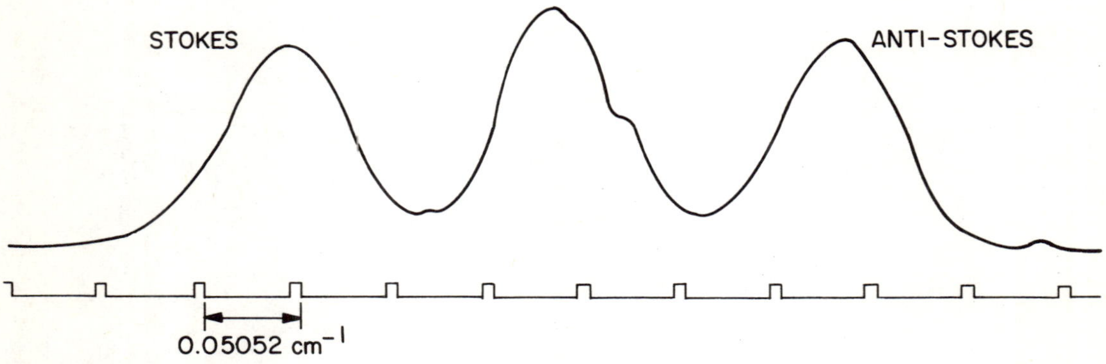

Fig. 2. Brillouin scattering of water taken from Ref. 9 showing the two Brillouin components (outside lines) and the Rayleigh line of water. The Rayleigh line for water shown in the picture is almost all coming from scattering of particles suspended in the sample since for pure water the Rayleigh scattering should be very small.

GENERALIZED RAMAN EFFECT

For the purposes of this paper we shall designate as Raman effects all those inelastic light scattering phenomena in which the scattering mechanism produces a change in the polarizability tensor associated with each molecule, as viewed in our laboratory frame of reference.

ROTATIONAL RAMAN EFFECT

To each molecule we can associate a polarizability tensor which is tied to the symmetry axis of the molecule: x, y, z. Let us diagonalize this tensor and call the new tensor diagonals α_1, α_2, α_3. In a completely spherical molecule $\alpha_1 = \alpha_2 = \alpha_3$ so if the molecule rotates, the tensor, viewed in the laboratory axis x', y', z' stays constant. No change or modulation of the polarizability occurs during the rotation and without a change no oscillating dipole develops so this rotation will be inactive in scattering.

If $\alpha_1 \neq \alpha_2$, for instance, as in the case of a linear molecule, the polarizability viewed in the laboratory system will change when the molecule rotates and the rotational Raman effect can be observed. The selection rules for the rotational Raman effect are $\Delta J = 0, \pm 2$ because each component of the tensor viewed in the laboratory system is equal to the sum of the components in the molecular system of reference multiplied by a factor containing two cosine functions. The rotational Raman effect is completely depolarized (P = .75) and the $\Delta J = 0$ selection rule predicts an undisplaced scattering (Rayleigh) which is polarized. The rotational frequencies are inversely proportional to the molecular moments of inertia and the rotational displacements are of the order of 1 cm^{-1}.

To observe the rotational Raman effect high resolution spectrographs have been used almost exclusively up to date. Due, however, to the fact that the Rayleigh scattering is so highly polarized while the rotation Raman effect is depolarized, one can foresee coupling a laser source to a single monochromator (a double monochromator will not be needed because in the right geometry the Rayleigh line is weak) and photoelectric techniques as the ideal way to observe rotational Raman effects[11].

ANISOTROPY RAMAN SCATTERING IN LIQUIDS

This kind of scattering is quite commonly known as the "Rayleigh wing" scattering and is observed in liquids. It is, in one way, very closely related to the rotational Raman effect and in another related to the Kerr effect in liquids. This anisotropy scattering is due to the fact that in a liquid the molecules sampled by the laser beam are rotating in a viscous medium and that one views a changing polarizability in the laboratory system of reference. This changing polarizability is due to this "overdamped rotation" and also to changes in instantaneous aggregation states of the molecules. "Overdamped rotation" gives rise to the same kind of polarizability changes responsible for the rotational Raman effect but instead of discrete levels, the resulting spectrum is a low frequency continuum centered around the laser exciting frequency.

Debye[12] has worked out details of this anisotropy scattering predicting a Lorentzian line shape for the scattering with a width that is dependent on the volume of the molecule, the temperature and the shear viscosity. For most liquids the half width of this anisotropy scattering is of the order of 5-10 cm^{-1} but recent measurements[13] show deviations from the predicted Lorentz shape.

This anisotropy scattering, even though hardly explored today, should generate much information on the angular correlation functions in liquids so badly needed to understand nonlinear optical effects, like self focusing, which are dependent on the Kerr effect.

VIBRATIONAL RAMAN EFFECT IN MOLECULES

This is the oldest kind of Raman effect known and can be simply understood as a modulation of the polarizability tensor components due to a vibration of the molecule. Classically, if the polarizability is modulated, or if it changes with a vibration of the molecule

$$\alpha = \alpha_o + \frac{\partial\alpha}{\partial q_m} q_m = \alpha_o + \alpha_1 \cos\omega_M t \quad \text{so}$$

$$P = (\alpha_o + \alpha_1 \cos\omega_M t) \quad (E_o \cos\omega_L t)$$

$$= E_o\alpha_o \cos\omega_L t + \alpha_1 E_o [\cos(\omega_L + \omega_M) t + \cos(\omega_L - \omega_M) t] \tag{8}$$

We see from Eq. (8) that the polarization P will radiate energy at the frequencies $(\omega_L - \omega_M)$ and $(\omega_L + \omega_M)$, the anti-Stokes and Stokes-Raman vibrational frequencies, besides the Rayleigh scattering discussed before.

Group theory predicts the number of frequencies which are Raman, or infrared active, for all molecules provided that the shape of the molecule or its "point group" is known. Group theory also predicts for each normal mode those polarizability tensor components which are changing during the vibrational motion measured by the Raman effect. By counting the number of modes which are Raman and infrared allowed, and measuring the depolarization of the Raman lines, one can gain considerable knowledge about the shape of the molecule under investigation. All completely symmetric molecular vibrations (such as the "breathing" motions) are characterized by changes in the diagonal components of the polarizability tensor and their scattering is polarized (depolarization ratio is close to zero); in all other normal modes changes are found mostly in the off-diagonal terms of the polarizability tensor or have the trace of the tensor equal to zero. Associated scattering is depolarized (depolarization ratio = .75).

It is interesting to mention here the "vibrational overtone" Raman effect. Selection rule for a vibrational Raman effect is $\Delta v = \pm 1$; an overtone Raman effect means that we are observing a process in which $\Delta v = \pm 2$. This new selection rule can arise from two different causes: the mechanical anharmonicities of the harmonic oscillator or a non-linear term in the polarizability i.e., $\partial^2\alpha/\partial Q^2 \neq 0$. In either case a sharp line corresponding to $\Delta v = \pm 2$ appears with the Raman displacement in general being a little less or equal to twice the Raman displacements for the $\Delta v = \pm 1$ transition.

RAMAN SCATTERING BY PHONONS

The main difference between the vibrational Raman effect in liquids and in solids is that in liquids light is scattered by the changes of polarizability associated with a normal mode in the molecule. In solids, such as NaCl, if a pair $Na^+ - Cl^-$ oscillates, the sodium is so tightly bound to all of its Cl^- nearest neighbors that the change of position of this Na^+ ion will induce a corresponding movement of all those Cl^- ions in a continuing chain reaction. The molecule NaCl thus loses its identity as the vibration becomes a wave propagating the whole crystal. Characterized by a discrete phase velocity v, a frequency ω and a wave propagation vector k, this wave is called a phonon. It is a normal mode of the crystal. By contrast, we can consider the Raman effect in liquids as one in which $|k| = 0$, i.e., no phonon propagation occurs.

In practice for liquids we have to conserve only energy in the Raman scattering process, for solids both energy and momentum or, more properly, energy and wave vector $\underline{k}$ must be conserved. Another difference which is very important between Raman spectroscopy of liquids and solids is that where the liquid molecules are randomly oriented in relation to the laboratory system of reference all the unit cells of a solid are oriented in the same manner. No difference exists in solids, between the laboratory and the crystal systems of reference. Imagine that we calculate from group theory that a primitive cell of a solid has a vibration in which the only changing components in the polarizability tensor are xy and yx. To observe the Raman effect of that phonon, we first arrange the polarization of the incoming laser radiation parallel to the x axis and observe the scattered light with its polarization in the y direction or vice-versa, since only those two geometries would give non-zero results for the equation:

$$\underline{P} = \alpha\underline{E} \quad \text{i.e.} \quad P_x = \alpha_{xy}E_y \text{ and } P_y = \alpha_{yx}E_x \tag{9}$$

So in very elegant ways we can, for a solid, determine all the Raman-active phonons and with each we can associate a polarizability tensor and a definite symmetry. Fig. (3) shows a typical example for MnF_2. MnF_2 belongs to the D_{4h} point group and group theory predicts four Raman active modes: $1A_{1g}$ (with α_{xx}, α_{yy}, $\alpha_{zz} \neq 0$), $1E_g$ (α_{yz}, α_{xz}, α_{zy}, $\alpha_{zx} \neq 0$), 1 B_{1g} (α_{xx}, $\alpha_{yy} \neq 0$) and $1B_{2g}$ (α_{xy}, $\alpha_{yx} \neq 0$). Fig. (3) shows the spectra observed for the different α_{ij} spectra, in complete agreement with the results of group theory[14].

The conservation of momentum plays a very important role in the understanding of the spectra of solids. First the frequency of a phonon--either in the acoustical or optical modes*--varies with momentum throughout the Brillouin zone (the maximum $k = 2\pi/\alpha$ where α is the crystal lattice constant). For example, Fig. (4) shows how the frequency varies with momentum for two acoustical modes and two optical modes. With a visible source of light, where $|k| = \nu/c \cong 10^5$ cm^{-1}, a 90° scattering will create or destroy phonons with $|k| \cong 10^5$ cm^{-1}; in the scale of Fig. (4) this $|k|$ is very small so we often refer to the Raman effect measuring the $|k| \cong 0$ phonons. Let us examine the influence of this conservation of momentum in another case. In ZnO a doubly degenerate phonon of symmetry E_1 is present. It is both Raman and infrared active. When infrared active in the x direction its polarizability tensor components xz and zx are different from zero. On the other hand, when the phonon is polarized in the y direction the yz and zy polarizability components are different from zero[15]. Imagine that we are looking at the xz polarizability component of this E_1 line in ZnO: if the light is incident in the z direction and the observation is made along the y direction, by conservation of momentum, a phonon is produced in the yz plane with x polarization (since we are measuring the xz component of the tensor). So this phonon has a propagation direction perpendicular to its polarization and is a transverse optical phonon (TO). If, still measuring the xz spectrum, the light is incident in the z direction and the observation is in the x direction we produce a phonon in the xz plane with x polarization. This observed phonon is both transverse (propagation perpendicular to polarization) and longitudinal (LO). Since the transverse and longitudinal phonons have different frequencies we observe two lines. So in the z (xz) y spectrum we see only one line, the TO phonon, while in the z (xz) x spectrum we see two E_1 lines, the TO and LO phonons. Figs. (5A) and (5B) show the z (xz) x and the z (xz) y spectra of ZnO with the conservation of momentum diagrams which explain them. [16]

*An acoustical phonon is like a sound wave in which all constituents in the unit cell vibrate in the same direction; in an optical phonon, they vibrate against one another. At k = 0 in solids the two are markedly different. At very large values of k, they may tend to merge.

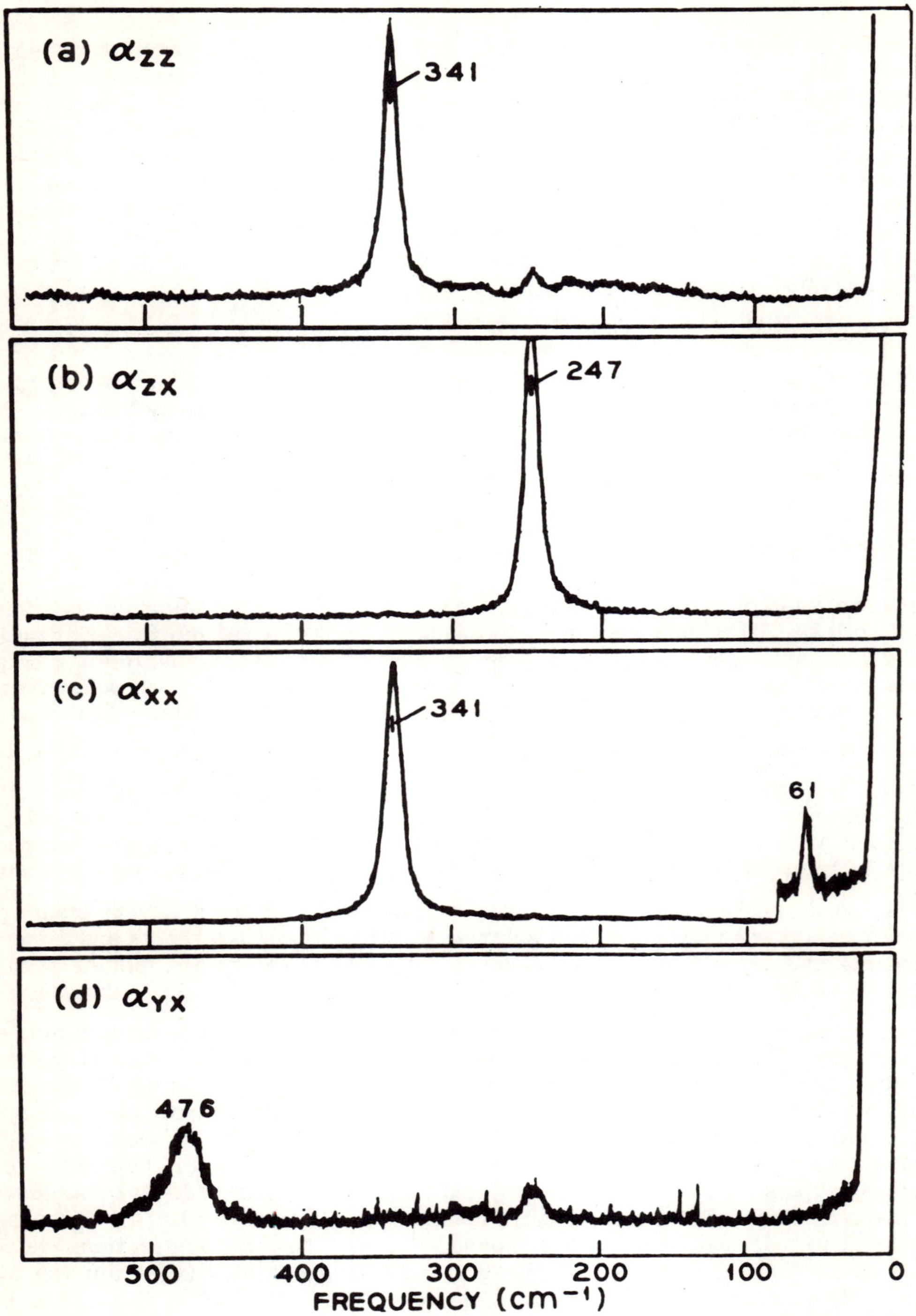

Fig. 3. Raman scattering of MnF_2. In this crystal there is an A_{1g} vibration at 341 cm^{-1} with $\alpha_{xx,yy,zz} \neq 0$, a B_{1g} vibration, with $\alpha_{xx,-yy} \neq 0$, at 61 cm^{-1}, a B_{2g} line with $\alpha_{xy,yx} \neq 0$ at 476 cm^{-1} and an Eg vibration at 247 cm^{-1} with the $\alpha_{xz,yz} \neq 0$. From Ref. [14].

RAMAN SCATTERING BY MULTIPLE PHONON PROCESSES

If we expand the polarizability tensor as in Eq. (8) and keep higher order terms:

$$\alpha = \alpha_0 + \frac{\partial \alpha}{\partial q_i} q_i + \frac{\partial^2 \alpha}{\partial q_i \partial q_i} q_i q_j \tag{10}$$

The production of two or more phonons will create a modulation in the polarizability and will scatter in a Raman-like process. Similarly, if the force constants between atoms are not that of harmonic oscillators but include terms like ax^3, bx^4 etc. the selection rules are relaxed and $\Delta n = \pm 2, \pm 3$ processes are now allowed. Again, this scattering process must conserve both energy and momentum and for a two-phonon process:

$$\omega_L = \omega_R \pm \omega_{PH1} \pm \omega_{PH2}$$

$$k_L = k_R + k_{PH1} + k_{PH2} \tag{11}$$

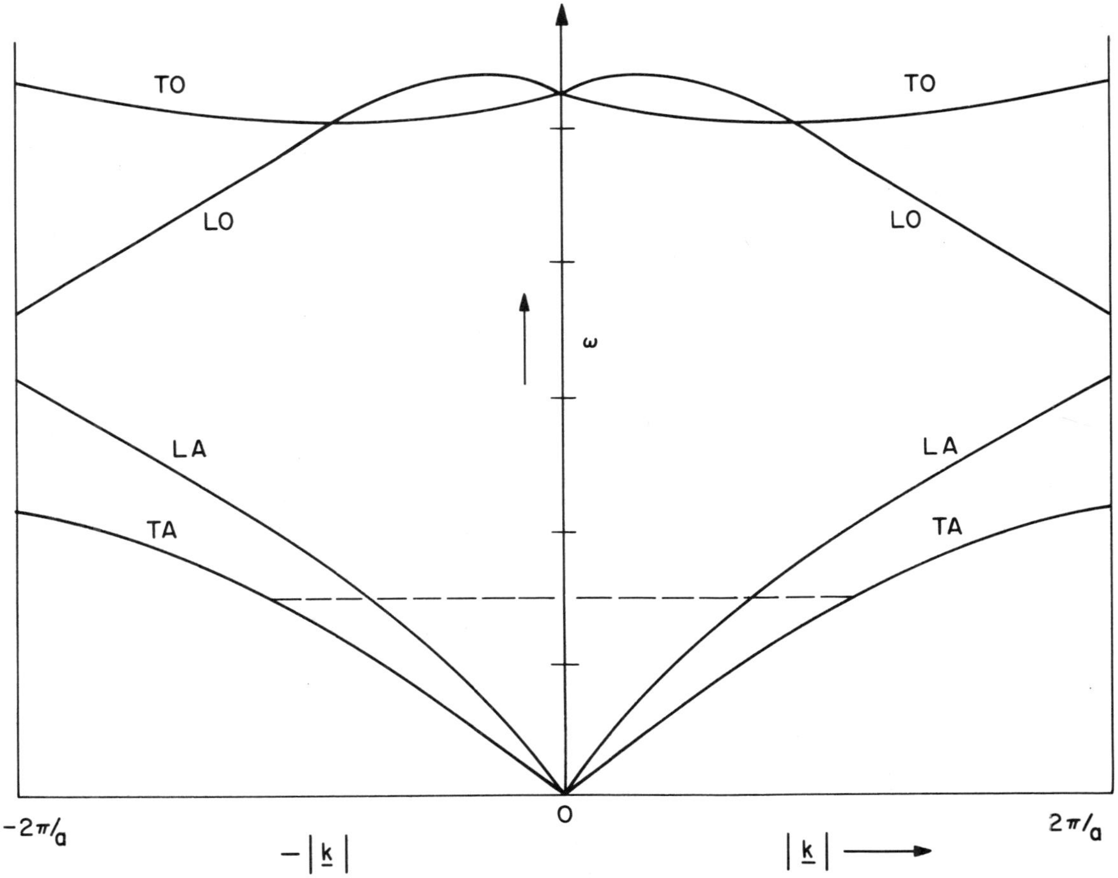

Fig. 4. Idealized dispersion curves of the acoustical and optical phonons of a crystal with two atoms per unit cell showing how the phonon frequencies change with wave vector.

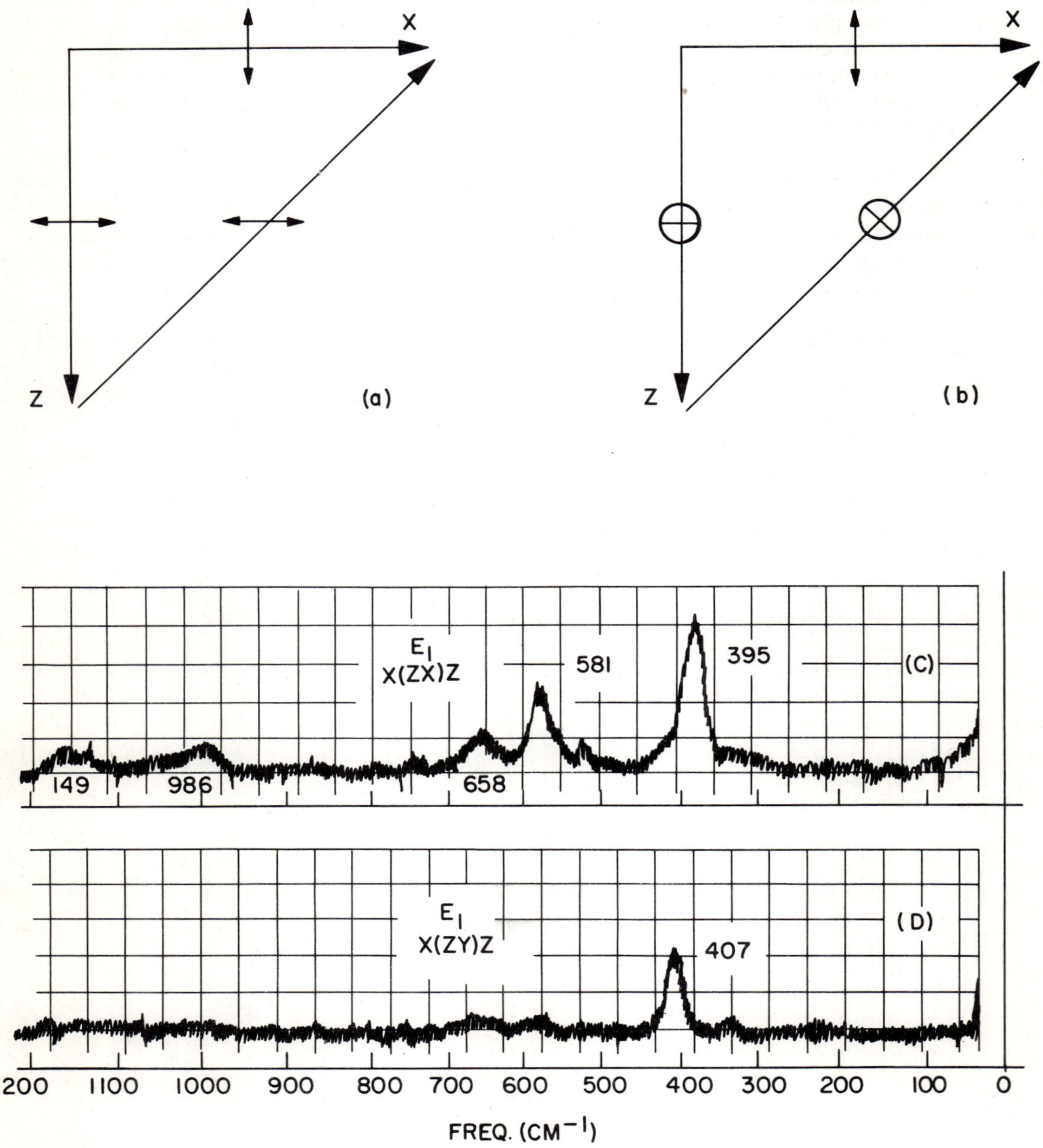

Fig. 5. Experimental demonstration of the conservation of momentum in an optical phonon scattering in ZnO. Fig. 5(a) is the conservation diagram to show that in 5(c) one obtains the scattering from the LO and TO phonons. Fig. 5(b) explains the data of Fig. 5(d) where the scattering is obtained from the TO phonon only. From Ref. [16].

In liquids, only phonons with k = 0 exist and an overtone Raman effect is produced which, as indicated before, consists of sharp lines. In the case of two-phonon scattering in solids, the two phonons can be produced throughout the whole Brillouin zone with only the conditions that energy and momentum are conserved in the scattering and that the process has the correct polarizability tensor of the experiment. Since, as seen in Fig. (4), the frequency of a phonon may vary drastically with momentum, in general multiple phonon scattering results in broad bands. We can imagine two successive Raman processes i.e., where when one phonon is produced, the Raman light is scattered and when another phonon is produced this will cause a sharp two-phonon or overtone line. Two phonon spectra are usually broad even though most of their intensity comes from large $|k|$ phonons, because here the density of states may be larger than that for phonons with small $|k|$.

Aside from the breadth, which sometimes can be misleading, there are two other ways of recognizing that a spectral feature arises from multiple phonon scattering. First, group theory will tell you that the symmetry of the two-phonon process is the product of the symmetries of the two phonons involved. The symmetry of the two-phonon process is thus more complex than that involving one phonon; the former may even appear not to obey the symmetries allowed for the specific point group under study. Since the product of two symmetries quite often contains the most symmetric representation, A_1, the second order Raman spectra may appear superimposed on the A_1 spectra. Another way to recognize a second order Raman process is from its temperature dependence. The one-phonon Stokes intensity decreases with temperature as (n + 1) while the intensity of the two-phonon spectra, occurring at the same frequency, will vary with temperature obeying a law like $(n + 1)^2$ where $n = (\exp h\omega/KT - 1)^{-1}$. The two-phonon process fades away quickly with decreasing temperature.

The old concept that in Raman spectra the strong lines represent one-phonon processes while the two-phonons give rise to broad and weak lines is very treacherous. In many substances like $BaTiO_3$, TiO_2, $KTaO_3$, some of the most prominent features of the Raman spectra are due to multiple phonon processes. Another fact to remember is that the $|k| \cong 10^5$ cm^{-1} acoustical phonon gives rise to the Brillouin spectrum with a frequency shift of the order of 1 cm^{-1}, while the two-phonon Raman scattering of the acoustical processes will extend to a few hundred cm^{-1} since most of the Raman effect is due to phonons with $|k|$ near the edge of the Brillouin zone. See Fig. (4).

Fig. (6) shows part of the (xx) spectrum of TiO_2 where the sharp line at 143 cm^{-1} cor-

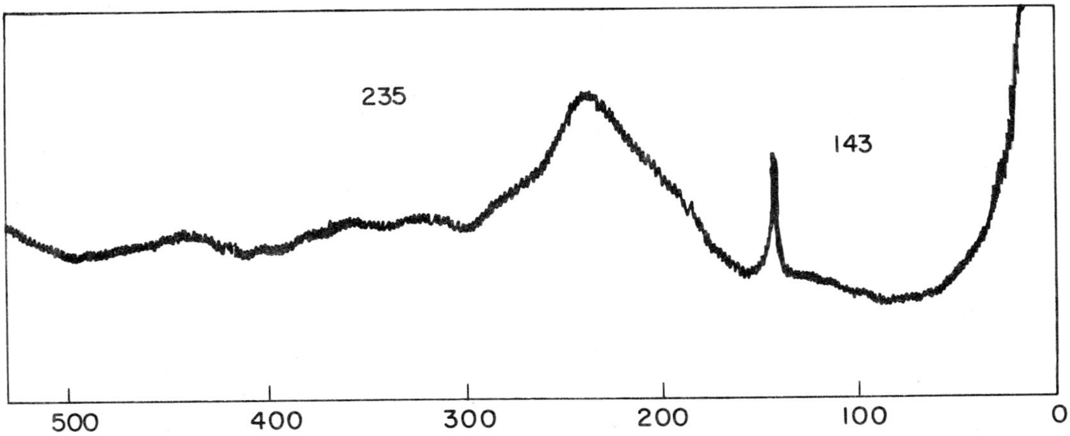

Fig. 6. Part of the xx spectrum of TiO_2 showing that the two phonon process at ~235 cm^{-1} can be stronger than the one phonon B_{1g} scattering at 143 cm^{-1}. From Ref. [14].

responds to a one-phonon process of symmetry B_{1g} (α_{xx}, $\alpha_{yy} \neq 0$) while the broad and strong band at ~ 234 cm^{-1} corresponds to the two-acoustical-phonon Raman scattering[14].

In principle, the two-phonon process should provide considerable information on the dispersion relation ($|k|$ vs ω) of phonons, critical points in the Brillouin zone where the population of those phonons is maximum, etc. However, due in part to the poor state of the experimental work on second order Raman processes, we feel that most of the original promises have not as yet been fulfilled.

RAMAN EFFECT OF F-CENTERS AND IMPURITIES IN CRYSTALS

Imagine a crystal like NaCl in which the first order Raman effect is forbidden for reasons of symmetry (each Na and Cl ion occupies a center of the cubic lattice). Should a Cl atom be removed and substituted by a vacancy in which an electron is trapped, an f-center is created. In the process the translational symmetry of the crystal is destroyed so that the Na^+ ions which are next-neighbors of the trapped electron are no longer at centers of cubes. First order Raman i.e., one-phonon interactions, are now allowed around the f-center. The same happens, for instance, when we substitute a Cl by Br or a Na by a K atom.

The problem of observing the Raman effect of f-centers is that we cannot introduce enough of them. By having just a few centers the Raman effect is distressingly weak. In NaCl, for instance, with 10^{17} f-centers/ml (one of every 100,000 Cl atoms) the first order Raman effect intensity of f-centers at 300°K is, at best, comparable to the weak two-phonon spectrum of the crystal. To observe the scattering from f-centers and to discriminate against the two-phonon processes, the temperature is lowered and the frequency of the laser excitation can then be chosen to be close to the strong electronic absorption of the f-center so that the resonant denominator will increase the cross section of the Raman process[17].

A substitutional impurity Raman effect, in alkalihalides--as distinguished from an f-center--cannot make ready use of this resonant denominator because unlike f-centers, visible light absorption of the impure crystal does not change. However, much larger concentrations of the impurity than f-centers can be substituted in the lattice without appreciably disturbing the crystal symmetry. The total cross section can then be made sufficiently large for observation[18].

An interesting characteristic of both the f-center and impurity Raman effect is that the Raman active centers, or impurities, disturb the translational symmetry characteristic of a solid; conservation of momentum then loses its meaning and the $|k|$ of the phonon can, in essence, assume any value within the Brillouin zone. This means that, like in the two-phonon processes, the scattering results from phonons with all values of $|k|$ and the spectrum is broad, reflecting the density of states functions for all the allowed phonons, instead of the normally sharp one-phonon processes obtained for a solid where only the $|k| \cong 0$ phonons are sampled.

RAMAN EFFECT OF POLARITONS

Electromagnetic radiation passing through a crystal is characterized by a frequency ω, a velocity (c/n) and a wave vector k. For low frequency light, the dispersion relation

(ω νs $|k|$) is a straight line passing through the origin. Imagine that in the same graph we plot the dispersion relation (ω νs $|k|$) of light and of an infrared active optical phonon with its LO and TO components. The k interval we are interested in is so small that we may consider that the phonon frequency itself is constant and independent of $|k|$. As seen from the dashed lines in Fig. (7) the two dispersion relations will cross where the phonon and the electromagnetic radiation have the same frequency and wave vector. If this phonon is infrared active there will be an interaction of the electromagnetic radiation with the mechanical vibration and the excitation, around the interaction region, will be partially phonon and partially light. This mixed excitation in the interaction region is called the polariton and its dispersion relation is also shown in Fig. (7) as the full lines. As can be seen from the figure we have two branches of the polariton: the upper or quasi-photon branch ω^+, which in the limit $|k| \rightarrow 0$ tends toward the frequency of the LO mode and which has escaped observation up to date, and the lower or quasi-phonon ω^- which has

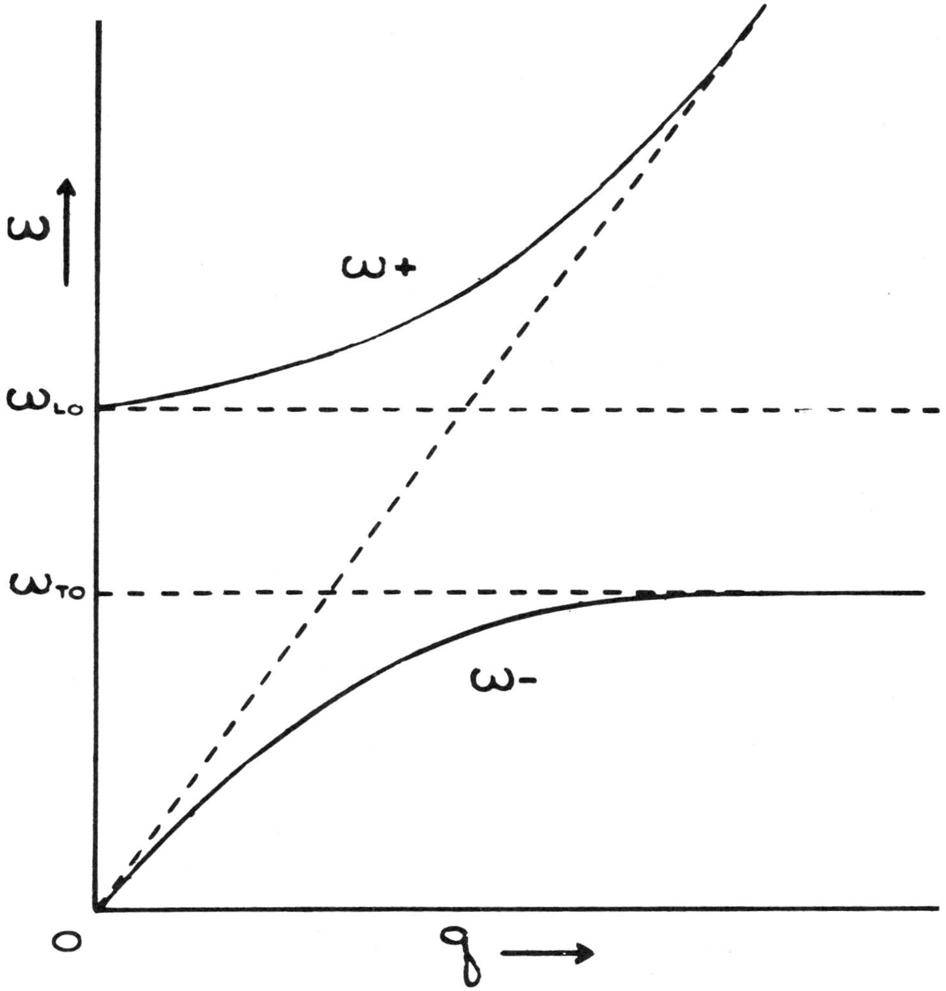

Fig. 7. Idealized dispersion curves of the coupled phonon - E.M. wave system (polariton) for the case of one infrared and Raman active phonon interacting with long wavelength light. The dashed lines are the dispersion of the uncoupled phonon and light waves. The solid curves are the dispersions of the coupled excitations.

been observed for GaP[19], ZnO[20] and quartz[21]. From Fig. (7) we see that the polariton exists only for very small values of $|k|$ so that in order to scatter from it we have to observe the Raman effect in the forward direction. Since this scattering process conserves momentum, in the forward direction a polariton with minimum $|k|$ will be produced. By observing the scattering let us say at 1°, 2°, 3° etc., from the forward region we can observe polaritons of higher and higher $|k|$ until the excitation becomes pure phonon for angles of the order of 10°. Fig. (8) shows the Raman effect of the polariton in ZnO[20] with the dispersion relation of the quasi-phonon polariton from 160 to 407 cm^{-1}.

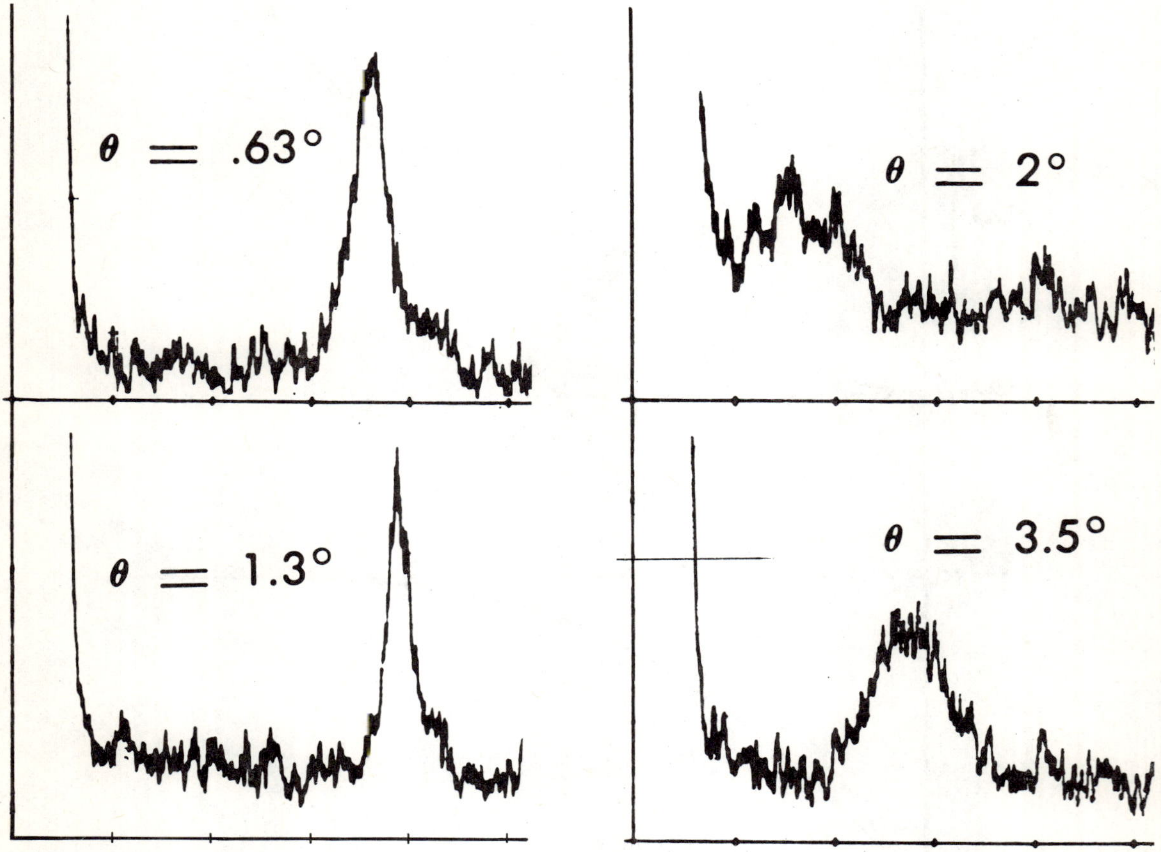

Fig. 8. Experimental observation of the scattering of polaritons. One can see easily that the frequency shift changes with the angle of observation from the forward direction. At 3.5° from the forward direction the polariton is almost completely the TO phonon whose frequency is 407 cm^{-1}. At .63° it is 160 cm^{-1}. From Ref. [20].

RAMAN EFFECT OF SPIN WAVES OR MAGNONS

Spin waves are excitations characterized by dispersion relations (ω vs $|k|$) very much like those of phonons. They occur in magnetic materials whose atoms have non-zero spins oriented in an ordered manner. The spins of all the atoms of ferromagnets (such as iron and nickel) are parallel with the same orientation. This, of course, gives rise to their very high degree of magnetization. In an antiferromagnet (MnF_2, FeF_2) the spin of one of the magnetic ions is pointed in a definite direction while the next ion is pointed in exactly the opposite direction; although the total magnetization of an antiferromagnet is zero all the spins are still oriented. In a ferrimagnet the spins of next neighbors are also antiparallel like in an antiferromagnet but they are of different magnitude so that cancellation is incomplete and a magnetization remains, see Fig. (9). Exactly as in the

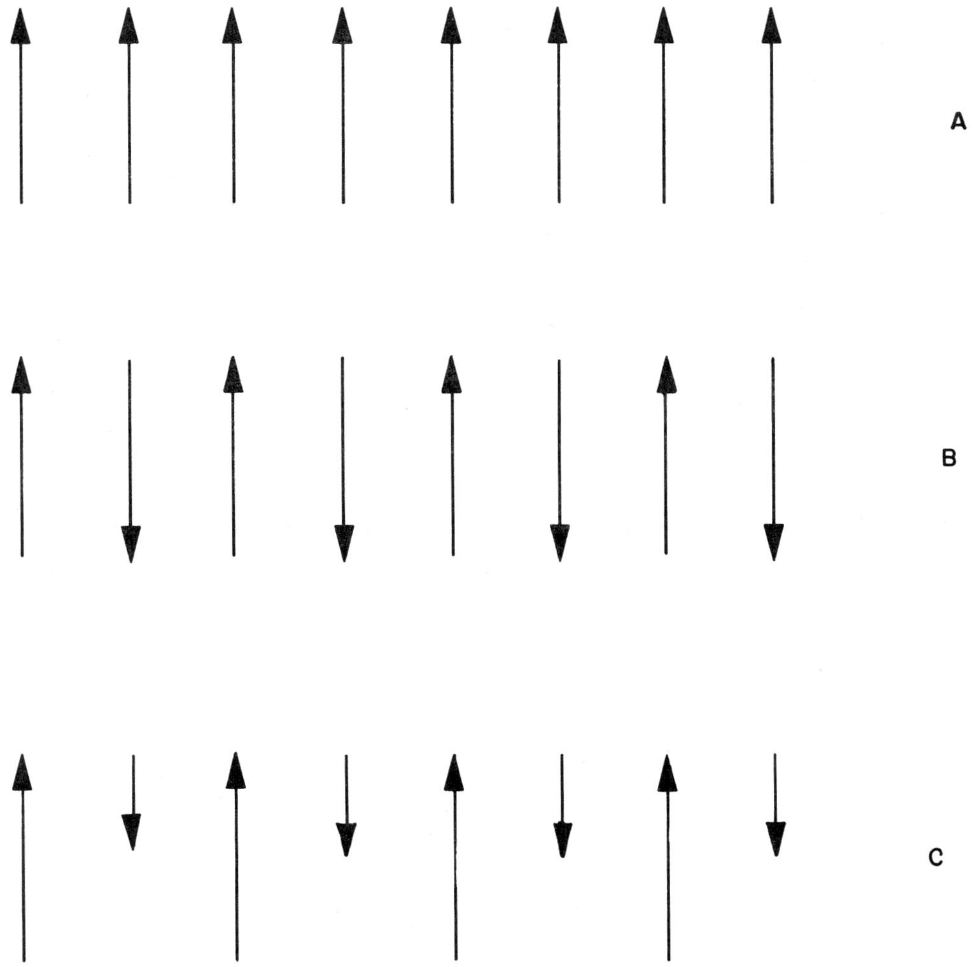

Fig. 9. Graphical representation of the spin alignment of (a) ferromagnetic materials, (b) antiferromagnets and (c) of ferrimagnets.

case of phonons, if we now disturb the orientation of one spin, since they are all coupled, this misorientation will be felt by the next neighbors creating again a wave which will travel through all the spins. This process, or this wave, is called a spin wave or magnon. If we heat a magnetic material we can break up the spin ordering and destroy its magnetic properties. When cooling the sample, the temperature at which magnetic ordering takes place is called the Neel-Curie temperature. In Fig. (3) we see the room temperature spectrum of MnF_2 without any trace of spin wave scattering; by cooling the crystal below its Neel temperature ($\cong 70\,°K$) Raman scattering by spin waves appears[22]. Fig. (10) shows the dramatic appearance of the one and two-magnon processes in FeF_2[22] as the temperature is lowered.

As in the case of phonons, the dispersion relations of magnons are not flat. The frequency of the broad two-magnon scattering does not, therefore, have to occur at twice the frequency of the one-magnon process since the two-magnon process consists of scattering by a pair of magnons throughout the Brillouin zone. The frequency distribution of the two-magnon process, like that of phonons, reflects the dispersion relation of the magnons in question and their density of states for the different points in the Brillouin zone[23].

ELECTRONIC RAMAN EFFECT

The electronic Raman effect is light scattering by a material in which conservation of energy is furnished by the quantum jump of an electron from one electronic state to another. This effect has been known for a long time, as far as Raman effects go; it was discussed by Placzek in 1934, who also discussed the experiments done up to that time[7]. More recently the electronic Raman effect has been observed in solids, using mercury excitation, in which electronic transitions occurred between the Stark split levels of a rare earth ion in a crystal field[24]. Even more recently, electronic Raman transitions were observed between the ground and impurity levels, in semi-conductors[25].

It is interesting to speculate on what new information one might obtain from the Raman spectra of rare earth Stark split levels. Let us imagine doping a crystal of $LaCl_3$ with Pr^{+3}. In free Pr^{+3} all electrons in the same orbit have identical energy levels--degenerate in quantum mechanical parlance. The anisotropy of the crystal field of $LaCl_3$ will remove the degeneracy of the Pr^{+3} levels. The ground state of Pr^{+3} ($^3H_4$), for instance will be split into six discrete levels each one of which can be characterized by a wave function ψ, an energy E, a crystal quantum number μ and a symmetry S in relation to all the operations of the crystal. So the electronic Raman effect of the Stark split levels of the $^3H_4$ state of Pr^{+3} will consist of five lines characterized by Raman displacements ΔE and polarizability tensors for each of the electronic transitions. These electronic Raman effect tensors, in general, will be more complicated than those arising from phonons in the same crystal but at the same time they will contain more information. From the form of the tensors for the different transitions we should be able to obtain many correlations such as energy levels with their respective quantum numbers and wave functions.

One might question the usefulness of the electronic Raman effect on the ground that most of the information can also be obtained from absorption and fluorescence spectroscopy. Perhaps so but the rebuttal is the same as can be given for phonons: in absorption or fluorescence arising from a dipole we measure but three possible components of a dipole vector; in Raman we measure nine components of a tensor and inherently we should extract much more information from the tensor.

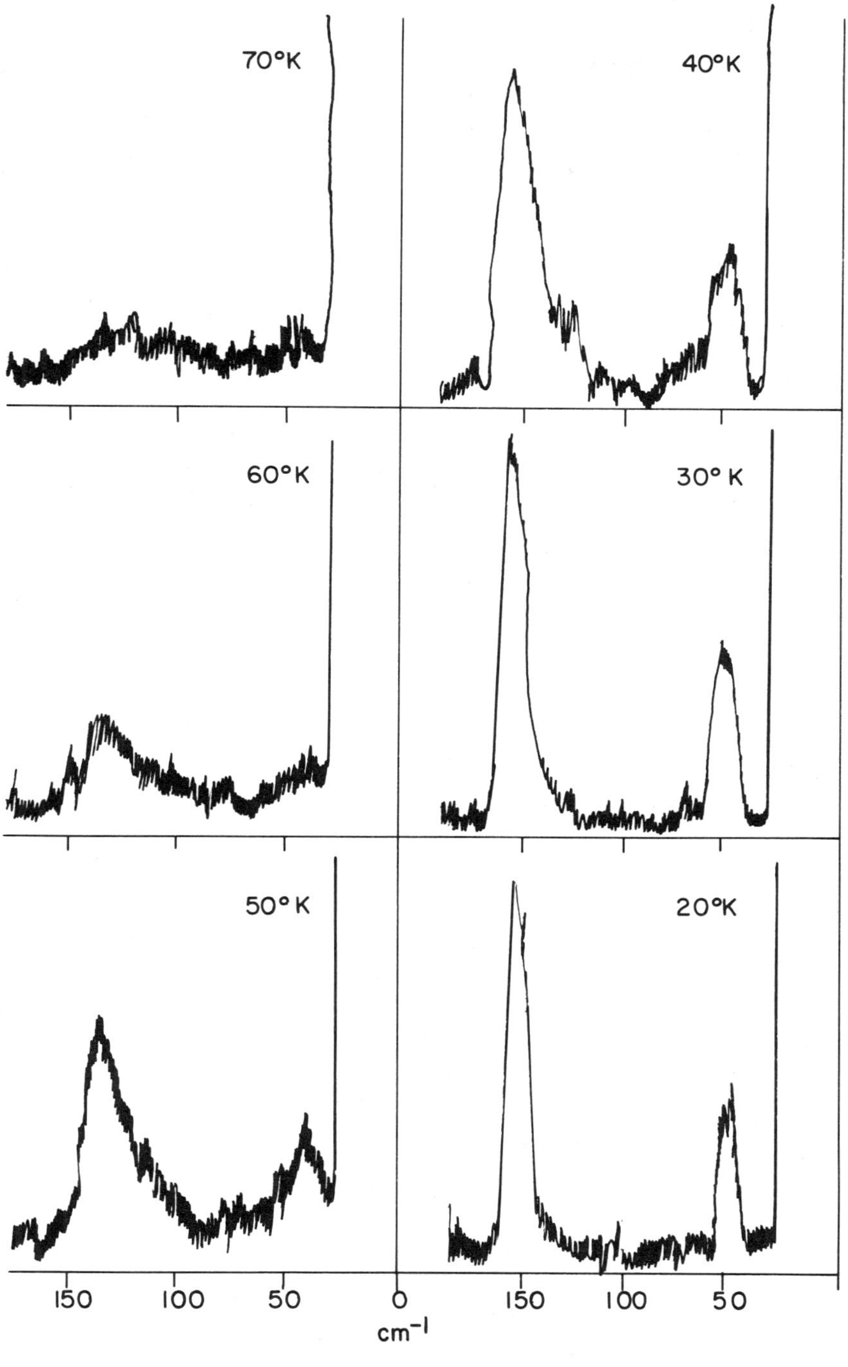

Fig. 10. The scattering from the one and two magnon states of FeF_2. FeF_2 becomes antiferromagnetic at ~70°K so the spins become oriented at that temperature and spin waves can propagate and scatter light. From Ref. [22].

PLASMON SCATTERING

Imagine N free electrons inside a cube of volume A^3. Coulombic repulsions taking place between the electrons will cause the electrons to congregate around configurations of minimum free energy. Cooled to 0°K these electrons organize themselves like atoms in a crystal. Any external disturbance will then propagate itself through the medium as a wave called a plasmon.

The plasmon, like the other excitations studied, can be characterized by its frequency ω_p and its momentum k given by:

$$\omega_P = \left(\frac{2\pi\rho e^2}{\epsilon M^*}\right)^{1/2} \quad \text{and} \quad |\underline{k}| = \left(\frac{4\pi\rho e^2}{kT}\right)^{1/2} \tag{12}$$

where ϵ is the dielectric constant, ρ the plasma density and M* is the effective mass of the electrically charged particle. For a gas, $\epsilon \cong 1$ and for a semiconductor, $\epsilon \cong 10$. So for a gas plasma $\omega \cong 10^1 - 10^4$ cm^{-1}; for a semiconductor plasma $\omega \cong 10^2 - 10^7$ cm^{-1}; for metals $\omega \cong 10^8$ cm^{-1}.

Let us scatter a laser beam from the plasma. We are going to transfer momentum in the scattering process from the light wave to the plasma; if the scattering experiment is done for instance at 90° the momentum transferred is $\cong 10^5$ cm^{-1} and if the scattering is observed in the forward direction the transferred momentum k drops towards zero. The important thing in the scattering process is that the momentum has to be transferred to the plasmon and we have to have allowed plasmons with the required |k|. If the |k| to be transferred in a scattering experiment is larger than the allowed |k| for the plasmon, we have no scattering from those plasma waves.

We can imagine qualitatively that in a scattering process the |k| transfer measures the "lattice spacing" between electrons. Large |k| transferred corresponds to large frequencies or small interparticle distance intervals probed. If the collective system, like a gas plasma, involves particles far away from each other its frequency and momentum (wave-vector) are small. To detect it we have to probe with a small momentum transfer.

If in a certain system the momentum transfer is smaller than the maximum allowed momentum of the plasmon we can detect the collective or plasma excitations. But if the momentum transfer is larger, we are probing smaller volumes. In a plasma, for instance, we are probing the velocity distribution of the individual scattering charges. This is beautifully illustrated in Fig. (11a and 11b)[26] which shows the near forward and 90° scattering of ruby laser light by a flash-produced H_2 plasma.

In the nearly forward scattering case of Fig. (11a) we see the Rayleigh and plasma-shifted frequencies. This scattering is similar to Brillouin scattering where the Rayleigh and plasma frequencies can be understood as propagating (pressure) and nonpropagating (entropy) fluctuations in the density of charges; the plasma shift observed is a function of |k| similar to that in a Brillouin experiment. In Fig. (11b) scattering at 90°, we obtain a spectrum for high values of |k| from the Doppler-broadened scattering by the individual particles, in this case molecular or atomic hydrogen ions surrounded by an electron cloud.

In the above we have treated only gaseous or "free" plasmas, ones not coupled to the medium through which they flow. Let us now consider a plasma in a solid, say a semiconductor. Here the same "Brillouin-like" spectrum discussed above will appear, the electronic mass, M, substituted by an effective mass M*. But a coupling of the host lattice with the internal electric field of the oscillating plasma will also take place. An externally or internally applied electric field will couple to the lattice and is, in effect, responsible for the LO-TO splitting of the optical vibrations. This coupling affects the dielectric constant of the material. In a plasma-free medium we can write the dielectric constant ϵ at low frequencies as:

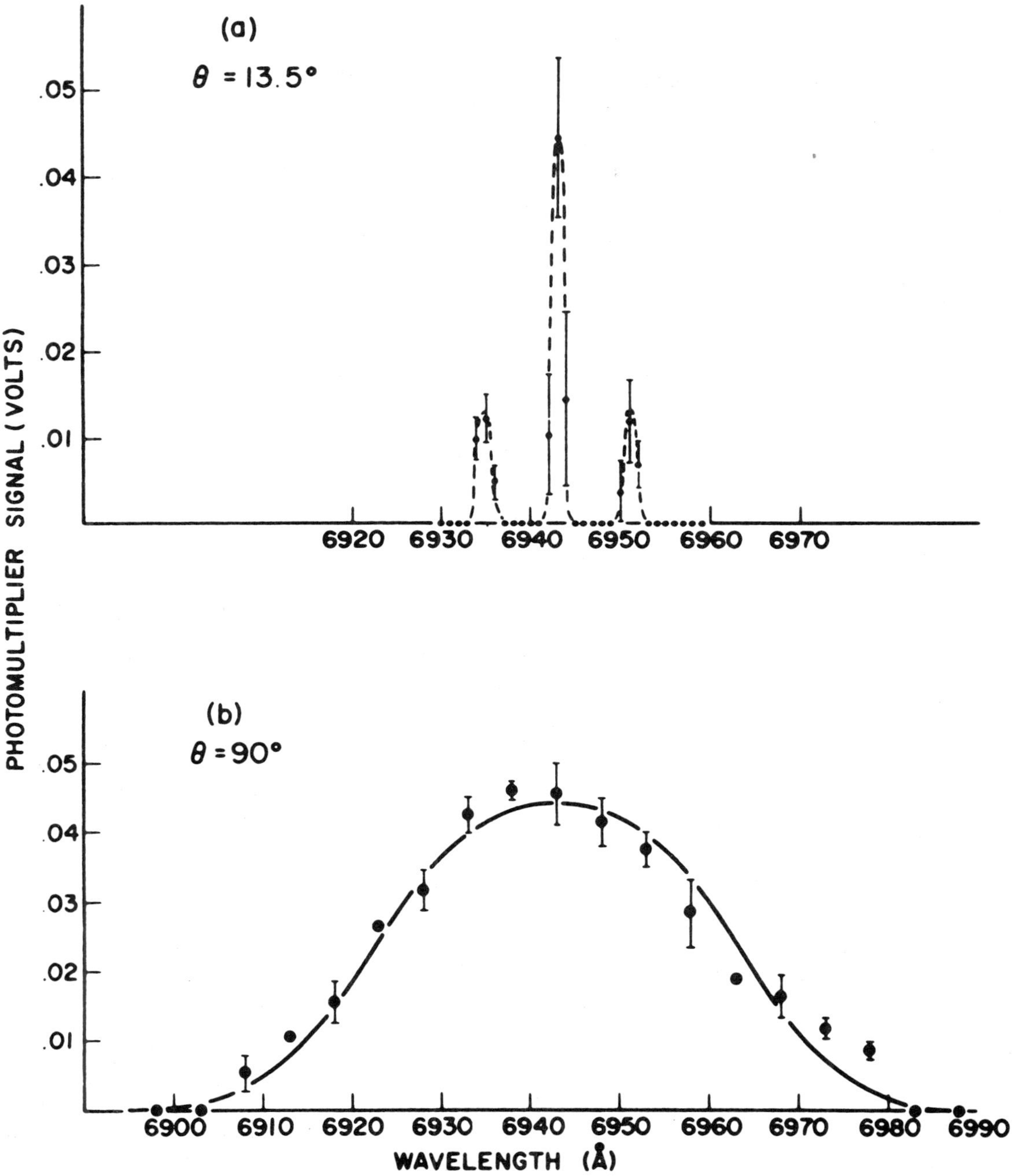

Fig. 11. Light scattering from a gaseous hydrogen plasma at the ruby laser frequency. On 11 (a) the scattering was observed at small angles for the plasma as well as the Rayleigh scattering. At large angle, 90°, and large $|k|$ transfer, we see the Doppler broadened scattering of the elementary scattering particles and not the collective excitations. From Ref. [26].

$$\epsilon(\omega) = \epsilon_\infty + \sum_i \frac{\omega_i^2 S_i}{(\omega_i^2 - \omega^2)} \tag{13}$$

where ϵ_∞ is the dielectric constant far away from the optical phonon frequencies, ω_i is the frequency of the infrared active optical phonons (transverse, or TO modes) and S_i is the infrared intensity of each i mode. If we now add free carriers with a plasma frequency ω_P, Eq. (12), the dielectric constant expression has to be modified to

$$\epsilon(\omega) = \epsilon_\infty + \sum_i \frac{\omega_i^2 S_i}{(\omega_i^2 - \omega^2)} - \frac{\omega_P^2 \epsilon_\infty}{\omega^2} \tag{14}$$

The infinities of the dielectric constant define the frequencies of the TO, or transverse optical modes and from expression (14) we see that the plasma does not change these TO frequencies, so we can say that the plasma does not couple to the TO modes. The zeroes of the dielectric constant are the LO (longitudinal) mode frequencies and those are definitely modified by the plasma. As a matter of fact we have now for each TO mode two coupled LO plasmon modes with their frequencies, line widths, etc. completely dependent on the plasma frequencies.

The scattering from this coupled plasmon-longitudinal mode-lattice can be classified as a generalized Raman effect. Fig. (12) shows the Raman effect of GaAs[27] with different carrier concentrations in which the TO mode appearance is hardly affected while the LO-plasma modes are very dependent on carrier concentration.

RAMAN SCATTERING OF LANDAU LEVELS

If a free electron is subjected to a magnetic field it will describe a circular trajectory with a radius and frequency determined by the magnetic field, electronic charge, mass of the particle and the dielectric constant of the medium. These "cyclotron orbits" inside a crystal are quantized and are called Landau levels. If we choose a crystal in which the effective electronic mass is small, for example in GaAs the effective electron mass is $\cong .07$ and in InSb it is $\cong .01$ of the rest electron mass, we can observe a large Landau splitting for relatively small magnetic fields.

The Landau levels of an electron in a magnetic field are equally spaced in energy, reflecting the fact that one can consider them as levels of an harmonic oscillator in which the fundamental frequency is equal to the cyclotron frequency ω_C. Carrying the harmonic oscillator approximation further we can predict the Raman selection rules and cross-section for a harmonic oscillator. The selection rules are found to be $\Delta n = 0, \pm 2$[28]. If we then include anharmonic terms, transitions with $\Delta n = \pm 1$ would be allowed but should naturally be weaker than the $\Delta n = \pm 2$ transitions. Recent experiments[29] in Raman scattering by Landau levels in InSb were performed and to a first approximation theoretical predictions of the properties of the scattering were observed. Disagreeing with predictions, the $\Delta n = \pm 1$ transitions were observed to be of the same strength as those with $\Delta n = \pm 2$. The harmonic oscillator theoretical approximation for the Landau levels has to be modified. In the same experiments[29] a "spin-flip" Raman transition, $\Delta s = \pm 1$, was also observed in these Landau levels.

ELECTRIC FIELD INDUCED RAMAN EFFECT

If we apply an electric field to a collection of coupled harmonic oscillators we can induce a dipole proportional to the polarizability; vibrations which were only Raman active without the field now become infrared active. This is the well known effect of Stark-induced infrared absorption. One can also show that the electric field will shift all the levels of the harmonic oscillator by the same amount, so the transition frequency does not change nor do the Raman effect $\Delta n = \pm 1$ selection rules for the oscillator change. The electric field may, however, change the symmetry of the unit cell in such a way that it may pull the center atom of a cubic crystal of symmetry 0_h away from the center position making the symmetry of the cell a tetragonal C_{4v}, for example.

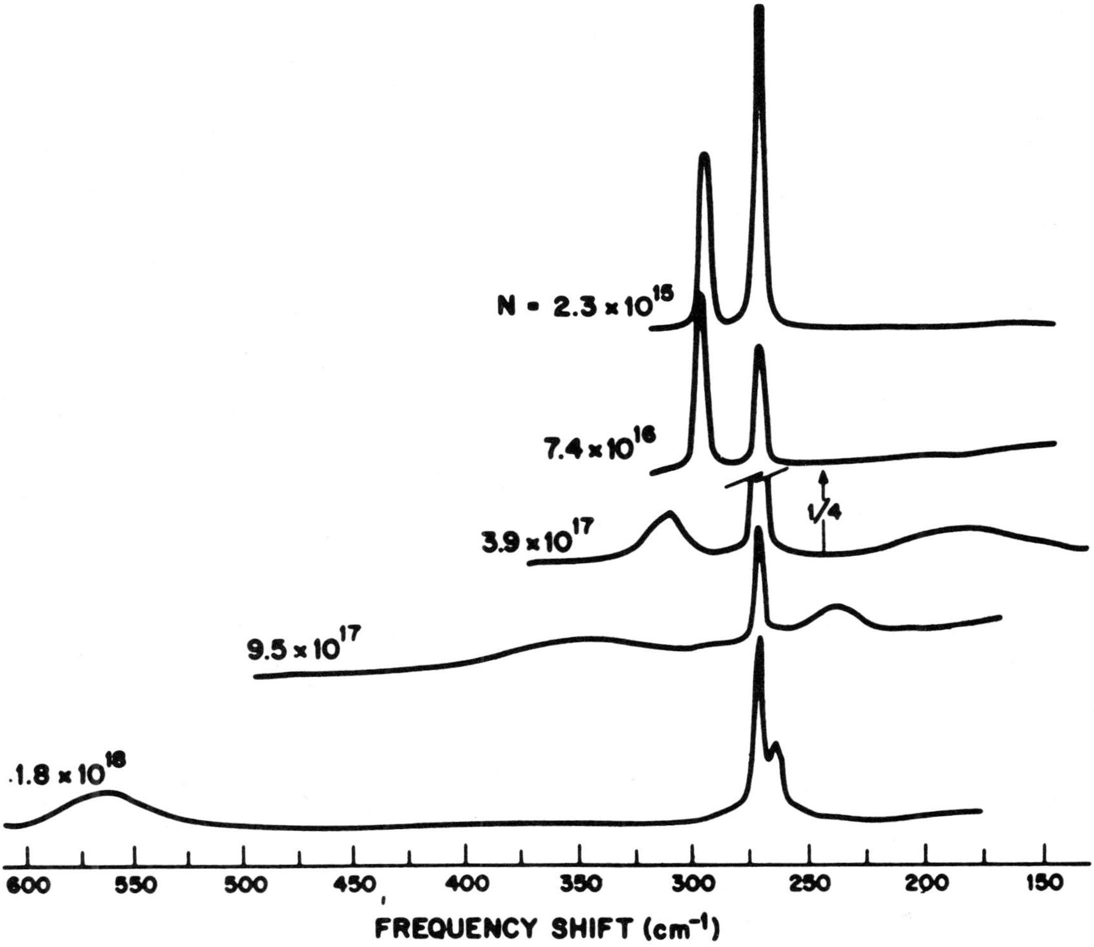

Fig. 12. Scattering from coupled LO-plasma modes in GaAs, from Ref.[27]. On top for a low carrier concentration we see the LO and TO modes of GaAs. As the concentration of carriers N is increased there is a coupling of the LO and plasma frequencies shown very clearly.

This small displacement of an atom in the unit cell may have drastic effects. Imagine that for reasons of symmetry a crystal has no first order Raman effect: by so slightly changing the symmetry the electric field makes the first order Raman effect allowed without affecting, in first order, the vibrational frequencies of the field-free crystal.

If the change of symmetry is small, the new one-phonon Raman effect at first will be as strong as the two-phonon process until, hopefully, it will become so strong as to dominate the spectrum. Assuming that the new, induced, one-phonon process is only as strong as the no-field-allowed, two-phonon process, a clever way of separating the two has been found, [30]. The laser is allowed to continuously shine upon the crystal while the applied electric field is oscillated in a square wave of frequency ω, let us say 100 cycles/sec. The detection of the Raman effect is made with a synchronous amplifier set at the electric field frequency ω or at 2ω; the electronics then measure the difference between the field-on and field-off spectra and, while the two-phonon spectra obligingly cancel out, the newly allowed field-induced one-phonon spectra rise.

This technique may presage a new way of looking at the forbidden modes and at a host of new crystals while per se the results will furnish information on the effect of electronic fields, on vibrational mode frequencies, etc.

We have tried to summarize today's research in the field of light scattering with laser sources. So extensive is the current work that all of us participating await eagerly each week's Phys. Rev. Letters to learn what new effects have been discovered or what unexplained details of the older ones have been clarified. It is an age of excitement in the light scattering field, brought about by the availability of the laser source.

TABLE I

Scattering Phenomena

	Instrumental Requirements		
Type of Scatter	Optical	Laser Excitation	Principal Applications
Rayleigh	Fabry-Perot; Photobeating	Single mode, single frequency, low power	Critical opalescence; phase transitions; atmospheric propagation
Brillouin	Fabry-Perot; Photobeating; High-resolution spectrograph	Single mode, single frequency, low power	Phase transitions; velocity of sound; damping processes in sound transmission
Rotational Raman	High-resolution ($0.0X\ cm^{-1}$) spectrograph	High-power ion (1 watt)	Molecular structure of gases
Vibrational Raman	Double Grating Spectrometer	50 mW He-Ne	Molecular structure; force constants; chemical analysis and identification
Phonon	Double Grating Spectrometer	High-power ion	Lattice properties of and phase transitions in solids
Multiple Phonon	Double Grating Spectrometer	High-power ion	Anharmonicities of force constants in solids; critical points in the Brillouin zone

TABLE I

Scattering Phenomena (cont)

Type of Scatter	Instrumental Requirements Optical	Instrumental Requirements Laser Excitation	Principal Applications
F-Center	Double Grating Spectrometer	High-power ion	Physics and analysis of lattice defects; phonon propagation in defective lattices
Polariton	Double Grating Spectrometer (forward scatter)	High-power ion	Coupling of phonons and light in the infrared
Spin Wave	Double Grating Spectrometer	High-power ion	Magnetic phase transition around Curie points; coupling of light with magnetic states
Electronic	Double Grating Spectrometer	High-power ion	Electronic energy levels of ions in solids
Plasmon	Double Grating Spectrometer	Ruby	Properties of gaseous, liquid, and solid plasmas; coupling of plasma with lattice modes
Landau levels	Single Grating Spectrometer	High-power CO_2	Cyclotron resonances and damping of nearly free electrons in solids
Field induced Raman	Double Grating Spectrometer	High-power ion	Electric field induced lattice deformation; "forbidden" Raman effects

REFERENCES

1. Lord Rayleigh, Phil. Mag. 47, 375 (1899), Coll. papers IV, 397.
2. L. Brillouin, Ann. Phys. (Paris) 17, 88 (1922).
3. A. Smekal, Naturwiss. 11, 873 (1923).
4. C.V. Raman, Indian J. Phys. 2, 387 (1928); C.V. Raman and K.S. Krishnan, Nature 121, 501 (1928).
5. G. Landsberg and L. Mandelstam, Naturwiss. 16, 57 (1928).
6. E. Gross, Nature 126, 201 (1930).
7. G. Placzek, Marx's Handbuch der Radiologie VI, 2, 209 (1934).
8. In particular the Indian work published in the Ind. Acad. Sc. by R.S. Krishnan and others and in this country by D.H. Rank et al., and more recently with laser sources by R.Y. Chiao and B. Stoicheff, J. Opt. Soc. Am. 54, 1286, (1964).
9. G. Benedek, J.B. Lastovka, and K. Fritch, J. Opt. Soc. Am. 54, 1284 (1964); G. Minichino, R. O'Brien, G.J. Rosaseo, and A. Weber, Bull. Am. Phys. Soc. 12,1132 (1967). Both groups used laser sources.

10. J.B. Lastovka and G.B. Benedek, "Proc. Phys. Quantum Electr. Conf.," Puerto Rico 1966; M.G. Cohen and E.I. Gordon, Bell Syst. Tech. J. 44, 693 (1965); and H.Z. Cummins, N. Knable and Y. Yeh, Phys. Rev. Letters 12, 150 (1964).
11. A. Weber, S.P.S. Porto, L.E. Cheesman, and J.J. Barrett, J. Opt. Soc. Am. 57, 19 (1967).
12. P. Debye, "Polar molecules," Dover Publ., N.Y. 1929 Chapt. V.
13. S.L. Shapiro and H.P. Broida, Phys. Rev. 154, 129 (1967).
14. S.P.S. Porto, P.A. Fleury, and T.C. Damen, Phys. Rev. 154, 522 (1967).
15. R. Loudon, Adv. in Phys. 13, 423 (1964).
16. T.C. Damen, S.P.S. Porto, and B. Tell, Phys. Rev. 142, 570 (1966).
17. J.M. Worlock and S.P.S. Porto, Phys. Rev. Letters 15, 697 (1965).
18. J. Hurrell, T.C. Damen, S.P.S. Porto, and S. Mascarenhas, Phys. Letters 26A, 194 (1968).
19. C.H. Henry and J.J. Hopfield, Phys. Rev. Letters 15, 964 (1965).
20. S.P.S. Porto, B. Tell, and T.C. Damen, Phys. Rev. Letters 16, 450 (1966).
21. J.P. Scott, L.E. Cheesman, and S.P.S. Porto, Phys. Rev. 162, 834 (1967).
22. P.A. Fleury, S.P.S. Porto, L.E. Cheesman, and H.J. Guggenheim, Phys. Rev. Letters 17, 84 (1966).
23. P.A. Fleury, S.P.S. Porto, and R. Loudon, Phys. Rev. Letters 18, 658 (1967).
24. J.T. Hougen and S. Singh, Phys. Rev. Letters 10, 406 (1963); Proc. Roy. Soc. (London) A277, 193 (1964).
25. C.H. Henry, J.J. Hopfield, and L.C. Luther, Phys. Rev. Letters 17, 1178 (1966).
26. S. Ramsden and W. Davies, Phys. Rev. Letters 16, 303 (1966).
27. A. Mooradian and G.B. Wright, Phys. Rev. Letters 16, 999 (1966).
28. P.A. Wolff, Phys. Rev. Letters 16, 225 (1966).
29. R.E. Sluscher, C.K.N. Patel, and P.A. Fleury, Phys. Rev. Letters 18, 77 (1967).
30. J.M. Worlock and P.A. Fleury, Phys. Rev. Letters 19, 1176 (1967).

A-2: POLARITONS, RAMAN SCATTERING, ELECTRO-OPTIC EFFECT AND PARAMETRIC AMPLIFICATION

Rodney Loudon
Physics Department, Essex University
Colchester, England

ABSTRACT

The properties of long-wavelength lattice vibrations are reviewed, with particular reference to polar phonons. The interaction of such phonons with the electromagnetic field, leading to the formation of polariton modes, is discussed. The theory of polaritons in cubic and multi-atomic uniaxial crystals is developed and illustrated by calculations for gallium phosphide and quartz. The theory of first-order Raman scattering by these excitations is reviewed and formulae are presented for the intensity of scattering by a given polariton. The connections between this theory and that of the electro-optic effect are described. The related theory of stimulated Raman scattering and parametric amplification using polaritons is outlined and the gain of the polariton wave is computed for a simple case. The possible practical application as a far-infrared tunable radiation source is emphasised.

INTRODUCTION

We consider a range of optical phenomena whose theories are closely related and which all involve in some way the vibrations of a crystal lattice. The vibrations of interest have wavelengths long compared with interatomic spacing, where the short-range interatomic forces contribute a restoring force independent of wavelength. For a homopolar crystal vibration the frequency of the long-wavelength phonons is therefore constant. Determination of the phonon frequencies is valuable for lattice-dynamical studies but is not of interest for the effects considered here. More interesting effects occur for polar lattice vibrations, where there are long-range electric-dipole forces which influence the phonon frequency ω. In addition, polar lattice vibrations interact with the transverse electromagnetic waves and cause absorption and dispersion of infrared radiation. In crystals which have no centre of inversion symmetry the polar phonons may be active in first-order Raman scattering.

The electric-dipole forces have two main consequences which can be investigated directly by Raman scattering. In a 90° scattering experiment, vibrations having wave vector k of order $10^5 cm^{-1}$ are produced. The frequency ω is independent of $|k|$, but depends on the phonon polarisation for cubic crystals, and also on the direction of k relative to the crystal axes for non-cubic crystals. These properties were first recognized in Raman spectra and explained theoretically by Poulet[1].

For the more difficult scattering experiments close to the forward direction one can observe phonons of much smaller k. For $|k|$ of order ω/c, where c is the velocity of light, the interactions between the polar phonons and the transverse electromagnetic waves are strong, leading to the formation of coupled phonon-photon waves called polaritons. Polaritons in cubic crystals were first treated theoretically by Huang[2]. Because of the constancy of the short-range restoring forces, one can describe the phonon modes very simply by means of Lorentz oscillators. The resulting polariton frequency varies rapidly with wave vector in a way which can be observed directly by Raman scattering.

The theory of these effects is developed in the following two sections for cubic and for uniaxial crystals respectively. Formulae for scattering intensities are derived in the penultimate section, where the relation of polar-mode scattering to the theory of the electro-optic effect is considered. In the final section we show how the electro-optic and Raman coefficients control the stimulated Raman effect and parametric amplification close to a lattice resonance.

POLARITONS AND LONG-WAVELENGTH PHONONS IN A DIATOMIC CUBIC CRYSTAL

For a long-wavelength optic phonon the vibrational frequency under the influence of the short-range forces has a constant value ω_0. We suppose the relative displacement of the positive and negative charges to be Q and the associated dipole moment in the unit cell to be e^*Q. The classical Lorentz model gives a very good description of the absorption and dispersion in the vicinity of such a lattice resonance. The Lorentz equation of motion is

$$M(\ddot{Q} + \Gamma\dot{Q} + \omega_0^2 Q) = e^*E \tag{1}$$

where M is the reduced mass of the atoms in the unit cell and Γ is a damping constant. Here E can be taken to be the <u>macroscopic</u> field if appropriate local-field corrections are included[3, 4] in ω_0 and e^*.

According to Maxwell's equations,

$$\nabla \times \nabla \times E - (\omega^2/c^2)E = (\omega^2/c^2)4\pi P \tag{2}$$

where P is the polarisation. Assuming that Q, E and P have space and time dependence $\exp i(k.r - \omega t)$, (2) has the solution

$$E = -4\pi \frac{k(k.P) - (\omega^2/c^2)P}{k^2 - (\omega^2/c^2)} \tag{3}$$

There are contributions to P from the lattice and the electrons,

$$P = (Ne^*/V)Q + \{(\epsilon_\infty - 1)/4\pi\}E \tag{4}$$

where N is the number of unit cells in the crystal, V is the crystal volume, and ϵ_∞ is the electronic contribution to the dielectric constant.

For a cubic crystal, (1), (3) and (4) have two types of solution[2, 4]. Firstly, there are <u>transverse</u> modes for which k is perpendicular to P and

$$\frac{k^2c^2}{\omega^2} = 1 + \frac{4\pi P}{E} = \epsilon_\infty + \frac{S\omega_0^2}{\omega_0^2 - \omega^2 - i\omega\Gamma} = \epsilon \tag{5}$$

where ϵ is the complex dielectric constant and

$$S = 4\pi Ne^{*2}/MV\omega_0^2 \tag{6}$$

Conventionally one sets

$$kc/\omega = \epsilon^{1/2} = n + i\varkappa \tag{7}$$

where n and $\varkappa$ are the refractive index and extinction coefficient. For many crystals Γ is small and can be neglected for some applications. In this case (5) becomes a real relationship between k and ω. This is plotted in Fig. 1 for GaP using the parameters measured by Kleinman and Spitzer[5]. The transverse dispersion curve has two branches. The excitations in the region shown in the figure are mixtures of electromagnetic wave and mechanical vibration of the lattice. These excitations are called polaritons. Henry and Hopfield[6] have measured part of the polariton dispersion curve for GaP by Raman scattering. The polaritons in GaP have been fully discussed by Barker[7].

Secondly there are longitudinal solutions for which k and P are parallel and,

$$E = -4\pi P \tag{8}$$

With Γ still neglected the longitudinal vibration frequency ω_ℓ obtained from (1), (4) and (8) is given by

$$\omega_\ell^2 = \omega_0^2 \{1 + (S/\epsilon_\infty)\} \tag{9}$$

The longitudinal branch is also shown in Fig. 1.

For zero damping there is a stop-band for frequencies between ω_0 and ω_ℓ, where n vanishes and $\varkappa$ is large. In the presence of damping n is not quite zero within the stop-band but $\varkappa$ remains large. In spontaneous Raman scattering experiments one is interested in the propagating waves outside or on the edge of the stop-band. However, for stimulated Raman scattering, where polariton waves of high intensity may be built up, it is worth considering the possibility of exciting waves within the stop-band.

The energy density of the polariton wave is[8, 9]

$$W = (NM/2V)(|\dot{Q}|^2 + \omega_0^2 |Q|^2) + (1/8\pi)(\epsilon_\infty |E|^2 + |H|^2) \tag{10}$$

Using (1), (5) and (7) this can be manipulated into the form

$$W = (|E|^2/4\pi)\{(2\omega n\varkappa/\Gamma) + n^2\} \tag{11}$$

In a similar way the Poynting vector is

$$G = (cn/4\pi)|E|^2 \tag{12}$$

In terms of these equations it is possible to define the velocity v_E with which the energy in the polariton is transported through the lattice as[8]

$$v_E = \frac{G}{W} = \frac{c}{n + (2\omega\varkappa/\Gamma)} \tag{13}$$

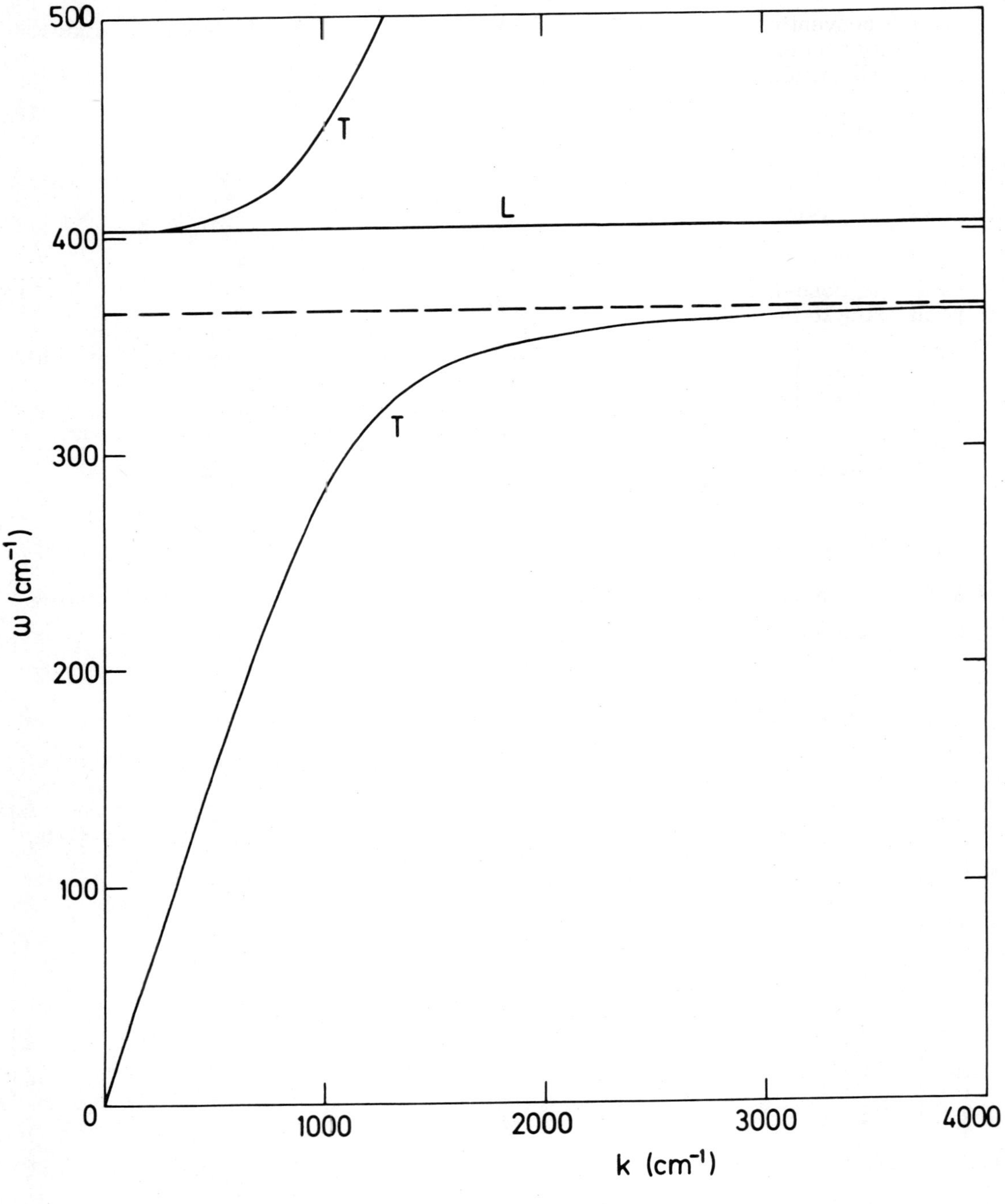

Fig. 1. The undamped transverse polariton dispersion curves (marked T) and the longitudinal branch (marked L) for GaP. The dashed curve shows the transverse lattice vibration frequency for the phonons at slightly larger wave vectors.

The energy velocity for GaP is plotted in Fig. 2. Outside the stop-band v_E becomes the same as the conventionally defined group velocity. However, within the stop-band the group velocity becomes a meaningless concept[9].

The electromagnetic field associated with the polariton is quantised by putting[10]

$$E = ig\eta\{\alpha_k \exp(ik.r) - \alpha_k^\dagger \exp(-ik.r)\} \tag{14}$$

where $\alpha_k^\dagger$ and α_k are creation and destruction operators for the polariton of wave vector k and η is a unit vector. The coefficient g is determined by the requirement

$$VW = \hbar\omega(\nu + 1/2), \tag{15}$$

where ν is the Bose-Einstein factor at the polariton frequency ω. Using (5), (7) and (11) and specialising to the case of zero damping

$$g^2 = \frac{2\pi\hbar\omega}{V}\left\{\epsilon_\infty + \frac{S\omega_o^4}{(\omega_o^2 - \omega^2)^2}\right\}^{-1} \tag{16}$$

The quantised lattice displacement Q can be obtained from (1), (14) and (16).

GENERALISATION TO MULTIATOMIC UNIAXIAL CRYSTALS

For a crystal having many atoms in the unit cell there are many modes of vibration and each of these can be described by a Lorentz equation of motion[3]. Let Q^m be the

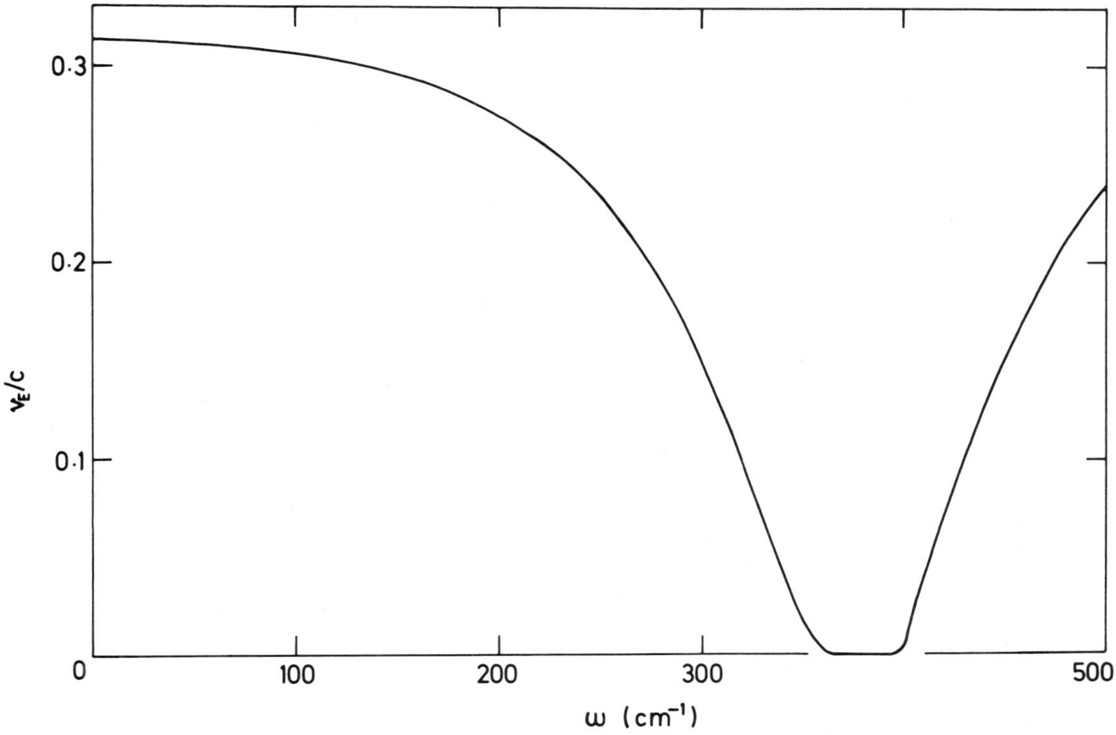

Fig. 2. The velocity of energy propagation for the polaritons in GaP as a function of frequency.

vibrational amplitude in the mth optic mode. For the polar modes Q^m will point along one of the principal axes of the dielectric constant tensor; if this axis is labelled j, then $Q_j^m = Q^m$ and the Lorentz equation for this mode is

$$M_m(\ddot{Q}_j^m + \Gamma_m \dot{Q}_j^m + \omega_m^2 Q_j^m) = e_m^* E_j \tag{17}$$

where E_j is the component of E along principal axis j, M_m is a suitably defined reduced mass, and the damping constant, natural frequency and effective charge now all depend on the particular normal mode under consideration.

The polarisation component is

$$P_j = \sum_m (Ne_m^*/V)Q_j^m + \{(\epsilon_{j\infty} - 1)/4\pi\}E_j \tag{18}$$

where $\epsilon_{j\infty}$ is the electronic contribution to the dielectric constant for principal axis j. The total dielectric constant for this axis is

$$\epsilon_j = \epsilon_{j\infty} + \sum_m^j \frac{S_m \omega_m^2}{\omega_m^2 - \omega^2 - i\omega\Gamma_m} \tag{19}$$

where the sum runs only over those modes m for which $Q_j^m = Q^m$ and the strength of the contribution of mode m is governed by the parameter

$$S_m = (4\pi N e_m^{*2}/M_m V \omega_m^2) \tag{20}$$

The damping constants Γ_m are often small and we ignore them for the remainder of this section.

For a uniaxial crystal[11] we denote the ordinary and extraordinary axes by symbols $\perp$ and $||$. Then

$$P_{||} = \{(\epsilon_{||} - 1)/4\pi\}E_{||} \tag{21}$$

$$P_{\perp} = \{(\epsilon_{\perp} - 1)/4\pi\}E_{\perp} \tag{22}$$

Eq. (3) remains valid for any crystal and can be combined with (21) and (22) to eliminate $E_{||}$, $E_{\perp}$, $P_{||}$ and $P_{\perp}$. There are two cases to consider:

P perpendicular to c-axis and to k

From (3),

$$E_{\perp} = \frac{4\pi P_{\perp}}{(kc/\omega)^2 - 1}, \quad E_{||} = 0, \tag{23}$$

and the dispersion relation for such waves is

$$(kc/\omega)^2 = \epsilon_{\perp} \tag{24}$$

These waves are the ordinary polaritons, being polarised perpendicular to the c-axis.

P in the plane containing the c-axis and k

Suppose that k makes an angle θ with the c-axis. In this case (3), (21) and (22) give

$$\frac{k^2c^2}{\omega^2} = \frac{\epsilon_{||}\epsilon_{\perp}}{\epsilon_{||}\cos^2\theta + \epsilon_{\perp}\sin^2\theta} \qquad (25)$$

This is the dispersion relation for the extraordinary polaritons. Except for $\theta = 0$ or $90°$ these polaritons do not have any simple polarisation. We note that (24) and (25) reduce to the cubic result for $\epsilon_{||} = \epsilon_{\perp} = \epsilon$.

The polariton dispersion relations for a uniaxial crystal are best appreciated by considering a particular example. Figs. 3, 4 and 5 show some data on the polariton curves for quartz. The ordinary dispersion relation (24) is independent of θ, but the extraordinary curves given by (25) are different for different values of θ. Figs. 3 and 4 show the calculated dispersion relations for $\theta = 0$ and $\theta = 90°$. The parameters required for the calculation are taken from the measurements of Spitzer and Kleinman[12] and of Russell and Bell[13]. The general properties of polariton dispersion curves in uniaxial crystals have been discussed by Loudon[11]. Note that the frequencies at k = 0 are independent of propagation direction and have the degeneracies predicted by group theory. At finite values of k the two-fold degeneracy of the E modes is lifted except for propagation parallel to the c-axis ($\theta = 0°$). Scott, Cheesman and Porto[14] have measured part of the polariton dispersion curves for quartz using small-angle Raman scattering.

For ordinary 90° Raman scattering the modes observed lie off the right-hand edges of Figs. 3 and 4 where

$$kc/\omega >> 1 \qquad (26)$$

The dispersion relations become independent of $|k|$

$$\omega = \omega_m \qquad (27)$$

for ordinary waves, where m refers to the ordinary vibrational modes, and

$$\epsilon_{||}\cos^2\theta + \epsilon_{\perp}\sin^2\theta = 0 \qquad (28)$$

for the extraordinary modes. Fig. 5 shows the calculated positions of the phonon frequencies as functions of θ.

We observe in Fig. 5 that two different types of behaviour occur for the extraordinary waves as k swings round from the c-axis to the a-axis. The wave can change its polarisation character from transverse to longitudinal or vice versa, as occurs for example with the 394-403 and 790-778 vibrations. Or the wave can display the same polarisation character at both ends of the range, as for example in the 450-495 and 1241-1237 vibrations. The reason for this is that there are two competing influences on the polarisation of a wave[11]. The electromagnetic forces try to make waves either longitudinal or transverse, and as we have seen they succeed completely in cubic crystals where they meet no opposition. However, in a uniaxial crystal the anisotropic short-range forces on atoms try to make the atoms move either parallel or perpendicular to the c-axis irrespective of the direction of k, at any rate for small wave vectors. The behaviour of the modes in Fig. 5 depends on which of these two constraining forces predominates. The ordinary waves are always polarised both transversely and perpendicular to the c-axis and no conflict arises. Unfortunately A_2 modes in quartz are not Raman active and it is not possible to measure the predicted curves of Fig. 5 completely by light-scattering experiments. The longitudinal mode frequencies at $\theta = 0°$ and $90°$ have previously been given by Elcombe[15] and by Scott and Porto[16] for $\theta = 90°$.

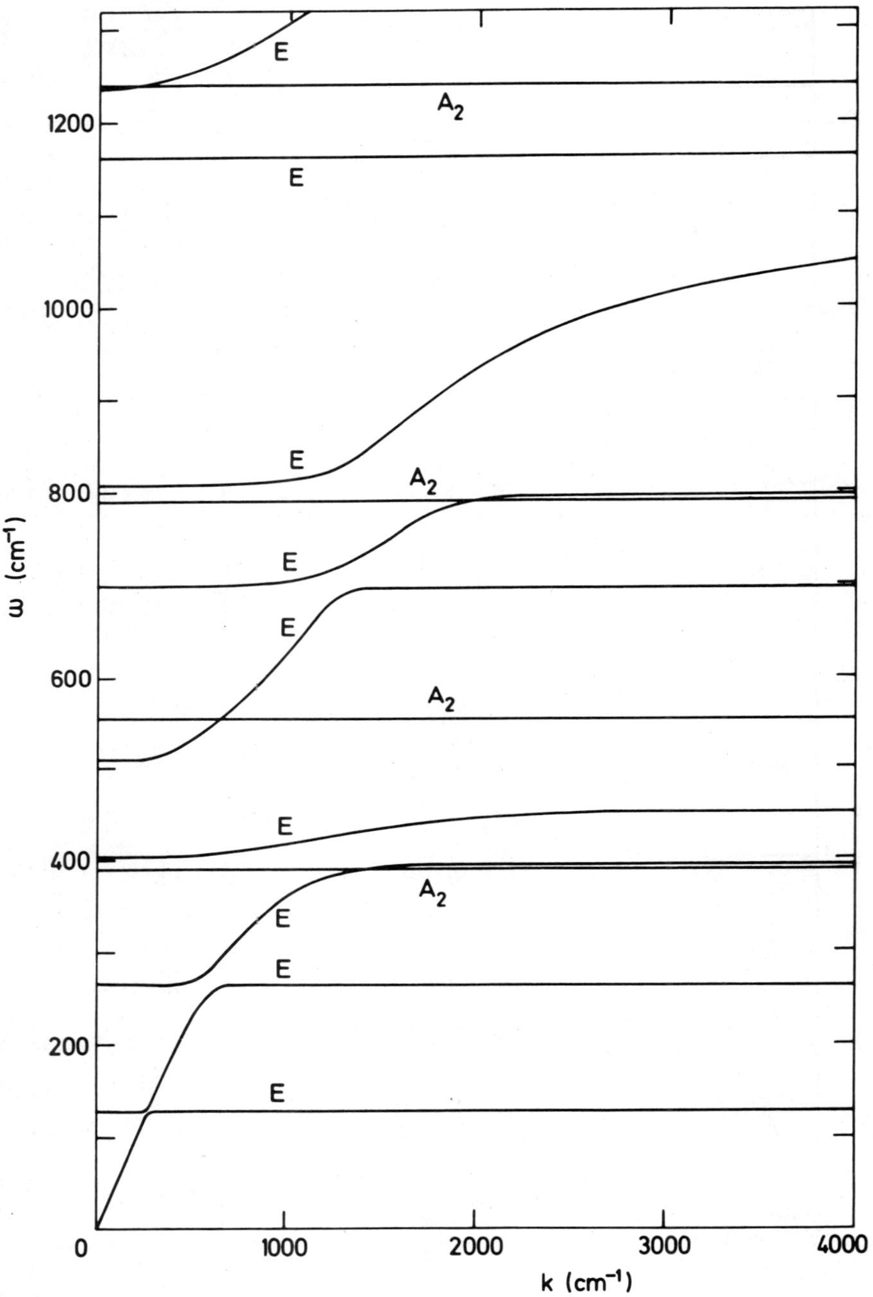

Fig. 3. Polariton dispersion curves for quartz. The direction of propagation is parallel to the c-axis. The modes labelled E are twofold degenerate transverse polaritons. The A_2 modes are longitudinal, being polarised parallel to c-axis.

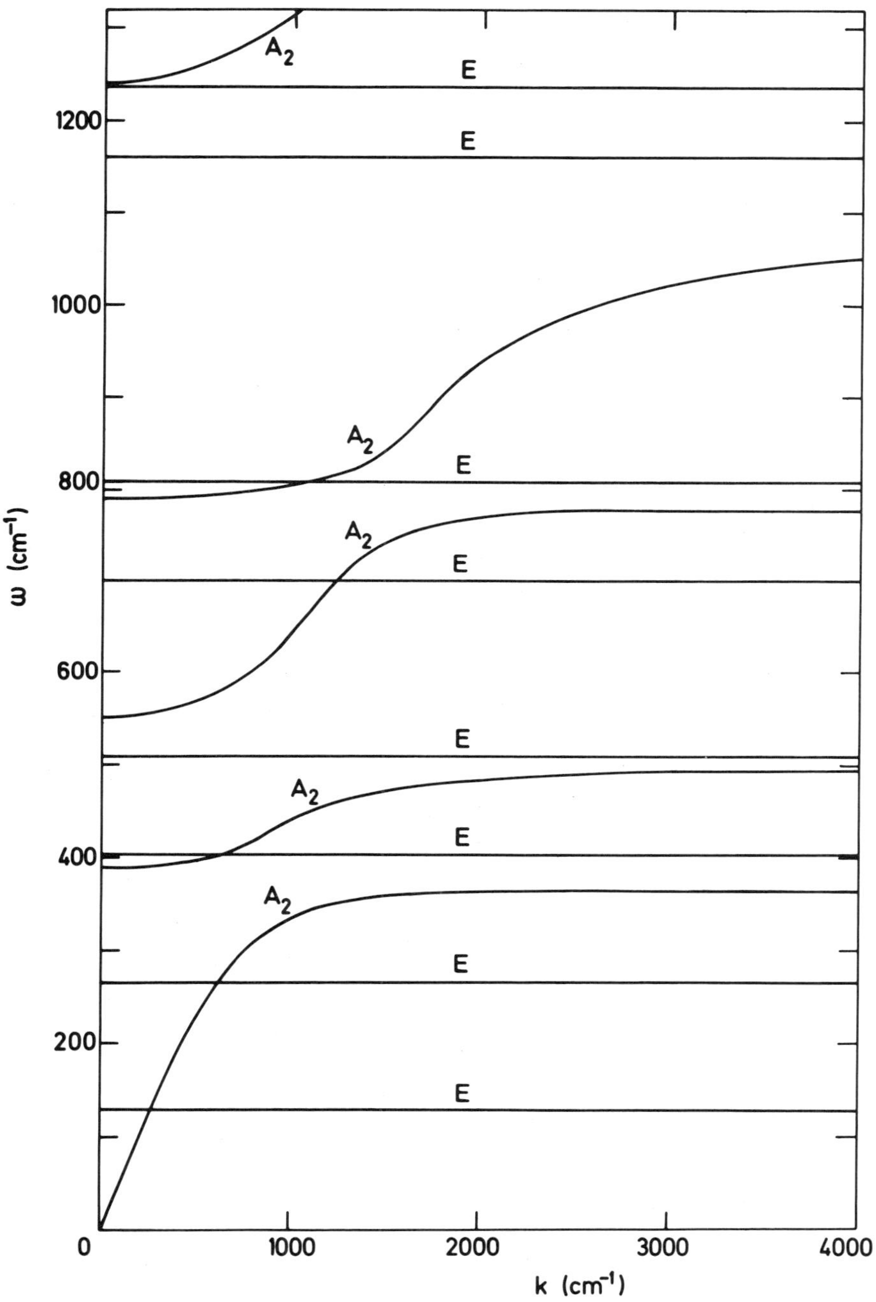

Fig. 4. Polariton dispersion curves for quartz for propagation perpendicular to the c-axis. The A_2 modes are transverse extraordinary polaritons. The E modes are non-degenerate longitudinal polaritons. Not included in the graph are the non-degenerate transverse ordinary polaritons; these have identical curves to those labelled E in Fig. 3.

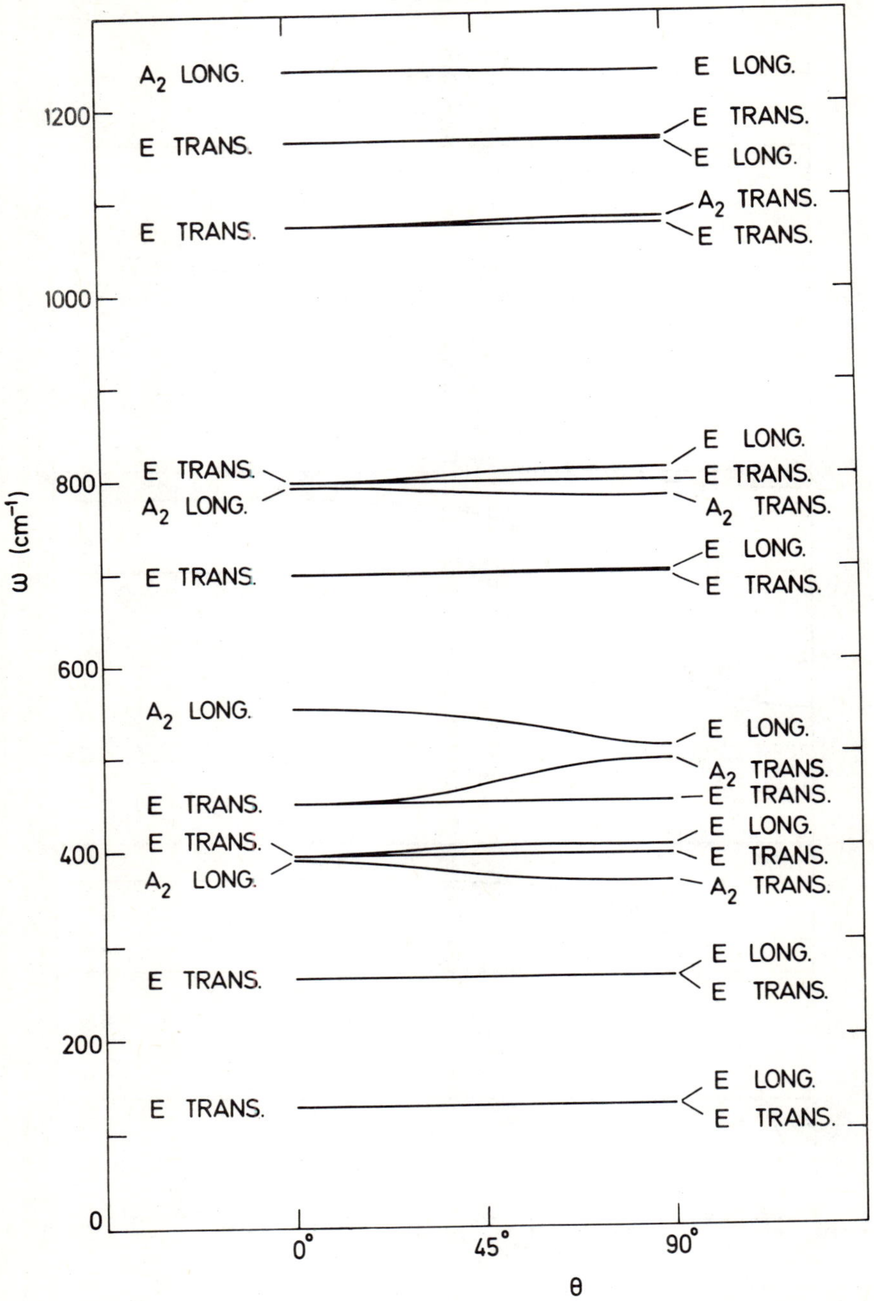

Fig. 5. Phonon frequencies in quartz in the region accessible to 90° first-order Raman scattering plotted against the angle of propagation relative to the c-axis.

For a uniaxial crystal the electric field quantisation is

$$E_j = ig_j \{\alpha_k \exp(ik.r) - \alpha_k^{\dagger} \exp(-ik.r)\} \tag{29}$$

similar to (14) but with g_j in general different for the components $||$ and $\perp$. The quantisation condition is

$$\hbar\omega = \frac{g_{||}^2 V}{2\pi} \left\{ \epsilon_{||\infty} + \sum_m^{||} \frac{S_m \omega_m^4}{(\omega_m^2 - \omega^2)^2} \right\} + \frac{g_{\perp}^2 V}{2\pi} \left\{ \epsilon_{\perp\infty} + \sum_m^{\perp} \frac{S_m \omega_m^4}{(\omega_m^2 - \omega^2)^2} \right\} \tag{30}$$

where

$$\frac{g_{||}}{g_{\perp}} = \frac{(\epsilon_{\perp} - 1)\,\epsilon_{||} \cos\theta}{(\epsilon_{||} - 1)\,\epsilon_{\perp} \sin\theta}$$

$$\text{or } -\frac{\epsilon_{\perp} \sin\theta}{\epsilon_{||} \cos\theta}$$

$$\text{or } 0 \tag{31}$$

for the two solutions of the extraordinary polariton dispersion relation, and for the ordinary polariton respectively.

The lattice displacement Q associated with the quantised polaritons now involves a linear superposition of the basic mode amplitudes Q^m. The coefficients in the sum are obtained from (17) with the Γ_m neglected and $\ddot{Q}_j^m$ set equal to $-\omega^2 Q_j^m$. The electric field components from (30) and (31) are substituted to complete the quantisation.

SPONTANEOUS RAMAN SCATTERING AND THE ELECTRO-OPTIC EFFECT

In any inelastic scattering experiment we have to consider three waves, an exciting light wave of frequency ω_p, usually obtained from a laser, a scattered light wave of frequency ω_s, and a wave of frequency ω excited in the crystal. For Stokes scattering,

$$\omega_p = \omega_s + \omega \tag{32}$$

Let the electric fields associated with the exciting and scattered waves be

$$E^p(r,t) = E^p \exp(ik_p . r - i\omega_p t) + c.c$$

$$E^s(r,t) = E^s \exp(ik_s . r - i\omega_s t) + c.c \tag{33}$$

The wave excited in the crystal is typified by a lattice displacement Q and an electric field E, which we now take in real form

$$Q(r,t) = Q \exp(ik.r - i\omega t) + c.c$$

$$E(r,t) = E \exp(ik.r - i\omega t) + c.c \tag{34}$$

where Q is a linear combination of the normal modes m with amplitude Q^m as discussed in the previous section.

The intensity of scattering is calculated as follows. The polarisation at frequency ω_s produced by mixing of the waves at frequencies ω_p and ω can be written,

$$P_h^s = \alpha_{hij}^m E_i^p Q_j^{m*} + \xi_{hij} E_i^p E_j^* \tag{35}$$

where

$$\alpha_{hij}^m = \partial \alpha_{hi} / \partial Q_j^{m*} \quad , \quad \xi_{hij} = \partial \alpha_{hi} / \partial E_j^* \tag{36}$$

the derivatives being evaluated at $Q_j^m = E_j = O$, and h, i and j represent principal axis components, repeated indices being summed. The linear polarisability tensor α_{hi} refers to the concentration N/V of unit cells in the crystal and is defined in such a way that it must be multiplied by the <u>macroscopic</u> field to give the polarisation; it is evaluated at frequency ω_p.

Now Q_j^m and E_j are related by (17) which we write

$$Q_j^m = \beta^m(\omega) E_j \quad ; \quad \beta^m(\omega) = \frac{e_m^*/M_m}{\omega_m^2 - \omega^2 - i\omega\Gamma_m} \tag{37}$$

Hence the non-linear polarisation is

$$P_h^s = \{\alpha_{hij}^m \beta^m(\omega)^* + \xi_{hij}\} E_i^p E_j^*$$

$$= 2\chi_{hij}^{(2)} (\omega_p - \omega) E_i^p E_j^* \tag{38}$$

in terms of the usual notation of non-linear optics[17, 18]. We note that α_{hi} is the polarisability at optical frequencies and is thus almost entirely due to the electronic states. We therefore expect α_{hij}^m and ξ_{hij} to vary slowly with ω, even as ω passes through lattice resonances, and there is a considerable advantage is dividing up $\chi_{hij}^{(2)}$ as in (38) since the main frequency dependence appears explicitly in the $\beta^m(\omega)$.

The scattered intensity is now obtained by calculating the energy radiated by the polarisation at frequency ω_s, taking the part of the quantised expression for E_j which corresponds to creation of an additional polariton. The result is conveniently expressed in terms of the attenuation length λ for the exciting beam due to scattering into unit solid angle by the process considered. A standard calculation (see for example ref. [4]) gives

$$1/\lambda = (\omega_p/c)^4 V(\nu + 1) \mid \eta_h^s \eta_i^p g_j \{\alpha_{hij}^m \beta^m(\omega)^* + \xi_{hij}\} \mid^2 \tag{39}$$

where η^s and η^p are unit vectors in the directions of E^s and E^p, g_j is defined in (29) - (31) and repeated indices are still summed.

We consider particular cases of (39) later in the section. It is of interest first to consider the relationship of the parameters introduced to the linear electro-optic coefficient r_{hij}. This is defined in terms of the coefficient in (38) for the case when ω is very small compared with the ω_m but is large enough not to excite sample acoustic resonances. We set $\omega = 0$, and the correspondence is

$$-(1/4\pi)\epsilon_{h\infty}\, r_{hij}\, \epsilon_{i\infty} = \alpha^m_{hij}\beta^m(O)^* + \xi_{hij} \tag{40}$$

This relation can be used to re-write the expression for the scattering length in a variety of forms. For example, ξ_{hij} can be eliminated to give

$$1/\lambda = (\omega_p/c)^4 V(\nu + 1)\,|\,\eta^s_h \eta^p_i g_j \{\alpha^m_{hij}\,[\beta^m(\omega)^* - \beta^m(O)^*] - (1/4\pi)\epsilon_{h\infty} r_{hij}\epsilon_{i\infty}\}|^2 \tag{41}$$

Note that as $\omega \to 0$ the term in α^m_{hij} goes out and the scattered intensity depends only on the electro-optic coefficient.

Relations similar to (40) and (41) have been derived by McGill and Yariv[19] and by Kaminow and Johnston[20]. The latter authors have used them to predict values of electro-optic coefficient using α^m_{hij} determined from 90° Raman scattering observations, $\beta^m(O)$ determined from infrared absorption, and the ξ_{hij} deduced from second harmonic generation coefficients. They find good agreement with values of the r_{hij} measured directly.

Returning now to the general Eq. (39) we note some simplifications which occur in particular cases. Consider 90° scattering by an excitation whose polarisation is purely transverse. For this case the polariton is in fact a purely vibrational excitation and its frequency must be equal to one of the normal mode frequencies, ω_m say. Then using (30) and (37), (39) becomes

$$\frac{1}{\lambda} = \frac{\hbar\omega_p^4\, V(\nu + 1)|\eta^s_h\, \eta^p_i\, \alpha^m_{hij}|^2}{2Nc^4 M_m \omega_m} \tag{42}$$

where j is the polarisation direction of normal mode m. There is no similar simplification in (39) for modes of any other polarisation.

For a cubic crystal having a single threefold degenerate normal mode of frequency ω_o, (39) becomes

$$\frac{1}{\lambda} = \frac{2\pi\hbar\omega\omega_p^4\,(\nu + 1)}{c^4[\epsilon_\infty(\omega_o^2 - \omega^2)^2 + S\omega_o^4]}\,|\eta^s_h \eta^p_i \eta_j\,[\alpha_{hij}(e^*/M) + \xi_{hij}\,(\omega_o^2 - \omega^2)]\,|^2 \tag{43}$$

This formula has been given previously by Loudon[10] and the special case of 90° scattering was treated some time ago[21]. It should be noted that the quantity z in these references corresponds only to the electronic contribution and does not therefore represent the whole of the electro-optic coefficient.

For GaP, Faust and Henry[22] have measured the relative intensities of the LO and TO Raman lines in 90° Raman scattering and use them to determine

$$\alpha_{hij}\, e^*/M\, \xi_{hij}\, \omega_o^2 = -0.53 \qquad (44)$$

It is therefore possible to plot the scattered intensity as a function of the polariton Raman shift for GaP and this is done in Fig. 6. Henry and Hopfield[6] have observed the polariton in GaP over the range 300cm^{-1} to ω_o by first-order scattering. They select different polaritons by varying the scattering angle from O° up to large angles, thus changing the polariton wave vector which satisfies the phase-matching requirement $k = k_p - k_s$.

Other experimental observations of polaritons by Raman scattering have been made on ZnO[23], ZnS, ZnSe[24] and $LiNbO_3$[25]. Finally we mention an important new development by Scott, Fleury and Worlock[26], who have shown by experiments on $SrTiO_3$ and $KTaO_3$ that polaritons in centro-symmetric crystals can be studied by electric-field induced Raman scattering.

STIMULATED RAMAN EFFECT AND PARAMETRIC AMPLIFICATION

It is now well known that, as the intensity of the exciting light in a Raman experiment is increased, a point is reached at which stimulated emission of scattered photons may take over from the spontaneous emission of the normal Raman experiment considered so far. An excellent review of the stimulated Raman effect has recently been published[27]. In this section we consider the theory as it applies to scattering by polaritons.

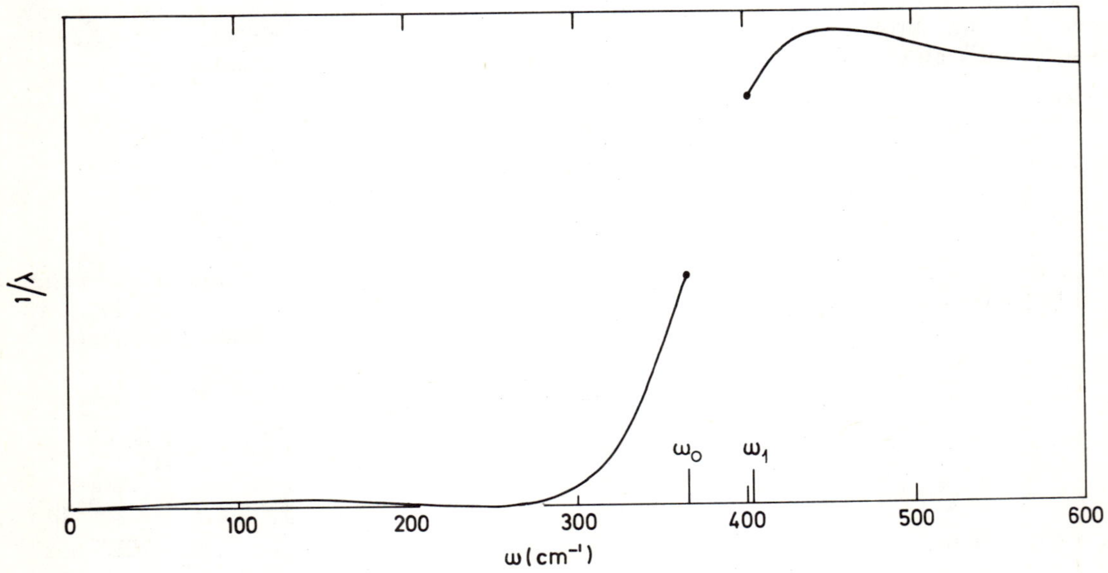

Fig. 6. Raman intensity in arbitrary units for GaP as a function of polariton frequency. The curve was constructed using data obtained by measuring the relative Raman intensity at the transverse and longitudinal optic frequencies[22] as indicated by the two blobs. Note that theoretically the longitudinal mode and the transverse polariton at frequency ω_ℓ should have the same Raman strength.

The particularly interesting feature of the polariton in this connection is the fact that part of its energy is carried in the form of an electromagnetic wave. Thus in any experiment which produces a beam of polaritons inside a crystal, the electromagnetic component of the energy can emerge as far-infrared radiation. Further, since the polariton produced in a scattering process can have its frequency varied by adjustment of the experimental geometry, the experiment leads in principle to a tunable source of radiation. This possibility was pointed out some time ago by Loudon[28] but no practical source has so far been constructed. However, it has been shown by Tannenwald and Weinberg[29] that the lowest frequency E-symmetry vibration in quartz can be produced by stimulated Raman scattering, and this crystal appears promising for obtaining observable emission in the far-infrared (see also a review by De Martini[30]).

It should be mentioned that, independently of the work discussed above, Chiao, Garmire and Townes[31] proposed a similar experiment using stimulated Raman scattering by molecular vibrational levels. Infrared radiation has recently been successfully generated in this way[32].

For the stimulated polariton scattering it is necessary to treat three travelling coupled waves; the incident pump beam, which we assume to pass through the crystal with negligible attenuation, the scattered light beam, and the polariton beam. The equations which describe the experiment are generalisations of (2) and (17) with some non-linear terms inserted on the right. Assuming that the polarisation vectors lie along principal axes and using (18), the equations are

$$M_m(\ddot{Q}_j^m + \Gamma_m \dot{Q}_j^m + \omega_m^2 Q_j^m) = e_m^* E_j + \alpha_{hij}^m E_h^{s*} E_i^p (V/N) \tag{45}$$

$$\nabla \times \nabla \times E_j - (\omega^2 \epsilon_{j\infty}/c^2) E_j = (4\pi\omega^2/c^2)\{(Ne_m^*/V) Q_j^m + \xi_{hij} E_h^{s*} E_i^p\} \tag{46}$$

The scattered wave is described by a similar equation

$$\nabla \times \nabla \times E_h^s - (\omega_s n_s/c)^2 E_h^s = (4\pi\omega_s^2/c^2)\{\alpha_{hij}^m E_i^p Q_j^{m*} + \xi_{hij} E_i^p E_j^*\} \tag{47}$$

where n_s is the refractive index for the scattered light. A summation over modes m is implied on the right-hand sides of (46) and (47).

The solution of the problem represented by Eqs. (45) to (47) has been attempted by a number of authors. The treatment is quite straightforward when the polariton frequency ω lies outside the stop-bands and linewidths of the infrared absorption lines. For such polariton frequencies the infrared dielectric constant at ω can be taken real (i.e. $n >> \varkappa$ in (7)). The problem can be solved in this region by time-dependent perturbation theory using a quantised electromagnetic field method. This was done by Loudon[28], who presented the results in a somewhat obscure form, and more recently by Henry and Garrett[33], who obtain essentially the same result but in a much more useful form.

A more straightforward approach is to solve the coupled wave Eqs. (45) - (47) by substitution of assumed plane-wave solutions. This method was used by Shen[34], who produced solutions of great generality which are difficult to apply to possible experimental situations. Independently Butcher, Loudon and McLean[35] used the same approach but made approximations tantamount to assuming $n >> \varkappa$, which produces slight errors within the stop-band region. Most recently Henry and Garrett[33] have given solutions valid in a general way for all values of n and $\varkappa$, but being interested in a particular experimental system they do not consider explicitly the situation within the stop-band. The particular difficulty within the stop-band occurs in choosing the experimental beam geometry which gives the largest gain for the stimulated polariton beam. Although it is in principle possible to produce a growing wave within the stop-band the gain is relatively small in this region and we do not consider this aspect in detail.

To show the form of solution obtained, define

$$\alpha_R = \frac{2\pi\omega\omega_s V|E_i^p|^2}{cn_s} \sum_m \frac{(\alpha_{hij}^m)^2\Gamma_m/M_m}{(\omega_m^2 - \omega^2)^2 + \omega^2\Gamma_m^2} \tag{48}$$

$$\alpha_P = \frac{2\pi|E_i^p|}{c}\left[\frac{\omega\omega_s}{n_s(n + i\varkappa)}\right]^{1/2}\left[\sum_m \frac{e_m^* \alpha_{hij}^m/M_m}{\omega_m^2 - \omega^2 - i\omega\Gamma_m} + \xi_{hij}\right] \tag{49}$$

We note that for a lattice vibration which is Raman-active but has zero dipole moment ($e_m^* = 0$), the system of equations reduces to that for the ordinary stimulated Raman effect[17, 27] with a gain constant α_R. On the other hand, for polariton frequencies well away from the ω_m (where α_R, $\omega\varkappa/c << \alpha_P$), the equations describe parametric amplification[17, 27] of the polariton and scattered beams with gain α_P.

For polariton frequencies closer to the ω_m, where the linear attenuation exceeds the parametric coefficient α_P, but where n is still larger than $\varkappa$, the gain has the simple form

$$\alpha = \mathrm{Re}(c\alpha_P^2/\omega\varkappa) \tag{50}$$

For frequencies where $\varkappa$ is greater than n the solution is more complicated and one must evaluate carefully the phase-matching conditions required for maximum growth of the polariton wave. We do not write down the equations here but illustrate the form of solution obtained in Fig. 7, which shows the gain constant α as a function of ω for GaP. The curve was constructed using the numerical data of Kleinman and Spitzer[5] and Faust and Henry[22]. It was assumed that the inequality $\omega\varkappa/c >> \alpha_P$ held throughout the range. The curve is similar to that of Henry and Garrett[33] except in the stop-band region where these authors assume an unphysical phase-matching condition. Note the very small gain in this frequency region.

For a crystal having a single lattice resonance like GaP, (50) can be written

$$\alpha = \frac{2\pi\omega_s|E^p|^2 V}{cn_s\omega\Gamma NM}\{\alpha_{hij} + (M\xi_{hij}/e^*)(\omega_o^2 - \omega^2)\}^2 \tag{51}$$

where (5), (6), (7) and (49) have been used. This result has been given in a variety of forms by all the authors who have published solutions to the problem. The zero in α for GaP close to $250\mathrm{cm}^{-1}$ is due to cancellation of the two terms in the bracket of (51). Phase matching for different frequencies is achieved by varying the angle between exciting and scattered beams.

Also included in Fig. 7 is a plot of the gain which would be obtained in a crystal with Raman scattering strength given by α_{hij} of equal magnitude to GaP but for which the optic mode was infrared inactive as in a crystal with a centre of inversion symmetry, or for the A_1 vibrations of quartz. Here α_P vanishes and the problem is that of the ordinary stimulated Raman effect[17, 27].

In a uniaxial crystal like quartz with many lattice vibrations one could construct curves for the growth constant as a function of polariton frequency similar to Fig. 7. Because the polariton spectrum changes with the propagation angle θ and since one has the choices of ordinary and extraordinary polarisation for the exciting and scattered light, there are more phase matching possibilities for a uniaxial crystal, and a greater spread of polariton frequencies can usually be covered. This may be a valuable feature in constructing a tunable far-infrared source.

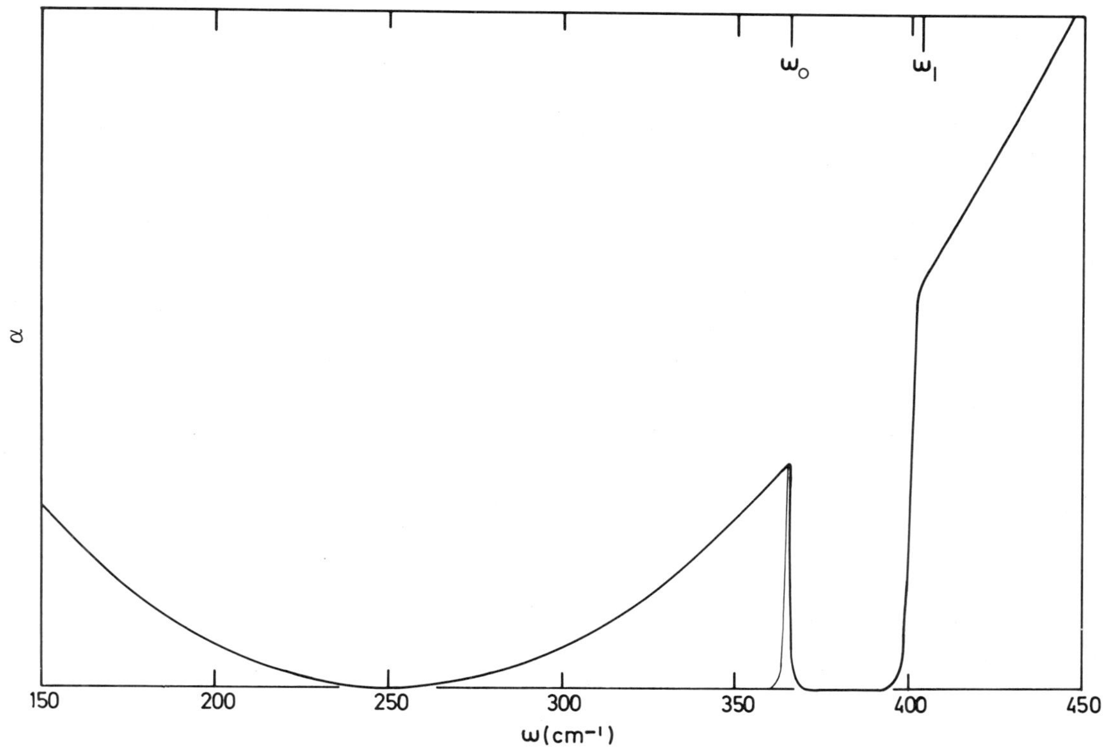

Fig. 7. Parametric gain of the polariton wave in GaP as a function of frequency. For a given pump frequency phase matching can be achieved for only part of the frequency range[33]. The Lorentzian curve centered at ω_0 shows the contribution of the stimulated Raman gain α_R.

REFERENCES

1. H. Poulet, Ann. Phys. Paris 10, 908 (1955).
2. K. Huang, Proc. Roy. Soc. A208, 352 (1951).
3. A.S. Barker, Jr., Phys. Rev. 136, A1290 (1964).
4. M. Born and K. Huang, "Dynamical Theory of Crystal Lattices," Clarendon Press, Oxford. 1954.
5. D.A. Kleinman and W.G. Spitzer, Phys. Rev. 118, 110 (1960).
6. C.H. Henry and J.J. Hopfield, Phys. Rev. Letters 15, 964 (1965).
7. A.S. Barker, Jr., Phys. Rev. 165, 917 (1968).
8. R. Loudon, R.R.E. Memorandum 2155 (1965).
9. L. Brillouin, "Wave Propagation and Group Velocity," Academic Press, New York, 1960.
10. R. Loudon, Proceedings of the International School in Physics "E. Fermi" Course 42, 1967, Academic Press, New York (to be published).
11. R. Loudon, Advan. Phys. 13, 423 (1964).
12. W.G. Spitzer and D.A. Kleinman, Phys. Rev. 121, 1324 (1961).
13. E.E. Russell and E.E. Bell, J. Opt. Soc. Amer. 57, 341 (1967).
14. J.F. Scott, L.E. Cheesman and S.P.S. Porto, Phys. Rev. 162, 834 (1967).
15. M.M. Elcombe, Proc. Phys. Soc. 91, 947 (1967).
16. J.F. Scott and S.P.S. Porto, Phys. Rev. 161, 903 (1967).
17. N. Bloembergen, "Nonlinear Optics," W.A. Benjamin Inc., New York, 1965.
18. P.N. Butcher, "Nonlinear Optical Phenomena," Ohio State Univ., Columbus, 1965.
19. T.C. McGill and A. Yariv, Phys. Letters 25A, 411 (1967).
20. I.P. Kaminow and W.D. Johnston, Jr., Phys. Rev. 160, 519 (1967).
21. R. Loudon, Proc. Roy. Soc. A275, 223 (1963).
22. W.L. Faust and C.H. Henry, Phys. Rev. Letters 17, 1265 (1966).
23. S.P.S. Porto, B. Tell and T.C. Damen, Phys. Rev. Letters 16, 450 (1966).
24. S. Ushioda, A. Pinczuk, W. Taylor and E. Burstein, "Proc. of the 1967 International Conference on II-VI Semiconducting Compounds," Benjamin, New York, 1967.
25. H.E. Puthoff, R.H. Pantell, B.G. Huth and M.A. Chacon, J. Appl. Phys. 39, 2144 (1968).
26. J.F. Scott, P.A. Fleury and J.M. Worlock, Phys. Rev. (to be published).
27. N. Bloembergen, Am. J. Phys. 35, 989 (1967).
28. R. Loudon, Proc. Phys. Soc. 82, 393 (1963).
29. P.E. Tannenwald and D.L. Weinberg, I.E.E.E. J. Quantum Electron. QE-3, 334 (1967).
30. F. De Martini, J. Appl. Phys. 37, 4503 (1966).
31. R.Y. Chiao, E. Garmire and C.H. Townes, Proceedings of the International School in Physics "E. Fermi", Course 31, 1963, Academic Press, New York, 1964.
32. J.P. Biscar, R. Braunstein and S. Gratch, Phys. Rev. Letters 21, 195 (1968).
33. C.H. Henry and C.G.B. Garrett, Phys. Rev. 171, 1058 (1968).
34. Y.R. Shen, Phys. Rev. 138, A1741 (1965).
35. P.N. Butcher, R. Loudon and T.P. McLean, Proc. Phys. Soc. 85, 565 (1965).

A-3: RAMAN SCATTERING BY POLARITONS IN POLYATOMIC CRYSTALS*

E. Burstein, S. Ushioda, and A. Pinczuk
Laboratory for Research on the Structure of Matter and Department of Physics, University of Pennsylvania, Philadelphia, Pennsylvania
and
J. F. Scott
Bell Telephone Laboratories, Incorporated
Holmdel, New Jersey

ABSTRACT

Polaritons participate in first order Raman scattering via atomic displacement and macroscopic electric field induced changes in the electric susceptibility, $\chi_{\pi}^{(1)}(\vec{u})$ and $\chi_{\pi}^{(1)}(E)$, respectively. In polyatomic crystals, polaritons correspond to photons coupled to all of the $q \approx 0$ IR active TO phonons. $\chi_{\pi}^{(1)}(\vec{u})$ is accordingly expressed in terms of the atomic displacement vector $\vec{u}_j(\omega_{\pi})$ and the atomic displacement susceptibility tensor a_j of the $q \approx 0$ TO phonons. $\chi_{\pi}^{(1)}(E)$ is expressed in terms of the macroscopic electric field of the polariton mode $\vec{E}_T(\omega_{\pi})$, and the macroscopic electro-optic coefficient, b, of the crystal. $\vec{u}_j(\omega_{\pi})$ is expressed in terms of $\vec{E}_T(\omega_{\pi})$ by means of the coupled equations of motion and $\vec{E}_T(\omega_{\pi})$ is obtained from the expressions for the energy density of electromagnetic radiation in a dispersive medium and the frequency dependent dielectric constant $\epsilon(\omega)$. The use of polariton scattering spectra to obtain ω vs q polariton dispersion curves and $\epsilon(\omega)$ is illustrated for the A_1 symmetry polaritons in tetragonal $BaTiO_3$.

INTRODUCTION

As a result of the interaction of EM radiation with the electric dipole excitations of a dielectric medium (e. g. TO phonons, excitons, free carrier cyclotron excitations, etc.)

*Research supported in part by the U.S. Army Research Office - Durham.

the propagating modes correspond to coupled photon-electric dipole excitation modes called polaritons. In crystals lacking a center of inversion the polaritons participate in first order Raman (R) scattering[1, 2], frequency mixing, [3] and light diffraction[4] via electric dipole excitation and via macroscopic electric field induced changes in the electric susceptibility[5].

Following the first observations of R scattering by phonon-polaritons e.g., coupled photon-TO phonon modes in GaP[1] and ZnO[2], the investigation of R scattering by polaritons has had a number of important applications. For example, data on Raman scattering by polaritons in GaP have been used to obtain polariton dispersion curves and thereby to obtain the contribution of the TO phonons to the dielectric constant[6]. Data on the relative scattering intensities of LO phonons, TO phonons and polaritons have been used to obtain information about the relative magnitudes and relative signs of the atomic displacement and the electro-optic contributions to the Raman scattering tensor (RST) in zincblende type crystals[7]. More recently, R scattering by polaritons was used to establish the frequency of the infrared active $q \approx 0$ TO phonons of A_1 symmetry in tetragonal $BaTiO_3$, over which there has been some controversy, and to obtain polariton dispersion curves and a value for the low frequency dielectric constant along the c-(ferroelectric-) axis[8].

The theory of the R scattering efficiency of phonon-polaritons has been given for crystals (such as zincblende and wurtzite crystals) having only a single set of infrared (IR) active $q \approx 0$ TO phonons[5, 9]. In the present paper we extend the theory to Raman scattering by polaritons in (polyatomic) crystals having several sets of IR active $q \approx 0$ TO phonons in which the polaritons correspond to photons coupled to all the IR active TO phonons. We also present a discussion of the ω vs q dispersion curves of A_1 symmetry polaritons in tetragonal $BaTiO_3$ and the frequency dependent dielectric constant which are derived from (small and large angle) R scattering data.

THE POLARITON MODES IN POLYATOMIC CRYSTALS

The wave vector dependent frequencies of the polariton modes $\omega_\pi(q)$ in a polyatomic crystal (assumed optically isotropic) are given by the dispersion relation[10]

$$q^2c^2 = \omega_\pi^2\,\epsilon(\omega) = \omega_\pi^2\left\{\epsilon_o(\omega) + \sum_j \frac{4\pi\beta_j\,\omega_{oj}^2}{\omega_{oj}^2 - \omega_\pi^2 - i\,\omega_\pi\,\gamma_j}\right\} \tag{2.1}$$

$\epsilon(\omega)$ is the frequency dependent complex dielectric constant; $\epsilon_o(\omega)$ is the dielectric constant in the absence of the photon-TO phonon coupling; β_j is the zero frequency contribution to the dielectric susceptibility from the jth IR active TO phonon; ω_{oj} is the frequency of the $q \approx 0$ j type TO phonon (in the absence of the photon-TO phonon coupling); and γ_j is the damping constant of the TO phonons assumed for simplicity to be independent of frequency. The parameter β_j which measures the strength of the photon-TO phonon coupling can also be expressed in terms of the dynamic (effective) ionic charge, e_j[11], and the reduced mass, m_j as follows[11]:

$$\beta_j = \frac{Ne_j^2}{Vm_j\omega_{oj}^2} \tag{2.2}$$

where N/V is the number of unit cell per unit volume. The e_j[11] is defined by the relation

$$\vec{M}_j(\omega_{oj}) = \left(\frac{\partial M}{\partial u_j}\right)_{E_T} \vec{u}_j(\omega_{oj}) = e_j \vec{u}_j(\omega_{oj}) \tag{2.3}$$

where $\vec{M}_j$ is the electric moment which is set up by the atomic displacements, $\vec{u}_j$, of the j type TO phonon, and $\vec{E}_T$ is the transverse electric field[11].

The polariton dispersion curves, $\omega_{\pi n}(q)$, in the limit $\gamma_j \to 0$, and the corresponding LO phonon branches are shown in Fig. 1 for a crystal having three sets of IR active TO phonons. We designate the lowest frequency polariton branches by $\omega_{\pi 1}(q)$, the next highest by $\omega_{\pi 2}(q)$, etc and we similarly designate the lowest LO phonon branch by $\omega_{\lambda 1}(q)$ etc.

We note from the dispersion relation for polaritons (Eq. (2.1)) that the frequencies of the polariton modes at $q = 0$, $\omega_{\pi n}(q = 0)$, are equal to the frequencies of the long wavelength LO phonon modes, $\omega_{\lambda j}$, since the latter correspond to frequencies at which $\epsilon(\omega) = 0$.

The TO phonon (mechanical) character of the polariton modes, which is measured by $u_j^2(\omega_\pi)$ and the photon (electromagnetic) character which is measured by $E_T^2(\omega_\pi)$, are functions of $\omega_\pi(q)$. They can be expressed in terms of the phonon strengths, $\mathcal{S}_j(\omega_\pi)$, and the photon strengths $\mathcal{S}_E(\omega_\pi)$ of the coupled modes which are defined by

$$u_j^2(\omega_\pi) = \mathcal{S}_j(\omega_\pi)\, u_j^2(\omega_{oj}) = \mathcal{S}_j(\omega_\pi)\left(\frac{\hbar}{NVm_j\omega_{oj}}\right) \tag{2.4}$$

$$E_T^2(\omega_\pi) = \mathcal{S}_E(\omega_\pi)\, E_{To}^2(\omega_\pi) = \mathcal{S}_E(\omega_\pi)\left(\frac{4\pi\hbar\omega_\pi}{\epsilon_0 V}\right) \tag{2.5}$$

where $\vec{u}_j(\omega_{oj})$ is the atomic displacement amplitude of the j type TO phonons at ω_{oj}, and $\vec{E}_{To}(\omega_\pi)$ is the EM electric field amplitude of the photon in the absence of photon-TO phonon coupling. The $\vec{u}_j(\omega_\pi)$ can also be expressed in terms of $\vec{E}_T(\omega_\pi)$ by means of the relation

$$\vec{u}_j(\omega_\pi) = \frac{e_j \vec{E}_T(\omega_\pi)}{m_j(\omega_{oj}^2 - \omega^2 - i\omega\gamma_j)} \tag{2.6}$$

which is obtained from the coupled photon-TO phonon equations of motion. We note (in the limit $\gamma_j \to 0$), that in the case of polariton modes having $\omega_\pi \approx \omega_{oj}$, $\vec{E}_T(\omega_\pi)$ is zero and, correspondingly, $\mathcal{S}_E(\omega_j)$ is zero and $\mathcal{S}_j(\omega_{oj})$ is unity. We also note that the relative sign of $\vec{u}_j(\omega_\pi)$ and $\vec{E}_T(\omega_\pi)$ is determined by the sign of e_j[11] and by the sign of $(\omega_{oj}^2 - \omega_\pi^2)$.

The polariton modes are damped waves whose damping constants are determined by the damping constants and phonon strengths of the TO phonons involved. In the case of frequency independent TO phonon damping constants, the damping of the polariton modes will peak at the frequencies of the TO phonons, ω_{oj}, and will be small at frequencies

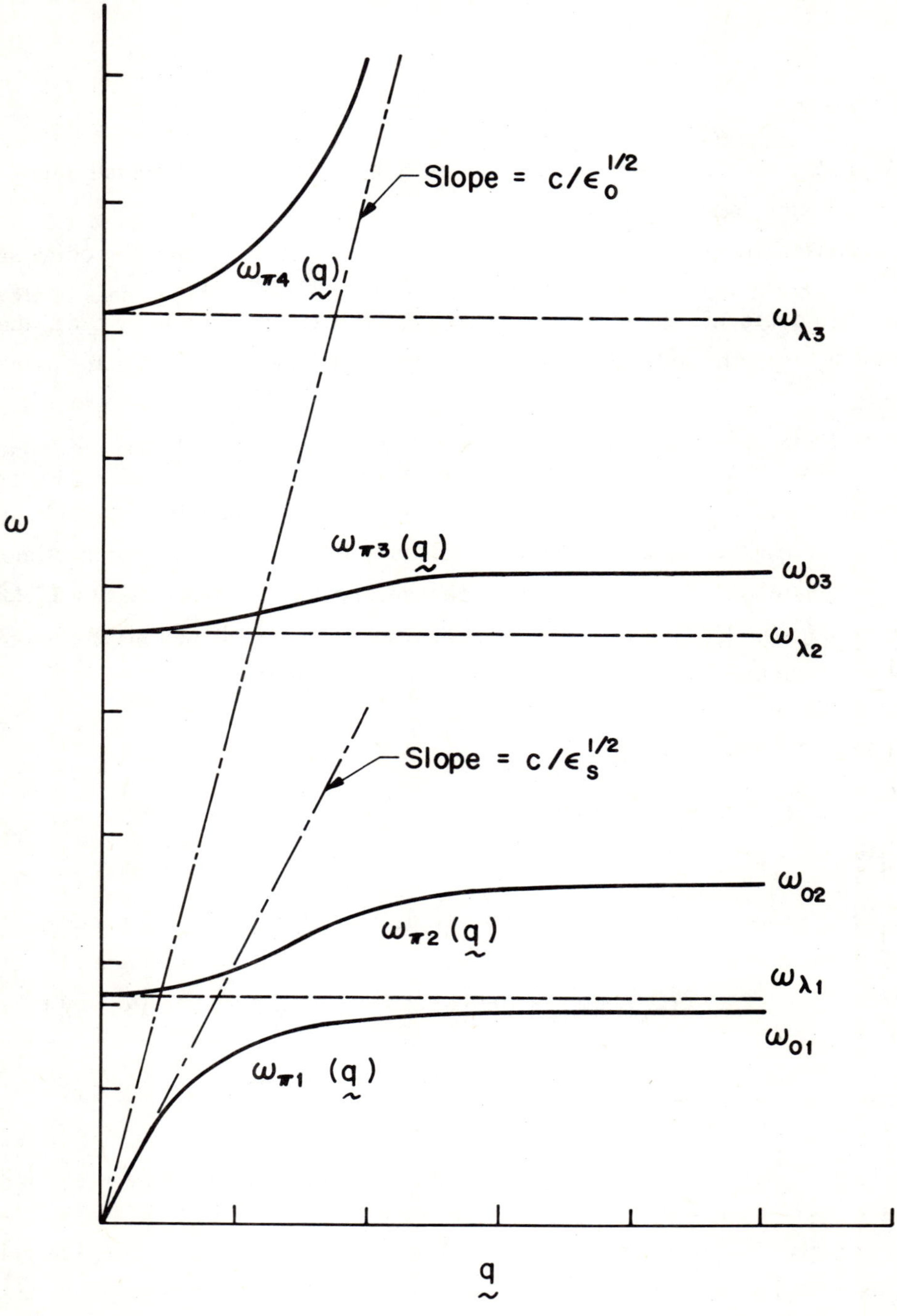

Fig. 1. The form of the polariton dispersion curves, in the limit $\gamma_j \to 0$, for cubic crystals having three IR active TO phonon (doubly degenerate) branches.

where the photon strength is large. Furthermore the damping of the TO phonons, apart from introducing an imaginary part to $\epsilon(\omega)$, sets an upper limit to the real part of $\epsilon(\omega)$ at $\omega_\pi = \omega_{o1}, \omega_{o2}, \ldots$ which in turn sets an upper limit to the real part of the wave vector of the photons which can couple with the TO phonons to form the polariton modes in the various polariton branches[12]. The real part of the polariton wave vector generally dominates the imaginary part in the frequency regions where $\epsilon(\omega)$ is positive. However, the imaginary part is generally larger than the real part, and the polariton modes are non-propagating in the frequency regions where $\epsilon(\omega)$ is negative. In actual practice, the anharmonic coupling of the TO phonons to other phonons leads to damping constants which are frequency dependent[13] and thereby to more complicated damping constants for the polariton modes.

THE RAMAN SCATTERING EFFICIENCY OF POLARITONS

The Stokes R scattering efficiency (per unit volume of crystal per unit solid angle) of a polariton mode $\omega_\pi(q)$ is given by[14]

$$S_\pi(\omega_\pi) = \frac{\omega_s^4 L}{2c^4} \mid \hat{e}_s(k_s) \cdot \chi_\pi^{(1)}(\omega_\pi) \cdot \hat{e}_o(k_o) \mid^2 (n_\pi + 1) \tag{3.1}$$

where $\omega_s = \omega_o - \omega_\pi(q)$ is the frequency of the scattered radiation; ω_o is the frequency of the incident radiation; $\vec{k}_o$, $\vec{k}_s$, and $\hat{e}_o$, $\hat{e}_s$ are the wave vectors and polarization vectors respectively of the incident radiation; $\vec{k}_o - \vec{k}_s$ is the "scattering wave vector" which determines the wave vector q and thereby the frequency, $\omega_\pi(q)$, of the polariton modes involved in the scattering; and $\chi_\pi^{(1)}(\omega_\pi)$, the RST, corresponds to the polariton induced first order change in the electric susceptibility. As a result of damping, the wave vector of the propagating polariton modes has an imaginary part. This results in a decrease in the scattering coherence length and, thereby, to a decrease in the scattering efficiency[3, 15]. In frequency regions where $\epsilon(\omega)$ is negative and the wave vector is predominantly imaginary, the polariton modes are non-propagating and, correspondingly, the scattering length and scattering efficiency are very small.

The RST of the polariton modes can be expressed in the form

$$\chi_\pi^{(1)}(\omega_\pi) = \left(\frac{\partial \chi(\omega_o)}{\partial u_\pi}\right)_{E_T} \vec{u}_\pi(\omega_\pi) + \left(\frac{\partial \chi(\omega_o)}{\partial E_T}\right)_{u_\pi} \vec{E}_T(\omega_\pi)$$

$$= a_\pi(\omega_o)\, \vec{u}_\pi(\omega_T) + b(\omega_o)\, \vec{E}_T(\omega_\pi) \tag{3.2}$$

where $\chi(\omega_o)$ is the electronic polarizability per unit volume. The coefficient $(\partial\chi/\partial\vec{u}_\pi(\omega_\pi))_{\vec{E}} = a_\pi(\omega_o)$ is the atomic displacement, or deformation potential susceptibility tensor of the polariton mode which depends on the distribution of the phonon strengths of the TO phonons involved and, therefore, on ω_π, and $(\partial\chi/\partial\vec{E}_\pi(\omega_\pi))_{\vec{u}} = b(\omega_o)$ is the electro-optic tensor which is, essentially, independent of ω_π. The first term of Eq. (3.2) represents the contribution of the phonon (mechanical) part of the polariton mode and the second term represents the contribution of the photon (electromagnetic) part of the polariton mode.

Since $a_\pi(\omega_o)$ is a function of ω_π and, therefore, has a different value for each polariton mode, $\omega_\pi(q)$, we take a different point of view, and treat the R scattering by the polariton modes as a superposition of the scattering by the different TO phonon parts of the coupled modes. Accordingly, we write the RST in terms of $\vec{u}_j(\omega_\pi)$ and a_j of the TO phonons involved as follows:

$$\chi_\pi^{(1)}(\omega_\pi) = \sum_j \left(\frac{\partial \chi}{\partial u_j}\right)_{E_T} \vec{u}_j(\omega_\pi) + b\,\vec{E}_T(\omega_\pi) \tag{3.3}$$

On introducing the expression of $\vec{u}_j(\omega_\pi)$ in terms of $\vec{E}_T(\omega_\pi)$ Eq. (2.6), in the limit $\gamma_j \to 0$, into Eq. (3.5), we obtain $\chi_\pi^{(1)}(\omega_\pi)$ in terms of $\vec{E}_T(\omega_\pi) = \mathcal{S}_E^{1/2}(\omega_\pi)\,\vec{E}_{TO}(\omega_\pi)$,

$$\chi_\pi^{(1)}(\omega_\pi) = \left\{\sum_j a_j \frac{e_j}{m_j(\omega_{oj}^2 - \omega_\pi^2)} + b\right\}\left(\frac{4\pi\hbar\omega_\pi}{\epsilon_o V}\,\mathcal{S}_E(\omega_\pi)\right)^{1/2}\hat{d}_E\,, \tag{3.4}$$

or alternatively, in terms of $u_j(\omega_\pi) = \mathcal{S}_j^{1/2}(\omega_\pi)\,u_j(\omega_{oj})_j$

$$\chi_\pi^{(1)}(\omega_\pi) = \sum_j \left\{a_j + \frac{b}{r}\,\frac{m_j(\omega_{oj}^2 - \omega^2)}{e_j}\right\}\left(\frac{\hbar}{NVm_j\omega_{oj}}\,\mathcal{S}_j(\omega_\pi)\right)^{1/2}\hat{d}_j \tag{3.5}$$

where $\hat{d}_E$ and $\hat{d}_j$, are the polarization vectors of $\vec{E}_T$ and $\vec{u}_j$, and r is the number of TO phonons that are coupled to the photon. When ω_π is equal to ω_{oj}, the frequency of one of the $q \approx 0$ TO phonons, $E_T(\omega_\pi = \omega_{oj})$ is zero and, therefore, $\mathcal{S}_E(\omega_\pi = \omega_{oj})$ is zero. The phonon strengths of the other TO phonons are also equal to zero and that of the type j TO phonon is equal to unity. The RST of the polariton mode $\omega_\pi(q) \approx \omega_{oj})$ is accordingly given by

$$\chi_\pi^{(1)}(\omega_\pi = \omega_{oj}) = a_j\,\vec{u}_j(\omega_{oj}) = \chi_j^{(1)}(\omega_{oj}) \tag{3.6}$$

We note that although m_j, the reduced mass of the type j TO phonons is a useful parameter for characterizing the optical phonons, it is not experimentally accessible in the case of polyatomic crystals. (It is even difficult to obtain m_j theoretically since it is sensitive to the force constant model used). For that matter, e_j, [11] the dynamic ionic charge of the TO phonons is also not experimentally accessible. The parameters which are determined from infrared (lattice vibration) data are ω_{oj}, $\omega_{\lambda j}$, and $\beta_j = Ne_j^2/m_j\omega_{oj}^2$. [11] We therefore combine a_j and $1/m_j^{1/2}$ into a single parameter $a_j' = a_j/m_j^{1/2}$ and write the expressions for $\chi_j^{(1)}(\omega_j)$ and $\chi_\pi^{(1)}(\omega_\pi)$ as follows:

$$\chi_j^{(1)}(\omega_{oj}) = a_j\left(\frac{\hbar}{NVm_j\omega_{oj}}\right)^{1/2}\hat{d}_j = a_j'\left(\frac{\hbar}{NVm_j\omega_{oj}}\right)^{1/2}\hat{d}_j \tag{3.7}$$

$$\chi_{\pi}^{(1)}(\omega_{\pi}) = \left\{ \sum_j a_j^{\cdot} \left(\frac{e_j}{|e_j|}\right)\left(\frac{\beta_j \omega_{oj}^2 V}{N(\omega_{oj}^2 - \omega_{\pi}^2)}\right)^{1/2} + b \right\} \left(\frac{4\pi\hbar\omega_{\pi}}{V\epsilon_o} \mathcal{S}_E(\omega_{\pi})\right)^{1/2} \hat{d}_E \quad (3.8)$$

where the factor $e_j/|e_j|$[11] is included in Eq. (3.9) to take into account the sign of the dynamic ionic charge of the type j TO phonons.

To complete the picture we note that the RST of the $q \approx 0$ LO phonons, $\chi_{\lambda}^{(1)}(\omega_{\lambda j})$ is given by

$$\chi_{\lambda}^{(1)}(\omega_{\lambda}) = \sum_j \left(\frac{\partial \chi}{\partial u_j}\right)_{E_L} \vec{u}_j(\omega_{\lambda j}) + b\, \vec{E}_L(\omega_{\lambda j}) \quad (3.9)$$

Since $\omega_{\lambda j}$ is equal to $\omega_{\pi n}(q = 0)$, it follows (from the fact that the shift in the frequency of the type j LO phonons from ω_{oj} to $\omega_{\lambda j}$ is due to the macroscopic field) that $\vec{E}_L(\omega_{\lambda j})$ is equal to $\vec{E}_{\pi}(\omega_{\pi} = \omega_{\lambda j})$. Therefore the expression for the RST for the LO phonons at $\omega_{\lambda j}$ is the same, apart from the polarization vectors, as that for the polariton modes (at $q = 0$) having the same frequency, i.e., $\chi_{\lambda}^{(1)}(\omega_{\lambda j}) = \chi_{\pi}^{(1)}(\omega_{\pi} = \omega_{\lambda j})$.

Expressions for $\mathcal{S}_j(\omega_{\pi})$ and $\mathcal{S}_E(\omega_{\pi})$ can be derived by the procedures used to obtain the phonon and plasmon strengths of coupled plasmon-LO phonon modes[15]. They can be obtained more simple and directly from the expression for the energy density of EM radiation in a dispersive medium and from the relation between $\vec{u}_j(\omega_{\pi})$ and $\vec{E}_T(\omega_{\pi})$ Eq. (2.6) which comes from the coupled photon-TO phonon equations of motion[5, 9].

The energy density of electromagnetic radiation in a dispersive medium is given by[16].

$$\langle W \rangle = \frac{c\,\epsilon(\omega)^{1/2} \langle E_T^2(\omega_{\pi})\rangle}{v_g(\omega_{\pi})\, 4\pi} = \left(\frac{\bar{n}_{\pi} + 1/2}{V}\right) \hbar\omega_{\pi} \quad (3.10)$$

where $v_g(\omega_{\pi}) = (\partial\omega/\partial q)_{\omega_{\pi}}$ is the group velocity of the polariton mode; $v_p(\omega_{\pi}) = \omega/q = c/\epsilon^{1/2}(\omega_{\pi})$ is the phase velocity of the polariton mode; and $\bar{n}_{\pi}/V$ is the average number of polaritons per unit volume. On solving for $\langle E_T^2(\omega_{\pi})\rangle$ per polariton mode per unit volume we obtain

$$\langle E_T^2(\omega_{\pi})\rangle = \frac{v_g(\omega_{\pi})\, 4\pi\hbar\omega_{\pi}}{\epsilon(\omega_{\pi})^{1/2}\, c} = \frac{v_g(\omega_{\pi})\,\epsilon_o}{v_p(\omega_{\pi})\,\epsilon(\omega)} \langle E_{To}^2(\omega_{\pi})\rangle \quad (3.11)$$

The group velocity of the polariton mode is obtained from the dispersion relation $q^2 c^2 = \omega^2 \epsilon(\omega)$. It is (in the limit $\gamma_j \to 0$) given by

$$v_g(\omega_{\pi}) = \frac{v_p(\omega_{\pi})\,\epsilon(\omega)}{\epsilon_o} \left\{ 1 + \sum_j \frac{4\pi\beta_j \omega_{oj}^4}{\epsilon_o(\omega_{oj}^2 - \omega_{\pi}^2)^2} \right\}^{-1} \quad (3.12)$$

On introducing the expression for $v_g(\omega_\pi)$ into Eq. (3.15), we obtain

$$<E_T^2(\omega_\pi)> = \left\{ 1 + \sum \frac{4\pi\beta_i\omega_{oj}^4}{\epsilon_0(\omega_{oj}^2 - \omega_\pi^2)^2} \right\}^{-1} <E_{To}^2(\omega_\pi)> \tag{3.13}$$

$\mathcal{S}_E(\omega_T)$ is accordingly given by

$$\mathcal{S}_E(\omega_\pi) = \frac{v_g(\omega_\pi)\epsilon_0}{v_p(\omega_\pi)\epsilon(\omega)} = \left\{ 1 + \sum \frac{4\pi\beta_j\omega_{oj}^4}{\epsilon_0(\omega_{oj}^2 - \omega_\pi^2)^2} \right\}^{-1} \tag{3.14}$$

$<u_j^2(\omega_\pi)>$ is obtained from Eq. (3.18), using the relation between $\vec{u}_j(\omega_\pi)$ and $\vec{E}_T(\omega_\pi)$ given in Eq. (2.6), as follows:

$$< u_j^2(\omega_\pi)> = <u_j^2(\omega_{oj})> \frac{4\pi\beta_j\omega_{oj}^3\omega_\pi}{(\omega_{oj}^2 - \omega_\pi^2)^2} \mathcal{S}_E(\omega_\pi) \tag{3.15}$$

$\mathcal{S}_j(\omega_\pi)$ is accordingly given by

$$\mathcal{S}_j(\omega_\pi) = \frac{4\pi\beta_j\omega_{oj}^3\omega_\pi}{\epsilon_0(\omega_{oj}^2 - \omega_\pi^2)^2} \mathcal{S}_E(\omega_\pi) \tag{3.16}$$

The magnitude of $\chi_\pi^{(1)}(\omega_\pi)$ which determines the scattering efficiency of the polariton mode, ω_π, can be seen to depend on $a_j' = a_j/m_j^{1/2}$, b and $e_j/|e_j|$[11] which may have positive or negative signs, and on β_j and ω_{oj} which are positive quantities. The scattering efficiency of the coupled modes of a given branch, $\omega_{\pi n}(q)$, may be expected to vary with frequency, and even exhibit maxima and minima, as a result of the relative signs of a_j', b and $e_j/|e_j|$[11] and of the variation of $\mathcal{S}_j(\omega_\pi)$ and $\mathcal{S}_E(\omega_\pi)$ with frequency. (See for example the paper by Scott and Ushioda, [17] this conference, on the scattering intensities of the multiphonon-polariton modes in quarts.

RAMAN SCATTERING BY POLARITIONS IN TETRAGONAL $BaTiO_3$

The room temperature R spectra of A_1 symmetry TO phonons in tetragonal $BaTiO_3$ exhibit three peaks, a narrow peak at 170 cm^{-1}, and two relatively broad peaks at 270 cm^{-1} and 520 cm^{-1}[18, 19, 20]. On the basis of observed shifts in the frequency of all three bands with direction of the scattering wave vector, $q = k_o - k_s$, relative to the c axis, Pinczuk et al[18] have concluded that the bands were first order. DiDimenico et al, [20] on the other hand, consider the two broad bands at 270 cm^{-1} and 520 cm^{-1} to originate in higher order scattering processes. More recent data on forward R scattering by polaritons of A_1 symmetry[8] show that the two bands at 270 cm^{-1} and 520 cm^{-1} shift to lower frequencies and approach the frequencies of the A_1 symmetry LO phonons at 185 cm^{-1} and 475 cm^{-1} as the scattering wave vector is decreased to small values, in precisely the manner expected for scattering by polariton modes. These results demonstrate conclusively that the two broad bands correspond to IR active $q \approx 0$ TO phonons and are therefore first order bands.

The forward scattering experiments were carried out on a single domain crystal, with the c axis in the plane of the surface, which showed "complete" extinction between crossed polarizers. The spectra were obtained using a double grating spectrometer designed by A. Filler, and a He-Ne laser operating at 6328 Å, with the incident and scattered light polarized along the axis. On the basis of the polarization selection rules[14] only A_1 symmetry modes contribute in this configuration. The forward scattering measurements were carried out for scattering angles $\theta(k_o, k_s)$ ranging from 0.6° to 8.5° with k_o along the x axis and the scattering wave vector, $q = k_o - k_s$, along either the y or z axis.

The polariton ($x(zz)x + \Delta y$) spectra obtained for $\theta(k_o, k_s) > 8°$ correspond to q values for which $\omega_\pi \approx \omega_{oj}$ and, therefore, involve scattering by TO phonons, i.e., $S_E(\omega_{oj}) \approx 0$ and $S_j(\omega_{oj}) \approx 1$. Thus the spectrum for $\theta(k_o, k_s) = 8.5° \pm 0.5°$ shown in Fig. 2 is essentially identical to that obtained for TO phonons in the back scattering configuration ($\theta(k_o, k_s) = 180°$). The spectrum exhibits a narrow peak at 170 cm^{-1} and two broad peaks at 270 cm^{-1} and 520 cm^{-1}. The corresponding LO phonon ($x(zz)x + \Delta z$) spectrum also shown in Fig. 2 exhibits three narrow peaks at 185 cm^{-1}, 475 cm^{-1} and 725 cm^{-1} and two broad peaks at 265 cm^{-1} and 515 cm^{-1} (labeled α and β). Data obtained with the scattering vector inclined at an angle to the y and z axes, which involve scattering by coupled A_1 and E modes have shown that the three narrow bands in the LO phonon spectra shift in frequency with change in inclination angle, whereas the two broad bands do not shift. One may therefore conclude that the narrow bands involve scattering by LO phonons and are first order bands, and that the broad bands are second order bands. The frequencies of the TO and LO phonons derived from these spectra are summarized in Table I.

The polariton scattering spectra for various small scattering angles are shown in Fig. 3. One sees that the three bands located at 170 cm^{-1}, 270 cm^{-1}, and 520 cm^{-1} in the large angle scattering spectra show a dependence of frequency on scattering angle, as expected for polaritons, and therefore correspond to one phonon processes. One also sees two broad bands at 270 cm^{-1} and 515 cm^{-1} labeled α and β which do not shift with scattering angle. These are, in part, due to back scattering by TO phonons and in part to second order scattering processes. It appears therefore that the broad bands at 270 cm^{-1} and 520 cm^{-1} which are observed in the large angle (TO phonon) scattering spectra involve a superposition of first and second order bands.

In Fig. 4 we show the ω vs q curves for the A_1 symmetry polaritons which are derived from the experimental data using the relation between q, ω, and θ (for q and ω expressed in cm^{-1}, and $v_g(\omega_o) \approx v_p(\omega_o) = c/\epsilon_{co}^{1/2}$)

$$q^2 = \omega^2 \epsilon_{co} + \omega_o^2 (\omega_o - \omega) \epsilon_{co} \theta^2 \tag{4.1}$$

to determine the scattering wave vector. On extrapolating the low frequency polariton branch, ω_π (q), to the origin, we obtain from the slope at $\omega \approx 0$, a value of 38 ± 6 for ϵ_{cs}, the low frequency dielectric constant along the c axis.

A value for ϵ_{cs} can also be obtained from the Lyddane, Sachs, Teller (LST) relation

$$\frac{\epsilon_{cs}}{\epsilon_{co}} = \prod_{i=1}^{i=3} \left(\frac{\omega_{\lambda j}}{\omega_{oj}} \right)^2 \tag{4.2}$$

using the frequencies of the TO and LO phonons, ω_{oj} and $\omega_{\lambda j}$ respectively, (Table I) and $\epsilon_{co} = 5.07$. We obtain by this procedure a value of 37 ± 5, in agreement with the value determined from the ω vs q dispersion data, which again serves to confirm the assignment of the two broad bands in the TO phonon spectra as first order bands. The values

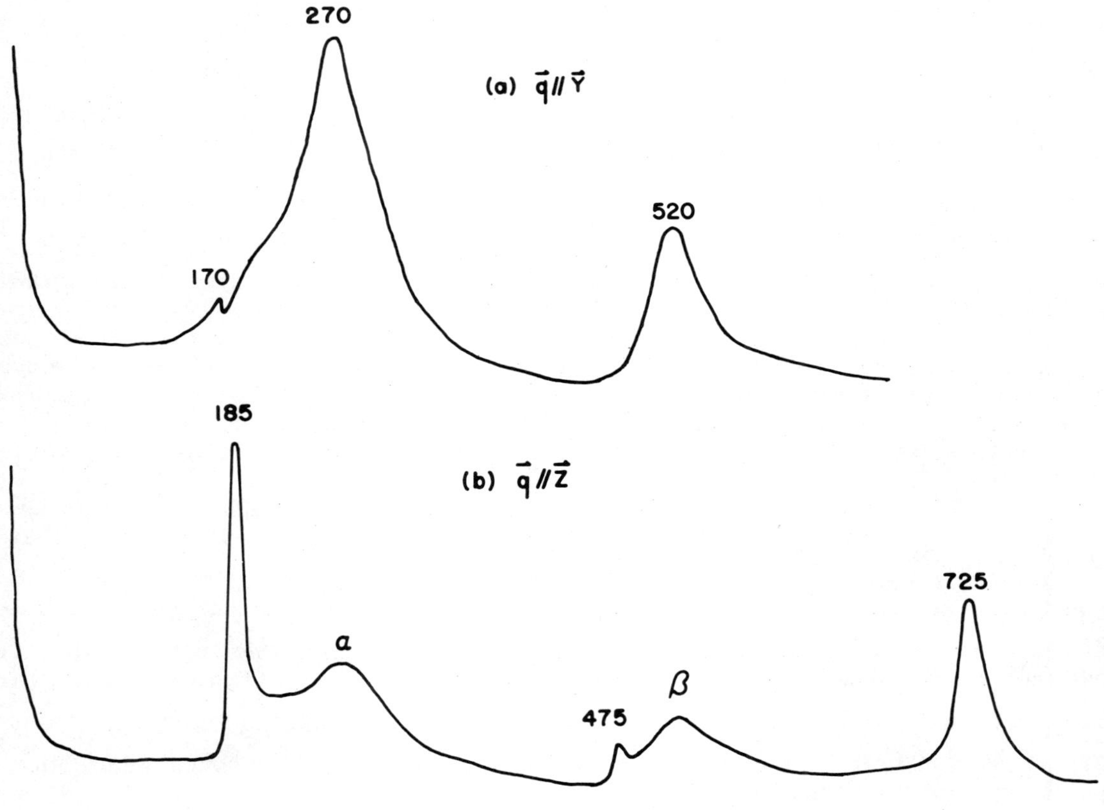

Fig. 2. Room temperature R spectra of A_1 symmetry TO phonons (x(zz)x + Δy) and LO phonons (x(zz)x + Δz) in tetragonal $BaTiO_3$ for $\theta(k_s, k_o) = 8.6^o \pm 0.5^o$.

TABLE I

Optical Phonons of A_1 Symmetry in $BaTiO_3$

ω_{oj}(TO)	$\omega_{\lambda j}$(LO)	$4\pi\beta_j$
cm^{-1}	cm^{-1}	
170	185	8
270	475	22
520	725	1

for ϵ_{cs} which we obtain are considerably lower than the value ϵ_{cs} = 80 determined by electrical measurements at microwave frequencies. [21] The reason for the large difference in the ϵ_{cs} values obtained by the two experimental methods is not yet known.

The contributions of the j type TO phonons to the low frequency dielectric constant, $4\pi\beta_j$, can be evaluated using the expression for the real part of the dielectric constant given in Eq. (2.1) in which the damping constants γ_j are assumed to be frequency independent.

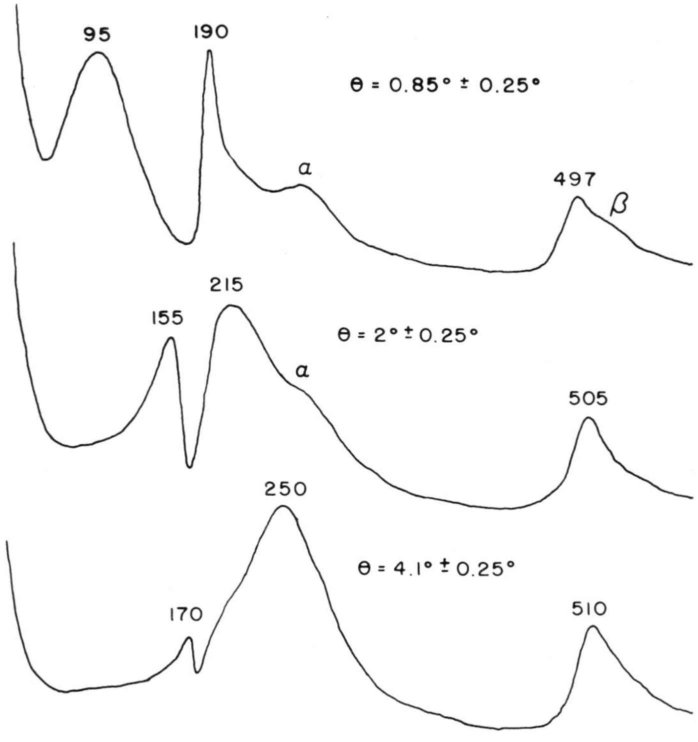

Fig. 3. Room temperature forward spectra of A_1 symmetry polaritons in tetragonal $BaTiO_3$.

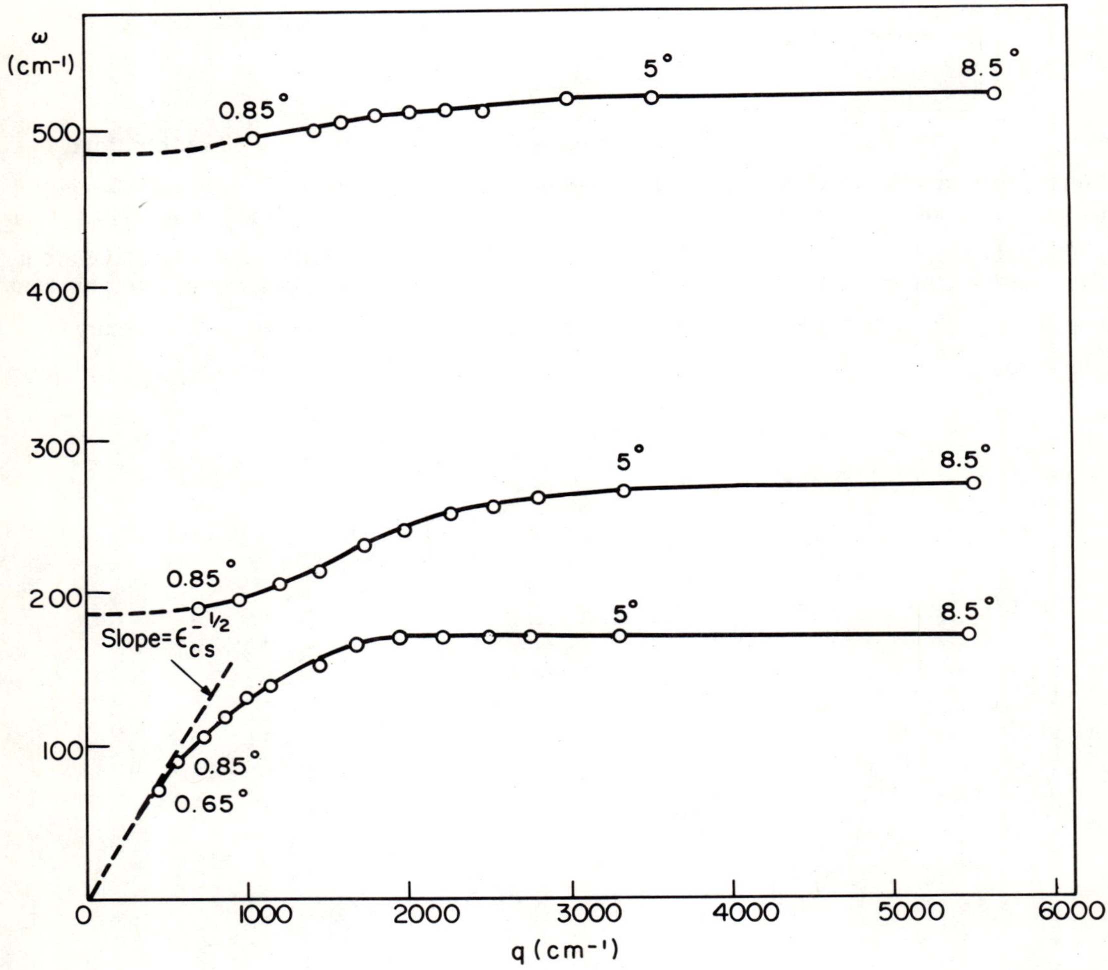

Fig. 4. Room temperature ω vs q data for polaritons of A_1 symmetry in tetragonal $BaTiO_3$ derived from forward scattering measurements.

On introducing the values for the TO phonon frequencies ω_{oj} and assuming that the Re $\epsilon(\omega)$ goes to zero at the LO phonon frequencies, $\omega_{\lambda j}$, an assumption which is reasonable since the LO phonons exhibit well defined with relatively small damping, we obtain values for $4\pi\beta_j$, given in Table I, which are insensitive to the values of γ_j. It is of interest to note that the mode at 270 cm^{-1} makes the largest contribution to ϵ_{cs}.

A satisfactory fit of the experimental polariton dispersion (ω vs q) curves cannot be obtained using an expression for the dispersion relation involving frequency independent damping factors. Efforts to fit the data using frequency dependent damping factors (and frequency dependent TO phonon frequencies) are now under way. As shown by Maradudin and Fein[13], the frequency dependent damping factors for a given optical phonon are determined by the strengths of the anharmonic coupling parameter and by the combined density of states of the phonons coupled by anharmonicity. The neutron scattering data for $BaTiO_3$[22] indicate that a large combined density of two phonon states occurs at the Brillouin zone boundary in the region between 230 cm^{-1} and 280 cm^{-1} due to acoustical phonons. A large combined density of states also occurs in the region of 520 cm^{-1} due to processes which involve two TO phonons. This can readily account for the large width of the first order bands.

ACKNOWLEDGEMENT

We wish to acknowledge valuable discussions with A. A. Maradudin and with D. L. Mills.

REFERENCES

1. C.H. Henry and J.J. Hopfield, Phys. Rev. Letters 15, 964 (1965).
2. S.P.S. Porto, B. Tell, and T.C. Damen, Phys. Rev. Letters 16, 450 (1966).
3. W.L. Faust and C.H. Henry, Phys. Rev. Letters 17, 1265 (1966).
4. E. Burstein, A. Pinczuk and Y. Sawada, Bull. Am. Phys. Soc. 12, No. 3, 421 (1967).
5. R. Loudon, Proc. Phys. Soc. 82, 393 (1963). This paper was the first to derive expressions for the Raman scattering efficiency of polaritons. See also the paper by Loudon, A-2, this conference.
6. A.S. Barker, Jr., Phys. Rev. 165, 917 (1968).
7. S. Ushioda, A. Pinczuk, W. Taylor, and E. Burstein, "Proc. Int. Conf. on II-VI Compound Semiconductors," 1185, D.G. Thomas, (ed.) W.A. Benjamin, 1968.
8. A. Pinczuk, E. Burstein, and S. Ushioda, Solid State Communications (in press).
9. E. Burstein, S. Ushioda, and A. Pinczuk, Solid State Commun. 6, 407 (1968).
10. Similar expressions apply for polaritons propagating along the principal axes of $\epsilon(\omega)$ in optically anistropic crystals having trigonal, tetragonal, hexagonal, and orthorhombic structures. The dispersion relations are somewhat more complicated in crystals of lower symmetry in which the orientations of two (or all three) principal axes are functions of frequency and the polaritons are quasi-transverse modes.
11. The TO phonons are characterized by the normal coordinates u_j. The definitions of e_j and m_j follow from the expressions for the electric moment, $M_j = e_j u_j$, and the energy density, $\langle W \rangle = (N/V) m_j \omega_j^2 \langle u_j^2 \rangle$.
12. H.E. Puthoff, R.H. Pantell, B.G. Huth, and M.A. Chacon, J. Appl. Phys. 39, 2144 (1968)
13. A.A. Maradudin and A.E. Fein, Phys. Rev. 128, 2568 (1962).

14. R. Loudon, Adv. in Phys. 13, 423 (1964).
15. E. Burstein, A. Pinczuk, and S. Iwasa, Phys. Rev. 157, 611 (1967).
16. H. Pelzer, Proc. Roy. Soc. (London) 208, 365 (1961).
17. J.F. Scott and S. Ushioda, paper A-4 this conference.
18. A. Pinczuk, W. Taylor, E. Burstein, and I. Lefkowitz, Solid State Commun. 5, 429 (1967).
19. J.L. Parsons and J.L. Rimai, Solid State Commun. 5, 423 (1967).
20. M. DiDomenico, Jr., S.H. Wemple, S.P.S. Porto, and R.P. Bauman, Phys. Rev. (in press).
21. G. Shirane, B.C. Frazer, V.J. Minkiewzcz, J.A. Leake, and A. Linz, Phys. Rev. Letters 19, 234 (1967).

A-4: POLARITION INTENSITIES IN α - QUARTZ

James F. Scott
Bell Telephone Laboratories, Incorporated
Holmdel, N.J.
and
S. Ushioda
Department of Physics, University of Pennsylvania
Philadelphia, Pennsylvania

The initial studies of light scattering from polaritons emphasized kinematic aspects [1-4]. Polariton dispersion was detailed experimentally for several materials, and the effect of birefringence, damping, and index of refraction variation near the laser frequency were examined. However, interest in polariton cross-sections was dormant until Faust and Henry [5] discovered a zero in the nonlinear electric susceptibility of GaP and showed how Poulet's [6] two-component phenomenological model could be used to explain their optical mixing data. Poulet's model has been elaborated upon by Loudon [7] and by Kleinman [8] and has recently been used by Kaminow and Johnston [9] and by Ushioda et al. [10] to analyze the influence of displacive and electro-optic contributions upon the Raman susceptibility. Extensions of the earlier work have most recently been made by Henry and Garrett [11] and by Burstein et al. [12] In the present analysis of quartz the Burstein formalism [12] will be used, because it is already cast into a form suited to computer solution of the many-mode case. In quartz we shall consider eight modes of E symmetry and obtain numerical solutions of the equations developed by Burstein et al. The formalism of Henry and Garrett can be shown to be equivalent, if damping constants are handled carefully, however their calculations are developed only for crystals having a single optical mode and contain approximations at various stages which are not suitable for quartz. Their conclusion that resonant absorption exactly cancels resonant gain near IR- and Raman-active phonon frequencies [11] is not true for quartz, nor for other crystals where two or more IR- and Raman-active modes are present, one of which has large Raman gain and very small IR oscillator strength.

The equations leading to the frequency-dependent polariton scattering intensity $I(\omega)$ are given below from Ref. [12].

$$C_{\mu\nu\lambda}(j) = \frac{b_{\mu\nu\lambda}}{a_{\mu\nu\lambda}(j)} = \left\{ \pm\left(\frac{\omega_{sj}}{\omega_{sjL}}\right)^2 \left(\frac{\omega_j^2 - \omega_{jL}^2}{\Omega_j}\right) \left[\frac{I_{LO}(j)}{I_{TO}(j)} \left(\frac{n_{jT}+1}{n_{jL}+1}\right) \frac{1 + \Sigma\, \Omega_i^2 \omega_i^2 (\omega_i^2 - \omega_{jL}^2)^{-2}}{\omega_j \omega_{jL}}\right]^{1/2} - 1 \right\} \left\{\left(\omega_j^2 - \omega_{jL}^2\right)\right\}^{-1} \tag{1}$$

$$a'_{\mu\nu\lambda}(j) = \left(\frac{8\pi\hbar}{V\epsilon_0 \Omega_j^2 \omega_j}\right)^{-1/2} \chi_{\mu\nu}^{(1)}(\omega_j) = \left(\frac{e_j^*}{m_j}\right) a_{\mu\nu\lambda}(j) \tag{2}$$

Here $\underset{\sim}{b}$ is the electroptic susceptibility tensor $\frac{\partial \chi}{\partial E}$; $\underset{\sim}{a}'$ (j) is proportional to the atomic displacement susceptibility tensor $\frac{\partial \chi}{\partial u_j}$, where u_j is the displacement vector for the j-th normal mode; ω_{jL} are the frequencies of the j-th transverse and longitudinal mode; $I_{LO}(j)$, $I_{TO}(j)$ are the integrated Raman scattering intensities of the j-th LO and TO mode; and Ω_j is proportional to the IR oscillator strength of the j-th mode.

$$\chi_{\mu\nu}^{(1)}(\omega) = \sum_j a'_{\mu\nu\lambda}(j) \left[1 + C_{\mu\nu\lambda}(j)(\omega_j^2 - \omega^2)\right] \left(\frac{8\pi\hbar\omega}{V\epsilon_0}\right)^{1/2} \prod_{i \neq j} (\omega_i^2 - \omega^2) \left\{ \prod_i (\omega_i^2 - \omega^2)^2 + \sum_i \Omega_i^2 \omega_i^2 \prod_{k \neq i} (\omega_k^2 - \omega^2)^2 \right\}^{-1/2} \tag{3*}$$

$$I_{\mu\nu}(\omega) = \left(\frac{\omega_s}{C}\right)^4 \frac{V}{2} \left[\chi_{\mu\nu}^{(1)}(\omega)\right]^2 \left[n(\omega) + 1\right] \tag{4}$$

*In calculating the curves in Figs. 2 and 3 we have approximated the term

$$\left\{ \prod_{i \neq j} \left(\omega_i^2 - \omega^2\right) \left[\prod_i \left(\omega_i^2 - \omega^2\right)^2 + \sum_i \Omega_i^2 \omega_i^2 \prod_{k \neq i} \left(\omega_k^2 - \omega^2\right)^2\right]^{-1/2} \right\}$$

in Eq. (3) by

$$\left[\left(\omega_j^2 - \omega^2\right)^2 + \Omega_j^2 \omega_j^2\right]^{-1/2}$$

to eliminate computational difficulties. Physically, this treats the phonon branches as independently coupled to the radiation field; in reality, the electric field also couples the phonon modes to each other.

Other parameters in Eqs. (1-4) are defined in Ref. [12] and may be evaluated from conventional right-angle phonon scattering data. Several approximations are made in applying these equations to quartz. First, the equations were developed by using the macroscopic electric field rather than the local field produced by the polar phonon. Nozieres and Pines have shown [13] that this is a better approximation for semiconductors than for materials of low electrical conductivity where electrons are highly localized; the use of E_{MAC} here is justified empirically. Since E_{LOC} would contribute to TO scattering also, any error introduced by E_{MAC} would be experimentally equivalent to error in the measured LO/TO intensity ratio and absorbed by our adjustable e_j^*/m_j coefficient. Second, the basic question of how to pair off LO and TO phonons arises. The modes of E symmetry in quartz are generally of low oscillator strength and hence have only a few cm^{-1} frequency separation between the LO and TO of a given mode, compared to say $100cm^{-1}$ between modes (i.e. between TO's); consequently, we have compared intensities of adjacent LO's and TO's -- with one exception -- the LO phonon at $1235cm^{-1}$ must be associated with the TO at $1072cm^{-1}$, despite the $1163\text{-}1164cm^{-1}$ pair between them. Where LO/TO doublets in the Raman spectrum are not resolved we have calculated LO frequencies from TO's and IR oscillator strengths [14]. In such cases the actual LO/TO intensity ratio is unimportant for polariton intensity calculations, because that mode does not contribute to the cross-sections of polaritons whose frequencies are more than about a linewidth from that of the TO phonon. For computational convenience we have set $C_j = 0$ for such modes.

Quartz is uniaxial and exhibits highly non-cubic optical properties. Polariton intensities have previously been analyzed only for single-mode cubic crystals [5, 10], for which scalar calculations could be used throughout. Hence in the present case, in addition to complications inherent in coupling photons to a whole system of phonons, we have to consider the nonlinear susceptibility in all its tensorial glory. The displacive and electrooptic tensors a_j and b each have two independent elements [15]; that they are very different numerically is shown in Fig. 1 for the $394\text{-}401cm^{-1}$ TO-LO pair. If a_j and b were cubic tensors, the ratio of LO/TO scattering for α_{xy} and α_{xz} scattering would be the same, contrary to that shown in Fig. 1. Note that the phonon propagation is the the same in each trace.

While most of the polariton data on quartz collected to date [3] are for α_{xz} polarization, some α_{xy} measurements are reported here. In α quartz the four E modes which have strong α_{xy} scattering have very low oscillator strength and hence couple weakly to photons; polaritons at frequencies near those modes will scatter weakly. This situation

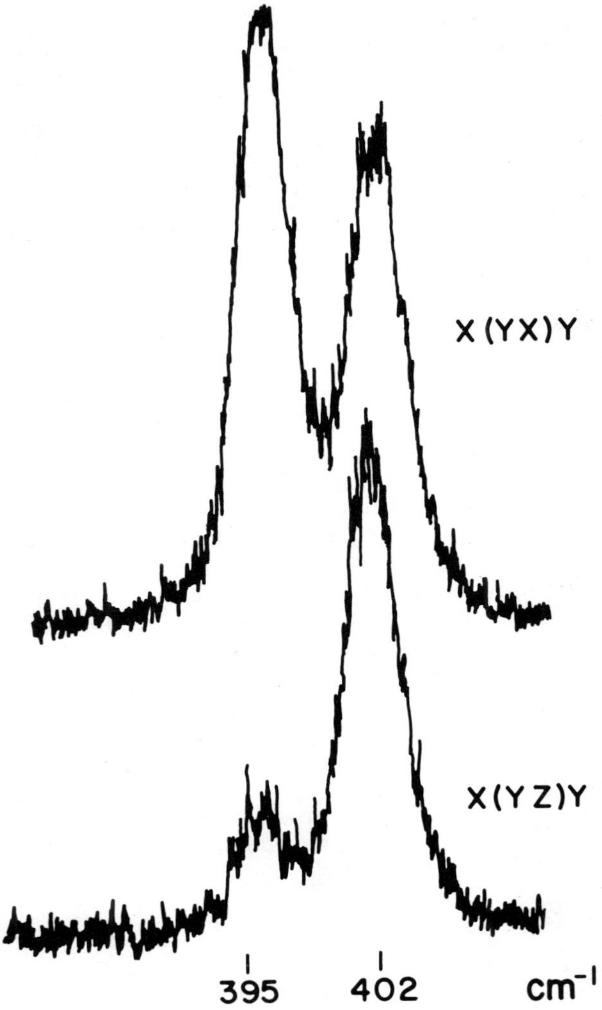

Fig. 1. Raman intensities for TO/LO pair at 394/403 cm^{-1} for α_{xy} and α_{yz} scattering and propagation in the xy-plane.

is not accidental. In β-quartz, which is only slightly different structurally from α-quartz, these E xy-polarized modes become E_2 and must have zero oscillator strength [16].

Figs. 2 and 3 show the calculated polariton intensity for α_{xy} and α_{xz} scattering, along with experimental data points. Fig. 4 shows data in the 800-100cm^{-1} region, illustrating the signal-to-noise ratio obtained.

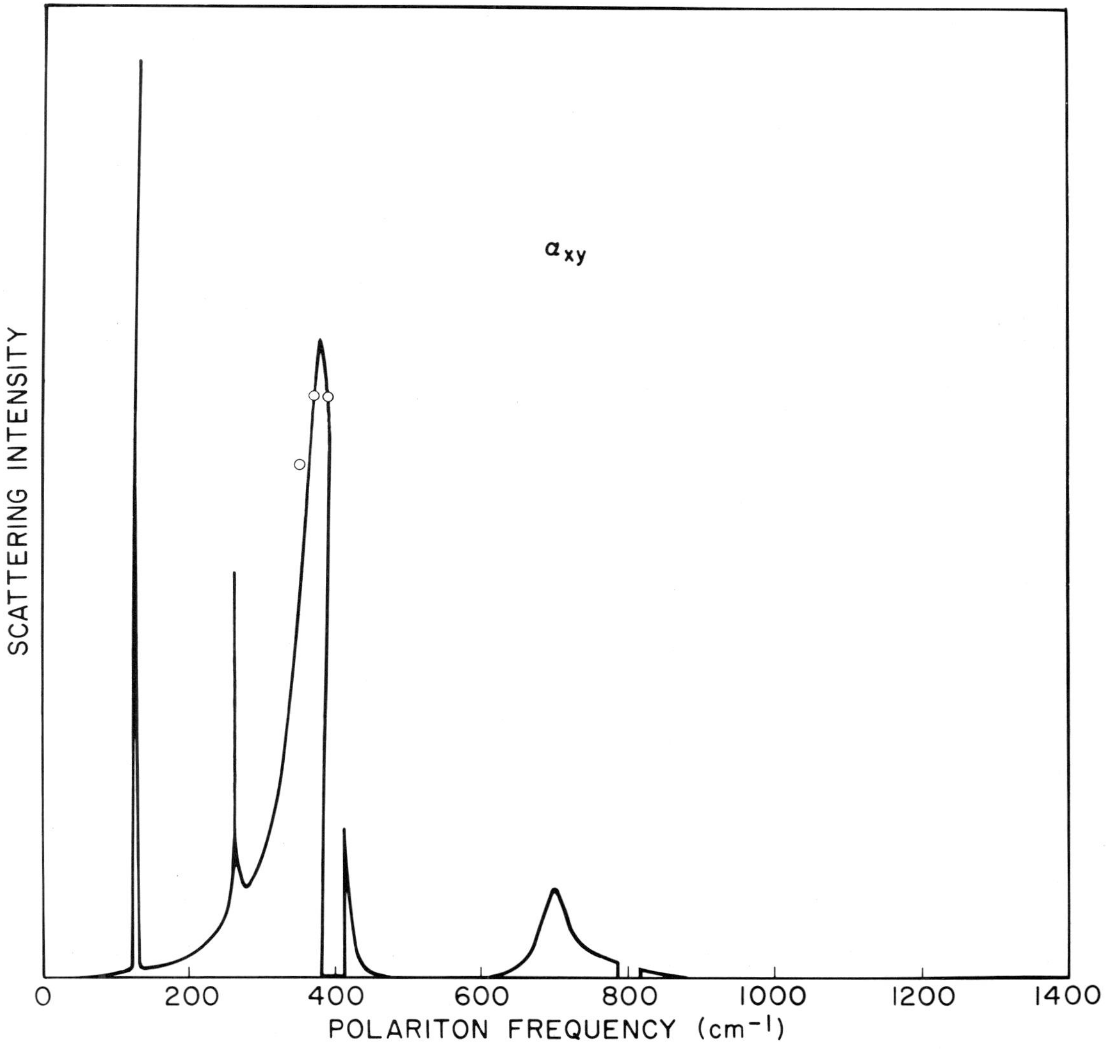

Fig. 2. Calculated polariton scattering intensities for α_{xy} scattering. Circles are data points; curves are constrained to fit peak intensities at LO and TO phonon frequencies.

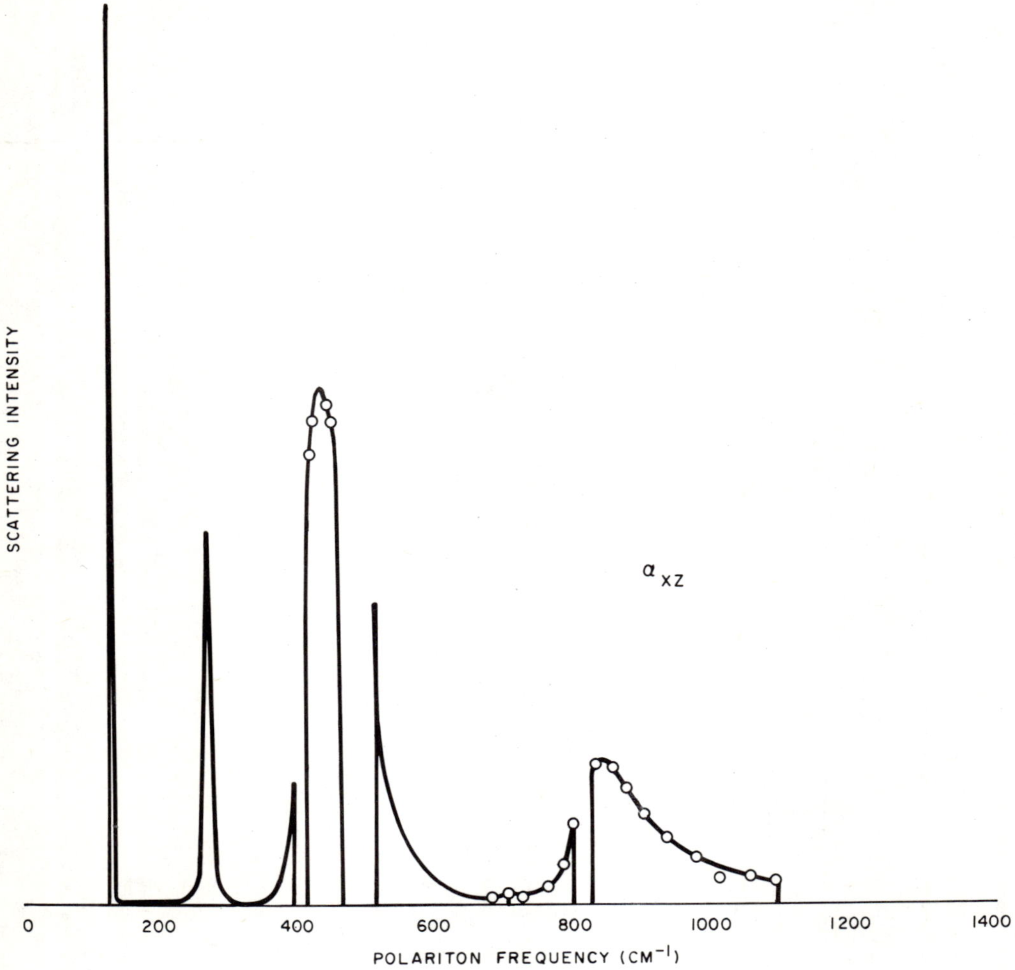

Fig. 3. Calculated polariton scattering intensities for α_{xz} scattering.

Experimentally, x(yz) x and z(xy)z geometries were employed. Both natural and synthetic quartz samples (~ 1cc) were used, with no discernible differences. Excitation and detection were via a 1.4W argon ion laser at 4880Å and a double spectrometer and photon counting electronics.

It is shown in Figs. 3 and 4 that fairly good agreement between theory and experiment is obtained. Note the zero at ~735cm^{-1} for α_{xz} scattering. It is also important that substantial intensity is present at 945cm^{-1} (10.6μ); this shows that optical mixing of Ar II and CO_2 lasers should be possible, thus providing up-conversion of CO_2 signals, just as Faust and Henry's experiment up-converted H_2O laser signals. Since 10.6μ is at an absorption minimum in the reststrahlen, the gain obtainable may be adequate for cw operation. This experiment will be attempted by one of us shortly.

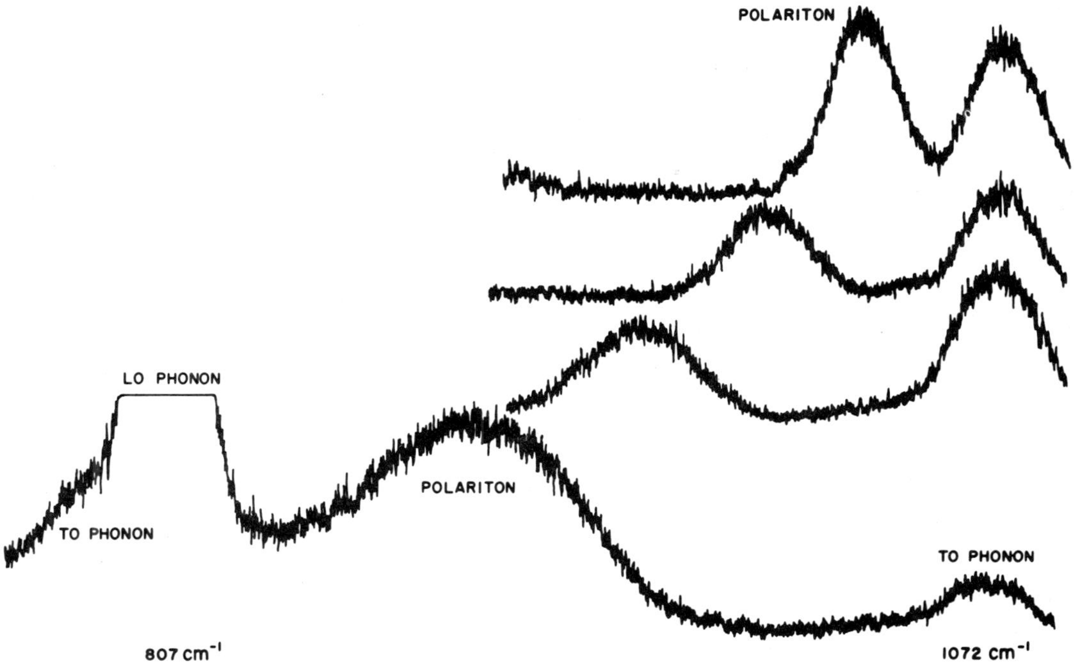

Fig. 4. Data in the 800-1100cm^{-1} region of the quartz spectrum for α_{xz} scattering.

Of special interest are the numerical results near 128cm^{-1}. Figs. 3 and 4 show that the gain drops essentially to zero just off resonance. In fact, the gain is non-negligible only within a linewidth, just as concluded earlier by one of us from semi-qualitative arguments [17]. This shows that while the 128cm^{-1} mode has sufficient gain to produce a Raman laser [18], such a laser will not be tunable by varying the scattering angle. The predictions here of gain near the 128cm^{-1} resonance may be compared with those of Henry and Garrett [11] and of Butcher, Loudon and MacLean [19].

As a final comment, we wish to compare polaritons and phonons in the 0-200cm^{-1} region. Just as the polariton can be "tuned" through this frequency range by varying wave vector (or scattering angle), a soft optic phonon can be tuned through the region by varying the temperature [20, 21]. One of us has shown [22] that the soft optic mode couples anharmonically to a two-phonon excitation. The question: why do polaritons in the region not also couple? The answer lies in a symmetry argument. The two phonon Raman scattering peak experimentally exhibits only α_{xx} and α_{zz} scattering, hence A_1 symmetry. While the soft A_1 optic mode can couple to the two-phonon excitation, producing mixed un-phonolike anharmonic excitations, such excitations are still of A_1 symmetry. No anharmonic coupling of polaritons, which have E symmetry in quartz, to the two-phonon peak will be allowed. Consequently, no "anticrossings" or linewidth anomalies are expected in the quartz polariton spectrum.

ACKNOWLEDGMENTS

We would like to acknowledge technical assistance from L.E. Cheesman and Mrs. M.L. Thomson and helpful discussions with Professor E. Burstein and Dr. C.H. Henry. The natural quartz sample was provided by Dr. W. J. Brya.

REFERENCES

1. J.J. Hopfield and C.H. Henry, Phys. Rev. Letters 15, 964 (1965).
2. B. Tell, S.P.S. Porto, and T.C. Damen, Phys. Rev. Letters 16, 450 (1966).
3. J.F. Scott, L.E. Cheesman, and S.P.S. Porto, Phys. Rev. 162, 834 (1967).
4. H.E. Puthoff, R.H. Pantell, B.G. Huth, and M.A. Chacon, J. Appl. Phys. 39, 2144 (1968).
5. W.L. Faust and C.H. Henry, Phys. Rev. Letters 17, 1265 (1966).
6. H. Poulet, Ann. Phys. (Paris) 10, 908 (1955).
7. R. Loudon, Advan. Phys. 13, 423 (1964).
8. D.A. Kleinman, Phys. Rev. 126, 1977 (1962).
9. I.P. Kaminow and D. Johnston, Jr., Phys. Rev. 160, 519 (1967).
10. S. Ushioda, A. Pinczuk, W. Taylor, and E. Burstein, "Proc. Int. Conf. on II-VI Semiconducting compounds", p. 1185, D.G. Thomas (ed.), W.A. Benjamin, 1968.
11. C.H. Henry and C.G.B. Garrett, Phys. Rev. 171, 1058 (1968).
12. E. Burstein, S. Ushioda, A. Pinczuk, and J.F. Scott, paper A-3 this conference.
13. P. Nozieres and D. Pines, Phys. Rev. 109, 762 (1958).
14. W.G. Spitzer and D.A. Kleinman, Phys. Rev. 121, 1324 (1961).
15. F.N. Butcher, "Nonlinear Optical Phenomena", Engineering Experiment Station Bulletin, p. 200, Ohio State University, 1965.
16. J.F. Scott and S.P.S. Porto, Phys. Rev. 161, 903 (1967).
17. J.F. Scott, J. Quant. Elec. 3, 693 (1967).
18. P.E. Tannenwald and D.L. Weinberg, J. Quant. Elect. 3, 334 (1967).
19. P.N. Butcher, R. Loudon and T.P. MacLean, Proc. Phys. Soc. 85, 565 (1965).
20. S.M. Shapiro and H.Z. Cummins, Bull. Am. Phys. Soc. 12, 588 (1967).
21. S.M. Shapiro, D.C. O'Shea and H.Z. Cummins, Phys. Rev. Letters 19, 361 (1967).
22. J.F. Scott, Phys. Rev. Letters 21, 907 (1968).

A-5: RAMAN SCATTERING FROM THE SOFT OPTIC MODE IN FERROELECTRIC CRYSTALS

S. H. Wemple and M. DiDomenico, Jr.
Bell Telephone Laboratories, Incorporated
Murray Hill, New Jersey

ABSTRACT

A microscopic description of Raman scattering from the soft ferroelectric mode in the class of oxygen-octahedra ferroelectrics is presented. By relating energy-band shifts produced by critical point polarization fluctuations to fluctuations in the optical dielectric constant tensor we are able to calculate the magnitude of the soft mode Raman scattering efficiency. The polarization-induced energy-band shifts are described using a polarization potential tensor which relates these shifts to the square of the total crystal polarization. The magnitude of the polarization potential is found to be essentially the same in all oxygen-octahedra ferroelectrics based on a straightforward connection between this quantity and the clamped quadratic electro-optic coefficients. Theoretical estimates are also available from energy-band calculations from which we conclude that modulation of the pdπ energy overlap integral by the soft mode polarization fluctuations leads to the electron-phonon interaction observed in Raman scattering. We find further that the strength of the interaction depends only on the dielectric quantity $\varkappa^S P_s^2$, where $\varkappa^S$ is the relevant static dielectric constant, and P_s is the crystal spontaneous polarization. The polarization fluctuation model can also be used to calculate the soft mode Raman lineshape. Based on the Nyquist theorem combined with a classical damped harmonic oscillator description of the soft mode, we obtain the Raman lineshape in the presence of phonon damping.

INTRODUCTION

The lattice dynamical theory of displacive ferroelectric phase transitions ascribes the paraelectric-ferroelectric transition to an instability of the lowest-frequency transverse optic (TO) vibrational mode at the Brillouin zone center[1]. Overwhelming experimental evidence in support of this viewpoint has been obtained from infrared reflectivity, inelastic neutron scattering, and Raman scattering experiments. These data reveal a strongly temperature dependent soft TO mode whose frequency tends towards zero as the temperature approaches the transition temperature. The fundamental interrelationship between optical phonons and ferroelectricity suggests that Raman scattering experiments should provide a valuable technique for the investigation of ferroelectric phenomena. In this paper we present a microscopic description of Raman scattering from the soft TO mode in the class of oxygen-octahedra ferroelectrics. This class includes perovskite (e.g., $BaTiO_3$), tungsten bronze (e.g., $Ba_2NaNb_5O_{15}$), and $LiNbO_3$-type structures. A unique property of oxygen-octahedra ferroelectrics, which enables us to obtain a unified description of the Raman scattering process in terms of a single specific electron-phonon interaction, is that they all contain the same basic BO_6 structural building block (B is a transition metal ion: Ti, Nb, or Ta). The electron-phonon interaction relates perturbations in the electronic energy band structure to thermodynamic fluctuations in the soft ferroelectric TO phonon mode via a polarization potential analogous to the deformation potential associated with acoustic phonons. Using this interaction we compute the absolute magnitude of the Raman scattering efficiency tensor coefficients as well as the lineshape of the Raman scattered light. We describe the thermally induced soft phonon mode vibrations in terms of its associated polarization fluctuations. These critical point fluctuations are composed of soft TO phonon wave packets and are generally quite complex to describe since they require knowledge of short-range spatial correlation between microscope lattice polarization amplitudes[2]. Raman scattering experiments using visible light are, however, essentially independent of spatial correlation and measure the long wavelength TO phonon components connected with low frequency dielectric behavior.

MICROSCOPIC DESCRIPTION OF RAMAN SCATTERING

The fluctuations in optical dielectric constant which lead to Raman scattering can be viewed as arising from fluctuations in the energy separation between interband critical points induced by lattice polarization fluctuations. Thus, the uv oscillators which give rise to the optical dielectric constant fluctuate in position (and possibly strength) due to an interaction between the microscopic polarization amplitude and the electronic energy-band structure. Calculation of this electron-phonon interaction is unusually simple for the soft mode in oxygen-octahedra ferroelectrics for the following reasons: (1) the soft mode normal coordinates involve a simple displacement of a transition metal ion relative to its oxygen-octahedron cage; (2) because all these materials contain the same basic BO_6 unit the important energy-bands are essentially the same and consist of d-like conduction-bands derived from the transition metal d-orbitals and p-like valence bands derived from the oxygen 2p-orbitals; and (3) the major energy-band perturbation associated with the soft mode nuclear displacement is a simple shift of interband energy spacings rather than a distortion of bands and an alteration of transition probabilities.

In order to describe the energy-band shifts at critical points produced by soft mode polarization fluctuations, we introduce a polarization potential tensor $\sigma_{ij}^{n}(k)$ defined by the relation[3]

$$\Delta \varepsilon^{n}(k) = \sum_{i,j} \sigma_{ij}^{n}(k) P_i P_j \tag{1}$$

where $\Delta\varepsilon^{n}(k)$ is the energy shift of the nth band at the critical point k in the Brillouin zone, and P_i is the total crystal polarization consisting in general of a static spontaneous part and a fluctuating component connected with the soft mode. In writing Eq. (1) we have taken the energy-bands in the centrosymmetric paraelectric phase to be the energy reference and therefore arrive at the indicated quadratic dependence on total polarization. In general, a different set of polarization potential tensor coefficients applies for the normal coordinates applicable to each of the optical modes. For acoustic modes, an analogous set of deformation potentials can be defined by replacing $P_i P_j$ in Eq. (1) by the second-rank strain tensor.

The magnitudes of the σ-coefficients for the soft TO mode can be estimated from Brews' modifications[4] of the Kahn and Leyendecker[5] LCAO energy-band calculations for $SrTiO_3$. Brews finds that $\sigma \sim 1$ eV-m^4/C^2 for the important $p \rightarrow d\epsilon$ critical transitions situated near 5 eV and $\sigma \sim 0$ for the $p \rightarrow d\gamma$ transitions situated near 9 eV. Electroreflectance data in $KTaO_3$ reported by Frova and Boddy[6] confirm this estimate giving $\sigma \approx 2$-8 eV-m^4/C^2 in the vicinity of the 5 eV $p \rightarrow d\epsilon$ transitions. The LCAO calculations show further that the physical origin of the σ-coefficients lies in a modulation of the $pd\pi$ energy overlap integral by a displacement of the transition metal ion relative to its oxygen-octahedron cage, i.e., by the soft TO mode nuclear motions. This can be seen clearly in Fig. 1 where we show schematically the important p and d orbitals responsible for the $p \rightarrow d\epsilon$ transitions and their region of overlap.

Unfortunately, at the present time energy-band calculations are not sufficiently precise to permit a complete calculation of all the σ-coefficients and all the optical selection rules required to give an accurate first principles calculation of Raman scattering efficiencies in oxygen-octahedra ferroelectrics. It is convenient, therefore, to introduce an effective polarization potential tensor β which is an average of the more fundamental σ-coefficients weighted by the appropriate optical selection rules. This parameter β then relates the shifts in a single effective uv oscillator near 5 eV associated with $p \rightarrow d\epsilon$ transitions to the polarization diadic $P_i P_j$. The next higher oscillator near 9 eV associated with $p \rightarrow d\gamma$ transitions is assumed to remain unshifted by the polarization. We emphasize that β is of the same order of magnitude as σ so that, based on the energy-band calculations, we expect β to fall in the 1-10 eV-m^4/C^2 range.

We now relate shifts in the 5 eV oscillator described by the polarization potential tensor β to fluctuations in the optical dielectric constant. These fluctuations will, in turn, be related to the Raman scattering efficiency. Our two-oscillator model is shown schematically in Fig. 2 where we indicate the shifts in the 5 eV $S_\epsilon, \lambda_\epsilon$ oscillator associated with particular geometrical configurations. Because all energy shifts are measured relative to a crystal containing centrosymmetric oxygen-octahedra, the only tensor components of β, in reduced index notation, are β_{11}, β_{12}, and β_{44} as indicated in Fig. 2. The fluctuations in the relative optical dielectric constant $\Delta\epsilon_{ij}$ may be written in terms of fluctuations in the impermeability $\Delta\left(\frac{1}{\epsilon}\right)_{kl}$, i.e.,

$$\Delta\epsilon_{ij} = - \sum_{k,l} \epsilon_{ik} \Delta\left(\frac{1}{\epsilon}\right)_{kl} \epsilon_{lj} \tag{2}$$

In a polarized crystal, which gives first-order Raman scattering, we can relate $\Delta\left(\frac{1}{\epsilon}\right)_{kl}$ to polarization fluctuations ΔP_m through a third-rank tensor f_{klm} as follows:

$$\Delta\left(\frac{1}{\epsilon}\right)_{kl} = \sum_m f_{klm} \Delta P_m \tag{3}$$

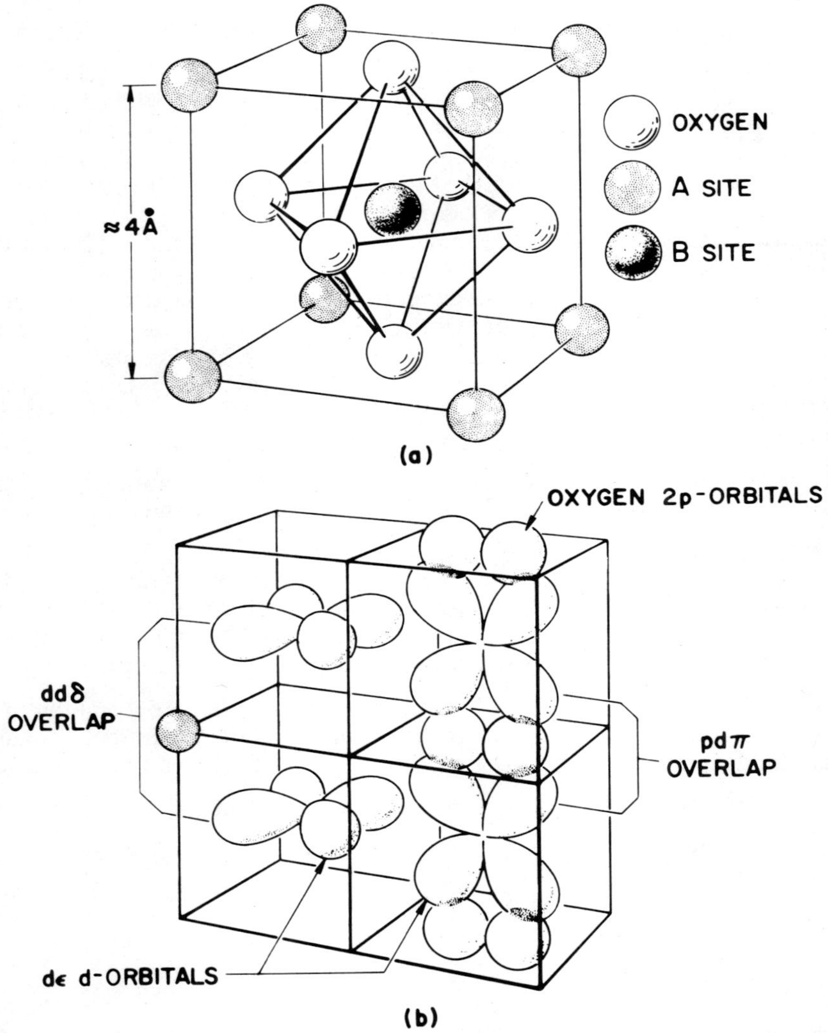

Fig. 1. (a) Unit cell of perovskite oxide ferroelectrics. B is a transition metal ion Ti(3d), Nb(4d), or Ta(5d). The BO_6 octahedron structure is common to the class of oxygen-octahedra ferroelectrics. (b) Schematic of important p and d orbitals in oxygen-octahedra ferroelectrics showing the physical location of pdπ energy overlap integral.

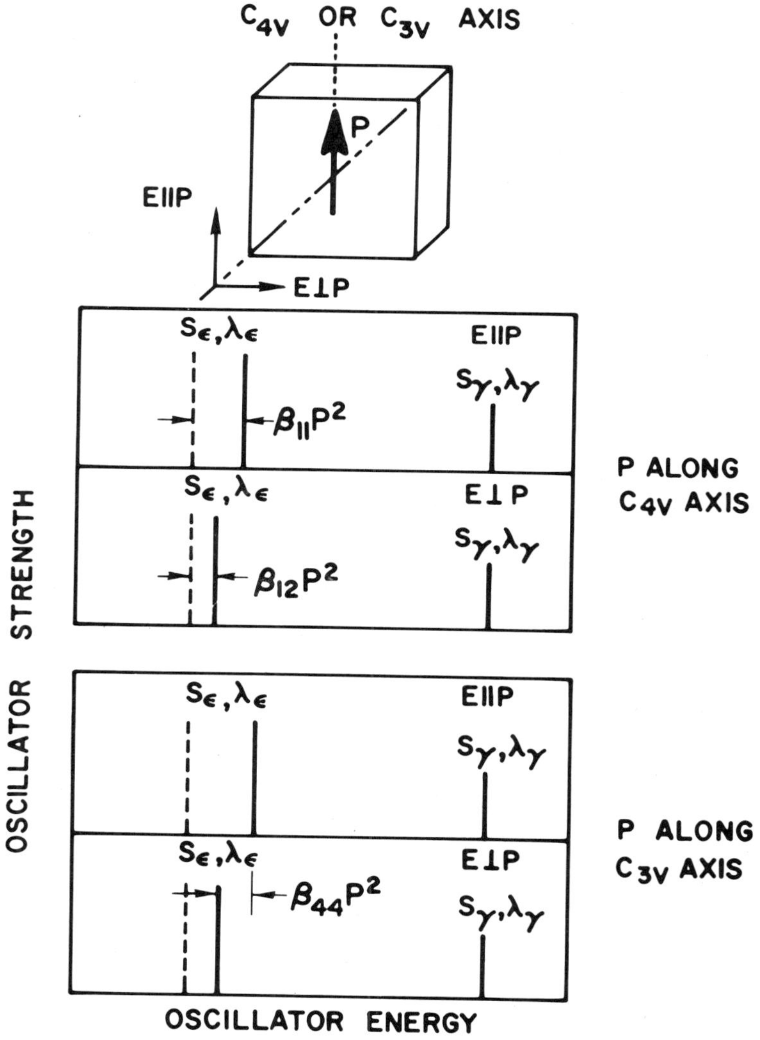

Fig. 2. Schematic representation of oscillator shifts produced by polarization P. The notation for the polarization potential β_{ij} is consistent with the usual contracted index notation for the fourth-rank quadratic electro-optic g-tensor. P along C_{4v} axis and P along C_{3v} axis refer respectively to P along the four-fold and three-fold octahedral axes. E | | P and E ⊥ P refer to the directions of light polarization.

We restrict the analysis to crystals having (1) C_{4v} or nearly C_{4v} symmetry (i.e., perovskite or tungsten bronze ferroelectrics), and (2) nearly C_{6v} symmetry (i.e., $LiNbO_3$-type ferroelectrics). In either case the only nonzero f_{klm} tensor components are $f_{13} = f_{23}$, f_{33}, and $f_{42} = f_{51}$ in reduced index notation. As a result, we should observe fluctuations in $\Delta\epsilon_{11}$, $\Delta\epsilon_{22}$, $\Delta\epsilon_{33}$, $\Delta\epsilon_{23}$, and $\Delta\epsilon_{13}$. The calculation of the f-tensor components in terms of energy band perturbations is simplified considerably if we use the two oscillator model shown in Fig. 2 to calculate the optical dielectric constant[7]. Thus, in the centrosymmetric phase we have

$$\epsilon - 1 = \frac{S_\epsilon \lambda_\epsilon^2}{1 - (\lambda_\epsilon/\lambda)^2} + \frac{S_\gamma \lambda_\gamma^2}{1 - (\lambda_\gamma/\lambda)^2} \tag{4}$$

where S_ϵ and S_γ are the strengths of the $p \to d\epsilon$ and $p \to d\gamma$ transitions at 5 and 9 eV respectively, and λ_ϵ and λ_γ are their corresponding oscillator positions in wavelength units. As discussed above and indicated in Fig. 2 only λ_ϵ is affected strongly by lattice polarization. In the presence of a spontaneous or field-induced polarization P_s directed along the z-axis and polarization fluctuation components ΔP_m, the f-tensor and β-tensor components can be related as discussed in detail in previous publications[3, 7]. Substituting those results into Eqs. (2) and (3) gives the expressions shown in Table I.

TABLE I

Fluctuations in Optical Dielectric Constant Induced in Polarized Ferroelectric Crystals by Lattice Polarization Fluctuations Connected with the Soft TO Phonon Mode

ϵ_{ij}	C_{4v} crystals	C_{6v} crystals
$\Delta\epsilon_{11}$	$-2\epsilon_{11}^2 AP_s\beta_{12}\Delta P_3$	$-\frac{2}{3}\epsilon_{11}^2 AP_s(\beta_{11}+2\beta_{12}-\beta_{44})\Delta P_3$
$\Delta\epsilon_{22}$	$-2\epsilon_{22}^2 AP_s\beta_{12}\Delta P_3$	$-\frac{2}{3}\epsilon_{22}^2 AP_s(\beta_{11}+2\beta_{12}-\beta_{44})\Delta P_3$
$\Delta\epsilon_{33}$	$-2\epsilon_{33}^2 AP_s\beta_{11}\Delta P_3$	$-\frac{2}{3}\epsilon_{33}^2 AP_s(\beta_{11}+2\beta_{12}-\beta_{44})\Delta P_3$
$\Delta\epsilon_{23}$	$-\epsilon_{22}\epsilon_{33}AP_s\beta_{44}\Delta P_2$	$-\frac{1}{3}\epsilon_{22}\epsilon_{33}AP_s(2\beta_{11}-2\beta_{12}+\beta_{44})\Delta P_2$
$\Delta\epsilon_{13}$	$-\epsilon_{11}\epsilon_{33}AP_s\beta_{44}\Delta P_1$	$-\frac{1}{3}\epsilon_{11}\epsilon_{33}AP_s(2\beta_{11}-2\beta_{12}+\beta_{44})\Delta P_1$

The constant A appearing in Table I is given by[7]

$$A = 2(1+R)^{-1/2}(e/hc)^2(\mathcal{E}_o/S_o)(1-1/\epsilon)^2 \tag{5}$$

where $R = S_\gamma\lambda_\gamma^2/S_\epsilon\lambda_\epsilon^2 \approx 0.5$, e is the electronic charge, h is Planck's constant, and c is the speed of light. The quantity $(\mathcal{E}_o/S_o)$ in Eq. (5) is the ratio of <u>average</u> oscillator energy $\mathcal{E}_0 = hc/e\lambda_0$ to average oscillator strength defined by a single term Sellmeier fit to optical dielectric constant dispersion data, i.e., $\epsilon - 1 = S_0\lambda_0^2\left[1 - (\lambda_0/\lambda)^2\right]^{-1}$. The dispersion parameter $\mathcal{E}_0/S_0$ is found to be remarkably constant for all oxygen-octahedra ferroelectrics and has the value $6.0 \pm 0.5\times10^{-14}$ eV-m^2 [7]. Substituting this value into Eq. (5) and taking $\epsilon \approx 5$, we find that $A \approx 0.04$ eV^{-1}.

We now have known relationships, given in Table I, between the dielectric tensor fluctuations $\Delta\epsilon_{ij}$ and the polarization fluctuations ΔP_i in terms of the static polarization and the polarization potential β-tensor. Since the Raman and electro-optic effects stem basically from the same nuclear motions[8], the β-coefficients which we have defined are directly related to the clamped quadratic electro-optic g-coefficients. We have shown[3, 7], for example, that for C_{4v} crystals $g_{ij} \approx A\beta_{ij}$. As we have previously reported[9], the g_{ij} coefficients are essentially the same in all oxygen-octahedra ferroelectrics. We thus obtain from the g-coefficients[7, 9] the following values of β: $\beta_{11} \approx 4/\zeta^3$, $\beta_{12} \approx 1/\zeta^3$, and $\beta_{44} \approx 3/\zeta^3$ eV-m^4/C^2 where ζ is the volume density of oxygen-octahedra relative to that in a simple perovskite structure with a 4 Å lattice constant[7], (e.g., $\zeta \approx 1$ in $BaTiO_3$ and $Ba_2NaNb_5O_{15}$, and $\zeta \approx 1.2$ in $LiNbO_3$).

To determine the third-rank Raman scattering efficiency tensor components we relate these quantities to the appropriate optical dielectric constant fluctuations given in Table I. This can be done using the general scattering theory discussed by Landau and Lifshitz[10]. In crystals having C_{4v} or C_{6v} symmetry the only nonzero tensor coefficients are S_{33} and S_{13} for A_1 modes and S_{42} for E modes in the reduced notation of Kaminow and Johnston[8]. We thus obtain the following relations for a 90° scattering geometry:

$$dS_{33}/d\omega = \left(\frac{1}{4\pi}\right)^2\left(\frac{\omega}{c}\right)^4 \langle|\Delta\epsilon_{33}(\omega)|^2\rangle V$$

$$dS_{13}/d\omega = \left(\frac{1}{4\pi}\right)^2\left(\frac{\omega}{c}\right)^4 \langle|\Delta\epsilon_{12}(\omega)|^2\rangle V$$

$$dS_{42}/d\omega = \left(\frac{1}{4\pi}\right)^2\left(\frac{\omega}{c}\right)^4 \langle|\Delta\epsilon_{23}(\omega)|^2\rangle V \tag{6}$$

where ω is the incident light frequency, V is the interaction volume, and the angular brackets denote an ensemble average over the volume V. The absence of S_{22} in Eqs. (6) reflects our neglect of small distortions from C_{4v} or C_{6v} symmetry. Combining Eqs. (6) with Table I then yields

$$dS_{33}/d\omega = 4\left(\frac{1}{4\pi}\right)^2\left(\frac{\omega}{c}\right)^4 (\epsilon_{33}A\beta_{11}P_s)^2 \langle|\Delta P_3(\omega)|^2\rangle V$$

$$dS_{13}/d\omega = 4\left(\frac{1}{4\pi}\right)^2\left(\frac{\omega}{c}\right)^4 (\epsilon_{11}\epsilon_{22}A\beta_{12}P_s)^2 \langle|\Delta P_3(\omega)|^2\rangle V$$

$$dS_{42}/d\omega = \left(\frac{1}{4\pi}\right)^2\left(\frac{\omega}{c}\right)^4 (\epsilon_{22}\epsilon_{33}A\beta_{44}P_s)^2 \langle|\Delta P_2(\omega)|^2\rangle V \tag{7}$$

By integrating Eqs. (7) over all frequencies, we obtain expressions for the total Raman scattering efficiency S, i.e., $S = (1/2\pi)\int_0^\infty (dS/d\omega)d\omega$. Using the classical thermodynamic theory of dielectric fluctuations[11] it can be shown that the frequency integrals over $<|\Delta P_3(\omega)|^2>$ and $<|\Delta P_2(\omega)|^2>$ are $kT\epsilon_0\varkappa_3^s/V$ and $kT\epsilon_0\varkappa_2^s/V$, respectively, where k is Boltzmann's constant, T is the absolute temperature, ϵ_0 is the free space permittivity, and $\varkappa_3^s$ and $\varkappa_2^s$ refer to the static relative dielectric strength of the soft mode along the indicated axes. This strength is somewhat less than the measured clamped dielectric constants which include all the TO modes as well as the electronic contribution of approximately five. Substituting the above expressions into Eqs. (7) along with values of A and β given earlier, taking $\epsilon_{11} \approx \epsilon_{22} \approx \epsilon_{33} \approx 5$ appropriate to the class of oxygen-octahedra ferroelectrics, and dividing by two to give the scattering efficiency appropriate to either Stokes or anti-Stokes components separately, we obtain at room temperature

$$S_{33} \approx 0.12\times10^{-6}\ \varkappa_3^s P_s^2 / \lambda_\mu^4 \zeta^6\ (\text{cm-sr})^{-1}$$

$$S_{13} \approx 0.007\times10^{-6}\ \varkappa_3^s P_s^2 / \lambda_\mu^4 \zeta^6\ (\text{cm-sr})^{-1}$$

$$S_{42} \approx 0.016\times10^{-6}\ \varkappa_3^s P_s^2 / \lambda_\mu^4 \zeta^6\ (\text{cm-sr})^{-1} \qquad (8)$$

where λ_μ is the incident light wavelength in microns, and P_s is in C/m^2. Eqs. (8) give the absolute Raman scattering efficiencies in terms of the material quantity $\varkappa^s P_s^2$. The temperature dependence of this quantity can be calculated from the Devonshire theory[12]. We point out that expressions relating Raman scattering efficiencies to the linear electro-optic r-coefficients can be obtained from Eqs. (8) by making use of previously derived relations[3, 9] between the β and r tensors. Our results appear to differ by a factor of two from those of Kaminow and Johnston[8].

An analysis along the above lines can also be given for Brillouin scattering where the strain-induced energy-band shifts are described by the usual deformation potential D-tensor. The ratio of Raman to Brillouin scattering efficiency is then easily shown to be given by the expression

$$S_R/S_B \approx P_s^2 (\beta/D)^2 (\varkappa^s/\epsilon_0 s) \qquad (9)$$

where s is an elastic compliance coefficient. Taking $s \approx 10^{-11} m^2/N$, we find that $S_R/S_B \approx (\beta/D)^2 \varkappa^s P_s^2$. Since the magnitudes of β and D are similar, and since $\varkappa^s P_d^2 \gtrsim 10$, we generally expect that Raman scattering from the soft mode will be more intense than Brillouin scattering in oxygen-octahedra ferroelectrics.

NUMERICAL CALCULATIONS

To give numerical examples, we let $\lambda_\mu = 0.488$ so that Eqs. (8) become $S_{33} \approx 2.1\ \varkappa_3^s P_s^2/\zeta^6$, $S_{13} \approx 0.13\ \varkappa_3^s P_s^2/\zeta^6$, and $S_{42} \approx 0.28\ \varkappa_2^s P_s^2/\zeta^6$ in units of 10^{-6}

$(cm\text{-}sr)^{-1}$. These relations, we emphasize, have been derived in the classical limit where $\hbar\omega_p \ll kT$ ($\hbar\omega_p$ is the phonon energy) and thus should be applicable to ferroelectrics not too far from their Curie temperatures. In the more general case where $\hbar\omega_p \gtrsim kT$ (e.g., $LiNbO_3$ and $LiTaO_3$) the calculated Stokes line scattering efficiencies should be multiplied by the factor $(\hbar\omega_p/kT)\,[1 - \exp(-\hbar\omega_p/kT)]^{-1}$ [8]. In Table II we list the results of our calculations. Also listed (in parentheses) are the measured values[8] for $LiTaO_3$ and $LiNbO_3$.

TABLE II

Calculated Scattering Efficiencies in Units of $10^{-6}(cm\text{-}sr)^{-1}$ for Several Oxygen-Octahedra Ferroelectrics at Room Temperature

Material	ζ	$P_s(C/m^2)$	$\varkappa_3^s$	$\varkappa_2^s$	S_{33}	S_{13}	S_{42}
$BaTiO_3$ (a)	1.0	0.25	75	2300	10	0.6	40
$SrTiO_3$ (b)	1.0	0.01	300	300	0.07	0.004	0.01
$Ba_2NaNb_5O_{15}$ (c)	1.03	0.40	25	195	7	0.4	8
$LiTaO_3$ (d)	1.20	0.50	≈ 25	≈25	15(10.4)	1.1(0.49)	2.1(1.43)
$LiNbO_3$ (e)	1.20	0.71	16	22	10(16)	0.6(1.76)	1.5(3.8)

(a) Wemple, S.H., DiDomenico, Jr., M., and Camlibel, I., J. Phys. Chem. Solids 29, 1797 (1968).

(b) The values listed for $SrTiO_3$ assume that a dc biasing field is applied so as to give $P_s = 0.01\ C/m^2$.

(c) See Ref. [9]. The values for $\varkappa_2^s$ and $\varkappa_3^s$ have been obtained from unpublished data of A.W. Warner.

(d) See Ref. [9]. The listed values for $\varkappa_2^s$ and $\varkappa_3^s$ are only estimates since accurate data are not available.

(e) See Ref. [9]. The listed values for $\varkappa_2^s$ and $\varkappa_3^s$ are taken from Barker, Jr., A.S. and Loudon, R., Phys. Rev. 158, 433 (1967).

Based on the reasonable agreement shown in Table II between measured and calculated Raman scattering efficiencies, we conclude that the energy band-polarization fluctuation model presented here provides a valid quantitative description of Raman scattering from the soft TO mode in oxygen-octahedra ferroelectrics.

RAMAN LINE SHAPE

We now turn our attention to the soft mode Raman line shape. It is easy to show using the Nyquist theorem that the spectral density of polarization fluctuations in the classical limit is given by[13]

$$< | P^2(\omega) | > = (4kT/V)\, [\epsilon''(\omega) / \omega] \tag{10}$$

Here $\epsilon''(\omega)$ is the imaginary part of the dielectric dispersion function, and V is the volume. Using a damped harmonic oscillator model for the soft mode[13, 14] and the results given above, we arrive at the following expressions at room temperature for the absolute spectral density $dS/d\omega$ of Raman scattered light

$$dS_{33}/d\omega = 2.7\times10^{-6} \frac{2\pi\varkappa_3^s P_s^2 \Gamma_A \omega_{TO,A}^3 / \zeta^6}{\left(\omega_{TO,A}^2 - \Delta\omega^2\right)^2 + 4\Gamma_A^2 \omega_{TO,A}^2 \Delta\omega^2} \quad (\text{cm-sr/sec})^{-1}$$

$$dS_{42}/d\omega = 0.36\times10^{-6} \frac{2\pi\varkappa_2^s P_s^2 \Gamma_E \omega_{TO,E}^3 / \zeta^6}{\left(\omega_{TO,E}^2 - \Delta\omega^2\right)^2 + 4\Gamma_E^2 \omega_{TO,E}^2 \Delta\omega^2} \quad (\text{cm-sr/sec})^{-1}$$

$$dS_{13}/d\omega = \frac{1}{16}\, dS_{33}/d\omega \tag{11}$$

In Eqs. (11) $\omega_{TO,A}$ and $\omega_{TO,E}$ are the undamped mode frequencies of A and E modes respectively; Γ_A and Γ_E are corresponding dimensionless damping constants ($\Gamma = 1$ for critical damping); and $\Delta\omega$ is the Stokes shift from the laser exciting line.

The Raman line shapes predicted by Eqs. (11) have been reported elsewhere[13]. Using this model good agreement between theory and experiment has been observed for the overdamped ($\Gamma > 1$) soft E mode in $BaTiO_3$[13, 14]. We observe, based on our analysis, that the Raman line shape is a measure of the spectral density of the polarization fluctuations, and that the total integrated intensity (scattering efficiency) is a measure of the mean square polarization fluctuations.

REFERENCES

1. W. Cochran, Advan. Phys. 9, 387 (1960).
2. S. H. Wemple, M. DiDomenico, Jr., and A. Jayaraman, Phys. Rev. (to be published).
3. M. DiDomenico, Jr. and S. H. Wemple, Appl. Phys. Letters 12, 352 (1968).
4. J.R. Brews, Phys. Rev. Letters 18, 662 (1967).
5. A. H. Kahn and A. J. Leyendecker, Phys. Rev. 135, A1321 (1964).
6. A. Frova and P. J. Boddy, Phys. Rev. 153, 606 (1967).
7. M. DiDomenico, Jr. and S. H. Wemple, J. Appl. Phys. 40 (February, 1969).
8. I. P. Kaminow and W. D. Johnston, Jr., Phys. Rev. 160, 519 (1968).
9. S. H. Wemple, M. DiDomenico, Jr., and I. Camlibel, Appl. Phys. Letters 12, 209 (1968).
10. L. D. Landau and E. M. Lifshitz, "Electrodynamics of Continuous Media," Chap. XIV, Addison-Wesley Publishing Co. Inc., Reading, Massachusetts, 1960.
11. See, for example, "Fluctuation Phenomena in Solids," R. E. Burgess (ed.), Academic Press, New York, 1965.
12. S. H. Wemple and M. DiDomenico, Jr., J. Appl. Phys. 40 (February, 1969).
13. M. DiDomenico, Jr., S. H. Wemple, S. P. S. Porto, and R. P. Bauman, Phys. Rev. 174, 522 (1968).
14. M. DiDomenico, Jr., S. P. S. Porto, and S. H. Wemple, Phys. Rev. Letters 19, 855 (1967).

A-6: RAMAN SPECTRA OF CADMIUM CHLORIDE AND CADMIUM BROMIDE*

D. J. Lockwood
Department of Physics, University of Canterbury
Christchurch, New Zealand

ABSTRACT

The first-order lattice Raman spectra of cadmium chloride and cadmium bromide crystals were measured at room temperature on a Raman system comprising a two watt argon laser and a double monochromator. The spectra were recorded photoelectrically using conventional phase sensitive detection techniques. The Raman lines observed were assigned on the basis of a factor group analysis of the crystal structure (D_{3d}^{5}). The assignments were verified from the polarization properties, and confirmed by measurements of the first and second order infrared absorption spectra. With these Raman results established, it is possible to examine transition metal ion doped cadmium chloride and bromide crystals for electronic Raman scattering effects and determine low-lying electronic levels. Preliminary measurements on such systems are discussed.

INTRODUCTION

Crystals with the $CdCl_2$ structure after a period of neglect are now being extensively studied. In particular, $CdCl_2$ and $CdBr_2$ are very useful hosts for studying iron group transition metal ions. The crystals are optically clear, can be heavily doped, and the crystal field is only a slight trigonal distortion from cubic. Optical [1], ESR [2] and Jahn-Teller effect [3] measurements have been reported on transition metal ions doped in these crystals. A knowledge of the lattice vibrations of the host is often necessary in explaining such results.

In this paper the lattice vibrations of the $CdCl_2$ structure are analysed group theoretically. The theoretical results are used to interpret the Raman spectra of $CdCl_2$ and

*This work was supported in part by the U. S. Air Force Office of Scientific Research, under AFOSR Grant No. 1275-67, and by the New Zealand University Grants Committee.

$CdBr_2$. An attempt to measure the electronic Raman spectra of Fe^{2+} and Co^{2+} in these crystals is reported.

THEORY

The $CdCl_2$ structure shown in Fig. 1 is trigonal, space group D_{3d}^5, with one molecule to the unit cell [4]. The crystal is made up of layers of chlorine ions which are nearly cubic close-packed, with cadmium ions sandwiched between alternate chlorine layers. The layers are perpendicular to the [111] direction, the c axis of the crystal. Each cation is located at the centre of an octahedron consisting of six anions which is compressed along the [111] direction [4].

TABLE I

Number and Symmetry of Vibrational Modes in the $CdCl_2$ Structure

Representations of D_{3d}	Modes at k = 0 Acoustic	Optic	Selection Rules
A_{1g}	0	1	Raman
A_{2g}	0	0	Inactive
E_g	0	1	Raman
A_{1u}	0	0	Inactive
A_{2u}	1	1	IR(z)
E_u	1	1	IR(x, y)

The results of a factor group analysis [5] of the crystal structure are summarized in Table I. Because the crystal possesses a centre of inversion, no optical modes of vibration are simultaneously Raman and infrared active. Thus both Raman and infrared measurements are necessary to determine the lattice vibrations. All the optical modes are either infrared or Raman active, and hence, in principle, all the fundamental lattice frequencies can be measured directly.

Group theory predicts an infrared spectrum comprising one band with z axis polarization (A_{2u}) and one band polarized in the x, y plane (E_u). The Raman spectrum is expected to consist of two lines, one with A_{1g} symmetry and the other of E_g symmetry. The polarizability tensors for these symmetries have the following form [6].

$$A_{1g}: \begin{bmatrix} a & 0 & 0 \\ 0 & a & 0 \\ 0 & 0 & b \end{bmatrix}; \qquad E_g: \begin{bmatrix} c & 0 & 0 \\ 0 & -c & d \\ 0 & d & 0 \end{bmatrix}, \begin{bmatrix} 0 & -c & -d \\ -c & 0 & 0 \\ -d & 0 & 0 \end{bmatrix}.$$

Simple polarization measurements should provide sufficient information to enable a symmetry assignment of the observed frequencies.

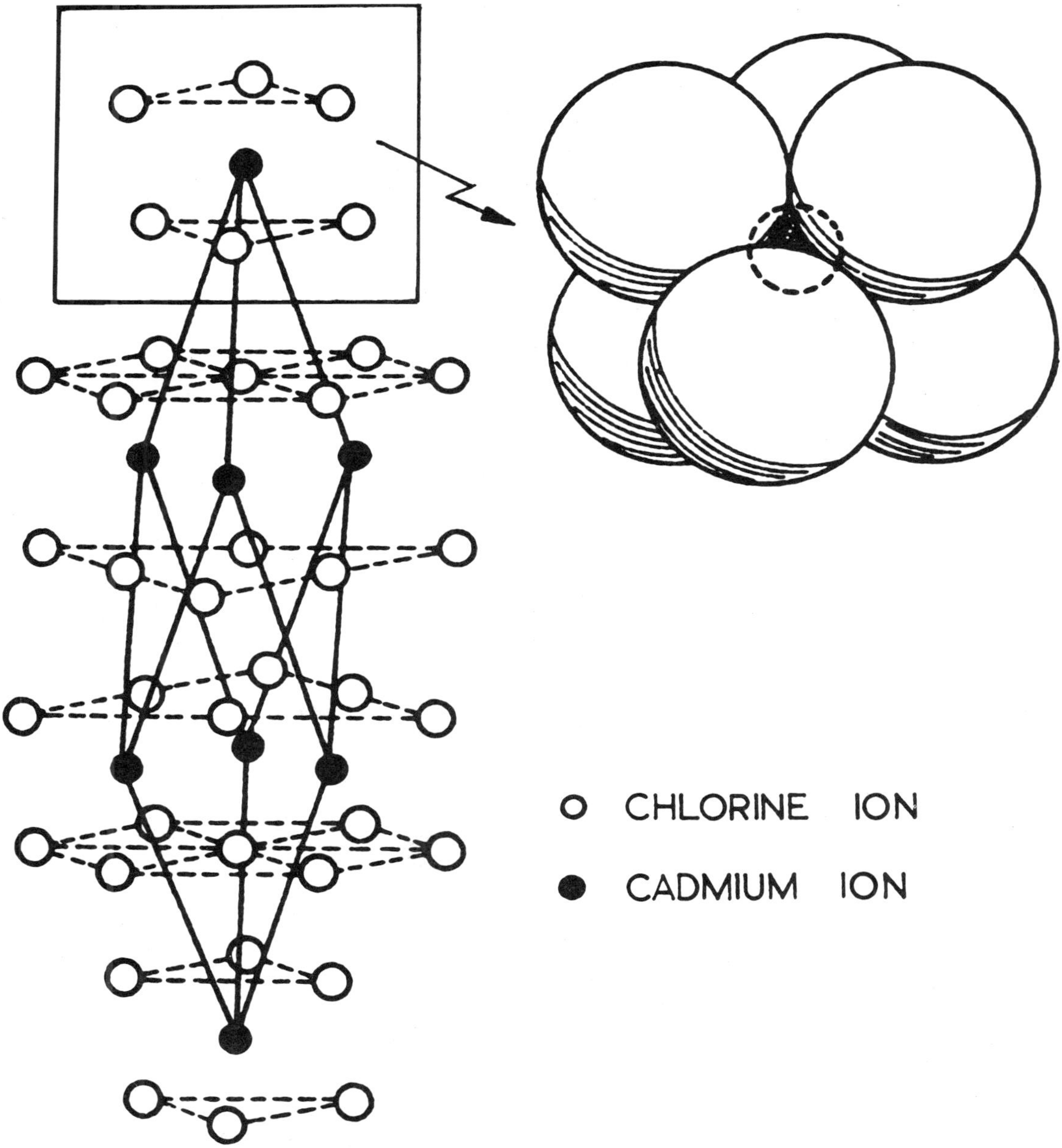

Fig. 1. Crystal structure of $CdCl_2$. (An adaption of Fig. 6 in K. Ôno, A. Ito and T. Fujita, J. Phys. Soc. Japan 19, 2119 (1964).)

Following Bhagavantam and Venkatarayudu [5], symmetry coordinates for the predicted lattice vibrations have been constructed. The results are shown pictorially in Fig. 2. It is immediately obvious that vibrations along the z axis are completely different from motions in the x, y plane. For example, in the Raman-active A_{1g} mode the planes of cadmium ions remain stationary while the planes of chlorine ions move directly against each other; whereas in the E_g mode the planes of chlorine ions move over each other with a sliding action as though they were acted upon by a shear force. Because of this difference, it would be reasonable to expect the A_{1g} mode to have a higher frequency than the E_g mode.

EXPERIMENTAL

Reagent grade powders of $CdCl_2$ and $CdBr_2 \cdot 4H_2O$ were used to grow the crystals. The powders were first dehydrated by heating in vacuum for several days, and then sealed into evacuated glass ampoules. The crystals were grown by slowly lowering the ampoules through a sharp temperature gradient. Crystals of $CdCl_2$ and $CdBr_2$ are deliquescent, the chloride more so than the bromide. To overcome rehydration problems, the ampoules in which the crystals were grown were also used as sample holders for the Raman measurements. This avoided exposing the crystals to atmospheric moisture. The ampoules were formed from 3-mm I. D. glass tubing, with the crystal-growing end flattened to form a window. Several single crystals, each at least 2.5 cm in length, were grown in this manner. The crystal orientation was determined from the direction of the cleavage planes, as these crystals cleave very readily along planes perpendicular to the c axis. The c-axis direction varied from crystal to crystal, and was usually inclined to the axis of the sample tube.

Raman spectra were recorded using conventional techniques. A Spacerays Model 5600 argon laser was used as the exciting source. Particular laser lines were chosen by means of a prism external to the laser. The 4800-Å laser line was predominantly used in this study; the power available in this line was normally at least 0.5 watt. A Jarrell-Ash double monochromator comprising a 1-m Czerny-Turner spectrometer coupled to a 0.33-m concave grating pre-monochromator was used to analyse the scattered light. The spectrometer output was detected using an EMI 9558QA photomultiplier, PAR HR-8 lock-in amplifier, and chart recorder. The laser beam was chopped inside the laser cavity to help discriminate against gas discharge lines. Plane-polarized laser light travelling parallel to the spectrometer slit was directed into the crystal through the window at the bottom of the sample tube. Scattered light was collected at 90° from incidence and focused on the spectrometer entrance slit. An analysing polarizer was placed between the sample and the spectrometer.

Typical room temperature spectra for $CdCl_2$ and $CdBr_2$ are shown in Fig. 3. The spectra were recorded under similar conditions, with a slit width of about 6 cm^{-1}. The $CdBr_2$ line is approximately five times stronger than the $CdCl_2$ line.

DISCUSSION OF RESULTS

As the spectra of Fig. 3 indicate, only one Raman line was found in both $CdCl_2$ and $CdBr_2$. The frequencies of the lines are 232.0 ± 1.2 cm^{-1} for the chloride and

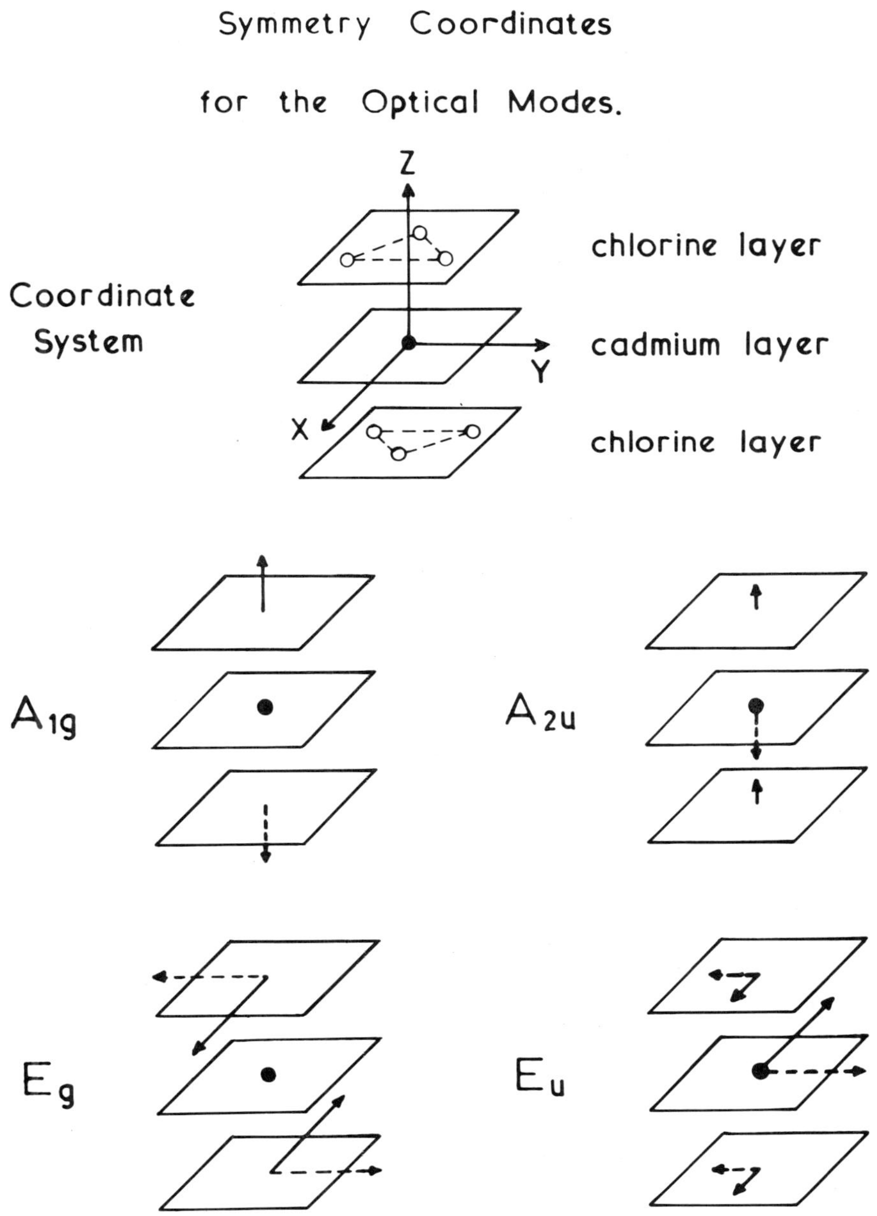

Fig. 2. Symmetry coordinates for the optical modes in $CdCl_2$. The z axis corresponds to the crystal c axis. The choice of the orthogonal x and y axes in the x,y plane is arbitrary.

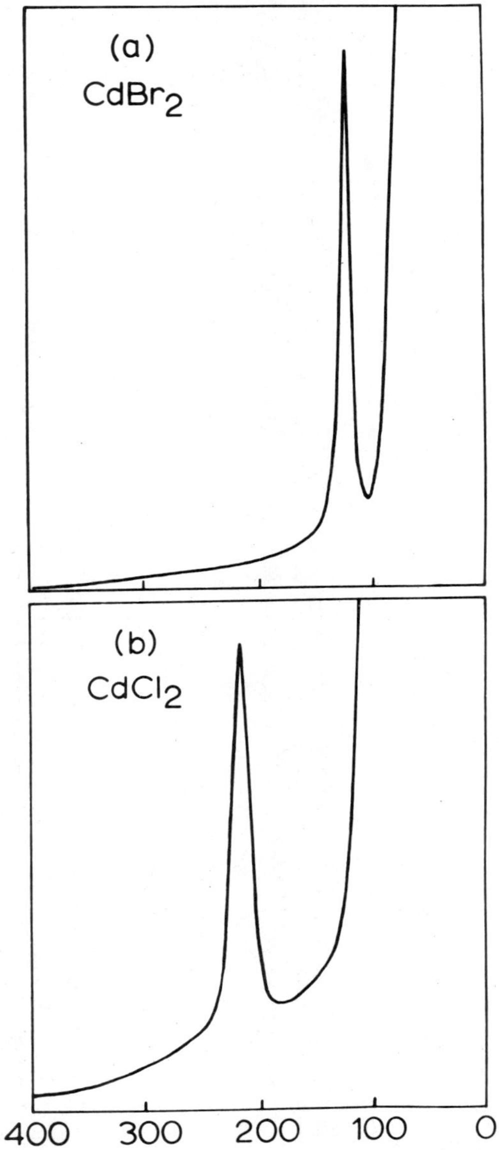

Fig. 3. Raman spectrum of (a) $CdBr_2$ and (b) $CdCl_2$ at room temperature.

147.3±1.0 cm^{-1} for the bromide, with linewidths of about 15 and 9 cm^{-1} respectively. Polarization studies of the scattered light indicate that the lines can be assigned to the E_g mode in each crystal. The chloride to bromide frequency shift is consistent with the fact that the E_g mode involves movement of the anions only. Electronic absorption spectra of Fe^{2+} in these crystals show vibronic side-bands such as those in Fig. 4. The energy shifts confirm that the Raman results are of the right order.

In order to facilitate the search for the missing lines, Raman measurements were made on a single crystal of $CdBr_2$ that was not enclosed in glass. The crystal was shaped into a cube of dimension 4 mm on a side, with two opposite faces being cleavage planes perpendicular to the crystal c axis. The other faces were arbitrarily cut at right angles to the x, y plane. Different orientations of the crystal relative to the polarization of the laser light and the scattered light were tried. With the experimental conditions z(yx)y and z(xz)y*, the frequency assigned to the E_g mode was observed. Under x(zz)y and z(xx)y conditions, where, according to the polarizability tensor, the A_{1g} mode should appear, no other frequency was obtained. This Raman line must be too weak to be seen.

The crystals were of indifferent optical quality. Cleavage planes and other crystal-growth imperfections were evident in all the samples. Because of this, the crystals scattered light by reflection and refraction, and the effect of this can be seen in Fig. 3 where the tail of the exciting line intrudes upon the spectrum. There is also a background of laser gas discharge lines at about one fifteenth the intensity of the $CdCl_2$ Raman line. These discharge lines were the limiting factor in finding the missing Raman lines. Better quality crystals are required to overcome these problems, and attempts to grow them are in progress.

Even though as yet the lines have not been observed, their frequencies can be predicted. First- and second-order infrared absorption spectra provide the information required. Some results of such measurements [8] are given in Table II.

TABLE II

Assignment of Peaks in the Second-Order Infrared Spectrum of $CdCl_2$ and $CdBr_2$

Crystal	Fundamentals (cm^{-1}) Infrared	+ Raman	= Allowed Combinations (cm^{-1}) Theoretical	Experimental
$CdCl_2$	205	232	437	430
	205	A_{1g}	-	555
$CdBr_2$	150	147	297	290
	150	A_{1g}	-	375

*The polarization notation of Damen et al. [7] is used here.

These results confirm the frequencies obtained for the E_g mode, and predict that the A_{1g} mode frequencies are 350±15 cm^{-1} for $CdCl_2$ and 225±15 cm^{-1} for $CdBr_2$.

ELECTRONIC RAMAN EFFECT

With the first-order lattice spectrum known, it is possible to examine the electronic Raman spectra of ions doped into $CdCl_2$ and $CdBr_2$ crystals. Several transition metal chlorides and bromides have the same structure as $CdCl_2$ [4], and thus high doping concentrations of these ions can be achieved. The ions chosen for detailed study were iron and cobalt. Cobalt doped $CdCl_2$ is blue in colour, and the bromide is green; the

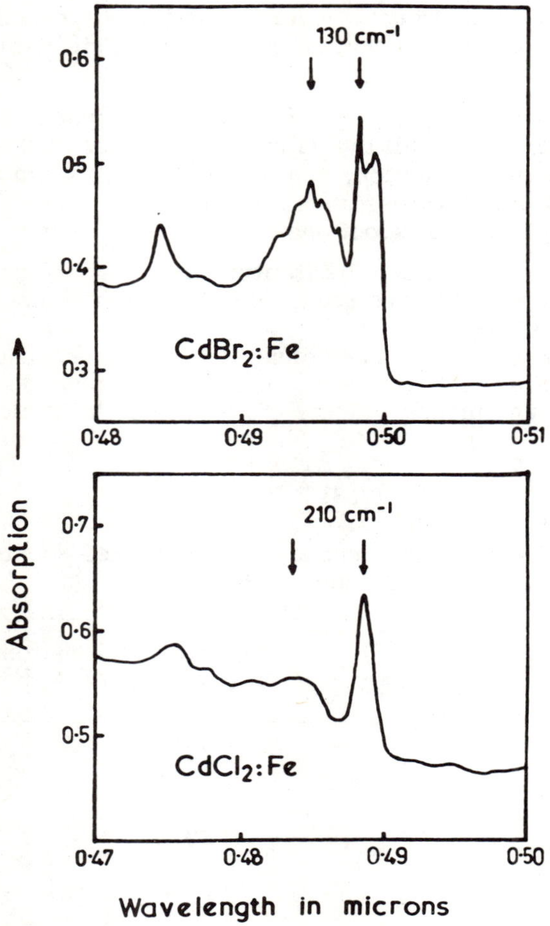

Fig. 4. Absorption spectra of Fe^{2+} in $CdBr_2$ and $CdCl_2$ at 4.2°K. The arrows mark the position of the electronic line and the first main peak in the vibronic side-band. The spacing between arrows is given in cm^{-1}.

iron doped crystals have a pale yellow colour. Thus these crystals are ideal for study using the blue-green argon laser lines. The Raman scattering in the case of the ferrous ion could be considerably enhanced under excitation by the 4880-Å line. As Fig. 4 shows, this line is close to weak Fe^{2+} absorptions.

Little is known about the low-lying energy levels of iron and cobalt in $CdCl_2$ and $CdBr_2$. The effect of a trigonal crystal field on the ground state of Fe^{2+} and Co^{2+} is shown in Fig. 5. The order of the lowest two levels is determined by the sign of the crystal field. ESR measurements on Co^{2+} in $CdCl_2$ and $CdBr_2$ [9] have shown that the orbital singlet ($^4A_{2g}$) is lowest. This means that for Fe^{2+} the doublet is lowest [10]. The energy levels as drawn are also split by the spin-orbit interaction, and additional complications arise from Jahn-Teller splitting of the degenerate E_g levels [3].

The energies of the ^5E_g-$^5A_{1g}$, $^4A_{2g}$-^4E_g transitions and Jahn-Teller splitting of the ^5E_g ground state were investigated by Raman scattering. Crystals of $CdCl_2$ and $CdBr_2$ doped with Fe^{2+} and Co^{2+} at five mole percent concentrations were examined at room temperature using the 4880-Å laser line. Careful examination of the 80-1200 cm^{-1} region at the limit of sensitivity of the present equipment failed to show any

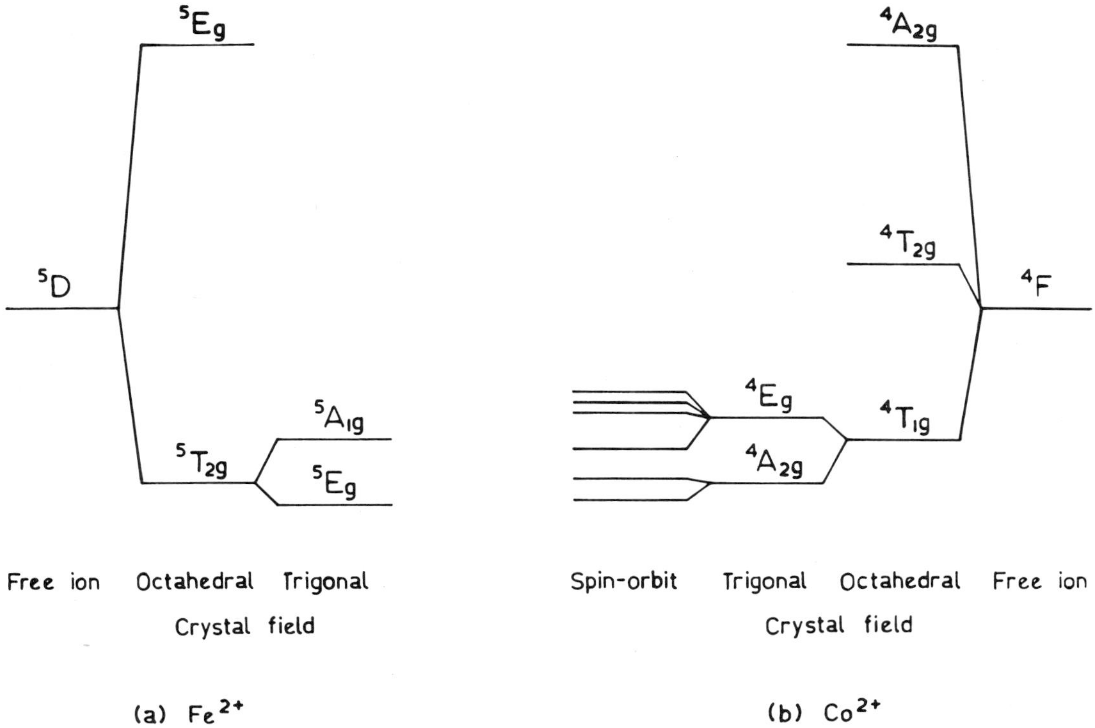

Fig. 5. Ground state energy-level splittings for (a) Fe^{2+} and (b) Co^{2+} ions in $CdCl_2$.

electronic Raman lines. Room temperature infrared measurements have shown that there is a broad electronic absorption centred at approximately 700 cm^{-1} in $CdCl_2:Fe^{2+}$ [3]. An infrared study of $CdCl_2:Co^{2+}$ has produced a similar band at 1160 cm^{-1}. Further low temperature Raman work is in progress to see if these transitions can be observed, and also to evaluate the Fe^{2+} ground-state splitting.

ACKNOWLEDGEMENTS

I am grateful to Professor A. G. McLellan for the opportunity to carry out this work and to Dr. G. D. Jones for bringing this crystal system to my attention. I wish to thank T. E. Freeman for providing his Fe^{2+} absorption spectra results prior to publication, J. H. Christie for assistance with the Raman measurements, and R. Ritchie for growing the crystals.

REFERENCES

1. H. Matsumoto, J. Phys. Soc. Japan 20, 1579 (1965).
2. T. Iri and G. Kuwabara, J. Phys. Soc. Japan 24, 127 (1968) and references therein.
3. T.E. Freeman and G.D. Jones (to be published).
4. R.W.G. Wyckoff, "Crystal Structures," 1, Interscience Publishers, New York, 1964.
5. S. Bhagavantam and T. Venkatarayudu, "Theory of Groups and Its Application to Physical Problems," Andhra University, Waltair, India, 1951.
6. R. Loudon, Advan. Phys. 13, 423 (1964).
7. T.C. Damen, S.P.S. Porto and B. Tell, Phys. Rev. 142, 570 (1966).
8. D.J. Lockwood (to be published).
9. K. Morigaki, J. Phys. Soc. Japan 16, 1639 (1961).
10. J.H. Van Vleck, Discussions Faraday Soc. 26, 96 (1958).

A-7: THE CROSS - SECTIONS OF THE RAMAN SCATTERING OF LIGHT IN CRYSTALS AND CRYSTALLINE POWDERS.

V. S. Gorelik, V. S. Rjazanov, M. M. Sushchinskii
P. N. Lebedev Physical Institute, Academy of Sciences
Moscow, USSR

ABSTRACT

The method of measuring cross-sections of the Raman effect of light in single crystals and powders is elaborated. The cross-sections are measured in stilbene, benzene, $NaClO_3$, GaP, CdS and other crystals under the excitation by different light sources. The temperature dependence of cross-sections is studied, and in GaP a sharp increase in the cross-section is detected with the increase of temperature. The cross-sections in stilbene and benzene are found to increase at the transition from liquid to crystal. The theoretical interpretation of the results obtained is given.

METHOD

The values of the cross-sections of the Raman scattering in crystals are of great interest; however, their direct measurement is difficult due to the existence of absorption and scattering in sample inhomogeneities. These difficulties can be overcome with the method[1-3] employing the theory of propagation of the exciting emission and the Raman effect in dispersion medium[4] and allowing to measure the cross-sections of the Raman scattering both in single crystals and polycrystalline powders. In the theory mentioned the dispersion medium is characterized by two parameters R and L connected with the dimensions of the powder particles and absorption of medium (R - the reflection coefficient of the infinitely thick layer, L - the effective absorption coefficient). The cross-section of the Raman scattering for one molecule in an angle 4π steradian in the experiment with a cuvette filled with polycrystalline powder in the approximation of the spherical indicatrice of scattering is to be determined from the expression:

$$X_o = \frac{4\pi Y_s}{x\, Y_o\, N\, f(R, Lx)\, \delta\Omega} \qquad (1)$$

where Y_s -the Raman scattering intensity, Y_o -the intensity of the exciting line, $\delta\Omega$ -the solid angle of a beam gathered in the cuvette; x -a layer thickness of powder in the cuvette, $f(R_1 Lx)$ -a certain function[2], N -a number of molecules in a unit of volume.

In (1) there are two unknown values - R and L. The value L can be determined by measuring the dependence of the intensity of the exciting light Y_e passed through the layer, upon the thickness of the latter. This dependence [4] is represented by the formula

$$Y_e = Y_o e^{-Lx} \tag{2}$$

(Y_o -the exciting light intensity at the cuvette's entrance). On the other hand by measuring the intensity of the Raman scattering lines under various thicknesses of layer one can calculate the value R.

The cross-section of the Raman scattering in the single crystal for the given solid angle $\delta\Omega$ can be calculated with the formula

$$X_o \delta\Omega = \frac{Y_s}{Y_o x e^{-Lx} N} \tag{3}$$

(X-the thickness of crystal, the other designations are the same as in (1)); this formula follows from (1) under the condition that the parameter R=0 for single crystals, which is confirmed experimentally.

By using the above-mentioned method we have measured the cross-sections of the Raman scattering for a number of crystals, crystalline powders and for liquids as well. The measurements were carried out with a double monochromator DFS-12; a mercury line was used as an exciting line $\lambda = 4358A$ and the line $\lambda = 6328A$ of the neon-helium laser with the capacity of 60 mw. A photomultiplier served as a detector.

To prove the above-described method we have measured the cross-sections of the Raman scattering for single crystals and polycrystalline powder of stilbene and GaP; the experiments showed that the cross-sections appeared the same both in a single crystal and in a powder within the limits of measurement error [2].

The results of measurements for solid and liquid stilbene and benzene, for the crystals of $NaC10_3$, GaP, and CdS are illustrated by Fig. 1, 2 and by the Table I. The results obtained for a liquid agree with the literature ones. It is worth mentioning a sharp difference of the cross-section values in a liquid and solid state discussed in paper [7]. The temperature dependence of cross-sections of small frequencies in benzene and $NaC10_3$ agrees well with the known theoretical dependence [8] except the line $\Delta\nu = 179\ cm^{-1}$ in $NaC10_3$; its anomalous properties are due to the phase transition in this crystal [9, 10].

A sharp increase in the cross-section of the Raman scattering with the temperature growth for GaP is, probably, explained by the increase of the absorption coefficient for the exciting line ($\lambda = 6328A$) connected with the intensity of lines of the Raman scattering[3].

The cross-section of the Raman scattering for the line $\Delta\nu = 207\ cm^{-1}$ CdS is considerably smaller in comparison with GaP which is due, in our opinion, to the fact that in GaP a covalent character of coupling is realized, and also by a proximity of the absorption band in the last crystal.

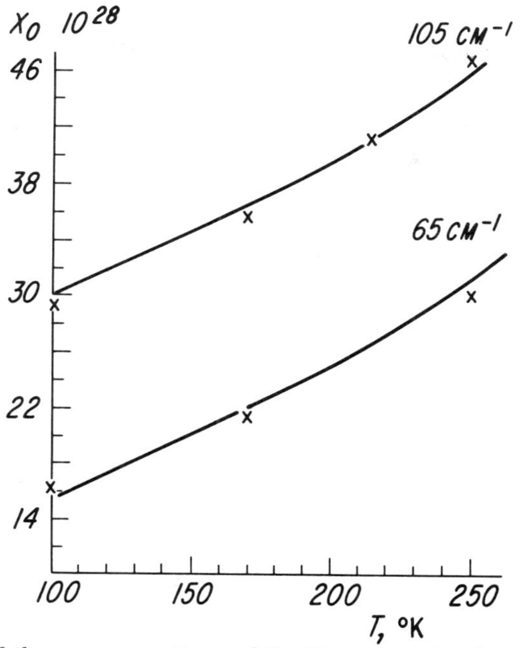

Fig. 1. The dependence of the cross-sections of the Raman scattering on temperature for low frequency benzene lines. Full lines- the theoretical dependence [8]; points - experimental data.

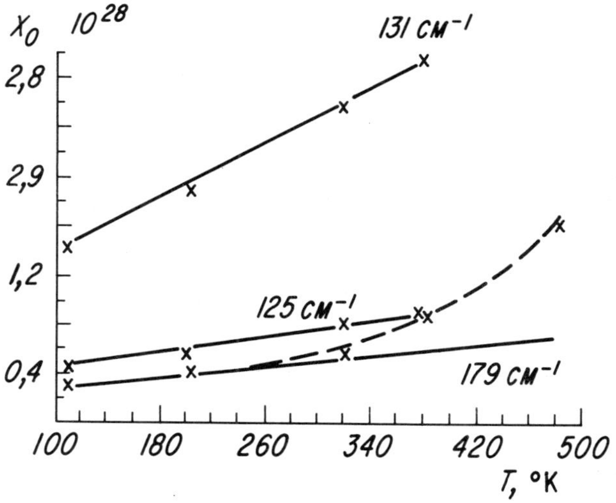

Fig. 2. The dependence of the cross-section of the Raman scattering on temperature for the lines of $NaClO_3$ of low frequencies. Full lines- the theoretical dependence [8]; points - experimental data; the temperature dependence for the line $\Delta\nu$=179 cm^{-1} is depicted by a dotted line.

TABLE I

Substance frequency	T°K	$10^{28} cm^2$
Benzene, crystalline powder $\Delta\nu = 922 cm^{-1}$	101 172 252	46 55 55
Benzene, liquid $\Delta\nu = 992 cm^{-1}$	293	2, 0
Stilbene, crystalline powder, single crystal $\Delta\nu = 1593 cm^{-1}$	293	460
Stilbene, liquid $\Delta\nu = 1593 cm^{-1}$	420°K	25
$NaClO_3$, crystalline powder $\Delta\nu = 936 cm^{-1}$	101 293 483	7, 7 7, 2 7, 9
GaP $\Delta\nu = 402 cm^{-1}$ cryst. powder	103 293 723	200 390 980
GaP $\Delta\nu = 402$ single crystal	293	370
CdS crystalline powder $\Delta\nu = 207 cm^{-1}$	293	0, 7

REFERENCES

1. V. S. Rjazanov, M. M. Sushchinskii Optika i spectroskopia (USSR) 23, 580 (1967).
2. V. S. Rjazanov, M. M. Sushchinskii JETP (USSR) 54, 1099 (1968).
3. V. S. Rjazanov, V. S. Gorelik, G. V. Peregudov, M. M. Sushchinskii and V. A. Chirkov Fizika Tverdogo Tela (USSR), 10, 1909 (1968).
4. R. E. Danil'tseva, V. A. Zubov, M. M. Sushchinskii, I. K. Shuvalov Proc. of the 15th conference on spectroscopy, USSR, VINITI, I, 696 (1964).
5. I. I. Kondilenko, P. A. Korotkov, V. L. Strizhevskii optika i spectroscopia (USSR), 8, 471 (1960); Physical Problems of Spectroscopy (USSR), I, 352 (1962).
6. I. I. Kondilenko, P. A. Korotkov Optika i spectroskopia (USSR) 17, 1051 (1964).
7. L. A. Shelepin JETP (USSR), 54, 5 (1968).
8. R. S. Landsberg, L. I. Mandelstam Zs. Phys., 58, 25 (1929).
9. V. S. Gorelik, I. V. Gavrilova, I. S. Zheludev, G. V. Peregudov, V. S. Rjazanov, M. M. Sushchinskii; JETF Letters (USSR) 5, 214 (1967)
10. V. S. Gorelik, G. V. Peregudov, V. S. Rjazanov, M. M. Sushchinskii, Modern Optics, New York, p. 349 (1967).

B-1: LIGHT SCATTERING FROM PHONONS

F. A. Johnson
Royal Radar Establishment
Malvern Worcestershire, England

ABSTRACT

In this paper we shall discuss the physics of light scattering from phonons and in particular the relationship between the electronic terms that define the phonon frequencies and the electronic terms that define the scattering cross-sections.

INTRODUCTION

It is well known that when light is scattered inelastically from a phonon the frequency ω_s and wave-vector k_s of the scattered photon is related to the frequency ω_i and wave-vector k_i of the incident photon by the equations

$$\begin{aligned} \omega_s &= \omega_i - \omega_o \\ k_s &= k_i - q \end{aligned} \tag{1}$$

where ω_o and q are the frequency and wave-vector of the emitted phonon and are related by the equation:

$$\omega_o = c_o |q| \tag{2}$$

where c_o is the phase velocity of the phonon. Further as ω_i is normally very much greater than ω_o we can write:

$$|q| = 2 |k_i| \sin \frac{\theta}{2} \tag{3}$$

where θ is the scattering angle. Experimentally one makes the distinction between Brillouin scattering where c_o is constant and thus ω_o varies with θ and Raman scattering where ω_o is a constant and thus c_o varies with θ.

More complex spectra result from scattering from two phonons since in this case we redefine the ω_o and q of Eqs. (1), (2) and (3) as

$$\begin{aligned} q &= q_1 + q_2 \\ \omega_o &= \omega_1 + \omega_2 \end{aligned} \tag{4}$$

for the case of two phonon emission and consequently phonons from the entire Brillouin zone can participate.

It is the purpose of this paper to discuss the origin of the scattering mechanism which leads to the simple results described above.

ELEMENTARY DIFFRACTION THEORY

The simplest and oldest explanation[1] is to assume that the phonon modulates the refractive index of the crystal so producing a moving diffraction grating as shown in Fig. 1.

Clearly we can write the path difference as

$$\frac{2\pi c_i}{\omega_i} = 2d \sin \frac{\theta}{2} \tag{5}$$

where c_i is the phase velocity of the incident light wave and for a single phonon

$$d = \frac{2\pi c_o}{\omega_o} \tag{6}$$

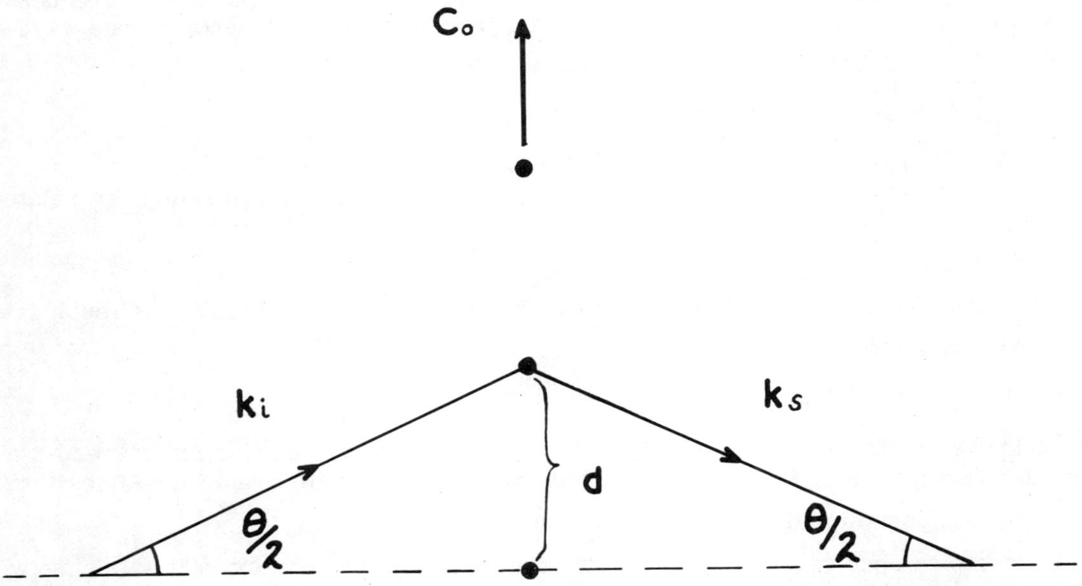

Fig. 1. The moving diffraction grating.

Now the doppler shift is given by

$$\frac{\Delta \omega_i}{\omega_i} = -2 \frac{c_o}{c_i} \sin \frac{\theta}{2}$$

$$= - \frac{\omega_o}{\omega_i}$$

thus

$$\Delta\omega_i = - \omega_o$$

$$= \omega_s - \omega_i \qquad (7)$$

Again in the case of two phonons the amplitude of the displacement is proportional to

$$\sin (\omega_1 t + q_1 \cdot r) + \sin (\omega_2 t + q_2 \cdot r)$$

$$= 2 \sin [\tfrac{1}{2}(\omega_1 + \omega_2)t + \tfrac{1}{2} q \cdot r] \cos [\tfrac{1}{2}(\omega_1 - \omega_2)t + (q_1 + \tfrac{1}{2} q) \cdot r]$$

where $q = q_1 + q_2$ and $|q| << |q_1|$ thus we have a rapid cosine function of r with a slow sine envelope whose length parallel to r is given by

$$d = 2\pi / |q| \qquad (8)$$

and whose phase velocity is

$$c_o = \frac{\frac{1}{2}(\omega_1 + \omega_2)}{\frac{1}{2}|q|} = \frac{\omega_1 + \omega_2}{|q|} \qquad (9)$$

It will be readily seen that Eqs. (5) through (9) are clearly equivalent to Eqs. (1) through (4), but we have now reduced the problem to the modulation of the refractive index of the crystal by one or more phonons.

PHENOMENOLOGICAL THEORIES

The classical approach[2] is to assume that the dielectric susceptibility tensor can be expanded as a power series in nuclear displacements from the equilibrium configuration thus

$$P^{(0)} + \lambda P^{(1)} + \lambda^2 P^{(2)} \ldots$$

$$= [X^{(0)} + \lambda X^{(1)} + \lambda^2 X^{(2)} + \cdots] E \qquad (10)$$

where P is the polarisation, E the electric field and X the susceptibility tensor. The terms $P^{(1)}$ and $P^{(2)}$ give rise to the one-phonon and two-phonon scattered radiation. In the special case of Brillouin scattering $X^{(1)}$ can be directly related to the elasto-optic tensor which can be measured independently and a reasonably satisfactory phenomeno-logical theory can be established. In the case of Raman scattering however this is not possible and one must resort to some oversimplified scheme to reduce $X^{(1)}$ and $X^{(2)}$ to a few adjustable parameters.

A more instructive approach is to write down Maxwells equations for the field in the crystal

$$\nabla\cdot \epsilon E = 0 \quad \nabla \times H - \frac{\partial}{\partial t} \epsilon E = 0$$

$$\nabla\cdot B = 0 \quad \nabla \times E + \frac{\partial}{\partial t} B = 0 \tag{11}$$

where ϵ is the dielectric permittivity tensor and expand all terms in powers of displacements of the nuclei from their equilibrium positions. In this way we can readily obtain a set of equations for all orders of the scattered field which can be written as follows:

$$\nabla\cdot \epsilon^{(0)} E^{(n)} = -\nabla\cdot P^{(n)} \quad \nabla \times H^{(n)} - \frac{\partial}{\partial t} \epsilon^{(0)} E^{(n)} = \frac{\partial}{\partial t} P^{(n)}$$

$$\nabla\cdot B^{(n)} = 0 \quad \nabla \times E^{(n)} + \frac{\partial}{\partial t} B^{(n)} = 0$$

where

$$P^{(n)} = \sum_{m=1}^{n} \epsilon^{(m)} E^{(n-m)} \quad \text{for } n \geq 1$$

$$= 0 \quad \text{for } n = 0 \tag{12}$$

Thus the radiation from n-phonon scattering processes $E^{(n)}$ obeys a set of Maxwells equations with an effective charge density $-\nabla\cdot P^{(n)}$ and an effective current density $\frac{\partial}{\partial t} P^{(n)}$. We note, in particular, that we can write

$$P^{(1)} = \epsilon^{(1)} E^{(0)}$$

and

$$P^{(2)} = \epsilon^{(1)} E^{(1)} + \epsilon^{(2)} E^{(0)} \tag{13}$$

thus the one-phonon scattered radiation arises from the direct interaction of the incident field with the first order permittivity but the two-phonon scattered radiation arises from two sources (i) the direct interaction of the incident field with the second order permittivity and (ii) the repeated first order scattering processes.

MICROSCOPIC THEORY OF SCATTERING

In the Hartree-Fock approximation the zero-order dielectric permittivity is determined by summing over all virtual transitions that result from the electron-photon coupling. This is illustrated diagramatically in Fig. 2 along with the diagrams appropriate to the various terms appearing in Eq. 13.

This result, which is identical with that derived by Loudon[3], is not very transparent since these perturbation terms must summed over all possible inter-band transitions and the prospect of starting with a calculated band structure and phonon spectrum and calculating the scattering cross-section is distinctly forbidding. However, it is important to note that while we have microscopic theories of the band structure and can formulate them for the dielectric constant we do not have a microscopic theory for the phonons themselves.

MICROSCOPIC THEORY OF PHONONS

In the adiabatic approximation[4] the effective nuclear potential is given by the expression:

$$V = V_{nn} + \mathcal{E} \tag{14}$$

where V_{nn} is the nuclear-nuclear electrostatic interaction and $\mathcal{E}$ is the total energy of the electron system for the particular nuclear configuration. The phonon frequencies are defined by the second order variation $V^{(2)}$ due to displacements of the nuclei from their equilibrium configuration thus we can write

$$V^{(2)} = V^{(2)}_{nn} + \mathcal{E}^{(2)} \tag{15}$$

In the Hartree-Fock approximation $\mathcal{E}^{(2)}$ is determined by summing over all virtual transitions that result from electron-one phonon and electron-two phonon coupling as illustrated in Fig. 3. Now we immediately see a striking similarity between these diagrams and those for the two-phonon scattering shown in Fig. 2. In fact we can regard the two-phonon scattering as an electromagnetic probe of the terms defining $\mathcal{E}^{(2)}$ i.e. the electronic contribution to the phonon spectrum.

Now in the case of the virtual transitions in Fig. 3 we can formally carry out the summation and the result is given by[5]

$$\mathcal{E}^{(2)} = \frac{1}{2}\int \rho^{(1)}(r)\, V^{(1)}_{en}(r)\, d^3r + \int \rho^{(0)}(r)\, V^{(2)}_{en}(r)\, d^3r \tag{16}$$

where $\rho^{(0)}(r)$ and $\rho^{(1)}(r)$ are the zero and first order electronic charge densities and $V^{(1)}_{en}(r)$ and $V^{(2)}_{en}(r)$ are the first and second order electron-nuclear electrostatic potential energies - the first integral comes from the repeated electron-one phonon matrix elements and the second from the electron-two phonon matrix elements. Again we see a formal similarity between $\mathcal{E}^{(2)}$ and $P^{(2)}$ although not quite so striking as before.

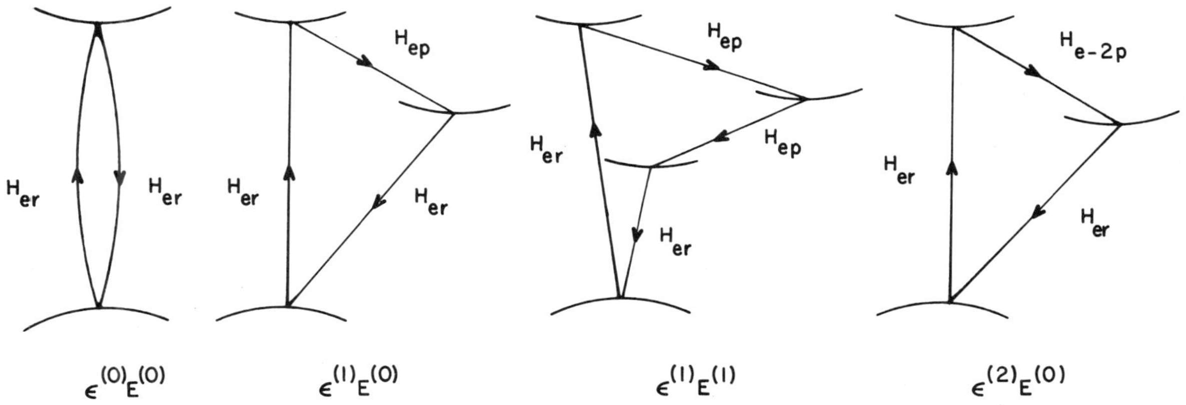

Fig. 2. Virtual transitions invalued in $\epsilon^{(0)}$, $\epsilon^{(1)}$ and $\epsilon^{(2)}$ Matrix elements: H_{er}, electron-radiation; H_{ep}, electron-one phonon; H_{e-2p}, electron-two phonon.

At this point the objection could be raised that excitons play an important role in determining Raman spectra[6, 7] and in determining the dielectric permittivity of a crystal and that this parallel between the microscopic theories of phonons and scattering theory is true only in the Hartree-Fock approximation. However, it can be shown[8] that the expansion for $\mathcal{E}$ based on the solutions of the exact many-electron Schrodinger equation can be written to all orders as:

$$\mathcal{E}^{(n)} = \sum_{m=1}^{n} \frac{m}{n} \int \rho^{(n-m)}(r)\, V_{en}^{(m)}(r)\, d^3r \tag{17}$$

Comparing this equation with Eq. 12 we can now state the physics of the problem as follows:

(a) the presence of a phonon in a crystal produces a modification in the electronic charge density which largely determines the frequency of the phonon.

(b) the phonon also produces a modification in the dielectric permittivity tensor-i. e. in the response of the electrons to an external electric field - which determines the scattering cross section.

Now the phonon modifies the charge density by modifying the wave functions and one consequence of this is that the optical absorption spectrum of the crystal is modified i. e. the imaginary part of the dielectric permittivity tensor is changed. As a consequence the real part of the dielectric tensor is also changed and we can write

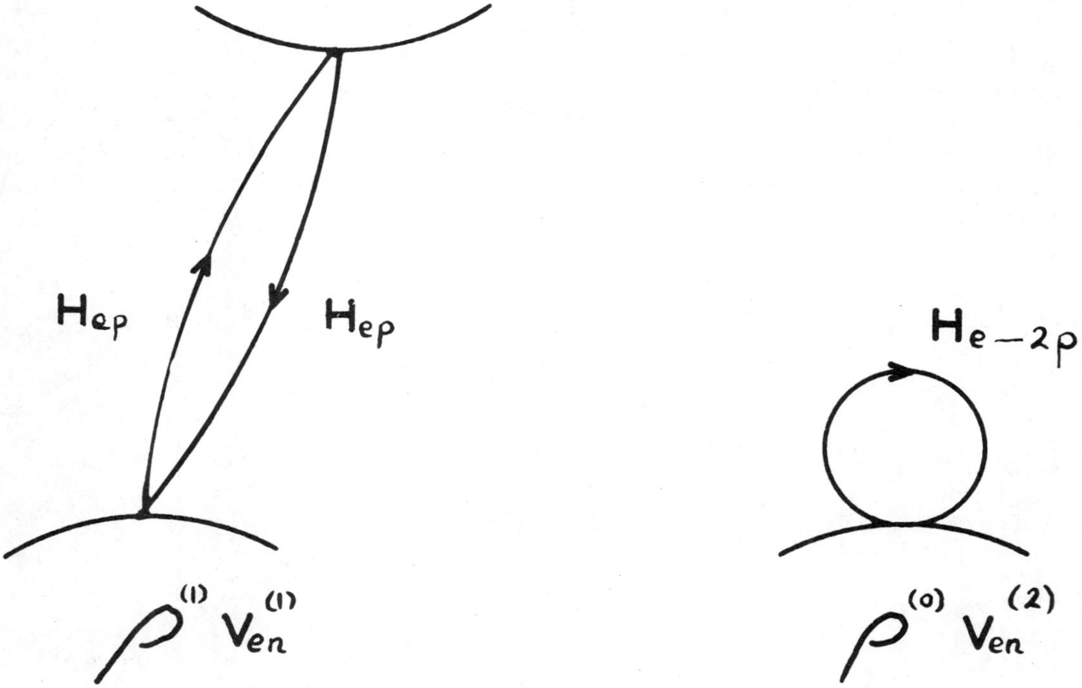

Fig. 3. Virtual transitions invalued in $\rho^{(1)}(r)\, V_{en}^{(1)}(r)$ and $\rho^{(0)}(r)\, V_{en}^{(2)}(r)$.

$$\epsilon_1^{(0)}(\omega_0) - 1 = \frac{2}{\pi}\int_0^\infty \frac{\omega\,\epsilon_2^{(0)}(\omega)}{\omega^2 - \omega_0^2}\,d\omega$$

and

$$\epsilon_1^{(n)}(\omega_0) = \frac{2}{\pi}\int_0^\infty \frac{\omega\,\epsilon_2^{(n)}(\omega)}{\omega^2 - \omega_0^2}\,d\omega \qquad \text{for } n \geq 1 \tag{18}$$

where $\epsilon_1^{(n)}(\omega_0)$ and $\epsilon_2^{(n)}(\omega_0)$ are the real and imaginary parts of the n'th order dielectric tensor at the frequency ω_0. We can now draw a number of conclusions from Eq. 18 namely:

(a) for a uniform translation of the crystal $\epsilon_2^{(n)}(\omega) = 0$ for $n \geq 1$ thus $\epsilon_1^{(n)}(\omega) = 0$ and as a consequence we expect $P^{(n)}$ to tend to zero for acoustic modes as q tends to zero.

(b) if $\epsilon_2^{(n)}(\omega)$ has a rapid onset at some frequency ω_0 then $\epsilon_1^{(n)}(\omega)$ will increase very rapidly as ω tends to ω_0 from the low frequency side and we shall see resonance enhancement of the scattering.

(c) $\epsilon_2^{(1)}(\omega)$ can be measured directly by stress modulation techniques and Eq. 18 could then be used to calculate $\epsilon_1^{(1)}(\omega)$ and hence Brillouin scattering cross-section.

To summarise we note that the phonon frequencies depend on modifications to the ground state wave functions and the scattering cross-sections depend on modifications to both the ground and excited state wave functions.

REFERENCES

1. L. Brillouin, Ann. d. Phys. 17, 88, (1922).
2. M. Born and M. Bradburn, Proc. Roy. Soc. A188, 161, (1947).
3. R. Loudon, Proc. Roy. Soc. A275, 218, (1963).
4. M. Born and K. Huang, "Dynamical Theory of Crystal Lattices," Oxford University Press, 1954.
5. F.A. Johnson, Proc. Roy. Soc. (to be published).
6. R.C. Leite and S.P.S. Porto, Phys. Rev. Letters 17, 10, (1966).
7. J.L. Birman and A.K. Ganguly, Phys. Rev. Letters, 17, 647, (1966).
8. P.D. DeCicco and F.A. Johnson, Proc. Roy. Soc. (to be published).

B-2: THEORETICAL INTERPRETATION OF THE SECOND-ORDER RAMAN SPECTRA OF THE ALKALI FLUORIDE CRYSTALS*

John R. Hardy
Behlen Laboratory of Physics, University of Nebraska
Lincoln, Nebraska
and
Arnold M. Karo
Lawrence Radiation Laboratory, University of California
Livermore, California

ABSTRACT

In two recent papers we have presented theoretical calculations of the second-order Raman spectra of CsF and NaF. In the case of the latter salt, the agreement with experimental measurements is extremely good. In the case of the former salt, no experimental data are available with which comparison can be made. The object of the present paper is to present systematic calculations of the second-order Raman spectra of the sequence of salts, NaF, KF, RbF, and CsF, so that we may investigate the effects of the different vibrational spectra of these crystals on the computed second-order spectra. In all four cases, we have used two alternative models for the polarizability tensor. These are designed to reduce the eight disposable parameters obtained when one uses a polarizability tensor which is assumed to depend only on the configuration of nearest-neighbor ions to a total of three. For the earlier calculation on NaF, we found that there was no marked difference between the results obtained from the two models, whereas in the case of CsF there was a drastic difference between the two sets of results. The object of the present calculations is to observe the manner in which this difference develops as one proceeds through the fluoride sequence.

DESCRIPTION OF THE CALCULATIONS

In the present paper we shall discuss an extension of our earlier work[1,2] on the second-order Raman spectra of the fluoride sequence of alkali halide crystals, with the objective of obtaining a systematic understanding of the spectra for the sequence from NaF through CsF. In the previous papers we described the theory in detail and applied it to NaF. Extremely good agreement was found between the computed and the experimentally measured spectra. The same theory was then applied to CsF with corresponding results. However, there is no known experimental test of our results for this crystal.

The basic model used in the calculations in this paper, as far as the lattice dynamics are concerned, is the deformation dipole model developed some time ago by one of us[3]

*Work performed in part under the auspices of the U.S. Atomic Energy Commission.

but wherein we have now included second-neighbor central forces between fluorine ions in the calculation for a "DDNNN model". [It is, of course, justifiable to argue that, in a crystal such as CsF, one should consider the possibility of second-neighbor positive ion forces. At this stage, however, we have chosen to make a consistent sequence of calculations for all four salts on the basis of the same model.] To determine the first and second derivatives of the non-Coulomb forces between first and second neighbor ions, we fit the static and high-frequency dielectric constants, the equilibrium lattice constant, the compressibility, the shear modulus C_{44}, and the infrared dispersion frequency ω_o. The high-frequency dielectric constants used in these calculations were derived from the TESSMAN, KAHN, and SHOCKLEY[4] polarizabilities, using the Clausius-Mosotti relation; thus, they may differ slightly from observed values.

As a first stage in the present calculations, we have derived the single-phonon vibrational spectra for these salts (as shown in Fig. 1), which were obtained using the interpolation method of GILAT and RAUBENHEIMER[5]. However, we did encounter a slight problem when using this technique; namely, one can obtain spurious spikes on the spectra in the vicinity of critical points. Nevertheless, if one allows for the presence of these spikes, the technique provides a speedy method of revealing most of the detail of the spectrum.

For the two-phonon spectra, where we require the eigenvectors, we have used a direct sampling procedure and have made calculations of the eigenfrequencies and eigenvectors for a mesh of 64, 000 points within the first zone. Using these data, the Raman spectrum is then computed in the same way as described in the earlier paper on sodium fluoride[1]. Thus, we consider the scattering geometry in which the light is incident along a [100] direction, polarized along a [010] direction, and viewed along a [001] direction. We make the assumption (which appears to be justified by the experimental results for sodium fluoride) that the polarizability tensor which determines the Raman scattered intensity is determined entirely by the configuration of first-neighbor ions. This gives a polarizability tensor which depends, at most, on eight parameters. For the specific geometry we are considering, the polarized and depolarized intensities are determined by the two components of the scattering tensor, $i_{xx,xx}$ and $i_{xy,xy}$, respectively[6]. The experimental results for NaF indicate that $i_{xy,xy}$ is approximately zero. This fact is consistent with the assumption that there is a central dependence of the polarizability tensor on the configuration of the first neighbors. This immediately reduces the number of disposable constants from eight to three, as described in detail in our NaF paper[1]. In the present paper we retain this assumption for each fluoride, and we can then present results for the same two extreme variations ("variations 1 and 2") given in that paper. As discussed there, we expect the relative magnitudes of the true constants to be somewhere between the bounds defined by the variations 1 and 2.

In Fig. 2 we show the polarized experimental spectrum, measured by J. P. Russell[1], for the geometry described previously. In Figs. 3 through 6 we predict the Raman spectra for the four salts using both variations 1 and 2. Thus, we show in Fig. 3 the computed Stokes and anti-Stokes components of the combination bands computed on the basis of variation 1 at room temperature. (We do not show here the results for absolute zero, which we have also obtained. We find the qualitative shape of the Stokes component is essentially the same as our room temperature results. The anti-Stokes components, of course, will vanish.) It can be observed that some of the spectra have very sharp features which would in practice be anharmonically broadened. In Fig. 4 we show the corresponding spectra computed for variation 2. In the case of NaF, we can indicate on the computed spectra the positions of the experimental features. It will be seen that, in both cases, there is remarkable agreement with experiment.

We can now comment on the manner in which differences (Figs. 3 and 4) between the various salts develop. One can see that for sodium fluoride, potassium fluoride, and

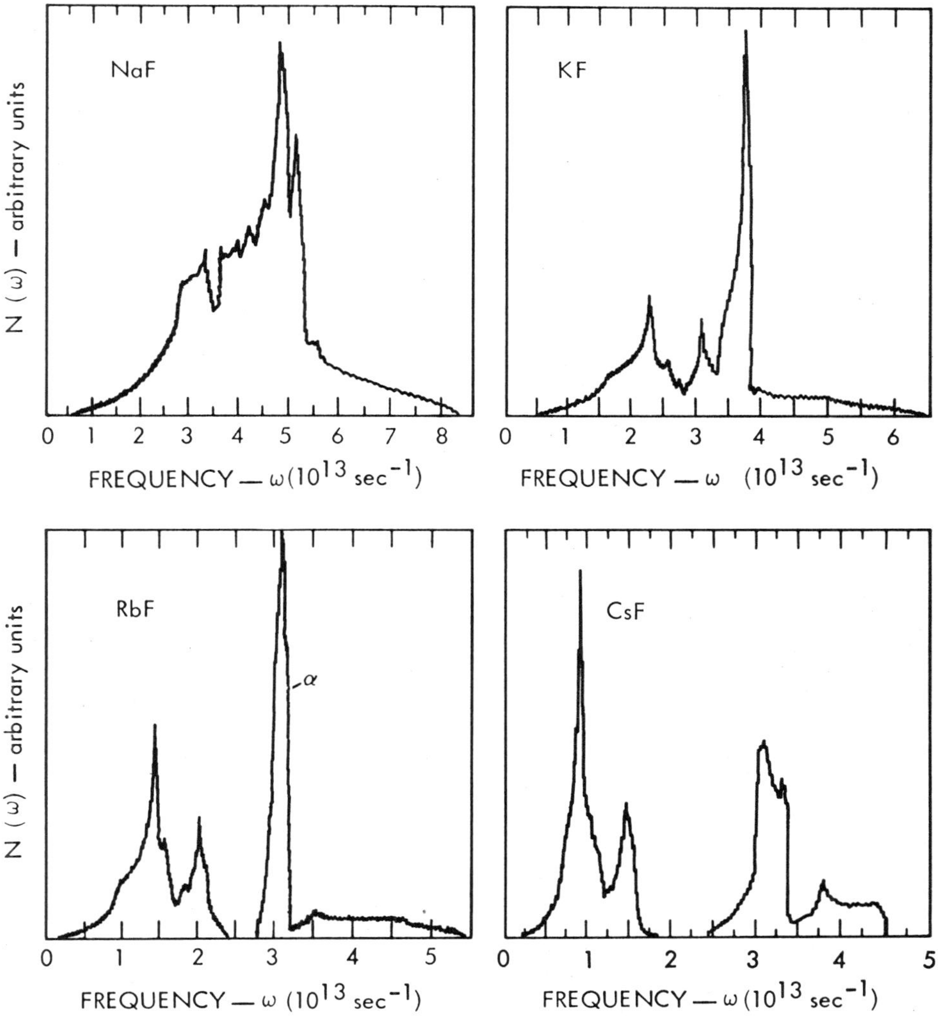

Fig. 1. The alkali fluoride frequency spectra for the deformation dipole model described in the text with room-temperature input parameters. (The main peak α for RbF is referred to in the text.)

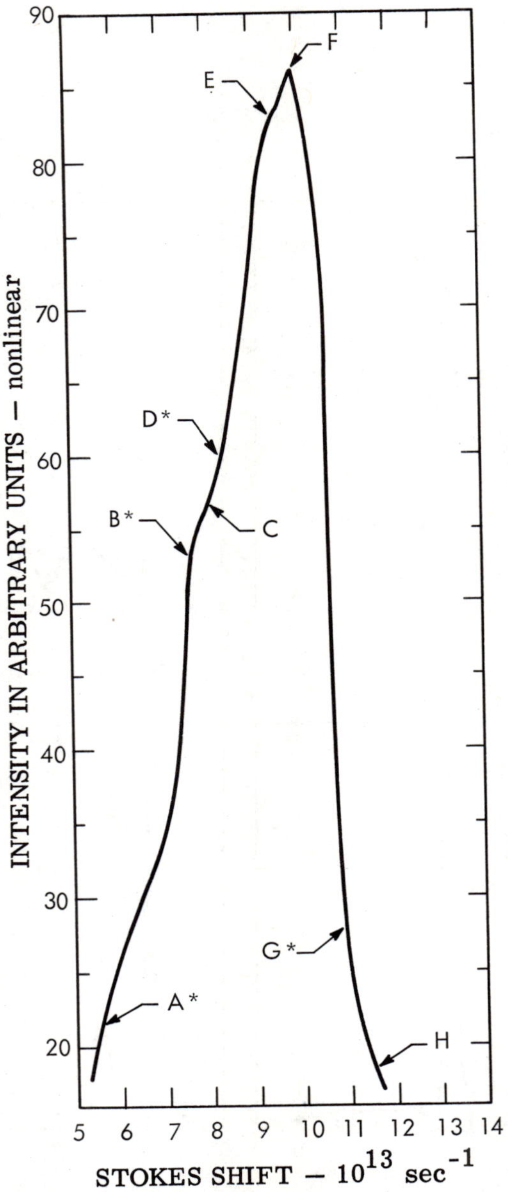

Fig. 2. Experimental 300°K second-order Raman spectrum of NaF (resolution ≈ 10 cm^{-1}). (The positions of the experimental features are marked A, B, etc. Asterisks refer to the edges of the main features; the small high-frequency bump H may be a third-order line.)

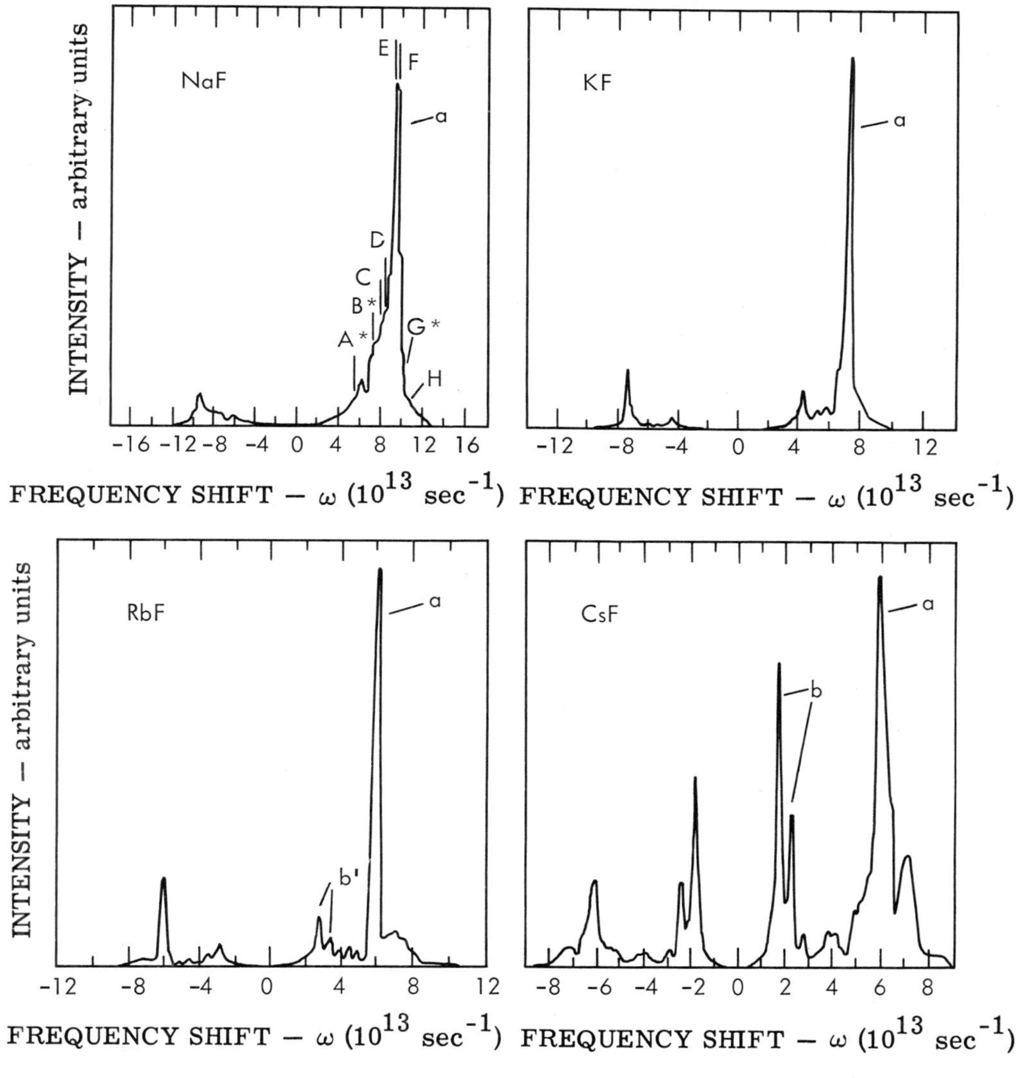

Fig. 3. Predicted polarized 300°K Stokes and anti-Stokes spectra for the DDNNN model described in the text using the polarizability option, variation 1. (Note that the intensity scales are linear, but differ for the various fluorides, so that the results can be conveniently displayed. For NaF the positions of experimental features are indicated by A, B, etc. Features indicated by a and b are referred to in the text.)

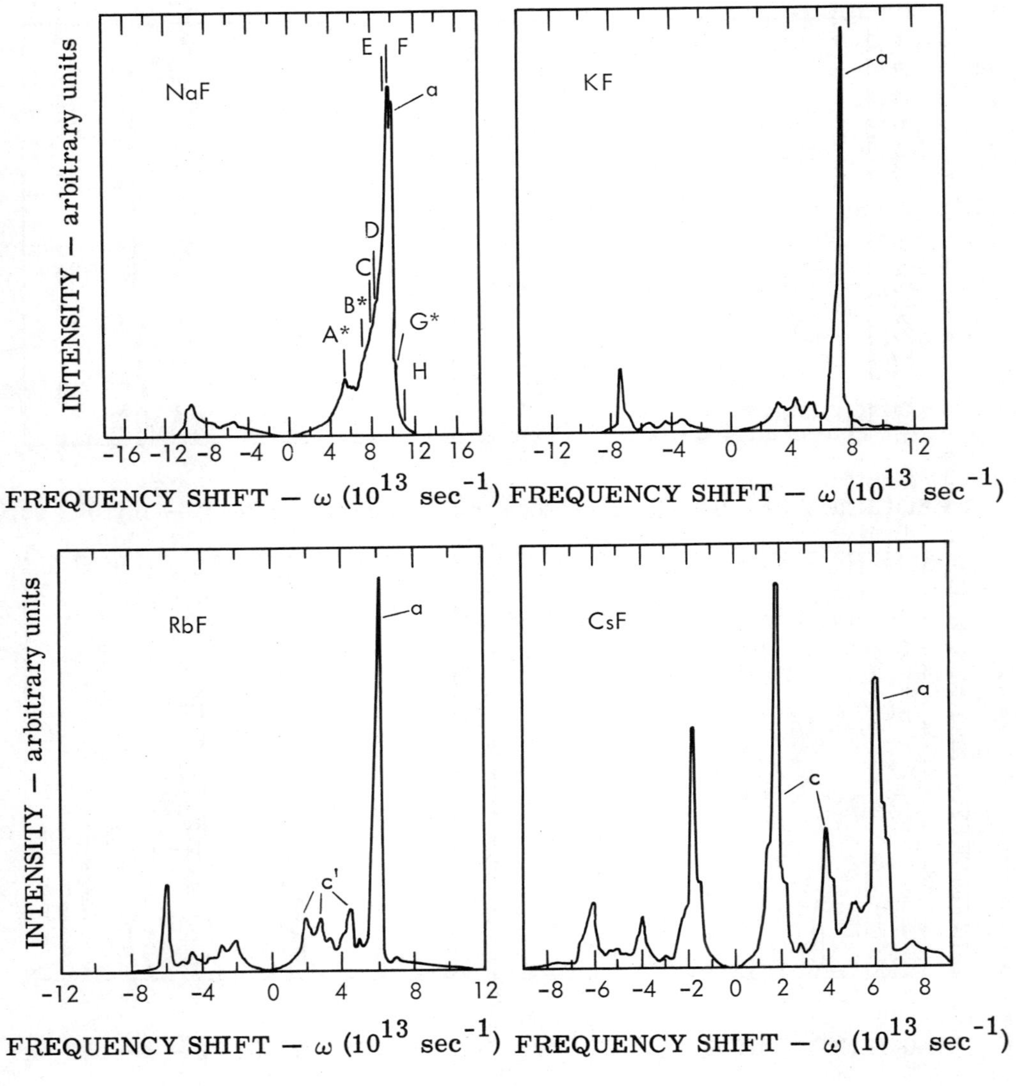

Fig. 4. Predicted polarized 300°K Stokes and anti-Stokes spectra for the DDNNN model described in the text using the polarizability option, variation 2. (Note that the intensity scales are linear, but differ for the various fluorides, so that the results can be conveniently displayed. For NaF the positions of experimental features are indicated by A, B, etc. Features indicated by a and c are referred to in the text.)

rubidium fluoride, the change from one salt to the next is relatively gradual and systematic. In all three cases the most striking feature is the sharp peak marked "a" on each spectrum. Then, from rubidium to cesium fluoride, the character of the computed spectra apparently changes for both variations. Though there is still a relatively sharp peak marked as "a" (Figs. 3 and 4), there has now appeared equally strong low-frequency structure, peaks "b" for variation 1 and "c" for variation 2. One can correlate these low-frequency peaks with those marked "b'" and "c'" on the corresponding spectra for RbF, although the latter are very much weaker than the main peak "a". The probable reason for this difference is the strikingly narrow transverse optical peak α in the RbF single-phonon spectrum.

Concerning the difference bands shown in Figs. 5 and 6, one can see a definite trend through the whole sequence, and in all cases, there is a marked difference between the results computed for variations 1 and 2. In variation 1 (Fig. 5) there appears a strong central maximum, which is suppressed in variation 2 (Fig. 6). As we previously pointed out for NaF, the distinction between variations 1 and 2 can be made experimentally if there are available detailed observations on the difference bands.

In conclusion, it should be pointed out that we have only discussed the two components of the scattering tensor $i_{\alpha\beta, \gamma\delta}$; namely, $i_{xx, xx}$ and $i_{xy, xy}$ – the second of which is, in fact, identically zero for the model we are using. There is, however, a third independent component of the scattering tensor, the $i_{xx, yy}$ component. If measurements were made for other scattering geometries[7], it would, in principle, be possible to determine the form of this third component, and one could then distinguish between the theoretical variations 1 and 2. This is because, for variation 2, $i_{xx, xx} = i_{xx, yy}$; whereas for variation 1, the two components differ.

We have presented a comprehensive, theoretical account of the second-order Raman spectra of the alkali fluoride sequence of crystals, excluding LiF. It remains for detailed experimental work on each crystal to determine just how valid are the models we have used in this calculation. We do not suggest that the validity of our model is equally good all through the sequence. However, the present calculations do represent a consistent systematic extension to the whole sequence of a model which has proved singularly successful in the case of NaF.

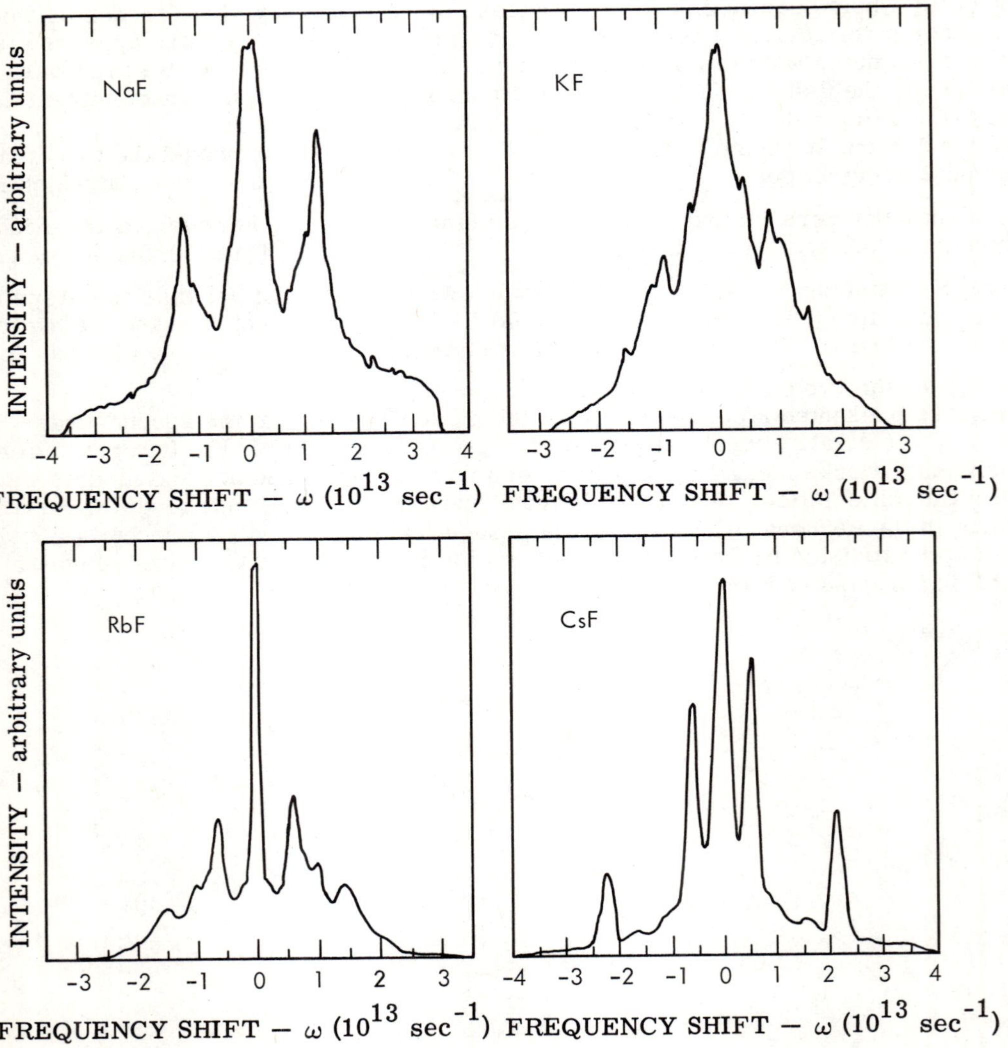

Fig. 5. The polarized 300°K difference bands predicted for the DDNNN model including the effect of variation 1 in the polarizability tensor. (Note that the intensity scales are linear, but differ for the various fluorides, so that the results can be conveniently displayed.)

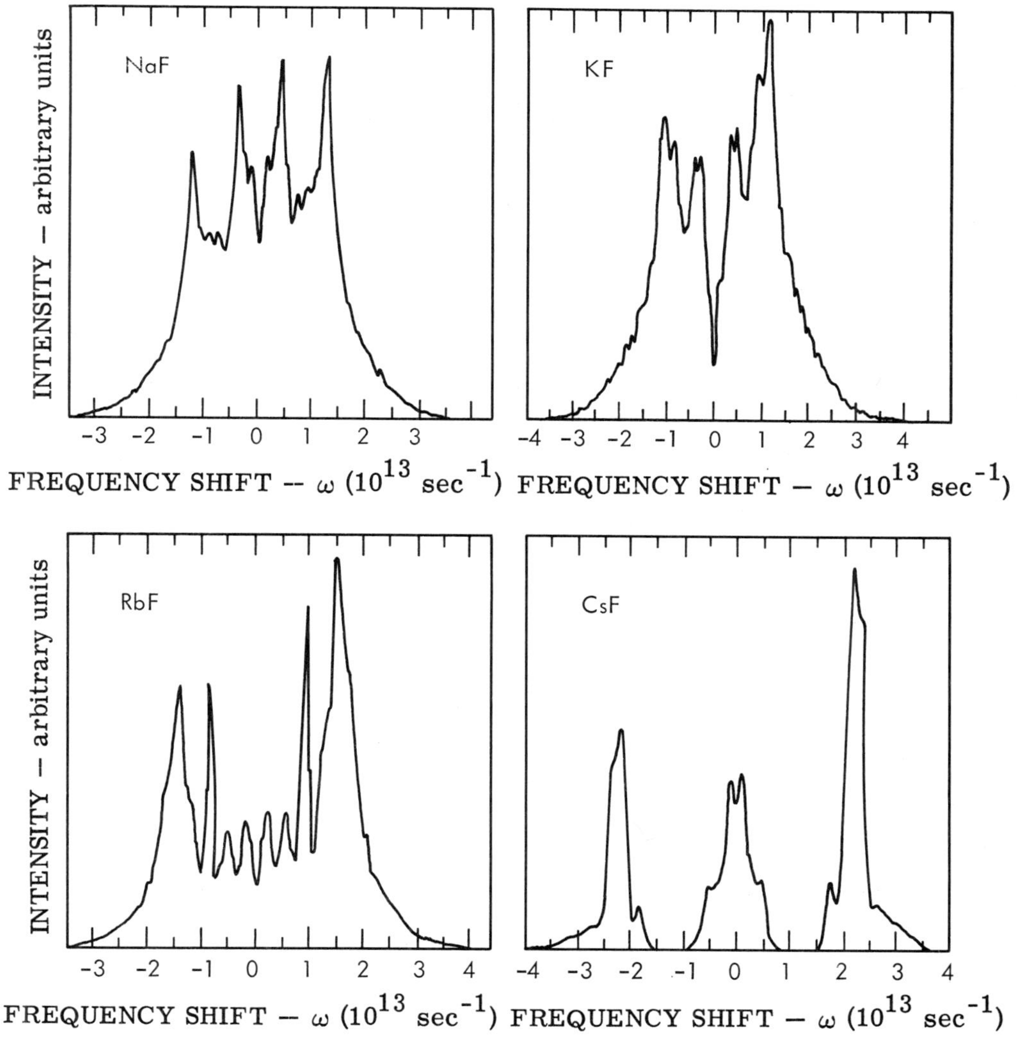

Fig. 6. The polarized 300°K difference bands predicted for the DDNNN model including the effect of variation 2 in the polarizability tensor. (Note that the intensity scales are linear, but differ for the various fluorides, so that the results can be conveniently displayed.)

REFERENCES

1. J.R. Hardy, A.M. Karo, I.W. Morrison, C.T. Sennett, and J.P. Russell, Lawrence Radiation Laboratory, Livermore, Report UCRL-70745 (to be published).
2. J.R. Hardy and A.M. Karo, Phys. Rev. 168, 1054 (1968). The relative intensities for "variation 2" shown in this reference differ from present results because of a possible compiler error at that time.
3. J.R. Hardy, Phil. Mag. 7, 315 (1962).
4. J.R. Tessman, A.H. Kahn and W. Shockley, Phys. Rev. 92, 890 (1953).
5. F. Gilat and L.J. Raubenheimer, Phys. Rev. 144, 390 (1966). (We would like to express our gratitude to these authors for communicating to us the details of their program.)
6. M. Born and K. Huang, "Dynamical Theory of Crystal Lattices," 368, Eq. 49.2, Oxford University Press, New York, 1954.
7. R.A. Cowley, Proc. Phys. Soc. 84, 281 (1964).

B-3: SECOND ORDER RAMAN - LASER SPECTRA OF SOME CUBIC BINARY SINGLE CRYSTALS

M. Krauzman
Departement des Recherches Physiques de la Faculte, des Sciences de Paris
Paris, France

ALKALI HALIDES

The Raman scattering spectra of polarized light by NaCl, KBr and KI (Fig. 1) have been studied elsewhere[1, 2]. The dispersion curves of frequencies obtained by neutrons scattering[3, 4, 5, 6] have the same shape as those calculated by Karo and Hardy[7] and can thus be approximately completed. This method allows a more full up interpretation than before. The assignments for NaCl, KBr and KI are given in Tables I, II and III and the principal phonon frequencies are gathered in Table IV for the three compounds. One remarks there that the gap between acoustical and optical frequencies increases with the ratio of masses of the two atoms. For instance: $[\mathrm{LO(L)/LA(L)}] \times (M_1/M_2)^{1/2} = 1,07 \pm 2,5\%$ and $[\mathrm{TO(L)/TA(L)}] \times (M_1/M_2)^{1/2} = 0,98 \pm 2,5\%$, ($M_1$ is the lighter atom). As experiments allow comparison with complete selection rules[2], these have been calculated by reducing, in the space group, the direct product (combinations) and the symmetrized square (overtones) of allowed representations[8]. The site of the lighter atom is taken as the origin of the coordinates.

The study of KCl in polarized light allows one to find the irreducible representation of the singularities which are never well marked, but no tentative assignments are given. The Fig. 2 shows how similar the NaCl and KCl spectra of well chosen tensor components are when one of the spectra is translated of about 55 cm^{-1}.

The spectra of RbI (Fig. 3) have not yet been analyzed. The intensities have been drawn to the same scale as well as on Fig. 1.

ZINCBLENDE

The Raman spectrum of zincblende is much more intense than the preceding ones and is mainly composed of lines which have been studied in Ref. [9]. Assignments have been confirmed by examination of temperature influence. The ratio of intensities of the lines at 80 K and 360 K has been plotted versus frequency (Fig. 4). The curve FG gives within the same temperature range, the theoretical variation of intensity for first order Raman lines[10]. The other curves of Fig. 4 give the same calculated variations for combination

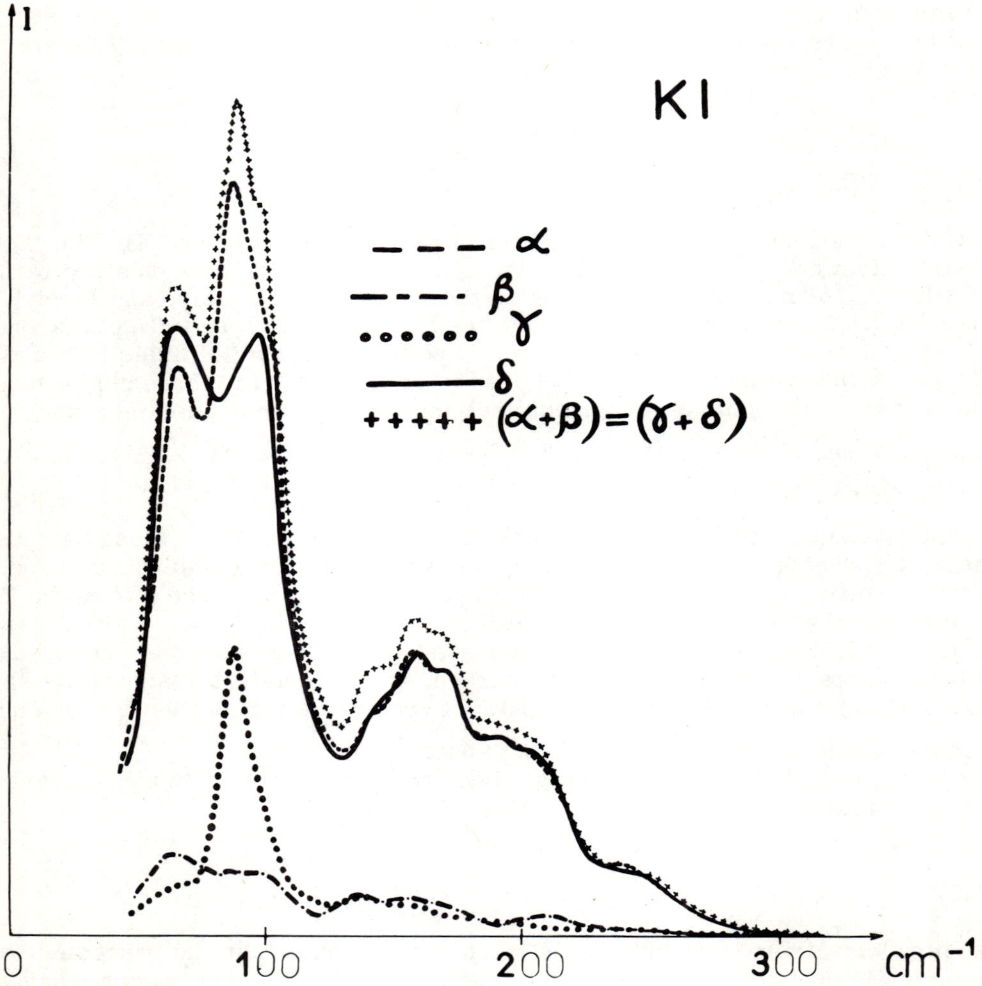

Fig. 1. Spectra of KI; $\alpha = \lambda_A + \lambda_E + \lambda_F$; $\beta = 3\lambda_E$; $\gamma = \lambda_F$; $\delta = \lambda_A + 4\lambda_E$. For comments, see Ref. [2].

TABLE I

Interpretation of NaCl Spectra

EXP.		CALC.		
$\nu\,cm^{-1}$	modes	Attributions	Selection rules	$\nu\,cm^{-1}$
		$TO\{z\} - TA\{x\bar{y}\}\ (\Sigma)$	F	53
		$LA - TA\ (X)$	F	55
55	F	$LA - TA\ (L)$	E F	55
		$Q_1 - Q_1$	A E F	57
		$W_3^{o} - W_3^{A}$	A E F	59
60, 5	F	$Q_1 - Q_2$	E F	59
		$LO - TO\ (L)$	E F	86
87	E	$TO - TA\ (X)$	A E F	86, 5
104	A E F	$LO - TO\ (\Gamma)$	A E F	102
174, 5	A E F	$2\ TA\ (X)$	A E F	175
		$2\ W_3^{A}$	A E F	230
		$2\ Q_2$	A E F	231
231	A	$Q_1 + Q_2$	E F	233
233	E			
		$LA + TA\ (X)$	F	230
235, 5	F			
		$2\ Q_1$	A E F	236
239	A E			
		$Z_3 + Z_4$	F	236, 5
		$2\ TA\ (L)$	A E F	236
		$TA\{z\} + TA\{x\bar{y}\}\ (\Sigma)$	F	238
248	A E	$2\ TA\{x\bar{y}\}\ (\Sigma)$	A E F	248
		$LA + TA\ (\Delta)$	F	250
251, 5	F	$W_1 + W_3$	F	253
258	A	$TO + TA\ (X)$	A E F	261, 5
259	F	$LO + TA\ (\Delta)$	F	260
266	A	$2\ TO\{x\bar{y}\}\ (\Sigma)$	A E F	266
		$LO + TA\ (X)$	F	270
		$TA\{x\bar{y}\} + TO\{x\bar{y}\}\ (\Sigma)$	A E F	271
273, 5	F	$W_{2'} + W_3^{A}$	F	273, 5
		$Z_1 + Z_3$	F	270
276	A E	$2\ W_1$	A E	276
280	F	$2\ TO\ (L)$	A E F	280
282	F	$TA\{z\} + LA\ (\Sigma)$	F	282
286	A	$2\ LA\ (X)$	A E	285
		$W_3^{A} + W_3^{o}$	A E F	289
288	F	$Q_1 + Q_2$	E F	289

TABLE I

Interpretation of NaCl Spectra (cont)

EXP.		CALC.		
$\nu\,cm^{-1}$	modes	Attributions	Selection rules	$\nu\,cm^{-1}$
		$Q_1 + Q_1$	A E F	293
294	A	$2\ Q_1$	A E F	294
300	F	$Q_1 + Q_2$	E F	300
307	A	?		
314	E	LO + $TA\{x\bar{y}\}\ (\Sigma)$	E	314
		LA + $TO\{x\bar{y}\}\ (\Sigma)$	E	314
316	F	TO + LA (X)	F	316, 5
317	A	$2\ W_{2'}$	A E	317
332	A	2 LA (Δ)	A E	331
		$W_{2'} + W_3^{o}$	F	332, 5
333	F	TO + LA (Δ)	F	336
343	E	LO + LA (Δ)	A E	343
		2 LA (L)	A F	346
		2 TO (X)	A E F	348
347 to 355	A	$2\ W_3^{o}$	A E F	348
		$2\ Q_1$	A E F	349
		LO + LA (Σ)	A E F	350
		2 LO (Δ)	A E	356
		LO + TO (Δ)	F	350
354	F	LO + $TO\{z\}\ (\Sigma)$	F	352
		LO + TO (X)	F	356, 5
360	E	2 LO (X)	A E	365
381	F	?		
394	A	?		
524 to 543	A	2 LO (Γ)	A E F	528

TABLE II

Interpretation of KBr Spectra

EXP.		CALC.		
$\nu\,cm^{-1}$	modes	Attributions	Selection rules	$\nu\,cm^{-1}$
		TO - LA (X)	F	45
46, 5	A E F	LO - TO (Λ)	E F	46, 5
		LO - TO (L)	E F	47, 5
		LA - TA (Δ)	F	47, 5
61	A E	LO - LA (X)	A E	60, 5
76	A E F	TO - TA (X)	A 2E F	76
86	A E (F)?	2 TA (X)	A 2E F	84
116	F	LA + TA (X)	F	115

TABLE II

Interpretation of KBr Spectra (cont)

EXP.		CALC.		
$\nu\,cm^{-1}$	modes	Attributions	Selection rules	$\nu\,cm^{-1}$
125	F	TA{z} + TA{$x\bar{y}$} (Σ)	F	124
		LA + TA (Δ)	F	125
135	E F	2 TA{$x\bar{y}$} (Σ)	A E F	135
138	A	2 TA (L)	A E 2F	140
145	A	2 LA (X)	A E	146
150	F	W_1 + W_3^A	F	150
157	F	LO + TA (Δ)	F	157
158	A	TO{$x\bar{y}$} + TA{$x\bar{y}$} (Σ)	A E F	157
		TO + TA (X)	A 2E F	160
165	E F	LA + TA (L)	E F	162
178	A (E) ?	2 LA (Σ)	A E F	177
		2 W_1	A E	178
182	F	2 TO{$x\bar{y}$} (Σ)	A E F	182
184, 5	A	2 LA (L)	A F	184
		TO + LA (X)	F	191
193	A (E) ? F	W_3^o + W_3^A	A 2E F	191, 5
		2 TO (L)	A E 2F	193
		$W_{2'}$ + W_1	E	200
200	E F	TO + LA (Δ)	F	200
207	F	TO{z} + LA (Σ)	F	208
208	A E	LO + LA (X)	A E	206, 5
		LO + LA (Δ)	A E	208
214	A E F	LO + LA (Σ)	A E F	215
221	E F	W_3^o + W_1	F	219, 5
		2 $W_{2'}$	A E	222
231	A E	2 TO (Δ)	A 2E F	232
235, 5	A F	2 TO (X)	A 2E F	236
		LO + TO (Δ)	F	236
242	A	2 LO (Δ)	A E	242
246	E (F)?	LO + TO{z} (Σ)	F	245
251, 5	F	LO + TO (X)	F	251, 5
259	A	2 W_3^o	A 2E F	261
290	A F	2 LO (L)	A F	288
End 316 to 336	A E F	2 LO (Γ)	A E F	326

TABLE III

Interpretation of KI Spectra

EXP.		CALC.		
$\nu\,cm^{-1}$	modes	Attributions	Selection rules	$\nu\,cm^{-1}$
62	A E	2 TA (X)	A 2E F	62
70	A E	TO - TA (X)	A 2E F	71, 5
		LO - TA (Δ)	F	75
		LO - TA (X)	F	77
77	F	$W_3^o - W_3^A$	A 2E F	77
		2 W_3^A	A 2E F	78
85	A E	2 TA{x$\bar{y}$} (Σ)	A E F	85
90	F	$W_1 + W_3^A$	F	90
		LA + TA (Δ)	F	91
97, 5	F	LA + TA{z} (Σ)	F	97
		TA{z} + TO{x$\bar{y}$} (Σ)	F	98
		2 W_1	A E	102
102	A E	2 LA (X)	A E	102
		2 TA (L)	A E 2F	103
		LA + TA{x$\bar{y}$} (Σ)	E	103
116 ?	F?	LA + TA (L)	E F	116
122, 5	E	2 LA (Σ)	A E F	122
126 ?	E?	?		
129	A F	2 LA (L)	A F	129
130	E	2 LA (Δ)	A E	132
134, 5	A	TA + TO (X)	A 2E F	133, 5
136, 5	E	$W_1 + W_{2'}$	E	137
139, 5	F	TA + LO (X)	F	139
144	E	TO{x$\bar{y}$} + LA (Σ)	E	144
150	A E	2 TO{x$\bar{y}$} (Σ)	A E F	150
153	F	TO + LA (X)	F	153, 5
158	E	LO + TA{x$\bar{y}$} (Σ)	E	158
		LO + LA (X)	A E	159
162	A F	LA + TO{z} (Σ)	F	163
		LA + TO (Δ)	F	166
166, 5	A (F)?	LA + LO (Δ)	A E	167
		$W_1 + W_3^o$	F	167
173, 5	A	2 $W_{2'}$	A E	172
178	E	LO + LA (Σ)	A E F	176
182	E	?		
186	A F	2 TO (L)	A E 2F	186
196	A E	2 TO (Δ)	A 2E F	198
199, 5	A	2 LO (Δ)	A E	200
205	A (E)? F	2 TO (X)	A 2E F	205
213	E	2 LO (X)	A E	216
216	A			

TABLE III

Interpretation of KI Spectra (cont)

EXP.		CALC.		
$\nu\,cm^{-1}$	modes	Attributions	Selection rules	$\nu\,cm^{-1}$
218	E F	TO + LO (L)	E F	218
229-232	A E F	2 W_3^o	A 2E F	232
250	A F	2 LO (L)	A F	250
270 to 305 end		2 LO (Γ)		278

TABLE IV

Main Phonon Frequencies (cm^{-1})

	Γ		L		X		W	
NaCl	LO	264	LO	226	LO	182,5	W_3^o	174
	TO	162	TO	140	TO	174	$W_{2'}$	158,5
			LA	173	LA	142,5	W_1	138
			TA	118	TA	87,5	W_3^A	115
KBr	LO	163	LO	144	LO	133,5	W_3^o	130,5
	TO	113	TO	96,5	TO	118	$W_{2'}$	111
			LA	92	LA	73	W_1	89
			TA	70	TA	42	W_3^A	61
KI	LO	139	LO	125	LO	108	W_3^o	116
	TO	101	TO	93	TO	102,5	$W_{2'}$	86
			LA	64,5	LA	51	W_1	51
			TA	51,5	TA	31	W_3^A	39

modes versus their frequencies: $\nu = \nu' \pm \nu''$ in the following limiting cases: curve AC: $\nu' = \nu''$ (overtones); curve AB: $\nu = \nu' \pm 0$; curve BC: $\nu = 350 + \nu''$. (We assume that the highest phonon frequency is that of LO (Γ)); curve DB: $\nu = 350 - \nu''$. Thus, the range of the additive combination modes is the area ABC, that of the substractive modes is ABD. If the thermal variation of line widths is taken into account (points with error bars), the positions of the points on the Fig. 4 corresponding to the best defined lines are in agreement with our assignments[9]. The variation of the line at 312 cm^{-1} due to iron impurities is intermediate between a first order and a second order one.

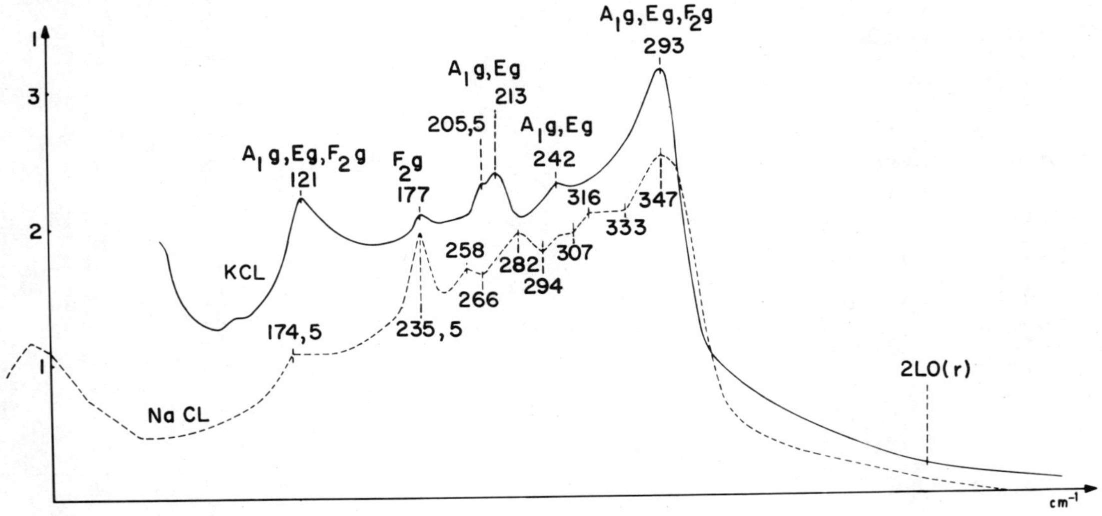

Fig. 2. $\lambda_A + 4\lambda_E + \lambda_F$ spectra of KCL and NaC1 (which is translated of about 55 cm^{-1}).

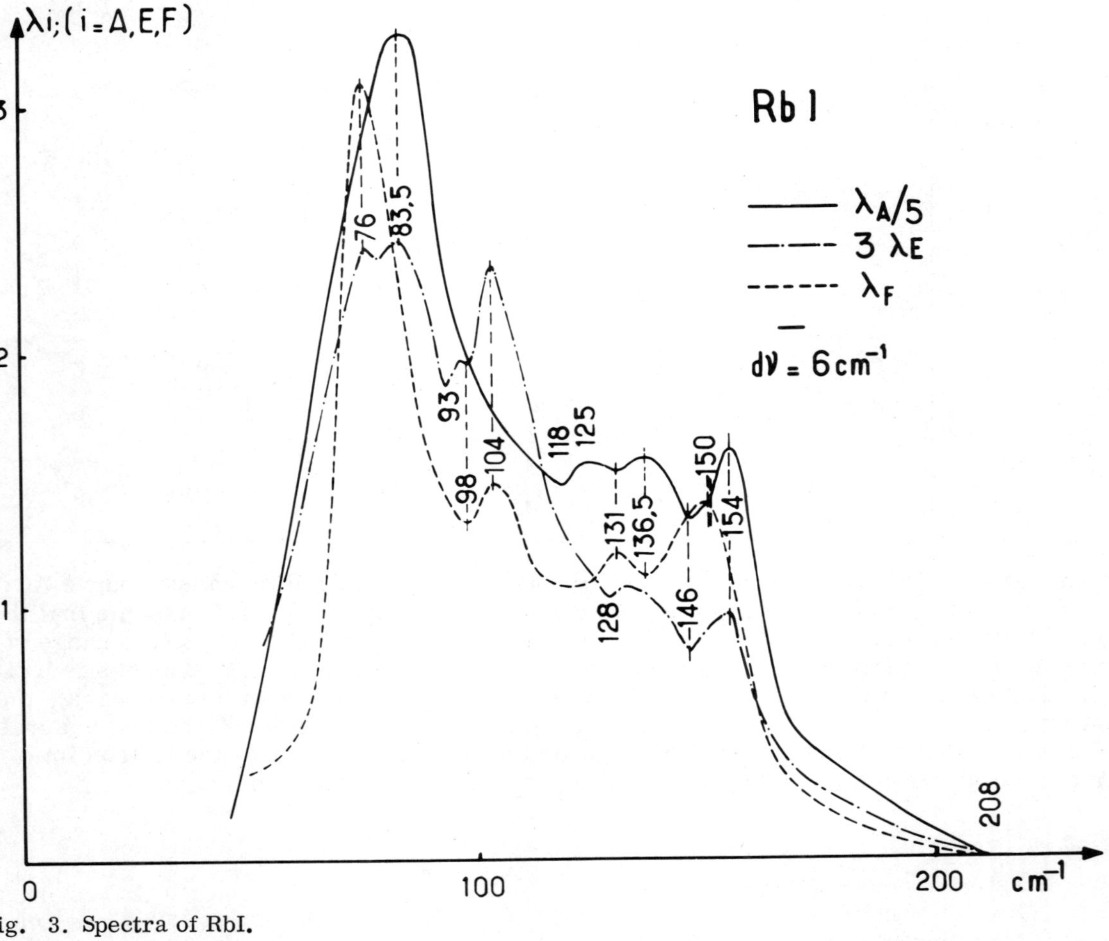

Fig. 3. Spectra of RbI.

CUPROUS CHLORIDE

A single CuCl crystal of unknown orientation gave the spectra of Fig. 5. At 300 K, the effect of the polarization of the incident beam (E_y and E_z curves) can be seen. At 90 K, all the lines are narrower (this agrees with rising of steep sides in the infrared reflection spectrum[11]) and their frequencies increase substantially. The two lines of highest frequencies are TO (Γ) and LO (Γ). The second order spectrum surprisingly does not extend farther than the first order. The ratio ν_L/ν_T remains near $(\epsilon_s/\epsilon_\infty)^{1/2}$ = 1,245 within 1%, between 300 K and 130 K. The Lyddane-Sachs-Teller law [12] is thus verified. At 90K, ν_L/ν_T = 1,22.

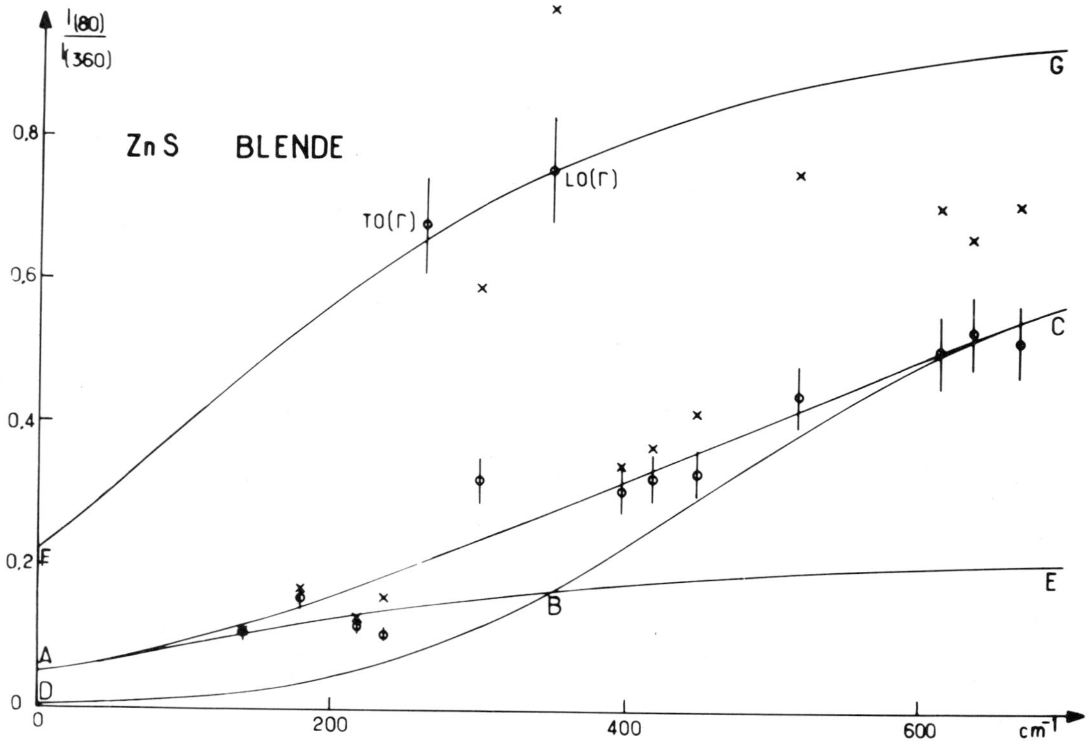

Fig. 4. Ratios of products (height x width) at 80 K and 360 K versus frequencies of the best defined lines of zincblende (points with error bars). The crosses with the same abscissas give the ratios of heights.

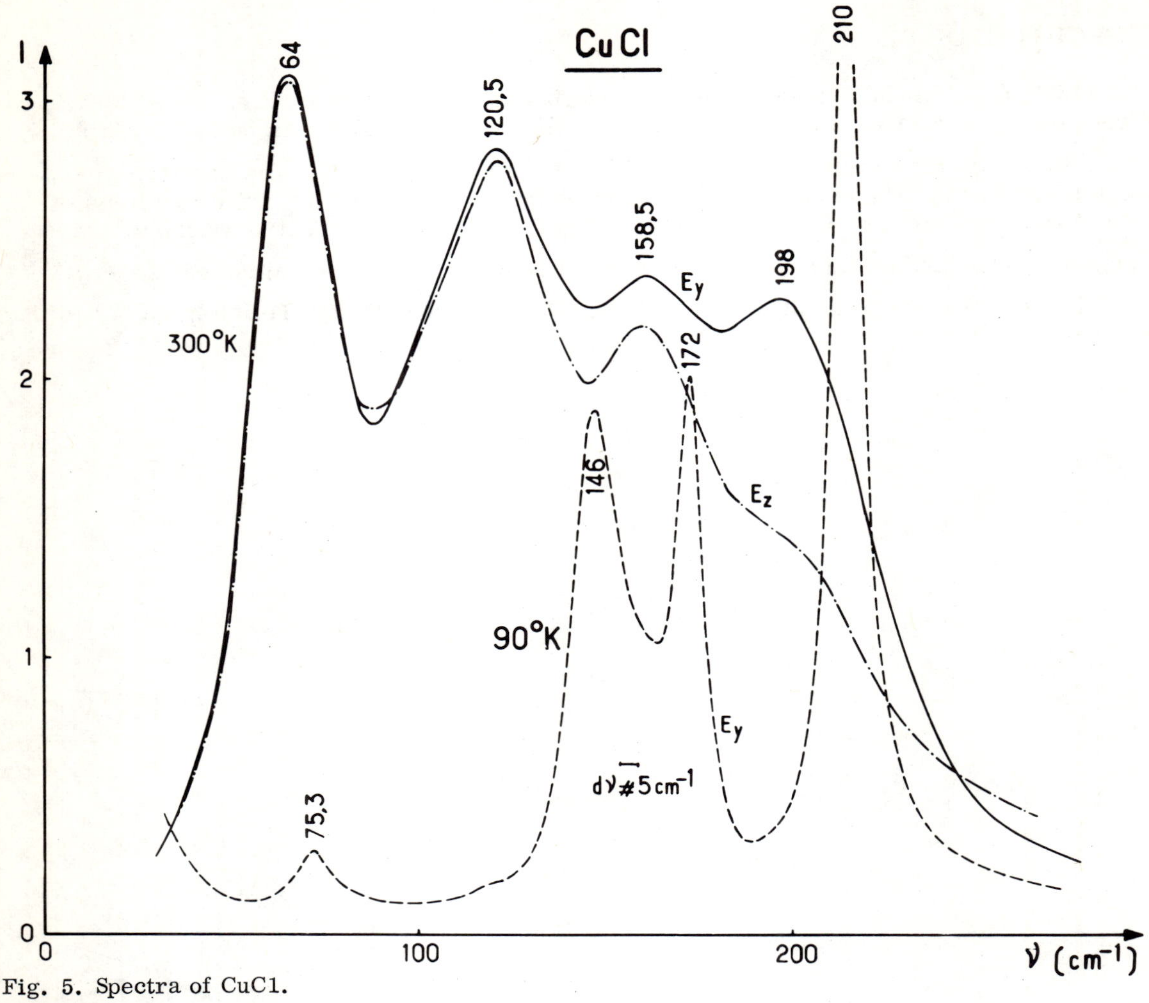

Fig. 5. Spectra of CuCl.

REFERENCES

1. M. Krauzman, Comptes Rendus Acad. Sc. (Paris) 265 B, 1029 (1967) and 266 B, 186 (1968).
2. M. Krauzman, Comptes Rendus Acad. Sc. (Paris) 265B, 689 (1967).
3. R.E. Schmunk, Bull. Am. Phys. Soc. 12, 281 (1967).
4. R.E. Schmunk has kindly communicated to us his latest results on neutrons scattering with more data than in Ref. [3] and with some modifications.
5. A.D.B. Woods, B.N. Brockhouse, R.A. Cowley, and W. Cochran, Phys. Rev. 131, 1025 (1963).
6. G. Dolling, R.A. Cowley, C. Schittenhelm, and I.M. Thorson, Phys. Rev. 147, 577 (1966).
7. A.M. Karo and J.R. Hardy, Phys. Rev. 141, 696 (1966).
8. H. Poulet, J. Phys. 26, 684 (1965).
9. M. Krauzman, Comptes Rendus Acad. Sc. (Paris) 266B, 1224 (1968).
10. M. Born and K. Huang, "Dynamical Theory of Crystal Lattices," 367, Oxford University Press, 1962.
11. J.N. Plendl, A. Hadni, J. Claudel, Y. Henninger, G. Morlot, P. Strimer, and L.C. Mansur, Applied Optics 5, 397 (1966).
12. R.H. Lyddane, R.G. Sachs, and E. Teller, Phys. Rev. 59, 673 (1941).

B-4: THE SECOND - ORDER RAMAN - SPECTRUM OF THE CRYSTAL NaCl FOR LOW TEMPERATURES

A. I. Stekhanov and A. P. Corolkov
AF Ioffe Physico-Technical Institute,
Academy of Sciences of the U. S. S. R.
Leningrad K-21, U. S. S. R.

The Raman spectrum of the crystal NaCl was investigated for the temperature 90°K. The spectrum was induced by the mercury line 2537 A. The registration was made by a spectrograph with dispersion 8 $\frac{A}{MM}$ in the mercury line region.

The investigations show that in the region of the resonance frequencies the spectrum practically coincides with one for 300°K. In the region of the sum frequencies, except for reasonable changes of the frequency maxima, there is also a broad band 420-520 cm^{-1}. For 300°K in this region, after abruptly decreasing for 360 cm^{-1}, there is a weak monotonic decrease of the intensity. We suppose that this effect is due to decreasing of the temperature factor more rapidly for the region 230-360 cm^{-1} than for the region 420-520 cm^{-1}.

An estimate shows that the ratio of the intensities of these frequencies for 90°K must be 1.5 times less than for 300°K, which corresponds well with the observed change of the intensity.

The theoretical calculations of Karo and Hardy show the second order NaCl spectrum has a very small intensity. It is possibly due to the calculation being made taking into account only the density of states. Besides, the calculations give a very abrupt maximum for the frequency 398 cm^{-1}. In the present experimental investigations such a maximum was not observed.

B-5: THE VIBRATIONAL SPECTRA OF MAGNESIUM OXIDE

Jean-Pierre Mon
Département de Recherches Physiques
Faculté des Sciences de Paris, Laboratoire associé au C. N. R. S.
Paris, France

INTRODUCTION

Magnesium oxide crystallizes in the rocksalt lattice structure with two ions per unit cell. It is a convenient host crystal for transition group elements which enter the crystal by substituting for Mg ions. The study of the fluorescence lines associated with these impurities shows a growing interest [1], but requires an accurate knowledge of the phonon spectrum.

With the development of laser sources the second order Raman effect has become a very used tool in the analysis of the phonon spectrum, although dispersion curves are still needed as a guide in phonon energy assignments; besides, the depolarization ratio ρ provides important information in the identification of the two phonon states

According to this introduction, after a brief theoretical survey, we shall look at the experimental results from second order Raman scattering for different polarizations of the incident light and for different orientations of the crystal.

The Raman spectrum and the vibronic structure associated with the $^2E \rightarrow {}^4A_2$ fluorescence line of Chromium in MgO will be compared to the two-phonon and one-phonon density of states curves, respectively.

THEORETICAL

First, we would like to point out the important role played by the second order raman effect in the knowledge of the phonon spectrum. In this process, one incident photon of energy h ν_o and wavevector $\vec{k}_o$ excites the crystal from an initial electronic and vibrational state to some intermediate virtual state. The crystal, then, makes a transition from this virtual state to a final electronic and vibrational state which differs from the initial state by two vibrational quanta, emiting in the process a secondary photon of freqquency ν_s and wavevector $\vec{k}_s$. This is observed as raman scattered radiation. As in one-phonon Raman scattering, energy and momentum must be conserved, leading to the equations:

$$h\nu_o = h\nu_s \pm h\nu_j \pm h\nu_{j'}, \quad (1)$$

$$\vec{k}_o = \vec{k}_s \pm \vec{k}_j \pm \vec{k}_{j'} + N\vec{k} \quad (2)$$

where $h\nu_j$ and $\vec{k}_j$ are respectively the energy and wave-vector of the phonons involved in the interaction, while $\vec{k}$ is any one of the reciprocal lattice vectors, and N is a positive or negative integer which may be zero. The plus sign corresponds to the creation of one phonon while the minus sign signifies the destruction of one phonon. Stokes frequency shifts thus involve the creation of two phonons or the creation of one phonon and the destruction of a second phonon, provided, of course, that the energy of the created phonon exceeds that of the destroyed one. In the following, we shall restrict our attention to the case where both phonons are created. The only requirement for the phonon wave-vectors is that their sum should balance the change in wave-vector of the scattered photon. Thus, the phonons wave-vectors can range over the entire Brillouin zone. In general, the photon wave-vectors are very small compared to the Brillouin zone dimension and Eq. (2) can be written

$$\pm \vec{k}_j \pm \vec{k}_{j'} + N\vec{k} = 0 \quad (3)$$

which shows that the wave-vectors of the two phonons should be equal and opposite. This condition being satisfied by a large range of phonons, the two phonon interaction will result in a band spectrum. In the scattering process considered, the vibrational lattice modes involved may be either optical or acoustical or even a combination of both. In optical processes, phonons which give appreciable contribution originate from regions of the Brillouin zone where there is a high density of states per unit wave-vector interval, i.e. regions where the dispersion curves are nearly flat. The maxima of the one and two-phonon density of state curves are well accounted for by these phonon energies. However, a more complete interpretation of the structure of the spectra is obtained if a critical point analysis is used. Critical points are points of the wave-vector space where every component of $\nabla_h(\vec{k})$ is either zero or changes sign discontinously. They have been classified by Van Hove [2] as minima (m), maxima (M) or saddle point (S_1 and S_2). Most of them occur at high symmetry points of the Brillouin zone, refered to as Γ, X, L, W and K [3]. In order to perform such a critical point analysis of the phonon spectrum the knowledge of the phonon dispersion curves is needed. Then, for the identification of the two-phonon states, it is useful to determine which of the Raman irreducible representations are present. In the cubic group the polarization operator responsible for Raman scattering is a second rank tensor which may be decomposed into the three symmetrical irreducible representations [4]:

$$\Gamma_1^{(+)} = a \begin{vmatrix} 1 & 0 & 0 \\ 0 & 1 & 0 \\ 0 & 0 & 1 \end{vmatrix} \; ; \; \Gamma_{12}^{(1)+} = b \begin{vmatrix} 1 & 0 & 0 \\ 0 & 1 & 0 \\ 0 & 0 & -2 \end{vmatrix} \; ; \; \Gamma_{12}^{(2)+} = b \begin{vmatrix} -\sqrt{3} & 0 & 0 \\ 0 & \sqrt{3} & 0 \\ 0 & 0 & 0 \end{vmatrix}$$

$$\Gamma_{25}^{(1)+} = d \begin{vmatrix} 0 & 1 & 0 \\ 1 & 0 & 0 \\ 0 & 0 & 0 \end{vmatrix} ; \quad \Gamma_{25}^{(2)+} = d \begin{vmatrix} 0 & 0 & 1 \\ 0 & 0 & 0 \\ 1 & 0 & 0 \end{vmatrix} \quad \Gamma_{25}^{(3)+} = d \begin{vmatrix} 0 & 0 & 0 \\ 0 & 0 & 1 \\ 0 & 1 & 0 \end{vmatrix}$$

Experimentally, this identification can be done by inserting a polarizer and an analyzer. Couture and Mathieu [5] have pointed out that some orientations give more information than others.

EXPERIMENTAL RESULTS

For Raman scattering the 5145 Å radiation of an ionized argon laser was used as the exciting line. The spectra were recorded with a coupled grating double monochromator. A half-wave plate allowed rotation of the plane of polarization of the incident light and an analyzer could be inserted between the crystal and the spectrometer slit. According to the results of [5], we used two single crystals, refered to as (I) and (II), respectively cut along: - (I) a parallelepiped, the sides of which were parallel to the Ox' (110), Oy' ($\bar{1}$10) and Oz (001) directions (Fig. 1a), - (II) a parallelepiped Ox''y''z'' having two faces normal to a three-fold axis Oz'' (Fig. 1b).

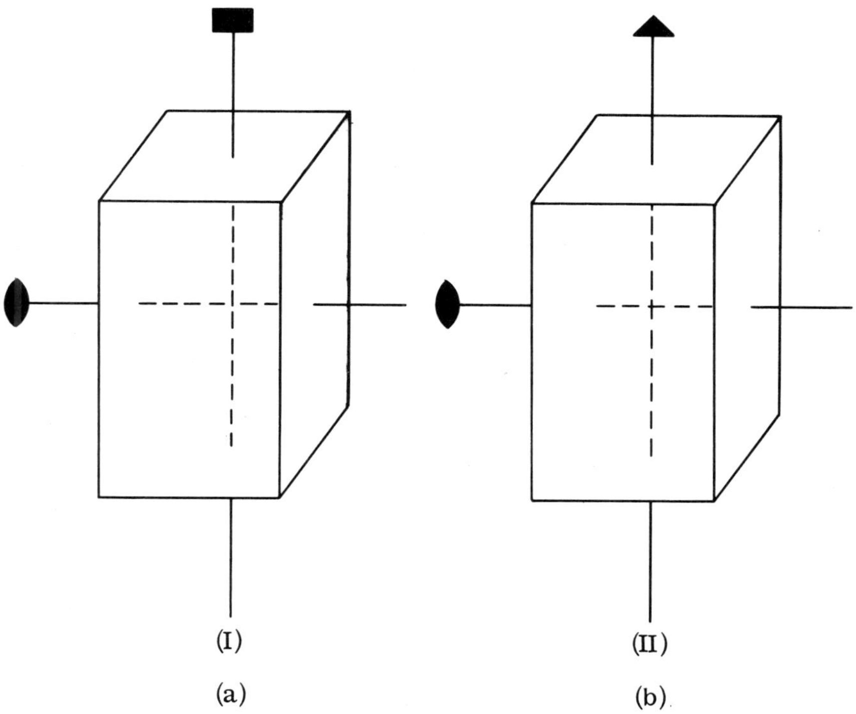

Fig. 1. (a) crystal orientation (I) (b) crystal orientation (II)

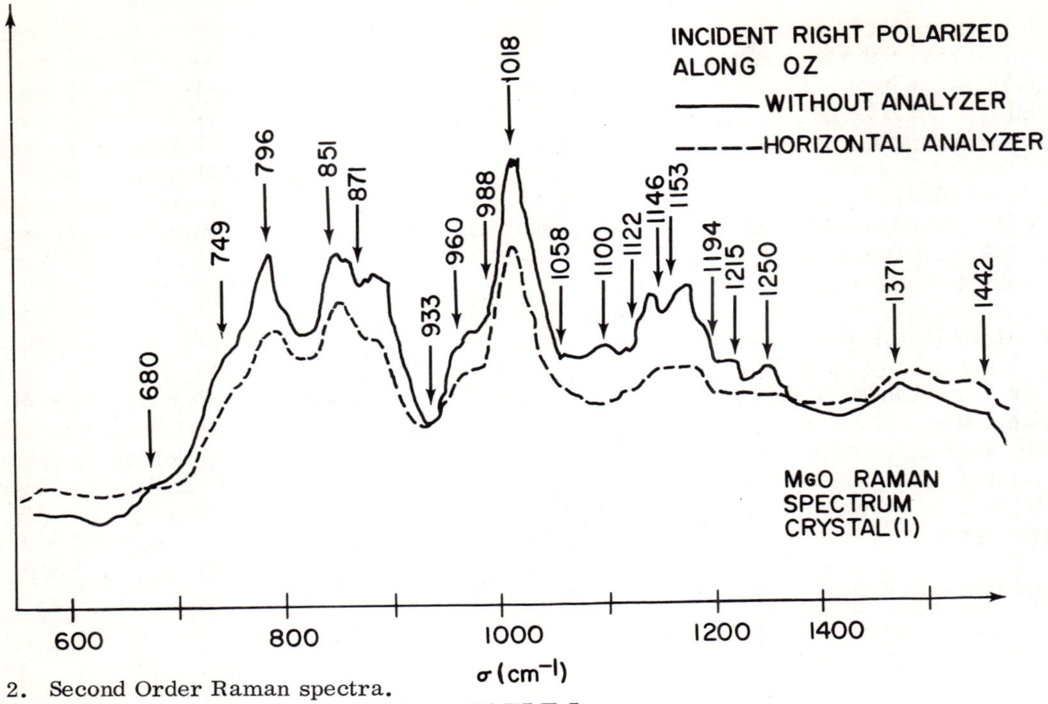

Fig. 2. Second Order Raman spectra.

TABLE I

Raman shift in cm^{-1}		Assignment
observed	calculated	
593	596	2TA (L)
617	610	2TA (X)
685	690	2TO (L)
727	727	TO+TA (X)
746	741	LA+TA (X)
796	796	2TO (Γ)
815	815	LA+TA (L)
826	827	LO+TA (X)
849	844	2TO (X)
857	858	LA+TO (X)
871	872	2LA (X)
942	944	LO+TO (X)
954	958	LO+LA (X)
1023	1029	LO+TO (L)
1371	1368	2LO (L)
1478	1480	2LO (Γ)

Several spectra were recorded for various orientations of the crystal and different polarizations of the incident light. The crystal (I) was oriented with its Oz axis parallel to the incident beam OX. The crystal (II) was set up with its Oz'' axis parallel to the direction of scattering OY. The incident light was polarized either along the OY axis or along an OZ axis, OZ being normal to the OX and OY directions. The scattered light was in some cases analyzed along OX. The most significant feature of these spectra is that they exhibit a great number of Van Hove's singularities and four main peaks between 700 and 1100 wavenumbers, as shown in Fig. 2.

The frequencies of some singularities and of some peaks are entered in Table I. An attempt at assignment has been made using the results of a shell-model calculation [6] and of infrared measurements [7]. Unfortunately, the laser intensity could not be kept very constant and therefore the depolarization ratios have not been measured.

The fluorescence spectrum of chromium doped MgO has been excited with the 6328Å radiation of an Helium-Neon laser and recorded with the spectrometer used for Raman scattering.

DISCUSSION

Using the results of Ref. [6], we have plotted the two-phonon combined density of state curves but the histogram which was so obtained was not sufficiently fine grained to reveal reliably the critical points. Fig. 3 shows this histogram and one of the Raman spectra. It is seen that the agreement is not quite good.

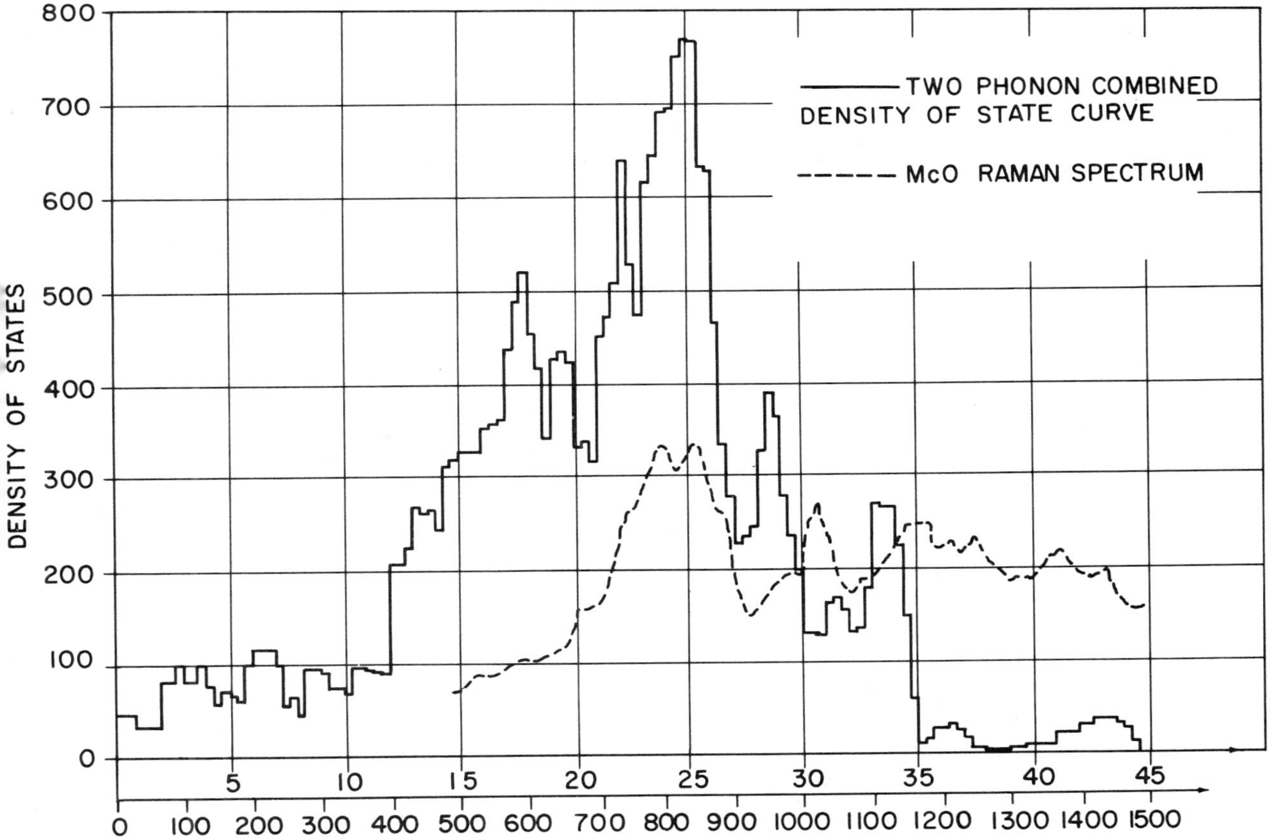

Fig. 3. Two-phonons combined density of state curve and second order Raman spectrum.

Some discrepancies appear also between the one-phonon density of state curve and the vibrational structure accompanying the zero phonon line of the fluorescence of chromium in MgO (Fig. 4). These discrepancies can be in some extent explained by the fact that the spectrum is obscured by the strong emission of the Cr ions at non-cubic sites.

We guess that a calculation involving a great number of points of the Brillouin zone could improve appreciably the one and two- phonon density of state curves.

We propose, in conclusion, the following phonon energies at symmetry points.

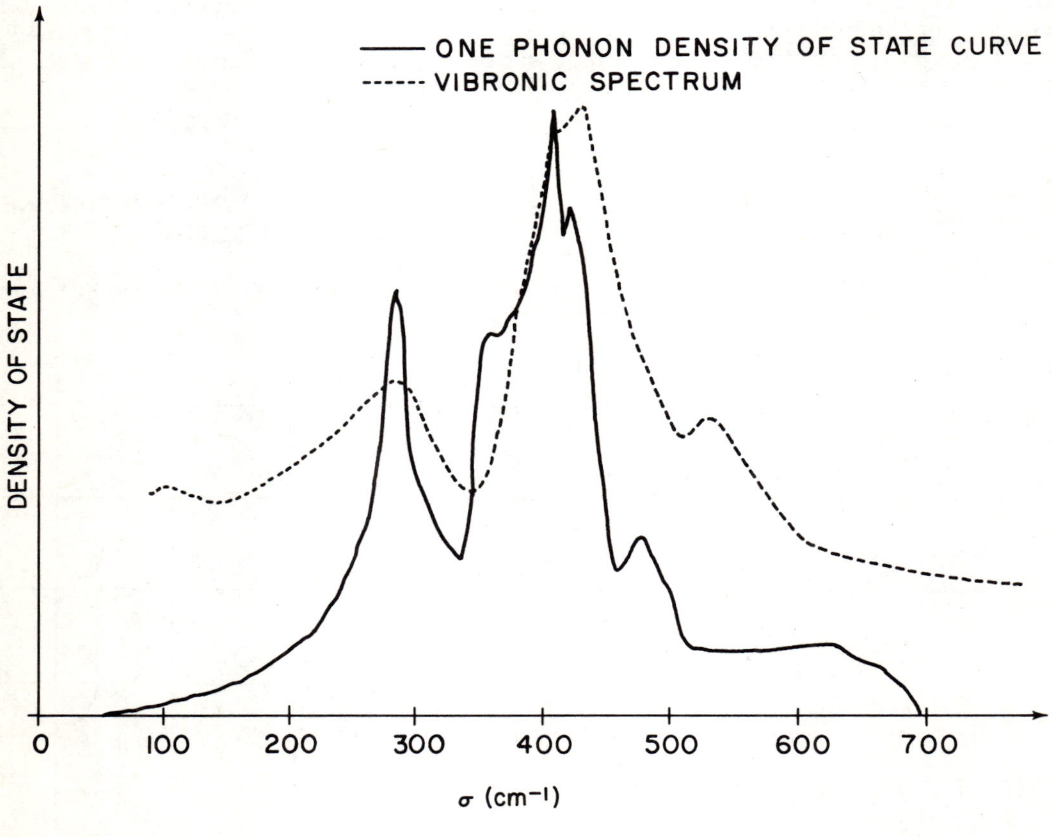

Fig. 4. One phonon density of state curve and vibronic spectrum.

Phonon Energies in CM^{-1}

Point	Mode	Energy
Γ	LO	740
	TO	398
L	LO	684
	TO	341
	LA	519
	TA	296
X	LO	520
	TO	422
	LA	434
	TA	306

REFERENCES

1. G.F. Imbusch, W.M. Yen, A.L. Schawlow, D.E. McCumber, and M.D. Sturges, Phys. Rev. 133, A 1029 (1964).
2. L. Van Hove, Phys. Rev. 89, 1189 (1953).
3. L.P. Bouckaert, R. Smoluchowski, and E. Wigner, Phys. Rev. 50, 58 (1936).
4. R. Loudon, Adv. in Phys. 13, 423 (1964).
5. L. Couture and J.P. Mathieu, Ann. de Phys. 12, 521 (1948).
6. G. Peckham, Proc. Phys. Soc. 90, 657 (1967).
7. B. Piriou and F. Cabannes, C.R. Acad. Sc. Paris 264, 630 (1967).

B-6: SECOND-ORDER RAMAN SPECTRA OF SOME ZINC BLENDE AND WURTZITE CRYSTALS

W. G. Nilsen
Bell Telephone Laboratories, Incorporated
Murray Hill, New Jersey

ABSTRACT

The second-order Raman spectra of several crystals with the zinc blende and wurtzite structure are investigated and the results interpreted in terms of the phonon energies of the individual crystals near the Brillouin-zone boundary. The Raman spectrum of cubic ZnS is examined in most detail and a reasonably complete interpretation of the spectrum is given. In addition, the calculated selection rules for second-order Raman processes at critical points X and L in the zinc blende structure agree fairly well with the observed polarization characteristics of the cubic ZnS spectrum. The second-order Raman spectrum of cubic ZnSe and cubic ZnTe are also interpreted largely by comparison with the ZnS spectrum. In the series ZnS to ZnSe to ZnTe, the individual phonon energies decrease throughout the Brillouin zone as does the difference in energy between differently-polarized optical and acoustical phonons. The second-order Raman spectrum of hexagonal ZnS and SiC are also discussed and the similarity between the cubic and hexagonal ZnS spectra is pointed out.

INTRODUCTION

Second-order Raman spectra has increased in interest and importance in recent years principally because of the ease and reliability with which such spectra can be measured using modern laser-Raman techniques. This type of Raman spectra promises to become a significant and unique source of information on the lattice dynamics of crystals especially where interest is centered on the Brillouin-zone boundary. At present, two interrelated things need further study and development if the second-order Raman effect is to become of value in studying phonons or other excitons in crystals. First, more experience and background is needed in interpreting the observed second-order spectra. This includes not only assigning the observed peaks to phonon branches but also determining the points of the Brillouin-zone from which the scattering originates and the extent to which scattering from different points in the Brillouin zone overlap one another. Second, a clearer idea is needed of the extent that the polarization properties of the second-order spectra can be used to identify the origin (both phonon branch and point in the Brillouin zone) of the scattering giving rise to a particular Raman peak and to determine the symmetry species of the individual phonons participating in the scattering event.

With these two points in mind, we undertook a study of the second-order Raman effect in a series of crystals in which the crystal structure (and therefore the phonon-mode structure) is relatively simple and in which the selection rules were known. A rather complete and detailed study has been made of the second-order Raman effect in cubic ZnS[1] and the results of this study are summarized here. Using these results as a model, we discuss the second-order effect in two crystals which are structurally similar to cubic ZnS; namely, cubic ZnSe and cubic ZnTe. The gross features of the second-order Raman spectra of these two crystals are rather easily interpreted when compared to the ZnS spectra. Next, the second-order Raman effect in hexagonal ZnS and SiC are examined and the close relation between cubic and hexagonal ZnS is pointed out.

We assume that the reader is familiar with the elementary principles of the second-order Raman effect. Specifically, momentum conservation does not restrict the range of phonon wave vectors taking part in the scattering as is the case with first-order scattering. The various structural characteristics of the second-order spectra are due largely to variations in the combined density of states and selection rules for the various scattering processes. We assume that most if not all of the observed spectra originates from near the Brillouin-zone boundary (critical points X and L for the zinc-blende crystals) where the combined density of states is very high. An argon- or krypton-ion laser was used as the exciting source and an in-tandem, double-grating spectrometer was used to disperse the scattered light. The experimental set-up is described in more detail in an earlier paper[2]. The spectra are shown in Figs. 1-5. Only right-angle scattering was examined and the direction of polarization of the exciting and detected scattered light are the same so that these spectra correspond to the diagonal elements of the scattering tensor.

THE CUBIC ZnS SPECTRUM

The second-order Raman spectrum of cubic ZnS is fairly easy to interpret principally because the various modes at the zone boundary are well separated and the overtones of these modes form a prominent part of the spectrum. Thus, the peak at 181 cm^{-1} is the transverse acoustic or TA overtone and the one at 222 cm^{-1} is the longitudinal-acoustic or LA overtone. The corresponding optical overtones are located at 621 cm^{-1} for the TO and 672 cm^{-1} for the LO modes. We assume that the second-order scattering comes predominantly from critical points X and L on the zone boundary and that phonon energies for the various branches are essentially the same at these two points. From the overtone energies, we derive the single-phonon energies at the zone boundary as follows: TA 88 cm^{-1}, LA 110 cm^{-1}, TO 306 cm^{-1} and LO 333 cm^{-1}. In deriving these values, we have used the energy shifts from a calibrated spectrum in which a helium-neon discharge spectrum is superimposed on the Raman spectrum rather than the values given in Fig. 1. The (LO+TO) combination band is at 644 cm^{-1}; the (LA+TA) band is not seen possibly because of overlap with the intense LA and TA overtones. The corresponding difference combination bands are detected only under polarization conditions and crystal orientations which measure the Γ_{15} irreducible representation of the polarizability. These two bands are superimposed on one another at an energy shift of 27.4 cm^{-1}. The (TO+TA) and (LO+LA) combination bands are seen at 401 and 457 cm^{-1} respectively and the two combination bands (TO+LA) and (LO+TA) are superimposed on one another at 433 cm^{-1}. On a calibrated spectrum, the above energy shifts come out to be 636, 386, 448, and 422 cm^{-1} The origin of the peaks at 304 and 525 cm^{-1} is not well established. They are possibly overtone bands that originate from points other than X and L in the Brillouin zone; for example, critical point W on the zone boundary. The sharp peak at 353 cm^{-1} is the first-order LO line which is forbidden in this polarization but appears with greatly reduced intensity (about 1/40) due to slight depolarization effects in the sample. The internal consistency in the energy shifts of the assigned modes and the agreement in selection rules given below gives us a reasonable amount of confidence in our assignments.

The polarization characteristics of the second-order Raman spectrum of cubic ZnS was also investigated and compared with the calculated selection rules for the zinc blende structure[3, 4]. We choose to discuss the polarization properties of the single-crystal Raman spectra in terms of the irreducible representations of the polarizability appropriate to the T_d point group. Such concepts as depolarization ratios and depolarized spectra apply more directly to liquid or polycrystalline samples but are less suitable for single-crystal work. These measurements were done on a crystal oriented so that one <110> crystallographic axis was parallel to the laser beam, another <110> axis pointed toward the spectrometer slits and a <100> axis pointed along the third orthogonal direction. The individual irreducible representations of the polarizability, Γ_1, Γ_{12} and Γ_{15} are obtained by measuring specific components of the scattering tensor for this crystal orientation[1]. For example, the diagonal component measures $(\Gamma_1 + \Gamma_{12})$ as is the case in Fig. 1. The results are given in Table I where they are compared with the calculated selection rules for the zinc blende structure.

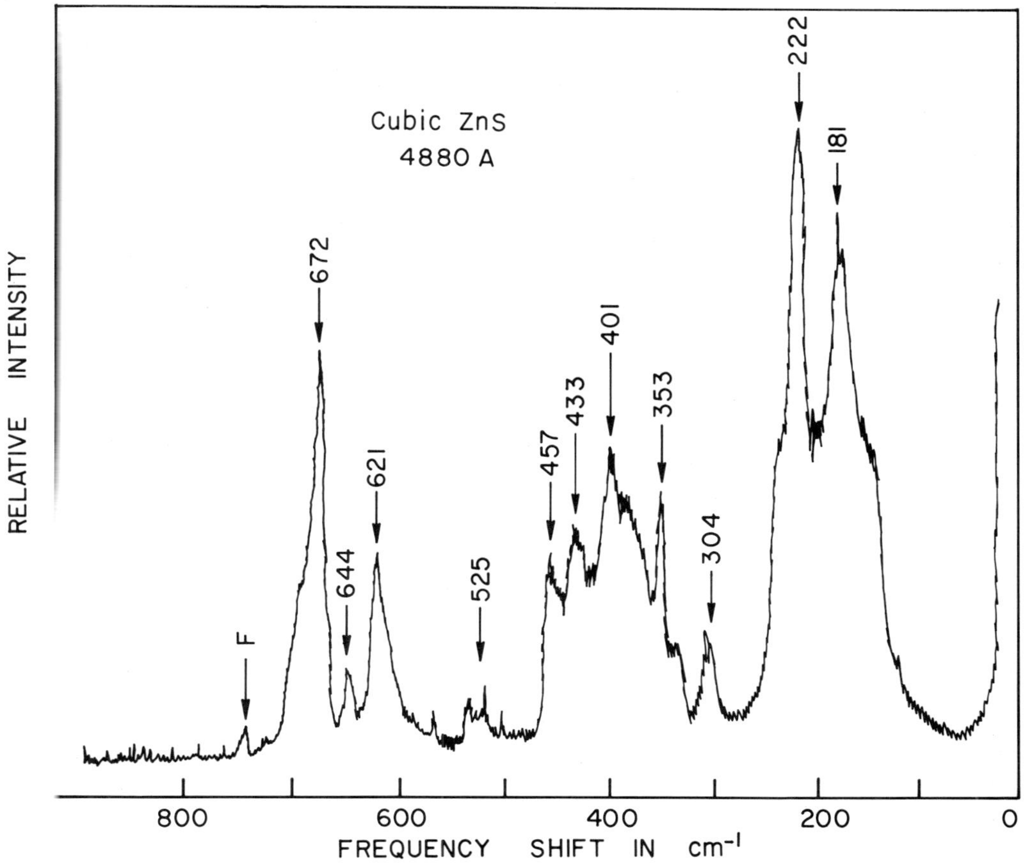

Fig. 1. Raman spectrum of cubic ZnS. The crystal is oriented so that <100> crystallographic axes are parallel to the three orthogonal axes defined by the direction of propagation of laser beam of the detected scattered light, and the third mutually-perpendicular direction.

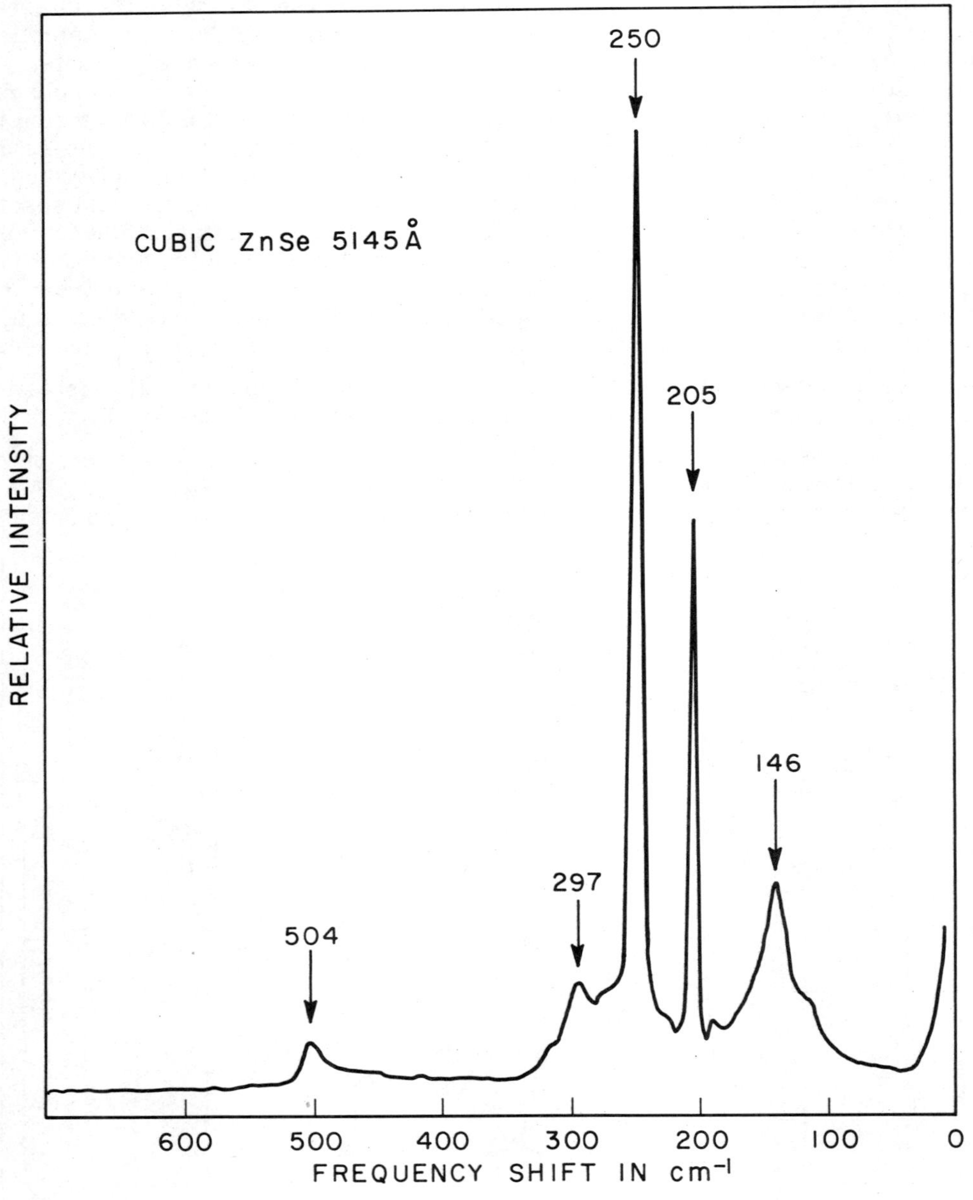

Fig. 2. Raman spectrum of cubic ZnSe. The laser propagates along a <121> crystallographic direction and the detected scattered light along a <111> direction. The third orthogonal direction is a <110> axis.

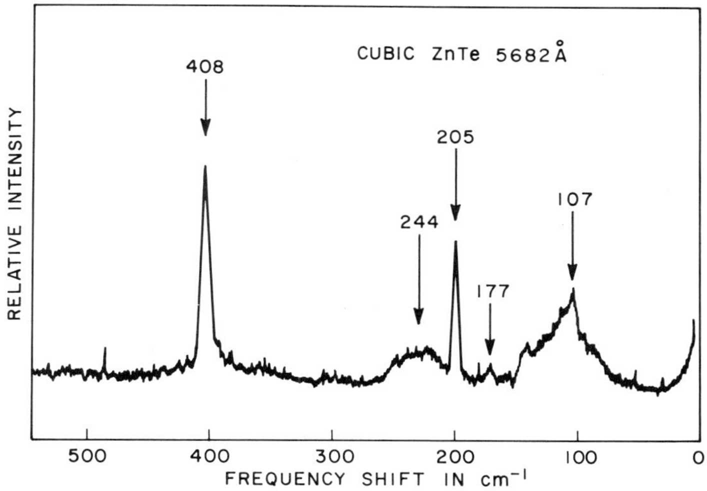

Fig. 3. Raman spectrum of cubic ZnTe. The crystal orientation is the same as in Fig. 1.

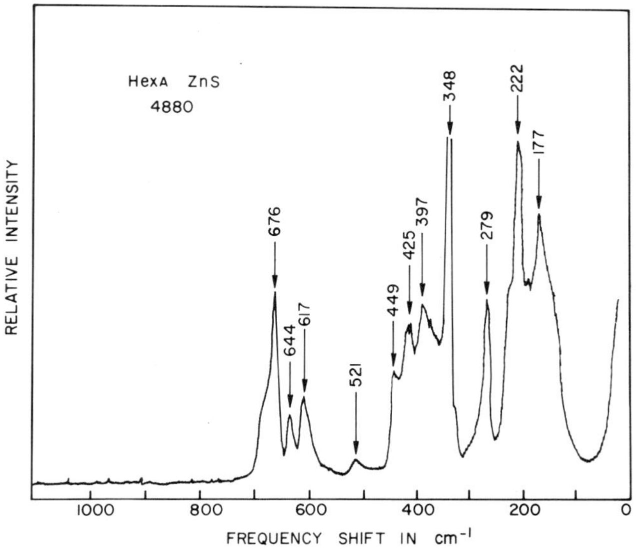

Fig. 4. Raman spectrum of hexagonal ZnS. The laser beam propagates along the C-axis and the detected scattered light along a <10· 0> direction.

TABLE I

Comparison of Predicted and Observed Polarization Characteristics of the Second-Order Raman Effect in Cubic ZnS.

Raman Process	Polarization Characteristic* Predicted† X	L	Observed
LA-TA	Γ_{15}	$\Gamma_{12}+\Gamma_{15}$	Γ_{15}
LO-TO	Γ_{15}	$\Gamma_{12}+\Gamma_{15}$	Γ_{15}
2(TA)	$\Gamma_1+2\Gamma_{12}+\Gamma_{15}$	$\Gamma_1+\Gamma_{12}$	Γ_1
2(LA)	$\Gamma_1+\Gamma_{12}$	$\Gamma_1+\Gamma_{15}$	$\Gamma_1+\Gamma_{12}+\Gamma_{15}$
TO+TA	$\Gamma_1+2\Gamma_{12}+\Gamma_{15}$	$\Gamma_1+\Gamma_{12}+2\Gamma_{15}$	$\Gamma_1+\Gamma_{12}+\Gamma_{15}$
TO+LA	Γ_{15}	$\Gamma_{12}+\Gamma_{15}$	Γ_1(?)$+\Gamma_{12}+\Gamma_{15}$
LO+TA	Γ_{15}	$\Gamma_{12}+\Gamma_{15}$	Γ_1(?)$+\Gamma_{12}+\Gamma_{15}$
LO+LA	Γ_{15}	$\Gamma_1+\Gamma_{15}$	Γ_1
2(TO)	$\Gamma_1+2\Gamma_{12}+\Gamma_{15}$	$\Gamma_1+\Gamma_{12}$	$\Gamma_1+\Gamma_{15}$
LO+TO	Γ_{15}	$\Gamma_{12}+\Gamma_{15}$	Γ_1(?)$+\Gamma_{15}$
2(LO)	$\Gamma_1+\Gamma_{12}$	$\Gamma_1+\Gamma_{15}$	$\Gamma_1+\Gamma_{15}$

*The polarization properties of the second-order spectra are specified in terms of the irreducible representations of the polarizability appropriate for the T_d point group.

†Obtain from the space-group selection rules for the zinc blende structure given by Birman J. L., Phys. Rev. 127, 1093 (1962). The polarization characteristics from both critical point X and L on the Brillouin-zone boundary are given separately in the table.

The comparison of calculated and observed selection rules for cubic ZnS may be summarized as follows. The Γ_1 spectra gives the most intense scattering and is observed experimentally in each combination band where it is allowed by symmetry. The Γ_{12} and Γ_{15} spectra is less intense and in some instances are not observed although allowed by symmetry. This is obviously not a violation of the selection rules but merely indicates that the scattering intensity from the Γ_{12} and Γ_{15} spectra for some second order Raman processes is too weak to be detected especially in the presence of the more intense Γ_1 spectra. It should be noted that our sensitivity for detecting Γ_{12} or Γ_{15} spectra in the presence of Γ_1 spectra is usually limited by depolarization effects in the sample (about 20-40:1) rather than the sensitivity of our instrument. For some combination bands, Γ_{12} is allowed only at critical point X and Γ_{15} only at L but both Γ_{12} and Γ_{15} spectra are observed. This result indicates that scattering from both critical point X and L contribute significantly to the observed spectra. In a few cases, Γ_1 appears to be present in a band even though it is forbidden by symmetry at critical points X, L, and W. This result is not firmly established since in one case (LO+TO), the corresponding difference band does not show Γ_1 polarization and in the other cases (TO+LA and LO+LA), overlap with adjacent peaks makes it difficult to distinguish Γ_1 from Γ_{12} spectra.

THE CUBIC ZnSe AND ZnTe SPECTRA

The second-order Raman spectrum of cubic ZnSe and ZnTe have not been as fully analyzed as in the case of cubic ZnS. Nevertheless, comparison of the ZnSe and ZnTe spectra with the ZnS spectra allows us to draw some conclusions immediately. In the case of ZnSe, the LO and TO modes at the zone boundary seem to be much closer together in energy than in the case of ZnS (see Fig. 2). Thus, the LO and TO overtone are superimposed on the peak at 504 cm^{-1}. A similar situation seems to occur with the acoustic modes at the zone boundary and the LA and TA overtones are superimposed at 146 cm^{-1}. The various optic-acoustic sum combination bands appear at 297 cm^{-1}. Much of the width of these peaks might be due to the variation in the mode frequencies with points (especially critical points) on the zone boundary. The sharp Raman lines at 250 and 205 cm^{-1} are the LO and TO phonon modes at the zone center.

The ZnTe spectrum shown in Fig. 3 can be interpreted in the same way as the ZnSe spectrum. The peak at 408 cm^{-1} is a superposition as the LO and TO overtones and the one at 107 cm^{-1} is a superposition of the LA and TA overtones. The various optic-acoustic sum combination bands are seen around 244 cm^{-1}; the corresponding difference

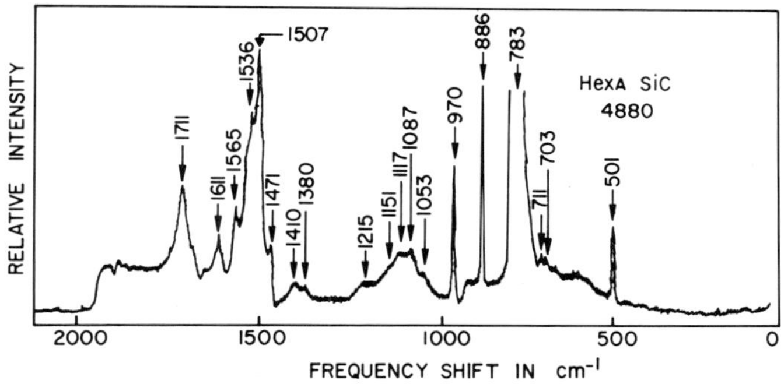

Fig. 5. Raman spectrum of hexagonal SiC. The laser beam propagates along a <10·0> direction and the detected scattered light propagates along a <12·0> direction.

bands are probably responsible for the spectra around 150 cm^{-1}. The first-order LO and TO lines are at 205 and 177 cm^{-1}. These results are summarized in Table II. As we go along the series from ZnS to ZnSe to ZnTe, the phonon energies throughout the Brillouin Zone decrease and the difference in energy between differently-polarized optical or acoustic phonons decreases.

TABLE II

Phonon Energies of Various Zinc Blende and Wurtzite Crystals at Zone Center and Zone Boundary

Crystal	Phonon Energies in cm^{-1}					
	Zone center		Zone boundary			
	TO	LO	TA	LA	TO	LO
Cubic ZnS	271	352	88	110	306	333
Cubic ZnSe	205	250	76	76	252	252
Cubic ZnTe	177	205	54	54	204	204
Hexa. ZnS	279	348	88	111	308	338
Hexa. SiC	783	970	?	?	753	856

THE HEXAGONAL ZnS AND SiC SPECTRA

The close relation between the crystal structure of cubic and hexagonal ZnS makes it of interest to compare their second-order Raman spectra. The Raman spectrum of hexagonal ZnS is shown in Fig. 4. As can be seen from Figs. 1 and 4, the second-order spectra of cubic and hexagonal ZnS are identical to within the precision and resolution of our measurement. The first-order LO and TO lines have diagonal scattering tensor components in the hexagonal structure so that these lines, located at 348 cm^{-1} and 279 cm^{-1}, form a prominent part of the spectrum.

The second-order Raman spectrum of hexagonal SiC is shown in Fig. 5. This spectrum has not been completely interpreted but by comparison with the ZnS spectrum some preliminary remarks can be made. The peaks at 1711 and 1507 cm^{-1} are optical overtones and the one at 1611 cm^{-1} is the corresponding sum band. The assignment of LO and TO overtones has not been made although we made an obvious assumption to obtain the data in Table II. Also, there seems to be more anisotropy in the energies of the phonon branches on the zone boundary as evidenced by the greater structure seen in the SiC spectra than the ZnS spectra. The peaks around 1087 cm^{-1} appear to be optic-acoustic sum bands and the acoustic overtones are probably hidden under the intense first-order line at 783 cm^{-1}.

CONCLUSIONS

As a source of information on the lattice dynamics of a crystal, second-order Raman measurements probably give less detailed information than neutron-diffraction measurements but probably more reliable data than infrared measurements. The main drawback with the second-order Raman measurements is that individual points in the Brillouin zone can not be examined. The second-order spectra is an average of scattering from several, at times unknown, points in the zone with the high-symmetry or critical points on the zone boundary contributing most of the scattering. In theory, the Raman measurements because of their polarization characteristics should give some information on the symmetry properties of the phonons at various points in the zone. A case in point is in assigning the X_1 or X_3 representation to the LA or LO phonon mode at critical point X in the zinc blende structure (3, 4). Unfortunately, the two species have exactly the same selection rules for second-order Raman and no assignment can be made. The main advantage of second-order Raman is that measurements can be made on small single crystals with relatively simple equipment. Ideally, second-order Raman measurements and neutron-diffraction measurements should supplement one another; the former giving symmetry information and the latter energy information on the phonons at various points in the Brillouin zone. Also, the value of second-order Raman spectra increases when a structurally related series of crystals are examined.

ACKNOWLEDGMENT

The author wishes to thank J. R. Potopowicz for his excellent technical assistance.

REFERENCES

1. W.G. Nilsen, Phys. Rev. (to be published).
2. W.G. Nilsen and J.G. Skinner, J. Chem Phys. 47, 1413 (1967).
3. J.L. Birman, Phys. Rev. 127, 1093 (1962).
4. J.L. Birman, Phys. Rev. 131, 1489 (1963).

B-7: THE BRILLOUIN, RAMAN AND INFRA-RED SPECTRA OF GALLIUM PHOSPHIDE

S. Fray, F.A. Johnson, R. Jones, S. Kay, C.J. Oliver, E.R. Pike, J. Russell, C. Sennett, J. O'Shaughnessy and C. Smith
Royal Radar Establishment
Malvern, Worcestershire, England

INTRODUCTION

The first purpose of this talk is to describe digital techniques used in our laboratory to obtain Brillouin spectra by Fabry-Perot interferometry. We have also used some of these techniques in Raman spectroscopy and this work will also be discussed. The spectra we have chosen to show are of GaP since we have also obtained high-resolution, infra-red two-phonon absorption bands for this material. A preliminary attempt at a shell-model calculation of the lattice-vibration bands has been made using the elastic constants found from the Brillouin spectra and compared with the second-order Raman and infra-red data. Structure near the central laser peak in the Raman spectrum has been found, which has not yet been explained. This extends to about 100 cm^{-1} and has two components.

BRILLOUIN SPECTRUM

The Brillouin spectrum of GaP (input direction 113 output $1\bar{1}0$) is shown in Fig. 1. The sample was a solution-grown platelet. The equipment used for obtaining this spectrum has been developed for rather more exacting applications. The novel features are concerned with obtaining high resolution at low light levels by signal averaging and servo control of drift. The system is used in such applications with a high-power "supermode" single-frequency laser and copes with both drifts of the laser frequency and of the Fabry-Perot cavity length. A block diagram is shown in Fig. 2. The laser drawn in the diagram is a modified Spectra Physics model 125 and gives up to 15 mW at a single frequency. It was used in the conventional way to obtain Fig. 1 since the finesse limitation due to lack of plate flatness ($\lambda/50$ plates were used) reduced the resolution below the point where the laser line-width became important. The reflection finesse was chosen to be 100 to impose little further degradation.

The piezo-electric element is a stack of six lead-zirconate annular discs and is driven by a sawtooth waveform of about 150V amplitude. A D.C. level is applied from a servo loop to stablize the spectrum. This servo-loop is controlled by obtaining a reference

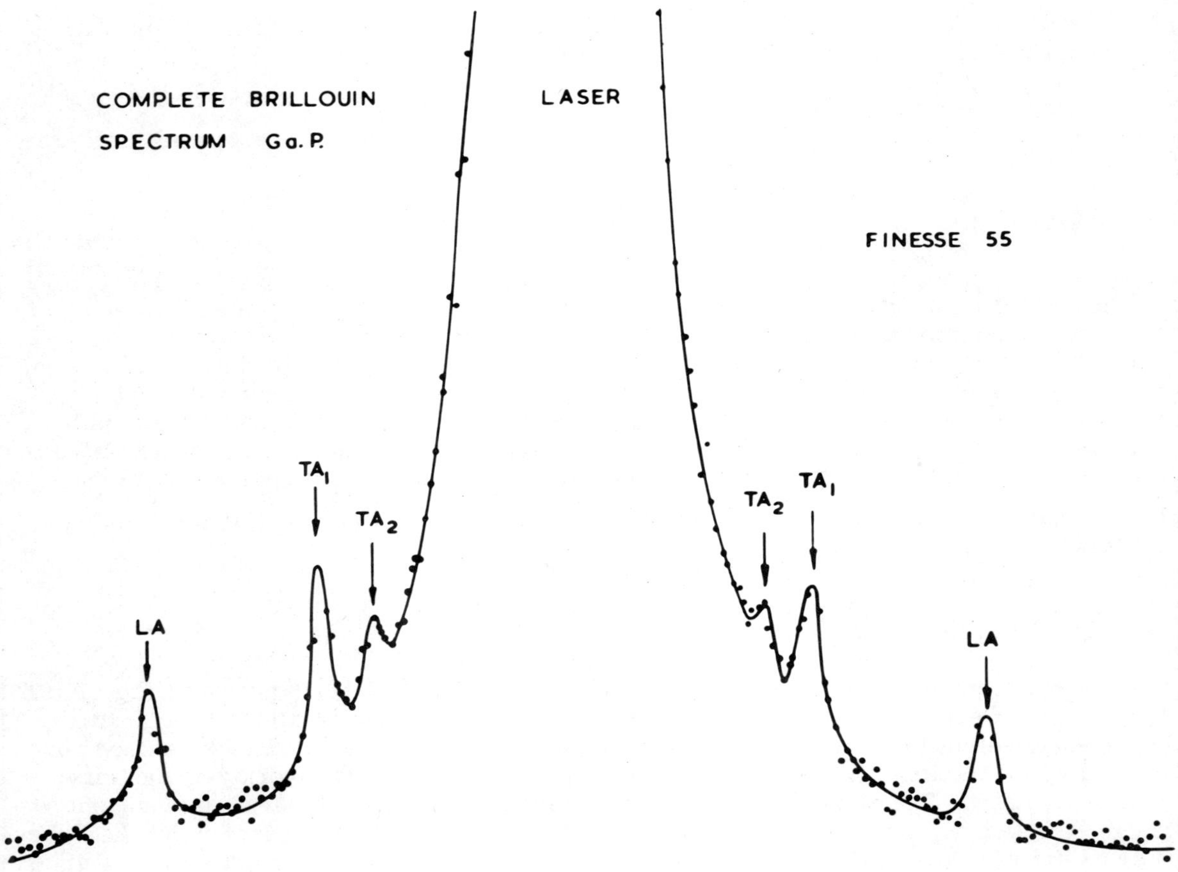

Fig. 1. Brillouin spectrum of GaP, input direction [113], output direction [1$\bar{1}$0].

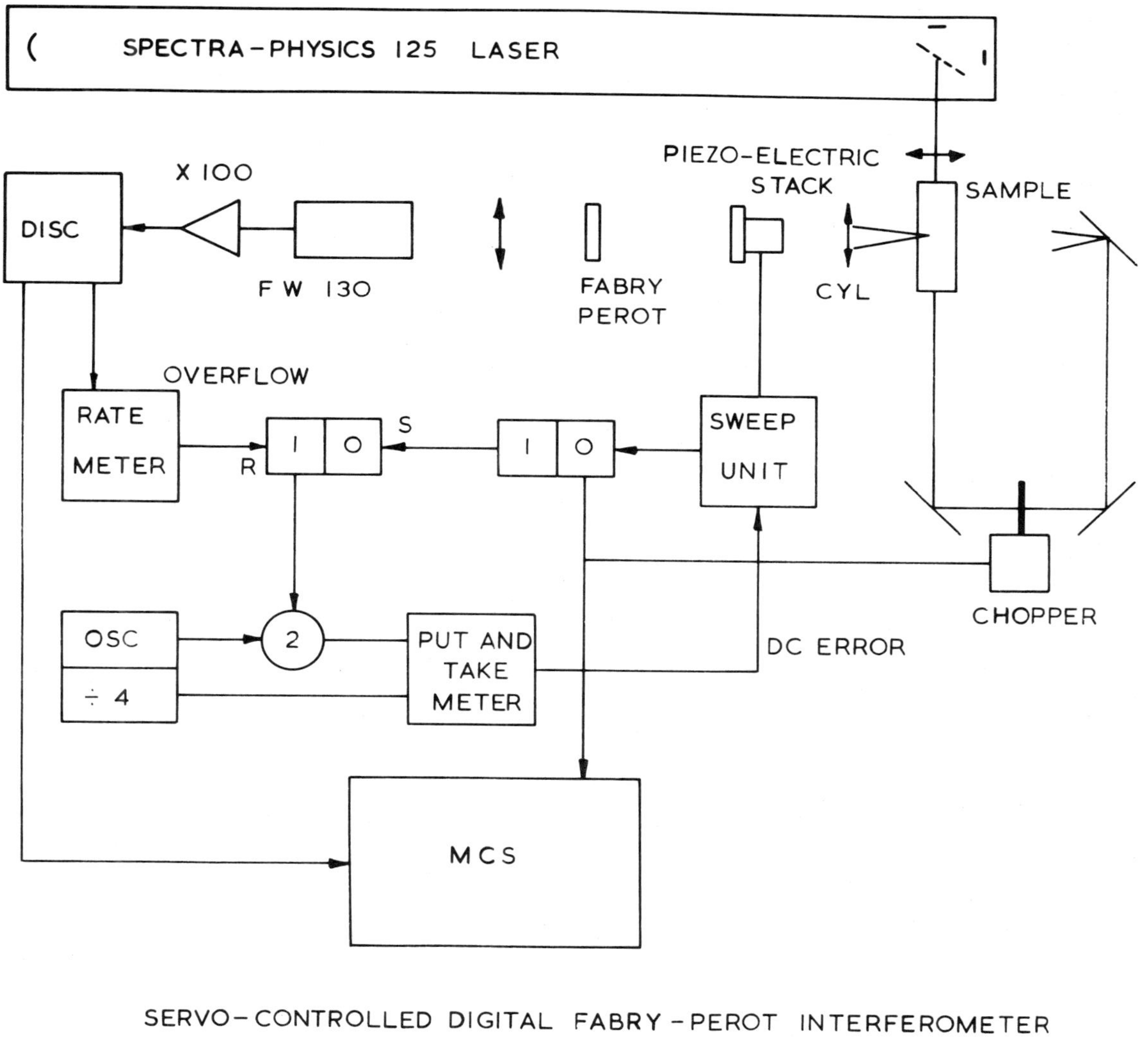

Fig. 2. Block schematic of scanning Fabry-Perot interferometer.

spectrum of the laser line on alternate cycles of the sawtooth. The laser beam, after passing through the sample, is returned back through the Fabry-Perot interferometer via a mechanical chopper every second cycle.

The rate-meter overflows at the point in each reference sweep where the laser line appears and shuts off a train of pulses which starts at the beginning of the sweep and enters one side of a "put and take" rate meter; a second train of pulses at one quarter the frequency is fed constantly into the other side. The D. C. output of this rate meter is integrated over a number of cycles and will move in one direction if the laser line appears later than half way across the reference sweep and in the other direction if the reference line appears earlier. This signal is amplified and fed back to the piezoelectric stack on top of the constant-amplitude sawtooth with the correct polarity to complete the servo loop. Drifts of the laser frequency and of the Fabry-Perot cavity length at rates slower than the integration time of the servo loop are thus removed. The spectrum is recorded by accumulating the photons detected in a multichannel scalar, following the signal-averaging technique described by Jackson and Pike[1], in such a way that each channel corresponds to a small preset range of frequency. The model used here was a 512-channel Hewlett-Packard model 5400A.

The multiplier tube used was an International Telephone and Telegraph Co. FW130 which has a far superior performance for this type of work than any other tube we have tried. The tube is cooled to about -25°C and gives dark-count rates of less than one per second. The performance characteristics of the tube are given in Fig. 3 which are essentially integral and differential-bias curves although obtained and plotted more

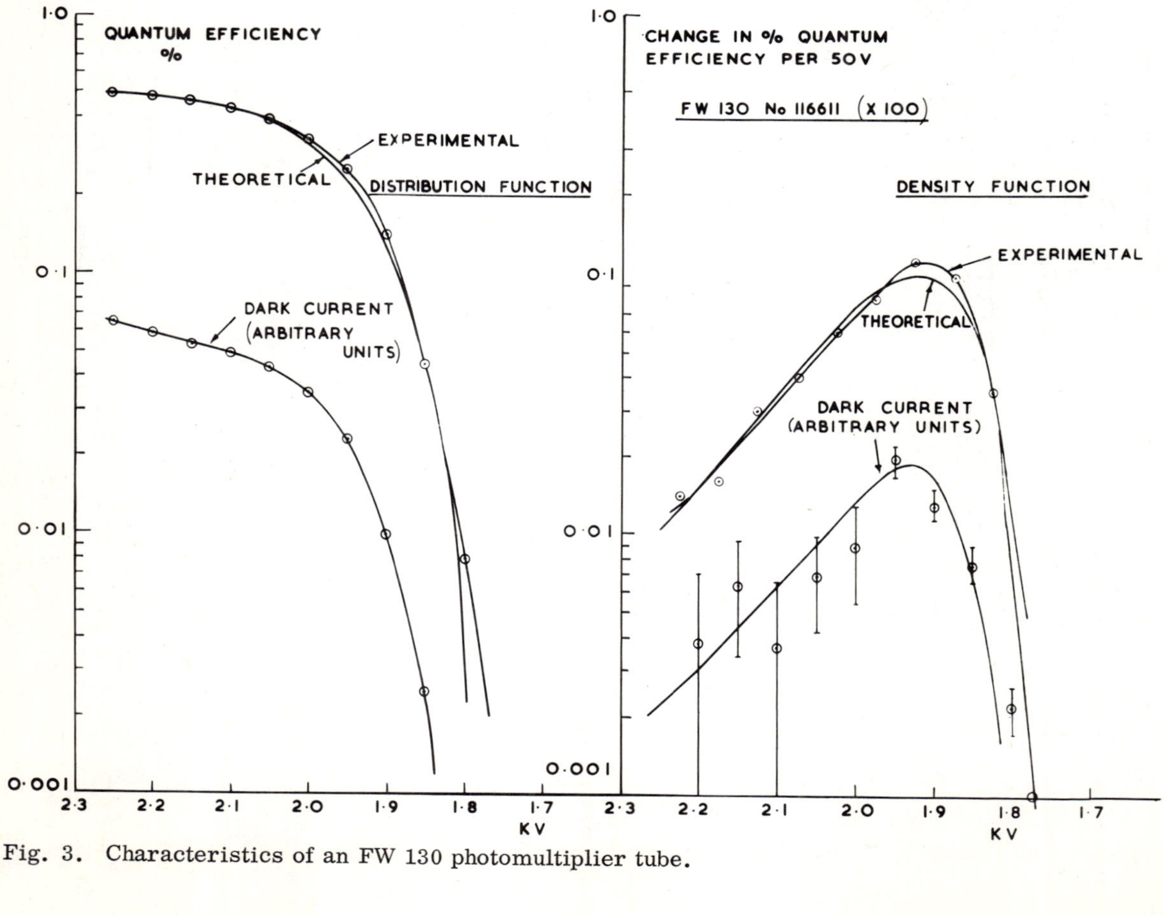

Fig. 3. Characteristics of an FW 130 photomultiplier tube.

conveniently for practical application by varying the tube voltage, using minimum amplification and maximum discriminator sensitivity. The absolute quantum efficiency is used in the ordinate scales. It will be noticed that these values, which are very easy to obtain using a laser and a counting system, are considerably lower than might be expected using the known values of S20 cathode quantum efficiency. This is due to loss of electrons in the first few stages of the tube and in our experience is common to many types of tube. It may also be noticed that the curves show no increase of pulses at low amplitude, which are a feature of many published distributions of this nature, right up to the highest voltage used. Many of these tails are undoubtedly due to poor electronic techniques and to obtain curves which fit the theoretical expectations[2] great care must be used in the selection of amplifiers and circuitry. If a tube is hooked up arbitrarily to a multichannel analyser through any commercial high-gain amplifier, such "tails" will usually be seen. A discussion in detail of the use of multipliers for photon counting is given by Foord, Jones, Oliver and Pike[3]. The general principles to be followed are good high-frequency design of component layout and wiring, operation into low resistance circuits, matching of lines at both ends and utilisation of amplifiers with both sufficiently wide-band performance and good overloading characteristics.

A further feature of interest is that the dark-current is similar to the light-current in its pulse-height distribution and can be safely attributed in the main to thermionic emission from the photocathode. We have checked its statistical behaviour also and found it to be Poisson which ensures the best performance for low-level light detection[4].

The GaP spectrum of Fig. 1 has been analysed to give the three elastic constants of the material using the value of refractive index at 6328 Å of 3.308 given by Bond[5]. The frequency shifts were measured to about one per cent for directions which were slightly off the nominal 113 and $1\bar{1}0$ and gave the elastic constants shown in Table I.

TABLE I

Elastic Constants in Units of 10^{11} dynes/cm^2

	c_{11}	c_{12}	c_{44}
Brillouin scattering	14.36 ± 1.1	6.26 ± 0.56	7.58 ± 0.45

RAMAN SPECTRUM

The use of a multichannel scaler for recording numbers of photons detected in given small wavenumber increments is advantageous also in Raman Spectroscopy. The well known benefits of digital recording are enhanced by the speed and convenience of the storage and display facilities of a multiscaler, and although stability is not such a severe problem, the spectrum may be "pulled out" of noise by repeated additions of identical runs if desired. Some signal averaging equipments divide the total numbers accumulated by the number of cycles completed so that the store never overflows; the capacity of a multiscaler is normally, however, sufficiently high to make this unnecessary. We show in Fig. 4 the Raman spectrum of GaP at three different magnifications and repeat details in Fig. 5. It can be seen that even the two orders of magnitude difference between peak intensities of the first- and second-order spectra are easily accommodated. The photon-counting considerations are the same as discussed in the section above and the same tube type and circuitry was used. A Spex Raman spectrometer was used with 600 line per mm gratings and with slit widths of 250 microns. The spectrum was recorded in 40 minutes at room temperature using about 30 mW of He-Ne laser power. The first-order shifts are given in Table II.

TABLE II

First-Order Raman Shifts in GaP

ω_{TO}	367.3 cm^{-1}
ω_{LO}	403.0 cm^{-1}

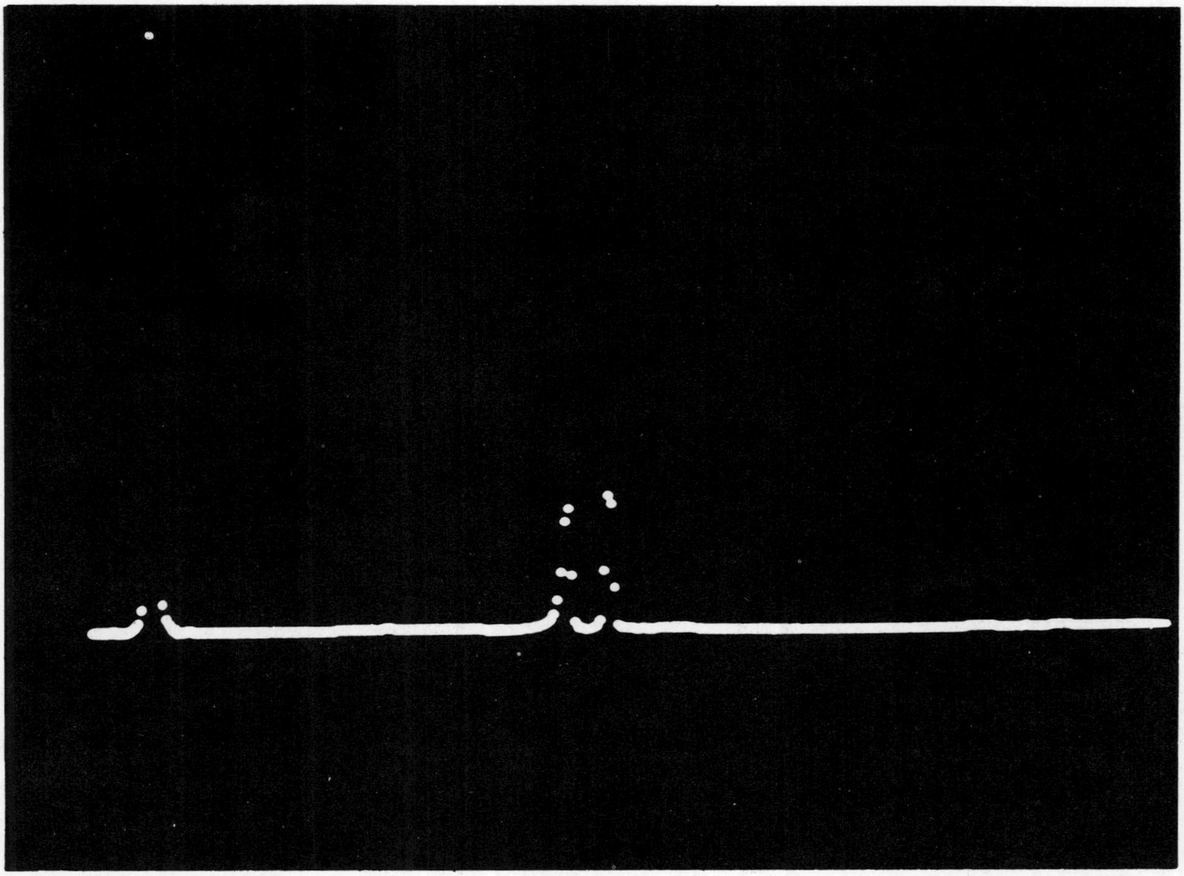

Fig. 4a. Raman spectrum of GaP 10^5 counts/channel full scale.

INFRA-RED SPECTRUM

The infra-red absorption spectrum of GaP at 77°K is shown in Fig. 6 together with a theoretical calculation of the two-phonon density of states. This work was done with an Ebert spectrometer developed in this laboratory shown schematically in Fig. 7. The resolution was 0.5 cm^{-1}. The optical constants were determined absolutely by the simultaneous measurement of transmission and reflection coefficients and the reduction procedure described by Fray, Goodwin, Johnson and Quarrington[6]. The strong similarity between the infra-red absorption spectrum and the second-order Raman spectrum will be seen. To obtain either, theoretically, one requires first a lattice band calculation of the phonon dispersion relations and we have used a type of shell model with some next-nearest-neighbour interactions and only one ion polarizable to obtain the two-phonon density of states shown. The calculation uses the five numbers of Tables I and II, and the value of ϵ_∞ of 9.084 obtained from Bond[5], as input parameters and follows the lines of a similar calculation performed by Dolling and Waugh[7]. We have made some attempt to put in the appropriate matrix elements to obtain the second-order Raman spectrum from the lattice-band calculations but the agreement is not good and further work is required to improve the theory.

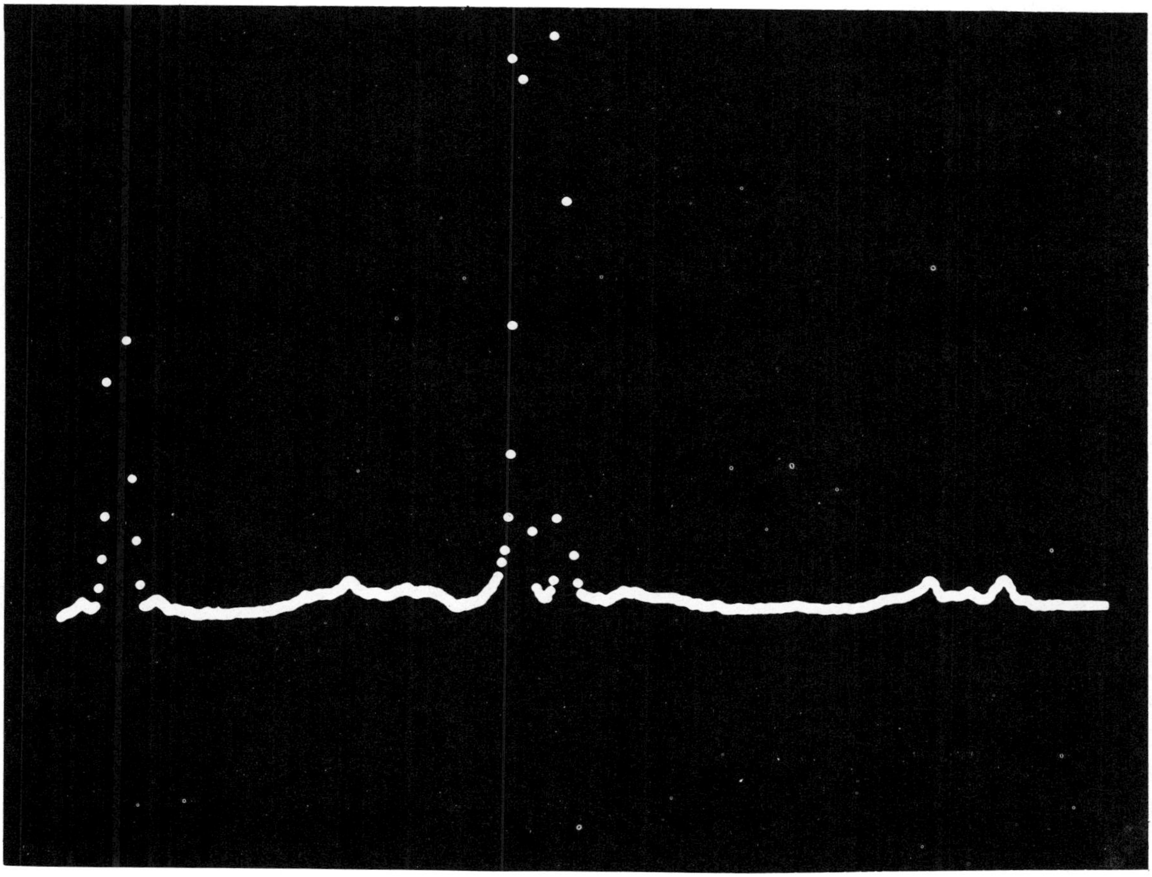

Fig. 4b. Raman spectrum of GaP 10^4 counts/channel full scale.

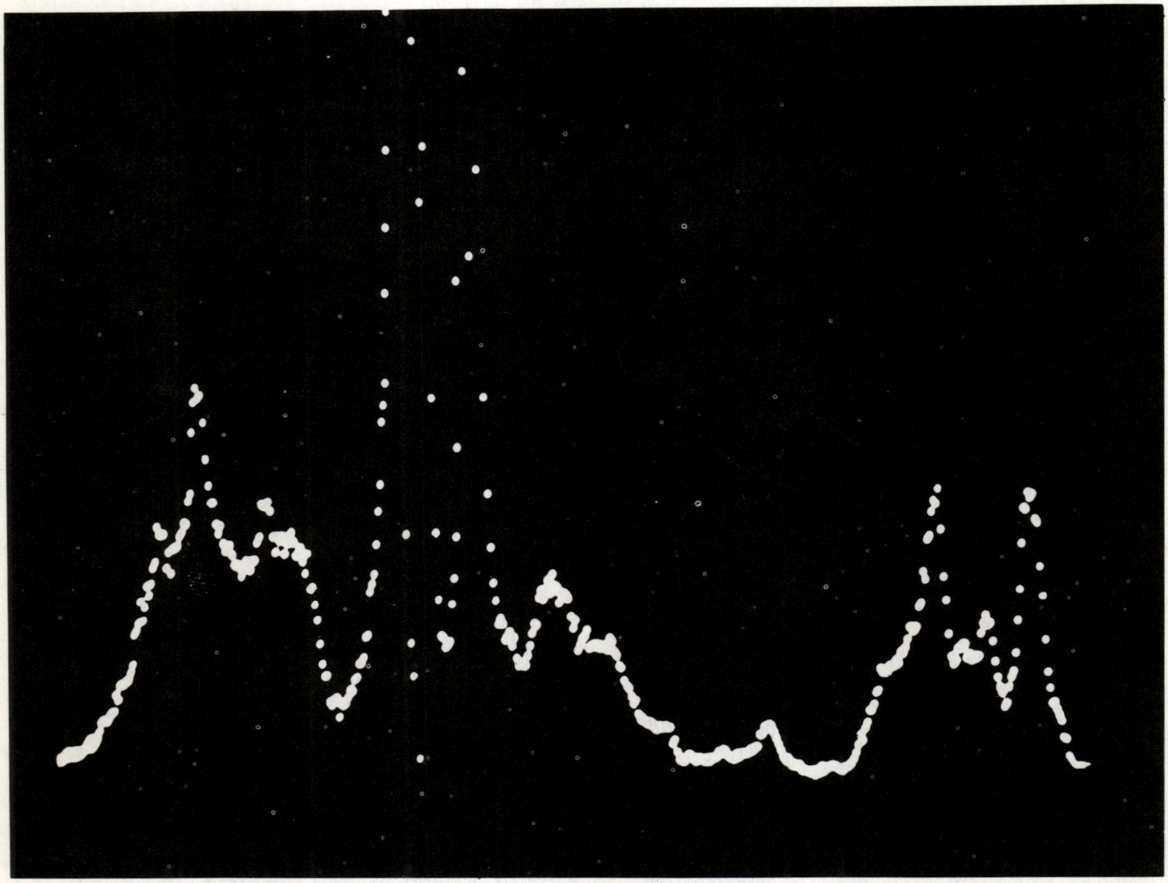

Fig. 4c. Raman spectrum of GaP 10^3 counts/channel full scale.

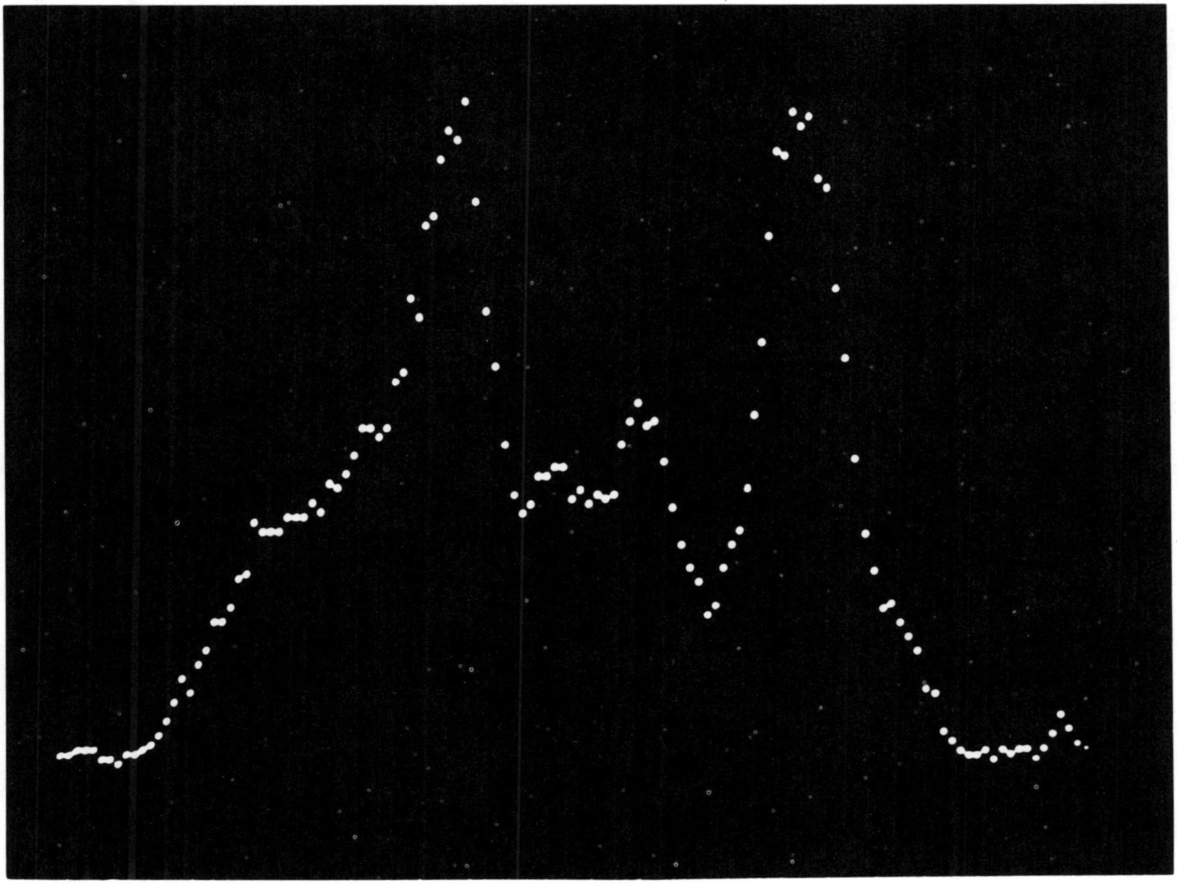

Fig. 5a. Details of second-order Raman spectrum of GaP optic + optic combination bands.

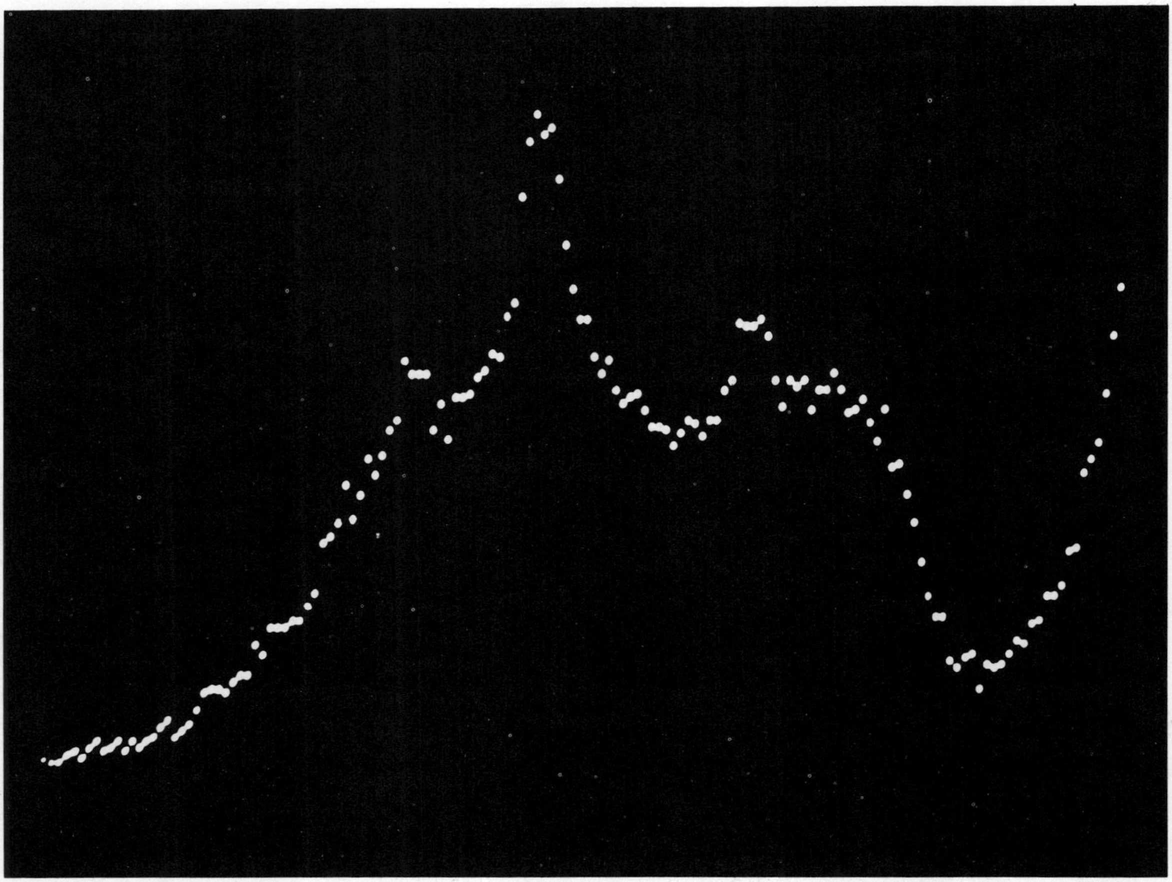

Fig. 5b. Details of second-order Raman spectrum of GaP transverse acoustic overtones and TA + LA combination bands.

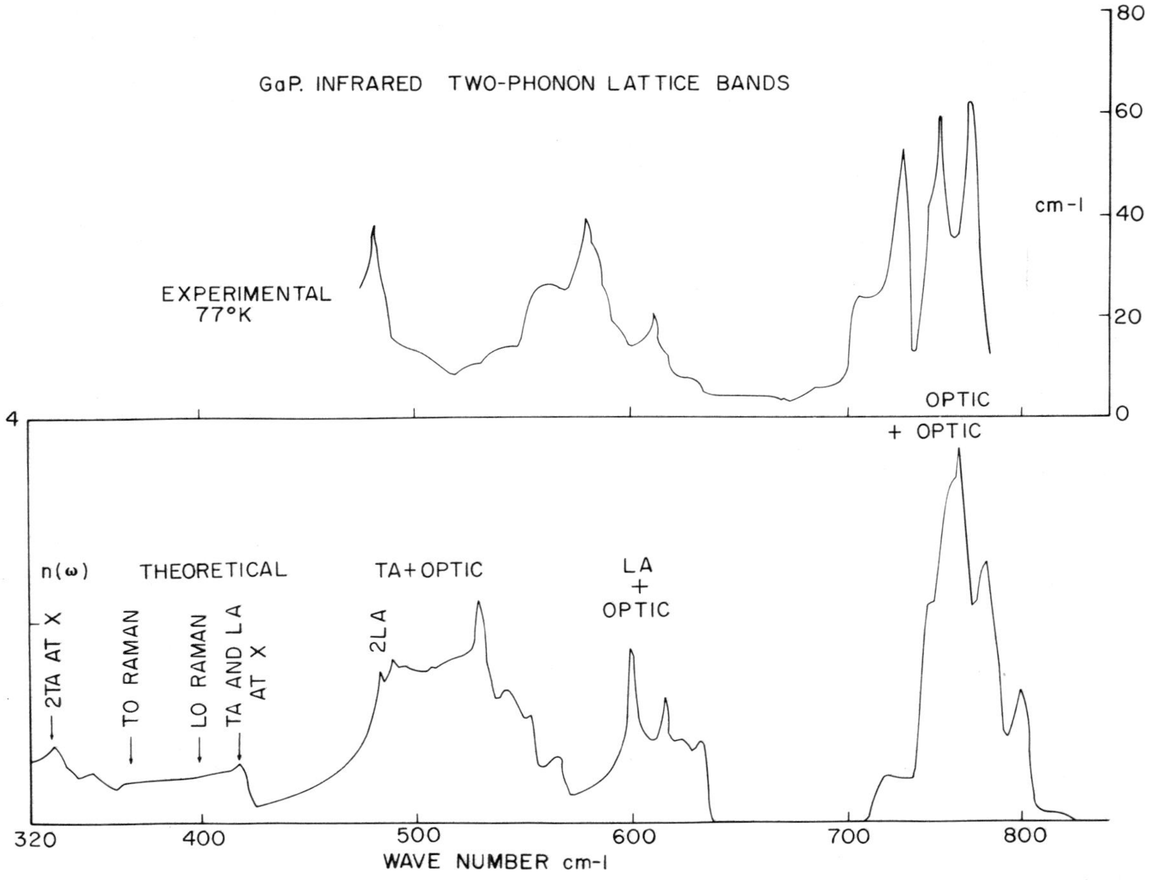

Fig. 6. Infra-red absorption spectrum of GaP together with shell-model calculation of two-phonon density of states.

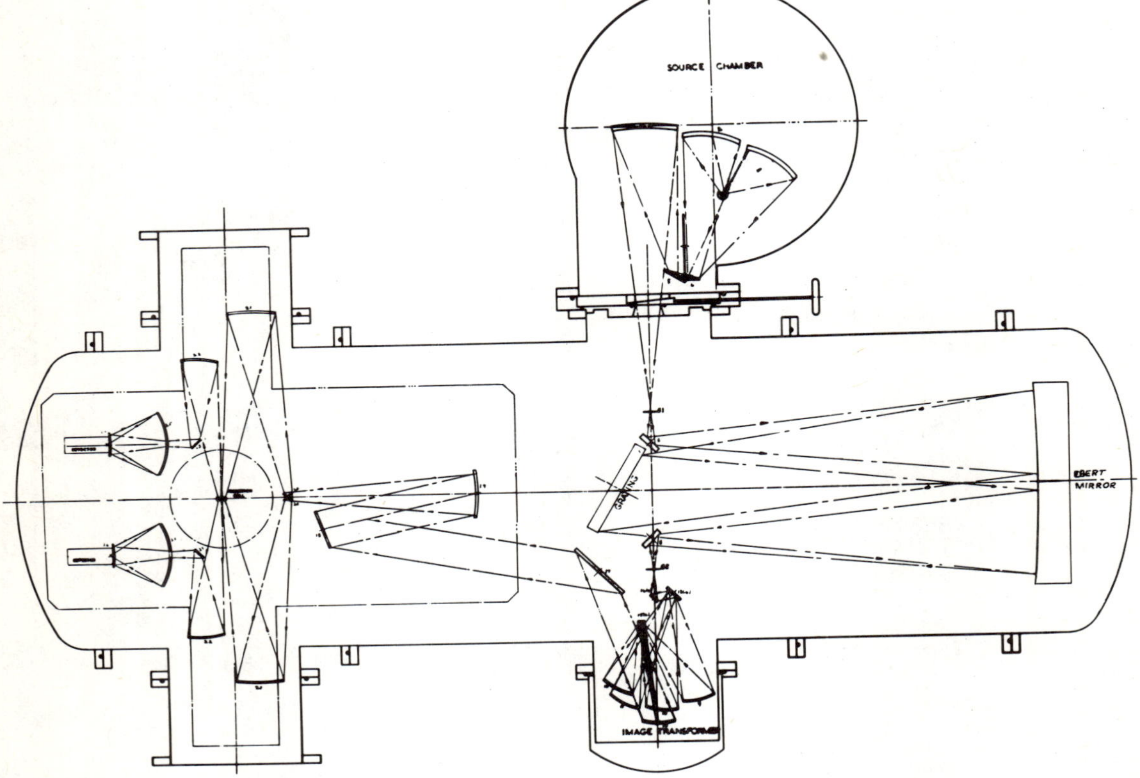

Fig. 7. Diagram of Ebert Infra-Red Spectrometer.

REFERENCES

1. D. Jackson, and E.R. Pike, J. Phys. E.1, 394 (1968).
2. P.M. Woodward, Proc. Camb. Phil. Soc. 44, 404 (1948).
3. R. Foord, R. Jones, C.J. Oliver, and E.R. Pike, (to be published).
4. C.J. Oliver, and E.R. Pike, J. Phys. D. 1, 1459 (1968).
5. W.L. Bond, J. Appl. Phys. 36, 1674 (1965).
6. S.J. Fray, A.R. Goodwin, F.A. Johnson, and J.E. Quarrington, Japanese Journal of App. Physics 4 Supp. 1, 594 (1965).
7. G. Dolling and J.L.P. Waugh, "Proceedings of International Conference on Lattice Dynamics," Pergammon Press, Copenhagen, 1963.

B-8: STUDY OF DYNAMICAL MODEL OF ICE LATTICE IN ORDER TO INTERPRET THE LOW-FREQUENCY RAMAN SPECTRUM

André Kahane and Pierre Faure
Laboratoire de Spectrométrie Physique de la Faculté des
Sciences de Grenoble, France

The structure of the oxygen atoms lattice in hexagonal ice corresponds to symmetry group D_{6h}^4 with four atoms A_1, A_2, A_3, A_4 in the unit cell (Fig. 1). The symmetry types of the principal normal modes of such a lattice are A_{1g}, B_{1g}, B_{2u}, E_g^-, E_g^+, and E_u^+. The active modes in the first-order Raman effect are A_{1g}, E_g^- and E_g^+.

In the low frequency Raman spectrum, the most intense line at 226 cm^{-1} (at 150°K) is polarized in a way corresponding to the superposition of two normal modes A_{1g} and E_g^+[1]. Beyond this line, the Raman spectrum presents a large number of lines and bands that are not well identified[2, 3].

In order to give an interpretation of the low-frequency optical spectra of ice (Raman and infra-red[4, 5]) we use a dynamical model of the crystal lattice of oxygen atoms. The protons that are responsible for the crystal cohesion by hydrogen bonding, are represented through force constants. This model depends on two constants:

K, the stretching constant of the bond between nearest neighbours and

G, the angle bending constant connecting three neighbouring atoms. In order to represent the 12x12 dynamical matrix of the crystal lattice, the Bright Wilson method, normally used for molecules, is applied[1]. The matrix terms are expressed as a function of K, G and the wave vector $\vec{k}$ defined in the first Brillouin zone.

The vibration frequencies are given as a function of K and G for the high symmetry points Γ, M and K of the first Brillouin zone (Fig. 2) and the dispersion curves of the model are determined for the three principal symmetry directions (Γ → A, Γ → M, Γ → K) (Fig. 3). The ratio G/K is taken equal to 0.034. This value is determined from the comparison between the experimental[6] and calculated[1] values of the elastic constants of ice.

The frequency spectrum is calculated for 275 points in the 1/24 th of the Brillouin zone. The value of K is taken so as to fit the principal peak of the spectrum with the main Raman frequency 226 cm^{-1}. This spectrum is compared (Fig. 4) with the experimental results of Prask, Boutin and Yip[7] obtained from slow neutron inelastic scattering.

This two-constant model seems to be more useful to represent low-frequency vibrations of ice than the Forslind[8] and Nakahara[9] six-constant model.

Experiments are underway to obtain the low-frequency Raman spectrum of ice at liquid helium temperature. We intend to use this two-constant model to interpret the first and second order Raman spectra[10].

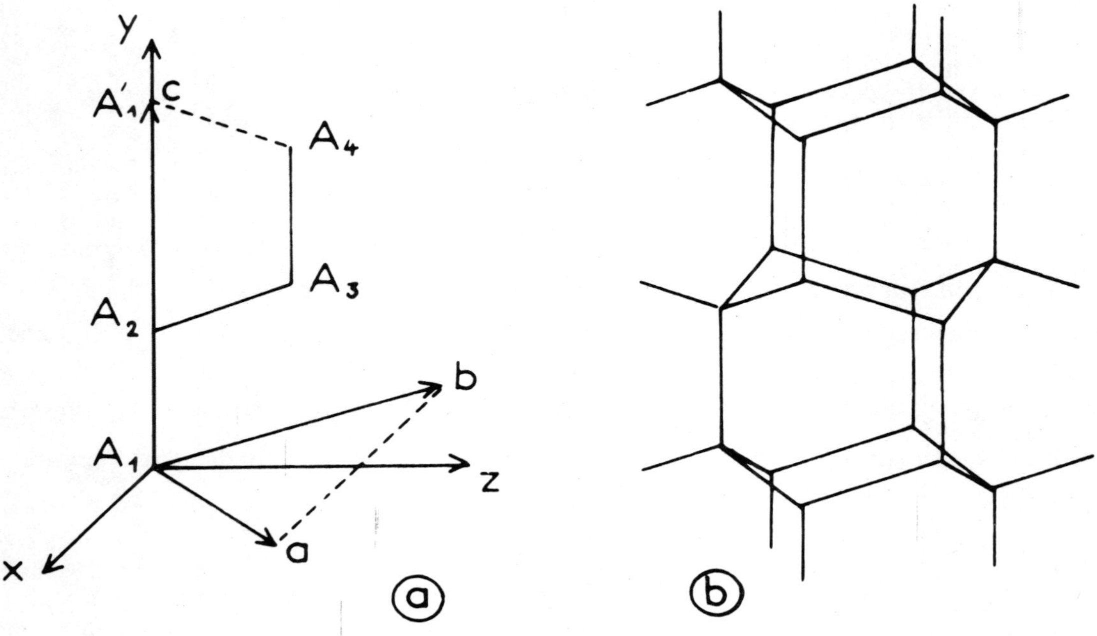

Fig. 1. Representation of the lattice of ice. a) Unit cell and Basis of the lattice b) Perspective view of a portion of the lattice

Unit cell:
- $\vec{a}$ $(a\sqrt{2}/\sqrt{3};\ a\sqrt{2};\ 0)$
- $\vec{b}$ $(-a\sqrt{2}/\sqrt{3};\ a\sqrt{2};\ 0)$
- $\vec{c}$ $(0;\ 0;\ 8a/3)$

Basis:
- A_1 (0; 0; 0;)
- A_2 (0; 0; a)
- A_3 $(0;\ 2a\sqrt{2}/3;\ 4a/3)$
- A_4 $(0;\ 2a\sqrt{2}/3;\ 7a/3)$

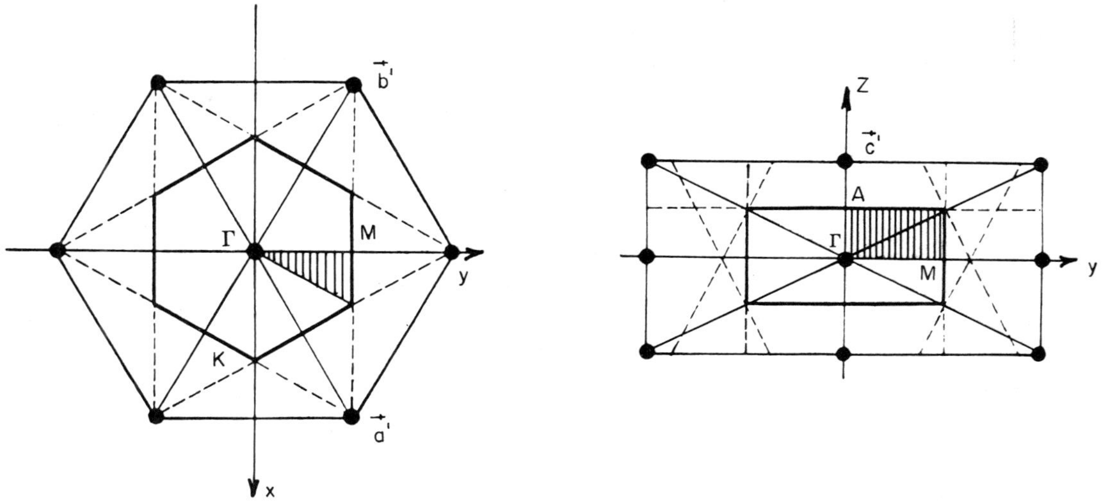

Fig. 2. Reciprocal lattice, Brillouin Zone, and points of high symmetry Γ, A, M, K.
●: Reciprocal Lattice
–: First Brillouin Zone
Cross hatched portion: 1/24 of the first zone

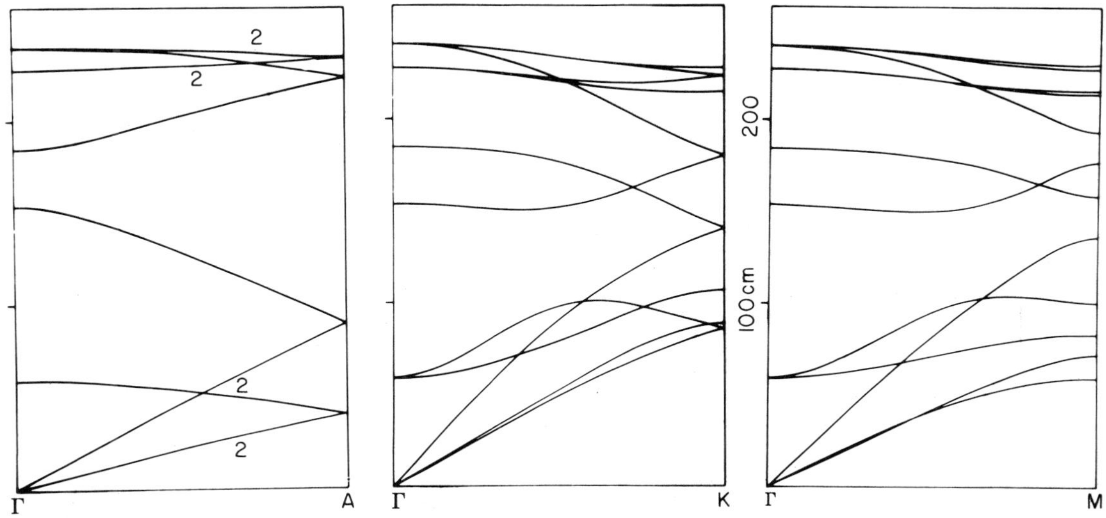

Fig. 3. Dispersion curves Γ→ A, Γ→ M, Γ→ K.

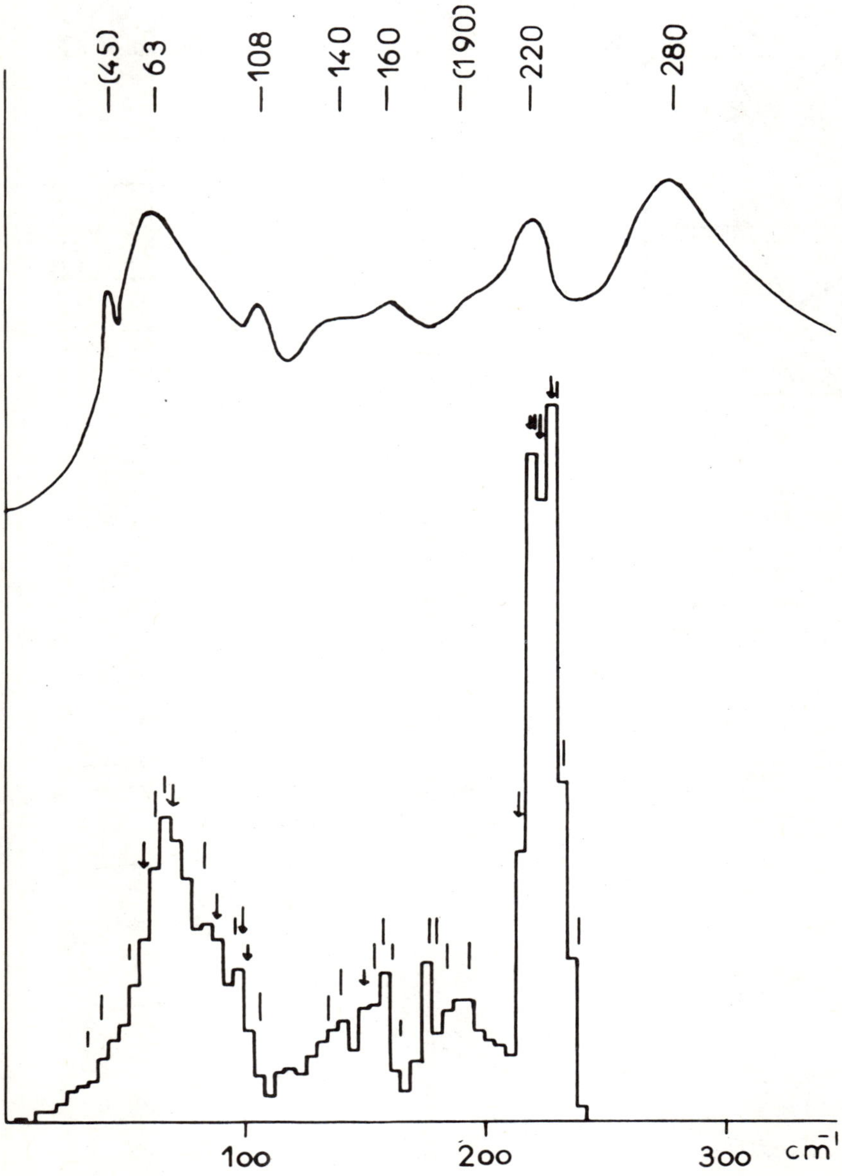

Fig. 4. Upper curve: experimental frequency spectrum. Lower curve: calculated frequency spectrum.

REFERENCES

1. A. Kahane, Thèse Paris (1962).
2. N. Ockman, Adv. in Phys. 7, 199 (1958).
3. V.I. Val'kov and G.L. Malenkova, Opt. i Spectros. 1, 881 (1956).
4. P.A. Giguere and J.P. Arraudeau, C.R. Acad. Sc. Paris 257, 1692 (1963).
5. J.E. Bertie and E. Whalley, J. Chem. Phys. 46, 1271 (1967).
6. F. Jona and P. Scherrer, Helv. Phys. Acta 29, 212 (1951).
7. H. Prask, H. Boutin and S. Yip, J. Chem. Phys. (to be published).
8. E. Forslind, Proc. Swedisch Cement and Concrete Res. Instr. 21, (1954).
9. Y. Nakahara, Jaeri - Memo n° 3108, Japan Atomic Energy Research Institute (1968).
10. P. Faure and A. Kahane, J. Phys. 28, 944 (1967).

B-9: RAYLEIGH AND RAMAN SCATTERING BY SURFACE MODES IN IONIC CRYSTALS

R. Ruppin and R. Englman
Soreq Nuclear Research Center
Yavne, Israel

SURFACE MODES IN IONIC CRYSTALS

In finite size diatomic cubic crystals there exist three types of long wave optical phonons: (a) Longitudinal bulk modes, (b) Transverse bulk modes, and (c) Surface modes. The features characterizing the surface modes in crystals of arbitrary shape are[1]: Their frequencies form a series lying in the range between ω_T and ω_L (the transverse and longitudinal frequencies) and converging to an intermediate frequency ω_s which satisfies the relation $\varepsilon(\omega_s) = -1$,
where ε is the frequency dependent dielectric constant

$$\varepsilon(\omega) = \varepsilon_\infty + \frac{\varepsilon_0 - \varepsilon_\infty}{1 - \omega^2/\omega_T^2} \tag{1}$$

The vibration amplitudes corresponding to the surface modes decay with increasing distance from the surface of the sample. Due to the existence of the long range Coulomb interaction, this decay is rather slow, so that the vibration usually penetrates to the center of the crystal. The number of surface modes is proportional to the number of surface unit cells. These properties of the surface modes still hold when retardation effects are included[2] (in which case the normal modes will be polaritons rather than phonons). The transverse bulk modes, which without retardation are all degenerate with the frequency ω_T, form a bulk polariton band lying below ω_T. The longitudinal modes are not affected by the inclusion of retardation and remain degenerate with the frequency ω_L.

Only few experiments (e. g. infrared absorption[3], electron scattering[4]) pertaining to the surface modes have been performed. We present here the results of some theoretical calculations which show how the surface modes will manifest themselves in Rayleigh and Raman scattering experiments.

RAYLEIGH SCATTERING

In the scattering of infrared radiation from small samples maxima should occur at frequencies of surface modes and of transverse bulk modes (the longitudinal modes do not scatter). The scattering cross section can be exactly calculated[5] provided that the dielectric constant is known. Using a dielectric constant of the form (1) (modified only by the addition of a small damping term), with $\varepsilon_o = 5.934$ and $\varepsilon_\infty = 2.328$[6], we have calculated the scattering cross sections of small NaCl spheres and cylinders. Fig. 1 shows the scattering cross section of a sphere of radius c/ω_T(~ 10 μ). The vertical lines near ω_T and ω_s denote the positions of the first few frequencies of the bulk and surface mode series, respectively. Fig. 2 shows the scattering width (scattering cross section per unit length) of a cylinder of the same radius for two different polarizations of the normally incident beam. In the case in which the electric field is perpendicular to the cylinder axis the spectrum is similar to that of a sphere. In both cases there occur both bulk and surface mode scattering of comparable magnitude. Figs. 3 and 4 refer to thinner samples, of radius 0.1 c/ω_T(~ 1 μ). Again the scattering from a cylinder for perpendicular polarization is similar to that from a sphere. In both cases only a surface mode scattering peak appears. It should be noted, however, that for a sphere the peak is near 222 cm^{-1} whereas for a cylinder it is at 237 cm^{-1}. These frequencies correspond to $\varepsilon(\omega) = -2$ and $\varepsilon(\omega) = -1$ respectively, as could be predicted from the depolarization factors of the samples. The scattering from a very thin sample will be peaked at the frequency which satisfies the relation

$$\varepsilon(\omega) = 1 - \frac{4\pi}{N} \tag{2}$$

where N is the depolarization factor. For a sphere and a cylinder (at perpendicular polarization) the depolarization factors are equal to $4\pi/3$ and 2π, respectively, which yield the correct frequencies.

RAMAN SCATTERING

We now discuss first order Raman scattering from surface modes. Since the displacement patterns of the different atoms in the unit cell which are produced by a surface mode are similar to those produced by the corresponding transverse and longitudinal modes, they will obey the same selection rules. Only crystals which lack a center of inversion will exhibit first order scattering from surface modes. We therefore restrict the discussion to crystals of the zincblende structure.

In the scattering from large samples both energy and momentum are conserved

$$\omega_1 = \omega_2 + \omega \quad ; \quad \underline{k}_1 = \underline{k}_2 + \underline{q} \tag{3}$$

where quantities referring to the incident and scattered radiation have subscripts 1 and 2 respectively; $\underline{q}$ and ω refer to an optic phonon. In scattering from small specimens (i.e., crystals which have at least one dimension comparable with the wavelength of the phonons involved in the scattering) only energy is conserved. In fact the long wave optical phonons cannot in this case be characterized by a wave vector $\underline{q}$. Instead of plane waves $e^{i\underline{q}\cdot\underline{r}}$ the spatial dependence of the displacements of the atoms will be described by some

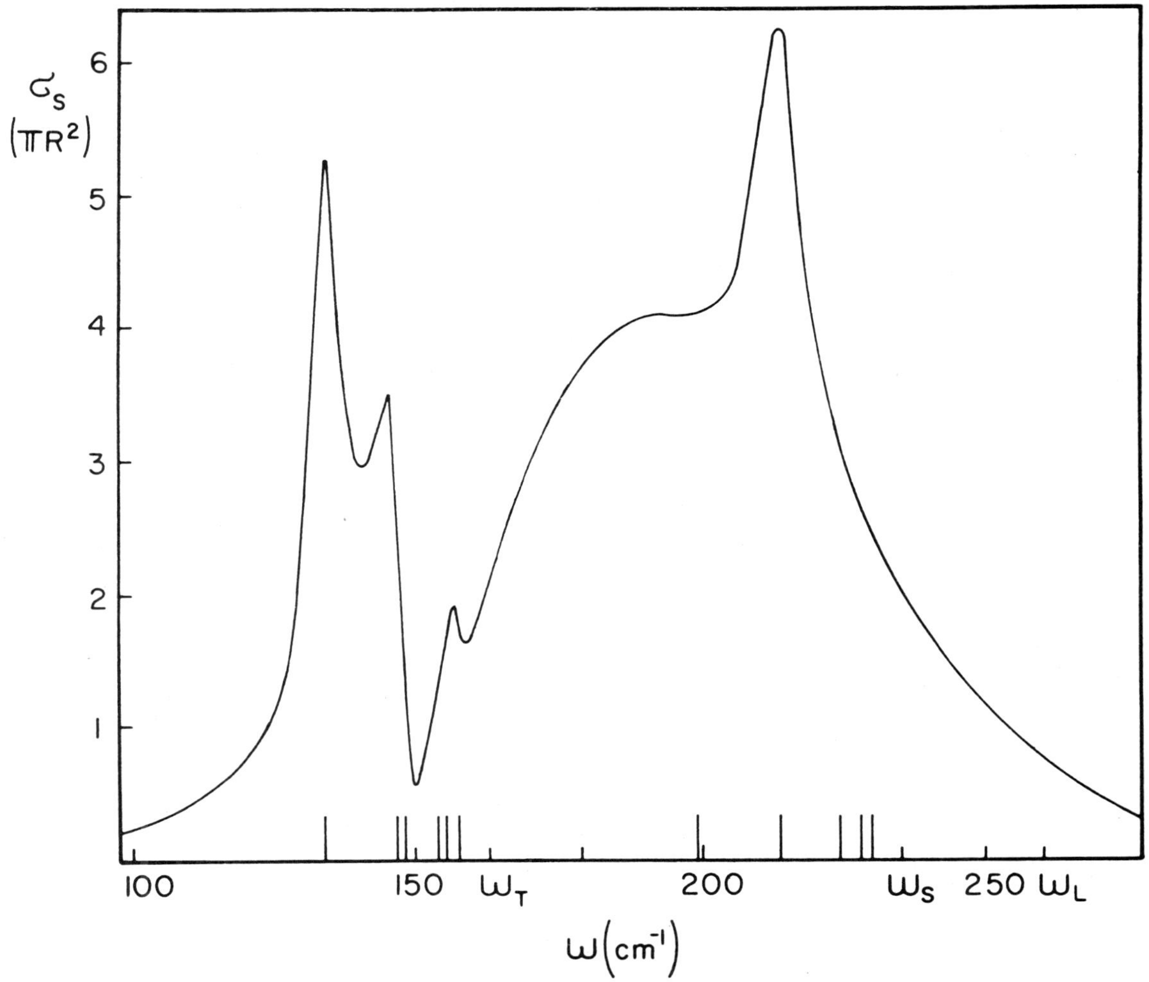

Fig. 1. Scattering cross section of a NaCl sphere of radius c/ω_T (~ 10 μ).

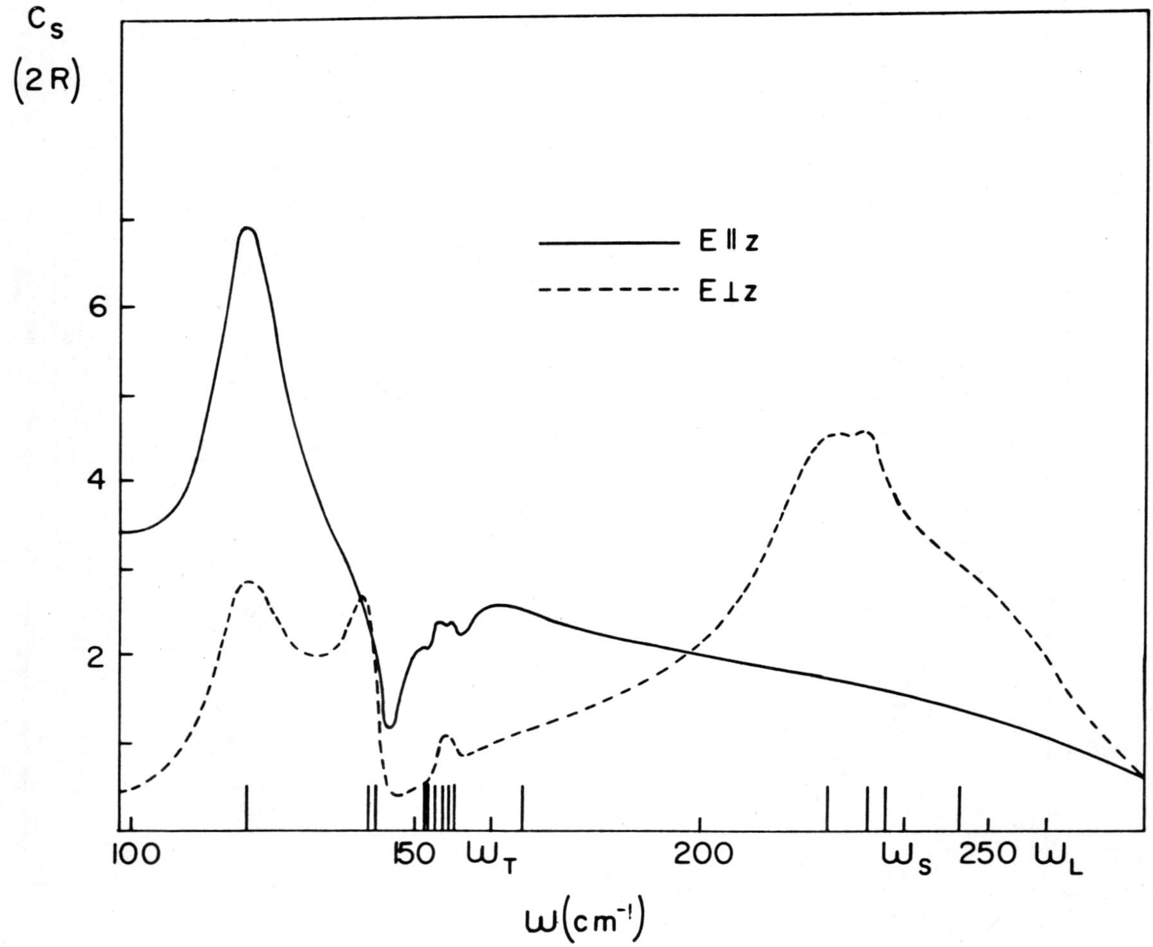

Fig. 2. Scattering width of a NaCl cylinder of radius c/ω_T(~ 10 μ).

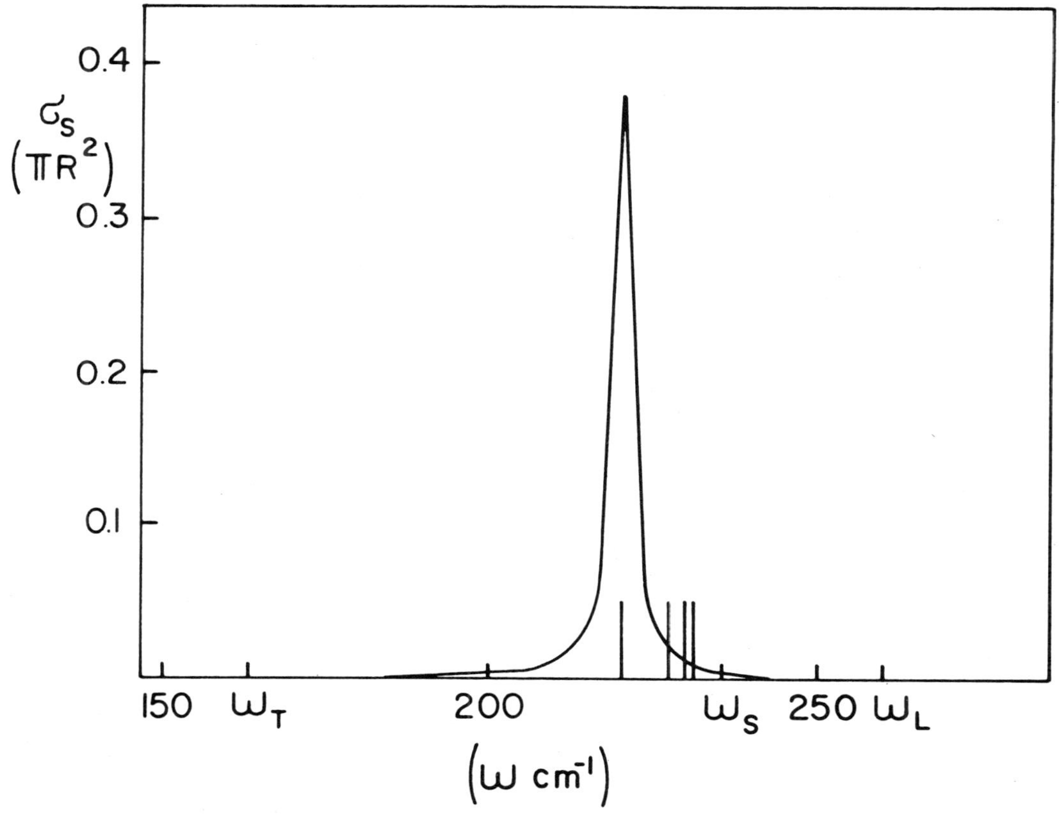

Fig. 3. Scattering cross section of a NaCl sphere of radius $0.1\ c/\omega_T$ ($\sim 1\ \mu$).

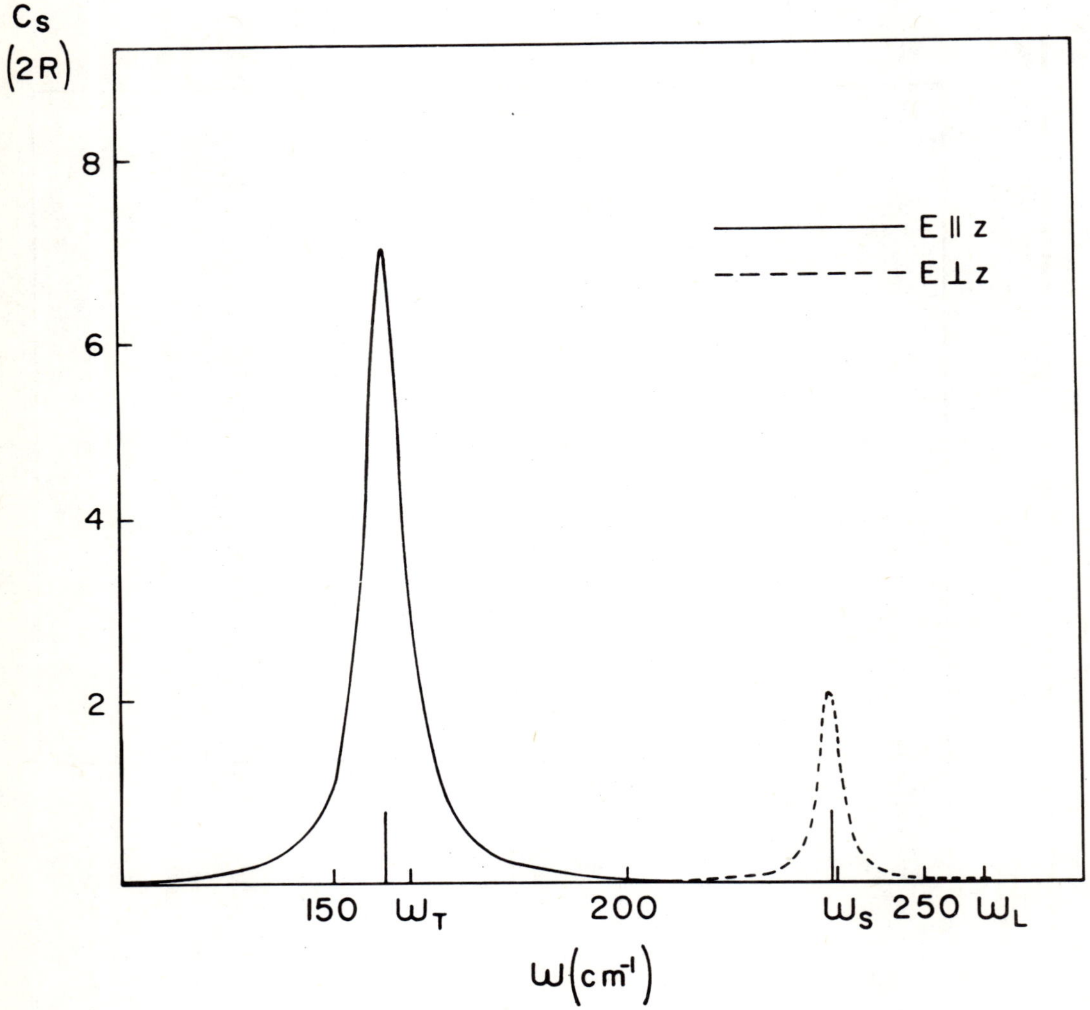

Fig. 4. Scattering width of a NaCl cylinder of radius 0.1 c/ω_T (~1 μ).

function $\underline{f}^p(\underline{r})$, where p stands for quantum numbers which characterize the phonon ($\underline{q}$ no longer being a good quantum number).

Taking the case of a slab and starting from Loudon's theory[7] one finds that the scattering efficiency for a process in which a phonon of type p is involved, is weighted by factors of the form

$$I_\tau = \int e^{i(\underline{k}_1 - \underline{k}_2)\cdot \underline{r}} \; f_\tau^p(\underline{r}) d^3 r \tag{4}$$

I_τ is an integral over the volume of the crystal which involves the τ component of the function $\underline{f}^p(\underline{r})$ which describes the displacements produced by the phonon p. It may be noted that for infinite crystals $\underline{f}^p(\underline{r})$ will follow an $e^{i\underline{q}\cdot \underline{r}}$ space dependence so that I_τ will be different from zero only if momentum is conserved.

Let the incident beam fall perpendicularly on an oriented slab and the scattering be observed at an angle θ (Fig. 5). In the two long directions of the slab cyclic boundary conditions can still be applied and as a result "two dimensional momentum" will be

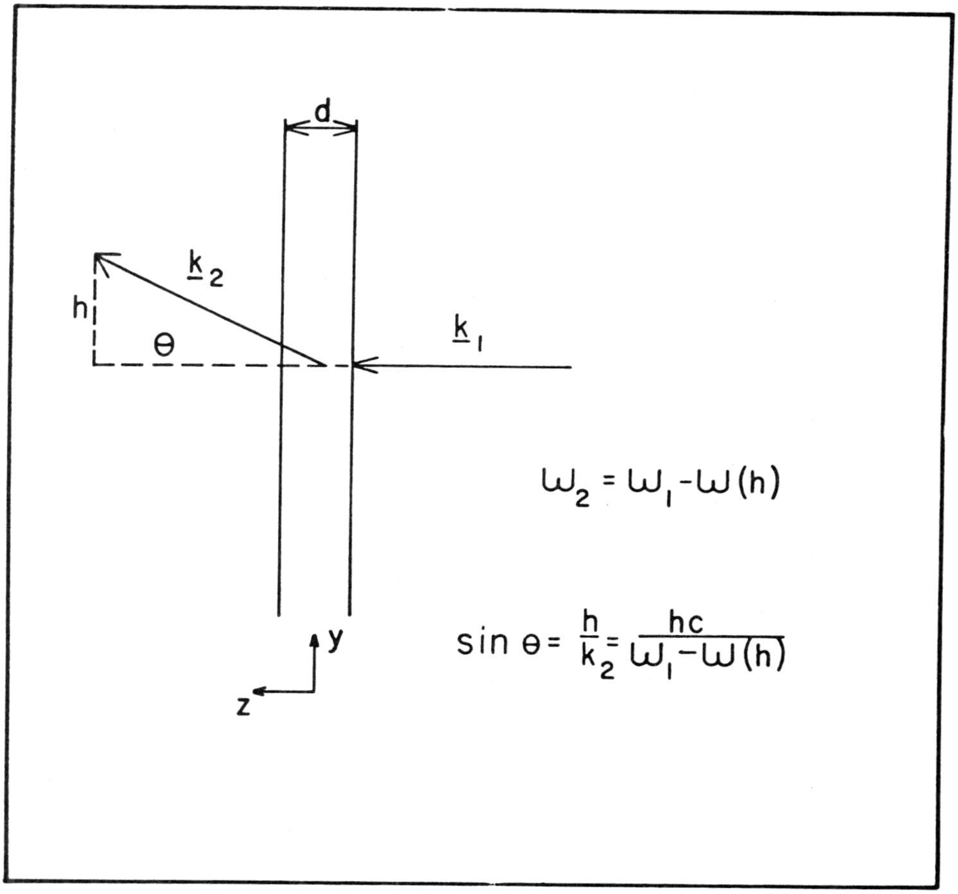

Fig. 5. Raman scattering geometry for a slab.

conserved in the scattering process. Let h be the magnitude of the phonon momentum parallel to the slab. Since the incident beam carries no momentum in the long direction it follows that the scattered photon will carry momentum h in that direction. It follows that $\sin\theta = h/|\underline{k}_2|$ and since $\omega_2 = \omega_1 - \omega$ we obtain

$$\sin\theta = \frac{hc}{\omega_1 - \omega(h)} \tag{5}$$

This relation between h and the angle of observation θ makes it possible to scan the dispersion curve ω(h) of the surface modes by varying θ. For each value of h there exist two surface modes whose frequencies we denote by ω_- and ω_+. Fig. 6 shows the dispersion curves for a SiC slab of thickness 0.4 μ. For ε_0, ε_∞ and ω_T the values given by Spitzer et al. [8] have been used. At any angle of observation, θ, four Raman lines (at ω_T, ω_-, ω_+, ω_L) should occur. The Raman frequencies for different angles can be calculated from (5). Some of these are shown in Fig. 6 (where for this calculation ω_1 was assumed to be the frequency of the 6328 Å He - Ne laser beam).

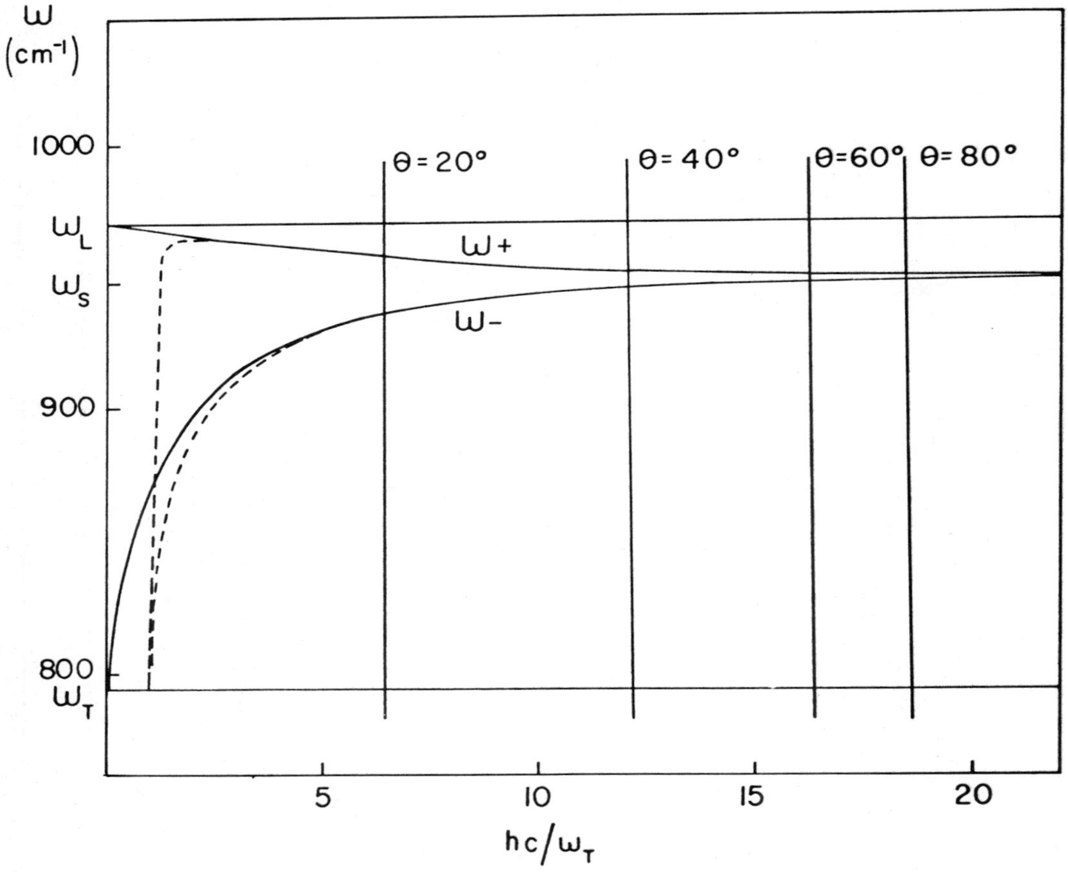

Fig. 6. The surface and bulk modes of a SiC slab of thickness 0.4 μ. The broken curves show the polariton lines. The frequencies observable at some angles are also shown.

Knowing the form of the functions $\underline{f}^p(\underline{r})$ [9] we can evaluate the relative intensity of the four Raman lines observable at any angle. Fig. 7 shows the intensities relative to that of the ω_T line as a function of the slab thickness. In this calculation the observation direction was assumed to be defined by $\theta = 30°$, $\varphi = 30°$ (where r, θ, φ are spherical coordinates whose origin is at the center of the slab). For thickness of the order of 1μ (or less) the surface mode scattering is comparable with (or stronger than) the bulk mode scattering. In this calculation only the deformation potential type of electron-lattice interaction was considered. The addition of the polar scattering mechanism will cause some quantitative changes in the relative intensities but the general behavior should be similar to that shown in Fig. 7.

In scattering from long cylinders momentum in the direction parallel to the cylinder axis will be conserved so that Eq. (5) will still apply. The dispersion curves of a SiC cylinder of diameter $0.4\ \mu$ are shown in Fig. 8. In this case there exists for every value of h a whole series of surface modes (of which only the first three are shown) which converge to the frequency ω_s. The curves obtained from Eq. (5) for some angles are also shown.

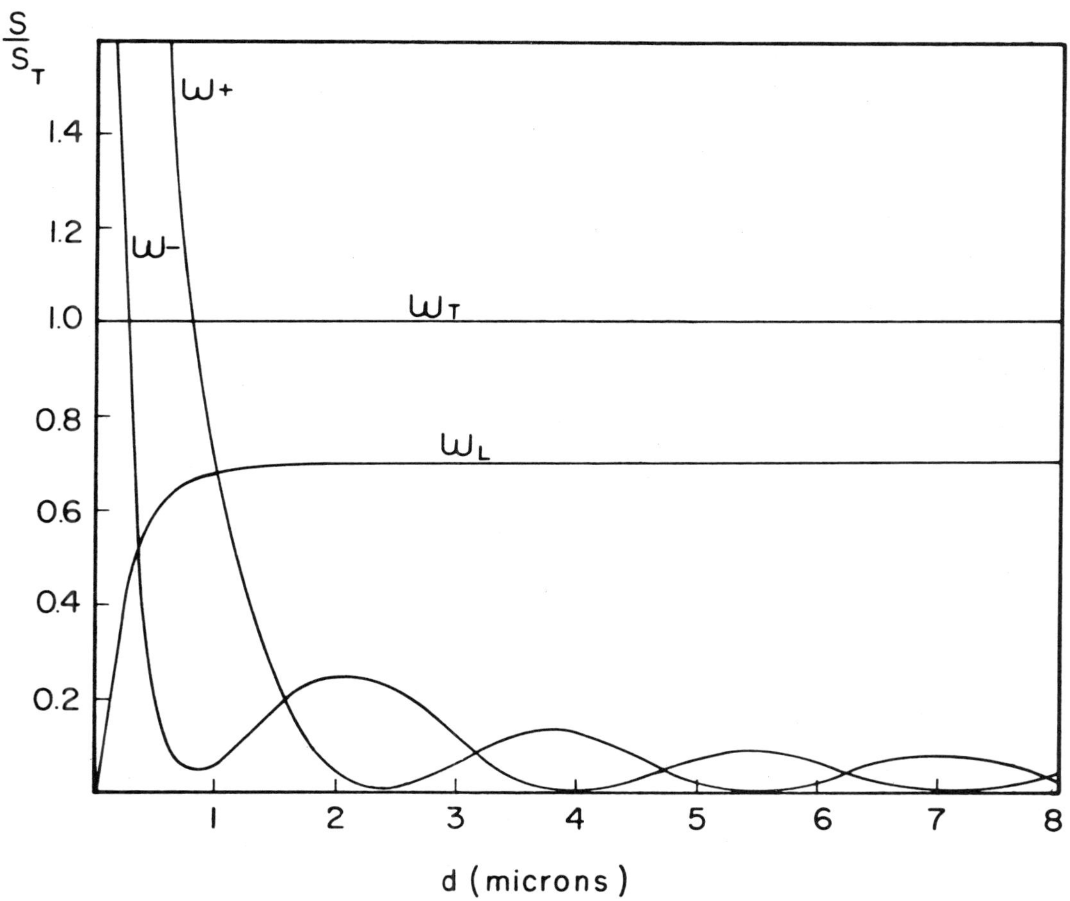

Fig. 7. Intensity of the four Raman lines relative to the ω_T line versus the SiC slab thickness (angle of observation $\theta = 30^o$, $\varphi = 30^o$).

Finally, we may mention that in the scattering from small spherical samples no component of the momentum is conserved so that the frequencies observed at different angles should be the same (but their relative intensities may vary).

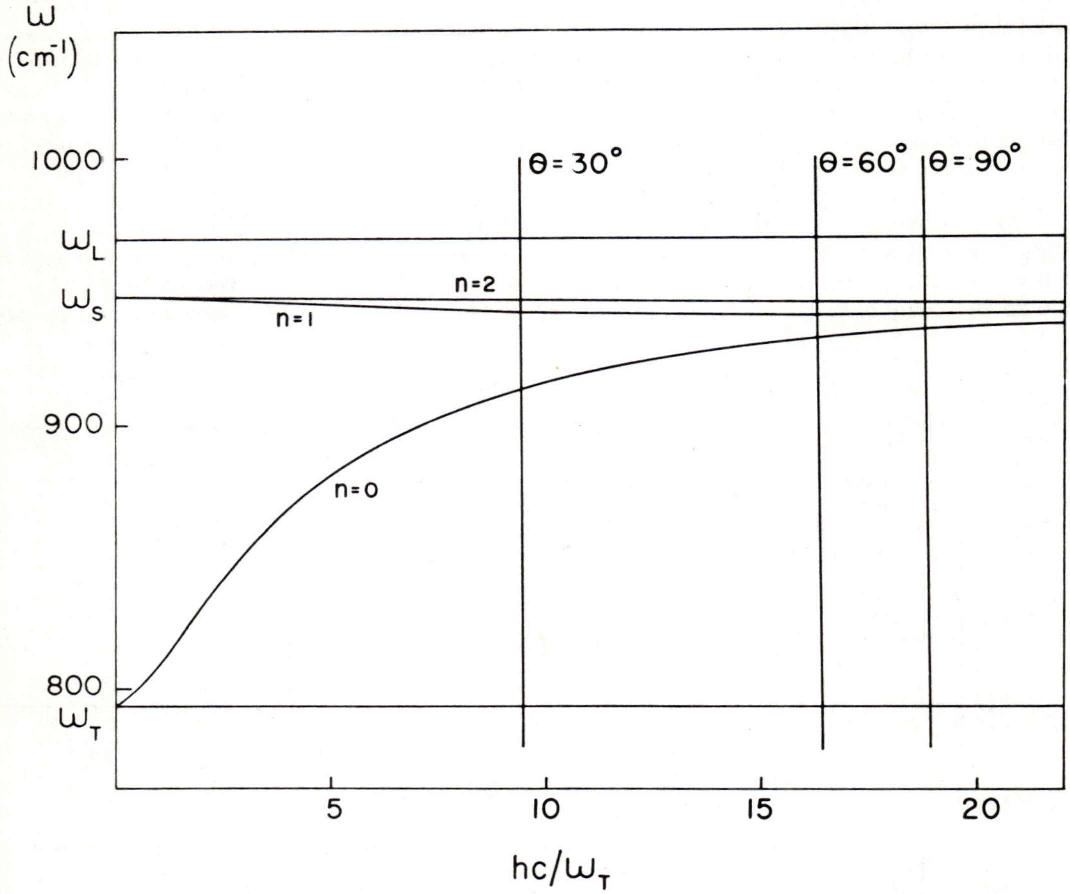

Fig. 8. The surface and bulk modes of a SiC cylinder of diameter 0.4 μ. The frequencies observable at some angles are also shown.

REFERENCES

1. R. Englman and R. Ruppin, J. Phys. C (Proc. Phys. Soc.) 1, 614 (1968).
2. R. Ruppin and R. Englman, J. Phys. C (Proc. Phys. Soc.) 1, 630 (1968).
3. W.W. Pultz and W. Hertl, Spectrochim. Acta 22, 573 (1966).
4. H. Boersch, J. Geiger and W. Stickel, Z. Physik 212, 130 (1968).
5. H.C. van de Hulst, "Light Scattering by Small Particles," Wiley, New York, 1957.
6. H. Bilz, L. Genzel and H. Happ, Z. Phys. 160, 535 (1960).
7. R. Loudon, Proc. Roy. Soc. A275, 218 (1963).
8. W.G. Spitzer, D.A. Kleinman, and C.J. Frosch, Phys. Rev. 113, 133 (1959).
9. R. Fuchs and K.L. Kliewer, Phys. Rev. 140, A2076 (1965).

B-10: LOW FREQUENCY RAMAN SPECTRA OF IONIC CRYSTALS

R. S. Krishnan, N. Krishnamurthy, T. M. Haridasan and J. Govindarajan
Department of Physics, Indian Institute of Science
Bangalore, India

INTRODUCTION

Several physical phenomena depend on the normal modes of vibration of the crystal. A ferroelectric phase transition in an ionic crystal is explained as due to a vanishing low frequency transverse optical mode at $q \to 0$ (soft mode)[1]. An antiferroelectric transition is due to a vanishing zone boundary phonon[2]. Phase transformations induced by pressure are similarly explained as due to a vanishing transverse acoustic phonon[3]. Order-disorder transformations in alloys are due to coupling of modes which are close in energy and at a phase difference $\pi/2$ [4]. These low frequency phonons can be easily observed in Raman spectra, though the far infra-red measurements are very difficult. The ferroelectric soft mode and its frequency variation have been observed by Perry et al[5] in Raman scattering.

The purpose of the present paper is to discuss the low frequency Raman spectra in a few ionic crystals in terms of the Born-von-Karman lattice dynamics and to discuss the results with the available experimental data.

CALCITE STRUCTURES

Sodium Nitrate and Potassium Nitrate: - Potassium nitrate exhibits polymorphic phase transitions. At room temperature, KNO_3 has aragonite structure[6]. On heating to about 130°C it goes over to calcite phase. The unit cell is an elongated rhombohedron and contains two molecules. The atomic positions are

$$K^+ \quad \ldots \quad \left(\frac{1}{4}, \frac{1}{4}, \frac{1}{4}\right); \quad \left(\frac{3}{4}, \frac{3}{4}, \frac{3}{4}\right)$$

$$NO_3^- \quad \ldots \quad (0, 0, 0); \quad \left(\frac{1}{2}, \frac{1}{2}, \frac{1}{2}\right)$$

and the space group is D_{3d}^6. The basis vectors of the lattice are

$$\vec{a_1} = s\vec{j} + r\vec{k}$$

$$\vec{a_2} = \sqrt{\frac{3}{2}}\, s\vec{i} - \frac{1}{2} s\vec{j} + r\vec{k}$$

$$\vec{a_3} = -\sqrt{\frac{3}{2}}\, s\vec{i} - \frac{1}{2} s\vec{j} + r\vec{k}$$

where $s = a(2(1 - \cos\alpha)/3)^{1/2}$

$r = a((1 + 2\cos\alpha)/3)^{1/2}$

and a, α are the trigonal cell dimensions.

The arrangement of the nearest neighbours around each atom is as follows: Each K^+ has six K^+ ions of the other type at a distance $(s^2 + r^2/4)^{1/2}$. Similarly, each NO_3^- ion has six NO_3^- ions of the other type at the same distance $(s^2 + r^2/4)^{1/2}$. Each K^+ ion is surrounded by three NO_3^- ions of one type and three NO_3^- ions of the other type at a distance of $(s^2 + r^2/16)^{1/2}$.

The dynamical matrix $D_{\alpha\beta}(KK')$ for this structure is of order 12 x 12 and the non-vanishing elements of the dynamical matrix, for a short-range potential[7]

where $\lambda = \beta_{AB}\, B_0 e^2 (r_A + r_B)^8$

β_{AB} is 1 for $(K\text{-}NO_3^-)$ interaction and 0.75 for $(NO_3^-\text{-}NO_3^-)$ interaction.

$B_0 = 0.02909$

r_A = Radius of K^+ ion

r_B = Radius of NO_3^- ion

are

$$\begin{bmatrix}12\\xx\end{bmatrix} = \begin{bmatrix}12\\yy\end{bmatrix} = \begin{bmatrix}34\\xx\end{bmatrix} = \begin{bmatrix}34\\yy\end{bmatrix} = -\frac{9\lambda\begin{pmatrix}K^+ - K^+\\ NO_3^- - NO_3^-\end{pmatrix}}{r_1^{11}}\left[(33r^2/r_1^2 - 6)\right]$$

$$\begin{bmatrix}12\\zz\end{bmatrix} = \begin{bmatrix}34\\zz\end{bmatrix} = -\frac{9\lambda\begin{pmatrix}K^+ - K^+\\ NO_3^- - NO_3^-\end{pmatrix}}{r_1^{11}}\left[(33r^2/r_1^2 - 6)\right]$$

$$\begin{bmatrix}13\\xx\end{bmatrix} = \begin{bmatrix}13\\yy\end{bmatrix} = \begin{bmatrix}14\\xx\end{bmatrix} = \begin{bmatrix}14\\yy\end{bmatrix} = \begin{bmatrix}23\\xx\end{bmatrix} = \begin{bmatrix}23\\yy\end{bmatrix} = \begin{bmatrix}24\\xx\end{bmatrix} = \begin{bmatrix}24\\yy\end{bmatrix}$$

$$= -\frac{9\lambda(K^+ - NO_3^-)}{r_2^{11}}\left[\frac{33s^2}{2r_2^2} - 3\right]$$

$$\begin{bmatrix}13\\zz\end{bmatrix} = \begin{bmatrix}14\\zz\end{bmatrix} = \begin{bmatrix}23\\zz\end{bmatrix} = \begin{bmatrix}24\\zz\end{bmatrix} = -9\left[\lambda(K^+ - NO_3^-)/r_2^{11}\right]\left[(33r^2/16r_2^2 - 3)\right]$$

where $r_1^2 = s^2 + r^2/4$ and $r_2^2 = s^2 + r^2/16$.

$\begin{bmatrix}KK\\ \alpha\beta\end{bmatrix}$ are obtained from translational invariance. The long range coupling coefficients for $q \to 0$ along [001] direction are calculated employing the Ewald method. The parameters λ's are calculated from the Goldschmidt radii. The frequencies of the normal modes at $q \to 0$ are obtained from their expressions as given by group-theoretical methods[8].

$$m_1 \omega^2_{A_{2g}} = -2(a + e)$$

$$m_1 \omega^2_{Eg} = -2(b + f)$$

$$\omega^2_{A_{1u}} = \frac{-2(m_1+m_2)}{m_1 m_2} e$$

$$\omega^2_{E_{1u}} = \frac{-2(m_1+m_2)}{m_1 m_2} f$$

$$m_2 \omega^2_{A_{2u}} = -2 (c + e)$$

$$m_2 \omega^2_{E_{2u}} = -2 (d + f)$$

where $a = \begin{bmatrix} 12 \\ zz \end{bmatrix}$, $c = \begin{bmatrix} 34 \\ zz \end{bmatrix}$, $e = \begin{bmatrix} 13 \\ zz \end{bmatrix}$, $b = \begin{bmatrix} 12 \\ xx \end{bmatrix}$, $d \begin{bmatrix} 34 \\ xx \end{bmatrix}$, $f = \begin{bmatrix} 13 \\ xx \end{bmatrix}$

and m_1, m_2 are the masses of K^+ and NO_3^- ions. The calculated frequencies for KNO_3 and $NaNO_3$ which is isomorphous with it are given in Table I.

TABLE I

Assignment	KNO_3	$NaNO_3$	
		Calculated	Experimental
$A_{2u}(LO_1)$	168	217	217
$A_{2g}(LO_2)$	119	146	140
$A_{1u}(LO_3)$	115	112	120
$E_u\ (TO_1)$	94	107	133
$E_g\ (TO_2)$	87	83	98
$E_u\ (TO_3)$	82	71	71

The agreement with experimental Raman[9] and Infrared frequencies[10] is fairly satisfactory for $NaNO_3$. No measurements are available for this phase of KNO_3.

On cooling from 130°C, the calcite phase KNO_3 transforms into a new phase at about 124°C. The structure of the γ-phase is also rhombohedral, but polar and contains only one molecule in the unit cell. The NO_3^- ion is slightly away from the line joining the two K^+ ions along the z-axis. Assuming short range forces between the origin ion and its six first and two second neighbours and employing the Group-theoretical expressions for $q \to 0$ longitudinal and transverse modes

$$\omega_\ell^2 = \frac{-(m_1 + m_2)}{m_1 m_2} \begin{bmatrix} 12 \\ zz \end{bmatrix}$$

$$\omega_t^2 = \frac{-(m_1 + m_2)}{m_1 m_2} \begin{bmatrix} 12 \\ xx \end{bmatrix}$$

Their frequencies turn out to be

$$\omega_\ell = 165\ \text{cm}^{-1} \quad \omega_t = 5\ \text{cm}^{-1}$$

This predicts that the ferroelectric phase transition in KNO_3 is due to a low frequency transverse optic mode and confirms the earlier predictions of Balkanski and Nusimovici[11] who have shown that KNO_3 phase III satisfies the condition

$$\Delta\omega \rightarrow 2\ \omega_o$$

where $\Delta\omega$ is the change in the frequency due to long range forces and ω_o is the frequency due to short range forces.

Ammonium Dihydrogen Phosphate: Ammonium dihydrogen phosphate is isomorphous with potassium dihydrogen phosphate and crystallises in the tetragonal system with the point group $\bar{4}2$ m. But the cell dimensions are markedly different from that of KH_2PO_4. For KH_2PO_4, a = 7.453 Å and c = 6.959 Å whereas for ADP, a = 7.479 Å and c = 7.516 Å. The $(NH_4)^+$ has 4 near neighbours of Oxygen atoms at distance of 2.97 Å and 4 oxygen slightly further away equal to 3.17 Å while K has 8 nearly equidistant oxygens (≈ 2.82 Å). The important property of ADP is that it exhibits antiferroelectricity at low temperature phase.

Taking into account short range interactions between the nearest-neighbouring oxygens of the phosphate groups and the $(NH_4)^+$ ions, the expressions for the short range coupling coefficients with the Pauling's potential become

$$\begin{bmatrix} 13 \\ xx \end{bmatrix} = \begin{bmatrix} 13 \\ yy \end{bmatrix} = \frac{-9\lambda}{r_{13}^{11}} \left[\frac{22\left[(\frac{1}{2} - x)^2\ a^2 + (\frac{1}{2} - y)^2\ a^2\right]}{r_{13}^2} - 4 \right]$$

$$\begin{bmatrix} 13 \\ zz \end{bmatrix} = \frac{-9\lambda}{r_{13}^{11}} \left[\frac{44\ z^2 c^2}{r_{13}^2} - 4 \right]$$

$$r_{13}^2 = \left[(\tfrac{1}{2} - x)^2 + (\tfrac{1}{2} - y)^2\right] a^2 + z^2 c^2$$

$$\begin{bmatrix}14\\xx\end{bmatrix} = \begin{bmatrix}14\\yy\end{bmatrix} = \frac{-9\lambda}{r_{14}^{11}} \left[\frac{22\left[\left(\frac{1}{2}-x\right)^2 + y^2\right] a^2}{r_{14}^2} - 4 \right]$$

$$\begin{bmatrix}14\\zz\end{bmatrix} = \frac{-9\lambda}{r_{14}^{11}} \left[\frac{44\left(\frac{1}{4}-z\right)^2 c^2}{r_{14}^2} - 4 \right]$$

$$r_{14}^2 = \left[\left(\frac{1}{2}-x\right)^2 + y^2\right] a^2 + \left(\frac{1}{4}-z\right)^2 c^2$$

where x, y, z are the oxygen parameters and 1, 2, 3, 4 designate respectively the positions of the two $(NH_4)^+$ ions and two $(H_2PO_4)^-$ ions. With the long range Coulomb coefficients, $\lambda_{(NH_4)-0}$ and the group theoretical expressions for the frequencies of the normal modes of the KDP structure as given by Schur[13] the corresponding frequencies of lattice modes for ADP were calculated and are compared with the experimental data by Chappelle[14] in Table II.

TABLE II

Symmetry species	Frequencies in cm^{-1}	
	Experimental	Calculated
B_1^1	69	64
E'	133	156
E''	72	73
E	172	180
B_1''	124	127
B_2'	---	218

In view of the interesting properties of the Scheelite structures whose vibration spectra have been thoroughly investigated by Porto and Scott[15] the computations are being extended to these structures also. An attempt is also being made to calculate the zone boundary phonons in ADP to detect any unstable normal mode.

In the case of sodium nitrite[16] however, calculations with the Paulings potential does not give any low frequency transverse modes confirming the origin of ferroelectricity as due to order-disorder transformation.

Detailed investigations on the infrared and laser excited Raman spectra of lanthanum trifloride and its isomorphs have been carried out by Porto and his group[17]. They have shown that the observed Raman spectra are consistent with a hexamolecular unit cell (D_{3d}^4) of the structure which is obtained from a slight distortion of the bimolecular unit cell of high symmetry (D_{6h}^4). In order to facilitate detailed calculations on the lattice vibration spectra of these structures the group theoretical expressions for $q \to 0$ normal modes for a general force constant scheme and the compatibility relations between the force constant tensors have been carried out and detailed numerical calculations are under progress.

The group theoretical expressions for the $q \to 0$ normal modes in Brucite type ionic crystals with the space group D_{3d}^3 also have been worked out. A short range model fails to explain the infrared and Raman frequencies and the work on the fundamental and multiphonon bands in Brucite with a long range model is under progress.

To conclude, it is hoped that the application of group theoretical methods and simple force constant models in ionic ferroelectrics and antiferroelectrics and a study of the low frequency Raman spectra with lasers will throw light on the mechanism of phase transitions and the interatomic forces in these crystals.

REFERENCES

1. W. Cochran, Adv. in Phys. 9, 387 (1960).
2. W. Cochran, Adv. in Phys. 10, 401 (1961).
3. J.R. Hardy and A.M. Karo, "Lattice Dynamics," (R.F. Wallis) p. 195, 1965.
4. G. Gilat, Phys. Lett. 24, 593 (1967).
5. C.H. Perry and D.W. Hall, Phys. Review Letters 15, 700 (1965).
6. F. Jona and G. Shirane, "Ferroelectric Crystals," p. 360, Pergamon Press, 1962.
7. L. Pauling, "Nature of the Chemical Bond," p. 523, Cornel University Press, 1960.
8. K. Huang, Z. Phy. 171, 213 (1962).
9. T.M.K. Nedungadi, Proc. Ind. Acad. Sci A8, 398 (1938)
10. R.K. Khanna, Ph.D. Thesis to I.I. Sc., p. 65, 1961.
11. M. Balkanski and M.A. Nuzimovic, Reports of GMR, p. 51, 1966.
12. H.D. Megaw, "Ferroelectricity in Crystals," p. 45, Methuen & Co., 1957.
13. M.S. Shur, Sov. Phys. Solid State 8, 43 (1966).
14. J. Chappelle, Jour. De. Chemi. Phys. 46, 30 (1949).
15. S.P.S. Porto and J.F. Scott, Phys. Rev. 157, 716 (1967).
16. M.L. Canut and J. Mendiola, Phys. Stat. Soli 5, 313 (1964).
17. R.P. Bauman and S.P.S. Porto, Phys. Rev. 161, 842 (1967).

B-11: THEORY FOR THE RAMAN SCATTERING TENSOR FOR COMBINATION AND OVERTONES

J. A. Koningstein
Department of Chemistry, Carleton University
E. W. R. Steacie Building
Ottawa, Canada

ABSTRACT

Theoretical expressions have been derived for the scattering tensor of Raman transitions which involve the excitation of one particular vibration by 2 quanta or the simultaneous excitation of two vibrations by one quantum each.

In order to incorporate the latter process it is necessary to expand [1] the electronic wave functions to second order in the nuclear coordinates; for the former process only the first order terms need to be considered. The electric dipole matrix element which appears in the expression for the Raman scattering tensor can be rewritten by introducing the expanded wave functions and selection rules. The symmetry of the scattering tensor and the theoretical relations of the intensity ratio of phonon bands of a solid or vibrational band of liquids and gases follow directly from this theoretical approach. It is found that asymmetry occurs in the scattering tensor for combination bands and the importance of this new phenomena for Raman bands, in particular those in solids, and the effect on the polarization properties shall be pointed out. It is now necessary to introduce in the character tables of the more familiar point groups the transformation properties of the antisymmetric part of the scattering tensor. This can be done in a relatively simple way [2].

The theory suggests also that the degree of antisymmetry increases if the Raman process involves the excitation of increasing numbers of vibrations.

REFERENCES

1. For the normal 1 quantum jump Raman effect see A. C. Albrecht, J. Chem. Phys., 34, 1476 (1961).
2. J.A. Koningstein and O. Sonnich Mortensen, Nature, 217, 445 (1968).

B-12: THEORY OF OPTICAL PROCESSES IN LIQUID HELIUM

J. Woods Halley
Department of Physics, University of California
Berkeley, California*

INTRODUCTION AND MODEL

In several many-particle systems, Raman scattering has been a useful tool for studying elementary excitations, complementing inelastic neutron scattering. In liquid helium, Brillouin scattering[1] has permitted the study of sound waves. It is of interest to ask if the rest (roton part) of the vibrational excitation spectrum observed by neutrons could also be observed in light scattering experiments. To investigate this possibility, we formulate a model for the study of optical processes in liquid helium and use it to study the two roton Raman scattering cross-section. The reasons for studying two roton scattering are first that one can produce excitations with larger wave vectors than in first order scattering and second that the matrix elements for scattering will be wave-vector dependent and will be larger for large wave-vector, so that one may hope to observe the two excitation scattering.

The model which we consider is described by the following Hamiltonian:

$$\mathcal{H} = \mathcal{H}_{\text{atoms}} + \mathcal{H}_{\text{atom-atom}} + \mathcal{H}_{\text{atom-field}} \tag{1}$$

where

$$\mathcal{H}_{\text{atoms}} = \sum_{\ell=1}^{N} \frac{-\hbar^2}{2M} \nabla_{\ell}^2 + \sum_{\ell=1}^{N} \left\{ \sum_{i=1}^{2} \frac{-\hbar^2}{2m_e} \nabla_{\ell i}^2 - \frac{2e^2}{|\underline{r}_{\ell i}|} + \frac{e^2}{|\underline{r}_{\ell 1} - \underline{r}_{\ell 2}|} \right\} \tag{2}$$

*Present address: School of Physics and Astronomy, University of Minnesota, Minneapolis, Minn.

$$\mathcal{H}_{\text{atom-atom}} = \sum_{\ell<m} \frac{4e^2}{|\underline{R}_{\ell m}|} + \sum_{i,j=1}^{2} \frac{e^2}{|\underline{R}_{\ell m} + \underline{r}_{\ell i} - \underline{r}_{mj}|}$$

$$-2e^2 \sum_{i=1}^{2} \left\{ \frac{1}{|\underline{R}_{\ell m} - \underline{r}_{\ell i}|} + \frac{1}{|\underline{R}_{\ell m} + \underline{r}_{mi}|} \right\} \tag{3}$$

$$\mathcal{H}_{\text{atom-field}} = \frac{-e}{mc} \sum_{\ell=1}^{N} \sum_{i=1}^{2} \underline{p}_{\ell i} \cdot \underline{A}\,(\underline{r}_{\ell i}, t)$$

Here M is the mass of the helium nucleus, m is the electron mass, $\underline{R}_{\ell m} = \underline{R}_{\ell} - \underline{R}_m$ where $\underline{R}_\ell$ is the position of the 1th nucleus, $\underline{r}_{\ell i} = \underline{x}_{\ell i} - \underline{R}_\ell$ and $\underline{x}_{\ell i}$ is the coordinate of the ith electron on the 1th nucleus (we neglect exchange effects). To express this Hamiltonian in terms of phonon and exciton variables we write an effective atom-atom interaction

$$\mathcal{H}_{\text{eff}}(\{R_\ell\}) = \langle \text{gnd} \mid \mathcal{H}_{\text{atom-atom}} \mid \text{gnd} \rangle \tag{4}$$

where $|\text{gnd}\rangle$ is the electronic ground state. We consider the part

$$\mathcal{H}_{\text{phonons}} = \sum_{\ell=1}^{N} \frac{-\hbar^2}{2M} \nabla_\ell^2 + \mathcal{H}^{\text{eff}}(\{R_\ell\}) \tag{5}$$

of the Hamiltonian. We assume this can be transformed to the form

$$\mathcal{H}_{\text{phonons}} = \sum_{\underline{k}} \hbar\omega_{\underline{k}}\, a_{\underline{k}}^{\dagger} a_{\underline{k}} \tag{6}$$

where $a_{\underline{k}}$ are phonon operators obeying boson commutation relations and related to the displacements $\delta\underline{R}$ by

$$\delta\underline{R}_\ell = \sum_{\underline{k}} \hat{k} \left(\frac{\hbar}{2\rho\,\omega_{\underline{k}} V} \right)^{1/2} \left(a_{\underline{k}}\, e^{-i\underline{k}\cdot\underline{R}_\ell^{(o)}} + \text{h.c.} \right) \tag{7}$$

where $\underline{R}_\ell^{(o)}$ is the position of the 1th nucleus in the state with no phonons excited and $\hbar\omega_{\underline{k}}$ is the observed phonon-roton energy at wave vector $\underline{k}$. We suppose that the $\underline{R}_\ell^{(o)}$ lie on a crystal lattice and later average over lattices. The Hamiltonian is then rewritten

$$\mathcal{H} = \mathcal{H}_o + \mathcal{H}_I$$

$$\mathcal{H}_o = \mathcal{H}_{phonons} + \mathcal{H}_{excitons}$$

$$\mathcal{H}_{excitons} = \sum_i \sum_\nu E_\nu \, b_\nu^{(\ell)\dagger} \, b_\nu^{(\ell)} \tag{8}$$

$$\mathcal{H}_I = \mathcal{H}_{atom-atom} - \mathcal{H}^{(eff)} \left(\{ \underline{R}_\ell \} \right)$$

$$+ \sum_{\ell=1}^{N} \sum_{i=1}^{2} \left\{ \frac{-2e^2}{|\underline{x}_{\ell i} - \underline{R}_\ell|} + \frac{2e^2}{|\underline{x}_{\ell i} - R_\ell^{(o)}|} \right\}$$

where $\mathcal{H}_{excitons}$ accounts for the electronic states of the helium atoms. The operator $b_\ell^{(\nu)\dagger}$ is defined by

$$b_\ell^{(\nu)\dagger} \; | \, gnd > \; = \; A \, \{ \psi_\nu^{(\ell)} \, (\underset{\sim}{x}_{\ell 1}, \sigma_{\ell 1} \, ; \underset{\sim}{x}_{\ell 2}, \sigma_{\ell 2}) \; x \tag{9}$$

$$\pi \; \psi_g^{(\ell')} \; (\underset{\sim}{x}_{\ell' 1}, \sigma_{\ell' 1} \, ; \underset{\sim}{x}_{\ell' 2}, \sigma_{\ell' 2}) \, \}$$

where $\psi_\nu^{(\ell)^{\ell' \neq \ell}}$ is the two electron wave function for the 1th helium atom in the $\nu^{\underline{th}}$ excited state and $\mathcal{A}$ antisymmetrises. Fourier transforming gives

$$\beta_{\underline{k}}^{(\nu)} = \frac{1}{\sqrt{N}} \sum_{\ell=1}^{N} e^{i\underline{k} \cdot \underline{R}_\ell^{(o)}} \, b_\ell^{(\nu)} \tag{10}$$

and

$$\mathcal{H}_{excitons} = \sum_{\underline{k}} \sum_\nu E_\nu \, \beta_{\underline{k}}^{(\nu)\dagger} \, \beta_{\underline{k}}^{(\nu)} \tag{11}$$

where E_ν is the energy of the $\nu^{\underline{th}}$ excited state. The remaining interactions can be written in terms of exciton and phonon variables. Expanding $\mathcal{H}_{\text{atom-atom}}$ in a series in $|\underline{r}_{\ell i}| / |\underline{R}_{\ell m}|$ in the usual way[2] gives a multipole moment expansion:

$$\mathcal{H}_{\text{atom-atom}} = \sum_{\ell<m} \sum_{i,j=1}^{2} \Bigg\{ \frac{-e^2}{R^3_{\ell m}} (2 Z_{\ell i} Z_{mj} - X_{\ell i} X_{mj} - Y_{\ell i} Y_{mj}) + \frac{3}{2} \frac{e^2}{R^4_{\ell m}} \Big[r^2_{\ell i} Z_{mj} - Z_{\ell i} r^2_{mj} + (2 x_{\ell i} x_{mj} + 2 y_{\ell i} y_{mj} - 3 Z_{\ell i} Z_{mj}) (Z_{\ell i} - Z_{mj}) \Big] \tag{12}$$
$$+ \frac{3}{4} \frac{e^2}{R^5_{\ell m}} \Big[r^2_{\ell i} r^2_{mj} - 5 r^2_{\ell i} Z^2_{mj} - 5 r^2_{mj} Z^2_{\ell i} - 5 z^2_{\ell i} z^2_{mj} + 2 (4 Z_{\ell i} Z_{mj} - x_{\ell i} x_{mj} - y_{\ell i} y_{mj})^2 \Big] + \dots \Bigg\}$$

Using this one can express $\mathcal{H}_I$ in terms of local exciton operators $b^{(\nu)}_{\ell}$ and local displacements $\delta\underline{R}_{\ell}$. Two sets of terms involve only exciton operators. When they are added to $\mathcal{H}_{\text{excitons}}$ the resultant Hamiltonian can be diagonalized by a canonical transformation and a Bogoliubov transformation. The modified exciton Hamiltonian will again be of the form (11) but E_ν will depend on $\underline{k}$. The width of the resultant exciton band will be small, however. There is a term in the exciton-phonon interaction which arises from the dipole-dipole interaction and which leads to Brillouin scattering. There are also terms bilinear in exciton and phonon operators which give two-roton Raman scattering. A term bilinear in phonon operators and linear in exciton operators and arising from the quadrupole-dipole interaction gives two-roton infrared absorption. As an illustration, we consider the term leading to two roton scattering. Transformed to exciton and phonon variables it is

$$\mathcal{H}_I^{(2\text{-}2)} = \sum_{\underline{k},\underline{k}'} \sum_{\underline{q}} \sum_{\nu,\nu'} \Big\{ \Big[\beta^{(\nu)\dagger}_{\underline{k}} \beta^{(\nu')}_{\underline{k}'} \mathfrak{M}^{(1)}(\nu, g; \nu', g) \times \tag{13}$$

$$\times \Big[\big(T(\underline{k}', \underline{q}, \underline{k}+\underline{k}'-\underline{q})\, a_{\underline{q}}\, a_{\underline{k}+\underline{k}'-\underline{q}} + T(\underline{k}', \underline{q}, -\underline{k}-\underline{k}'+\underline{q})\, a_{\underline{q}}\, a^{\dagger}_{\underline{k}+\underline{k}-\underline{q}} \big) + \text{h. c.} \Big]$$

$$+ \text{h. c.} \Big] + \Big[\beta^{(\nu)\dagger}_{\underline{k}}\, \beta^{(\nu')}_{\underline{k}'}\, \mathcal{M}^{(1)}(\nu, g; g, \nu')$$

$$\times \Big[\big(T(-\underline{k}', \underline{q}, \underline{k}-\underline{k}'-\underline{q})\, a_{\underline{q}}\, a_{\underline{k}-\underline{k}'-\underline{q}} + T(-\underline{k}', \underline{q}, -\underline{k}+\underline{k}'+\underline{q})\, a_{\underline{q}}\, a^{\dagger}_{-\underline{k}+\underline{k}'+\underline{q}} \big) + \text{h. c.} \Big]$$

$$+ \text{h. c.} \Big] \Big\} .$$

Here

$$T(\underline{k}', \underline{q}, \underline{q}') =$$

$$\frac{-e^2 \hbar}{4\zeta V} \sum_{i,j} \sum_{\ell-m} \Big(\Big[M^{(2;3)}_{ij}(\, , m) / (R^{(o)}_{\ell m})^5 \Big] \times \tag{14}$$

$$D(\ell, m; -\underline{k}', \underline{q}, \underline{q}') \Big[(\hat{q})_i\, (\hat{q}')_j / (\omega_{\underline{q}}\, \omega_{\underline{q}'})^{1/2} \Big] \Big)$$

in which

$$D(\ell, m; -k', \underline{q}, \underline{q}') = e^{-i\underline{k}' \cdot \underline{R}^{(o)}_m} \,(1 - e^{i\underline{q} \cdot \underline{R}^{(o)}_{m\ell}})\,(1 - e^{i\underline{q}' \cdot R^{(o)}_{m\ell}}$$

$$M^{(2;3)}(\ell, m) = \frac{3}{2} \left(\frac{5\, R^{(o)\, i}_{\ell m}\; R^{(o)\, j}_{\ell m}}{\left(R^{(o)}_{\ell m}\right)^2} - \delta_{ij} \right) \tag{15}$$

and $\mathcal{M}^{(1)}(\nu, \nu'; \nu'', \nu''')$ is a combination of electric dipole matrix elements whose explicit form is not needed. To calculate the Raman amplitude we also need to express $\mathcal{H}_{\text{atom-field}}$ in terms of photon operators $\alpha_{\underline{k}\lambda}$ and exciton operators $\beta^{(\nu)}_{\underline{k}}$. It has the form

$$\mathcal{H}_{\text{atom-field}} = \sum_{k, \nu, \lambda} \{ m(\underline{k}, \lambda, \nu)\, (\beta^{(\nu)\dagger}_{\underline{k}}\, \alpha_{\underline{k}\lambda} + \beta^{(\nu)\dagger}_{-\underline{k}}\, \alpha^{\dagger}_{\underline{k}\lambda}) + \text{h. c.} \} \tag{16}$$

in which

$$m(k, \lambda, \nu) = \frac{ie}{\sqrt{N/V}} \left(\frac{2\pi\hbar}{ck}\right)^{1/2} (\underline{X}(\nu, g) \cdot \hat{e}_{\underline{k}\lambda} \tag{17}$$

$$\underline{X}(\nu, g) = \int\int \psi_{\nu}^{(\ell)*} \sum_{i=1}^{2} \underline{x}_{\ell i} \, \psi_{g}^{(\ell)} \, d\tau_1 \, d\tau_2$$

CALCULATION OF THE SCATTERING RATE

We consider the scattering rate for the process

$$\begin{pmatrix} \text{1 photon} \\ \text{at } \underline{q} \end{pmatrix} \rightarrow \begin{pmatrix} \text{2 rotons} \\ \text{at} \\ \underline{k}, \underline{k}' \end{pmatrix} + \begin{pmatrix} \text{1 photon} \\ \text{at } \underline{q}' \end{pmatrix}$$

via the following mechanism: The incoming photon couples to an exciton. The exciton couples to two rotons and another exciton. The second exciton couples to the outgoing photon. One has

$$\sum_{\underline{k}, \underline{k}'} \frac{2\pi}{\hbar} \left| M(\underline{q};\underline{k}, \underline{k}', \underline{q}') \right|^2 \delta(\hbar c(|q| - |q'|) - \hbar\omega_{\underline{k}} - \hbar\omega_{\underline{k}'}) \tag{18}$$

where

$$M(\underline{q};\underline{k}, \underline{k}', \underline{q}) =$$

$$= \sum_{m', m''} \frac{\langle \underline{q} \,|\, \mathcal{H}' \,|\, m' \rangle \langle m' \,|\, \mathcal{H}' \,|\, m'' \rangle \langle m'' \,|\, \mathcal{H}' \,|\, \underline{k}, \underline{k}', \underline{q} \rangle}{(hc\,|\underline{q}| - E_{m'})\,(hc\,|\underline{q}| - E_{m''})} \tag{19}$$

and

$$\mathcal{H}' = \mathcal{H}_I^{(2-2)} + \mathcal{H}_{\text{atom-field}} \tag{20}$$

Writing out $M(\underline{q};\underline{k}, \underline{k}', \underline{q}')$ using the model then gives a complicated expression whose k-dependence is quite simple and can be written

$$M(\underline{q};\underline{k}, \underline{k}', \underline{q}) = \text{constant} \times T(\,|\underline{k}|\,) \tag{21}$$

where

$$T(k) = \frac{-e^2\hbar}{4\zeta V\omega_k} \sum_{\ell-m} \frac{3}{2}\left(\frac{5(\underset{\sim}{R}^{(o)}_{\ell m},\hat{k})^2}{\left(R^{(o)}_{\ell m}\right)^2} - 1\right) \frac{4}{R^{(o)\,5}_{\ell m}} \sin^2\left(\frac{\underline{k}\cdot \underset{\sim}{R}^{(o)}_{\ell m}}{2}\right) \tag{22}$$

Writing $\theta(k;\ell, m)$ for the angle between $\underline{k}$ and $\underline{R}^{(o)}_{\ell m}$ one has

$$T(k) =$$

$$\frac{-3e^2\hbar}{2\zeta V\omega_k} \sum_{\ell-m}\left(\frac{5\cos^2\theta(\hat{k};\ell, m) -1}{\left(R^{(o)}_{\ell m}\right)^5}\right)\sin^2\left(\frac{k\,R^{(o)}_{\ell m}\,\cos\theta(\hat{k};\ell, m)}{2}\right) \tag{23}$$

To evaluate this we average over lattices in a way consistent with the pair correlation function so that we can write

$$\sum_{\ell, m} (\ldots) \rightarrow 2\pi \int_{-1}^{+1} d\mu \int_{0}^{\infty} dR_{\ell m}\, g(R_{\ell m})\, (\ldots) \tag{24}$$

where $g(R_{\ell m})$ is the pair correlation function and $\mu = \cos\theta(\hat{k};\ell, m)$. One has after doing the angular integrals that

$$T(k) = -\frac{6\pi e^2\hbar}{\zeta V\omega_k}\int_{0}^{\infty}\frac{g(R_{\ell m})\,dR_{\ell m}}{R^5_{\ell m}} \tag{25}$$

$$\times\left(\frac{1}{3} - \frac{2\sin kR_{\ell m}}{kR_{\ell m}} - \frac{5\cos kR_{m}}{(kR_{\ell m})^2} + \frac{5\sin k\,R_{\ell m}}{(kR_{\ell m})^3}\right)$$

The transition rate is

$$W(\hbar\Delta\omega = \hbar\omega_q - \hbar\omega_{q'}) =$$

$$= \frac{2\pi}{\hbar}\sum_{\underline{k}} \left| M(\underline{k}, \underline{q}, \underline{q}') \right|^2 \delta(2\hbar\omega^P_{\underline{k}} - \hbar\omega_{\underline{q}} + \hbar\omega_{\underline{q}'}) \tag{26}$$

$$= \left(\frac{V}{\pi \hbar^2}\right) \sum_{i=1}^{3} \left\{ \frac{k_i^2}{\left| \frac{d\omega^{ph}_{k_i}}{dk_i} \right|} \; | M (\underline{k}_i, \underline{q}, \underline{q}') |^2 \right\}_{2\omega^p_{k_i} = \Delta\omega}$$

where k_1, k_2, k_3 are the three solutions to $2\hbar\omega^p_{\underline{k}} = \hbar\Delta\omega$. To evaluate this, one needs T(k) which can be found numerically from the experimental pair correlation function[3] and the spectrum[4] $\hbar\omega^p_{\underline{k}}$. The result is shown in Fig. 1 where the experimental data for ω_k and g(R) have been used. Using T(k) we compute $W(\Delta\omega)$ as shown in Fig. 2. $W(\Delta\omega)$ is related to the Raman scattering cross-section by the relation

$$\frac{d^2\sigma}{d(\hbar\omega_{q'})\, d\Omega_{q'}} = W(\Delta\omega) \frac{v^2 q'^2}{\hbar c^2 (2\pi)^3} \tag{27}$$

We estimate that the magnitude of the scattering rate is 10^{-18} photons/incident photon.

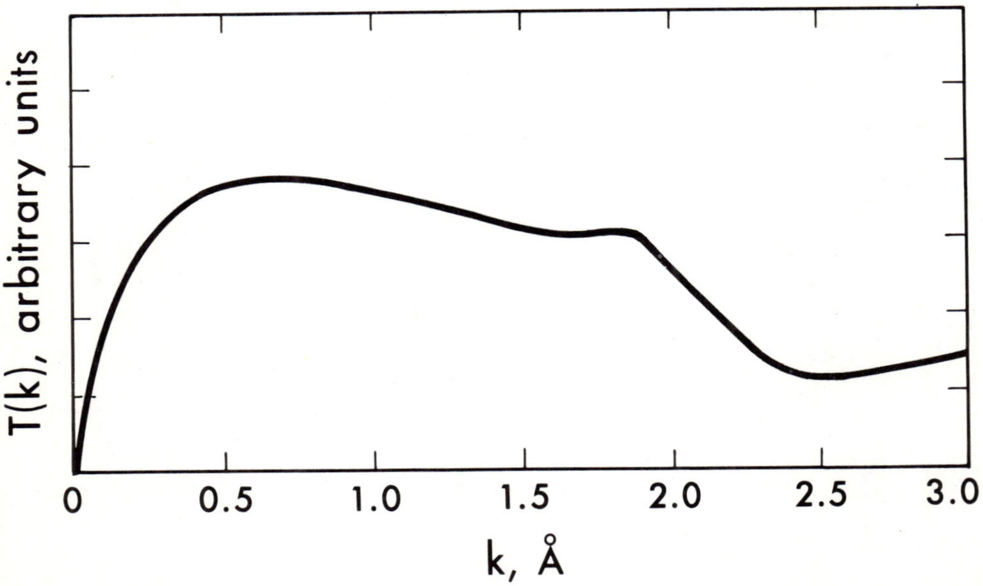

Fig. 1. The function T(k).

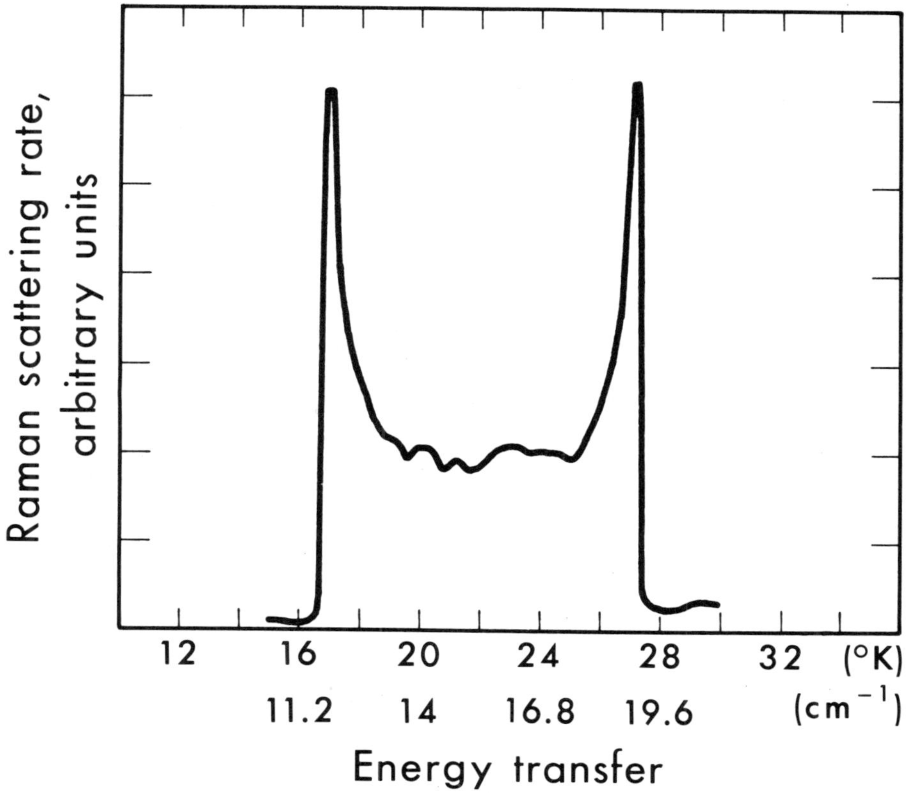

Fig. 2. Scattering rate W(Δω).

DISCUSSION

The model used only represents real liquid helium rather crudely. Because we treat the phonon-roton spectrum phenomenologically, however, this crudeness is not expected to affect the results very much. In particular, the dependence of the dipole-dipole interaction on the wave-vectors of the interacting excitations is not expected to depend in any important way on the approximations used in evaluating averages. It is this wave-vector dependence which gives the form of the predicted cross-section.

The excitons are only Fourier transformed for mathematical convenience and the question of whether the excitons in liquid helium are localized or band-like will not affect the conclusions reached here.

The present results do not seem to contradict any sum rules[5] on $S(\underline{q}, \omega)$ for the following reason. The effective coupling which we derive depends on the distance between the two displaced atoms involved. If we refer to the standard derivation[6] of the conclusion that the scattering cross-section for neutron or light scattering is proportional to $S(q, \omega)$ we see that it depends on the assumption of a coupling between exciting field and many-body system which is of form $\sum_i (v(\underline{x}-\underline{x}_i))$. If one tries to replace this by the form $\sum_{i,j} V(\underline{x}-\underline{x}_i, \underline{x}-\underline{x}_j, \underline{x}_i-\underline{x}_j)$ appropriate to this situation then the derivation fails.

The peak scattering cross-section for two roton scattering occurs when the two-excitation density of states is largest, i.e. when the energy transfer to the fluid is such that the rotons at one of the two flat places in the ω_k versus k curve for the excitations are produced. Because the liquid is isotropic, these peaks in the density of states occur at the same energy in all directions in $\underline{k}$ space and a larger peak is expected than in the analagous situation in most solids. The experimental Raman scattering peaks would come at convenient energy transfers of 12 cm^{-1} and 19 cm^{-1}. A Raman scattering experiment is also simpler to perform than the corresponding Brillouin scattering experiment because the energy and momentum conservation rules do not fix the angle of the scattered photon as a function of energy transfer in the two-excitation case. With regard to checking the present theory, it should be possible to divide the results of a two-roton scattering experiment by the roton-phonon density of states determined from neutron scattering and thus obtain a check on our form for T(k).

Finally we note that the present machinery can be used to study the mechanisms of uv absorption of light with the production of excitons, a process of some current experimental interest. [7]

REFERENCES

1. M. Woolf, P. Platzmann and M. Cohen, Phys. Rev. Letters 17, 294 (1966); T. Greytak, Bull. Amer. Phys. Soc.
2. F. Seitz, "The Modern Theory of Solids," p. 266, McGraw-Hill Book Co., New York, 1940.
3. L. Goldstein and J. Richie, Phys. Rev. 98, 857 (1955).
4. D.G. Henshaw and A.D.B. Woods, Phys. Rev. 121, 1266 (1961).
5. V. Ambegaokar, T. Conway, and G. Baym, "Proceedings of the 1963 Copenhagen Conference on Lattice Dynamics."
6. C. Kittel, "Quantum Theory of Solids," Ch. 19, John Wiley and Sons, Inc., New York, 1963.
7. C. Surko (private communication).

C-1: MAGNONS AND THEIR INTERACTIONS AS OBSERVED BY LIGHT SCATTERING

P. A. Fleury
Bell Telephone Laboratories, Incorporated
Holmdel, New Jersey

ABSTRACT

We describe results of inelastic light scattering experiments in transparent magnetic materials. Three kinds of information have been inferred from these experiments: (1) magnon dispersion relations in simply structured antiferromagnets like MnF_2 and FeF_2; (2) magnon-magnon interaction effects, most clearly illustrated in $RbMnF_3$; and (3) the existance of spin waves in the paramagnetic phase (paramagnons) observed in NiF_2 and $RbNiF_3$.

GENERAL INTRODUCTION AND OUTLINE

In this paper we review the past work on one- and two-magnon scattering of light, touching briefly on scattering mechanisms and on the inference of magnon dispersion relations. In addition we present some very recent results on magnon-magnon interactions and on the observation of spin waves in the paramagnetic phase. The theoretical aspects of the mechanisms for light scattering from magnons and of magnon interactions are treated in detail in the following papers by Halley and by Thorpe and Elliott respectively. Our concern in the three following sections of this paper will be with (a) reviewing the earlier results in the rutile antiferromagnets MnF_2 and FeF_2 (b) presenting the observation of magnon-magnon interactions - quantitatively in $RbMnF_3$ and (c) discussing the observation of paramagnetic spin waves in NiF_2 and $RbNiF_3$.

THEORETICAL AND HISTORICAL REVIEW

Inelastic light scattering from magnetic excitations was first considered theoretically by Bass and Kaganov[1], who calculated the scattering from a ferromagnet due to a direct magnetic dipole coupling between light and magnetization fluctuations. An alternative scattering mechanism involving a three step process which relies on spin-orbit coupling was proposed by Elliot and Loudon[2]. This mechanism was further explored theoretically by other authors[3]. The original observation of light scattering from magnons[4] bore out the predictions of the spin orbit mechanism for one magnon scattering and revealed a surprisingly strong two-magnon scattering, which proceeds by an altogether different mechanism.

These experiments were performed on the rutile structure antiferromagnets FeF_2 and MnF_2. They are among the most simply structured magnetic materials in that below their respective Neél temperatures of 78°K and 67°K the spins are aligned parallel and antiparallel to the tetragonal axis and are equally divided between two equivalent sublattices. The resulting magnon dispersion curves are doubly degenerate in the absence of external fields and have the general appearance indicated in Fig. 1. The simplicity of the magnon dispersion curves implies that they can be characterized by a very few parameters which can be deduced from measurement of magnon frequencies at a few key points in the Brillouin zone. Here is where the importance of two-magnon scattering becomes evident. Because the wave vector of visible light is so small ($\sim 10^5 cm^{-1}$) compared to Brillouin zone boundary ($\sim 10^8 cm^{-1}$), one is constrained by momentum conservation to examine excitations of essentially zero wave vector as seen in (a) of Fig. 1. This means a measurement of only the zone center magnon frequency in the one-magnon scattering. On the other hand a two-magnon excitation may have zero total wave vector even though the constituent magnons individually have wave vectors up to q_{max}. The shape of a two-magnon peak in a given experimental geometry depends on (a) the magnon density of states and (b) various q dependent weighting factors dictated by the symmetry of the magnetic crystal. (Magnon interactions are also important as discussed below.) Detailed calculations of such shapes indicate that for the rutile antiferromagnets (space group D_{2h}^{12}) xz and xy tensor components receive their dominant contributions from R and M point magnons respectively. Because the combination of one- and two-magnon scattering results yields magnon frequencies at the Brillouin zone center as well as at identifiable critical points on the zone boundary, we have been able to infer the magnon dispersion relations from light scattering experiments. See Ref. [5 and 6].

Fig. 2 illustrates the experimental observations in FeF_2 at $T = 0.2T_n$ of both one- ($52\ cm^{-1}$) and two-magnon ($154\ cm^{-1}$) scattering[4]. (Details of the experimental procedure are discussed in Ref. [5] and need not be repeated here.) Any difference in frequency between R and M point magnons can be attributed to the effects of J_1 and J_3 (exchange interactions between neighbors on the same sublattice in the [001] and [100] directions respectively). The similarity of xz and xy two-magnon peaks in FeF_2 indicates these constants are negligible. This however is not the case for MnF_2, whose two-magnon spectra appear in Fig. 3. The shift in peak frequency of some $10 cm^{-1}$ has been used to assign values of J_1 and J_3 which are in agreement with results of neutron scattering measurements. The sets of dashed curves are results of calculations based on the following picture of the two-magnon scattering mechanism[7]. Since the two-magnon peak shows no broadening or splitting in fields up to 52Koe, it was concluded that the excitation responsible for the scattering has zero spin, as well as even parity and zero wave vectory. (The state can be designated as $|0,+\rangle = |\uparrow q, \downarrow -q\rangle + |\downarrow q, \uparrow -q\rangle$.) This means one magnon is excited on each sublattice. Thus the most general form one may write for the scattering interaction Hamiltonian is[6].

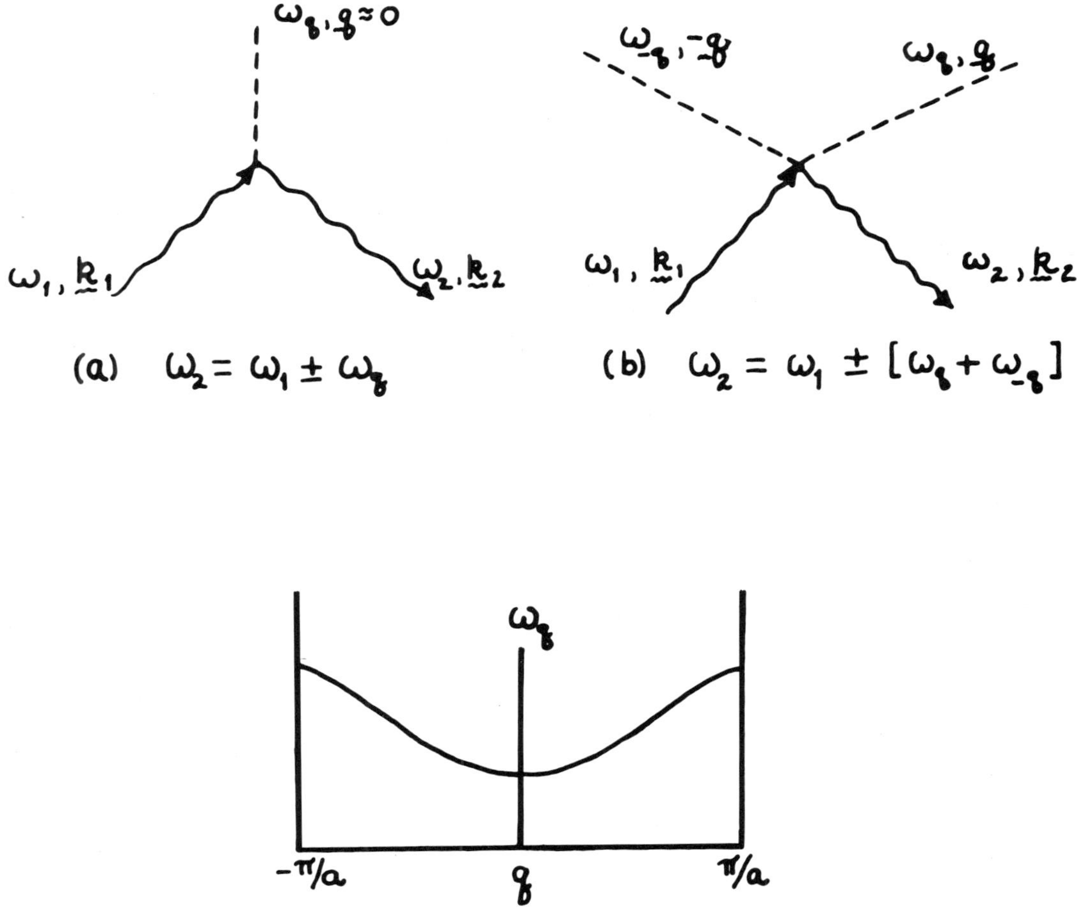

Fig. 1. Kinematics of one-magnon (a) and two-magnon (b) scattering of light. Since $|\vec{k}_1 - \vec{k}_2| \approx 0$ the two-magnon excitation is formed from magnons of opposite and nearly equal wave vectors. These, however, may lie anywhere along the dispersion curve illustrated at the bottom of the figure.

$$H^{II} = \sum_{i,j} \sum_{ab} J_{ab}' S_a^- S_b^+ E_1^i E_2^j G_{ij}^{ab} \qquad (1)$$

where $i, j = x, y, z$.

The fields E_1 and E_2 describe the incident and scattered radiation respectively J_{ab}' is an excited state exchange constant; S_a^- creates a magnon on the "a" (↓) sublattice; S_b^+ creates one on the "b" (↑) sublattice. The tenor G_{ij}^{ab} insures that H^{II} is invariant under the operations of the crystal's magnetic space group. The calculations depicted in Fig. 3 assumed the linearization of the Holstien Primakoff transformation (neglect of magnon-magnon interactions). The short dashed curve assumed J'^{1}_{ab} is nonzero for nearest neighbors only. Agreement is improved if one extends the range of J'_{ab} to include farther neighbors with an exponential weighting e^{-r/r_o}. The long dashed curves in Fig. 3 were obtained using $r_o = 0.4a$. A full discussion of these calculations appears in Ref. [5]. While the improvement in agreement is encouraging the necessity of having to introduce an additional parameter lessens the value of second-order light scattering measurements in determining magnetic parameters. It now seems however, that the extended range procedure is unnecessary provided one includes the effects of magnon-magnon interactions in the final state. Below we shall discuss these effects quantitatively with respect to $RbMnF_3$. Aside from these details of lineshape, all other aspects of the scattering of light by one- and two-magnon excitations - polarization selection rules, Raman tensor symmetry, magnetic field and temperature effects - in these antiferromagnets are rather well understood as described in detail in Ref. [5].

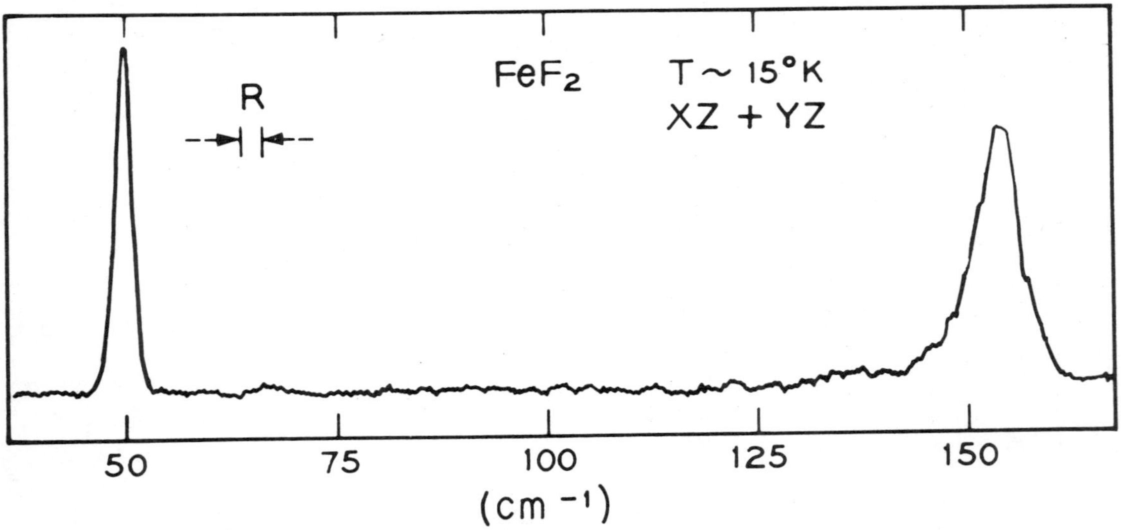

Fig. 2. One magnon (~50cm^{-1}) and two-magnon (~154cm^{-1}) scattering in FeF_2 at 15°K. The instrumental width R precludes measurement of the one-magnon width, but the true shape of the two-magnon peak is observed.

MAGNON-MAGNON INTERACTIONS

As mentioned above the previous discussion has assumed that the magnons created in the two magnon scattering process are independent of each other. This is equivalent to assuming that the transformation from spin to magnon variables is linear. Of course the Holstien-Primakoff transformation is not linear but it is given in a closed form so that all the nonlinear terms are known exactly. Thus it is possible to evaluate the effects of such nonlinear terms on the two-magnon spectrum. The details of such a calculation are described in the following paper by Thorpe and Elliott. Here we shall mention only those theoretical results which bear directly on our recent observations of magnon-magnon interactions[9] in $RbMnF_3$. It is often the case that for temperatures much less than T_n, effects of magnon magnon interactions are entirely negligible. This is true, for example, with regard to effects on magnon contributions to thermodynamic properties principally because a given magnon interacts with the whole thermal bath of other magnons resulting in only a small renormalization in its own frequency. As pointed out by Elliott et al[8] the two magnons created in a light scattering event are always in close proximity (in real space) and should interact strongly. This is a consequence of the local nature of the photon-magnon interaction. By retaining nonlinear terms in the spin-magnon transformation and by calculating the correlation function relevent to the two-magnon

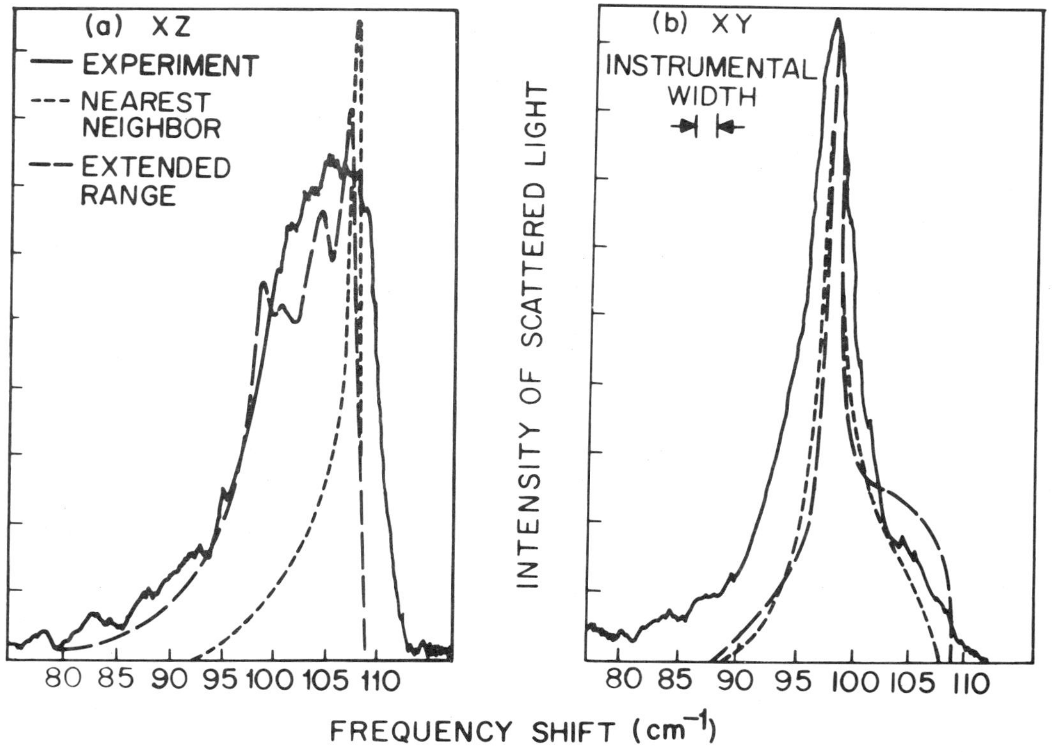

Fig. 3. Two-magnon spectra of MnF_2 at $10^{O}K$. (a) xz Raman tensor component primarily due to R point magnons (b) xy component due to M point magnons. Solid lines are experimental; dashed curves are calculated as described in the text.

scattering process Elliott et al[8] have accounted for magnon interaction effects on the spectra of simple antiferromagnets. They have applied their theory numerically to the simple cubic perovskite antiferromagnet, $RbMnF_3$, which has a Neél temperature of 82.5°K. $RbMnF_3$ is nature's best approximation to an isotropic Heisenberg antiferromagnet. Its anisotropy field is negligible (4.5 Gauss) and its magnetic properties are expressible quite well in terms of the single nearest-neighbor exchange[10] $J = 5 \pm 0.45\ cm^{-1}$.

As shown by Elliott et al[8] the symmetry of $RbMnF_3$ implies that the general form of Eq. (1) reduces to two contributions of the form

$$H^{II} \propto \sum_{ab} \{G_1(E_1 \cdot E_2) + G_3[(\vec{E}_1 \cdot \vec{r})(\vec{E}_2 \cdot \vec{r}) - \frac{1}{3}\vec{E}_1 \cdot \vec{E}_2]\} \vec{S}_a \cdot \vec{S}_b \qquad (2)$$

That the terms multiplying G_1 and G_3 transform as Γ_1^+ and Γ_3^+ respectively implies the following polarization combinations will govern the scattering:

$$\text{for } \Gamma_1^+ \quad (\vec{E}_1 \cdot \vec{E}_2)^2$$

$$\text{for } \Gamma_3^+ \quad (E_1^x E_2^x)^2 + (E_1^y E_2^y)^2 + (E_1^z E_2^z)^2 - E_1^x E_2^x E_1^y E_2^y - E_1^x E_2^x E_1^z E_2^z - E_1^y E_2^y E_1^z E_2^z$$

Since theory predicts different lineshapes for the two components it is necessary to distinguish experimentally between the Γ_1^+ and Γ_3^+ components. To do so, data were taken with (a) E_1 parallel to E_2 so that both contributions were present and (b) with E_1 parallel to [110] and E_2 parallel to [$\bar{1}$10] so that only Γ_3^+ appeared. Comparison of these spectra showed Γ_3^+ to be the strongly dominant contributor. Fig. 4 shows the Γ_3^+ component of the two magnon scattering in $RbMnF_3$ at 10°K. The dashed line represents the calculated shape ignoring magnon interactions; the dotted curve includes them. The agreement between the experiment (solid line) and the theory including magnon interaction effects provides striking confirmation of the importance of magnon interactions on two magnon spectra. It should be emphasized that there are no adjustable parameters in the theory of Elliott et al[8]. The theorectical curves were drawn using the value of $E_{max} = 2nSJ$ determined by[10] $n = 6$, $S = 5/2$ and $J = 4.7\ cm^{-1}$. The success of the theory for $RbMnF_3$ makes it quite likely that the previously described descrepancy in the MnF_2 data in attributable to failure to include magnon interaction effects.

There remains to be done some generalization of the theory of Elliott et al[8] both to more complex crystal structures and more important to high temperatures. With regard to the latter some motivation may be provided by the discussion of the next section on paramagnetic spin waves.

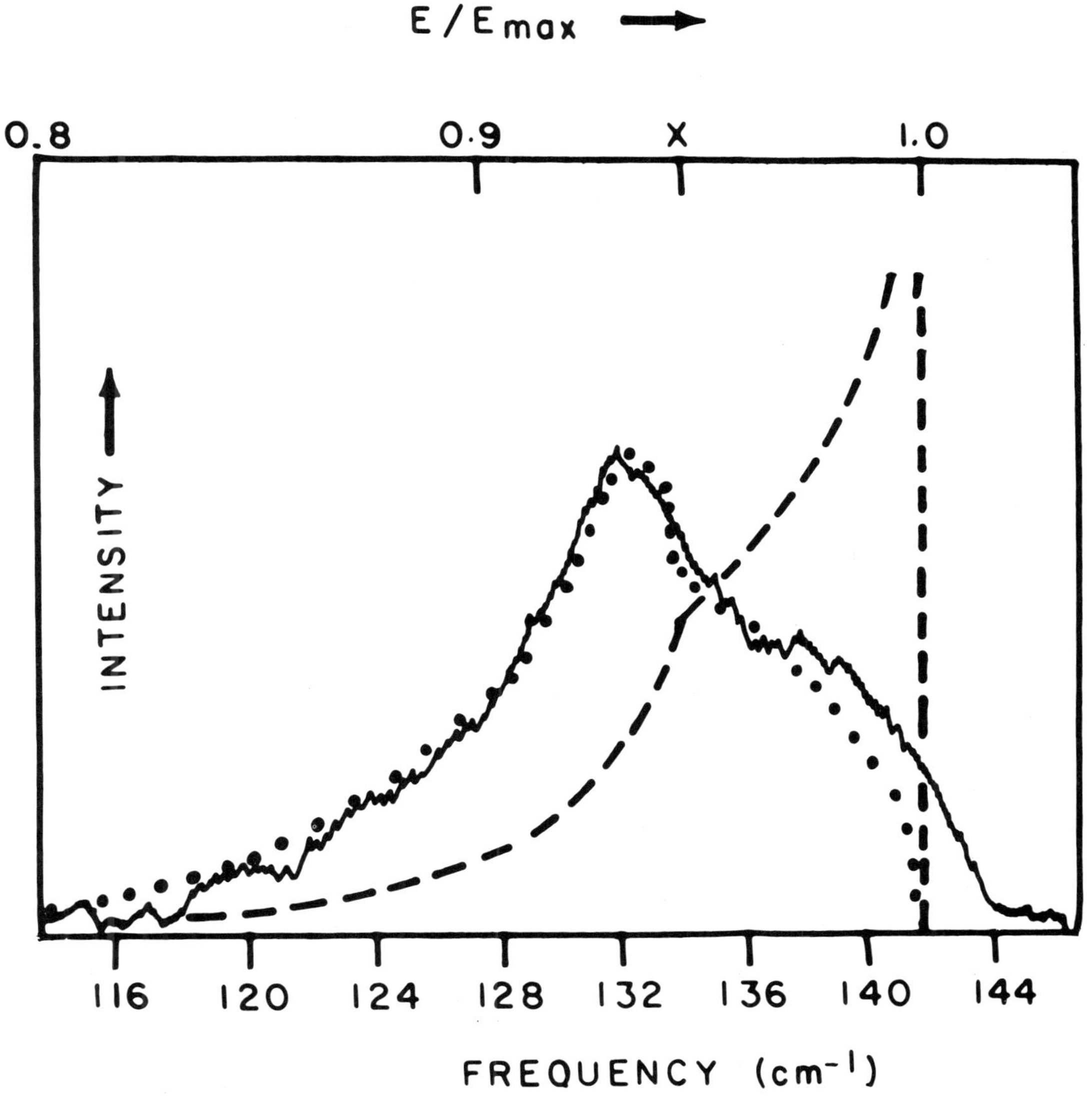

Fig. 4. Two-magnon spectrum (Γ_3^+ component) in $RbMnF_3$ at $10^{\circ}K$. Solid line is experimental. Dashed line is theoretical ignoring magnon interactions. Dotted line is theoretical including magnon interactions. X indicates magnon frequency at the X point of the Brillouin zone.

SPIN WAVES IN THE PARAMAGNETIC PHASE

The applicability of spin wave theory to magnetic systems is usually considered restricted to temperatures well below the magnetic transition temperature. As T is increased magnon lifetimes decrease because of magnon-magnon interactions so that at higher temperatures one does not expect the magnon to be a very well defined excitation. Certainly above T_n where there is no long-range order, there should be no well defined long-wavelength spin waves. However, it is possible that very short wavelength spin waves may remain well defined even into the paramagnetic phase because of their relative insensitivity to the disappearance of long-range magnetic order. As we have seen earlier, two-magnon light scattering probes magnons of very short wavelength (zone-boundary magnons). We will present in this section observations of spin waves in the paramagnetic phase (paramagnons) in two materials: NiF_2 and $RbNiF_3$.

NiF_2 is also a rutile structure antiferromagnet - but is slightly more complicated than MnF_2 or FeF_2. Below $T_n = 73\,°K$, the spins of NiF_2 align perpendicular to the c-axis such that the spins on the two sublattices are not quite antiparallel.

This results in two distinct magnon branches at small wave vector - which, however, become degenerate as the zone boundary is approached. Richards[11] in far infrared absorption has observed these zone center frequencies to be 3.3 cm^{-1} and 31 cm^{-1}. There have been no observations of two-magnon absorption in NiF_2 due to the presence of a strong IR active phonon[12] at 225 cm^{-1} - just about twice the frequency of a zone edge magnon estimated from theory and susceptability experiments. Recent neutron scattering work[13] shows the zone boundary magnon to have a frequency of 112 $cm^{-1} \pm 4$ cm^{-1}. As we see below the peak of the two magnon line in NiF_2 from light scattering is 202 cm^{-1}, indicating the importance of magnon-magnon interactions, in NiF_2. This point will not be pursued here.

The observation of paramagnons in NiF_2 is clearly illustrated in Fig. 5, where the magnon spectrum of NiF_2 is plotted for a wide range of temperatures in both the antiferromagnetic and the paramagnetic phases. In the former, two peaks are clearly visible, the first-order peak (I) due to a single zone center magnon (at ~ 31cm^{-1}) and the second-order peak (II) receiving in this geometry its major contributions from R and M point magnons. (We find no difference in the xy and xz spectra indicating that within experimental error R and M point magnons have the same frequencies [6]). As T is increased toward T_n the one-magnon line decreases in both frequency and intensity. Although it can be followed to $T = 0.92T_n$, it is clearly absent above T_n. That is, the scattering from the long wavelength magnon vanishes when long range order disappears. By contrast the two-magnon peak shows a continuous evolution as the crystal undergoes the transition from the antiferromagnetic to the paramagnetic phase. While the frequency and integrated intensity decrease as T is raised, the two-magnon scattering is clearly recognizable and well defined to at least twice T_n. Even at 4.1 T_n (300 °K) there is some magnetic scattering evident. Notice the shape of the instrumental background (dashed curve) in Fig. 5.

In Fig. 6 we show the temperature dependence of the peak of the two magnon scattering as well as that of the full width at half maximum. These curves indicate that the participating spin waves remain under-damped excitations up to at least 1.5 T_n.

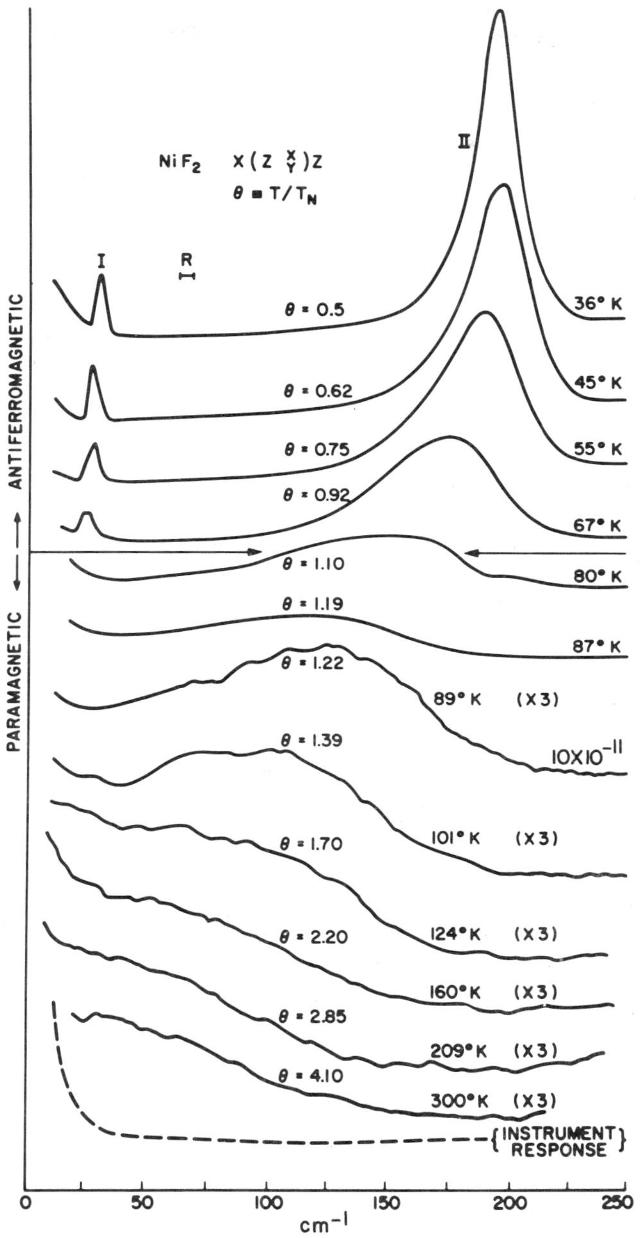

Fig. 5. Magnon spectra of NiF_2 in the antiferromagnetic and the paramagnetic phases. The one magnon peak (I) disappears for $T > T_n$. The two-magnon peak (II) persists indicating the existence of spin waves in the paramagnetic phase. Note the instrumental response (dashed curve).

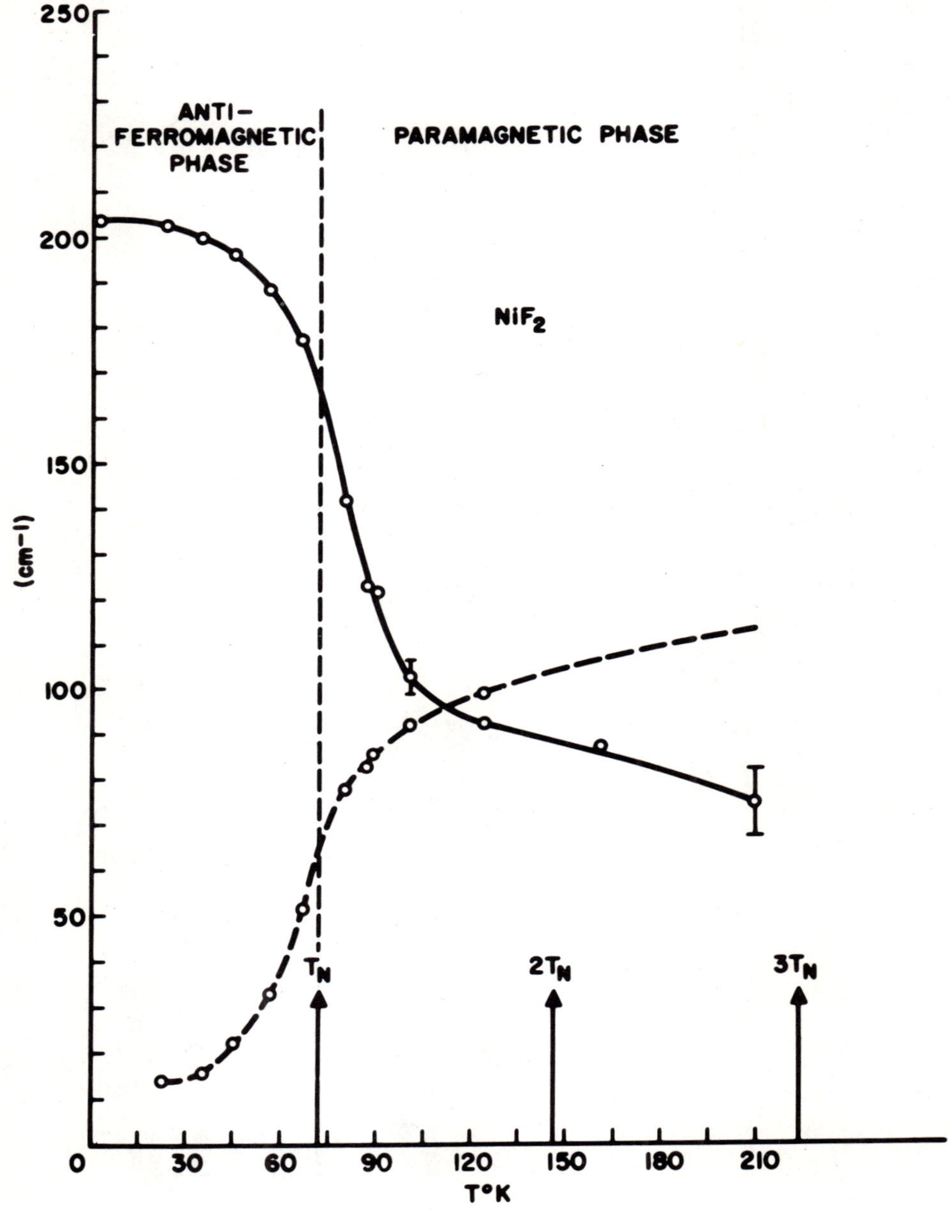

Fig. 6. Solid line indicates temperature dependence of the frequency of the peak of the two-magnon line in NiF_2. Dashed curve shows full width at half maximum of two-magnon peak. Note the peak is under damped up to ~1.5 T_n.

There is little theoretical work with which to compare these observations. (Reiter[14] has recently shown that for certain regions is the Brillouin zone paramagnons should exist. However, his calculations neglect lifetime effects). One might expect the temperature dependence of the magnon frequency to resemble that of some spin correlation function (say for nearest neighbor spins). Indeed the curve of Fig. 6 bears a qualitative similarity to that for the two spin correlation function with nearest neighbor exchange calculated by Callen and Callen[15]. However, spin correlation function at T_c drops to about 20 per cent of its zero temperature value - as compared with some 80 per cent in our experiment, so quantitative comparison is inappropriate[16].

Of course, the light scattering from two-magnons is really expressed in terms of a four spin correlation function[8] and the line shape receives contributions from weighted magnon densities of states[5], individual magnon lifetimes, and magnon-magnon interactions[8, 9]. So it is evident that more theoretical work is needed on the question of paramagnons in general and their interactions with light in particular.

Finally we mention similar observations of paramagnons in the hexagonal ferrimagnet, $RbNiF_3$. This is the most complicated crystal thus far studied by magnon-light scattering, possessing some six formula units per unit cell. The magnetic ions occupy two types of site - four ions on 'A' sites and two ions on 'B' sites[17]. Thus below $T_c = 139\,^\circ K$ the crystal exhibits a net magnetization due to the two extra 'A' site ions. An extensive study of the scattering from the thirty-three Raman active phonons as well as the two-magnon feature discussed below will be presented elsewhere[18]. In Fig. 7 is shown the temperature dependence of the xz spectrum of $RbNiF_3$ in the 200-600 cm^{-1} range. The sharp stationary peak is an E_{1g} phonon. The broad peak which decreases in intensity and frequency as T is increased, we attribute to a two-magnon scattering; one magnon each on the 'A' and 'B' sublattices. Space limitations preclude further discussion here, but the similarity between the results in $RbNiF_3$ and those in NiF_2 supports this identification, as do the observations that the line is uneffected by magnetic fields of 50 Koe parallel or perpendicular to the c-axis.

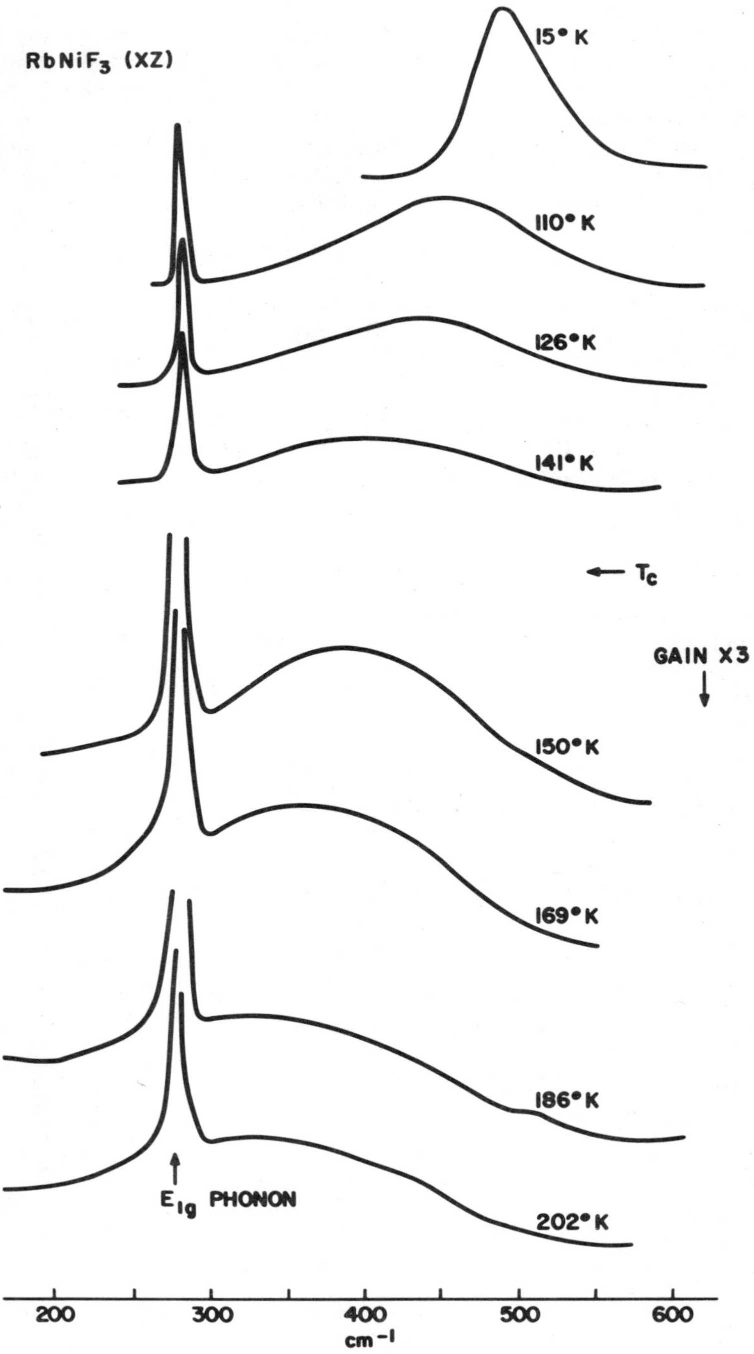

Fig. 7. xz component of the Raman spectrum of $RbNiF_3$ at various temperatures. The sharp line at $280cm^{-1}$ is an E_{1g} phonon. The broad peak is attributed to two-magnon scattering. Note the downward shift in its frequency with increasing T and its persistence well above $Tc = 139^{o}K$.

CONCLUSION

The utility of light scattering in the study of magnetic systems has been illustrated with regard to (1) determination of magnon dispersion relations (2) measurement of effects of magnon interactions and (3) observation of short range order in the paramagnetic phase. A coordinated program of both neutron and light-scattering experiments should provide valuable information for more sophisticated theories of magnetism.

ACKNOWLEDGMENTS

It is a pleasure to thank Professor R. Loudon for helpful discussions and H. J. Guggenheim for the excellent single crystal samples which made these experiments possible.

REFERENCES

1. F. G. Bass and M. I. Kaganov, Soviet Physics - JETP 10, 986 (1960).
2. R. J. Elliott and R. Loudon, Phys. Letters 3, 189 (1963).
3. Y. R. Shen and N. Bloembergen, Phys. Rev. 143, 372 (1966); T. Moriya, J. Appl. Phys. 39, 1042 (1968).
4. P. A. Fleury, S. P. S. Porto, L. E. Cheesman, and H. J. Guggenheim, Phys. Rev. Letters 17, 84 (1966).
5. P. A. Fleury and R. Loudon, Phys. Rev. 166, 514, (1968).
6. P. A. Fleury, Bull Am. Phys. Soc. 12, 420 (1967).
7. P. A. Fleury, S. P. S. Porto, and R. Loudon, Phys. Rev. Letters 18, 658 (1967).
8. R. J. Elliott, M. F. Thorpe, G. F. Imbusch, R. Loudon, and J. B. Parkinson, Phys. Rev. Letters 20, 147 (1968).
9. P. A. Fleury, Phys. Rev. Letters 20, 151 (1968).
10. C. G. Windsor and R. W. H. Stevenson, Proc. Phys. Soc. 87, 501 (1966).
11. P. L. Richards, J. Appl. Phys. 35, 850 (1964).
12. M. Balkanski, P. Moch, and G. Parisot, J. Chem. Phys. 44, 940 (1966).
13. R. J. Birgeneau (private communication).
14. G. F. Reiter, Phys. Rev. (to be published).
15. H. B. Callen and E. Callen, Phys Rev. 136, A1675 (1964).
16. A behavior similar to that in NiF_2 has recently been observed by neutron scattering in CoF_2 (P. Martel, R. A. Cowley, and R. W. H. Stevenson, J. Appl. Phys. 39, 1116 (1968)). However, significant differences also exist.
17. G. Zanmarchi and P. F. Bonglis, Solid State Comm. 6, 27 (1968), and references cited therein.
18. P. A. Fleury and J. M. Worlock (to be published).

C-2: TWO-MAGNON PAIRING EFFECTS ON THE OPTICAL SPECTRA OF ANTIFERROMAGNETS

M. F. Thorpe* and R. J. Elliott
Department of Theoretical Physics
Oxford, England

ABSTRACT

Recent experiments with light have made it possible to observe the interaction between pairs of magnons in antiferromagnets. We show that these interactions produce large effects on the observed spectra of $RbMnF_3$ and CoF_2.

INTRODUCTION

The interactions between magnons in magnetically ordered materials have been extensively studied. At low temperatures, where magnon theory holds, they lead to comparatively small changes in the magnon self-energy and in average thermodynamic properties. It has been pointed out by Wortis[1] and others, that in some circumstances the interaction can lead to the formation of bound pairs of magnons, but in ferromagnets there is no experimental data bearing on this prediction.

Recently, there has been great interest in the optical properties of antiferromagnets in which two magnons are created simultaneously by light absorption[2] or Raman scattering[3]. The interaction is such that two spin-deviations are created close together in real space, but with no change in the z component of the spin of the system. This is clearly impossible in a ferromagnet, but in an antiferromagnet two magnons are created in close proximity where they interact strongly. In this situation, the magnon interaction is expected to be much more important than for properties where the magnons are far apart on average. Indeed, it has become increasingly apparent that the two-magnon spectra could not be understood using simple spin wave theory, even at very low temperatures.

The lowest order interaction that we can write for the Raman process will be linearly proportional to the electric field of the incident radiation $\underline{E}$, the electric field of the scattered radiation $\underline{E}'$ and to the spins on neighboring sites belonging to different sublattices[4].

*Present address: Brookhaven National Laboratory, Upton, New York.

$$H'_{Ram} = \sum_{\substack{\underline{R}, \underline{r} \\ \alpha, \beta, \gamma, \delta}} E_\alpha E'_\beta B_{\alpha\beta\gamma\delta}(\underline{r}) S^\gamma_{\underline{R}} S^\delta_{\underline{R}+\underline{r}} \quad (1)$$

where α, β etc. are Cartesian components and B is a tensor whose symmetry is given by the group G' which leaves invariant the bond described by $\underline{r}$ in the lattice[5]. For a spin only magnet-like $RbMnF_3$, it is likely that terms in (1) $\propto \underline{S}_{\underline{R}} \cdot \underline{S}_{\underline{R}+\underline{r}}$ will be dominant, whereas in a magnet where the orbital contribution is important, like CoF_2, no such simplification can be made[6]. We assume that B is only finite for nearest neighbor pairs. The attempt to understand the lineshape in MnF_2 by assuming that this interaction acts over larger distances is probably not valid[3].

The most important terms in (1) for the creation of two magnons will be $\propto \sum_{\underline{R}} S^+_{\underline{R}} S^-_{\underline{R}+\underline{r}}$ although there will be small contributions from other terms like $\sum_{\underline{R}} S^z_{\underline{R}} S^z_{\underline{R}+\underline{r}}$. We shall however consider only the former terms and so the scattering will be related to the imaginary part of Green functions of the type

$$G_{rr'} = \langle\langle \sum_{R} S^+_{R} S^-_{R+r}; \sum_{R'} S^-_{R'} S^+_{R'+r'} \rangle\rangle_E \quad (2)$$

Before calculating these Green functions we will look at a system that can be solved exactly for the two magnon states at absolute zero.

ISING MODEL

We consider a two-sublattice antiferromagnet described by the Hamiltonian

$$H = \sum_{R'>R} J S^z_{\underline{R}'} S^z_{\underline{R}} \quad (3)$$

where the exchange acts only between nearest neighbors. If a single spin deviation is created, the excitation energy will be nJS, where n is the number of nearest neighbors. If there are two spin deviations in the lattice their energy will be 2nJS, unless they are on neighboring sites when their energy will be lowered to 2nJS-J, and the spin deviations can be thought of as a bound pair. We would expect a delta function response in the light scattering experiments, a fraction 1/2nS below the top of the 'two magnon band.' When terms $\frac{1}{2} \sum_{R'>R} J \left(S^+_{\underline{R}'} S^-_{\underline{R}} + S^-_{\underline{R}'} S^+_{\underline{R}} \right)$ are added to (3) to give a Heisenberg Hamiltonian, the flat band at 2nJS becomes a band stretching from 0 to 2nJS. For cubic antiferromagnets we find from the Green function calculations that the bound state/resonance is still about a fraction 1/2nS below the top of the two magnon band. This may be understood because cubic antiferromagnets with nearest neighbor Heisenberg exchange have a square root singularity at 2nJS in the density of states and so the Ising energy and the most probable energy coincide.

GREEN FUNCTION CALCULATION

In the ferromagnet at absolute zero[1], the two magnon states can be found exactly. This is not possible in the antiferromagnet, however, because the exact ground state is unknown. Let us consider a two sublattice antiferromagnet described by the Hamiltonian

$$H = \sum_{R'>R} J\underline{S}_{\underline{R}'} \cdot \underline{S}_{\underline{R}} + \sum_{R'>R} A S^z_{\underline{R}'} S^z_{\underline{R}} \tag{4}$$

with nearest neighbor interactions between the sublattices.

If we form an equation of motion for Green functions of the type (2), the inhomogeneous terms have to be approximated because the ground state is unknown. We have used both the Néel ground state and spin wave ground state for this purpose. A rather more serious problem is that higher order Green functions enter the problem and these must be decoupled[7]. Details of the procedure will be given elsewhere[8]. The essential point to notice is that the first spin deviation appears like an impurity to the second deviation and so we obtain a Dyson equation, typical of impurity problems[9] for the Green functions of the type (2), considered as matrices with indices $\underline{r}$ and $\underline{r}'$.

$$OG = C + \Delta G \tag{5}$$

where C is the inhomogeneous term and Δ is a localized matrix which describes the 'interaction' or improvement on simple spin wave theory. The method of solution for this type of equation is well known[9]. The solution may be written most conveniently in terms of symmetrized Green functions belonging to representations of the appropriate point group of the crystal.

The light scattering intensity, i.e. the imaginary part of G, may be written as a numerator over a denominator. The denominator vanishes when det $|1-0^{-1}\Delta| = 0$. Detailed calculations show that this occurs very close to the position as predicted by the simple Ising model, but is shifted slightly by symmetry effects. A true bound state will occur only if

$$(1 + A/J)^2 \gtrsim nS \tag{6}$$

No examples of true bound states have been found experimentally. The numerator depends on the ground state used and the detailed interaction (1). This will be a slowly varying function of frequency and so the scattering is dominated by the behavior of the denominator. This means that one can learn much about the two magnon states, but little about the detailed interaction between the light and the magnetic system.

APPLICATION OF RESULTS

We have applied the theory to $RbMnF_3$ where the magnon dispersion has been measured by inelastic neutron scattering[10] and its properties are well explained by a nearest neighbor Heisenberg Hamiltonian (i.e. Eq. (4) with A = 0, J = 4.7 cm^{-1} and S = 5/2).

The Raman Hamiltonian (1) for $RbMnF_3$ may be written[4]

$$H'_{Ram} = \sum_{R,r} \left[\frac{B_3(\underline{E}\cdot\underline{r})(\underline{E}'\cdot\underline{r})}{r^2} + (B_1 - 1/3B_3)(\underline{E}\cdot\underline{E}') \right] S_{\underline{R}} \cdot S_{\underline{R}+\underline{r}} \tag{7}$$

The Green functions (2) may be classified according to the cubic group[11] into Γ_4^-, Γ_1^+ and Γ_3^+ modes. The negative parity mode is responsible for the absorption and the Raman active modes are Γ_1^+ and Γ_3^+ which are plotted in Fig. 1. The intensity in the Γ_1^+ mode is proportional to B_1^2 and in the Γ_3^+ mode to B_3^2. The resonance in the Γ_3^+ mode is particularly interesting as it falls near to a critical point in the Brillouin zone. This mode has been observed by Fleury[12] and is plotted in Fig. 2 where the parameter B_3 has been adjusted to make the peak heights coincide.

We have also applied the theory to CoF_2, which is a complex magnetic system whose properties are not well understood. The magnon dispersion has been measured[13] and we describe the lowest branch with the nearest neighbor Hamiltonian (4) with $S = 1/2$, $J = 13.4\ cm^{-1}$ and $A = 2.9\ cm^{-1}$. This is very much a first approximation, but allows us to fit the lowest magnon branch fairly well except that with this approximation the points X, Z, A in the Brillouin zone become degenerate in energy which is incorrect[13]. Next nearest neighbor interactions would have to be included to resolve this difficulty.

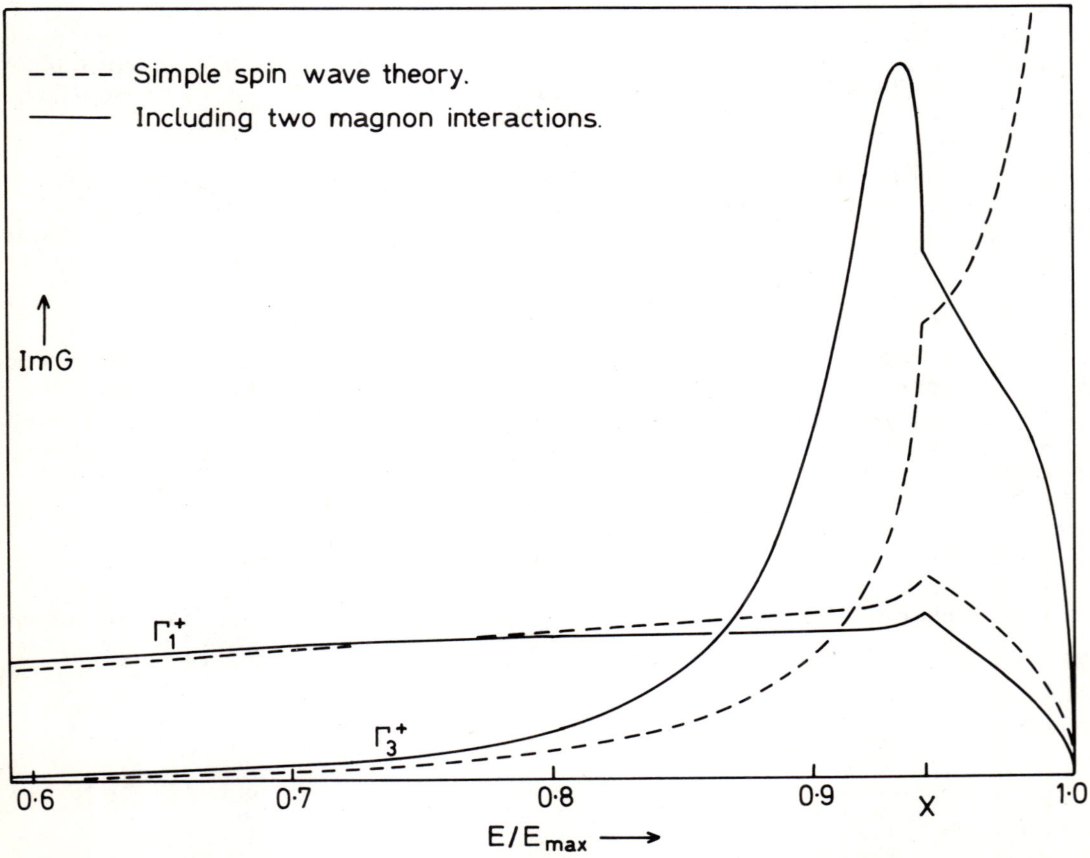

Fig. 1. Theoretical Raman modes in $RbMnF_3$.

The crystal becomes effectively cubic and the Raman active modes Γ_1^+ and Γ_5^+ are shown in Fig. 3. Unfortunately, there is no experimental data to compare these calculations with. The infrared absorption has been measured by Allen and Richards and the absorption with the electric field parallel to the c axis is shown in Fig. 4.

The predicted sharp resonance in the Γ_4^- mode can be clearly seen and its position is given reasonably well by this simple model. The sharp resonance is due to the low value of the density of states at resonance. The theoretical curves in Fig. 4 should be multiplied by a slowly varying function of frequency as the detailed interaction (1) contains many unknown parameters.

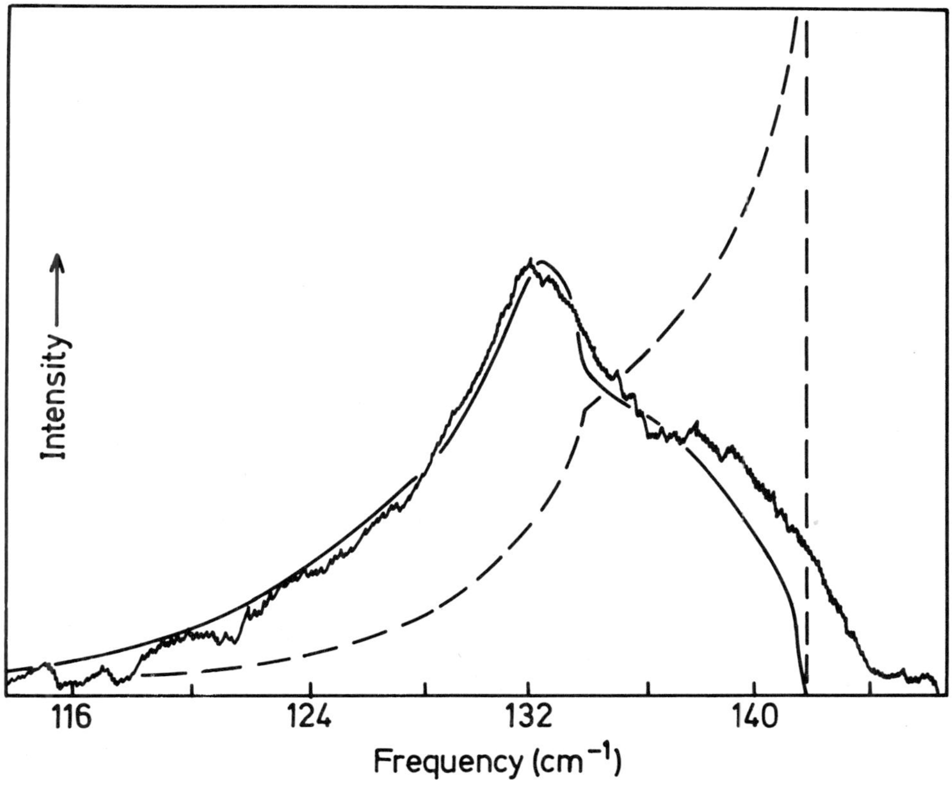

Fig. 2. Theoretical and observed Γ_3^+ mode in $RbMnF_3$. The dashed line is the simple spin wave theory and the solid line includes the interaction (from Fig. 1).

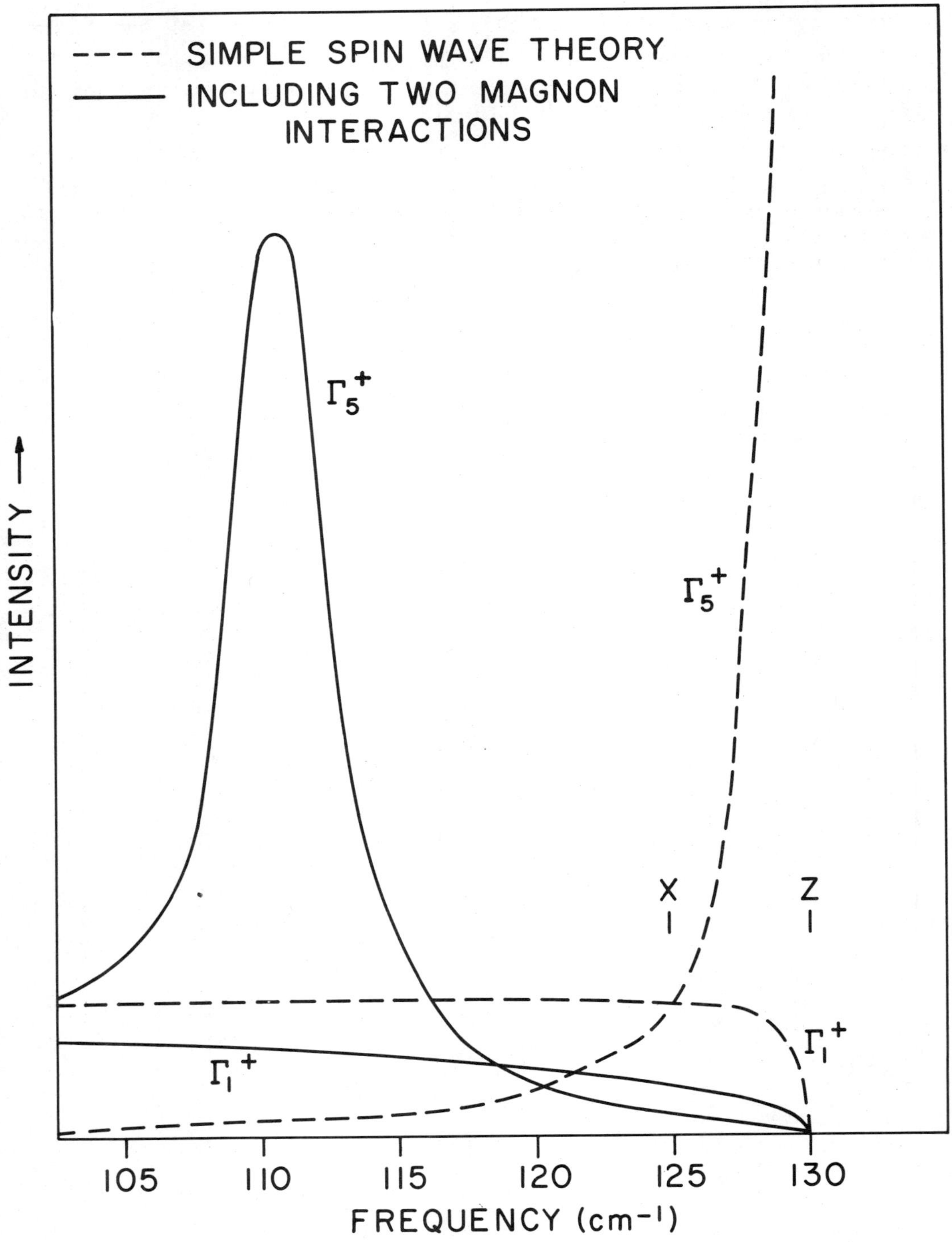

Fig. 3 Theoretical Raman modes in CoF_2. The X and Z points in the Brillouin zone are determined from the neutron data[13].

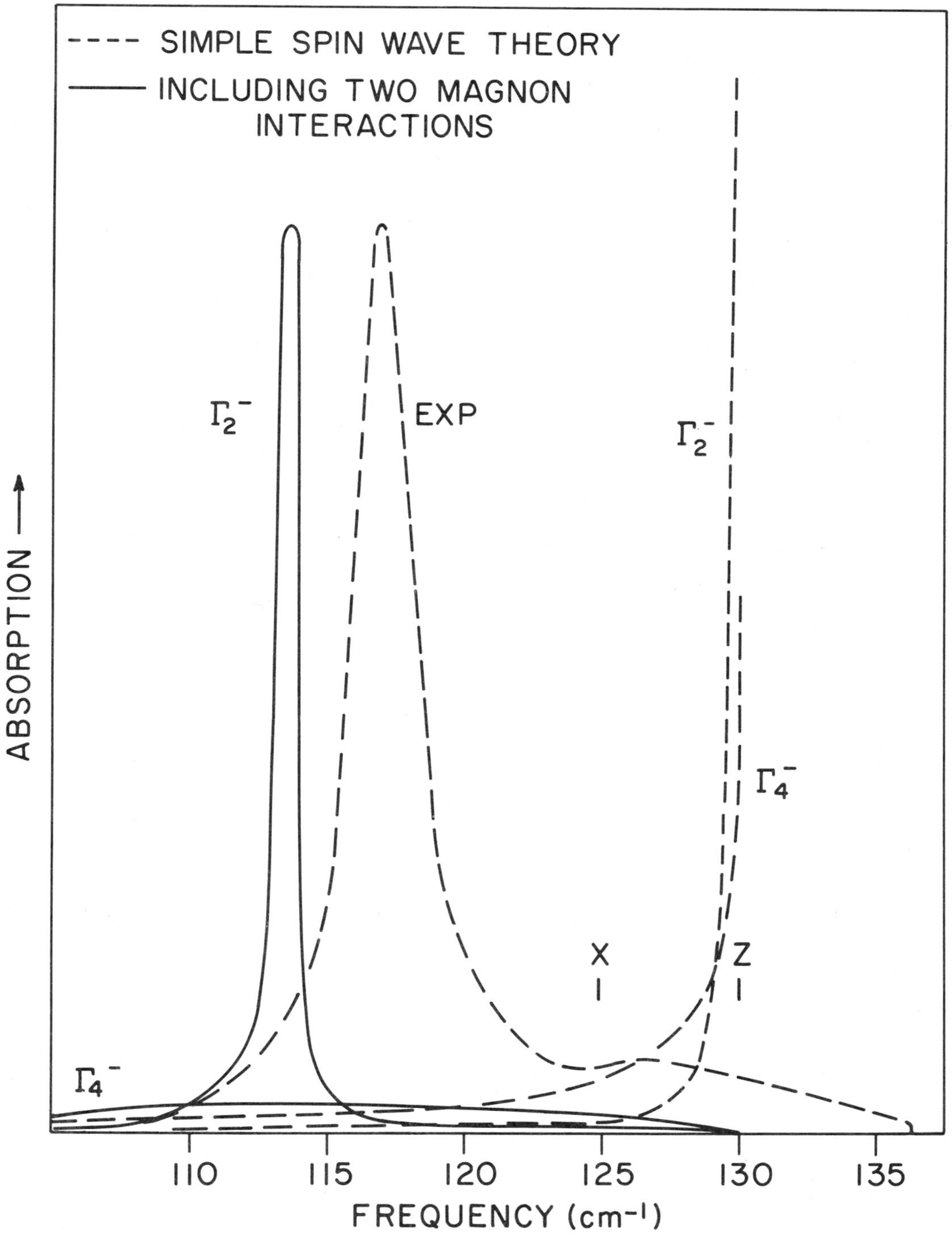

Fig. 4. Theoretical and observed infrared absorption modes in CoF_2.

CONCLUSIONS

We find that the interaction effects must be included to understand the absorption and Raman scattering experiments in antiferromagnets. The present calculations may be extended to finite temperatures, but it will be difficult to find a satisfactory decoupling scheme and diagrammatic techniques may be more satisfactory. Finally, we would like to thank R. Loudon and J. B. Parkinson for many interesting discussions on this topic.

REFERENCES

1. M. Wortis, Phys. Rev. 132, 85 (1963).
2. R. Loudon, Adv. Phys. 17, 243 (1968).
3. P.A. Fleury and R. Loudon, Phys. Rev. 166, 514 (1968).
4. R.J. Elliott, M.F. Thorpe, G. Imbusch, R. Loudon, and J.B. Parkinson, Phys. Rev. Letters 21, 147 (1968).
5. R.J. Elliott and M.F. Thorpe, J. Appl. Phys. 39, 802 (1968).
6. Microscopic theories for the B tensor are discussed by J. Woods Halley, paper C-3 this conference.
7. M.F. Thorpe, "D. Phil. Thesis," Oxford University (unpublished).
8. R.J. Elliott and M.F. Thorpe (to be published).
9. T. Wolfram and J. Callaway, Phys. Rev. 130, 2207 (1963).
10. C.G. Windsor and R.W.H. Stevenson, Proc. Phys. Soc. 87, 501 (1966).
11. The notation follows that of S.F. Koster, J.O. Dimmock, R.S. Wheeler and H. Statz, "Properties of the 32 Point Groups," M.I.T. Press, Cambridge, Mass., 1963.
12. P.A. Fleury, Phys. Rev. Letters 21, 151 (1968).
13. R.A. Cowley, P. Martel, and R.W.H. Stevenson, Phys. Rev. Letters 18, 162 (1967).
14. S.J. Allen and P.L. Richards (private communication).

C-3: SOME POSSIBLE EXPERIMENTS FOR STUDY OF MECHANISMS OF TWO SPIN-WAVE SCATTERING AND ABSORPTION

J. Woods Halley
Department of Physics, University of California
Berkeley, California*

INTRODUCTION

Recent interest[1] in two spin wave optical processes in rutile and cubic antiferromagnets has given rise to several proposals[2] for the origin of the coupling between the light and the spin waves. The question of which mechanism is likely to dominate in a given material does not seem easy to settle quantitatively on the basis of presently available experimental information. For this reason we point out here that certain of the proposed mechanisms have some striking qualitative features which other mechanisms don't have and which might be observable experimentally.

First we review the proposed mechanisms, using Table I for convenience of exposition. We do not consider here the related question of the mechanisms of coupling in spin-wave side bands in optical absorption by antiferromagnets in the ultraviolet. The two optical processes of interest are then two spin wave infrared absorption and two spin wave Raman scattering. For each process, three mechanisms have been proposed, based respectively on the spin orbit interaction plus multipole couplings between spins, on the electric field dependence of the exchange arising from mixing of high-lying electronic electronic states into the ionic wave functions and on the electric field dependence of the exchange arising from exchange-strictive coupling to optically active phonons. Factors entering the order of magnitude of the scattering and absorption amplitudes in each case are shown in Table I. It has been suggested[2] that in the case of infrared absorption, the multipole moment coupling may play a role in FeF_2 while it cannot be important in MnF_2. Studies of the Raman scattering mechanisms have been less detailed than those of infrared absorption mechanisms and we will have occasion to develop a more detailed consideration of one of them in the sequel.

For the purposes of exploiting the symmetry properties of the magnetic crystals involved, one can treat the infrared and Raman scattering processes phenomenologically without reference to a mechanism. For infrared absorption, one seeks a phenomenological perturbing Hamiltonian which is linear in the infrared electric field and bilinear

*Present address: School of Physics and Astronomy, University of Minnesota, Minneapolis, Minn.

TABLE I

Summary of Scattering and Absorption Mechanisms (key below)

Infrared	Raman Scattering
$\frac{\lambda^2 \text{ (Q-d) } (\underline{A} \cdot \underline{p})}{(\Delta E_e)^2 \; (\Delta E_o)}$	$\frac{\lambda^2 \text{ (d-d) } (\underline{A} \cdot \underline{p})^2}{(\Delta E_e)^2 \; (\Delta E_o)^2}$
$\frac{J_{od}}{(\Delta E_o)} \; \underline{A} \cdot \underline{p}$	$\frac{J_{od}}{(\Delta E_o)^2} \; (\underline{A} \cdot \underline{p})^2$
$\frac{\partial J}{\partial R} \; \delta R \; \frac{\underline{A} \cdot \underline{P}}{(\Delta E_{ph})}$	$\frac{\partial^2 J}{\partial R^2} \; (\delta R)^2 \; \frac{(\underline{A} \cdot \underline{P})^2}{(\Delta E_{ph})^2}$

λ	spin orbit interaction
Q-d	quadrupole dipole interaction
d-d	dipole-dipole interaction
$\underline{A} \cdot \underline{p}$	electric dipole coupling of electrons to photons
$\underline{A} \cdot \underline{P}$	electric dipole coupling of phonons to photons
J_{od}	off diagonal exchange
δR	phonon displacement
ΔE_o	energy to excited odd-parity electronic state
ΔE_e	energy to excited even-parity electronic state
ΔE_{ph}	energy to excited phonon state

in the spin operators. One knows that the Hamiltonian must be linear in the field because the absorption intensity depends experimentally on the direction of the electric field of the infrared light and that it must be bilinear in the spins because two spin waves are produced. Experimental independence of the two spin wave frequency on magnetic field imposes the condition that the two spin operators refer to opposite sublattices. Then requiring that the two spin operators be as close together as possible and that the Hamiltonian be invariant under the operations of the magnetic space group, one gets the following form[1]

$$\mathcal{H} = \sum_{i,j} \{ \pi_1 (S_i^x S_j^x + S_i^y S_j^y)(E_x \sigma_y + E_y \sigma_x)$$

$$+ \pi_2 (S_i^x S_j^y - S_i^x S_j^x)(E_x \sigma_y - E_y \sigma_x) \tag{1}$$

$$+ \pi 3 (S_i^x S_j^x + S_i^y S_j^y) E_z \sigma_x \sigma_y \sigma_z \}$$

Here $\underline{E}$ is the electric field, π_1, π_2, π_3 are real phenomenological constants and $\sigma_x = \mathrm{sgn}(x_i - x_j)$ where (x_i, y_i, z_i) is the position of the $i^{\underline{th}}$ site. A similar derivation leads to a phenomenological coupling describing Raman scattering with the production of two spin waves[2]

$$\mathcal{H} = \sum_{i,j} \Big\{ A(E_1^x E_2^x + E_1^y E_2^y) + B E_1^z E_2^z + C(E_1^x E_2^y + E_1^y E_2^x)\sigma_x\sigma_y + D\Big[(E_1^y E_2^z + E_1^z E_2^y)\sigma_y\sigma_z + (E_1^x E_2^z + E_1^z E_2^x)\sigma_x\sigma_y\Big] + F\Big[(E_1^y E_2^z - E_1^z E_2^y)\sigma_y\sigma_z - (E_1^z E_2^x - E_1^x E_2^z)\sigma_x\sigma_z\Big]\Big\}(S_i^x S_j^x + S_i^y S_j^y)$$

Conversion of these expressions to expressions in terms of spin wave operators then leads to expressions for the line shapes of scattering and absorption cross-sections as a function of photon energy and energy transfer which may be compared with experiment. Assuming that a) the restriction to nearest opposite-sublattice neighbors is valid and b) that final state two spin-wave interactions are not important, the task of a theory of the microscopic mechanism of scattering and absorption is then to calculate the constants π_1, π_2, π_3, A, B, C, D, F in terms of a microscopic model. It has been suggested[3] that a) is not valid and it now appears[4] that b) may not be valid. Nevertheless, we proceed here as if the task were in fact to find π_1, π_2, π_3, A, B, C, D, F. Most of our considerations would not be vastly changed by modifications of a) and b).

We concentrate on two results of microscopic studies of the coefficients

(1) For the phonon-modulated process[5], $\pi_2 = 0$

(2) The Raman coefficients A, B, C, D, F will have resonances at incident photon energies of hundreds of cm^{-1} if the phonon modulated process plays a role and at tens of thousands of cm^{-1} if one of the other two processes dominate the scattering[6].

We point out here that experiments can be done to find out if $\pi_2 = 0$ or not and if the Raman cross-section has resonances in the middle infrared. These experiments are respectively a Stark effect experiment on the two spin wave infrared absorption and a Raman scattering experiment with a laser in the middle infrared. We discuss these one at a time.

STARK EFFECT

The symmetry considerations leading to (1) are equally valid if the electric field is a static one. Microscopic considerations show that the coefficients π_i depend on the frequency of the electric field, but this dependence is weak except near a resonance of a sort analagous to the kind discussed in the next section. The same mechanisms give rise to finite π_i if the electric field is static and these π_i still obey the condition $\pi_2 = 0$ for the exchange strictive mechanism. One therefore expects that at least part of the effect of imposing a d. c. electric field on a rutile antiferromagnet can be accounted for by adding a Hamiltonian of the form (1) to the usual spin wave Hamiltonian, but with $\underline{E}$ the static d. c. field and that the π_i involved will be of the same order of magnitude and will obey the same selection rules as the π_i involved in (1) when $\underline{E}$ is an infrared field. There will, in addition, be other effects arising from the static field, but if the dominant two spin-wave effects arise from exchange through either of the second two mechanisms of Table I, then the terms of the form (1) will dominate. It is then of interest to note that this new term leads to a first order shift in the energy of the two spin wave state. An experimental study of the π_i is thus possible through measurement of the Stark shifts in spin wave frequencies. This appears to be a better way to measure the π_i than through the infrared absorption intensity, because frequency shifts are notoriously easier to measure than absorption intensities. To show that there is a first order shift we transform (1) using the relations[7]

$$S^+_{aj} = \sqrt{\frac{2S}{N}} \sum_{\underline{k}} e^{i\underline{k}\cdot\underline{x}j} \; c_{\underline{k}}$$

$$S^-_{aj} = \sqrt{\frac{2S}{N}} \sum_{\underline{k}} e^{-\underline{k}\cdot\underline{x}j} \; c^\dagger_{\underline{k}}$$

$$S^+_{b\ell} = \sqrt{\frac{2S}{N}} \sum_{\underline{k}} e^{-i\underline{k}\cdot\underline{x}_\ell} \; d^\dagger_{\underline{k}}$$

$$S^-_{b\ell} = \sqrt{\frac{2S}{N}} \sum_{\underline{k}} e^{i\underline{k}\cdot\underline{x}_\ell} \; d_{\underline{k}}$$

$$\alpha_{\underline{k}} = u_{\underline{k}} c_{\underline{k}} - v_{\underline{k}} d^\dagger_{-\underline{k}} \; ; \; \beta_{\underline{k}} = u_{\underline{k}} d_{\underline{k}} - v_{\underline{k}} c^\dagger_{-\underline{k}}$$

where the $u_{\underline{k}}$ and $v_{\underline{k}}$ are determined so that the usual spin wave Hamiltonian (cited below) is diagonal. Using (2), (1) becomes

$$\mathcal{H} = \sum_{\underline{k}} \Bigg[\Bigg\{ \Big((E_x M^x_{\underline{k}} + E_y M^y_{\underline{k}}) - \frac{i^{\pi}2}{\pi_1} (E^x_x M^x_{\underline{k}} - E_y M^y_{\underline{k}})$$

$$+ M^z_{\underline{k}} E^z \Big) \Big(u^2_{\underline{k}} \alpha_{\underline{k}} \beta_{-\underline{k}} + u_{\underline{k}} v_{\underline{k}} (\alpha_{\underline{k}} \alpha^\dagger_{\underline{k}} + \beta^\dagger_{\underline{k}} \beta_{\underline{k}}) +$$

$$v^2_{\underline{k}} \beta^\dagger_{-\underline{k}} \alpha^\dagger_{\underline{k}} \Big) \Bigg\} +$$

$$+ \left\{ \left(E_x M_{\underline{k}}^{x*} + E_y M_{\underline{k}}^{y*} + \frac{i\pi_2}{\pi_1} \left(E_x M_{\underline{k}}^{x*} - E_y M_{\underline{k}}^{y*} \right) \right.\right.$$

$$+ M_{\underline{k}}^{z*} E^z \Big) \left(u_{\underline{k}}^2 \alpha_{\underline{k}}^{\dagger} \beta_{-\underline{k}}^{\dagger} + u_{\underline{k}} v_{\underline{k}} \left(\alpha_{\underline{k}}^{\dagger} \alpha_{\underline{k}} + \beta_{-\underline{k}} \beta_{-\underline{k}}^{\dagger} \right) \right.$$

$$\left.\left. + v_{\underline{k}}^2 \alpha_{\underline{k}} \beta_{-\underline{k}} \right) \right\} \Bigg]$$

where

$$M_{\underline{k}}^x = 8\pi_1 \; i \; S \cos\left(\frac{k_x a}{2}\right) \sin\left(\frac{k_y a}{2}\right) \cos\left(\frac{k_z c}{2}\right)$$

$$M_{\underline{k}}^y = 8\pi_1 \; i \; S \sin\left(\frac{k_x a}{2}\right) \cos\left(\frac{k_y a}{2}\right) \cos\left(\frac{k_z c}{2}\right)$$

$$M_{\underline{k}}^z = 8\pi_1 \; i \; S \sin\left(\frac{k_x a}{2}\right) \sin\left(\frac{k_y a}{2}\right) \sin\left(\frac{k_y c}{2}\right)$$

where a and c are the lattice constants for the rutile structure and the fact that π_i are real has been used. By use of (4), the first order contributions to the Stark shift vanish for the π_1 and π_3 parts but remain for the π_2 part. We get a Stark shift in the energy of the two spin wave state $\alpha_{\underline{k}}^{\dagger} \beta_{\underline{k}}^{\dagger}$ | vacuum > of

$$\Delta(k, -k) =$$

$$32\, u_{\underline{k}} v_{\underline{k}} \pi_2 S \left[E_x \cos\left(\frac{k_x a}{2}\right) \sin\left(\frac{k_y a}{2}\right) - E_y \sin\left(\frac{k_x a}{2}\right) \cos\left(\frac{k_y a}{2}\right) \right]$$

$$\times \cos\left(\frac{k_z c}{2}\right)$$

The Stark shift thus depends on π_2 and will be zero if $\pi_2 = 0$ as predicted for the exchange strictive mechanism. There is no shift at the zone boundary (because $v_{\underline{k}} = 0$ there) or at the center of the zone (because the sines in the matrix element are zero there). The first order Stark shift arises only from a d. c. field normal to the c-axis. The two spin wave state has energy $2\hbar\omega_{\underline{k}} + \Delta(\underline{k}_1 -\underline{k})$. To calculate the Stark shifted two spin wave absorption line shape we use $u_{\underline{k}}$, $v_{\underline{k}}$ and $\hbar\omega_{\underline{k}}$ from the Hamiltonian[7]

$$\mathcal{H} = \sum_{i,j} J\, \underline{S}_{ai} \cdot \underline{S}_{bj} - 2\mu_0 H_A \sum_i S_{ai}^z + 2\mu_0 H_A \sum_i S_{bi}^z \qquad (6)$$

giving

$$h\omega_{\underline{k}} = \sqrt{(\omega_e + \omega_A)^2 - \omega_e^2 \gamma_{\underline{k}}^2}$$

$$\gamma_{\underline{k}} = \frac{1}{z} \sum_{\delta} e^{i\underline{k}\cdot\underline{\delta}} = \cos\left(\frac{k_x a}{2}\right) \cos\left(\frac{k_y a}{2}\right) \cos\left(\frac{k_z c}{2}\right) \tag{7}$$

$$\omega_e = 2Jz\ S\ , \quad \omega_A = 2\mu_o H_A$$

and

$$2u_{\underline{k}} v_{\underline{k}} = \frac{\gamma_{\underline{k}}}{\left(\left(1 + \frac{\mu_o H_A}{JzS}\right)^2 - \gamma_{\underline{k}}^2\right)^{1/2}}$$

Then the Stark shifted infrared two spin wave absorption rates are

$$a_{||}(\omega) \ \alpha \ \sum_{\underline{k}} |M_{\underline{k}}^z|^2 \ \delta(\hbar\omega - 2\hbar\omega_{\underline{k}} - \Delta(\underline{k}, -\underline{k}))$$

$$a_{\perp}(\omega) \ \alpha \ \sum_{\underline{k}} \left| M_{\underline{k}}^x - \frac{i\pi_2}{\pi_2} M_{\underline{k}}^x (u_{\underline{k}}^2 + v_{\underline{k}}^2) \right|^2 \ \delta(\hbar\omega - 2\hbar\omega_{\underline{k}} - \Delta(\underline{k}, -\underline{k}))$$

for the cases in which the infrared electric field is parallel to the c-axis and normal to the c-axis respectively. We have evaluated $a_{||}(\omega)$ for parameters appropriate to FeF_2 and a field of 10^6 volts/cm and find the result shown in Fig. 1. FeF_2 was chosen as an example because it seems easier to fit the experimental results without a d. c. field in this case without considering the complications arising from spin wave interactions and many-neighbor couplings.

The d. c. field needed to observe a shift is estimated as follows. Using the fact that the two spin wave absorption is of the same order as the antiferromagnetic resonance absorption one has

$$EM \sim \mu H$$

where E and H are the infrared electric and magnetic fields, M is the two spin wave matrix element and $\mu \sim 10^{-20}$ erg/gauss. Then because E = H in cgs units, one has $M \sim 10^{-20}$ erg/statvolt/cm. For E_{dc} $M \sim 1\ cm^{-1}$ one therefore needs

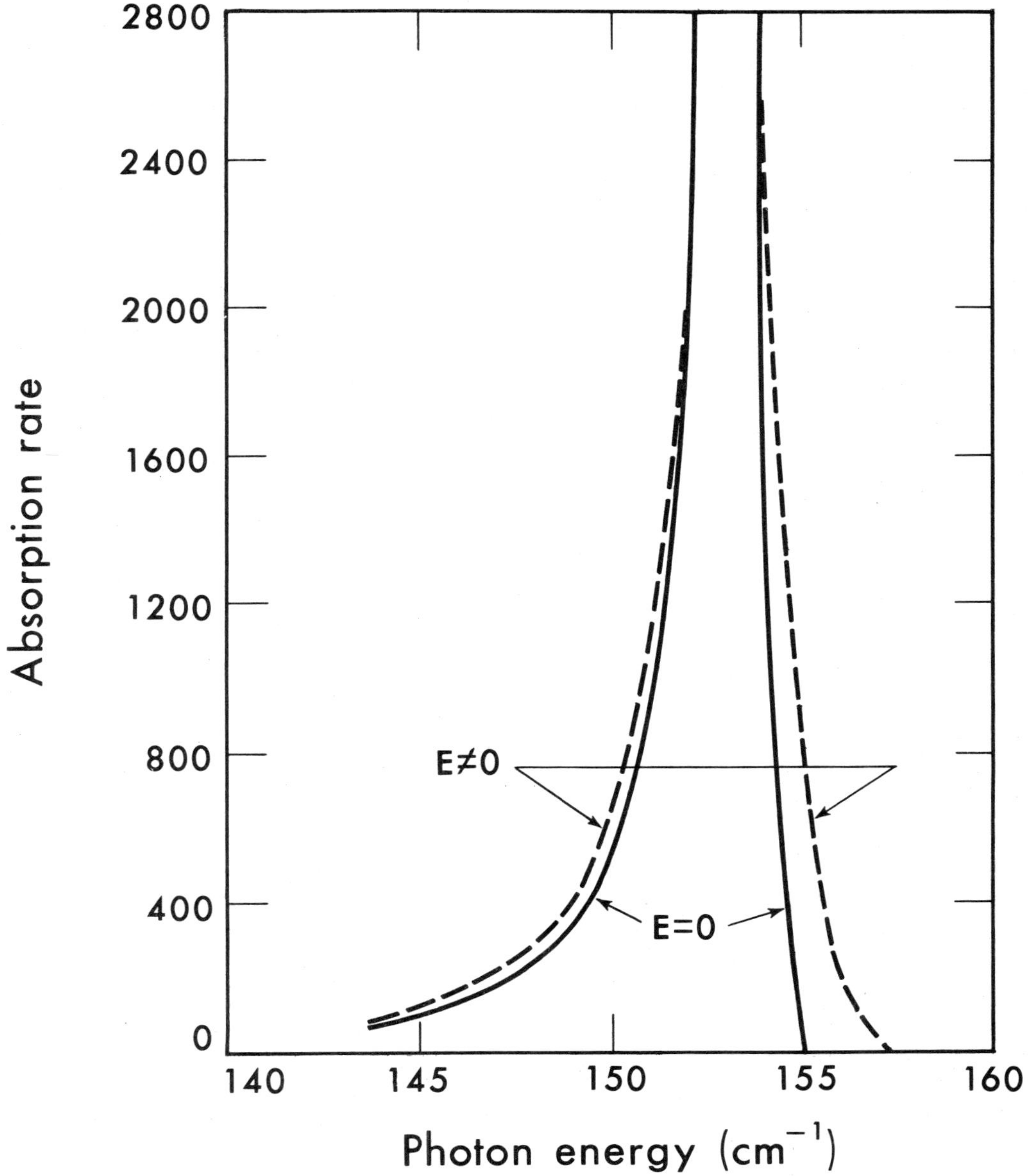

Fig. 1. Stark effect for infrared electric field parallel to c-axis in FeF_2. The d.c. field is normal to the c-axis.

$$E_{dc} \sim \frac{1\ \mathrm{cm}^{-1}}{10^{-20}\ \mathrm{erg/statvolt/cm}} \sim 10^{6}\ \mathrm{volts/cm}$$

RESONANCE RAMAN SCATTERING

In Raman scattering experiments one can investigate the scattering rate as a function, not only of energy transfer to the crystal, but also as a function of the incident photon energy. This should make it possible to check for the existence of resonances in A , B , C , D , F which are predicted by various Raman scattering mechanisms. To illustrate, we consider the case of the exchange strictive mechanism. We choose this mechanism to consider in detail because a) it has not been treated in detail at all before and b) the phonon frequencies and hence the resonance frequencies predicted are well known[8] and fall in a relatively convenient region of the spectrum. (The resonances for the other two mechanisms are in the ultraviolet.) In fact, the resonances turn out to be quite large, and one might hope to observe them even if the exchange strictive mechanism is not dominating the scattering off resonance.

To consider the exchange strictive mechanism, we take account of the dependence of J in Eq. (6) on the displacements $\underline{X}_{mc}$ of all the ions (nonmagnetic as well as magnetic) in the crystal. Here m labels the unit cell and c labels the ion within the unit cell. For the Raman scattering arising from exchange striction we need the second order term arising from expanding J as a function of the $\underline{X}_{mc}$:

$$\mathcal{H}^{(2)}_{\text{phon-spin}} = \frac{1}{2} \sum_{\ell,\ell'} \sum_{b,b'} \sum_{m,m'} \sum_{c,c'} \left[\underline{X}_{mc} \cdot \left\{ \nabla_{\underline{X}_{mc}} \nabla_{\underline{X}_{m'c'}} J_{\ell,b;\ell',b'} \right\}_{0} \cdot \underline{X}_{m',c'} \right] \underline{S}_{\ell,b} \cdot \underline{S}_{\ell',b'} \tag{12}$$

where the sums on ℓ, ℓ' are over unit cells and those on b, b' are on magnetic ions in the unit cells. The derivatives on J are evaluated at $\underline{X}_{mc} = \underline{X}_{m'c'} = 0$. We expand

$$\underline{X}_{m,c} = \sum_{\underline{k}',\mu} \left[\underline{C}^{(c)}(\mu,\underline{k}')\, e^{-i\underline{k}'\cdot \underline{X}^{(o)}_{mc}} A^{(\mu)}_{\underline{k}'} + \text{h. c.} \right] \tag{13}$$

where $A^{(\mu)\dagger}_{\underline{k}'}$ is a creation operator producing a phonon of symmetry type μ and wave vector $\underline{k}'$. The coefficients $\underline{C}^{(c)}(\mu,\underline{k}')$ are determined by solving the normal mode problem for the crystal and some relevant coefficients for this problem are determined by symmetry and are listed in Ref. [5]. To work with (12), we transform it to spin wave variables using (2) and neglecting the $v_{\underline{k}}$ which play no essential role here. The result is

$$\mathcal{H}^{(2)}_{\text{phon-spin}} = \sum_{\underline{k}} \sum_{\nu,\nu'} \Big\{ \Big[\mathfrak{M}^{(1)}_{\nu\nu'}(\underline{k})\, A^{(\nu)\dagger} A^{(\nu')} + \mathfrak{M}^{(2)}_{\nu\nu'}(\underline{k})\, A^{(\nu)\dagger} A^{(\nu')\dagger} + \mathfrak{M}^{(1)*}_{\nu\nu'}(-\underline{k})\, A^{(\nu)} A^{(\nu')\dagger} + \mathfrak{M}^{(2)*}_{\nu\nu'}(-\underline{k})\, A^{(\nu)} A^{(\nu')} \Big] c_{\underline{k}}\, d_{-\underline{k}} + \text{h. c.} \Big\} \tag{14}$$

where

$$\mathfrak{M}^{(2)}_{\nu\nu'}(\underline{k}) = \frac{N}{2} \sum_{\underline{\delta}} e^{-i\underline{k}\cdot\underline{\delta}} \sum_{c,c'} \Big[\underline{C}^{(c)}(\nu,o) \sum_{m,m'} (\nabla_{\underline{X}_{m,c}} \nabla_{\underline{X}_{m',c'}} J_{i,5\,;\,i+\underline{\delta},\,6}) \cdot \underline{C}^{(c')}(\nu',o) \Big] \tag{15}$$

$$\mathfrak{M}^{(2)}_{\nu\nu'}(\underline{k}) = \text{same with } \underline{C}^{(c)}(\nu,o) \rightarrow \underline{C}^{(c)*}(\nu,o)$$

Here the sum on $\underline{\delta}$ is over nearest magnetic neighbors on the opposite sublattice and the labels on the unit cell ions are those of Ref. [5].

The Raman scattering mechanism we consider is that in which the incident photon couples to an infrared active phonon which in turn couples to two spin waves and another infrared active phonon through (12). The second infrared active phonon couples in turn to the outgoing photon. In addition to (12) we therefore also need the phonon-photon coupling. The wavelength of the photons of interest in the resonance region is much longer than the lattice spacing, so we restrict attention to the infrared active phonons at the center of the zone. Then the phonon-photon coupling of interest is [5].

$$\mathcal{H}_{\text{phon-phot}} = \sum_{\nu=E_u(1)_1 \cdot\cdot E_u(6),\, A_{2u}} \sum_{\lambda} \Big[\mathfrak{M}^{*}(\underline{k}, \nu, \lambda)\, (A^{(\nu)} a_{\underline{k}\lambda} + A^{(\nu)} a^{\dagger}_{\underline{k}\lambda}) + \text{h. c.} \Big] \tag{16}$$

where

$$\mathfrak{M}^*(\underline{k}, \nu, \lambda) =$$

$$\Bigg[\Big(\delta_{\nu, A_{2u}} \, \delta_{\lambda, z} \, h\omega_{A_{2u}} \sum_b Z_b \, C_z^{(b)} \, (A_{2u}, 0)\Big)$$

$$+ \sum_{i=1}^{6} \Big(\delta_{\nu, E_u(i)} \, \delta_{\lambda, x} \, h\omega_{E(i)} \sum_b Z_b \, C_x^{(b)} \, (E\,(i), 0)\Big) \qquad (17)$$

$$+ \sum_{i=1}^{6} \Big(\delta_{\nu, E_u(i)} \, \delta_{\lambda, y} \, h\omega_{E_u(i)} \sum_b Z_b \, C_y^{(b)} \, (E_u(i), 0)\Big) \Bigg] \; x$$

$$x \; i \left(\frac{2\pi N}{h\omega_{\underline{k}\lambda \, \Omega}} \right)^{1/2}$$

Here we label the infrared active phonons at the zone center by their usual symmetry labels A_{2u}, E_u (1), . . . , E_u (6) the last six being three pairs of degenerate modes excited when the electric field is normal to the c-axis and the first being the only mode excited when the electric field is along the c-axis. λ gives the polarization of the photon and takes values x, y, z and $a_{\underline{k}\lambda}$ is a creation operator for a photon with wavevector $\underline{k}$ and frequency $\omega_{\underline{k}\lambda}$. N is the number of unit cells and Ω is the volume per unit cell.

To calculate the Raman scattering amplitude we do standard time dependent perturbation theory in third order with the perturbations (14) and (16), having as zero order Hamiltonian the free photon, phonon and spin wave fields. The result for the Raman scattering rate is

$$\mathfrak{w} \left(\left\{ \begin{matrix} \text{photon at} \\ \hbar\omega_{\underline{q}\lambda} \end{matrix} \right\} \rightarrow \left\{ \begin{matrix} \text{photon at } \hbar\omega_{\underline{q}'\lambda'} \\ + \text{ spin waves at } \underline{k}, -\underline{k} \end{matrix} \right\} \right) =$$

$$= \frac{2\pi}{\hbar} \; | \, M(0, \lambda; \underline{k}, -\underline{k}, 0, \lambda') \, |^2 \; \delta(\hbar\omega_{q\lambda} - \hbar\omega_{q'\lambda'} - 2\hbar\omega_{\underline{k}}^{s.\,w.})$$

where

$$M(o, \lambda; \underline{k}, -\underline{k}; o, \lambda') =$$

$$= \sum_{\nu, \nu'} \Bigg\{ \mathfrak{M}^{*}(q, \nu, \lambda)\, \mathfrak{M}^{(1)}_{\nu\nu'}(\underline{k})\, \mathfrak{M}^{*}(\underline{q}', \nu', \lambda') \times$$

$$\Bigg[\frac{1}{(\hbar\omega^{\nu} + \hbar\omega^{\nu'} + 2\hbar\omega^{sw}_{\underline{k}})\,(\hbar\omega_{q\lambda} + \hbar\omega^{\nu})} +$$

$$\frac{1}{(\hbar\omega^{\nu} + \hbar\omega^{\nu'} + 2\hbar\omega^{sw}_{\underline{k}})\,(-\hbar\omega_{q\lambda} + (\hbar\omega^{\nu} + 2\hbar\omega^{sw}_{\underline{k}}))} \Bigg]$$

$$+ \mathfrak{M}(\underline{q}, \nu, \lambda)\, \mathfrak{M}^{(1)}_{\nu\nu'}(\underline{k})\, \mathfrak{M}(\underline{q}', \nu', \lambda') \times$$

$$\times \Bigg[\frac{}{(\hbar\omega_{q\lambda} + (-2\hbar\omega_{\underline{k}} + \hbar\omega^{\nu}))\,(\hbar\omega^{\nu} + \hbar\omega^{\nu'} + 2\hbar\omega^{sw}_{\underline{k}})}$$

$$+ \frac{1}{(-\hbar\omega_{q\lambda} + \hbar\omega^{\nu})\,(-2\hbar\omega^{sw}_{\underline{k}} + \hbar\omega^{\nu} + \hbar\omega^{\nu'})} \Bigg] +$$

$$\mathfrak{M}(\underline{q}, \nu, \lambda)\, \mathfrak{M}^{(1)}_{\nu\nu'}(\underline{k})\, \mathfrak{M}^{*}(q, \nu', \lambda') \times$$

$$\times \frac{1}{(\hbar\omega_{\underline{q}\lambda} - \hbar\omega^{\nu})\,(\hbar\omega_{q\lambda} - (\hbar\omega^{\nu'} + 2\hbar\omega^{sw}_{\underline{k}}))}$$

$$+ \mathfrak{M}(\underline{q}, \nu, \lambda)\, \mathfrak{M}^{(1)}_{\nu\nu'}(\underline{k})\, \mathfrak{M}^{*}(\underline{q}, \nu', \lambda') \times$$

$$\times \frac{1}{(\hbar\omega_{\underline{q}\lambda} + (-2\hbar\omega^{sw}_{\underline{k}} + \hbar\omega^{\nu'}))\,(\hbar\omega_{q\lambda} + \hbar\omega^{\nu})} \Bigg] \Bigg\}$$

Here $\hbar\omega_{\underline{k}}^{sw}$ is the spin wave energy (7) and $\hbar\omega^\nu$ is the energy of a phonon with symmetry ν. Inspection of Eq. (19) shows that there are resonances in the second, fourth and fifth denominators at the photon energies $\hbar\omega_\nu + 2\hbar\omega_{\underline{k}}^{sw}$, $\hbar\omega^{\nu'}$ and $\hbar\omega^\nu$ and $\hbar\omega^{\nu'} + 2\hbar\omega_{\underline{k}}^{sw}$ respectively. The resonances at $\hbar\omega^\nu$ will be difficult to deal with experimentally because a big direct absorption is expected which will heat up the crystal and reduce the scattering. We therefore concentrate attention on the resonances at $\hbar\omega^\nu + 2\hbar\omega_{\underline{k}}^{sw}$ and $\hbar\omega^{\nu'} + 2\hbar\omega_{\underline{k}}^{sw}$. The photon energy ranges at which these resonances occur are known for several materials for which both the spin wave spectrum and the infrared active phonons have been experimentally determined. We show them for various polarizations of the incoming and outgoing light for the case of FeF_2 in Table II.

TABLE II

Range of Photon Energies for Exchange
Strictive Raman Scattering Resonance in FeF_2

E_{in}	E_{out}	Range of Resonance (cm^{-1})
PARALLEL TO C-AXIS	PARALLEL TO C-AXIS	546-594
PARALLEL TO C-AXIS	NORMAL TO C-AXIS	546-594 519-567
NORMAL TO C-AXIS	PARALLEL TO C-AXIS	426-474 306-354
NORMAL TO C-AXIS	NORMAL TO C-AXIS	519-567 426-474 306-354

For illustrative purposes we calculate the form of the resonant scattering rate for FeF_2 for the situation in which the incoming and outgoing polarizations are both parallel to the c-axis of the crystal. Only the A_{2u} phonon is involved and the matrix element is determined from symmetry to have the form

$$\text{constant} \times \left(\cos\left(\frac{k_x a}{2}\right) \cos\left(\frac{k_y a}{2}\right) \cos\left(\frac{k_z c}{2}\right) \right)$$

in agreement with the result of Loudon and Fleury[9]. It is interesting to note however that the k-dependence arising from the energy denominators in (19) has been omitted in the purely phenomenological treatment. The total Raman scattering rate given by

$$\mathfrak{w}\,(\hbar\Delta\omega = \hbar\omega_{q\lambda} - \hbar\omega_{q',\lambda'}\,,\ \hbar\omega_{q\lambda})$$

$$= \sum_{\underline{k}} \{\text{ W of equation (18) }\}$$

can be calculated numerically using (7), known properties of the FeF_2 spin wave spectrum and the fact that $\hbar\omega_A = 440\ \mathrm{cm}^{-1}$ for FeF_2. With $\hbar\omega_{q\lambda}$ in the region of the resonance we get the results shown in Fig. 2. The intensity is 10^4 to 10^6 times larger than that anticipated away from the resonance and the line shape is completely different from that expected for nonresonant two spin wave scattering. Similar calculations can easily be made for other polarizations.

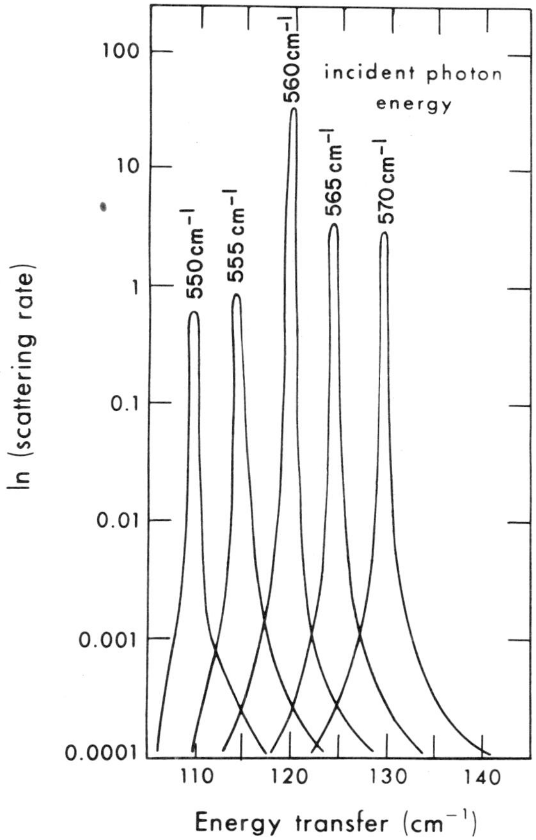

Fig. 2. Resonance Raman scattering for ingoing and outgoing photons polarized along the c-axis in FeF_2.

CONCLUSIONS AND REMARKS

The two possible experiments discussed here would particularly contribute to a conclusion of the question of whether exchange striction is involved in the two spin wave processes. Caution would be needed in interpretation of negative results, however, since the exchange strictive Raman mechanism is of higher order in the displacements than the corresponding infrared mechanism. If a Stark effect can be observed, then the interesting possibility of seeing an electric field induced phase transition might arise.

REFERENCES

1. R. Loudon, Advances in Physics 17, 243 (1968).
2. J.W. Halley, Phys. Rev. 149, 423 (1966) and reference to Tanabe et al. contained therein. Also discussed in Ref. [1].
3. S. Allen, R. Loudon and P. Richards, Phys. Rev. Letters 16, 463 (1966).
4. R.J. Elliott and J.M. Thorpe, preceding paper.
5. J.W. Halley, Phys. Rev. 154, 458 (1967).
6. Similar resonances have been observed in semiconductors.
7. C. Kittel, "Quantum Theory of Solids," J. Wiley and Sons Inc., New York, 1963.
8. M. Balanski et al., J. Chem. Phys. 44, 940 (1966).
9. R. Loudon and Fleury, Phys. Rev. 166, 514 (1968).

C-4: RAMAN SCATTERING IN FERROMAGNETIC $CdCr_4Se_4$

Günther Harbeke and Edgar F. Steigmeier
Laboratories RCA Ltd.
Zürich, Switzerland

$CdCr_2Se_4$ has the spinel crystal structure and is a ferromagnetic semiconductor with a Curie temperature $T_c = 129.5°K$[1]. The band gap which is about 1.3 eV at room temperature decreases by 0.2 eV below the Curie-Weiss temperature $\theta = 204°K$ due to magnetic ordering[2].

We have measured[3] the Raman spectra of $CdCr_2Se_4$ at different temperatures and polarization configurations using a He-Ne laser, a double grating spectrometer and a photon counting detection system. A back-scattering arrangement was used since the band gap is smaller than the laser photon energy of 1.96 eV. In the paramagnetic state ($T > 130°K$) we observe Raman scattering due to phonons. The analysis of the normal modes of vibration in the spinel lattice shows that one $\Gamma_1^+(A_{1g})$, one $\Gamma_{12}^+(E_g)$ and three $\Gamma_{25}^+(F_{2g})$ modes are Raman active. From the polarization dependence of the two strongest Raman lines at 154 and 239 cm^{-1} we conclude that they are of Γ_{12}^+ and Γ_1^+ symmetry, respectively.

At temperatures below the Curie point a new line appears at 168 cm^{-1} which is weakly present also somewhat above T_c. The line does not show any observable shift with temperature and has a half-width less than our instrumental resolution of 4 cm^{-1}. The intensity of this line, normalized to the intensity of the 154 cm^{-1} phonon line, increases strongly with decreasing temperature and resembles very closely the temperature dependence of the spin correlation function. This line obviously is related to the magnetic properties of the material and is interpreted as due to scattering from magnetic excitations.

The spin wave spectrum of $CdCr_2Se_4$ having four magnetic atoms per unit cell (magnetic space group $I\,4_1/a\,d'm'$) consists of one acoustical and three optical branches. The 168 cm^{-1} line could be caused by either (a) one-magnon scattering involving an optical zone-center magnon or (b) two-magnon scattering involving zone-boundary magnons. The zone-center magnons are characterized by irreducible representations of the unitary subgroup C_{4h} of the magnetic space group. These are Γ_1^+, Γ_2^+ and Γ_3^+, respectively for the three optical branches. Measurements with different polarization configurations of incident and scattered light showed that the components of the Raman scattering tensor of the 168 cm^{-1} line α_{yz} and α_{zy} are unequal but of the same sign. The same relation has been found to hold for α_{xz} and α_{zx}. These results favor the interpretation of the 168 cm^{-1} line as due to an optical zone-center magnon since the obtained scattering tensor components are in agreement with the calculated ones for a Γ_3^+ magnon.

REFERENCES

1. P.K. Baltzer, H.W. Lehmann and M. Robbins, Phys. Rev. Letters 11, 493 (1965).
2. G. Harbeke and H. Pinch, Phys. Rev. Letters 17, 1090 (1966).
3. G. Harbeke and E.F. Steigmeier, Solid State Communications 6, 747 (1968).

C-5: RAMAN SCATTERING FROM LOCALIZED MAGNONS IN Ni^{2+} AND Fe^{2+} DOPED MnF_2 *

A. Oseroff† and P. S. Pershan‡
Division of Engineering and Applied Physics,
Harvard University
Cambridge, Massachusetts

We have studied Raman scattering at low temperatures from MnF_2 doped with Ni^{2+} and Fe^{2+}. Three temperature and polarization dependent lines have been observed for each impurity. For each dopant we have identified one line as a two magnon excitation of the impurity and some linear combination of the host spins. When corrections are made for magnon-magnon interactions, the predicted energy is in excellent agreement with the experimental results.

INTRODUCTION

We have observed Raman scattering from localized magnon impurity modes in Ni^{2+} and Fe^{2+} doped MnF_2. Our measurements complement such other techniques as neutron diffration[1], fluorescence[2], IR absorbtion[3], and ESR[4] that have previously been used to identify localized magnons.

EXPERIMENTAL PROCEDURE

Polarized light from an argon ion laser was focused through oriented single crystals of MnF_2 containing either Ni^{2+} (0.13% or 0.98%) or Fe^{2+} (0.2% or 2.1% ∮) which were mounted in a helium vapor cooled, variable temperature dewar. Light scattered by

*This work was supported in part by the Advance Research Project Agency, and by the Division of Engineering and Applied Physics, Harvard University.
†National Science Foundation Predoctoral Fellow.
‡Alfred P. Sloan Foundation Fellow.
∮The Fe content of this sample was originally reported to be 0.65%. Further analysis has shown that the impurity distribution is quite inhomogeneous, and that the portion of the crystal transversed by the laser beam actually contains about 2.1% Fe.

90° was analyzed with a Spex double monochrometer and detected by photon counting techniques. The experimental geometry was chosen to allow the study of different components of the Raman scattering tensor α_{ij}, and care was taken to avoid depolarization of the scattered radiation by the birefringence of the crystal. Since MnF_2 is not completely transparent at 4880A°, there is a possibility of sample heating in the vicinity of the focussed laser beam. At the lower temperatures the incident power was varied over a 10 - 1 range to guard against such heating; no changes were observed in the Raman energies.

EXPERIMENTAL RESULTS

In addition to the previously reported two magnon peaks in pure MnF_2, [5] we find three extra lines for each of the impurities when $T < .8T_N$. Above this temperature the lines were too broad to be easily measured. In each case we have identified one of these lines as a two magnon excitation of the impurity and the nearest antiferromagnetically coupled host spins. These signals have integrated intensities roughly proportional to impurity concentration, and are about seven to ten times smaller than the host (Mn-Mn) two magnon line. In the Ni^{2+} doped samples they appear at 162.5 cm^{-1} and 165 cm^{-1} for α_{xy} and α_{xz} respectively, while for Fe^{2+} the xy component is at 140 cm^{-1}, and the xz at 143 cm^{-1}. In addition, both systems exhibit a weaker line at 185 cm^{-1} with α_{xx} polarization. The Ni^{2+} doped crystals also give a strong α_{xz} polarized line at 26.5 cm^{-1}, while the Fe^{2+} samples give a line at 164 cm^{-1} with α_{xx} and α_{xz} polarizations. The uncertainty in the position of the lines is about $\pm$ 1 cm^{-1}. Representative spectra are shown in Fig. 1.

DISCUSSION

If one considers an impurity spin at the body center of a rutile type lattice and assumes that it interacts principally with nearest antiferromagnetically coupled neighbors, it is evident that among these nine spins there are nine possible independent spin excitations, consisting of linear combinations of spin deviations on the impurity and on the eight neighbors. These linear combinations are most conveniently taken so as to transform as the irreducible representations of the impurity site point group[6, 7, 8]. The combinations are not necessarily single frequency eigenmodes corresponding to elementary excitations of the impurity-host spin wave system. In some cases one has true single frequency local modes, which lie outside the spinwave manifold for the host lattice. Often, however, one has "resonance modes" which can be approximately represented as "damped" eigenmodes, the damping resulting from the resonant interaction between the impurity centered excitation and the host magnons. As the damping increases, for example at frequencies near those of zone edge magnons, even this approximation fails and one can't really speak of either "local" or "resonance" modes.

In an analogous situation to the vibrational impurity problem, Green's function calculations show that when the impurity-host exchange J' and the impurity spin S' are sufficiently different from the intrinsic exchange and spin of the host, one of the excitations, conventionally denoted s_o, is a true local mode with a well-defined energy

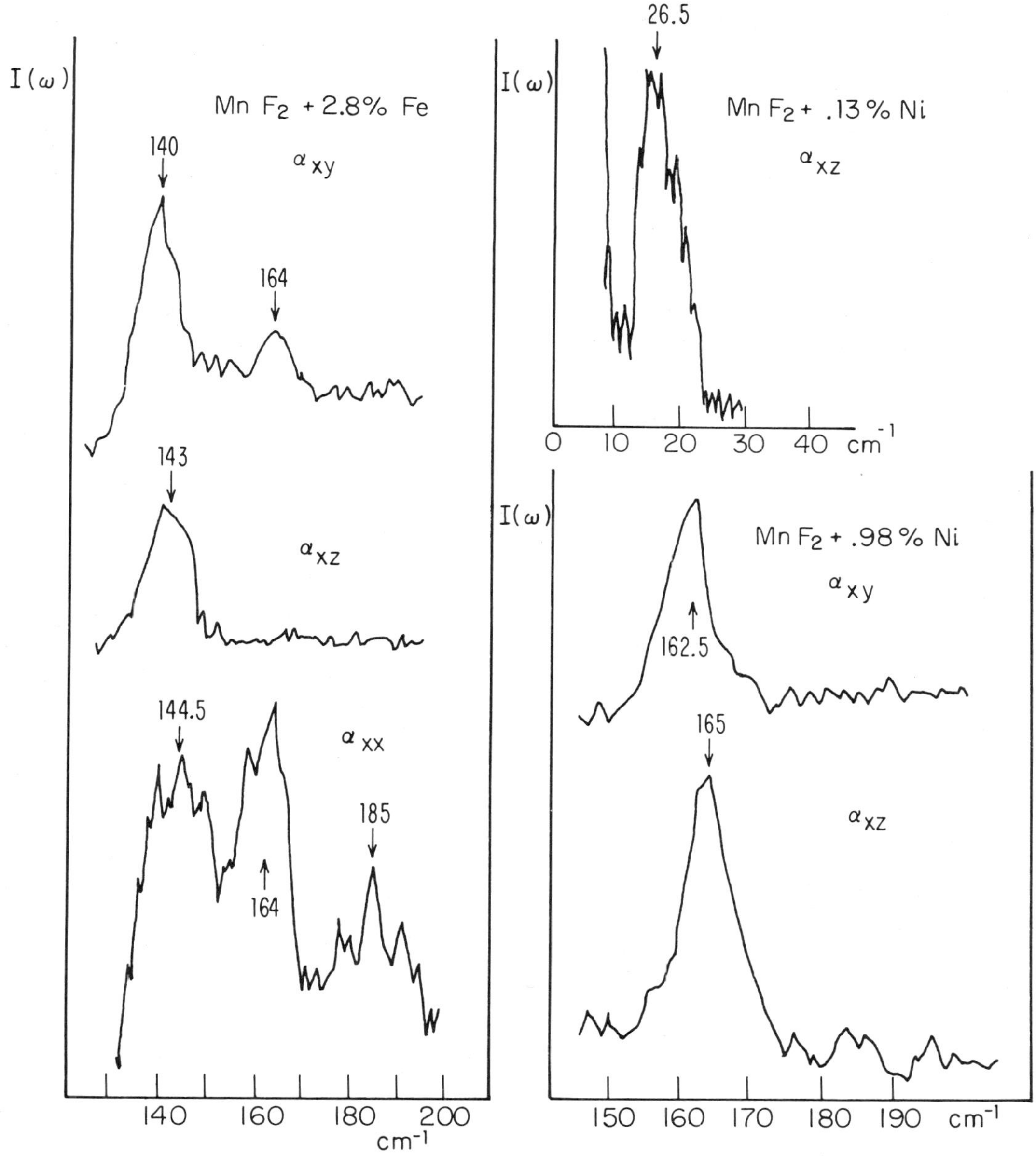

Fig. 1. Typical Raman spectra from Fe^{2+} and Ni^{2+} doped crystals at 10° K. Resolution is about 3 cm^{-1}. All intensities are arbitrary. The α_{xy} and α_{xz} Fe lines have about twice the intensity of the corresponding Ni lines.

that lies well outside the host-spin manifold, and with a spin wave amplitude localized primarily on the impurity site[6, 7]. In this case, the s_o mode is well approximated by a molecular field model in which the impurity spin processes in the effective magnetic field 16J' S of the eight antiferromagnetically coupled neighbors[2, 3]. With MnF_2 as a host, the Ni^{2+} and Fe^{2+} s_o modes have been observed at 120.4 [2] and 94.5 [3] cm^{1}, compared with the 50.4 cm^{-1} (X-point)/548 cm^{-1} (Z-point) boundary of the Brillouin zone for the host spin waves. The other eight excitations are basically linear combinations of the host spins. Only one of these eight has any admixture of the impurity spin[6, 7]. Green's function calculations demonstrate that for MnF_2 the major spectral content of these excitations occurs very near to energies corresponding to the Brillouin zone boundary for the pure crystal.

By analogy with the known Raman spectra from pure MnF_2[5, 9] we might anticipate either one or two magnon scattering from the various impurity excitations. However, we find no appreciable scattering at the s_o mode energies or within ± 30% of the Brillouin zone edge. This apparent absence of one magnon scattering is not completely unexpected, since previous experimental and theoretical work indicates a rather small cross section for the process[9, 10]. *

If the same excited state exchange mechanism holds in the doped as in the pure crystal, we would expect two magnon impurity scattering to be due to adjacent spin deviations on opposite sublattices. The excitation would consequently involve the s_o and one of the Mn^{2+} modes. The simplest model for this process ignores the details of the various Mn^{2+} modes and arrives at the energy of the two magnon line by summing the s_o energy and the exchange plus anisotropy energy necessary to excite one near neighbor Mn spin. Since the latter excitation corresponds approximately to the magnon energy at the Brillouin zone boundary, and since different points on this non-cubic boundary have different energies, we also expect small shifts in the two magnon spectrum for different polarizations of the incident and scattered light. From this model we would compute the Ni-Mn and Fe-Mn two magnon lines to be centered in the vicinity of 170-175 cm^{-1} and 146-151 cm^{-1} respectively, depending on which polarization components are involved.

Because the model ignores magnon-magnon interactions, the calculated values are too large. These effects are quite important, since we are describing the simultaneous creation of two spin waves that are physically near to one another[12]. We can obtain a rough estimate of the size of these interactions by considering an Ising calculation of of the energy difference between the creation of a pair of adjacent and a pair of distant magnons.

The creation of simultaneous spin deviations on an impurity and on a distant host requires an antiferromagnetic exchange energy of $2nS(J+J')$, while a pair of deviations on neighboring spins requires $2nS(J+J') - 2(JS-J'S') - 2J'$. The difference between these expressions is $-2(JS-J'S') - 2J'$. The first term just accounts for the fact that the Mn ion has the impurity and seven host spins as neighbors rather than eight host spins, while the second term gives the magnon-magnon interaction energy. In MnF_2, $J = -1.24$ and $S = 5/2$, while $J'_{Ni-Mn} = 3.11$ [2] and $S'_{Ni} = 1$. J_{Fe-Mn} has been estimated to be equal to -1.9 [3], while $S'_{Fe} = 2$. With these values, we compute the magnon-magnon correction to be 6.2 cm^{-1} for Ni^{2+} and 3.2 cm^{-1} for Fe^{2+}. The corresponding

*The s_o mode in Fe: MnF_2 has since been observed at 95 cm^{-1}.

two magnon modes should then occur at 163.8 cm^{-1} and 168.8 cm^{-1} for Mn^{2+} magnons at X and Z points in the Ni^{2+} doped crystals; and at 142.6 cm^{-1} and 147.6 cm^{-1} in the Fe^{2+} doped crystals. The calculated positions are in good agreement with the experimental data, particularly for the α_{xy} Ni^{2+} line.

In all cases the estimated energies are higher than those which are actually observed. This is to be expected, since we have approximated the peaks of the spectral distributions of the Mn^{2+} excitations by zone edge magnons, although they actually lie a few cm^{-1} inside the Brillouin zone.

Additional confirmation of the two magnon assignment is provided by the temperature shift of the two impurity lines (shown in Fig. 2 for the α_{xy} component) which closely follows that of the Mn-Mn two magnon line.

A more sophisticated model that includes the details of the Mn modes can be obtained from a Green's function calculation that accounts for the nonlinear magnon-magnon interaction in terms of the actual excitations. For such a calculation to have any quantitative significance it has to also include the anisotropic intra-sublattice ferromagnetic coupling. These computations are in progress, and will be reported in the future.

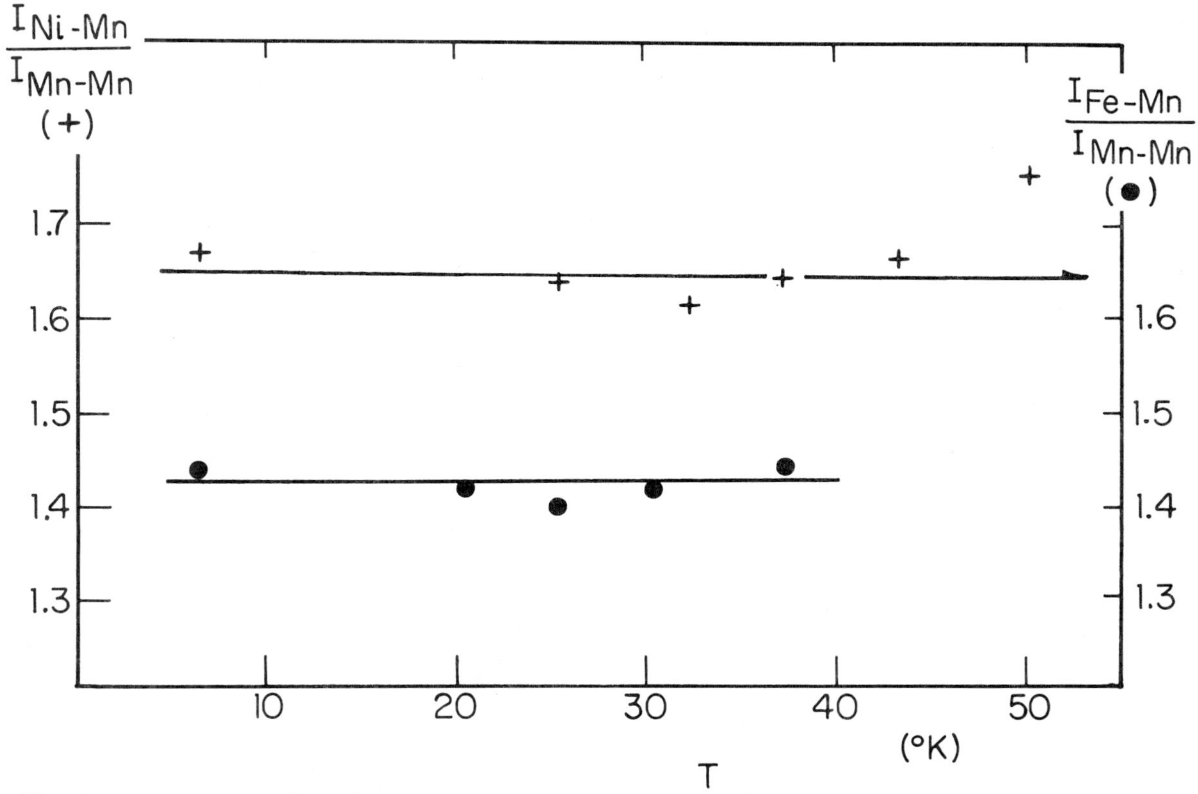

Fig. 2. Temperature dependence of the ratios of the energies of the xy components of the Ni-Mn/Mn-Mn (+) and the Fe-Mn/Mn-Mn (●) lines. The Fe-Mn line was much narrower at higher temperatures than the Ni-Mn line. It could not be followed above 40°K because it merged with the very broad 164 cm^{-1} line.

Independently of the Green's function results, we can draw some conclusions about the symmetry of the scattering tensors purely from group theoretical considerations. Let us consider only those two magnon combinations which result from the simultaneous excitation of the local s_o mode and one of the other eight impurity-centered modes. On the basis of previous work[2, 9] we can argue that the dominant Raman activity results from only those four of the eight possible combinations which have even parity. Employing the existing notation[7], these have approximately s, d_{xy}, d_{xz}, and d_{yz} like symmetry. The latter two (i. e. s_o-d_{xz} and s_o-d_{yz}) each contribute to both α_{xz} and α_{yz}, while the first two contribute to α_{xx}, α_{yy}, and α_{xy}. In addition, s_o-s_1 contributes to α_{zz}. Thus xy polarized scattering can be due to both s_o-s_1 and s_o-d_{xy} combinations. The two contributions can be separated by studying the anisotropy of the observed Raman spectra. Let x' and y' be two axes, perpendicular to the c axis of MnF_2. If x' makes an angle θ with the (100) crystal axis, one can obtain an expression for $\alpha_{x'y'}$ as a function of θ. The s_o-s_1 contribution to $\alpha_{x'y'}$ is independent of θ while the s_o-d_{xy} contribution is of the form $\alpha_{x'y'}(s_o\text{-}d_{xy}) - \alpha \sin(2(\theta - \theta_o))$, where θ_o is some angle that can be determined by symmetry arguments alone. We are in the process of making these measurements.

The origin of the 185 cm^{-1} α_{xx} line is not understood at this time, nor is it clear why the Fe^{2+} doped sample has α_{xx} and α_{xy} polarized lines at 164 cm^{-1}. They do not appear to be related to the lines of similar energy that occur in Ni^{2+} doped MnF_2, since in the latter samples we find an α_{xz}, but no significant α_{xx} component. Also, the Fe^{2+} doped crystals have been analyzed and found to contain less than 0.02% Ni.

The 26.5 cm^{-1} line in the Ni^{2+} doped crystals has the proper polarization components to be a one magnon line, but it does not seem to shift with temperature. Instead, it simply broadens with increasing T, becoming unmeasurable around 40°K. In addition, its intensity does not appear to increase with increasing impurity concentration, as do the 140/143 cm^{-1} (Fe^{2+} : MnF_2) and 162.5/165 cm^{-1} (Ni^{2+} : MnF_2) lines. Measurement of the possible antisymmetry of the scattering tensor for this line has been precluded so far by the small size of the available samples. This measurement should be done since a one magnon line would have an antisymmetric scattering tensor ($\alpha_{xz} + \alpha_{zx} = 0$).

ACKNOWLEDGMENT

We would like to acknowledge the free and open discussion with Dr. R. E. Dietz of pre-publication data which were obtained, completely independently of our own, in Professor Balkanski's laboratory at La Faculté des Sciences de Paris. We also wish to thank Dr. L. R. Walker for the information in Ref. [6], and for graciously making tables of the pure crystal Green's function available prior to publication

REFERENCES

1. W.J.L. Buyers, R.A. Cowley, T.M. Holden, and R.W. Stevenson, J. App. Phys. 39, 1118 (1968).
2. L.F. Johnson, R.E. Dietz, and H.J. Guggenheim, Phys. Rev. Letters 17, 13 (1966); A. Misetich and R.E. Dietz, Phys. Rev. Letters 17, 392 (1966).
3. R. Weber, Phys. Rev. Letters 21, 1260 (1968).
4. M. Motokawa and M. Date, J. Phys. Soc. Japan 23, 1216 (1967).
5. P.A. Fleury, S.P. Porto, and R. Loudon, Phys. Rev. Letters 18, 658 (1967).
6. T. Tomegawa, "Thesis," Osaka University Graduate School of Science, Japan, 1968 (unpublished).
7. S.E. Lovesay, J. Phys. Chem. (Proc. Phys. Soc.) 1, 102 (1968); 1, 118 (1968).
8. H. Callen, D. Hone, and L.R. Walker (to be published).
9. P.A. Fleury and R. Loudon, Phys. Rev. 166, 514 (1968).
10. See, for example, R.J. Elliott and R. Loudon, Phys. Letters 3, 189 (1963); Y.R. Shen and N. Bloembergen, Phys. Rev. 143, 372 (1966); T. Moriya, J. Phys. Soc. Japan 23, 490 (1967); P.S. Pershan, J. App. Phys. 38, 1482 (1967).
11. Y. Tanabe, T. Moriya, and S. Sugano, Phys. Rev. Letters 15, 1023 (1965).
12. R.J. Elliott, M.F. Thorpe, G.F. Inbush, R. Loudon, and J.B. Parkinson, Phys. Rev. Letters 21, 147 (1968).

C-6: OBSERVATION OF LOCALIZED MAGNONS BY RAMAN SCATTERING AND FAR INFRA-RED ABSORPTION IN NI DOPED MnF_2

P. Moch and G. Parisot
Laboratoire de Physique des Solides de la Faculté des Sciences*
Ecole Normale Superieure, Paris, France
and
R. E. Dietz and H. J. Guggenheim
Bell Telephone Laboratories, Incorporated
Murray Hill, New Jersey

INTRODUCTION

Localized magnons have already been identified employing various experimental techniques as fluorescence[1] or neutron scattering studies[2]. However, their observation by Raman scattering or far infra-red absorption has not yet been reported. We present below experimental results showing evidence of both processes in Ni doped MnF_2.

In antiferromagnetic materials, besides the well known antiferromagnetic resonance absorption which is of magnetic dipole character, one may also find 2 magnon absorption[3-4] of electric dipole character arising from the combined effects of the electric dipole coupling and the exchange interaction[5-7]. On the other hand, Raman scattering of light by magnons occurs by an indirect electric dipole interaction via a spin-orbit coupling when it involves a one magnon excitation and via an exchange interaction when it involves 2 magnons[8-9]. The latter process was found to be generally more efficient in pure materials[9-12].

For Ni doped MnF_2, a localized magnon mode due to the spin deviation of a Ni ion surrounded by 8 Mn second neighbors has been found at about 120 cm^{-1}[1], while impurity modes involving spin deviations of the second neighbor Mn ions lie near 50 cm^{-1}[13]. For a simultaneous excitation of these modes, one expects a frequency of approximately 170 cm^{-1}. Our experimental results, as discussed below, concern such a process.

EXPERIMENTAL TECHNIQUES AND RESULTS: RAMAN SCATTERING

We employed a double grating CODERG monochromator associated with a linearly polarized Argon laser providing between 30 and 100 mW of 4880 Å radiation. It was

*Work supported in part by funds supplied by D.R.M.E.

necessary to cool the photocathode of an EMI 9558 photomultiplier by flowing cold Nitrogen and to reduce its effective diameter with a permanent magnet. Because of the weakness of the observed signals measured by photon counting, the resolving width was kept slightly smaller than 3 cm^{-1}. In some experiments the sample was immersed in superfluid Helium (T ≃ 2°K). In other cases, a gas exchange cryostat allowed temperature dependence studies from about 8°K. The temperature regulation was generally inside a 0.3°K interval, but its absolute value was probably not known within better than 2°K.

The sample, an X-ray oriented slab, was cut from a 1% Ni doped MnF_2 single crystal. It was possible to study all polarized spectra except α_{zz} (here the first index refers to the incident polarization, the second to the scattered one[10]).

At low temperatures, in addition to the phonon and intrinsic 2 magnon scattering also observed in pure MnF_2, the Ni doped material scatters near 170 cm^{-1} as shown in Fig. 1. At 2°K or 8°K, for the α_{xy} and α_{zx} spectra, the integrated intensities are approximately equal and 5 or 6 times weaker than the corresponding integrated intensities of the intrinsic 2 magnon spectra. An α_{xy} line is centered near 164.5 cm^{-1} with a half-width of 6.5 cm^{-1}. The α_{zx} line is centered near 167 cm^{-1} with a half-width of 7 cm^{-1}. In the α_{xx} spectrum 2 lines appear: the 164.5 line observed in α_{xy} and another line centered at 185.5 cm^{-1} with a half-width of 6 cm^{-1}; both lines have integrated intensities 3 times weaker than those in α_{xy} and α_{zx}.

Temperature dependence measurements were performed for (α_{zx}, α_{zy}) which disappears near T_N (68.3°K); it is still observable at 60°K. 3 effects result from increasing the temperature: a shift towards smaller wavenumber, a broadening and an intensity decrease. The relative frequency $\frac{\omega(T)}{\omega(0)}$, shown on Fig. 2, rapidly decreases above 25°K and reaches 0.8 near 60°K. The half width increases to more than 30 cm^{-1} at 55°K (Fig. 3). Finally, as shown on Fig. 4, the relative integrated intensity, is nearly constant until 35°K, then falling rapidly: at 60°K it is less than 0.2.

EXPERIMENTAL TECHNIQUES AND RESULTS: FAR INFRARED ABSORPTION

Preliminary spectra were recorded using an evacuated grating spectrometer. Since the sample was mounted on the "cold finger" of a Helium cryostat, the crystal temperature was not exactly known but was estimated at 15°K when the cryostat was filled with liquid Helium, and at 85°K when filled with liquid Nitrogen. The light could be polarized and the same mounting allowed reflection and transmission spectra. The resolving width was about 3 cm^{-1}.

Transmission measurements could only be made with E||c because of the very strong absorption of the ordinary ray due to an infra-red "active" phonon near 160 cm^{-1} (as observed in pure MnF_2 at 300°K[14]). Reflectivity measurements for the ordinary polarization at 15°K indicated that the E_{u1} TO phonon mode lies at about 159-160 cm^{-1}.

Fig. 5 shows an absorption band at 15°K for E||c: the absorption maximum is found at 167 cm^{-1}, which is also the frequency of the Raman scattering line in α_{zx}. The maximum absorption coefficient is about 5 times smaller than that of the intrinsic

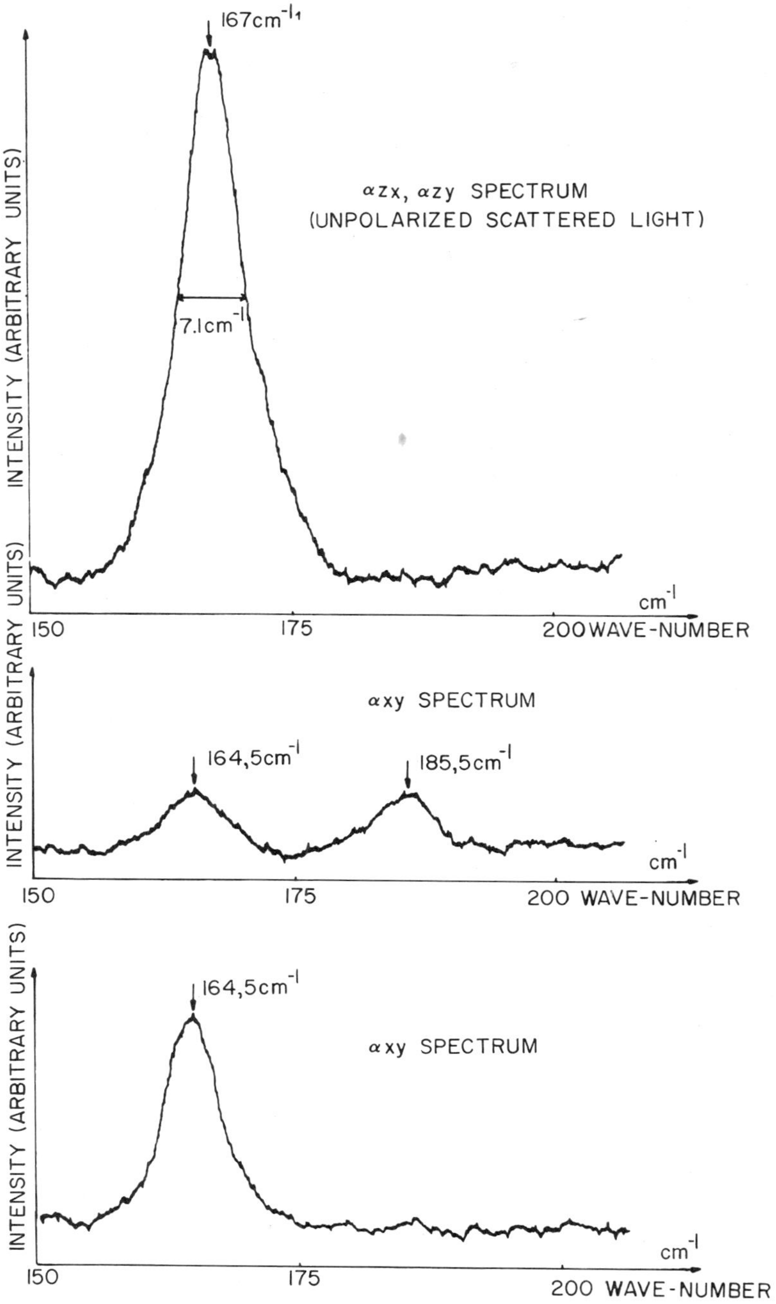

Fig. 1. Experimental Raman scattering spectra in 1% Ni doped MnF_2 at $8^{\circ}K$ between 150 and 200 cm^{-1} for different polarizations. Corrections due to instrumental polarization have not been made on the relative intensities of the different polarizations, but are estimated in the text.

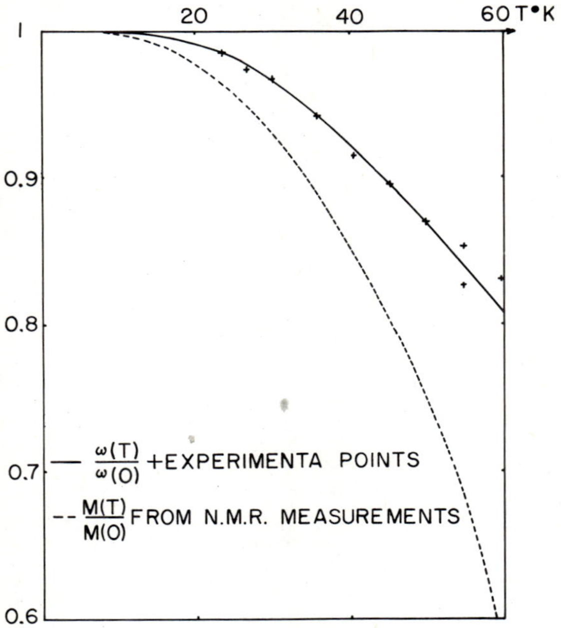

Fig. 2. Temperature dependence of $\frac{\omega(T)}{\omega(O)}$ for the α_{zx} spectrum. Comparison is made with $\frac{M(T)}{M(O)}$.

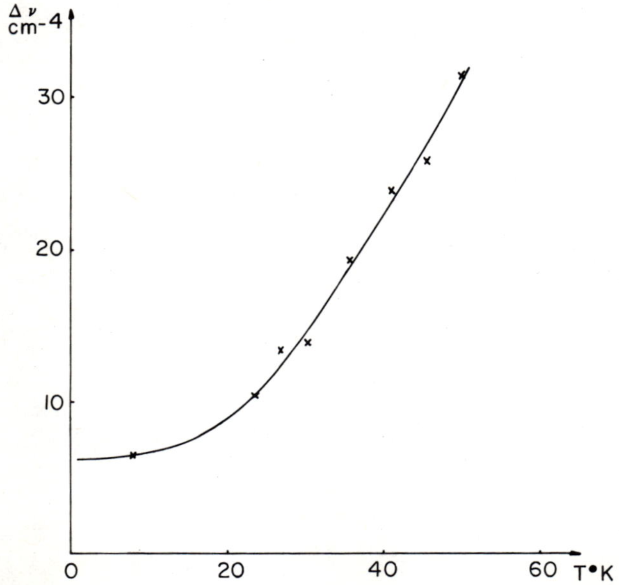

Fig. 3. Temperature dependence of the integrated Raman Intensity for the α_{zx} spectrum.

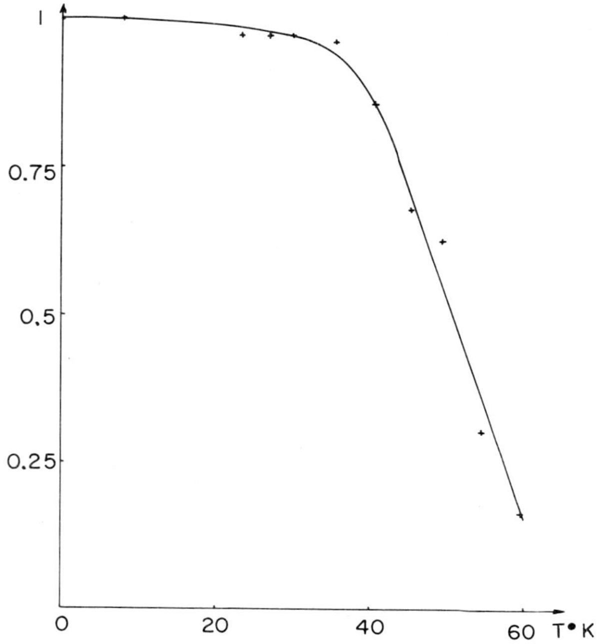

Fig. 4. Temperature dependence of the half width of the α_{zx} line.

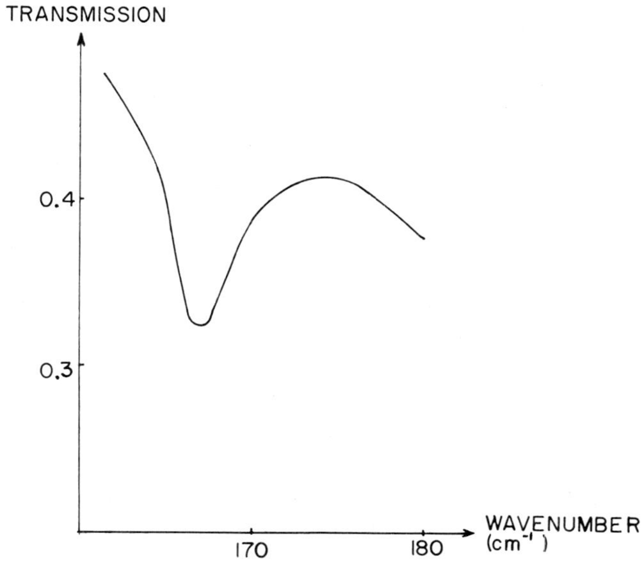

Fig. 5. Experimental far infra-red transmission spectrum in 1% Ni doped MnF_2 at $15^{\circ}K$, near 170 cm^{-1}, light polarization parallel to the Z axis.

2 magnon absorption in the same polarization. With the cryostat filled with liquid Hydrogen, the absorption is not significantly different; when it is filled with liquid Nitrogen, it disappears.

INTERPRETATION

The gross features of the observed spectra (temperature dependence of frequency, intensity, linewidth, and magnitude of the frequency) argue that the Raman scattering and infra-red absorption derive from the processes suggested in the introduction.

In the paramagnetic phase, the site group of the nickel ion is D_{2h}, but in the ordered, antiferromagnetic phase this is lowered to the unitary magnetic site group C_{2h} by the restrictions placed on the anti-unitary time reversal operator. Since Raman scattering proceeds without a change of parity, and since we may take the ground state to be A_g, then the excited states reached by Raman scattering must also be of even parity, and therefore must belong to either A_g or B_g. Polarizations α_{xx} or α_{xy} will involve two magnons having A_g as their direct product, while for α_{zx} or α_{zy} the two magnons will have B_g. Assuming that one of the scattered modes is the s_o (120.4 cm^{-1}) mode observed in fluorescence[15] (this is the only impurity mode residing mainly on the impurity sublattice) which has the representation A_g, then the representations given for the product states are also the representations for the impurity mode on the sublattice opposite to that of the impurity. The s_o mode has been shown to be highly localized on the Ni ion[15]. Similarly, modes coupling to the s_o mode reached by the infra-red absorption must be of odd parity; A_u modes will be active for $E \parallel c$, and B_u for $E \perp c$.

Although we may expect the s_o mode to couple to an infinite number of even parity impurity modes on the other sublattice which may be constructed from the complete set of MnF_2 wavevector states, only those low order modes which have their spin deviations concentrated on the second neighbors to the nickel ion will have appreciable Raman or infrared intensity. Theoretical calculations recently reported for MnF_2[13, 16, 17] show that one s (called s_1), three p, three d, and one f mode can be constructed from the eight second neighbors (these considerations neglect the first neighbor exchange, J_1, and the possible inequivalence of exchange between the impurity ion and the two classes of second neighbors). Of these, only s_1 and the three d modes, d_{xy}, d_{xz}, and d_{yz} have even parity. Since s_1 may also comprise spin deviations on the nickel ion, its energy is expected to be smaller than that of the d modes. The ratio of the energy of the d to that of the s_o mode of Ni in MnF_2 has been estimated[13] by a Green's function calculation (neglecting J_1) to be 0.426. From the experimental value for s_o, we then estimate a value of 51.3 cm^{-1} for the d mode. The 164 cm^{-1} mode (observed in α_{xx}, α_{xy}) transforms as A_g, and we assign it to d_{xy} while the 167 cm^{-1} (α_{zx}) mode transforms as B_g, or d_{xz}, d_{yz}. The splitting between these modes arises from the first neighbor exchange, which affects, again, d_{xy} differently from d_{xz}, d_{yz}. Since the d_{xy}

mode does not depend on J_1, its energy should agree well with the Green's function estimate. The α_{xy} peak frequency should equal the sum of the s_o and d modes.

However, this sum, 171.7 cm^{-1}, is 7.2 cm^{-1} larger than the observed α_{xy} scattering peak. This is to be expected, since we have neglected the interaction between the s_o and d modes which, like that observed in the intrinsic two-magnon scattering[12], is significant. Since a Green's function calculation of the interaction between impurity magnons is not yet available, we present a simpler, Ising model estimate of the s_o - d_{xy} scattering frequency.

If i and j represent second neighbors (on opposite sublattices), the interaction between two Mn ions is $-2J_{MnMn}\underset{\sim}{S}_{i\,Mn}\cdot\underset{\sim}{S}_{jMn}$ while the interaction between a Ni and a Mn ion is $-2J_{NiMn}\underset{\sim}{S}_{iMn}\cdot\underset{\sim}{S}_{jNi}$. For 1 Ni in MnF_2, the most probable situation for a Ni-Mn pair corresponds to the Ni ion surrounded by 8 Mn second neighbors with the Mn ions each surrounded by an additional 7 Mn second neighbors. In the scattering process, $M_{s_{Ni}}$ changes from -1 to 0, while the $M_{s_{Mn}}$ changes from +5/2 to +3/2.

For large molecular fields, the s_o mode energy can be represented[15] by an exchange term and a crystal field term:

$$E_{s_o} = -40J_{NiMn} - D = 120.4\ cm^{-1}$$

where $J_{NiMn} = -3.11\ cm^{-1}$, and $D = 4.05\ cm^{-1}$. The d_{xy} mode created separately from the s_o mode is

$$E_d = -35J_{MnMn} - 2J_{NiMn} = 48.9\ cm^{-1}$$

where J_{MnMn} has been taken[18] equal to $-1.22\ cm^{-1}$. The energy for both modes simultaneously at the same impurity is

$$E_{s_o-d} = -40J_{NiMn} - D - 35J_{MnMn} = 163.1\ cm^{-1}$$

Thus the interaction energy ΔE is

$$\Delta E = E_{s_o-d} - E_{s_o} - E_d = +2J_{NiMn} = -6.22\ cm^{-1}$$

Note that E_d compares favorably with the Green's function result (which includes anisotropy), and E_{s_o-d} is in excellent agreement with the α_{xy} peak at 164.5 cm^{-1}. In fact, if we add[11] to the d-mode energy 0.8 cm^{-1} which corrects for the anisotropy energy of the Mn ions, we obtain 163.9 cm^{-1}, even closer to the experimental value. This agreement may be considered evidence that the d modes are mainly localized on the second neighbors to the nickel impurity as assumed in the above calculation. However, this conclusion applies strictly to the d mode while interacting with the s_o mode since the interaction may significantly contribute to its localization.

It is not possible at this time to rationalize the frequencies of the α_{zx} scattering, or of the infra-red absorption. The odd parity states reached by the absorption may be of p_z or f symmetry, and are known from Green's function calculations[13] to have energies close to the d modes.

Further analysis of the shape and intensity of the scattering lines and infra-red absorption, and of their temperature dependence must await a more sophisticated model. At the present time the 185 cm^{-1} scattering is not understood.

ACKNOWLEDGMENTS

We are grateful to L. R. Walker for many stimulating discussions, and for permission to publish results of calculations in advance of publication. We have also benefitted from discussions with M. Balkanski, A. Misetich, D. Hone, J. P. van der Ziel, and A. Oseroff. It is a pleasure to thank C. Dugautier for technical assistance.

REFERENCES

1. L. F. Johnson, R. E. Dietz, and H. J. Guggenheim, Phys. Rev. Letters 17, 13 (1966).
2. T. M. Holden, R. A. Cowley, W. J. L. Buyers, and R. W. H. Stevenson, Sol. St. Com. 6, 145 (1968).
3. J. W. Halley and I. Silvera, Phys. Rev. Letters 15, 654 (1965).
4. S. J. Allen, R. Loudon, and P. L. Richards, Phys. Rev. Letters 16, 463 (1966).
5. Y. Tanabe, T. Moriya, and S. Sugano, Phys. Rev. Letters 15, 1023 (1965).
6. T. Moriya, J. Phys. Soc. Jap. 21, 926 (1966).
7. R. Loudon, Adv. Phys. 17, 243 (1968).
8. T. Moriya, J. Phys. Soc. Jap. 23, 490 (1967).
9. P. A. Fleury and R. Loudon, Phys. Rev. 166, 514 (1968).
10. P. A. Fleury, S. P. Porto, L. F. Cheesman, and H. J. Guggenheim, Phys. Rev. Letters 17, 84 (1966).
11. P. A. Fleury, S. P. Porto, and R. Loudon, Phys. Rev. Letters 18, 658 (1967).
12. P. A. Fleury, Phys. Rev. Letters 21, 151 (1968).
13. H. B. Callen, D. N. Hone, and L. R. Walker (to be published).
14. G. Parisot, C. Rend., Ac. Sc. Paris 265, 1192 (1967).
15. A. Misetich and R. E. Dietz, Phys. Rev. Letters 17, 392 (1966).
16. S. W. Lovesey, J. Phys. (C) 1, 102 (1968).
17. T. Tonegawa and J. Kanamori, Phys. Letters 21, 130 (1966).
18. S. J. Pickart, M. F. Collins, and C. D. Windsor, J. Ap. Phys. 37, 1054 (1966).

C-7: OBSERVATION AND INTERPRETATION OF ELECTRONIC AND VIBRATIONAL RAMAN EFFECTS OF RARE EARTH DOPED GARNETS

J. A. Koningstein
Carleton University, Ottawa, Canada
and
O. Sonnich Mortensen
University of Copenhagen, Denmark

INTRODUCTION

During the last three years we have been working on a program aimed at the detection and interpretation of laser excited electronic Raman scattering of trivalent rare earth ions in numerous inorganic crystals. Experimentally, we have been able to detect electronic lines for several different crystal systems. As far as the theory is concerned, we have been particularly interested in the question of antisymmetry in the scattering tensor and absolute intensities and we have been able to demonstrate by a priori calculations, that the scattering tensor for electronic transitions in these systems often is very highly antisymmetric. Just recently we have extended the theory of the electronic Raman effect to general molecules and the theory of this effect is then brought on a fairly quantitative basis.

SHORT OUTLINE OF THE THEORY

Let us start with a short outline of what is essential in the theory of the electronic Raman effect. Suppose we have the experimental situation shown in Fig. 1. The direct Raman transition is between states k and n, but the expression for the Raman intensity

$$I_s = I_i \times \frac{16\pi^4(\nu + \nu_{kn})^4 e^4}{c^3} |\alpha_{zy}|^2$$

$$\alpha_{zy} = \frac{1}{h} \sum_t \frac{\langle \Psi_n | z | \Psi_t \rangle \langle \Psi_t | y | \Psi_k \rangle}{\nu_{tk} - \nu - i\gamma_t} + \frac{[z \rightleftarrows y]}{\nu_{tn} + \nu + i\gamma_t}$$

Fig. 1. Raman scattering forces between two electronic states k and n. γ is the frequency of the exciting radiation.

involves a summation over all excited states of the system. The wave functions, in the expressions for the intensity are the total wave functions, depending on both the nuclear and electronic coordinates. By direct expansions and by using the Born-Oppenheimer approximation, it is readily seen that provided the electronic states are non-degenerate, and provided the laser frequency is not almost coincident with a strong absorption of the system, we can neglect the vibrational problem and deal only with the electronic states. [1] Now of course, for nearly all the systems with which we are dealing here, one or both of the electronic states are actually degenerate. Have we then reason for despair? No! We know from numerous experiments that the interaction between the rare earth ion in the crystal and the vibrations of the crystal are extremely small so that even though the electronic wave functions have a nuclear dependence, this dependence is so small as to be negligible, provided the transition under consideration is not forbidden for the pure direct electronic transition. The latter is of course the case in fluorescence and absorption spectroscopy and is the reason that we cannot neglect the vibronic coupling for these processes. What is left in the expression for the intensity is then, just the summation over excited electronic states. To break this summation down, we shall make the assumption that the wave functions of those states that are important in the summation are of an essentially free ion nature. That does not mean that we neglect crystal field effects, but it means a neglect of ligand type intermixing in the rare earth wave function. We have reason to believe that this assumption is reasonable. However, it should be made clear that little experimental information about the nature of the excited states is available.

Once this approximation is made, it is straightforward, though somewhat tedious, to break down the above mentioned summation by purely group theoretical means. We shall not go into the details of this procedure, most of which can be found in the literature, but shall just point to one important difference between our treatment and that of Judd[2], Ofelt[3] and Axe[4].

The group theoretical procedures make use of certain summations that greatly condense the final expressions. In the ordinary treatment these summations cannot be directly carried out, since the states involved in the summations have different energies so as to give different denominators in the equation for the scattering tensor. Previous authors have made the not very reasonable assumption that these differences in the energy denominators could be neglected. However, by using the Heisenberg equation of motion, one can transfer the energy dependence to an operator dependence and the following summations are then exact. Not only does this give a more reliable theory in itself, but the operator approach permits - as it has been demonstrated[5] - a straight-forward check on the goodness of the radial functions used.

When all this mathematics has been done, we end up with the expression shown in Fig. 2. Here α_Q^k are the irreducible tensor components, and the other symbols more or

$$(\alpha_Q^K)_{kn} = F(K,\nu) \times \sum_{\gamma SLJM} \sum_{\gamma' SL'J'M'} a^*(n;\gamma' SL'J'M')\, a(k;\gamma SLJM) \times (\gamma' SL'J'M' | U_Q^K | \gamma SLJM)$$

$$F(K,\nu) = (-1)^{K+1} \sqrt{2K+1} \sum_{\chi} (\ell || c^{(1)} || \ell')^2 \left\{ \begin{matrix} 1 & K & 1 \\ \ell & \ell' & \ell \end{matrix} \right\} \times \Bigg[(n\ell | r | n'\ell')^2 \left[\frac{1}{\bar{\nu}_{\chi k} - \nu} + \frac{(-1)^K}{\bar{\nu}_{\chi n} + \nu} \right] +$$

$$(n\ell | r | n'\ell')^2 \left[\frac{\bar{\nu}_{\chi k}}{\bar{\nu}_{\chi k - \nu}} + \frac{(-1)^K \bar{\nu}_{\chi n}}{\bar{\nu}_{\chi n} + \nu} \right] - \frac{(n\ell | r | n'\ell')(n'\ell' | [\mathcal{H} r]^r | n\ell)}{\bar{\nu}_{\chi k} - \nu} + (-1)^K \frac{(n\ell | [\mathcal{H} r]^r | n'\ell')(n'\ell' | r | n\ell)}{\bar{\nu}_{\chi n} + \nu} \cdots \Bigg]$$

Fig. 2. The expression for the irreducible tensor component $(\alpha_Q^k)_{kn}$.

less speak for themselves. From this information it is then straightforward to construct the theoretical Raman spectrum. Again we shall not go into the details, but will just mention one important feature. It has been pointed out already by Placzek[5], in 1932, that although the scattering tensor for ordinary phonon transitions is symmetric, this is not necessarily the case for electronic transition. The formalism presented here permits a straightforward calculation of the degree of antisymmetry in the tensor since the k=1 components are purely antisymmetric and the k=2 components purely symmetric. We have found that in many cases the antisymmetric part is dominant over the symmetric part and that this dominance increases with increasing laser frequency. For exact derivations of the theory, for examples of synthetic Raman spectra and for the experimental observation of a completely antisymmetric Raman tensor, we may refer to work already published by this group[6-18].

EXPERIMENTAL RESULTS ON GARNETS

In Fig. 3 the 80°K laser excited Raman spectra of YGaG, Yb in YGaG and the compound YbGaG are shown. The crystals investigated were rather small and the orientation differs somewhat. Generally speaking, the three spectra are rather similar but not identical. Even greater differences are found if these spectra are compared with the spectra of YbAℓG and YAℓG. These differences in the Raman spectra can be accounted for by the different site group splittings and factor group interactions that are to be

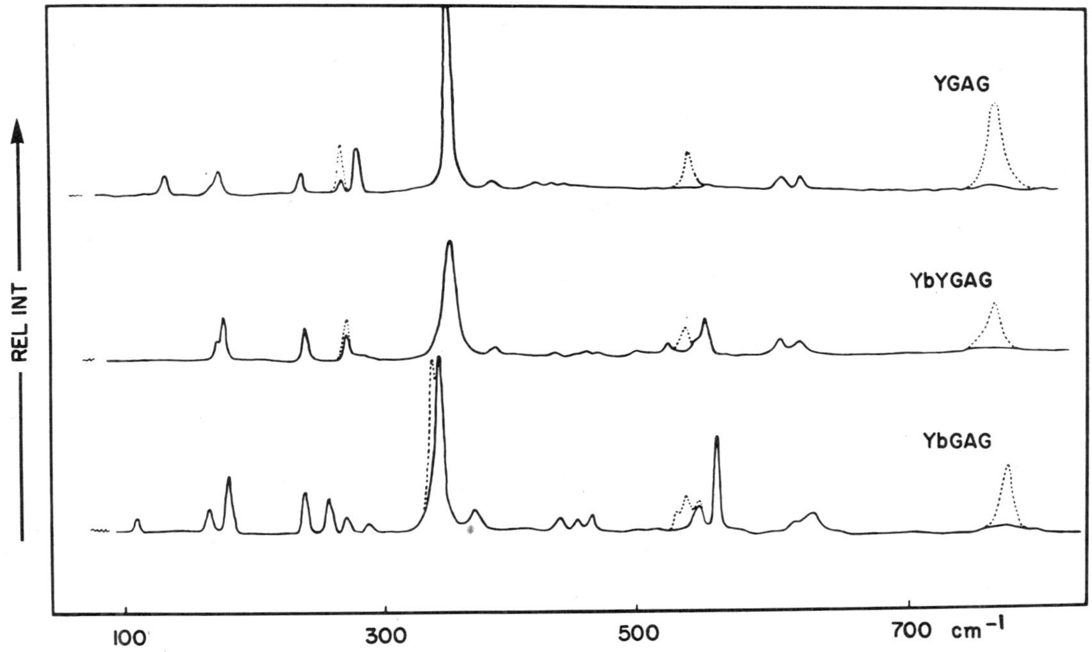

Fig. 3. The 80°K spectra of Garnets.

expected in such complicated systems, with the very low site symmetries. We do not want to go into detail here, but instead focus our attention to the appearance of electronic Raman lines in the YbGaG spectrum. The lines at ~ 550 cm^{-1} are dependent on the concentration of the rare earth ion and in particular the strong line at 552 cm^{-1} in YbGaG is distinctly weaker in the doped crystal and is completely absent in YGaG crystal. This line and the weaker line at 539 cm^{-1} must therefore be attributed to electronic Raman transitions in the trivalent Ytterbium ion. This is also clearly brought forward by the comparison shown in Fig. 4 between the TmGaG and YbGaG spectra in this region.

Apart from the electronic lines shown here, we have also found evidence for such effects of TmGaG, ErGaG, EuYGaG, TbAℓG, Eu in YaℓG, Eu in YVO_4 and Nd in YGaG.

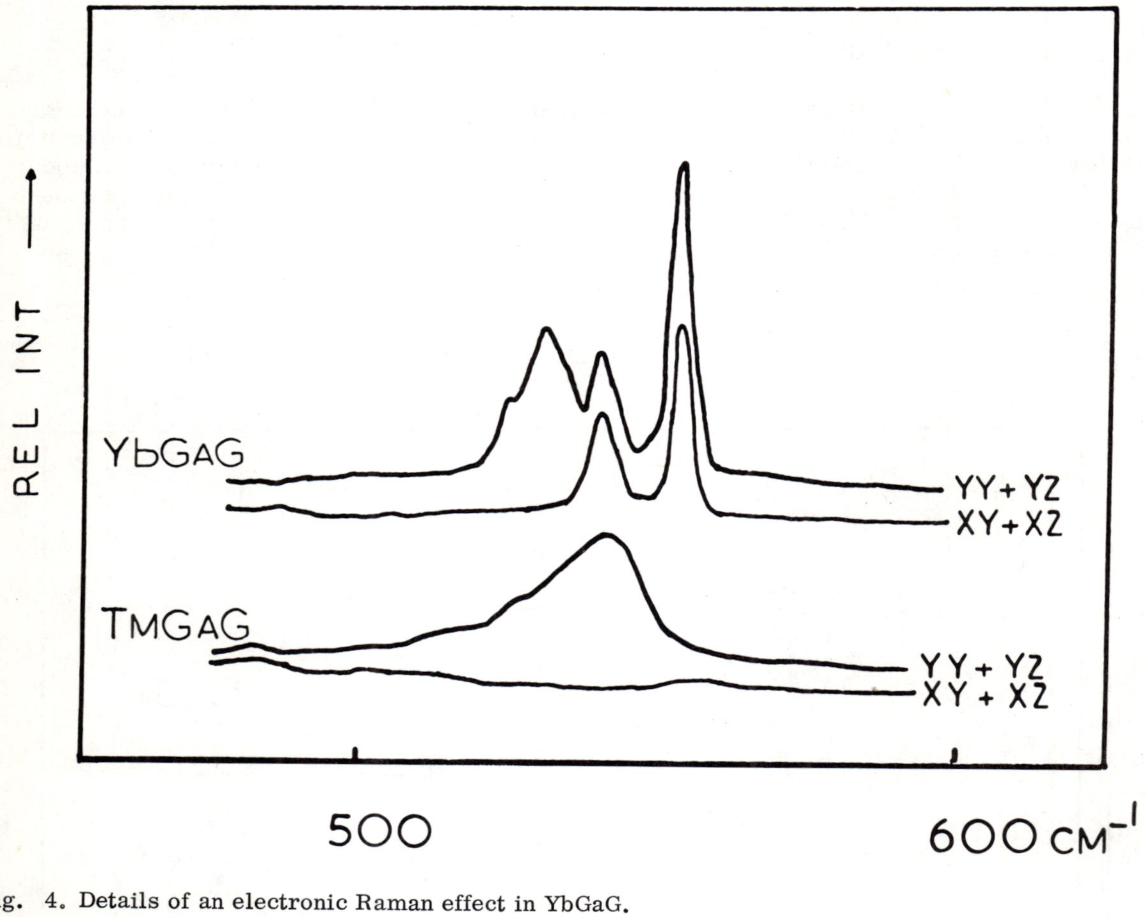

Fig. 4. Details of an electronic Raman effect in YbGaG.

CONCLUSIONS

We have found examples of numerous electronic Raman transitions both in Garnets and other crystal systems and have been able to relate the experiments to a convenient, and we think reliable, model. The electronic Raman effect, particularly in the doped crystals, is still a very weak effect and great care must be taken in the interpretation of the spectra. The Raman spectra of these systems change a great deal by substitution of one rare earth ion by another and Raman spectra of a whole series of systems is therefore needed for safe interpretation and identification of the electronic Raman transitions.

REFERENCES

1. O. Sonnich Mortensen and J.A. Koningstein, Chem. Phys. Letters 1, 409 (1967).
2. B.R. Judd, Phys. Rev. 127, 750 (1962).
3. G.S. Ofelt, J. Chem Phys. 37, 511 (1962).
4. J.D. Axe, Phys. Rev. 136, AY2 (1964).
5. G. Placzek, "Handbuch der Radiologie," VI, Ser. II, p. 205, Leipzig, 1934.
6. O. Sonnich Mortensen and J.A. Koningstein, J. Chem. Phys. 48, 3971 (1968).
7. J.A. Koningstein and O. Sonnich Mortensen, Phys. Rev. 160, 75 (1968).
8. J.A. Koningstein and O. Sonnich Mortensen, Nature 217, 447 (1968).
9. J.A. Koningstein and O. Sonnich Mortensen, Phys. Rev. Letters 18, 831 (1967).
10. J.A. Koningstein and O. Sonnich Mortensen, Chem. Phys. 1, 693 (1967).
11. J.A. Koningstein, J. Chem. Phys. 46, 2811 (1967).
12. J.A. Koningstein, J. Opt. Soc. 56, 1405 (1966).
13. J.A. Koningstein and O. Sonnich Mortensen, J. Opt. Soc. Am. 58, 1208 (1968).
14. J.A. Koningstein, Chem. Phys. Letters
15. J.A. Koningstein and Ng. Toaning, J. Opt. Soc. Am. 58, 1462 (1968).
16. J.A. Koningstein, Phys. Rev. 174, 477 (1968).
17. J.A. Koningstein, Appl. Spectr. 22, 438 (1968).
18. J.A. Koningstein and O. Sonnich Mortensen, J. Mol. Spectr. 28, 309 (1968).

C-8: ELECTRONIC RAMAN EFFECT IN RARE EARTH CHLORIDES

A. Kiel
Bell Telephone Laboratories, Incorporated
Holmdel, New Jersey

ABSTRACT

$CeCl_3$ provides a useful case for simultaneously testing the theory of the electronic Raman effect and of proving the utility of this technique for investigating energy levels in the far infrared. We have succeeded in determining the complete electronic structure of the f^1 configuration of Ce^{3+} in the trichloride. In this system we have seen pure antisymmetric transitions in certain polarizations and, in addition, have observed large asymmetries in the intensity of all lines allowed in xz polarization. The intensities of all lines are in reasonable agreement with theory and we can predict the orientations giving maximum scattering as well as the magnitude of the asymmetry for all those lines showing asymmetric scattering intensity. Some new results concerning selection rules and a number of practical considerations will be discussed in some detail.

INTRODUCTION

Since the early work of Hougen and Singh[1] there have been a number of interesting experiments[2-6] related to the electronic Raman effect. The basic theory of this effect is contained in the Heisenberg-Kramers dispersion relation and has been elaborated upon in a number of recent papers[6-8] of which I would single out Axe's as being most fundamental. In view of this I shall be rather brief with the theory and try to emphasize some anomalies in the selection rules, our experimental results on $CeCl_3$, and a number of practical considerations related to the general utility of the electronic Raman effect as a spectroscopic tool. We shall see that a number of surprises (at least to the author) occur when one considers relevant symmetries and the linear polarization selection rules in real crystals. Much of the work discussed in this paper was done in conjunction with T. Damen, S. P. S. Porto, S. Singh, F. Varsanyi, and J. Scott.

THEORY OF THE ELECTRONIC RAMAN EFFECT

As stated previously, standard second order perburbation theory is sufficient to describe all the main features of the pure electronic Raman effect. Skipping over the repetition of these equations (Kramers-Heisenberg relation), we may write the transition rate as

$$W(\nu) = (64\pi^5\nu^3 I/hc^4)\,|P(f, g; \nu, \epsilon_\beta, \epsilon_\alpha)|^2 G(\nu') \tag{1}$$

where the arguments of P represent a transition from the ion ground state g to final state f with exciting radiation of polarization $\vec{\epsilon}_\alpha$ and scattered radiation polarized $\vec{\epsilon}_\beta$. ν is the scattered light frequency (we shall ignore the difference in energy between initial and final electronic states which is certainly valid in the nonresonant case). I is the total intensity in the input laser beam and $G(\nu')$ is the line width function of the excited state ($\nu' = \nu - \nu_\beta$). By our convention, the initial electronic state and polarization of the excitation always appear as the right side of a pair. The function P may be written as ($|i\rangle$ are the intermediate states)

$$P(f, g; \nu, \epsilon_\beta, \epsilon_\alpha)$$

$$\equiv \frac{e^2}{2} \sum_i \left[\left(\frac{1}{E_i - h\nu} + \frac{1}{E_i + h\nu} \right) \left(\langle f|r_\beta|i\rangle\langle i|r_\alpha|g\rangle + \langle f|r_\alpha|i\rangle\langle i|r_\beta|g\rangle \right) \right.$$

$$\left. + \left(\frac{1}{E_i - h\nu} - \frac{1}{E_i + h\nu} \right) \left(\langle f|r_\beta|i\rangle\langle i|r_\alpha|g\rangle - \langle f|r_\alpha|i\rangle\langle i|r_\beta|g\rangle \right) \right] \tag{2}$$

The second term in (2) transforms like a cross-product or pseudovector but is not a simple $\vec{r}\times\vec{r}$ term (and therefore is, in general, nonzero). The first term in (2) includes the conventional symmetric terms in the Raman tensor.

If we reverse the polarizations of incident and scattered beams, we find that $P(f, g; \nu, \epsilon_\beta, \epsilon_\alpha) = P(f, g; -\nu, \epsilon_\alpha, \epsilon_\beta)$. This result is due to the nonzero pseudovector terms. If ν is negligibly small compared to E_i, the Raman scattering is rigorously symmetric (i. e., the energy part in the second term in (2) vanishes). It is well known that very near resonance the antisymmetric terms can become quite large. However, one has to be very far from resonance before the energy becomes negligible. For example, if $E_i = 45,000\ cm^{-1}$ and $h\nu = 20,000\ cm^{-1}$, the ratio of the two energy factors in (2) is 0.44. Hence, for many cases of interest, the relative importance of the antisymmetric terms is determined solely by the magnitude of the matrix elements $r_\beta r_\alpha - r_\alpha r_\beta$ (in the sense used in Eq. (2)).

It is useful to divide P into symmetric and antisymmetric terms, $P \equiv P_S + P_A$ (see Eq. (2)). The pseudovector part transforms exactly the same as the magnetic dipole moment of the system. We therefore state the general rule that in the electronic Raman effect involving any pair of states, antisymmetric (pseudovector) terms occur in the scattering tensor whenever magnetic dipole transitions are allowed between those states.

It is important to note, that in the electronic Raman effect, the initial and final electronic states are never identical. This is the crucial difference between this effect and the phonon Raman effect. In the latter case, we are usually in the situation where the initial and final electronic states are identical and nondegenerate. In this case it is easy to see that axial terms are not possible[9], i.e.,

$$\langle j|r_\alpha|i\rangle \langle i|r_\beta|j\rangle - \langle j|r_\beta|i\rangle \langle i|r_\alpha|j\rangle \equiv 0.$$

An important situation arises when, for given incident and scattered polarizations, both symmetric and pseudovector terms occur in the electronic Raman effect. This is almost always true for the case of yz or zx polarizations in any symmetry. Since $P(f, g; \nu, \epsilon_\alpha \epsilon_\beta) \neq (P(f, g; \nu, \epsilon_\beta, \epsilon_\alpha)$, the intensity of the scattering for one setting of the polarizers ($I \simeq |P|^2$) can be different in magnitude from that obtained with the reversed settings[6, 10]. To see this more explicitly, let us apply closure over the lowest excited manifold[6]. Then Eq. (1) may be rewritten as

$$P(f, g; \nu, \epsilon_\beta, \epsilon_\alpha) = P_S + P_A \quad \text{and} \quad P(f, g; \nu, \epsilon_\alpha, \epsilon_\beta) = P_S - P_A$$

where we have used Eq. (2) in the second relation.

Then for $\alpha\beta$ polarization, the scattered intensity is proportional to $|P_S + P_A|^2$ while for reversed polarization, $\beta\alpha$, $I \simeq |P_S - P_A|^2$. (The asymmetric scattering was predicted independently by the author [6] and in Ref. [10].)

Since the intensity is proportional to P^2, pure antisymmetric scattering will show no asymmetry in intensity. The maximum asymmetry occurs when $P(g, f; \nu, \epsilon_\alpha, \epsilon_\beta) = 0$ or $P(g, f; \nu, \epsilon_\beta, \epsilon_\alpha) = 0$. This situation occurs when

$$\frac{\langle f|r_\beta r_\alpha + r_\alpha r_\beta|g\rangle}{\langle f|r_\beta r_\alpha - r_\alpha r_\beta|g\rangle} = \pm \left(\frac{1}{E-h\nu} - \frac{1}{E+h\nu}\right) \Big/ \left(\frac{1}{E-h\nu} + \frac{1}{E+h\nu}\right)$$

The reader may wonder about the thermodynamic stability of this system since it appears to convert polarizations. However, it is easy to show that $P(g, f; \nu, \epsilon_\alpha, \epsilon_\beta) = P(f, g; \nu, \epsilon_\beta, \epsilon_\alpha)$, where the term on the right applies to normal Stokes scattering and the one on the left to anti-Stokes radiation with reversed polarization. This general reciprocity relation assures that normal equilibrium is maintained even in asymmetric systems.

SELECTION RULES AND ZEEMAN EFFECTS

The electronic Raman selection rules are only slightly more complicated than in the phonon case. We reduce the polarization tensor to the irreducible representations of the symmetry group, Γ_P. If the ground state belongs to the irreducible representation Γ_g and the final state to Γ_f, we must take the direct product $\Gamma_g \times \Gamma_f$ and reduce this, i.e., $\Gamma_g \times \Gamma_f = \Sigma_h \Gamma_h$. Then if a particular Γ_P occurs among the $\Sigma\Gamma_h$, we expect the polarization components to Γ_P to appear in the Raman scattering. The major differences in the electronic scattering case from the phonon case is the appearance of pseudovector terms in Γ_P and the fact that the Γ_g, Γ_f may be double-valued representations. In fact, these are very simple to deal with and complete tables of the electronic Raman effect selection rules for all states and symmetries appear in Ref. [11].

In principle, when no magnetic field is applied, rectangular coordinates are adequate for describing the polarization selection rules.

Consider for example scattering from a T_2 state to an E state in cubic symmetry; $E \times T_2 = T_1 + T_2$. The polarization components may be reduced as $A_1(xx+yy+zz)$, $E(xx-yy, 2zz-xx-yy)$, $T_2(xy+yx, xz+zx, yz+zy)$, $T_1(S_x, S_y, S_z)$ where $S_y = zx - xz$, $S_x = yz - zy$ and $S_z = xy - yx$. Therefore, for $E \leftrightarrow T_2$ Raman transitions, we expect both symmetric and antisymmetric terms of xy, xz and yz; no terms like xx, yy, zz should occur. For a $T_2 \to T_2$ electronic transition $T_2 \times T_2 = A_1 + E + T_1 + T_2$ and we therefore would expect all possible polarizations. An inspection of the tables in Ref. [11] shows that the occurrence of antisymmetric terms are indeed very common.

Experiments using a "Raman-Zeeman" effect are quite feasible. In rare earths, for example, the ground state of a J-manifold at liquid helium temperatures is often less than 1 cm^{-1} in good crystals. Hence, fairly modest magnetic fields (i. e., $H > 20{,}000$ gauss) will often be adequate for Zeeman work. To analyze Zeeman data we must use circular polarizations, i. e., $z \to x_0$, $x \to -(x_+ - x_-)/\sqrt{2}$, $y \to -i(x_+ + x_-)/\sqrt{2}$. In determining selection rules in the case where the electronic levels are split by a magnetic field, one determines the direct products of individual components of the representations Γ_g, Γ_f; that is, we need the product of the ith basis vector of Γ_g and j^{th} bases vector of Γ_f, $\Gamma_g(i) \times \Gamma_f(j)$, expressed in terms of a circular component of Γ_P. For example, if the basis functions of a T_2 electronic state is (η_+, η_0, η_-) and the basis of an E state is (u_+, u_-) we have for the products and selection rules, $\eta_- u_+ \to x_+ x_+ (T_2)$, $\eta_+ u_- \to x_- x_- (T_2)$; $\eta_0 u_- \to x_0 x_- (T_1, T_2)$, $\eta_0 u_+ \to x_0 x_+ (T_1, T_2)$; $\eta_- u_- \to x_+ x_- (T_1)$, $\eta_+ u_+ \to x_- x_+ (T_1)$. In this 6-line spectrum, the first two are pure symmetric, the third and fourth are mixed symmetric and antisymmetric, and the last two are pure antisymmetric. If one observes this spectrum with linear polarizers, all linear polarizations occur.

This last point has important consequences in interpreting linearly polarized spectra from samples without magnetic fields. Consider a real cubic crystal where the degeneracy is partly or entirely removed because of strains. This will cause individual components of T_2 and E to scatter independently. The linear selection rules are based on coherent scattering of all the sublevels, which leads to cancellation of certain terms. Therefore, one cannot expect to see the ideal linear selection rules in real crystals if strains are large enough to destroy the correlations of the magnetic sublevels. For example, in strained crystals, an "apparent" xx component would appear in a $T_2 \to E$ transition contrary to the previous linear selection rules. In concentrated crystals, the magnetic moments of the ions can lead to the same result. Some further important consequences of these considerations will be discussed in the next section.

$CeC\ell_3$ EXPERIMENTS

The electronic Raman effect experiments on $CeC\ell_3$ were performed with what is, by now, a conventional Raman spectrometer using several lines of an argon laser. More detailed description of the apparatus and experimental details are contained in Ref. [6]. This work was done in conjunction with T. Damen, S. P. S. Porto, S. Singh, and F. Varsanyi.

Fig. 1 shows our results using the 5145°A line of the argon laser. The four different polarizations are shown in the column on the right. The group on the right in Fig. 1 were all the lines detected within 300 cm^{-1} of the laser excitation. This group includes the phonon lines as well as the two lines we interpret as electronic Raman lines; the line at 45 cm^{-1} in yx and yy polarizations and the very strong line at 117 cm^{-1} in yz polarization. The group at the left were found in the region 2168 cm^{-1} to 2349 cm^{-1} below the laser line. This is the region where the ${}^2F_{7/2}$ states of Ce^{3+} should lie. Three lines definitely appear and we believe the broad line at 2349 cm^{-1} in yz polarization is real, accounting for the four lines expected from this manifold. We have compared these results with those of Varsany for Ce^{3+} in $LaC\ell_3$ [12]. We find that with only modest changes of the crystal field we can fit Varsanyi's results. We are therefore confident that we are observing electronic states. In addition the phonon lines seen in Fig. 1 right, are close to those in $LaC\ell_3$ and $PrC\ell_3$ so there is no difficulty in separating phonon and electronic lines.

Since the ground state in $CeC\ell_3$ is $\Gamma_{7,8}$ ($\mu_{odd} = \pm 5/2$), we can predict the polarizations (in order of energy of the excited states as

45 cm^{-1}	$\Gamma^*_{7,8} \to \Gamma^*_{9,10}$	xx, yy, xy
117 cm^{-1}	$\Gamma^*_{7,8} \to \Gamma^*_{11,12}$	xz, yz(S_x, S_y), xx, yy, xy
2168 cm^{-1}	$\Gamma^*_{7,8} \to \Gamma^*_{7,8}$	zz, xx + yy, S_z, xz, yz(S_x, S_y)
2222 cm^{-1}	$\Gamma^*_{7,8} \to \Gamma^*_{9,10}$	xx, yy, xy
2287 cm^{-1}	$\Gamma^*_{7,8} \to \Gamma^*_{7,8}$	zz, xx + yy, S_z, xz, yz(S_x, S_y)
2368 cm^{-1}	$\Gamma^*_{7,8} \to \Gamma^*_{11,12}$	xz, yz(S_x, S_y), xx, yy, xy

We have, in addition, calculated the relative intensities of all lines in the spectrum. In general the agreement of theory and experiment is good. Some of the more interesting features of Fig. 1 are: The lines in xy polarizations at 2168 and 2287 cm^{-1} are <u>pure antisymmetric transitions</u>, and based on calculations using only symmetric terms the lines at 2168 and 2287 should have been much stronger. This latter anomaly led us to search for asymmetry in the intensity. In Fig. 2 we show most of the scattering features but now using zy polarization. Note that the relative sizes of the 2168 and

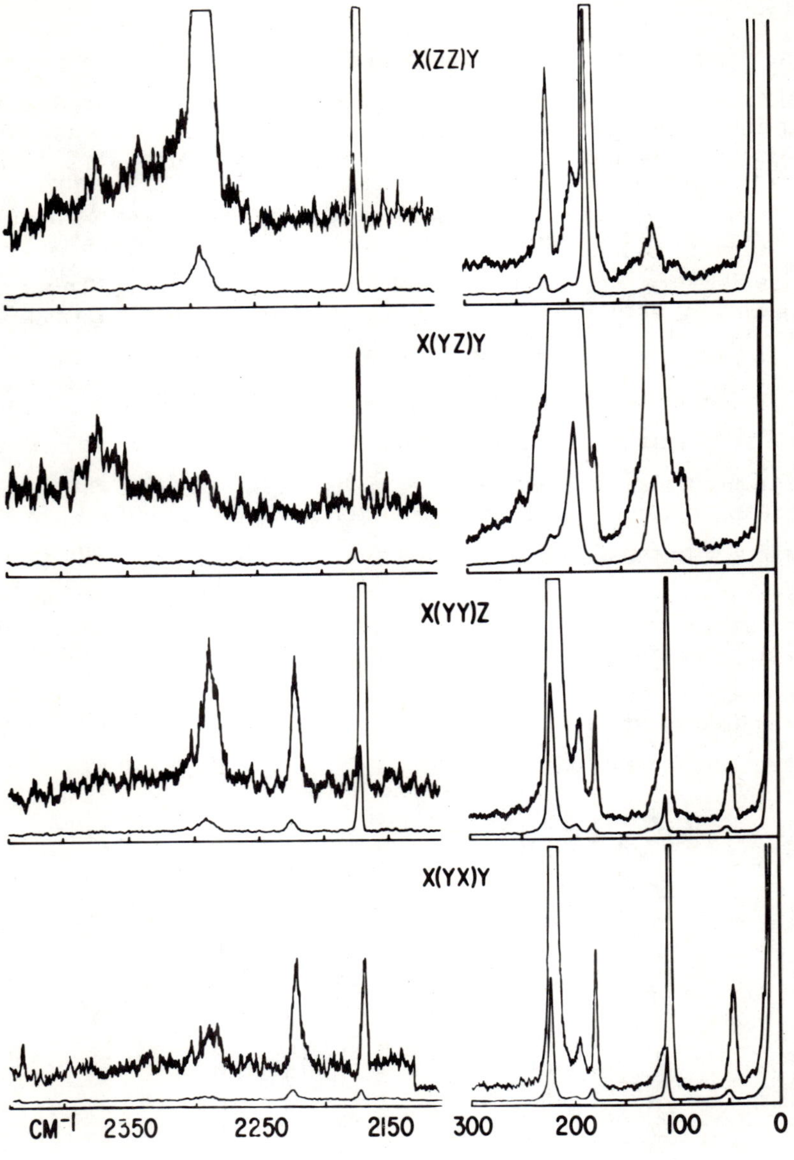

Fig. 1. Phonon and electric Raman lines in $CeC\ell_3$.

The polarizations of the lines in each row are within the parentheses, propagation directions outside the parentheses with letters to the left applying to excitation (4965Å line of Argon laser), on the right to scattering. Each chart has a low gain and high gain (x10) traces; wave length should be taken from the high gain trace. The lines on the left side are all electronic (J = 7/2); electronic lines at the right are at 45 cm^{-1} (xx and xy polarization) and 117 cm^{-1} (yz) polarization. Intensity of the lines at left and right cannot be compared directly.

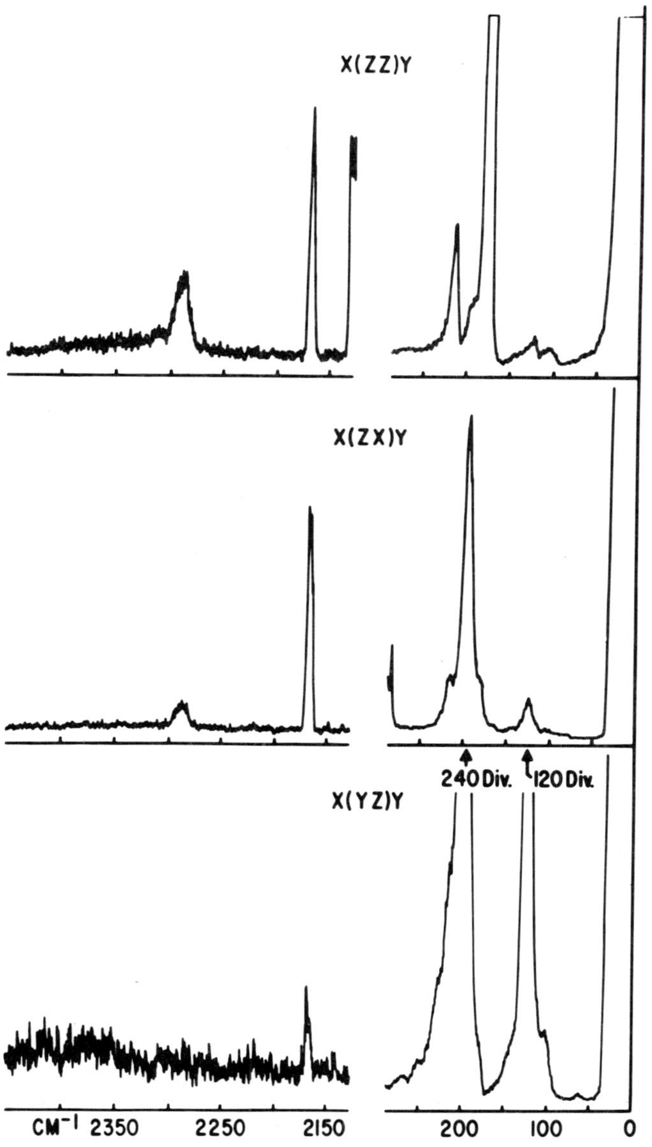

Fig. 2. Asymmetry in the electronic Raman effect.
The gain of the lowest trace at the left is three times as great as the upper two at the left.
The 4765A° Argon laser line was used in this case.

2287 cm^{-1} lines have greatly increased and the 117 cm^{-1} line is much weaker. This asymmetry effect was discussed in Section II. In every case, the calculated and observed relative intensities agreed to within a factor of two. Especially gratifying is the fact that for the lines which should show asymmetric intensities in xz/zx polarization we predict the correct orientations for the maxima and minima in every case.

There was a major problem in trying to understand the strength of the xx components of the 2168 and 2287 cm^{-1} lines. As pointed out in Section III, the rectangular selection rules only hold if the Kramers pair of states are degenerate and are coherent. In $CeC\ell_3$, however, there is a large dipole magnetic which is rapidly varying. The phase memory time of a state in this system is less than $\simeq 10^{-10}$ sec and the dipolar width of the ground state wil be $\sim 10^{-1}$ cm^{-1}. Therefore each state of a Kramer's pair scatters independently although we cannot resolve them. In all cases except the xx scattering, this leads to a reduction in absolute intensity of a factor of two. However, for the xx scattering, the incident beam propagates along the y axis, the scattered beam along the z axis. Hence in the scattered beam we are observing circular components belonging to both the xx scattering and the xy. The result is an increase of a factor of 3 in the intensity of the observed xx scattering.

This property of magnetic crystals will have a very important consequence in many cases of concentrated crystals. For in a majority of cases the crystal has more than one ion per unit cell with the result that the local site symmetry is lower than the crystal symmetry. If the ions do not interact, we expect the scattering to be characteristic of the space group of the crystal, i. e., coherent scattering. However, if the magnetic noise in the crystal breaks this coherence, the ions scatter independently and the site symmetry is displayed in the Raman scattering. Strains can cause similar effects when the states have non-Kramers degeneracy.

PRACTICAL CONSIDERATIONS

In this section we shall (affirmatively) answer the question, "Is Raman spectroscopy of low lying electronic states a generally useful technique." It is clear that this method will be especially useful when the crystal does not fluoresce. Even when there is fluorescence, the Raman data may be far easier to interpret, since one must know the complete set of energy levels before one can definitely establish the initial and terminal states in fluorescence. Therefore, at the very least, electronic Raman spectroscopy should be a useful subsidiary to fluorescence studies.

There is an additional advantage of Raman data over fluorescence results. The optical transitions involved in the latter are very often "forbidden." As a result, vibronic lines may have comparable intensity to the competing no-phonon transitions, leading to additional structure in an already complex spectrum. Since the pure electronic transitions are "allowed" in the Raman case, this competition does not exist, and we will usually see only no-phonon states.

The advantages of electronic Raman techniques over far infrared absorption are more striking. We have seen that one can observe electronic states even in the region of strong phonon absorption. This would not be possible with far infrared instruments where phonon lines tend to "swamp" everything. In any case the electronic Raman effect can be more sensitive than infrared absorption. For example, in our $CeC\ell_3$ results, the calculated differential cross-sections of the observed lines ranged from 4×10^{-31} cm^2 to 10^{-33} cm^2. The ion concentration was about 10^{22} grams/cm^3. A reasonably useful figure for the limiting sensitivity is about 10^{-11} (i.e., absolute cross-section per ion times ions/cm^{-3} divided by linewidth in wave numbers). We saw there were transitions in $CeC\ell_3$ that were several hundred times greater than this limit and Ce is not a particularly strong scatterer. Pr^{3+}, for example, has cross-sections much greater than Ce. It is important to note that very small crystals may be used for the Raman scattering since we can focus the laser beam to a very small area with no great loss in sensitivity.

In absorption[13], we can take the ion cross-section as $10^{-17} \cdot f\ cm^2$, where f is the oscillator strength (integrated). The absorption coefficient is approximately equal to $10^{-17} \cdot \eta f/\Delta$ where η is the number of ions per cubic centimeter and Δ is the linewidth. In the infrared we expect f to be no larger than 5×10^{-6}, and about the smallest absorption one can expect to see with a scanning instrument is 1 per cent). Taking $\Delta = 3\ cm^{-1}$ (rather less than an average linewidth) and a crystal 0.5 cm long, we find that the minimum detectible ion concentration is about 1.2×10^{21} ions/cm^3. Thus, even in these very favorable conditions (large cyrstal, small linewidth, fairly high oscillator strength), the limiting concentration is quite high. In the $CeC\ell_3$ case, where the concentration was 10^{22}, the best signal-to-noise ratio one expects in absorption is therefore 8:1 while in the electronic Raman work, using small crystals, we had a maximum S/N of several hundred.

The electronic Raman effect is particularly advantageous if one is able to take advantage of resonance enhancement. The sensitivity of the electronic Raman effect can then be increased by factors of 10^2 - 10^5 in favorable cases. Note that there is no disadvantage in having the laser frequency coincident with an absorption band so long as the scattered light is not greatly absorbed. Using this technique J. F. Scott has observed electronic Raman scattering[14] of 6×10^{17} ions/cm^3 of Ce^{2+} in CaF_2. This crystal did not fluoresce and no other technique could have been used to observe these low lying states of the fd manifold.

REFERENCES

1. J.T. Hougen and S. Singh, Proc. Roy. Soc. (London) A277, 193 (1964).
2. J.Y.H. Chau, J. Chem. Phys. 44, 1708 (1966).
3. J.A. Koningstein, J. Opt. Soc. Amer. 56, 1405 (1966).
4. J.A. Koningstein and O.S. Mortensen, Nature 217, 5127, 445 (1968).
5. J.A. Koningstein and O.S. Mortensen, J. Chem. Phys. 46, 2811 (1967).
6. A. Kiel, T. Damen, S.P.S. Porto, S. Singh, and F. Varsanyi, IEEE, Journal of Quantum Electronics 4, 318 (1968); Phys. Rev. (to be published).
7. J.D. Axe, Phys. Rev. 136, A42 (1962).
8. O.S. Mortensen and J.A. Koningstein, J. Chem. Phys. 48, 3971 (1968).
9. This is not exact since small antisymmetric terms arise due to the small difference in incident and scattered light frequencies.
10. J. A. Koningstein and O.S. Mortensen, Chem. Phys. Letters 1, 693 (1968).
11. A. Kiel and S.P.S. Porto, J. Mol. Spectroscopy (to be published).
12. F. Varsanyi and B. Toth, Bull. Am. Phys. Soc. 11, 242 (1966).
13. We consider here only the common situation where transitions to the low lying states are, to first order, forbidden in electric dipole radiation.
14. A. Kiel and J. Scott, Bull. Am. Phys. Soc. 13, 1438 (1968).

C-9: SPIN FLIP RAMAN SCATTERING IN CADMIUM SULFIDE

J. J. Hopfield
Palmer Physical Laboratory, Princeton University
Princeton, New Jersey
and
D. G. Thomas
Bell Telephone Laboratories, Incorporated
Murray Hill, New Jersey

INTRODUCTION

The Raman scattering with which this paper deals[1] is due to neutral donor and acceptor impurities in CdS. In this hexagonal crystal, both the usual neutral donors and acceptors have a spin degeneracy of 2 which is lifted by the application of a magnetic field. In the presence of a magnetic field H, an electronic Raman scattering can take place in which the electron or hole begins in one magnetic state, and is left in the other, the change of photon energy being $\pm \mu H$. We will indicate why this unlikely-sounding process has the observed large scattering cross section, and examine some details of the geometry, field, and energy dependence of the cross section.

EXPERIMENTAL

The crystals of CdS used were in the form of thin platelets grown from the vapor by passing a stream of argon over Eagle-Picher high purity CdS held at 1100°C. Doping was carried out by adding impurities to the gas stream. The main faces of the crystals were perpendicular to the $11\bar{2}0$) axis[2]. For the Raman experiments, it was important to make observations with light traveling parallel to the hexagonal (c) axis. For this purpose, crystals were found which had small (0001) faces as grown. A typical platelet was 0.2 x 0.1 x 0.003 cm wide. The samples were immersed in liquid helium, and the dewar placed in the gap of an electromagnet producing fields up to 30,400 Gauss.

An argon ion laser which could be tuned to different wavelengths was used for the scattering experiment. The laser beam was focused to a small spot. For the highest resolution and most accurate determinations of line separations, the back-scattered light was photographed with a Bausch and Lomb 2-m spectrograph having a resolution of 2Å/mm. Intensity measurements were made using a scanning photomultiplier.

THEORY - DONORS

In the calculation of Raman scattering, sums over intermediate states must be made. The weight of an intermediate state in the Raman scattering amplitude is proportional to the reciprocal of the amount by which the state fails to conserve energy. For the 4880 Å line, the optically allowed intermediate state, "exciton from top valence band bound to neutral donor", has such a small energy denominator compared to other paths expected to have similar matrix elements, that a good quantitative approximation should result if only this state is included. Besides giving an estimate of the absolute cross section, this simple model shows in detail how the selection rules originate.

In CdS, the valence band spin-orbit splitting is only about 0.07 eV[3]. The spin-orbit coupling of the valence band provides the dominant mechanism for spin-flip scattering, which would, of course, vanish without spin-orbit coupling. If typical energy denominators in intermediate state energy denominators are large enough that the .07 eV splitting is insignificant, an intermediate state sum over all valence bands will produce a complete cancellation between the nonzero contributions of each of the (spin-orbit split) valence bands. On the other hand, when one intermediate state energy denominator is small compared to this splitting, the cancellation does not occur, and the magnitude of the spin-orbit coupling is effectively infinite. This is the case for scattering at 4880 Å from donors in CdS.

Quantize all spins along the c-axis. The ground state of the system has an electron with spin up $\uparrow$ (or down $\downarrow$): the excited state has a hole with spin up $(P_x + iP_y)\uparrow$ [or spin down $(P_x - iP_y)\downarrow$][4]. Let M be the optical matrix element for linearly polarized light between the state $|\uparrow\rangle$ and the state $|(P_x + iP_y)\uparrow\rangle$. Then in terms of M, the optical decay lifetime τ of this excited state is given by

$$\frac{1}{\tau} = \frac{2\Omega M^2 k^2 n^2}{3\pi h^2 v_g} \tag{1}$$

where Ω is the volume of the crystal, k is the vacuum wave-vector at the transition energy, v_g is the group velocity of the ordinary ray, and n the ordinary index of refraction at the transition energy. The oscillator strength f can be defined in terms of the integrated absorption by

$$f(N/\Omega) = \frac{nmc}{2\pi^2 e^2} \int_0^\infty \alpha(\omega)\, d\omega = \frac{nmc}{2\pi^2 e^2} \left(\frac{2\, NM^2}{h^2 v_g}\right) \tag{2}$$

The matrix elements connecting the ground and excited states for the two polarizations of light perpendicular to the c-axis are shown in Fig. 1.

The basic unit of all cross sections is the Rayleigh cross section for light propagating along the c-axis and scattered either backward or forward. This differential cross section[5] can be approximately written, for ω near ω_0, as

$$\frac{d\sigma}{d\Omega} = \frac{1}{4} f^2 \cdot \left(\frac{e^2}{mc^2}\right)^2 \left(\frac{\omega}{\omega - \omega_0}\right)^2 \equiv \left(\frac{d\sigma}{d\Omega}\right)_0 \tag{3}$$

where ω and ω_0 are photon and bound state frequencies. This cross section, for donors, is the same for scattered light polarized either parallel or perpendicular to the incident light, if both beams are polarized $E \perp c$.

For a magnetic field in the z direction, there are no cross matrix elements, and no spin-flip scattering can occur. If the incident light is polarized in the x-direction, the differential cross section for backward Rayleigh scattering is $(\frac{d\sigma}{d\Omega})_0$ for both the x and the y polarization. For a magnetic field in any other direction, the ground state spins will not be resolved along the c-axis, and the same matrix elements given in Fig. 1 will then be capable of giving transitions between the spin states. For a magnetic field making an angle θ with the c-axis, the two ground spin eigenstates are

(a) $\uparrow \cos\theta/2 + \downarrow \sin\theta/2$
(b) $- \uparrow \sin\theta/2 + \downarrow \cos\theta/2.$

The matrix element for going from (a) to (b) via intermediate state (P_x+iP_y) with an incident "x" polarized photon and an outgoing "y" polarized photon is,

$$-iM^2 \cos\theta/2 \sin\theta/2.$$

The matrix element for going from (a) to (b) via intermediate state $(P_x-iP_y)\downarrow$ with the same polarization conditons is also

$$-iM^2 \cos\theta/2 \sin\theta/2.$$

Adding and squaring, the spin-flip differential scattering cross section for backward (or forward) scattering and crossed polarizations and propogation along the c-axis is

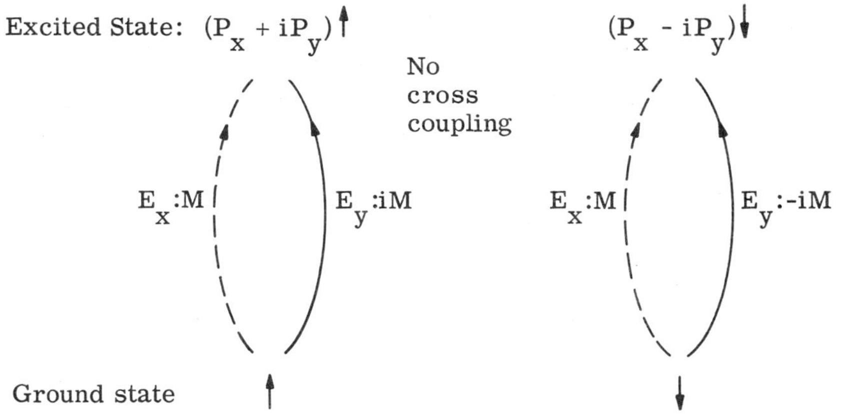

Fig. 1. The optical matrix elements between the ground and excited states of both spins for light polarized with $E \perp C$ in the x or y directions. The matrix elements are written immediately after the corresponding E-vectors. All other matrix elements vanish.

$$\left(\frac{d\sigma}{d\Omega}\right)_{\substack{\text{spin flip, H at}\\ \text{angle } \theta;\, k \| c\\ \text{crossed polarizations}}} = \left(\frac{d\sigma}{d\Omega}\right)_0 \sin^2\theta \tag{4}$$

RESULTS ON DONORS

The spectrum of light scattered from donors in two different crystals is shown in Fig. 2. The geometry used was that in which the cross section should be maximal. The line width of the scattered light is limited by instrumental resolution. The ratio of the Stokes to anti-Stokes scattering is approximately the thermal value in the doped crystal, but is extremely non-equilibrium in the undoped crystal. Reducing the laser power resulted in a slightly more nearly thermal ratio in the undoped crystal.

The relative sizes of the observed cross sections as a function of geometry are qualitatively as expected on the basis of the one-level model. For example, the experimental cross section drops by a factor of 130 in going from the 4880 Å line to the 4965 Å line. The ratio of the squared energy denominators is 100. While the signals in the geometry $K \perp C$, $E \perp$, $E \|$, $H \| C$ are not simply comparable to those in the (0001) face geometry, the qualitative experimental conclusion that the Raman scattering here was

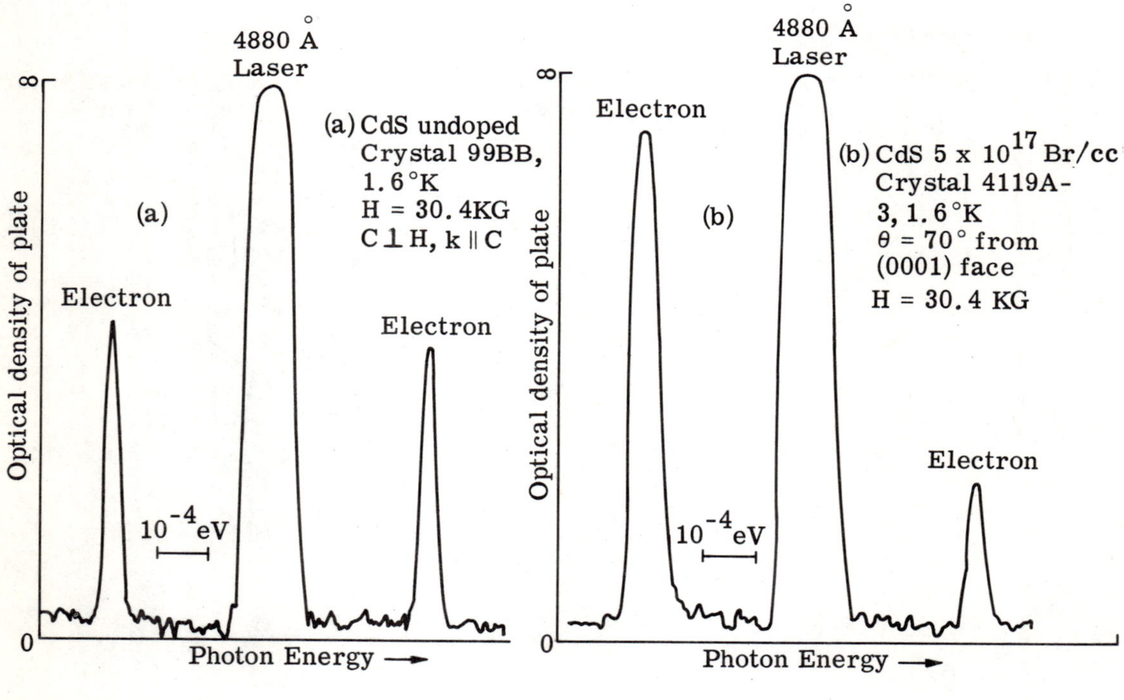

Fig. 2. The spin-flip Raman spectrum due to bound electrons in a "pure" crystal of CdS. a) The data is taken from a photographic plate using an exposure time of one second; b) is similar except a bromine doped crystal was used.

weak compared to the (0001) face geometry is also expected, for the squared energy denominator is here larger by an order of magnitude. Finally, for E ⊥, E ⊥, H || C, no spin-flip scattering could be detected (to an accuracy of ~ 1 in 300 of the E ⊥, E ⊥, H ⊥ geometry). This is in agreement with the effective mass estimate and of the one-level model that the cross section vanish in this geometry. All observed cross sections for donors are independent of the magnitude of the magnetic field, in accordance with theoretical expectations.

THE ABSOLUTE CROSS SECTION FOR DONORS

In the two level model, the absolute Raman cross section can be derived from the oscillator strength of the nearby bound exciton transition.

To measure the Raman differential cross section, several "pure" crystals were examined in the geometry k || C, C ⊥ H. For this geometry the Raman light typically has an intensity of about 1% of the laser light transmitted through the crossed polarizer. With a crossed polarizer the Raman intensity (the sum of both lines) was observed through the spectrograph. This intensity was found to be independent of H. A quarter wave plate was then inserted in the laser beam so that the beam was circularly polarized, and in place of the crystal was placed a surface covered with 3 M Velvet Coating No. 202 white paint which uniformly scattered and depolarized the light. The intensity of the 4880 Å laser light was now measured. From the ratio of crystal scattering to paint scattering, the Raman cross section can be directly calculated. The only additional information required is the mean depth of penetration (believed to be about 300 microns in a typical crystal.) and the impurity concentration. The donor concentration was determined by Hall measurements by H. C. Montgomery. The observed donor Raman cross section was 4×10^{-18} cm^2/steradian.

To check this cross section against the theoretical expression for the one level model, the oscillator strength was determined for several samples by integrating the area under the absorption curve. Such a measurement yielded $f = 9 \pm 2$. A calculated Raman cross section for such an f - value is 4×10^{-19} cm^2/steradian. The disagreement between these two numbers for theory and experiment is at least partly due to the uncertainty in the experimental value of the scattering cross section, the systematic error in which is perhaps a factor of 5 from the penetration depth alone.

OSCILLATOR STRENGTHS OF BOUND EXCITONS

It is extremely difficult to construct a wave function for bound excitons. A first estimate might be that the electrons and holes are in a product wave function which looks like a stationary exciton. Such a model would yield an oscillator strength of the order of that of the exciton. Another estimate could put an electron and hole in an exciton state moving as a unit around the impurity. If such an exciton is weakly bound to an impurity potential of short range, the distributed exciton resembles a larger "antenna" and radiates more effectively. Rashba and Gurgenishvili[6] have shown that in a one exciton band model the oscillator strength possible from such a wave function is the excition oscillator strength multiplied by the number of unit cells in a sphere whose radius is the size of the bound state. For an exciton whose binding energy to an impurity is only ~ .007 eV and whose mass is the order of the electron mass, this enchancement factor can be as large as 10^5. The experiments reported here give an oscillator strength for excitons bound to donors of about 10, whereas the intrinsic exciton oscillator strength[3] is only about $2x10^{-3}$ per unit cell. These numbers qualitatively confirm the predictions of Rashba and Gurgenishvili.

There exists an alternative means of calculating Rayleigh and spin-flip Raman scattering cross sections based upon polaritons. It treats the bound and free intermediate states simultaneously, and finds Rashba's giant oscillator strength as a natural outcome of a polariton point of view.

RAMAN SCATTERING FROM HOLES ON ACCEPTORS

Two types of acceptors are known in CdS. The simplest of these has a symmetry such that its g-value vanishes for $H \perp c$. The fluorescence line "I", at 4886 Å represents such a transition. The vanishing of the g-value in this geometry prevents conventional spin resonance observation of such holes, and for related reasons the Raman scattering with spin-flip is proportional to H^2. The magnitude of the Raman scattering is given by a formula similar to that which gives the magnitude for the neutral donor, but multiplied by the additional factor

$$\left(\frac{\mu H}{h\omega - h\omega_0} \right)^2$$

For the 4880 argon ion laser transition, this factor is about $\frac{1}{500}$ for the maximum magnetic field available. It is therefore not surprising that weak spin-flip Raman scattering was seen from acceptors, as shown in Fig. 3. That this scattering was due to acceptors could be demonstrated from its magnetic splitting energy and anisotropy. Fig. 3 clearly also shows that in the same crystal at the same time, the hole spins have thermalized while the electron spins have not.

Acceptors having a symmetry such that $g \neq 0$ for $H \perp c$ were also observed in spin-flip Raman scattering. One such acceptor lies so close to the laser frequency that a component of its absorption (or emission) spectrum can be magnetically tuned across the laser frequency.

The huge Raman cross sections and narrow lines produce huge theoretical Raman gains for CW argon ion pumping.

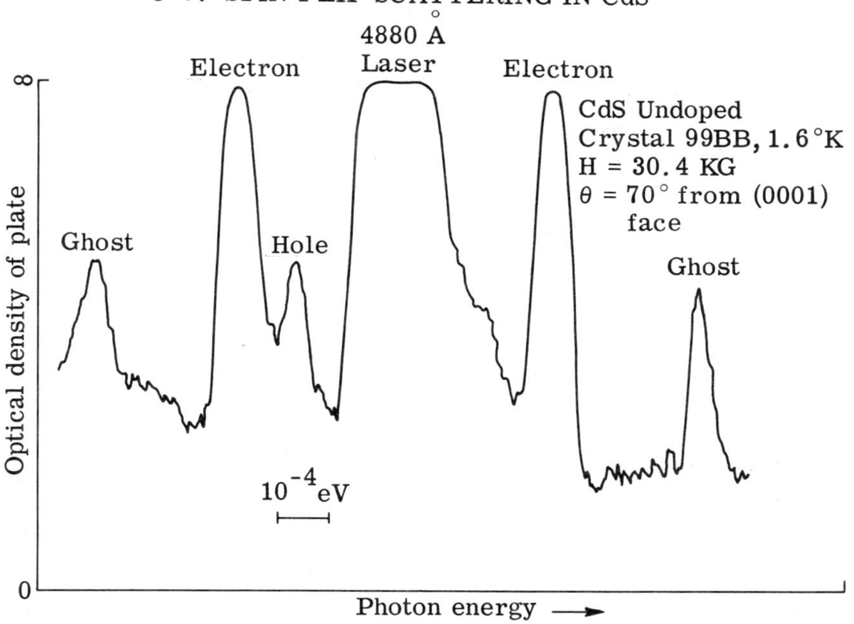

Fig. 3. A Raman spectrum which shows a line due to hole spin-flip scattering. Unlike the electron, in this "pure" crystal the hole thermalizes between its two levels, and at 1.6°K the anti-Stokes line is very weak. For this exposure requiring 30 seconds, grating ghosts of the 4880 Å line become visible.

CONCLUSION

As a tool for studying impurities in solids, spin-flip Raman scattering has several obvious assets. The small size of the sample volume and small number of spins needed, and the fairly direct relation to optical transitions due to the impurities, are all useful in semiconductor studies. The fact that the impurity optical transition itself is broad from strain effects or phonon broadening does not appreciably broaden the Raman lines, (which come from Kramers doublets), and thus makes the study of the ground state magnetic structure more readily performed by Raman scattering than by fluorescence and absorption studies. The particular selection rules CdS are not necessary, and the spin-flip Raman process should be easily studied in most direct band gap semiconductors.

The use of the Raman scattering to detect microwave resonance seems a promising line of approach to the chemical and geometrical nature of some of the yet unidentified shallow optical transitions in those semiconductors for which available laser transitions make cross sections favorable.

REFERENCES

1. A more complete version of this paper will be available in D.G. Thomas and J.J. Hopfield, Phys. Rev. (in press).
2. J.D. Levine and P. Mark, Phys. Rev. 144, 751 (1966).
3. D.G. Thomas and J.J. Hopfield, Phys. Rev. 116, 573 (1959).
4. J.J. Hopfield, J. Phys. Chem. Solids 15, 97 (1960).
5. W. Heitler, "Quantum Theory of Radiation", p. 132, Clarendon Press, Oxford, 1936.
6. E.I. Rashba and G.E. Gurgenishvili, Fiz. Tverd. Tela 4, 1029 (1962). (Translation Soviet Physics - Solid State 4, 759 (1962).

C-10: THEORY OF MAGNETIC FIELD EFFECTS ON THE RAMAN SCATTERING OF SHALLOW IMPURITY STATES IN SEMICONDUCTORS

P. J. Lin-Chung and R. F. Wallis
Naval Research Laboratory
Washington D. C.

INTRODUCTION

The recent upsurge of interest in electronic Raman scattering in semiconductors[1, 2] has prompted us to extend our calculations[3, 4] of magnetic field effects on infrared impurity spectra to the case of Raman scattering. Two cases have been investigated: the effect of small magnetic fields on shallow acceptors in germanium and the effect of large magnetic fields on shallow donors in indium antimonide.

SHALLOW ACCEPTORS IN GERMANIUM

An external magnetic field removes the degeneracies of electronic impurity states in semiconductors and should therefore produce splittings of the Raman lines associated with transitions between such states. At small magnetic fields the effect is analogous to the ordinary Zeeman splitting of optical absorption lines. We have investigated in detail the case of small magnetic fields for shallow acceptor impurities in germanium. Attention is given to the electronic transitions which are forbidden by one-photon selection rules. The effective mass wave functions of Mendelson and James[5] have been used to calculate the magnetic splittings of the ground state, (8+ 00), and the lowest excited state of even parity, (8+ 01), by means of first order perturbation theory[3]. The four-fold degeneracy of each of these states in the absence of a magnetic field is fully removed by the application of the field.

We have calculated the effective g-factors, $g_{|M_J|}$, characterizing the splittings of each state. The g-factors we have determined are $g_{3/2} = 0.2879$, $g_{1/2} = -0.9720$ for the (8+00) and $g_{3/2} = -0.0827$, $g_{1/2} = -1.5260$ for the (8+01) state. The shift in energy of the M_J-th component of a multiplet from its unperturbed value is given by

$$\Delta E = g_{|M_J|} \beta H M_J \qquad (1)$$

where H is the magnetic field and β is the Bohr magneton. Raman transitions are allowed between pairs of components of the ground state multiplet. Also allowed are Raman transitions between the components of the ground state multiplet and various components of the excited state multiplet. Selection rules and polarization properties for these two types of transitions are displayed in Table I.

TABLE I

Selection rules for strong Raman transitions between the (8+ 00) and (8+ 01) multiplets and between the components of the (8+ 00) ground state for Ge. $\vec{\epsilon}_I$ and $\vec{\epsilon}_F$ refer to the polarization vectors of the incident and outgoing radiations respectively.

(8+00) → (8+01)			(8+00) → (8+00)		
Δm_j	$\vec{\epsilon}_I$	$\vec{\epsilon}_F$	Δm_j	$\vec{\epsilon}_I$	$\vec{\epsilon}_F$
0, ±2	$\perp\vec{H}$	$\perp\vec{H}$	±2	$\perp\vec{H}$	$\perp\vec{H}$
±1, ±3	$\|\vec{H}$	$\perp\vec{H}$	±1, ±3	$\|\vec{H}$	$\perp\vec{H}$
	$\perp\vec{H}$	$\|\vec{H}$		$\perp\vec{H}$	$\|\vec{H}$
0	$\|\vec{H}$	$\|\vec{H}$			

In the case of Raman induced transitions among the ground state components, one line with energy 0.05 βH is produced by Raman scattering for $\vec{\epsilon}_I \perp \vec{H}$, $\vec{\epsilon}_F \perp \vec{H}$, where $\vec{\epsilon}_I$ and $\vec{\epsilon}_F$ are unit vectors in the directions of the electric vectors of the incident and scattered radiation, respectively. For $\vec{\epsilon}_I \| \vec{H}$, $\vec{\epsilon}_F \perp \vec{H}$ or $\vec{\epsilon}_I \perp \vec{H}$, $\vec{\epsilon}_F \| \vec{H}$ three lines with energies 0.918 βH, 0.972 βH, 0.864 βH, respectively, appear. A schematic diagram of these transitions is given in Fig. 1.

The transition between (8+ 00) and (8+ 01) i.e. the E line, is forbidden for strong electric dipole transitions but is Raman-active. The splittings of the Raman E line in a low magnetic field are shown in Fig. 2.

No studies of Raman scattering from acceptors in germanium appear to have been reported. Wright and Mooradian[1], however, have observed a Raman line in boron-doped silicon in the absence of a magnetic field which probably corresponds to the (8+ 00) → (8+ 01) transition.

SHALLOW DONORS IN INDIUM ANTIMONIDE

At high magnetic fields such that $\hbar\omega_c >> E_I$ (ω_c is the cyclotron frequency and E_I is the binding energy of the current carrier to the impurity in zero magnetic field), the effect on the Raman transitions may be regarded as an impurity-shifted Raman scattering from Landau levels. The high magnetic field wave functions and energy levels of the shallow donor impurities in indium antimonide determined by Wallis and Bowlden[4] were employed in this study. The energy levels are characterized by the quantum numbers ℓ, m, λ, and the wave functions are of the following form.

$$F_{\ell m \lambda}(\sigma, \Phi, z) = c e^{im\Phi} e^{-(1/2)\sigma} \sigma^{(1/2)|m|} L_{\ell+|m|}^{|m|}(\sigma) P_\lambda(z) \exp(-\tfrac{1}{4}\gamma\epsilon^2 z^2) \quad (2)$$

The symbols all have their previously[4] defined meanings.

Several types of Raman transitions are possible. In Table II, we list the selection rules and polarization properties of the various cases. The first type involves transitions between bound states associated with the L = 0 Landau level. Here L = ℓ + 1/2 (m+|m|).

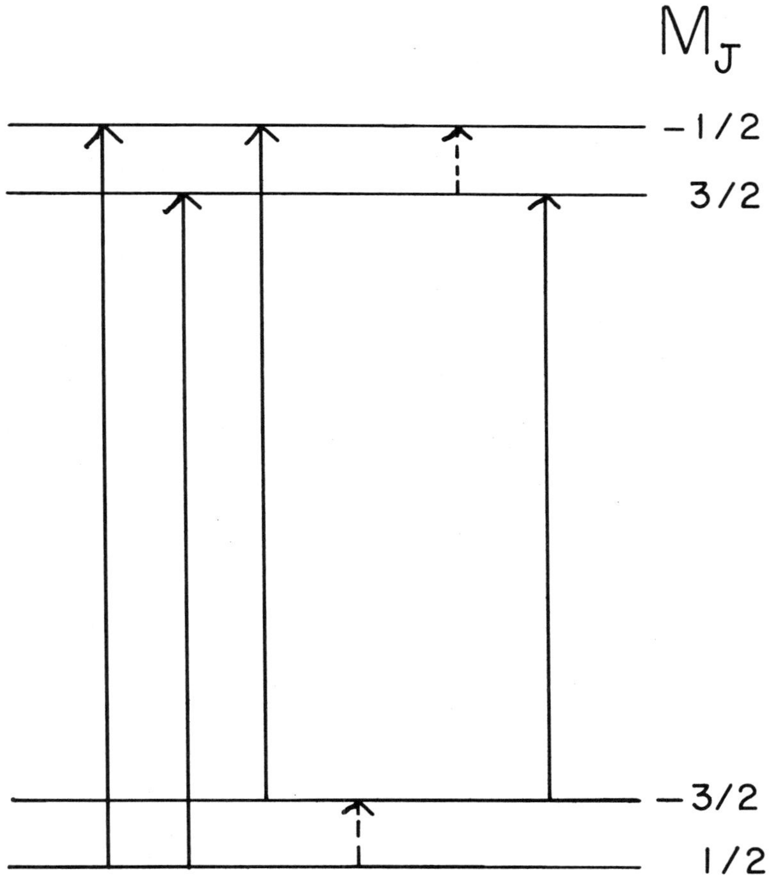

Fig. 1. Raman transitions between components of the (8+00) ground acceptor state in germanium. Solid lines represent transitions with $\vec{\epsilon}_I \perp \vec{H}$, $\vec{\epsilon}_F \parallel \vec{H}$ or $\vec{\epsilon}_I \parallel \vec{H}$, $\vec{\epsilon}_F \perp \vec{H}$; broken lines represent transitions with $\vec{\epsilon}_I$ and $\vec{\epsilon}_F$ perpendicular to $\vec{H}$.

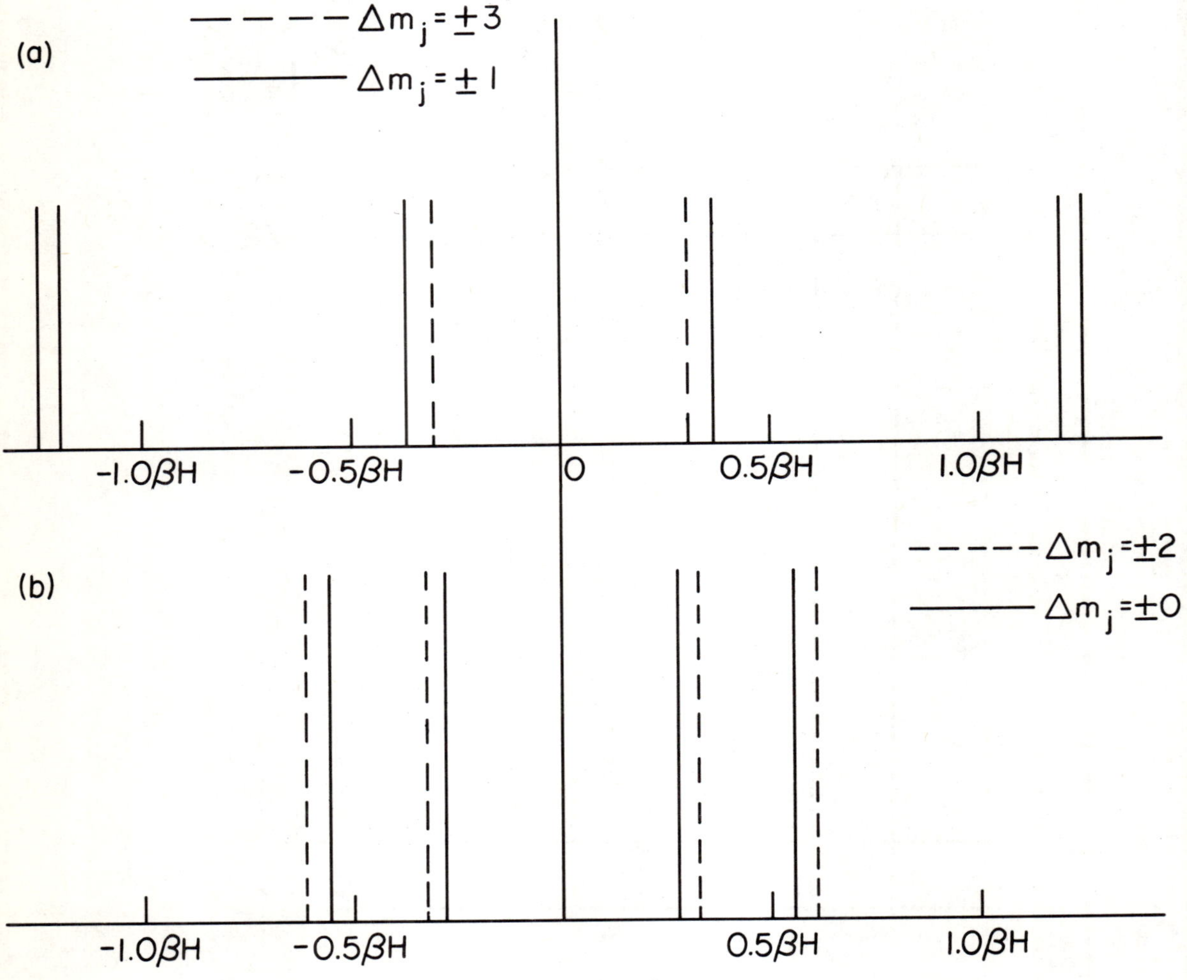

Fig. 2. (a) The splitting of the Raman E line in germanium in magnetic fields, when $\vec{\epsilon}_I \parallel \vec{H}$, $\vec{\epsilon}_F \perp \vec{H}$ or $\vec{\epsilon}_I \perp \vec{H}$, $\vec{\epsilon}_F \parallel \vec{H}$. Both $\Delta m_j = \pm 1$ (solid lines) and $\Delta m_j = \pm 3$ (broken lines) are allowed transitions.
(b) The splittings of the Raman E line in magnetic fields, when $\vec{\epsilon}_I \perp \vec{H}$, $\vec{\epsilon}_F \perp \vec{H}$ (both broken and solid lines); and when $\vec{\epsilon}_I \parallel \vec{H}$, $\vec{\epsilon}_F \parallel \vec{H}$ (solid lines only). The zero position is the position of the unperturbed E line.

TABLE II

Selection rules for Raman transitions between L=0, L=1, L=2 levels in InSb ($L = \ell + 1/2\ (m+|m|)$).

$\Delta\lambda$	$\Delta\ell$	Δm	$\vec{\epsilon}_I$	$\vec{\epsilon}_F$
even	0	-2	$\perp\vec{H}$	$\perp\vec{H}$
		2	$\perp\vec{H}$	$\perp\vec{H}$
even	0	0	$\perp\vec{H}$	$\perp\vec{H}$
			$\parallel\vec{H}$	$\parallel\vec{H}$
odd	0	1	$\perp\vec{H}$	$\parallel\vec{H}$
			$\parallel\vec{H}$	$\perp\vec{H}$
		-1	$\perp\vec{H}$	$\parallel\vec{H}$
			$\parallel\vec{H}$	$\perp\vec{H}$
even	1	0	$\perp\vec{H}$	$\perp\vec{H}$
odd	1	-1	$\perp\vec{H}$	$\parallel\vec{H}$
			$\parallel\vec{H}$	$\perp\vec{H}$
		1	$\perp\vec{H}$	$\parallel\vec{H}$
			$\parallel\vec{H}$	$\perp\vec{H}$

Let us introduce the quantity $\gamma = 1/2\ \hbar\omega_c/R_y^*$ as a measure of the magnetic field (R_y^* is the effective Rydberg, 0.00069 eV for InSb, and ω_c is the cyclotron frequency). The Raman transition diagram for $\gamma = 35$ (~ 55 kOe) is given in Fig. 3. We have also determined the frequency shifts as functions of magnetic field for these transitions. The result is shown in Fig. 4.

A second type of transition differs from the first in that the final states are bound states associated with the L = 1 Landau level. A third type involves final states consisting of bound states associated with the L = 2 Landau level. These two types of transition are shown in Fig. 5 for $\gamma = 35$. Their frequency shifts relative to the corresponding free carrier shifts are given as functions of magnetic field in Figs. 6 and 7. It is noticed in Figs. 4, 6, and 7 that, apart from the transitions involving both initial and final states with $\lambda = 0$, all other transitions have significant magnetic field dependence of the frequency shifts. From the coexisting features of the Rydberg series and the Landau levels[6], the transitions $(0\bar{1}0) \rightarrow (010)$ and $(0\bar{1}0) \rightarrow (001)$ pass continuously into the transitions between the 2p states in the low field limit. On the other hand, the transitions $(000) \rightarrow (100)$ and $(000) \rightarrow (0\bar{2}0)$ pass into those between 1s, 2s and 1s, 3d states.

We have not attempted to calculate scattering cross-sections; however, for certain of the transitions considered they should be comparable to those for the corresponding free carrier transitions.

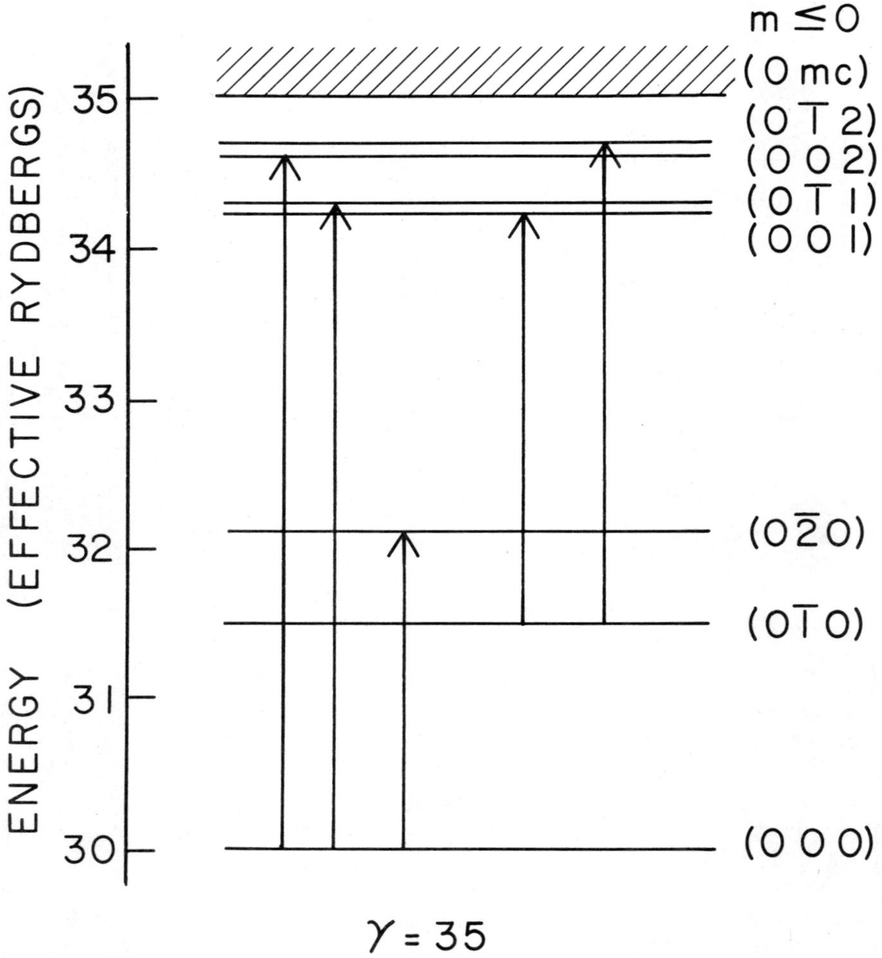

Fig. 3. Raman transitions between states associated with the L = 0 Landau level for shallow donor impurities in indium antimonide when $\gamma = 35$.

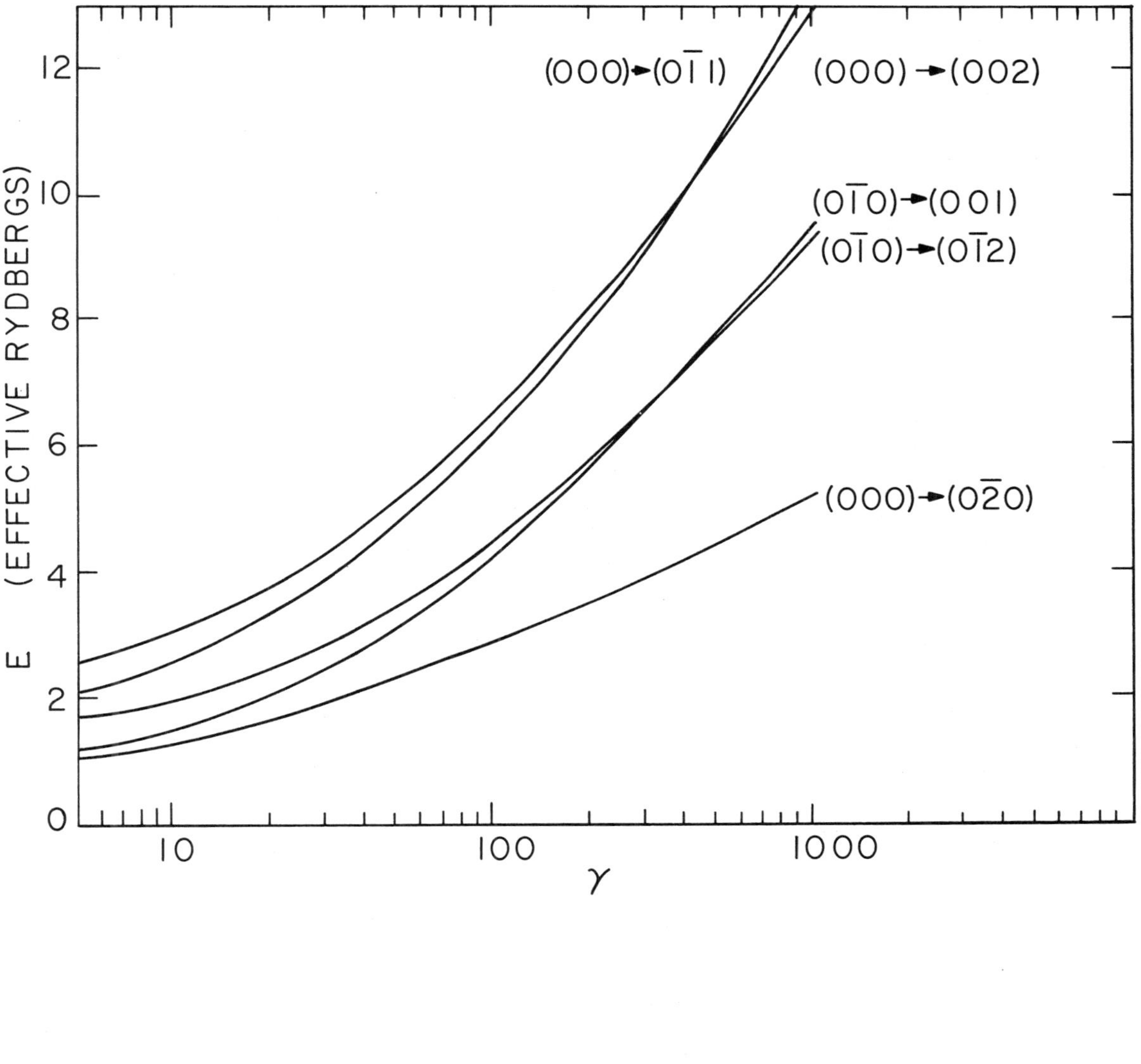

Fig. 4. Frequency shifts of transitions between states associated with the L = 0 Landau level as a function of γ.

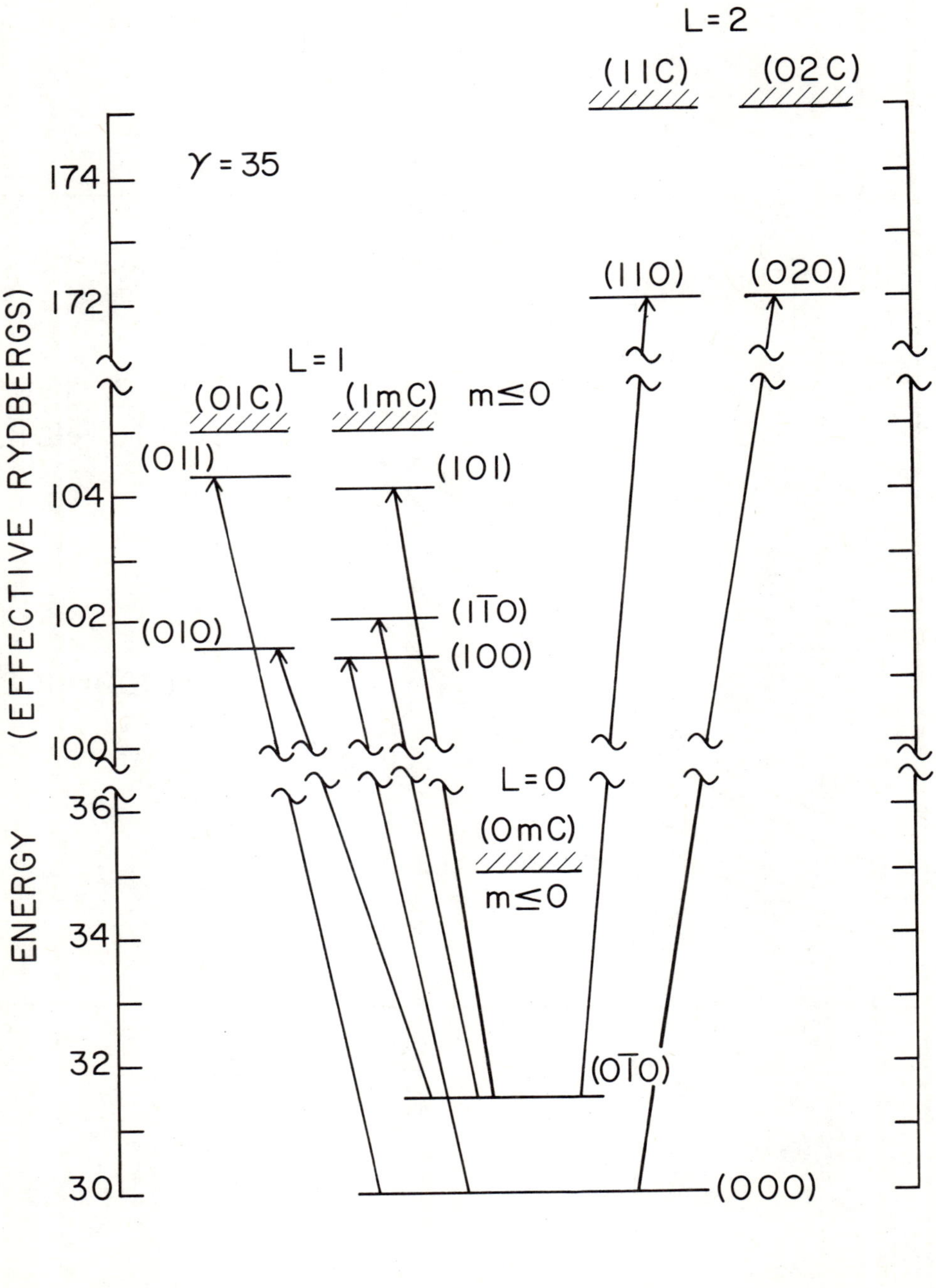

Fig. 5. Raman transitions between states associated with the L = 0 and the L = 1, L = 2 Landau levels for shallow donor impurities in indium antimonide when γ = 35.

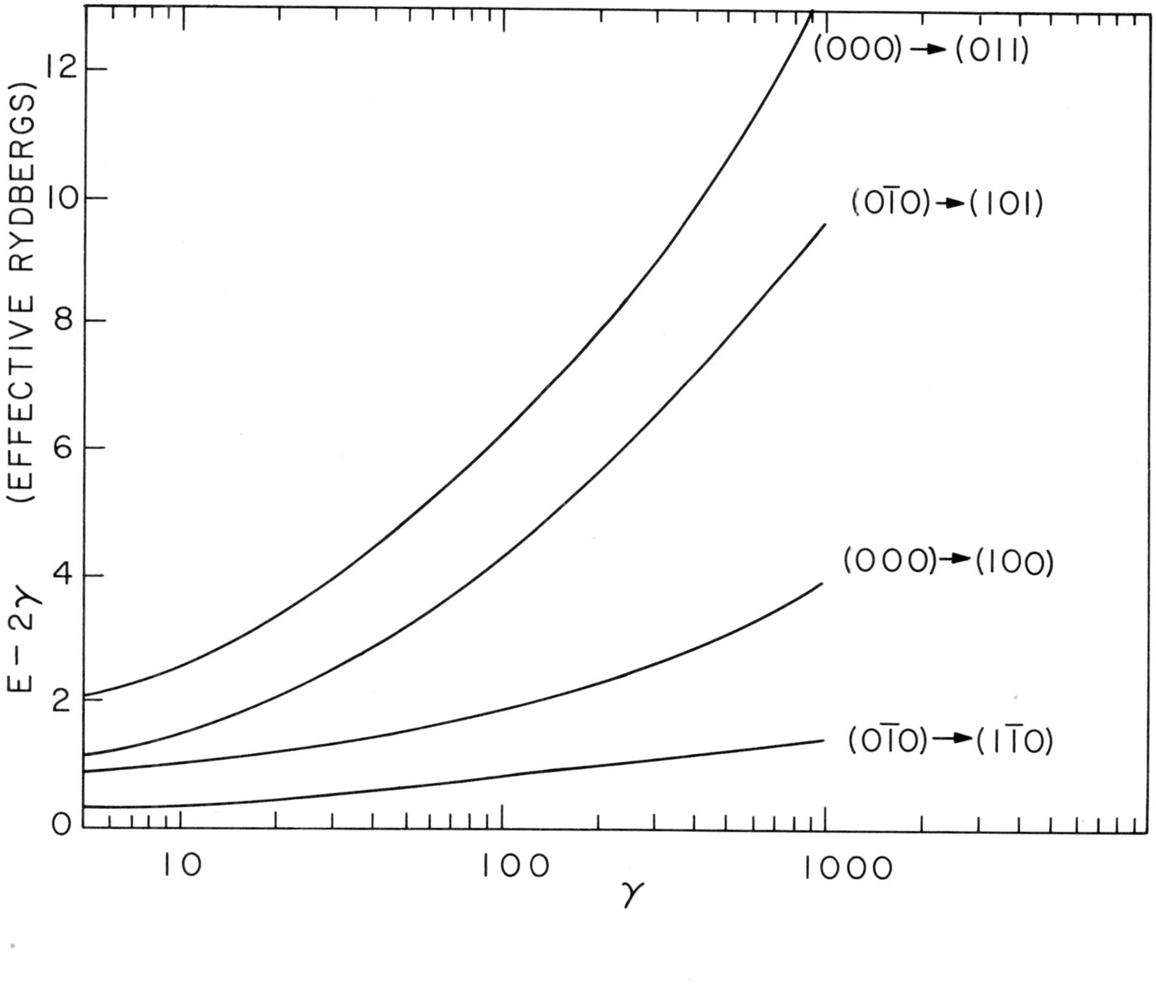

Fig. 6. Deviation from 2γ of the frequency shifts of transitions between states associated with the L = 0 and L = 1 Landau levels as a function of γ. The line $E - 2\gamma = 0$ is the line for the free carrier Raman transition.

The experimental observation of impurity-shifted Raman lines in high magnetic fields is most favorable for donors in InSb but faces a number of difficulties. To get well-defined impurity lines at convenient fields, the donor concentration should be $\sim 10^{14}$ cm^{-3} which is two orders of magnitude less than the concentrations used by Slusher et al. [2] in their work on free carriers. There thus may be some problem of sufficient intensity. Also, the experiments must be done at liquid helium temperature, and one must worry about the laser beam heating the sample and ionizing the impurities. Finally, the difference in frequency shifts of the bound carriers and free carriers is quite small and may be difficult to detect.

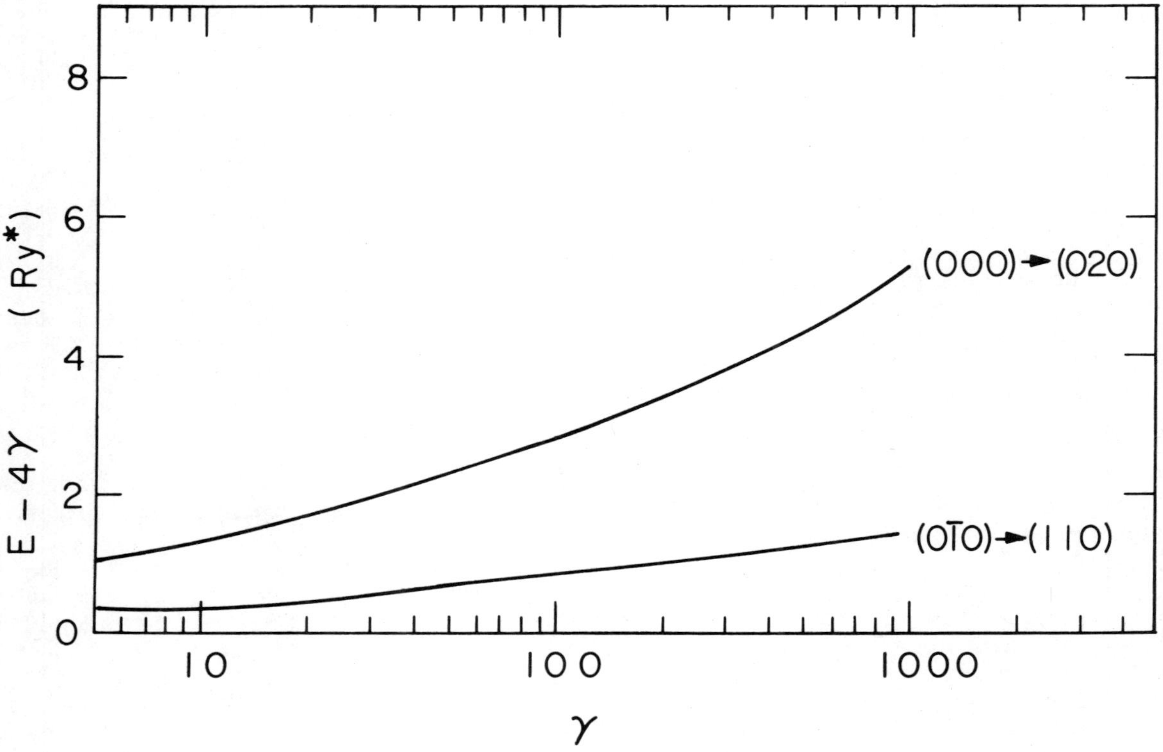

Fig. 7. Deviation from 4γ of the frequency shifts of transitions between states associated with the L = 0 and L = 2 Landau levels as a function of γ. The line E - 4γ = 0 is the line for the free carrier Raman transition.

REFERENCES

1. G.B. Wright and A. Mooradian, Phys. Rev. Letters 18, 608 (1967).
2. R.E. Slusher, C.K.N. Patel, and P.A. Fleury, Phys. Rev. Letters 18, 77 (1967).
3. P.J. Lin-Chung and R.F. Wallis, "Proceedings of the IXth International Conference on the Physics of Semiconductors" (to be published).
4. R.F. Wallis and H.J. Bowlden, J. Phys. Chem. Solids 7, 78 (1958).
5. K.S. Mendelson and H.M. James, J. Phys. Chem. Solids 25, 729 (1964).
6. H. Hasegawa (to be published).

D-1: LIGHT SCATTERING FROM SOLID STATE PLASMAS

P. A. Wolff
Bell Telephone Laboratories, Incorporated
Holmdel, New Jersey

INTRODUCTION

Light scattering from plasmas has been extensively studied during the past decade. Initially, this work was concerned with gaseous plasmas, but more recently the technique has been used to study plasmas in solids. Detailed theories of the scattering phenomenon have been developed for the case of the classical plasma[1]. This work indicates that the spectrum of radiation scattered from a plasma should consist of two distinct pieces; a single particle portion and a collective part. Single particle scattering is caused by individual moving electrons in the plasma, and is nearly elastic. This portion of the spectrum directly mirrors the electron velocity distribution, and can be used to determine it. In addition, there is collective mode scattering due to plasma waves in the electron gas.

Both the single particle and the collective mode scattering have been observed in experiments on ionospheric plasmas and laboratory gas discharges[2]. Such experiments are powerful tools for probing the structures of these plasmas because they determine two of the most important features which one wishes to know about a plasma; the frequencies and other properties of collective modes, and the velocity distribution of electrons in the plasma.

Light scattering experiments have also been used to study the properties of plasmas in semiconductors. The possibility of observing collective modes in solids by this method was suggested independently by McWhorter[3] and Platzman[4] and the first successful experiments were carried out by Mooradian and Wright[5]. They used light scattering to observe the plasma mode in n-type GaAs samples having various levels of doping. In such a semiconductor, the frequency of the plasma mode is modified because it couples to the lattice vibrations, but in all other respects the Mooradian-Wright experiments were understandable in terms of the classical theory developed earlier to explain light scattering from gaseous plasmas. Application of this theory to the solid state plasma case indicated that the collective mode scattering should have reasonable cross sections (as observed), but that under the experimental conditions employed by Mooradian and Wright the single particle scattering would be too weak to see. This was disappointing because, as we have seen, the single particle scattering can be used to determine electron velocity distributions.

Lately this situation has changed quite drastically. Mooradian[6] has actually observed the single particle portion of the spectrum in n-type GaAs and other semiconductors. In addition, a variety of theoretical calculations[7] have indicated that the scattering of light from solid state plasmas is a considerably more complicated phenomenon than the scattering of light from a classical gas plasma. The reason for this greater complexity is the more complicated dynamics of electron motion in crystals. Electrons in solids often have energy-momentum relations which are not the same as those of a classical free electron. These band structure effects produce important qualitative changes in the light scattering spectrum. In particular, they can greatly enhance the single particle portion. Thus, it now seems possible that single particle scattering can be used to determine electron velocity distributions in semiconductors. Several of the papers in this section are related in one way or another to this problem. To understand these new, and quite exciting, developments, it is absolutely essential to have some knowledge of the theory of light scattering from a classical plasma. The first part of this paper will review this theory, emphasizing the assumptions and the physical ideas involved. In discussing it, we will treat the plasma as a free electron gas, neutralized by a background of fixed positive charges. This model is the simplest one from the theoretical point of view and, fortunately, is also quite a good one for the semiconductors, such as n-type GaAs, InP, and CdTe, which have been investigated to date. We will discuss the predictions of this classical theory in various important limiting cases.

The latter part of the paper will describe recent work[7] concerning the effect of band nonparabolicity on light scattering from solid state plasmas. We will see that nonparabolicity produces new terms in the formula for the spectral distribution of scattered radiation, terms which quite strongly enhance the single particle portion of the spectrum. In materials such as InSb and InAs, these contributions to the single particle scattering are comparable in intensity to the collective mode scattering, which has been observed[8]. Thus there is a reasonable chance that one can measure velocity distributions in such materials via light scattering.

CLASSICAL THEORY OF LIGHT SCATTERING FROM A PLASMA

An important requirement of any good scattering experiment is that the incoming beam interacts weakly with the target. In the plasma case, this condition is insured by using primary radiation whose frequency is high compared to the electron plasma frequency. This condition is satisfied in all the experiments we have mentioned, and will be assumed from now on. When the frequency is high, the primary beam penetrates the target plasma essentially undeviated and unattenuated. Indeed, if the plasma were a perfectly homogeneous medium, this would be the only thing that happened. It is not, however, but consists of particles moving in space and time which can occasionally cluster together to produce a density fluctuation. These fluctuations give rise to light scattering. The spectrum of the scattered radiation directly measures the spectrum (in frequency and wave vector) of the density fluctuations within the plasma.

To develop a theory of this scattering process it is probably easiest to proceed from a Hamiltonian point of view. In Eq. (1) we have written the Hamiltonian for the electron gas coupled to an external electromagnetic potential, A.

$$H = \sum_i \left[\frac{\left(p_i - \frac{e}{c} A_i\right)^2}{2m} \right] + \frac{1}{2} \sum_{i \neq j} \left[\frac{e^2}{r_{ij}} \right] \tag{1}$$

We will assume that the frequencies of all the waves involved in this potential are large compared to the electron plasma frequency, so that they interact weakly with the plasma, and their coupling to it can be treated by perturbation theory. Notice that there are two types of electron-photon coupling terms in Eq. (1); terms of the form p· A, and terms of the form of A^2. It can easily be shown[9] that the p· A terms make an exceedingly small (of order v/c) contribution to the light scattering cross section. The whole light scattering cross section arises from the A^2 terms. For a scattering process in which an incident photon of wave vector q_0 scatters to a final state of wave vector q_1, these terms take the form shown in Eq. (2):

$$H_2 = \left(\frac{e^2}{mc^2}\right) \frac{2\pi hc^2}{\sqrt{\omega_0 \omega_1}} \quad (\epsilon_0 \cdot \epsilon_1) n(-q) \tag{2}$$

where

$$n(q) = \sum_i (e^{iq \cdot r_i})$$

$$q = q_0 - q_1$$

The photons couple directly to the electron density operator, n(q), thus confirming our earlier statement that electron density fluctuations are responsible for the light scattering.

It is now a straightforward matter, using first order perturbation theory, to calculate the scattering rate due to the A^2 terms in the electron-photon coupling. We will not discuss the details of this calculation, which is well known and quite standard[10], but just present the final results:

$$\frac{d^2\sigma}{d\Omega\, d\omega} = \left(\frac{e^2}{mc^2}\right)^2 \left(\frac{\omega_1}{\omega_0}\right) \int_{-\infty}^{\infty} e^{i\omega t} < n(q,t) n(-q,o) > \frac{dt}{2\pi} \tag{3}$$

where $\quad n(q,t) = e^{iH_0 t}\, n(q)\, e^{-iH_0 t}$

Here we see that the differential cross section for light scattering, into the element of solid angle dΩ and frequency interval dω, is directly proportional to the Fourier transform, in space and time, of the electron density-density correlation function. The light scattering experiment provides a direct measure of the density fluctuation spectrum in the plasma. One can also easily see that the frequency shifts in the scattering are Doppler shifts, due to the fact that the light is scattering from moving fluctuations. Eq. (3) is known as the Booker-Gordon formula[11] and is the starting point for all calculations of light scattering from classical plasmas.

Eq. (3) is an exact result, but the correlation function which appears in it must always be evaluated approximately. Actually, it is usually more convenient to evaluate a closely related function, the response function G(ω) of the electron gas:

$$G(t) = -i\theta(t) < [n(q,t),\ n(-q,0)] > \tag{4}$$

The two functions are related by the fluctuation dissipation theorem[12] (Eq. 5) which is a generalization of the Nyquist theorem.

$$\int_{-\infty}^{\infty} e^{i\omega t} < n(q,t) n(-q,o) > dt = \frac{-2}{(1-e^{-\beta\omega})} \ \mathrm{Im}\ [G(\omega)] \tag{5}$$

This equation is exact when the plasma is in thermodynamic equilibrium, and is approximately correct under other circumstances in weakly coupled plasmas.

One advantage of working with the response function is that it has a very direct physical interpretation. Let us imagine that we perturbed the plasma by a weak, external electrostatic potential, which couples to the electron density as shown in Eq. (6):

$$H' = e \int n(r)\varphi_{ext}(r, t)d^3r \tag{6}$$

This perturbation will produce a corresponding density fluctuation in the plasma. The induced electron density is related to the perturbing potential through the correlation function. Assuming an external potential of the form $\varphi_{ext}(rt) = \varphi_{ext}\, e^{i(q\cdot r-\omega t)}$ one finds

$$< n\,(q,\ \omega)> \;=\; eG(q, \omega)\, \varphi_{ext} \tag{7}$$

This is an exact result.

To calculate $G(\omega)$ one cannot, even in weakly interacting plasmas, treat the system as a gas of free, noninteracting electrons. The reason is quite simple. When a perturbation φ_{ext} is applied to the plasma, it induces an electron density as shown in Eq. (7). This induced electron density will, in turn, produce an induced potential in the plasma (φ_{ind}) which is comparable in magnitude to the external potential. One must always consider the response of the plasma to the total potential, $\varphi = \varphi_{ext} + \varphi_{ind}$. In weakly coupled plasmas it is permissible to treat the response to this total potential by perturbation theory. This is the basis for a very well-known approximation[13] in plasma work, the random phase approximation (RPA). To calculate $G(\omega)$ in this approximation we determine the response of a noninteracting electron gas to the total perturbation φ. This is a straightforward problem in quantum mechanical perturbation theory and the result is given in Eq. (8):

$$< n(q, \omega) > \;=\; -e\varphi\ \ F(q, \omega)$$

where

$$F(q, \omega) = \sum_k \left[\frac{f(k+q)-f(k)}{\omega + \frac{k^2}{2m} - \frac{(k+q)^2}{2m}} \right] \tag{8}$$

and f(k) is the electron momentum distribution.* The induced potential is related to the electron density through Poisson's equation:

$$q^2 \varphi_{ind} = 4\pi e\ \ <n(q, \omega)> \tag{9}$$

Combining Eqs. (7, 8 and 9) we may now obtain an expression for $G(\omega)$ valid in the random phase approximation. Finally, we may use Eq. (3) and Eq. (5) to obtain an expression for the light scattering cross section. This is our final formula:

*We use units such that $\hbar = 1$.

$$\frac{d^2\sigma}{d\Omega\, d\omega} = \left\{ \left(\frac{e^2}{mc^2}\right)^2 \left(\frac{\omega_1}{\omega_o}\right) \frac{1}{\pi(1-{}^{-\beta\omega})} \frac{\mathrm{Im}(F^*)}{\left| 1 + \frac{4\pi e^2}{q^2} F \right|^2} \right\} \tag{10}$$

If we had entirely neglected electron-electron interactions, a very similar formula would have resulted except that the square term in the denominator, which is the square of the longitudinal electron dielectric constant, $\epsilon(q, \omega) = \left(1 + \frac{4\pi e^2}{q^2} F\right)$, would not have been present. Thus, within the random phase approximation, the sole effect of electron-electron interactions is to reduce the differential cross-section for light scattering by the factor $| \epsilon (q, \omega) |^{-2}$.

Let us first consider a situation in which the effects of this dielectric constant screening are relatively unimportant. This occurs in the limit where the wave vector (q) transferred to the plasma is large compared to the characteristic plasma wave vector (the Debye wave vector in a classical plasma, or the Fermi-Thomas wave vector in a degenerate one). In this large q limit the dielectric constant which appears in Eq. (10) is nearly unity. The function IM(F*) in the numerator has the form shown in Eq. (11):

$$\mathrm{IM}(F^*) \approx \pi \sum_k \left[\frac{\partial f}{\partial \epsilon}\, \omega\, \delta\, \left(\omega - \frac{k \cdot q}{m}\right) \right] \tag{11}$$

The delta function appearing in this equation expresses energy-momentum conservation in a process in which a single electron scatters a light quanta. The total spectrum mirrors the electron velocity distribution in the plasma, and can be used to measure it. This limit of the spectrum is shown schematically by the solid line in Fig. 1.

Now let us consider the opposite limit, in which the wave vector q transferred to the plasma is small compared to the characteristic plasma wave vector, q_D. Under these conditions, the dielectric constant appearing in Eq. (10) is very large and greatly changes the predictions of the theory. For values of q and ω in the single particle range (see Fig. 1) $| \epsilon |^2$ is of order $(q_D/q)^4$. This factor enormously reduces single particle scattering in the limit $q << q_D$. In addition, the dielectric constant has a pole at the plasma frequency. This singularity contributes to the scattering cross section, giving rise to the collective mode scattering that we mentioned previously. Thus, the total spectrum in the small q limit has the form shown by the dotted curve in Fig. 1. The single particle portion of the spectrum is enormously reduced, and in addition, there is a peak at the plasma frequency. Notice that the total intensity is $(e^2/mc^2)^2 (q/q_D)^2$. It so happens that most light scattering experiments involving solid state plasmas are carried out in the limit $q << q_D$ described here. From Fig. 1 we see why, on the basis of the classical theory, one did not expect to see single particle scattering from such plasmas.

It is interesting to inquire into the physical reason for the reduction of the scattering cross section by electron-electron interactions. To investigate this point we integrate Eq. (3) to determine the total scattering cross section. This result is given in Eq. (12) and shows that the total cross section is proportional to the equal time, density correlation function.

$$\frac{d\sigma}{d\Omega} = \left(\frac{e^2}{mc^2}\right)^2 \iint \langle n(r)n(r')\rangle \, e^{iq\cdot(r-r')} d^3r d^3r' \qquad (12)$$

The form of this function is well known in the classical equilibrium case and is indicated in Eq. (13):

$$\langle n(r)n(r')\rangle = n_o \left[\delta(r-r') - \frac{q_D^2 e^{-q_D|r-r'|}}{4\pi|r-r'|}\right] \qquad (13)$$

An important point is that there are two terms in the correlation function; the first delta function represents correlation of any given electron with itself, the second term correlation of this electron with other electrons in the plasma. In plasmas with other sorts of velocity distributions the second term in this formula will have different forms, but the essential feature will remain that each electron carries with it a screening cloud which contains a net deficiency of exactly one electron. We now see what happens when one calculates the light scattering cross section. The second term in the Eq. (13) contributes negatively to Eq. (12). The reason is quite simple. In the vicinity of any given electron there is, because of correlation, a deficiency of electrons. There is less scattering from plasma in the vicinity of a given electron than there would be if one

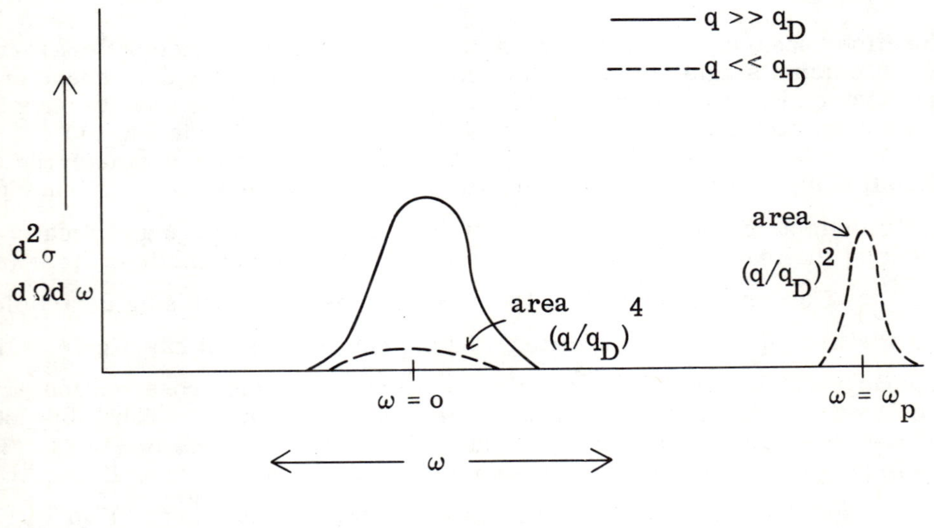

$$\omega_p^2 = \frac{4\pi ne^2}{\varepsilon m^*} \; ; \quad (q_D)^{-1} = \text{Debye length or Fermi-Thomas length}$$

Fig. 1. Spectrum of light scattered from a classical, single-component plasma in the limiting cases $q >> q_D$ and $q << q_D$.

calculated from the average electron density. This deficiency in scattering adds to the scattering from the given electron, but with a negative sign, and thus tends to reduce the light scattering cross section. It is this effect that is represented by the factor $(q/q_D)^2$ which reduces the total scattering cross section.

SCATTERING FROM A NONPARABOLIC PLASMA

In the proceeding section we have outlined the theory of light scattering from a classical, single component plasma. We now wish to see how these calculations are modified when the electrons have energy-momentum relations other than those of the classical form[7]. As we will see presently, band nonparabolicity enables light to couple to fluctuations other than electron density fluctuations in the plasma - for example, energy density fluctuations. Such fluctuations need not carry a charge, and thus are relatively less affected by Coulomb interactions than are the density fluctuations. As a consequence, nonparabolic plasmas can give rise to single particle scattering which is unscreened and relatively strong, even in the long wavelength (small q) limit. This scattering is expected to be particularly important in materials such as InSb or InAs, whose conduction bands are quite nonparabolic. There is a reasonable chance that light scattering experiments in these materials could be used to determine electron velocity distributions.

To make estimates of the strength of this new form of scattering, we proceed as before from the Hamiltonian. We will use the effective mass approximation to describe the plasma - an approach which can be shown to be valid in the limit where photon frequencies are small compared to the direct bandgap. In this approximation the Hamiltonian of the plasma coupled to the electromagnetic field is given by Eq. (14):

$$H = \sum_i \left[E\left(p_i - \frac{e}{c} A_i\right)\right] + \frac{1}{2}\sum_{i\neq j}\left(\frac{e^2}{r_{ij}}\right)$$

$$\cong H_0 - \frac{e}{c}\sum_i \left(A_i \cdot \frac{\partial E}{\partial p_i}\right) + \frac{e^2}{2c^2}\sum_i \left(A_i \cdot \frac{\partial^2 E}{\partial p_i \partial p_i} \cdot A_i\right) + \ldots \qquad (14)$$

where E(p) is the energy-momentum relation. Strictly speaking, this Hamiltonian should be symmetrized in the operators A_i and p_i, which do not commute. However, if we are considering the scattering of light of relatively large wavelength, this effect is not important and can be ignored. The expansion of the Hamiltonian in powers of the vector potential yields terms that are linear in A and quadratic in A. As in the classical plasma, it is the latter which make the important contributions to the scattering of cross sections.

It is important to notice that the A_i^2 term in Eq. (14), besides coupling light to density fluctuations, also couples it to other sorts of fluctuations of the plasma. To illustrate this point it is perhaps easiest to consider an example. In Eq. (15) we have written the Hamiltonian appropriate to electrons in the conduction band of materials such as InSb, and below it the coupling term which arises from such an energy momentum relation.

$$E(p) \simeq \frac{p^2}{2m^*} - \left(\frac{1}{E_G}\right)\left(\frac{p^2}{2m^*}\right)^2$$

$$H_2 = \left(\frac{e^2}{2m^*c^2}\right)\sum_i \left[A_i^2\left(1 - \frac{p_i^2}{m^*E_G}\right) - \frac{2(A_i \cdot p_i)(A_i \cdot p_i)}{m^*E_G}\right] \tag{15}$$

Notice that the last two terms couple light to energy density fluctuations of the plasma. These need not be screened and thus can give rise to fairly strong single particle scattering.

From this coupling it is a straightforward manner to calculate, by perturbation theory, the differential scattering cross section. The result is given in Eq. (16):

$$\frac{d^2\sigma}{d\Omega\, d\omega} = \left(\frac{\omega_1}{\omega_0}\right)\int_{-\infty}^{\infty} e^{i\omega t} \langle \mathcal{H}_2^+(t)\, \mathcal{H}_2(o)\rangle \frac{dt}{2\pi}$$

where

$$\mathcal{H}_2 = \left(\frac{e}{c}\right)^2 \sum_i \left[\left(\epsilon_0 \cdot \frac{\partial^2 E}{\partial p_i \partial p_i} \cdot \epsilon_1\right) e^{iq\cdot r_i}\right]$$

and

$$\mathcal{H}_2(t) = e^{iH_0 t}\, \mathcal{H}_2^+\, e^{-iH_0 t}. \tag{16}$$

As before, we may relate the correlation function appearing in Eq. (16) to a corresponding response function through the fluctuation dissipation theorem. These relations are shown in Eq. (17).

$$\bar{G}(t) = -i\theta(t) \langle [\mathcal{H}_2^+(t), \mathcal{H}_2(o)]\rangle$$

$$\int_{-\infty}^{\infty} \langle \mathcal{H}_2^+(t)\, \mathcal{H}_2(o)\rangle e^{i\omega t} \frac{dt}{2\pi} = \frac{-2\,\mathrm{Im}\,[\bar{G}(\omega)]}{(1-e^{-\beta\omega})} \tag{17}$$

$\bar{G}$ represents the response of the electron gas to a fictitious perturbation which couples to the operator $\mathcal{H}_2$ for the electron system, as shown in Eq. (18):

$$\langle \mathcal{H}_2^+(q, \omega)\rangle = \bar{G} V_{ext} \tag{18}$$

It is important here to realize that V_{ext} is not an electrostatic potential.

To determine $\bar{G}$, and hence the scattering cross section, we use the random phase approximation. The perturbation V_{ext} induces a charge density in the electron system which in turn, via Poisson's equation, will create an electrostatic potential. We may treat the plasma as a noninteracting electron gas in calculating its response to this total perturbation, shown in Eq. (19).

$$H'_{total} = \mathcal{H}_2(-q)V_{ext}(q,\omega) + en(-q)\,\varphi_{ind}(q,\omega) \qquad (19)$$

Two quantities are required, the induced value of $\mathcal{H}_2^+$ and the induced electron density. Finally, the induced potential is related to the electron density by Poisson's equation. These relations are shown in Eq. (20).

$$\begin{aligned} \langle \mathcal{H}_2^+(q,\omega)\rangle &= -V_{ext}(q,\omega)F_2(q,\omega) - e\,\varphi_{ind}(q,\omega)F_1(q,\omega) \\ \langle n(q,\omega)\rangle &= -V_{ext}(q,\omega)F_1(q,\omega) - e\,\varphi_{ind}(q,\omega)F_0(q,\omega) \\ q^2\,\varphi_{ind}(q,\omega) &= 4\pi\langle n(q,\omega)\rangle \end{aligned} \qquad (20)$$

where

$$F_\alpha(q,\omega) = \left(\frac{e}{c}\right)^2 \sum_k \left\{ \frac{\left(\epsilon_c \cdot \frac{\partial^2 E}{\partial p\partial p} \cdot \epsilon_1\right)^\alpha [f(k+q)-f(k)]}{\omega + E(k) - E(k+q)} \right\}$$

The last step in the analysis is the elimination of φ_{ind} and $\langle n(q,\omega)\rangle$ from these equations to obtain a relation between the induced value of $\mathcal{H}_2$ and the external perturbation φ. This relation determines the response function $\bar{G}$ and thereby the scattering cross section. The final result is given in Eq. (21).

$$\frac{d^2\sigma}{d\Omega\,d\omega} = \left(\frac{\omega_1}{\omega_0}\right)\left[\frac{1}{\pi(1-e^{-\beta\omega})}\right] \times \left\{ \mathrm{Im}\left[\frac{-F_2(q,\omega)}{1+\frac{4\pi e^2}{q^2}F_0(q,\omega)}\right] + \left(\frac{4\pi e^2}{q^2}\right)\mathrm{Im}\left[\frac{F_1^2(q,\omega)-F_0(q,\omega)F_2(q,\omega)}{1+\frac{4\pi e^2}{q^2}F_0(q,\omega)}\right]\right\} \qquad (21)$$

We have broken this formula into two terms: the first corresponds to the classical result in the limit where the bands are parabolic; the second vanishes for parabolic bands, but gives an important contribution when the bands are nonparabolic. In the limit $q \ll q_D$, the factors $4\pi e^2/q^2$ cancel in the numerator and denominator of the second term, yielding a result which is independent of q.

We have evaluated[7] Eq. (21) in detail for the Hamiltonian appropriate to n-type InSb. These calculations have been performed for both Fermi-Dirac and Maxwellian distributions. The results are illustrated in Figs. 2 and 3. It is important to realize that the area under

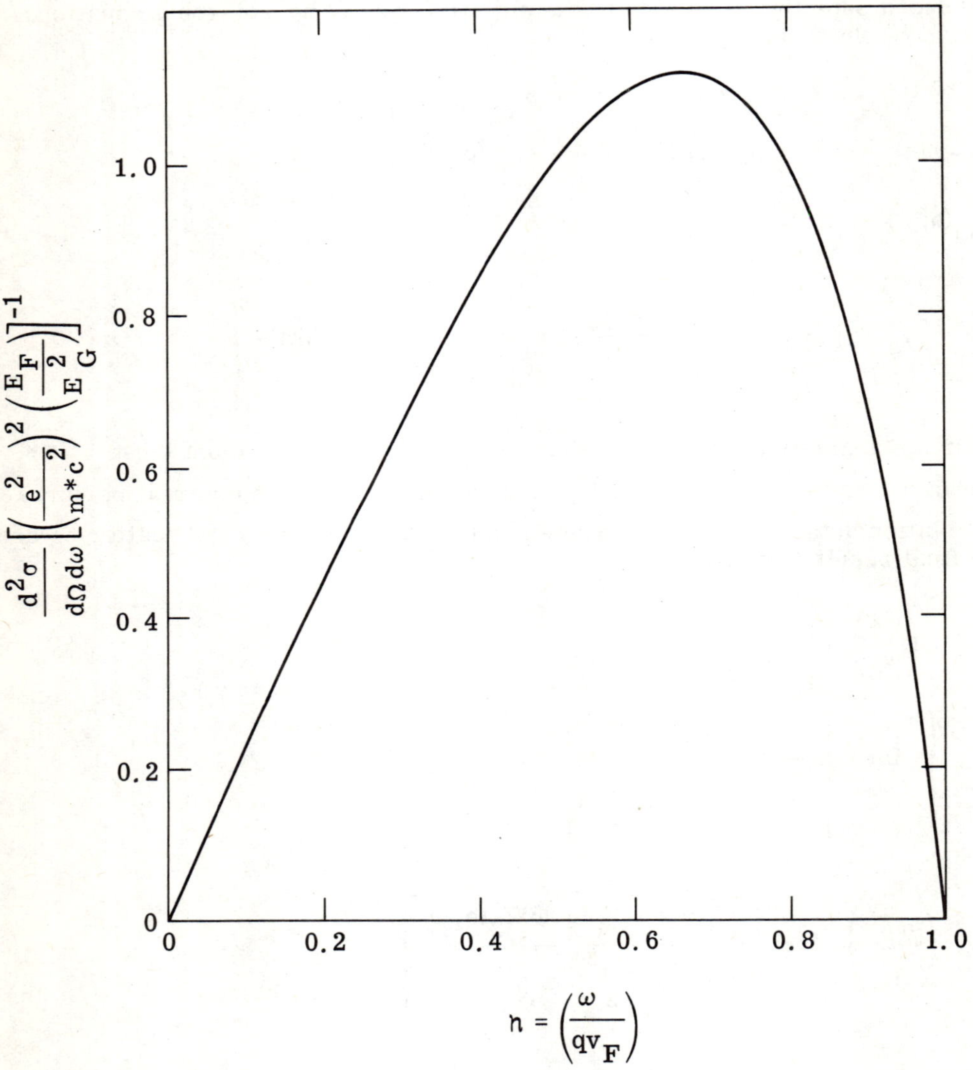

Fig. 2. Single particle (quasi-elastic) spectrum for a degenerate plasma in n-type InSb.

these spectra does not contain the factor $(q/q_D)^2$ which so greatly reduces the single particle scattering in the classical case. This is because, as we indicated earlier, the scattering in these plasmas is caused by energy density fluctuations, which are unscreened. In the Maxwellian case the area under the single particle line is several times $\left(\frac{e^2}{m^*c^2}\right)^2 (kT/E_G)^2$. For electron temperatures of the order of 100°K this scattering intensity is larger than that in the plasma line. The width of the single particle spectrum directly measures the electron temperature. If this portion of the light scattering spectrum could be observed it would be useful as a monitor of average electron energy. In cases where the velocity distribution is not of the Maxwellian type it might also be possible to use the light scattering spectrum to determine some of its properties.

The Fermi-Dirac case is somewhat less interesting because the exclusion principle introduces a factor which reduces the total cross section. Nevertheless, here too, single particle light scattering is unscreened and might be observable.

Finally, it should be mentioned that this calculation does not appear to be capable of explaining the single particle scattering that Mooradian[6] has observed in n-type

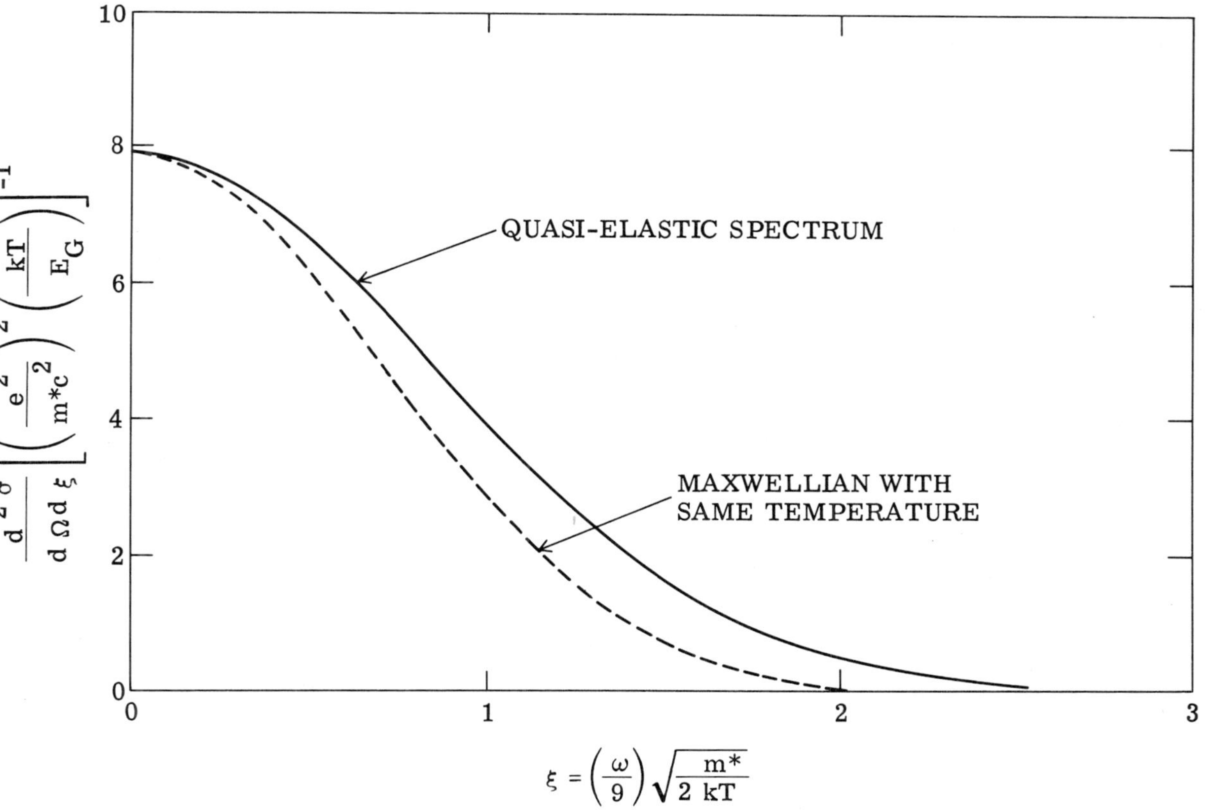

Fig. 3. Single particle (quasi-elastic) spectrum for a Maxwellian plasma in n-type InSb. The corresponding Maxwellian distribution is indicated for comparison.

gallium arsenide. For the gallium arsenide case our theory predicts a single particle scattering cross section considerably larger than that predicted by the classical theory, but still not big enough to explain the observations. The gallium arsenide band structure is nearly parabolic so the mechanism proposed here does not play an important role.

CONCLUSION

In this paper we have reviewed the classical theory of light scattering from plasmas, and suggested a new mechanism by which light may couple to fluctuations other than those of electron density in plasmas having nonparabolic energy momentum relations. The classical theory indicates that, under most experimental conditions, it will be difficult to observe the single particle portion of the light scattering spectrum. Its intensity is reduced because of screening. On the other hand, in nonparabolic plasmas, light couples to fluctuations which are unscreened. As a consequence, the single particle scattering can be quite strong. Detailed calculations indicate that in InAs or InSb its strength should be comparable to, or possibly greater, than that of the plasma line. Since the plasmon scattering has been observed in InAs there seems a good chance that one might also be able to see the single particle scattering and use it as a tool to measure electron temperatures and even velocity distributions in such crystals.

REFERENCES

1. J.P. Dougherty and D.T. Farley, Proc. Roy. Soc. (London) A259, 79 (1960); E.E. Salpeter, Phys. Rev. 120, 1528 (1960); J.A. Fejer, Can J. Phys. 38, 1114 (1960); M.N. Rosenbluth and N. Rostoker, Phys. Fluids 5, 776 (1962).
2. K.L. Bowles, Phys. Rev. Letters 1, 454 (1958); J.V. Evans and M. Lowenthal, Planetary Space Sci. 12, 915 (1964); W.E.R. Davies and S.A. Ramsden, Phys. Letters 8, 179 (1964); H.J. Kunze, E. Fünfer, B. Krossast, and W.H. Kegel, Phys. Letters 11, 42 (1964); A.W. DeSilva, D.E. Evans, and M.J. Forrest, Nature 203, 1321 (1964); B. Krossast, H. Rohr, E. Glock, H. Zwicher, and E. Fünfer, Phys. Rev. Letters 16, 1082 (1966); S. Ramsden et al, J. Quant. Elects. QE-2, 8, 267 (1966).
3. A.L. McWhorter, "Proceedings of the International Conference on the Physics of Quantum Electronics, Puerto Rico 1965," McGraw-Hill, New York, 1965.
4. P.M. Platzman, Phys. Rev. 139, A379 (1965).
5. A. Mooradian and G.B. Wright, Phys. Rev. Letters 16, 999 (1966).
6. A. Mooradian, Phys. Rev. Letters 20, 1102 (1968).
7. P.A. Wolff, Phys. Rev. 171, 436 (1968). Also see paper D-4 this conference.
8. C.K.N. Patel and R.E. Slusher, Phys. Rev. 167, 413 (1968).
9. P.M. Platzman and N. Tzoar, Phys. Rev. 136, A11 (1964).
10. L. Van Hove, Phys. Rev. 95, 249 (1954). Also see Ref.[9].
11. W.E. Gordon, Proc. Inst. Radio Engrs. 46, 1824 (1958).
12. D.N. Zubarev, Sov. Phys. Uspekhi 3, 320 (1960).
13. See, for example, David Pines and Philippe Nozieres, "The Theory of Quantum Liquids," I, W.A. Benjamin Inc., New York, 1966.

D-2: LIGHT SCATTERING FROM SINGLE - PARTICLE ELECTRON AND HOLE EXCITATIONS IN SEMICONDUCTORS

A. Mooradian
Lincoln Laboratory, * Massachusetts Institute of Technology
Lexington, Massachusetts

ABSTRACT

Light scattering from single-particle electron excitations has been observed in GaAs, InP, CdTe, and AlSb, as well as from holes in p-type GaAs and n-type epitaxial GaAs. A YAG:Nd^{3+} laser operating at 1.06 microns with a CW output of up to 20 W was used to excite the single-particle excitations. GaAs samples with electron concentrations from 4×10^{14} to $3 \times 10^{18} cm^{-3}$ were measured. The single-particle scattering could be observed in samples with electron concentrations as low as $10^{12} cm^{-3}$. For electron concentrations of a few times $10^{16} cm^{-3}$ and less, the single-particle excitation spectrum overlaps and Landau damps the collective plasmon excitations at room temperature. The intensity of the single-particle scattering in single-crystal GaAs was measured as a function of polarization and concentration to determine the dominant scattering mechanisms. In p-type GaAs at room temperature the scattering from both light and and heavy holes was observed, the line shape being characterized by two superimposed Maxwellian velocity distributions having different thermal velocity cutoffs. The effect of an applied electric field on the velocity distribution of electrons is discussed.

INTRODUCTION

The inelastic scattering of light from a solid state plasma provides a great deal of information concerning not only the excitation spectrum of the electron gas[1] but details of the electron-photon interaction in the medium itself. Because of recent developments in the area of high-power continuous lasers which operate in the infrared, study of the light scattering spectra in previously opaque solids has become possible. The particular solids of interest here are semiconductors such as GaAs, InP and CdTe which have their gaps in the infrared and are thus studied by the 1.06 micron YAG:Nd^{3+} laser.

*Operated with support from the U.S. Air Force.

EXPERIMENTAL

The experiments were performed using a 1.064 micron YAG:Nd^{3+} laser which could operate with a continuous output in excess of twenty watts. Most of the samples were rectangular parallelepipeds cut to dimensions of 3x3x5 mm. These samples were mounted in a dewar where the temperature could be varied from 1.5°K to 300°K by adjusting the flow of cold helium gas. The polarization measurements were carried out on oriented single-crystal samples. A polarized laser beam was incident along a (100) axis of the crystal while the scattered light was collected at 90 degrees along another (100) axis. In the best samples, the elastically scattered laser light was sufficiently low in intensity to permit measurements to within 1.5 cm^{-1} of the laser line at low temperatures. Spectra were recorded with incident powers less than 10 milliwatts to make certain that local heating of the sample did not affect the spectra. A check on the sample temperature was made from the intensity of the anti-Stokes components and was found to be in quite good agreement with the value obtained from direct measurement.

In most cases, the electron concentrations of each sample or an adjacent slice was measured by the Hall or Van Der Pol method. The entire optical system response was calibrated using a standardized quartz iodine lamp. The S-1 response phototube sensitivity drops off almost exponentially towards longer wavelength, which accounts for anti-Stokes components being about equal in intensity to Stokes components at room temperature.

RESULTS

The single-particle excitation spectrum at 0°K arises from electrons (or holes) with momentum $\vec{p}$ that are excited from occupied states below the Fermi surface to unoccupied states just above the Fermi surface with momentum $\vec{p} + \vec{q}$, where $\vec{q}$ is the momentum transferred to the electrons by the scattered light. When the momentum transfer is much smaller than twice the Fermi momentum p_F, the excitation spectrum increases linearly with frequency and cuts off sharply at qV_F, where V_F is the Fermi velocity. For q still less than $2p_F$, but now an appreciable fraction of it, the spectrum has a maximum that shifts toward a lower frequency. As the temperature of the electron gas increases, the Fermi surface becomes smeared and more electrons can participate in the scattering process. In the high temperature limit of a Maxwellian velocity distribution, the single-particle excitation spectrum has a finite value at zero frequency and a Gaussian line shape with a tail determined by kT, where T is the electron temperature. When q is larger than the screening momentum q_D in the plasma, the excitations of the system are the collective modes. However, when q is small compared to q_D, then the excitations of the system correspond to those of single particles. In semiconductors such as GaAs the screening momentum can vary from much less to much greater than the momentum imparted by a 1 eV laser. In such a material we can study then the entire range of the electron excitation spectrum from the single-particle to the collective regime.

In Fig. 1 we show the spectra of scattered light at room temperature from a series of n-type GaAs samples with electron concentration varying from 3 x $10^{15} cm^{-3}$ to 1.4 x $10^{18} cm^{-3}$. The top trace, which is for a semi-insulating sample (room temperature electron concentration of about $10^8 cm^{-3}$), shows the Stokes and anti-Stokes scattering

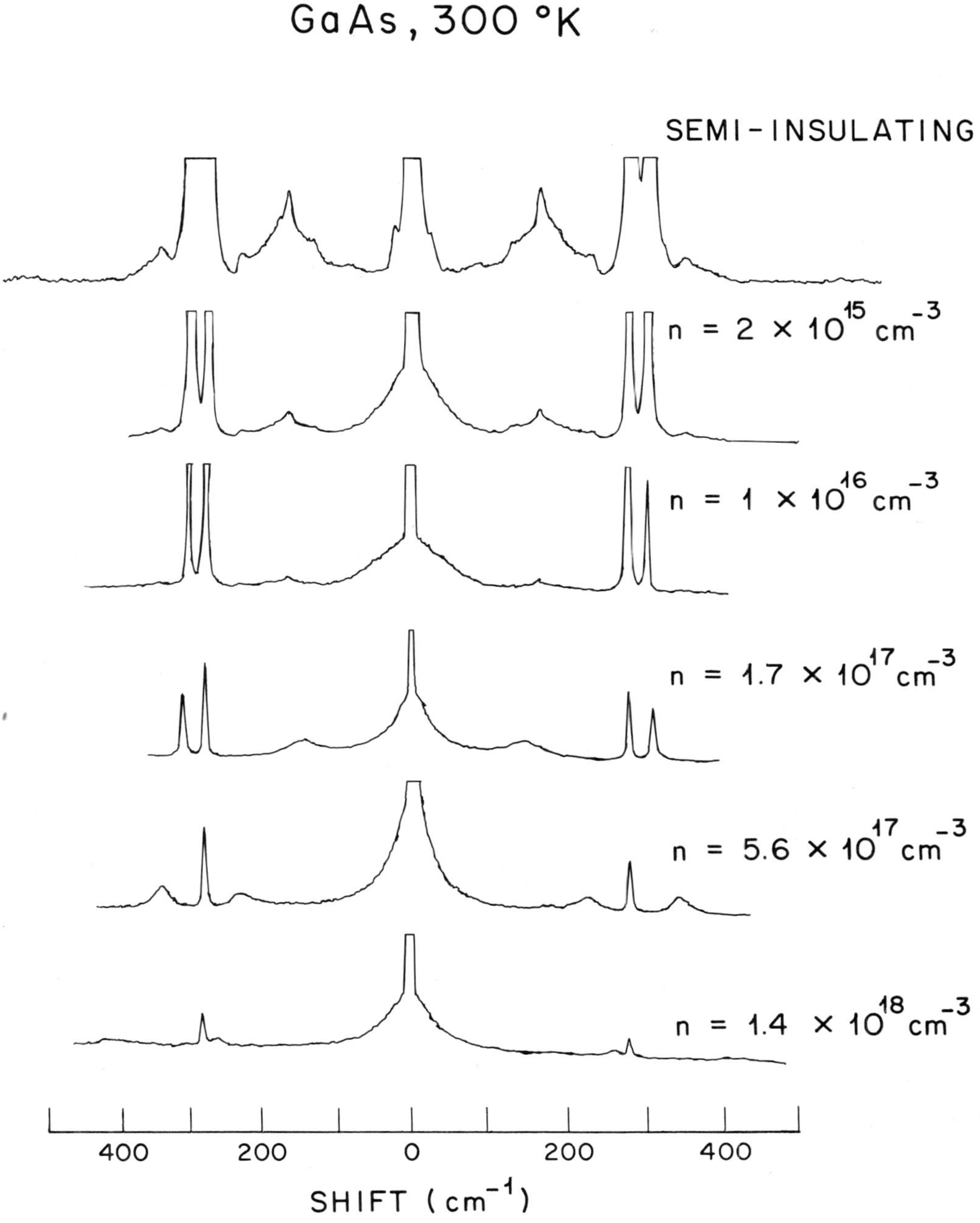

Fig. 1. Raman spectra of n-type GaAs samples at room temperature.

from one and two phonon processes. The two strong peaks at 272 cm^{-1} and 296 cm^{-1} are from first-order scattering of zone center transverse and longitudinal optical phonons, respectively. The remaining structure comes from scattering involving two phonon processes. The strong peaks around 20 cm^{-1} in the semi-insulating spectrum are due to grating ghosts and come from the large amount of laser light that is scattered into the spectrometer from occlusions of metallic chromium. Chromium has a tendency to precipitate in such closely compensated samples of semi-insulating GaAs. For samples with electron concentrations of up to $10^{16} cm^{-3}$, an exponential continuum spectrum is observed with a finite intensity at zero frequency. This scattering arises from the single-particle excitations of the conduction electrons. For n less than $10^{16} cm^{-3}$, the plasma frequency is less than qV_{th}, where V_{th} is the thermal velocity of the electrons. Most of the modes of the electron gas are then contained in the single-particle excitations rather than the collective excitations, which are damped out. The sample with $n = 1 \times 10^{16} cm^{-3}$ just barely shows the plasma line emerging from the single-particle continuum. The increasingly higher concentration samples show the plasmon mode coupled to the longitudinal optic (LO) phonon as the plasma frequency passes through the LO phonon frequency. The presence of the one- and two-phonon Raman bands serve as a reference intensity marker for the electron scattering.

Fig. 2 shows the temperature dependence of the single-particle spectrum in GaAs with $n = 1 \times 10^{16} cm^{-3}$. At room temperature, q/q_D is about unity and most of the modes in the electron gas are single-particle excitations as the plasmon is heavily Landau damped. At low temperature the plasma line emerges from the single-particle tail as the screening momentum increases from the Debye to the Fermi-Thomas momentum, and most of the modes in the electron gas become collective plasma oscillations. At low temperature the line width is dominated by electron collisions with impurities, although there is still a small amount of overlap of the single-particle tail with the collective mode due to the lifetime smearing of the single-particle cutoff. This overlap was quite prominent in the very heavily doped samples studied. The plasmon mode, while heavily damped, was observed at 5°K for electron concentration down to $3 \times 10^{15} cm^{-3}$. Some high-mobility epitaxial GaAs samples having thicknesses of 120 microns or less were studied in the range of concentration of around $1 \times 10^{16} cm^{-3}$. The mobilities were in excess of 20,000 cm^2/Vsec, which gave a collision lifetime almost four times greater than that in a comparably doped boat-grown sample. The plasmon line width at 80°K was consistent with the difference in the mobility lifetime in the two samples. At liquid helium temperature the plasma line width in the epitaxial samples showed a considerable broadening, which is attributed to a decrease in mobility lifetime as there is a strong onset of ionized impurity scattering at low temperature. This behavior shows that the dominant contribution to the plasmon and single-particle line shape in the low temperature limit comes from collisions of electrons with impurities. This degraded lifetime at the low temperatures may preclude use of such otherwise desirable samples for say magnetic field studies. Epitaxial samples with electron concentrations less than a few times $10^{15} cm^{-3}$ exhibit carrier freeze-out at low temperature and would therefore present some further problems. Light scattering from the single-particle and collective modes of electrons in InP, CdTe, and AlSb has also been observed.

The cross-section for scattering from electron density fluctuations has been calculated within the effective mass approximation by Platzman[2] and McWhorter[3] for parabolic bands. Wolff[4] has extended these calculations to include the effects of the interband

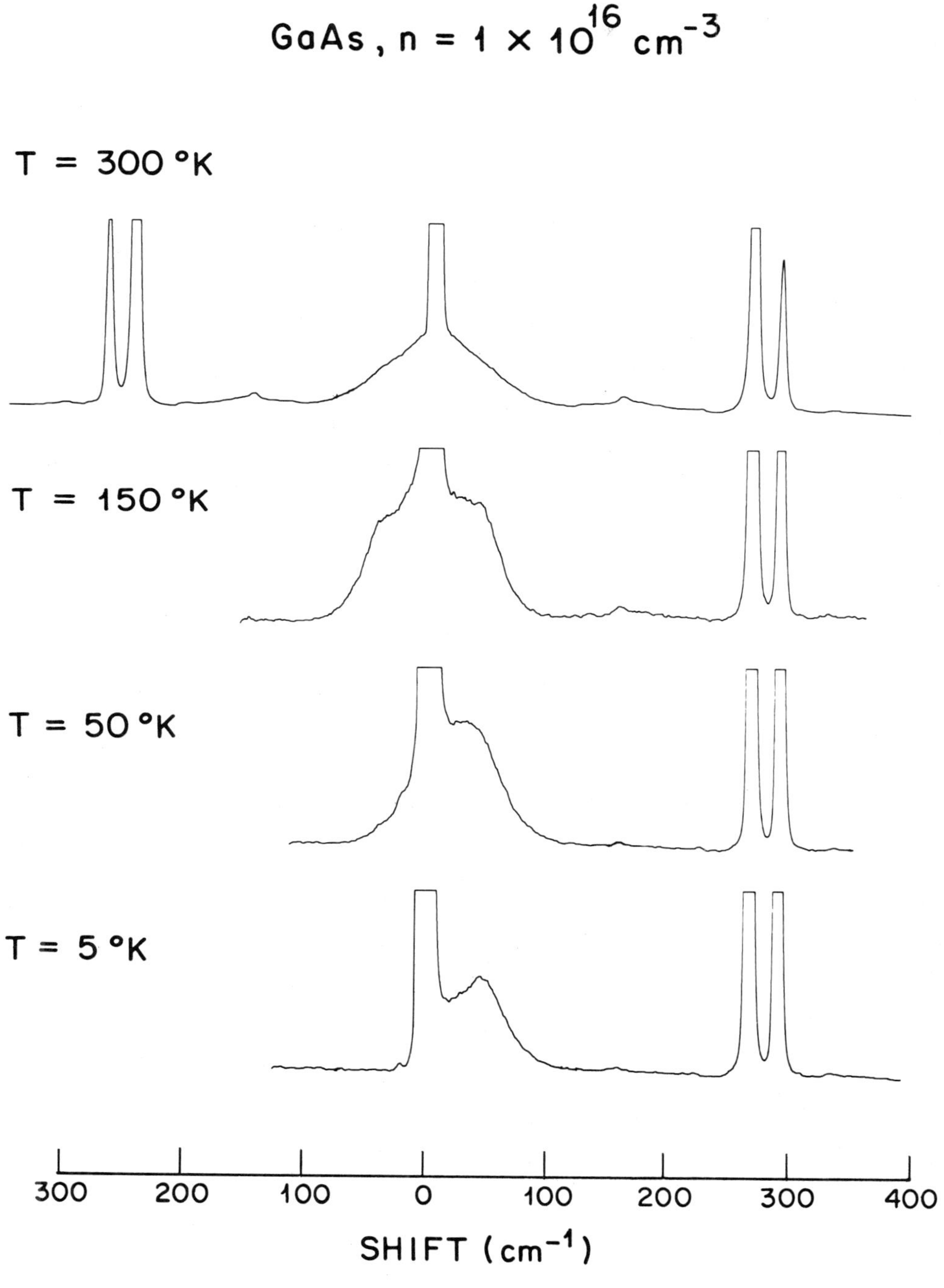

Fig. 2. Single particle spectrum in GaAs, $n = 1x10^{16} cm^{-3}$ as a function of the temperature.

contribution to the single-particle cross section. The single-particle cross section for scattering from electron charge-density fluctuations is given by

$$\frac{d\sigma}{d\omega} = (n_\omega + 1) \left[\frac{E_g^2}{E_g^2 - \hbar^2 \omega_i^2} \right]^2 \left(\frac{e^2}{m^* c^2} \right)^2 \frac{1}{\pi} (\vec{e}_i \cdot \vec{e}_s) \frac{\mathrm{Im}\ Q}{\left| 1 - \frac{4\pi e^2}{\epsilon(\omega) q^2} Q \right|^2} \quad (1)$$

where for $q << p_F$

$$Q = 2 \int \frac{d^3 \vec{p}}{h^3} \frac{\vec{v} \cdot \vec{p} \frac{\partial f}{\partial \epsilon}}{\vec{v} \cdot \vec{q} - \omega - i/\tau}$$

$(e^2/m^*c^2)^2$ is the effective Thompson cross section, m^* is the electron effective mass, n_ω is the Boson occupation factor, $f(\epsilon)$ is the electron distribution function, ω is the frequency of the excitation, τ is a phenomenological lifetime, E_g is the energy gap, ω_i the laser frequency, $\vec{e}_i$ and $\vec{e}_s$ are the polarization vectors of the incident and scattered light, respectively, and $\epsilon(\omega)$ is the lattice contribution to the dielectric constant. The latter is given by

$$\epsilon(\omega) = \epsilon_\infty + \frac{(\epsilon_o - \epsilon_\infty)\omega_t^2}{\omega_t^2 - \omega^2} \quad (2)$$

where ω_t is the TO phonon frequency, ϵ_o the static dielectric constant, and ϵ_∞ the optical dielectric constant.

The integrated single-particle scattering intensity was measured as a function of the electron concentration for samples with carrier concentrations ranging from $4 \times 10^{14} \mathrm{cm}^{-3}$ to $1.4 \times 10^{18} \mathrm{cm}^{-3}$ in order to determine the nature of the scattering mechanism. The results are shown in Fig. 3 where the scattering intensity is normalized to the TO phonon intensity. In the limit of low concentrations, i.e. q/q_D greater than unity, the scattering is dominated by electron charge-density fluctuations. The theoretical curve for the charge-density fluctuation scattering given by Eq. (1) is the solid line peaking at $q/q_D = 1$. This curve, which varies as $[1 + (q/q_D)^2]^2$ was normalized to the lowest concentration single-particle intensity. At higher concentrations where q/q_D becomes less than unity, the single-particle cross section contribution from the charge-density fluctuations becomes screened out, as most of the excitations revert to the collective modes. The incident and scattered light from the charge-density fluctuations should both be polarized in the same direction, which is consistent with the experimental observations, i.e. $(\perp, \perp)$. Despite the fact that the charge-density fluctuations are screened out in the higher concentration samples, the single-particle scattering not only increases with increasing electron concentrations but all possible polarization contributions appear. Wolff[5] has calculated a contribution to the single-particle cross section which arises from a nonparabolic conduction band. This mechanism would not be screened out in the region where $q/q_D < 1$. A previous[6] comparison of the ratio of the integrated single particle intensity to the plasmon intensity has shown that

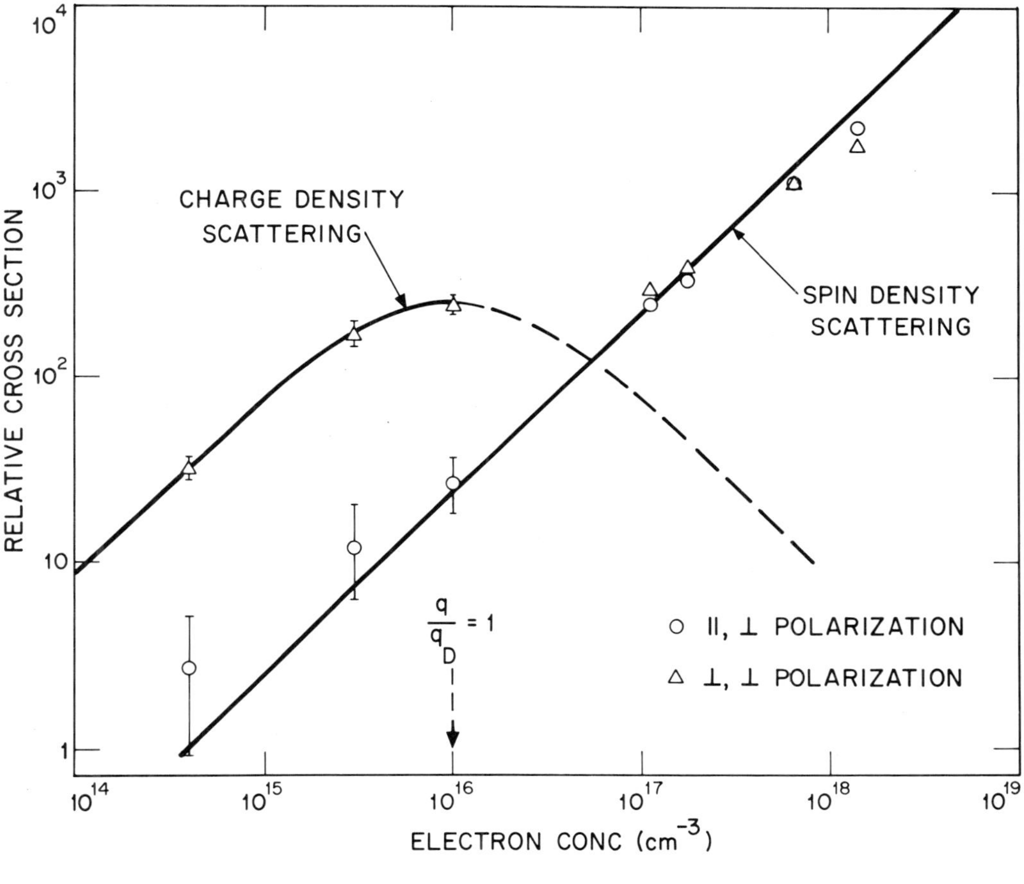

Single Particle Raman Cross-Section Relative to TO Phonon
n-type GaAs, 300 °K

Fig. 3. The integrated single particle Raman cross section in n-type GaAs at 300 $^{\circ}$K as a function of the electron concentration showing both the charge density and the spin density scattering.

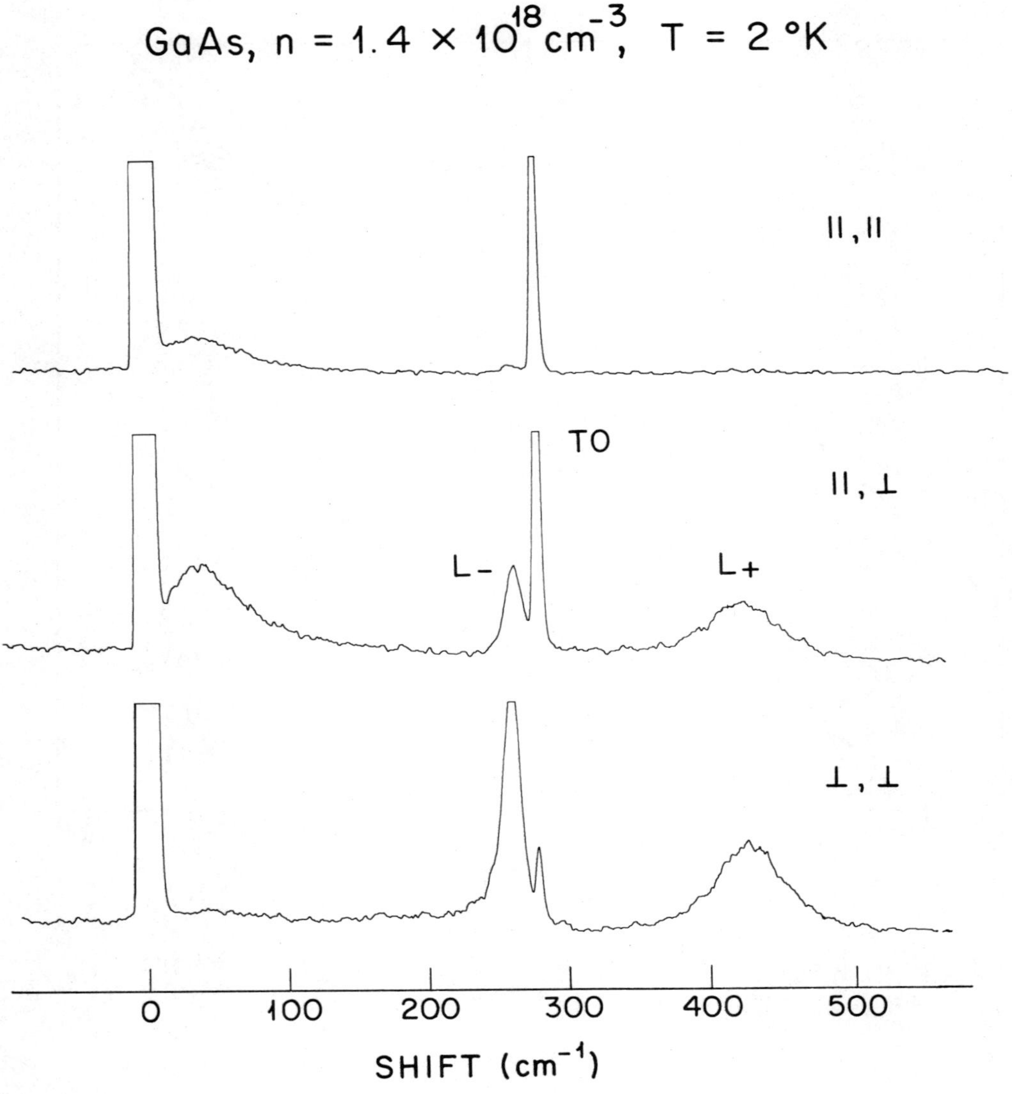

Fig. 4. Polarized Raman spectrum at 2°K in GaAs, n = $1.4x10^{18}$ cm^{-3}. The || and ⊥ signs refer to the polarization of the incident and scattered light in the scattering plane.

the nonparabolic contribution to the single particle intensity is between one and two orders of magnitude too small to explain the observed intensity. Recently, Hamilton and McWhorter[7] have calculated the contribution of the spin-orbit split valence band to single-particle scattering cross section. The interband $p \cdot A$ matrix elements between the conduction band and the spin-orbit split valence band give rise to spin-density scattering. This contribution appears to account for the observed single-particle intensity and the variation with electron concentration for the $(\|, \perp)$ and $(\|, \|)$ components at 300°K. The straight line in Fig. 3 was drawn by calculating the ratio of the charge-density scattering to the spin-density scattering in the low-concentration limit (a theoretical factor of 32.4) and passing a straight line with unity slope through this point. The fit at high concentrations is seen to be remarkably good. The spin-density scattering theory predicts the existence of all polarization components except $\vec{e}_i \cdot \vec{e}_s$ at low temperature. Fig. 4 shows the polarization of a high concentration ($n = 1.4 \times 10^{18} cm^{-3}$) sample at 2°K. The crystal was oriented so that the incident and scattered light propagated along the (100) directions in order to use the TO phonon for a check of the alignment. The $\vec{e}_i \cdot \vec{e}_s$ polarization component $(\perp, \perp)$ is at least an order of magnitude smaller than the other polarization components. As the temperature is increased, however, the size of this polarization component relative to the $(\|, \perp)$ and $(\|, \|)$ components increases almost linearly with T up to room temperature. This variation is accounted for in a forthcoming publication[8].

Eq. (1) in the limit $\omega_p << qV_{th}$ and infinite relaxation time has the following shape for a Maxwellian velocity distribution:

$$\frac{d\sigma}{d\omega} \rightarrow \exp\left(-\frac{\omega^2}{q^2 V_{th}^2}\right) \tag{3}$$

This line shape is plotted in Fig. 5 along with the experimentally determined line shape for a sample of GaAs at 300°K having 3×10^{15} electrons per cm^3. The fit between experiment and theory is good to within a few per cent over most of the frequency range. The experimental uncertainty at low-frequency is due to the presence of the laser line while the uncertainty at high-frequency is due to smaller single-particle intensity coupled with subtracting out the two-phonon combination bands. This line shape, which is characteristic of a Maxwellian velocity distribution, occurred in samples with concentrations up to $10^{18} cm^{-3}$, where at room temperature the electron distribution became degenerate.

Scattering from both light and heavy holes in p-type GaAs at room temperature was also observed and could be characterized by two superimposed Maxwellian distributions. The line shape could be fitted quite well by using the values of light and heavy hole masses as determined from magneto-optical studies.

An interesting aspect of scattering from the single-particle excitations is the possibility of determining experimentally the carrier velocity distribution functions under the influence of an applied electric field. Deviations from a Maxwellian would be quite evident. For electric fields in the region where Gunn oscillations occur, electrons are involved in inter-valley transfer with the production of large numbers of phonons having momenta not at $q = 0$. Because the two-phonon Raman bands are easily observed, it should be possible to determine some of the details of the electron kinetics from the nature of the driven two-phonon spectra. Both the types of phonons as well as their symmetry might be determined.

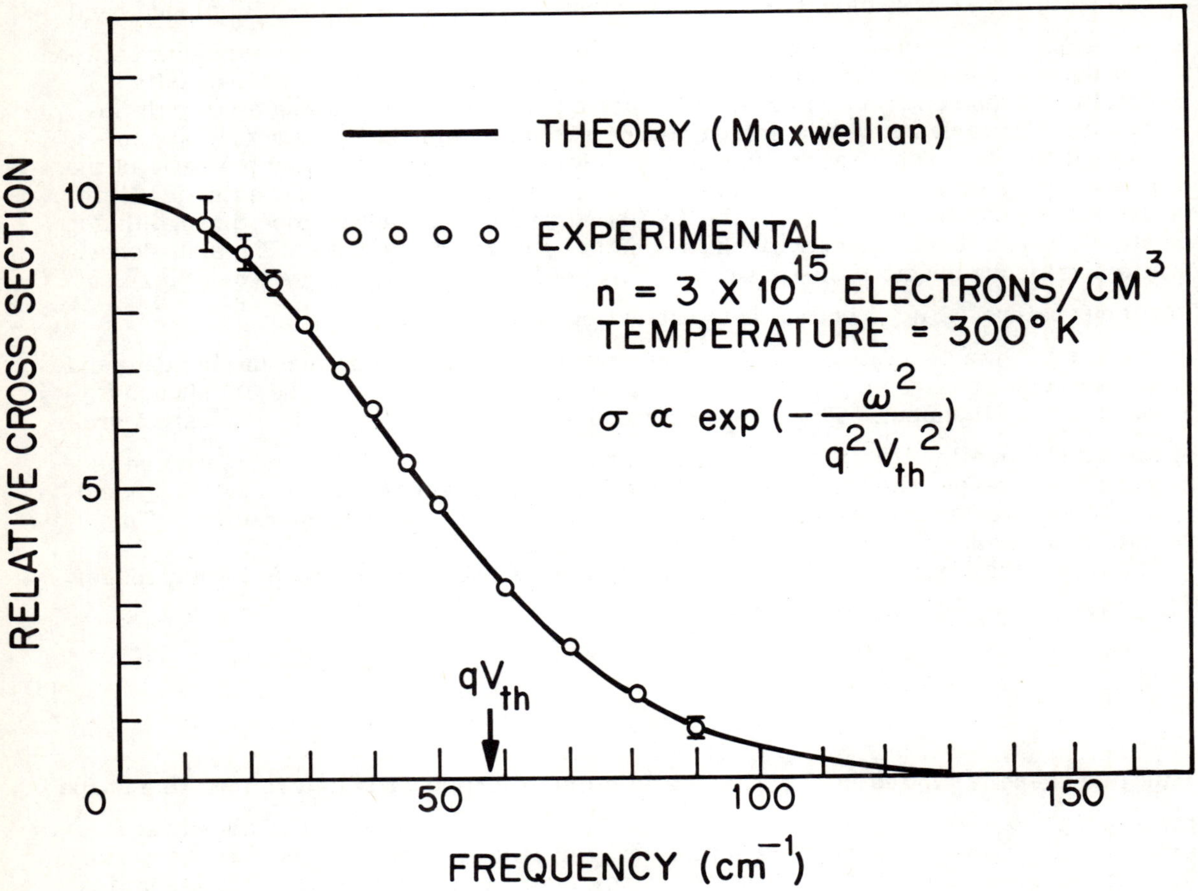

Fig. 5. Single particle line shape theory vs. experiment for GaAs, $n = 3 \times 10^{15}\,cm^{-3}$ at room temperature showing the Maxwellian profile.

ACKNOWLEDGEMENTS

The author would like to thank A. L. McWhorter, D. C. Hamilton, and P. A. Wolff for helpful discussions. He would also like to thank D. J. Wells for assistance with the measurements. Also, W. Laswell for polishing and K. Nearen for cutting the number of crystals used.

REFERENCES

1. An extensive treatment of the properties of a solid state electron gas is contained in "The Theory of Fermi Liquids" D. Pines and P. Nozieres, W.A. Benjamin Co. Inc., New York, 1966; and "Elementary Excitations in Solids", D. Pines, W.A. Benjamin Co. Inc., New York, 1963.
2. P.M. Platzman, Phys. Rev. 139, A379 (1965).
3. A.L. McWhorter, "Physics of Quantum Electronics," p. 111, P.L. Kelley, B. Lax, and P.E. Tannenwald (eds.) McGraw-Hill Book Company Inc., New York, 1966.
4. P.A. Wolff, Phys. Rev. Letters 16, 225 (1966).
5. P.A. Wolff, Phys. Rev. 171, 436 (1968).
6. A. Mooradian, Phys. Rev. Letters 20, 1102 (1968).
7. D.C. Hamilton and A.L. McWhorter, Intl. Conf. on Light Scattering in Solids, paper D-4 New York, 1968
8. A.L. McWhorter, A. Mooradian, and D.C. Hamilton (to be published).

D-3: LIGHT SCATTERING FROM PLASMONS AND PHONONS IN Ga As

A. Mooradian and A. L. McWhorter
Lincoln Laboratory*, Massachusetts Institute of Technology
Lexington, Massachusetts

ABSTRACT

Previous studies of the properties of the scattered light from mixed plasmon-phonon modes in n-type GaAs have been extended to cover a wider range of electron concentration. Oriented single-crystal samples with electron densities from 4×10^{17} to 3×10^{18} cm^{-3} were investigated both at room temperature and liquid helium temperature. The polarization properties and relative intensities of the scattered light from the mixed plasmon-phonon modes are in satisfactory agreement with cross-section calculations based on contributions from charge-density fluctuations as well as from deformation potential coupling and electro-optic coupling. Improved values are obtained for the Raman scattering tensor and electro-optic coefficient in GaAs, partly as a result of treating more carefully the effect of resonant enhancement.

INTRODUCTION

Light scattering from a solid state plasma was first observed by Mooradian and Wright[1] in n-type GaAs, where the conduction electron plasmons were seen to couple to the longitudinal optical (LO) phonons as the plasma frequency was swept through the LO phonon frequency. A brief description of the polarization and intensity dependence of the scattered light from the coupled plasmon-phonon modes in GaAs at liquid helium temperature has previously been reported[2]. The present work extends these measurements to cover a wider range of concentrations and in addition includes some results at room temperature. The theory has also been extended to treat more carefully the effect of resonant enhancement. Reasonably good agreement is obtained between theory and experiment, although there may be a small systematic discrepancy at the higher concentrations.

*Operated with support from the U.S. Air Force.

EXPERIMENTAL

The scattered light was generated using a 1.06 micron YAG:Nd^{3+} laser. Most of the experimental techniques are described elsewhere[2, 3]. Polished single-crystal samples with (100) faces and dimensions 3x3x5 mm were used. The polarized laser beam was incident along a <100> axis of the crystal while the scattered light was collected at 90° along another <100> axis. The solid angle of collection was about 0.08 sr. The system response was calibrated as a function of polarization and wavelength by using a standardized quartz-iodine light source.

The two most difficult experimental problems encountered in making these quantitative measurements were luminescence and internal reflections. Most of the GaAs samples luminesced quite strongly under the intense 1.06 micron laser excitation, thereby introducing phototube shot noise and a sloping baseline. Samples were carefully selected for their low luminescent intensity at all temperatures in the broad 1 micron emission band characteristic of most GaAs. Despite careful optical alignment, multiple internal reflection in the samples prevented the rejection of unwanted polarization components in the spectra. This was partly overcome by using larger samples; the stronger absorption in the more heavily doped samples also helped by damping out the internal reflections.

The range of electron concentrations accessible for quantitative measurements of the plasmon-phonon modes in GaAs lies between the limits of about 4×10^{17} cm^{-3} and 5×10^{18} cm^{-3}. The high-concentration limit is set by the solubility of donors in GaAs. Above a few times 10^{18} cm^{-3} the electron density becomes quite non-uniform throughout the crystal. Even before this high-concentration limit is reached, however, it becomes impossible with the present apparatus to carry out polarization studies on the upper-frequency plasmon-phonon branch because of the rapid falloff of the phototube response. For concentrations much below about 4×10^{17} cm^{-3}, the low-frequency plasmon-phonon mode merges with the tail of the single-particle spectrum[3] and becomes Landau damped.

COUPLED PLASMON-PHONON MODES

The general properties of the coupled plasmon-phonon modes have already been described[1, 2], but will be summarized here for completeness. Fig. 1 is a plot of the Raman frequency shifts as a function of the square root of the electron concentration and shows the mixing between the LO phonon mode of frequency ω_ℓ and the longitudinal plasma mode of frequency $\omega_p = (4\pi ne^2/\epsilon_\infty m^*)^{1/2}$. Here, n is the electron concentration, m* the conduction band effective mass, and ϵ_∞ the optical dielectric constant. The eigenfrequencies of longitudinal collective modes are given by the zeros of the total dielectric response function, which in the long wavelength limit can be written as

$$\epsilon(\omega) = \epsilon_\infty \left[1 - \frac{\omega_p^2}{\omega(\omega + i/\tau)} \right] + \frac{(\epsilon_0 - \epsilon_\infty)\,\omega_t^2}{\omega_t^2 - \omega^2} \qquad (1)$$

where ϵ_0 is the static dielectric constant, ω_t is the transverse optical phonon frequency, and τ is a phenomenological collision time. The solid lines in Fig. 1 are the zeros of $\epsilon(\omega)$ with $\tau \to \infty$.

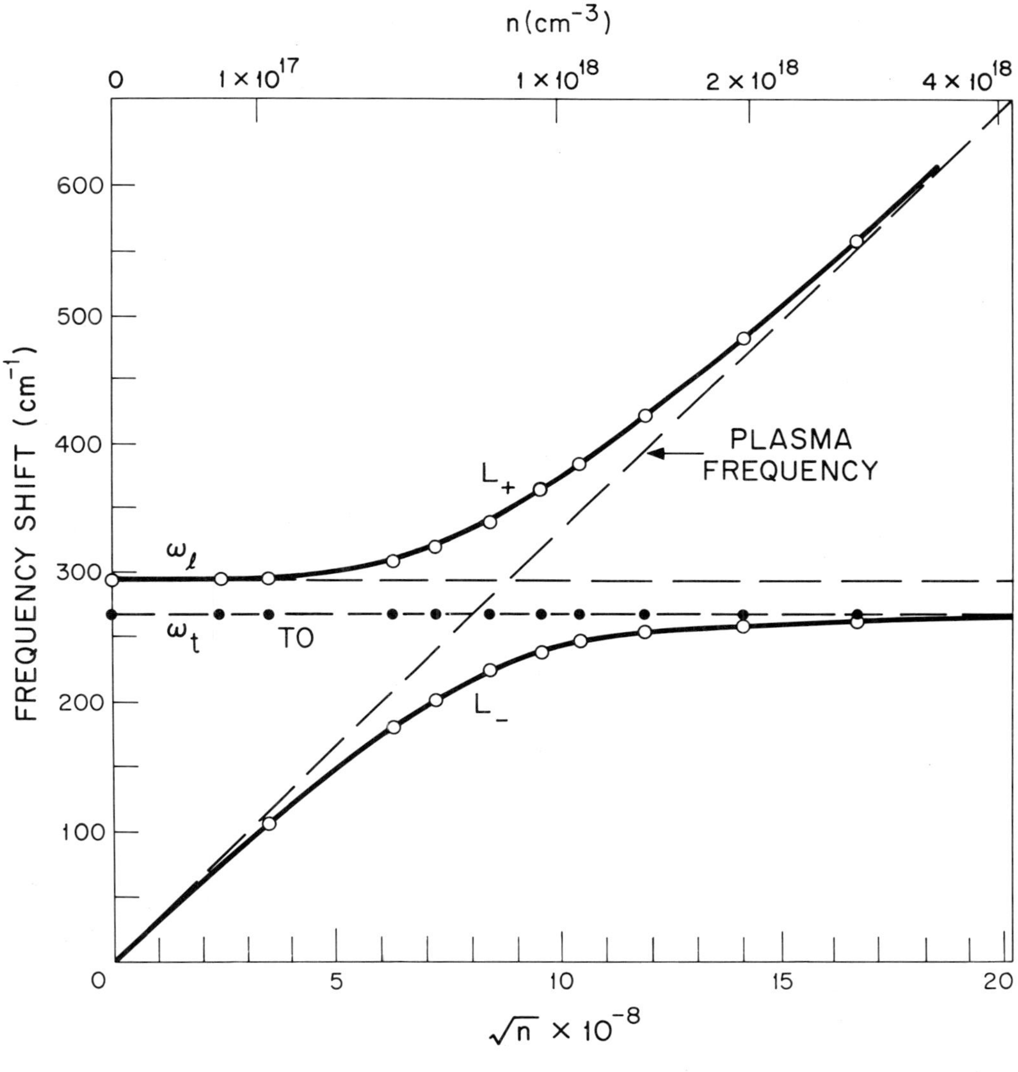

Fig. 1. Frequency shift of the Raman-scattered light in GaAs at room temperature as a function of the square root of the electron concentration. The solid curves labeled L_+ and L_- give the calculated frequencies of the mixed longitudinal plasmon-phonon modes.

The transverse optic (TO) mode at frequency ω_t is unaffected by the presence of free carriers and exhibits the polarization properties predicted by group theory for a zinc blende insulator[4]. The polarization properties of the upper and lower modes, L_+ and L_-, differ from those found for a pure longitudinal optic mode in semi-insulating GaAs. The latter, in accordance with standard group-theory predictions, shows zero LO scattering at 90° when the incident and scattered beams are both polarized parallel to the plane of scattering (||, ||) or both perpendicular to the plane of scattering (⊥ , ⊥). Fig. 2 shows a representative polarization trace for a sample with $n = 1.9\times10^{18}$ cm ^{3} taken at a temperature near that of liquid helium. Good optical alignment is assured by the strong suppression of the TO mode in the (⊥ , ⊥) polarization configuration.

THEORY

We shall assume that the Raman scattering by the free carrier fluctuations can be adequately described within the framework of the Kane two-band model[5]. Omitting spin-density fluctuations, which are unimportant for the collective modes, we then have from[6] for the free electron scattering

$$\frac{d^2\sigma}{d\Omega d\omega} = \frac{\omega_2}{\omega_1}\left(\frac{e^2}{mc^2}\right)^2 (n_\omega + 1)\frac{1}{\pi}\ \mathrm{Im}\left[L_2 - \frac{4\pi e^2}{q^2}\frac{L_1^2}{\epsilon(\omega)}\right](\hat{e}_1\cdot\hat{e}_2)^2 \tag{2}$$

where ω_1 and ω_2 are the incident and scattered frequencies, n_ω is the Bose-Einstein factor, q is the momentum transfer, $\epsilon(\omega)$ is given by (1), and $\hat{e}_1/\hat{e}_2$ are the incident and scattered polarization vectors. Unity scattering volume is assumed throughout and h is not written explicitly. For $q<<k_F$ and with collisions neglected,

$$L_n = \frac{2}{(2\pi)^3}\int d^3k A^n(k,\omega_o)\ \frac{v\cdot q\ \partial f_o/\partial\epsilon}{\omega - v\cdot q} \tag{3}$$

where f_o is the equilibrium distribution function of the electrons and

$$A(k,\omega_o) = 1 + \frac{2P^2}{3m}\sum_i \frac{E_{gi}}{E_{gi}^2 - \omega_o^2} \qquad (i = 1, 2, 3) \tag{4}$$

In (4) the electron wave functions have been evaluated at k = 0, a small anisotropic term involving $\hat{k}\hat{k}$ has been dropped[6], and as a better approximation for finite ω we have used $\omega_o = (\omega_1 + \omega_2)/2$ rather than ω_1 in the energy denominators. Also E_{gi} is the energy difference between the conduction band and the ith valence band evaluated at momentum k (1 = heavy hole band, 2 = light hole band, 3 = split-off band) and P is the interband matrix element of the momentum.

For scattering from the plasmon-phonon modes the second term in the square brackets in (2) will dominate, since at those frequencies $\epsilon(\omega)$ becomes very small. As long as the collective mode resonances lie outside the single-particle excitation spectrum, we may approximate (2) for our purposes by

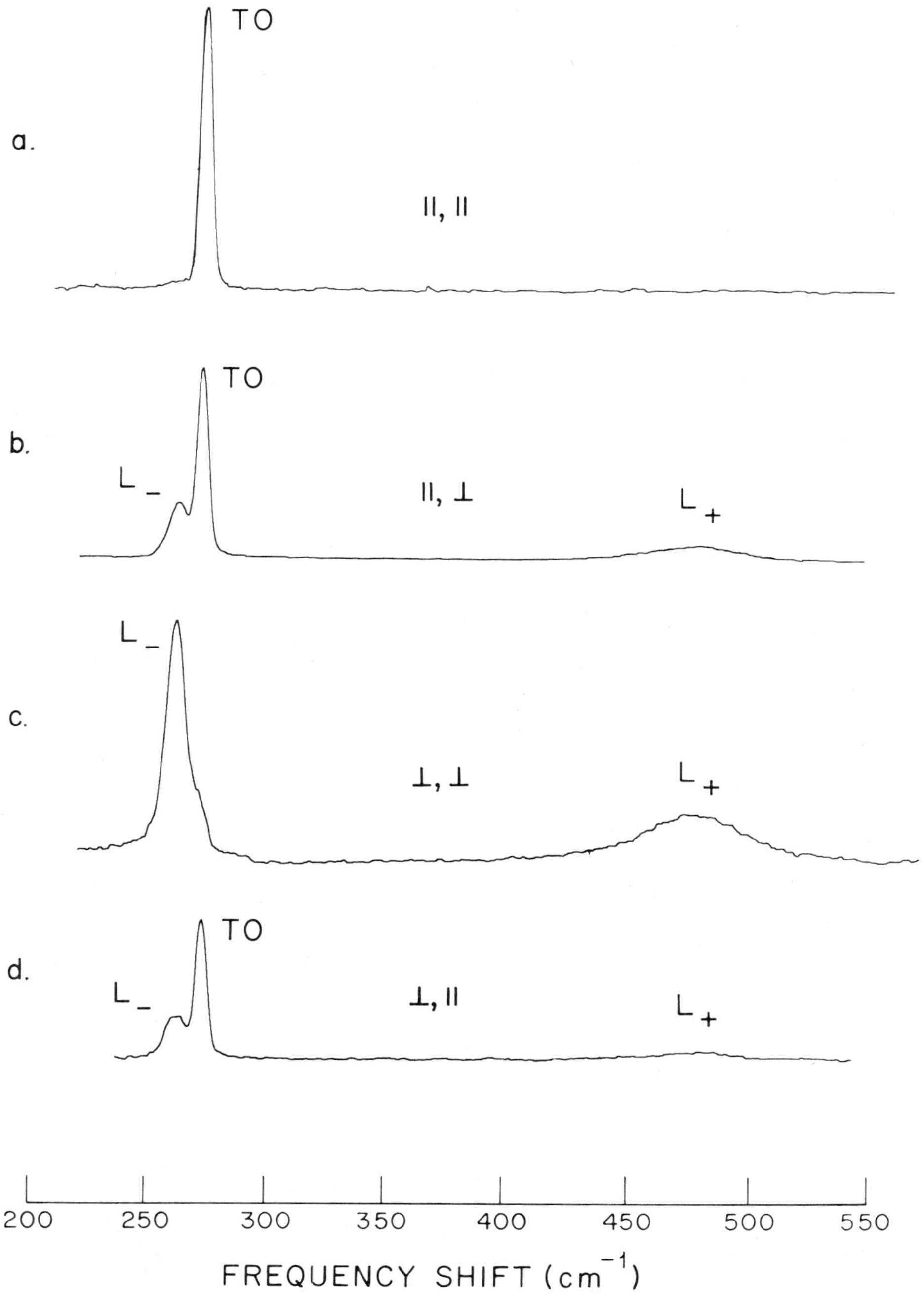

Fig. 2. Polarization recorder traces of the plasmon-phonon coupled modes for GaAs near liquid helium temperature with $n = 1.9 \times 10^{18}$ cm^{-3}. The scattering angle is 90°, with propagation along <100> directions. The polarization of the incident (scattered) light with respect to the plane of scattering is indicated by the first (second) symbol by each trace.

$$\frac{d^2\sigma}{d\Omega d\omega} = -\frac{\omega_2}{\omega_1}\left(\frac{e^2}{mc^2}\right)^2 (n_\omega + 1)\frac{4e^2 L_1^2}{q^2} (\hat{e}_1\cdot\hat{e}_2)^2 \, \mathrm{Im}\,\frac{1}{\epsilon(\omega)} \tag{5}$$

For $\omega_p^2, \omega_\ell^2 >> q^2 v_F^2$ and degenerate electron statistics

$$L_1 \approx -\frac{q^2\epsilon_\infty}{4\pi e^2}\frac{\omega_p^2}{\omega^2} A(k_F, \omega_o) \tag{6}$$

and hence we find for 90° scattering from the longitudinal collective modes

$$\frac{d^2\sigma_\ell(\perp, \perp)}{d\Omega d\omega} \approx -\frac{\omega_2}{\omega_1}\left(\frac{e^2}{mc^2}\right)^2 (n_\omega + 1)\frac{q^2}{4\pi^2 e^2}\left(\frac{\epsilon_\infty \omega_p^2}{\omega^2}\right)^2 A^2(k_F, \omega_o) \times \mathrm{Im}\,\frac{1}{\epsilon(\omega)} \tag{7}$$

with zero cross section for the free carrier scattering in all other polarizations. The same expression may be obtained by specializing the results of[7] to the Kane two-band model. Note the effect of resonant enhancement contained in the factor $A^2(k_F, \omega_o)$.

By using either phenomenological arguments directly[8] or a microscopic treatment like that in[7], one can show that for $\omega << \omega_1$ the scattering due to deformation potential and electro-optic coupling is

$$\frac{d^2\sigma}{d\Omega d\omega} \approx \left(\frac{\omega_1}{c}\right)^4 \left| e_2^\alpha e_1^\beta \left(\frac{\partial \chi_{\alpha\beta}}{\partial u_\gamma}\delta u_\gamma + \frac{\partial \chi_{\alpha\beta}}{\partial E_\gamma}\delta E_\gamma\right)\right|^2 \tag{8}$$

where $\chi_{\alpha\beta}$ is the electron susceptibility evaluated at frequency ω_1, δu is the thermal fluctuation in the optical mode lattice displacement, and δE is the fluctuation in the electric field. In the long wavelength limit and for frequencies near the coupled mode resonances

$$\delta \underset{\sim}{u} = \left(\frac{\epsilon_o - \epsilon_\infty}{4\pi M}\right)^{1/2} \frac{\omega_t}{\omega_t^2 - \omega^2}\delta \underset{\sim}{E} \tag{9}$$

where M is the reduced mass density of the two sublattices. (For ω very near ω_t, one must use the more general expression in[9], which include the spontaneous fluctuations in δu.) From the fluctuation-dissipation theorem we have for the longitudinal fluctuations in electric field

$$\langle \delta E \delta E^+ \rangle_\omega = -4(n_\omega + 1)\, \mathrm{Im}\,\frac{1}{\epsilon(\omega)} \tag{10}$$

Hence for 90° scattering with propagation along <100> axes, deformation potential and electro-optic coupling give for the longitudinal modes

$$\frac{d^2\sigma_\ell(\perp,\|)}{d\Omega d\omega} = \frac{d^2\sigma_\ell(\|,\perp)}{d\Omega d\omega} = -2(n_\omega + 1)\left(\frac{\omega_1}{c}\right)^4 \left(1 + C_1 \frac{\omega_t^2}{\omega_t^2 - \omega^2}\right)^2 \times |b_{41}|^2 \operatorname{Im}\frac{1}{\epsilon(\omega)} \tag{11}$$

where

$$C_1 = \left(\frac{\epsilon_o - \epsilon_\infty}{4\pi M}\right)^{1/2} \frac{1}{\omega_t}\frac{a_{41}}{b_{41}} \tag{12}$$

with $a_{41} = \partial\chi_{zy}/\partial u_x$ and $b_{41} = \partial\chi_{zy}/\partial E_x$. We shall treat C_1 as a constant to be determined from the data. By introducing both the Raman tensor[4]

$$R^x_{yz} = -(m^2\omega_1^2 a/e^2)(\partial\chi_{zy}/\partial u_x) \tag{13}$$

where a is the lattice constant, and the electro-optic coefficient

$$z_{41} = -4\pi\epsilon_\infty^{-2}(\partial\chi_{zy}/\partial E_x) \tag{14}$$

we can also write (11) with the aid of (1) in the form

$$\frac{d^2\sigma_\ell(\perp,\|)}{d\Omega d\omega} = -\frac{n_\omega+1}{2\pi^2c^4}\left|\frac{e^2\epsilon_\infty}{\omega_t m^2 a}\left(\frac{\pi/M}{\epsilon_o - \epsilon_\infty}\right)^{1/2}\left(1 - \frac{\omega_p^2}{\omega^2}\right)R^x_{yz} - \frac{1}{2}\omega_1^2\epsilon_\infty^2 z_{41}\right|^2 \operatorname{Im}\frac{1}{\epsilon(\omega)} \tag{15}$$

This expression more closely resembles that given by Loudon[4] for insulators and was the form previously used[2]. The quantity z_{41} is not the total electro-optic coefficient as measured experimentally, but involves only the direct electronic contribution.

Finally, for the TO modes the integrated $(\|,\|)$ cross section is[4]

$$\frac{d\sigma_t(\|,\|)}{d\Omega} = \left(\frac{\omega_1}{c}\right)^4 \frac{n_t+1}{2M\omega_t}|a_{41}|^2 \tag{16}$$

with 1/2 this amount for the $(\perp,\|)$ and $(\|,\perp)$ scattering, and zero for the $(\perp,\perp)$ scattering.

Since only the relative scattering cross sections are measured experimentally, it is convenient to work with the integrated cross sections normalized to the TO $(\|,\|)$ cross section. We then find for $\omega\tau \gg 1$.

$$\frac{d\sigma_\ell^{\pm}(\perp, ||)/d\Omega}{d\sigma_t(||, ||)/d\Omega} \approx \frac{n_{\omega_\pm}+1}{n_t+1} \frac{\epsilon_0 - \epsilon_\infty}{\omega_t(\partial\epsilon/\partial\omega_\pm)} \left(\frac{1}{C_1} + \frac{\omega_t^2}{\omega_t^2 - \omega_\pm^2}\right)^2 \tag{17}$$

$$\frac{d\sigma_\ell^{\pm}(\perp, \perp)/d\Omega}{d\sigma_t(||, ||)/d\Omega} \approx \frac{n_{\omega_\pm}+1}{n_t+1} \frac{\epsilon_0 - \epsilon_\infty}{\omega_t(\partial\epsilon/\partial\omega_\pm)} \left(\frac{\omega_p}{\omega_\pm}\right)^4 \frac{A^2(k_F, \omega_{0\pm})}{C_2^2} \tag{18}$$

where

$$C_2 = \left(\frac{\epsilon_0 - \epsilon_\infty}{4\pi M}\right)^{1/2} \frac{2^{3/2}\pi m\,\omega_1^2}{\epsilon_\infty q e\,\omega_t}\, a_{41} \tag{19}$$

is a second constant which will also be determined by fitting the data. Note that except for the Bose-Einstein and the $A^2(k_F, \omega_{0\pm})$ factors, the relative scattering intensities really involve only two frequency ratios, ω_p/ω_t and ω_ℓ/ω_t, which can be evaluated directly from the Raman data, and the two dimensionless adjustable parameters C_1 and C_2.

EXPERIMENTAL RESULTS AND DISCUSSION

In all samples the observed line shapes for the two longitudinal modes could be fitted by Lorentzians consistent with (1), (7) and (11) in the limit $\omega\tau >> 1$. The value of τ was found to be of the order of 10^{-13} sec for the samples studied, which agrees with the collision time determined from dc mobility measurements.

In Fig. 3 the integrated scattering intensities of the longitudinal modes at liquid helium temperature are shown for the ($\perp$, $||$) and ($\perp$, $\perp$) polarization configurations as a function of electron concentration, with all cross sections normalized to the TO($||$, $||$). The cross sections for $L_\pm$ ($||$, $\perp$) are not shown as they were always equal to those for $L_\pm$ ($\perp$, $||$) within the experimental uncertainty of about 10%; no scattering was observed for $L_\pm$ ($||$, $||$), as expected. Also $\sigma_t(\perp, ||) \approx \sigma_t(||, \perp) \approx 0.5\sigma_t(||, ||)$ for all samples within the same 10% uncertainty.

The curves in Fig. 3 were determined from (17) and (18) for T = 0°K with $C_1 = -0.51$ and $|C_2| = 7.0$. The fit is reasonably good, but the discrepancy at the high concentration end appears to lie outside experimental error. In the $A^2(k_F, \omega_{0\pm})$ factor in (18) we took $E_g = 1.51$ eV for the k = 0 gap, $2P^2/3mE_g = 4.94$, $m_e = 0.067$ m, $m_{h1} = 0.45$ m and $m_{h2} = 0.082$ m from Vrehen[10], and computed $m_{h3} = 0.33$ from the Kane model with a spit-orbit splitting $\Delta = 0.33$ eV. If we further put $\omega_t = 273$ cm^{-1} and $\omega_\ell = 296$ cm^{-1} from the Raman data, and take $\epsilon_\infty = 11.1$ and $\epsilon_0 = (\omega_\ell/\omega_t)^2\epsilon_\infty = 13.1$, we obtain from (12) and (19).

$$|a_{41}| = 6.4 \pm 1.0 \times 10^{7}\ \mathrm{cm}^{-1}$$
$$|b_{41}| = 8.4 \pm 1.2 \times 10^{-7}\ \mathrm{esu.} \qquad (T = 4^\circ K) \qquad (20)$$

A similar comparison between theory and experiment was made for the room temperature data. The analysis is much more complicated at 300°K because the single-particle excitations overlap the collective modes. Also there is some second-order Raman scattering which must be subtracted out. The best fit with experiment was achieved with $C_1 = -0.46$ and $|C_2| = 7.3$. In obtaining this result the $A^2(k_F, \omega_0)$ factor in (18) was replaced with the appropriate thermal average and E_g was taken as 1.43 eV. Using $\omega_t = 269\ \mathrm{cm}^{-1}$ and $\omega_\ell = 292\ \mathrm{cm}^{-1}$, we find

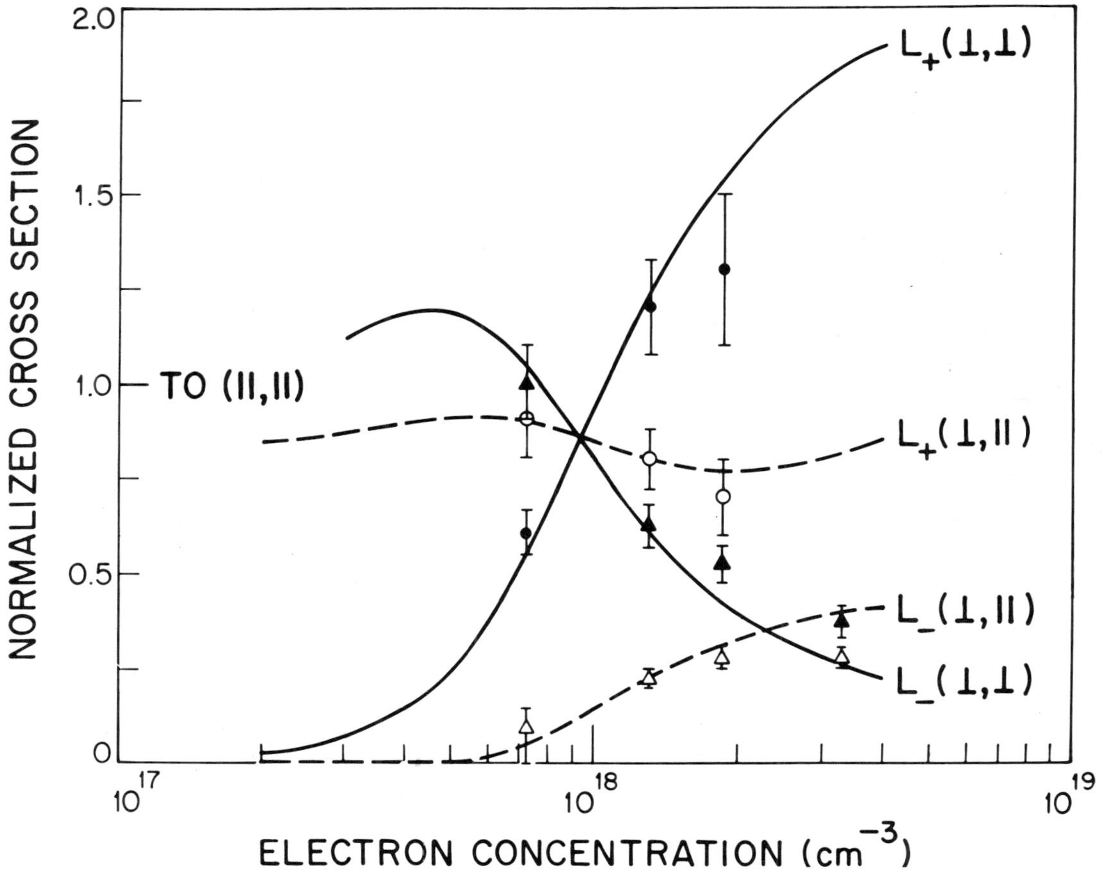

Fig. 3. Integrated plasmon-phonon cross sections relative to the TO (||, ||) cross section as a function of electron concentration. The points are experimental for GaAs near liquid-helium temperature; the curves are theoretical.

$$|a_{41}| = 6.5 \pm 1.0 \times 10^{7}\ \mathrm{cm}^{-1}$$
$$(T = 300°K) \qquad (21)$$
$$|b_{41}| = 9.7 \pm 1.4 \times 10^{-7}\ \mathrm{esu}$$

Comparison of (20) and (21) shows that a_{41} remains almost constant between 4° and 300°K, whereas on the basis of a simple two-band model one would expect[4] a variation as $(E_g - \omega_1)^{-1/2}$. However, the Kane two-band model gives no deformation coupling if the k = 0 wave functions are used. Thus the dominant contributions to a_{41} must arise either from three-band virtual transitions, as in the electro-optic coupling, or from two-band transitions elsewhere in the Brillouin zone.

Walsh[11] has made a direct measurement of the electro-optic coefficient in semi-insulating GaAs over a range of wavelengths. At 1μ he obtains $r_{41} = 1.2 \times 10^{-10}$ cm/V, with no difference between the clamped and unclamped values. From (11) we predict at 300°K a clamped (constant strain) value of

$$|r_{41}| = \frac{4\pi}{\epsilon^2(\omega_1)} |b_{41}| (1 + C_1) = 4.6 \pm 1.0 \times 10^{-8}\ \mathrm{esu}$$

$$\approx 1.5 \pm 0.3 \times 10^{-10}\ \mathrm{cm/V}$$

if we use Walsh's value of $\epsilon(\omega) \approx 12$ at 1.06μ. The agreement would be improved if k-dependent wave functions were used in (4), since $A(k_F, \omega_0)$ would be decreased (about 10% for $n = 1.5 \times 10^{-18}\ \mathrm{cm}^{-3}$), resulting in a corresponding reduction in $|a_{41}|$ and $|b_{41}|$. From the 10.6μ second-harmonic-generation experiments[12] one would deduce[13] a value of $|b_{41}| = 2|d_{123}| = 3.5 \pm 1.2 \times 10^{-6}$ esu, much larger than that found here. The discrepancy is not due to the difference in wavelength because r_{41} increases only to about 1.6×10^{-10} cm/V in the 3-10μ range[11, 14].

Using the value of C_1 obtained above, we find from (11) that at $n \approx 4 \times 10^{17}\ \mathrm{cm}^{-3}$ the electro-optic coupling just cancels the deformation potential coupling for the L_- mode, giving zero scattering for the ($\perp$, $||$) and ($||$, $\perp$) configurations. As shown in Fig. 3, the computed L_-($\perp$, $||$) intensity has a very broad minimum in this concentration region and does not recover appreciable strength at lower concentrations. Figure 4 shows the polarization traces for a sample with $n = 7 \times 10^{17}\ \mathrm{cm}^{-3}$ taken at room temperature where the best alignment could be achieved. The ($||$, $\perp$) trace indicates almost a complete absence of the L_- mode. Lower concentration samples also show no significant L_-($||$, $\perp$) scattering, in agreement with the theoretical curve in Fig. 3. As an additional check, the ratio of the integrated LO ($\perp$, $||$) to TO ($||$, $||$) intensity was measured at 300°K on a semi-insulating GaAs sample. A value of 0.88 was found, consistent within experimental error with that obtained by extrapolating the normalized L_+($\perp$, $||$) intensity to zero concentration. Finally, we note that the L_-($\perp$, $\perp$) mode scattering which persists in the high concentration samples arises from the conduction electrons that heavily screen the LO phonon mode in this limit.

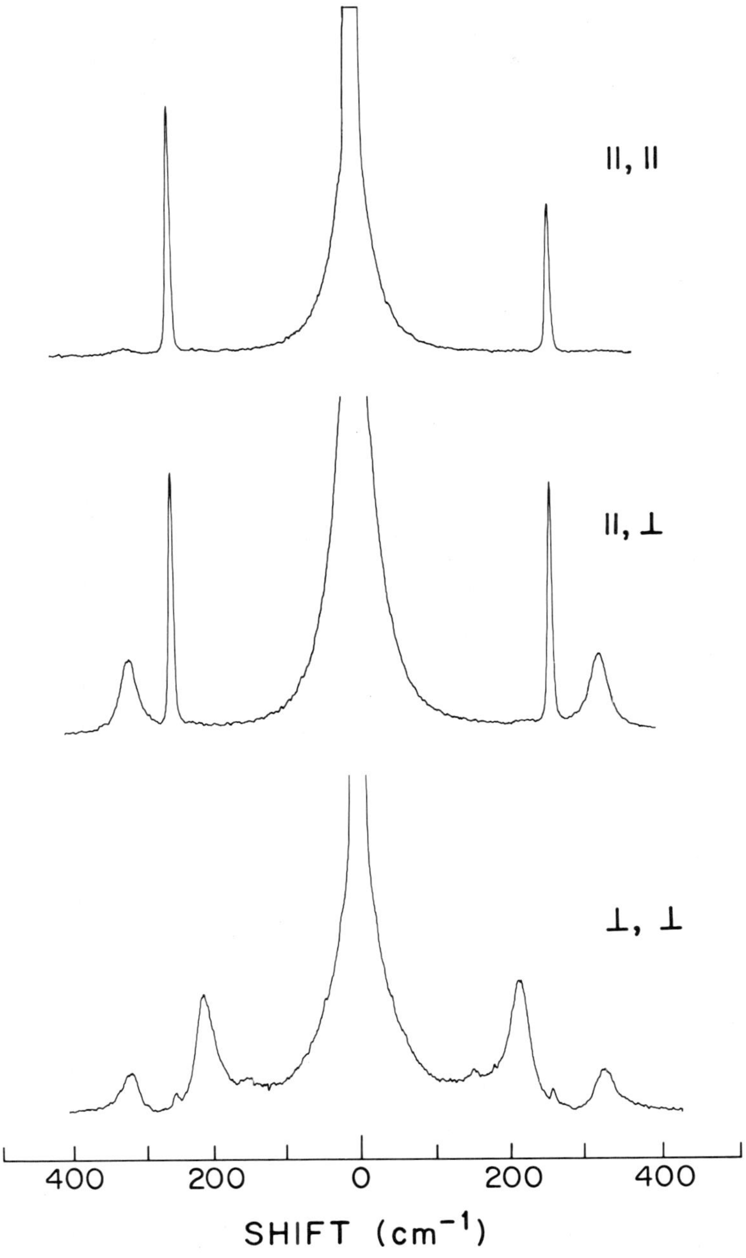

Fig. 4. Polarization recorder traces for GaAs at 300°K with n = 7 x 10^{17} cm^{-3}.

ACKNOWLEDGEMENTS

We would like to thank D. J. Wells for assistance with the measurements, and K. Nearen and W. Laswell for cutting and polishing the samples.

REFERENCES

1. A. Mooradian and G.B. Wright, Phys. Rev. Letters 16, 999 (1966).
2. A. Mooradian and A.L. McWhorter, Phys. Rev. Letters 19, 849 (1967).
3. A. Mooradian, paper D-2 this conference.
4. R. Loudon, Advan. Phys. 13, 423 (1964).
5. E.O. Kane, J. Phys. Chem. Solids 1, 249 (1957).
6. D.C. Hamilton and A.L. McWhorter, paper D-4 this conference.
7. A.L. McWhorter and P.N. Argyres, paper D-6 this conference.
8. E. Burstein, A. Pinczuk, and S. Iwasa, Phys. Rev. 157, 611 (1967).
9. N.D. Strahm and A.L. McWhorter, paper F-2 this conference.
10. Q.H.F. Vrehen, J. Phys. Chem. Solids 29, 129 (1968).
11. T.E. Walsh, RCA Review 27, 323 (1966). See also E.H. Turner and I. Kaminow, J. Opt. Soc. Am. 53, 523 (1963).
12. C.K.N. Patel, Phys. Rev. Letters 16, 613 (1966).
13. F.N.H. Robinson, Bell System Tech. J. 46, 913 (1967), particularly appendix.
14. A. Yariv, C.A. Mead, and J.V. Parker, J. Quantum. Electron. QE-2, 243 (1966); I. Kaminow, J. Quantum Electron. QE-4, 23 (1968).

D-4: RAMAN SCATTERING FROM SPIN - DENSITY FLUCTUATIONS IN n - Ga As

David C. Hamilton and A. L. McWhorter
Lincoln Laboratory, * Massachusetts Institute of Technology
Lexington, Massachusetts

ABSTRACT

The anomalously large cross sections observed for scattering from single-electron excitations in GaAs can be explained in part as scattering from spin-density fluctuations. The electromagnetic field is coupled to the spins through second-order p·A perturbation terms involving the spin-orbit splitting of the valence band. At high carrier densities, where the screening wave vector is much greater than the momentum transfer, the spin-density fluctuations are not screened out as the charge-density fluctuations are, and hence the cross section can be much larger than that due to charge-density fluctuations alone. The polarization properties of the scattered radiation are also different for the two processes. For spin-density scattering the matrix element is proportional to $\vec{\sigma} \cdot (\hat{\epsilon}_1 \times \hat{\epsilon}_2)$, whereas for charge-density scattering it varies as $(\hat{\epsilon}_1 \cdot \hat{\epsilon}_2)$, where $\hat{\epsilon}_1$ and $\hat{\epsilon}_2$ are the incident and scattered polarization vectors. The magnetic field is taken to be zero throughout.

INTRODUCTION

Mooradian [1] has recently observed unexpectedly large Raman scattering in GaAs at high carrier densities where charge-density fluctuations should be screened out. For two scattering angles (90° and 180°) and hence two values of momentum transfer q, the scattered light intensity was measured as a function of energy transfer ω. Besides the peaks due to the transverse optical phonons and the mixed longitudinal plasmon-phonon modes, a broad peak was seen extending from $\omega = 0$ to $\omega \lesssim qv_F$. This broad peak was attributed to single-electron excitations [2].

In Mooradian's data there was a glaring discrepancy between theory and experiment, in that the single-electron scattering was over two orders of magnitude larger relative to the plasmon than expected on the basis of scattering from charge-density fluctuations.

*Operated with support from the U. S. Air Force.

Furthermore, scattering was observed when the polarization vectors of the incident and scattered light were perpendicular to each other, which should not be the case for charge-density fluctuations. Although an increase in the single-electron scattering and a change in the polarization properties would be expected on the basis of Wolff's calculations [3], which showed that nonparabolicity of the conduction band would result in scattering from fluctuations in the angular distribution of the electrons, this effect is not large enough to account for the data.

In the following we will show that Mooradian's results can be partly explained in terms of scattering from electron spin-density fluctuations, which are coupled to the light waves via spin-orbit effects. Since spin-density fluctuations are not screened, the cross section can be much larger than that due to charge-density fluctuations. Spin-orbit coupling has been used by Elliot and Loudon [4] to provide a mechanism for scattering from spin waves in magnetic crystals. In semiconductors the cross section for spin-flip Raman scattering does not go to zero in the limit of zero magnetic field [5]. This is just the limit in which our spin-density fluctuations occur, and some of the expressions given by Yafet [5] are quite similar to ours. However, in Yafet's treatment qv/ω is implicitly set equal to zero at the outset, which is not permissible in zero field. In other words, Yafet's treatment implies that the frequency shift for spin-flip Raman scattering should go to zero as μH when H goes to zero, while actually it goes to qv_F.

CALCULATION

The Raman scattering cross section involves two terms to be treated in perturbation theory, the A^2 term in first order and the $p \cdot A$ term in second order. The vector potential is introduced into a Hamiltonian which is the sum of the usual one-electron kinetic and potential energies, the Coulomb interaction between the electrons and the interaction between the electrons and the LO phonons. The interband matrix elements of the $p \cdot A$ term give a coupling to the spins of the conduction band due to the spin-orbit splitting of the valence band, which we shall treat in the Kane two-band model [6]. For simplicity, all interband matrix elements are evaluated at the center of the zone. Finally, the conduction band correlation function is evaluated in the random-phase approximation (RPA).

The differential cross section for scattering from $k_1\omega_1$ to $k_2\omega_2$ is given by

$$\frac{d^2\sigma}{d\omega d\Omega} = \frac{\omega_2}{\omega_1}\left(\frac{e^2}{mc^2}\right)^2 \left\langle \sum_f \delta(E_f - E_i - \omega)\, | M_{fi} |^2 \right\rangle \tag{1}$$

where $\omega = \omega_1 - \omega_2$, $q = k_1 - k_2$, M_{fi} is the matrix element for a transition between the many-electron initial state of energy E_i and final state of energy E_f, and the angular brackets denote a thermal average over the initial state. Here and elsewhere we do not write $\hbar$ or k_B explicitly. It can be shown that

$$M_{fi} = (\hat{\epsilon}_1 \cdot \hat{\epsilon}_2) \langle f | n_q | i \rangle + \frac{1}{mi} \int_{-\infty}^{0} d\tau\, e^{i\omega_2 \tau} \langle f | j_1 j_2(\tau) | i \rangle + \frac{1}{mi} \int_{0}^{\infty} d\tau\, e^{i\omega_2 \tau} \langle f | j_2(\tau) j_1 | i \rangle \tag{2}$$

where

$$j_1 = \vec{j}(k_1)\cdot\hat{\epsilon}_1\ ,\quad j_2 = \vec{j}(-k_2)\cdot\hat{\epsilon}_2 \tag{3}$$

$$\vec{j}(k) = \sum_{\substack{pnn'\\ \alpha\alpha'}} a^{\dagger}_{p+kn'\alpha'}\, a_{pn\alpha} \langle p+kn'\alpha' \,|\, e^{ik\cdot r}\vec{p} \,|\, pn\alpha\rangle$$

$$n_q = \sum_{\substack{pn\alpha\\ n'\alpha'}} a^{\dagger}_{p+qn'\alpha'}\, a_{pn\alpha} \langle p+qn'\alpha' \,|\, e^{iq\cdot r} \,|\, pn\alpha\rangle$$

Here $\hat{\epsilon}_1$, $\hat{\epsilon}_2$ are incident and scattered polarization vectors, and $a^{\dagger}_{pn\alpha}$, $a_{pn\alpha}$ are the creation and destruction operators for a Bloch state $|np\alpha\rangle$ of band n, momentum p and spin α. This form for M_{fi} has been noted by Jha [7].

We now specialize to the Kane two-band model and assume an n-type semiconductor with full valence band, which is the situation of experimental interest [1]. Since we will not be concerned here with electro-optic coupling [8], which is small for the single-electron excitations, we will omit interband terms in the Coulomb interaction between the electrons and in the electron-phonon interaction. Then to a good approximation we can write for $n \neq n'$

$$a^{\dagger}_{pn\alpha}(t)\, a_{p'n'\alpha'}(t) = e^{i(\epsilon_{pn\alpha} - \epsilon_{p'n'\alpha'})t}\, a^{\dagger}_{pn\alpha}\, a_{p'n'\alpha'} \tag{4}$$

where $\epsilon_{pn\alpha}$ is the energy of Bloch state $|pn\alpha\rangle$. An alternate approach has been given without this restriction [9].

This approximation determines the time evolution of j and yields

$$M_{fi} = \langle f |\sum_{p\alpha\beta} \gamma_{\alpha\beta}(p) a^{\dagger}_{p+q\alpha}\, a_{p\beta} | i\rangle \tag{5}$$

with

$$\gamma_{\alpha\beta} = (\epsilon_1\cdot\epsilon_2)\,\delta_{\alpha\beta} + \frac{1}{m}\sum_{np\sigma} \left(\frac{\langle\alpha|\vec{p}\cdot\hat{\epsilon}_1|n\sigma\rangle\langle n\sigma|p\cdot\epsilon_2|\beta\rangle}{\epsilon_{pc} - \epsilon_{pn} - \omega_1} + \frac{\langle\alpha|\vec{p}\cdot\hat{\epsilon}_2|n\sigma\rangle\langle n\sigma|\vec{p}\cdot\hat{\epsilon}_1|\beta\rangle}{\epsilon_{pc} - \epsilon_{pn} + \omega_1} \right) \tag{6}$$

The matrix elements have been evaluated at p = 0; we have also let $k, q \to 0$ and taken $\omega_1 \approx \omega_2$. The summation in (6) runs over the three valence bands; the conduction band index c is suppressed except in the energy denominators. Wolff [10] has given a similar expression for the case of a single electron. The difference is that here spin indices and creation operators are introduced.

Once we evaluate γ, the cross section can be written in terms of a correlation function:

$$\frac{d^2\sigma}{d\omega d\Omega} = \left(\frac{e^2}{mc^2}\right)^2 \frac{1}{1 - e^{-\omega/T}} \left(-\frac{1}{\pi}\right) \operatorname{Im} \sum_{p\alpha\beta} \gamma^{\dagger}_{\alpha\beta}(p)\, g_{\alpha\beta}(p\omega q) \tag{7}$$

where

$$g_{\alpha\beta}(\omega) = \int_{-\infty}^{\infty} dt\ e^{i\omega t}\ g_{\alpha\beta}(t)$$

$$= -i \int_{0}^{\infty} dt\ e^{i\omega t} < [a^{\dagger}_{p\beta}(t)\ a_{p+q\alpha}(t),\ \sum_{p\alpha\beta} \gamma_{\alpha\beta}\ a^{\dagger}_{p+q\alpha}\ a_{p\beta}] > \tag{8}$$

This type of transformation of the square of the matrix element is due to Van Hove [11].

If we use the wave functions given by Kane [6] evaluated at the center of the zone, we find

$$\gamma_{\alpha\beta}(p) = \hat{\epsilon}_1 \cdot \overleftrightarrow{A}_p \cdot \hat{\epsilon}_2 \delta_{\alpha\beta} + i\,(\hat{\epsilon}_1 \times \hat{\epsilon}_2) \cdot \overleftrightarrow{B}_p \cdot \vec{\sigma}_{\alpha\beta} \tag{9}$$

where

$$\overleftrightarrow{A}_p = \overleftrightarrow{I} \left[1 + \frac{2P^2}{3m} \left(\frac{E_{g1}}{E_{g1}^2 - \omega_1^2} + \frac{E_{g2}}{E_{g2}^2 - \omega_1^2} + \frac{E_{g3}}{E_{g3}^2 - \omega_1^2} \right) \right]$$

$$- \overleftrightarrow{R}\, \frac{P^2}{m} \left(\frac{E_{g1}}{E_{g1}^2 - \omega_1^2} - \frac{E_{g2}}{E_{g2}^2 - \omega_1^2} \right) \tag{10}$$

and

$$\overleftrightarrow{B}_p = \overleftrightarrow{I}\, \frac{P^2 \omega_1}{3m} \left(\frac{1}{E_{g1}^2 - \omega_1^2} + \frac{1}{E_{g2}^2 - \omega_1^2} - \frac{2}{E_{g3}^2 - \omega_1^2} \right)$$

$$+ \overleftrightarrow{R}\, \frac{P^2 \omega_1}{m} \left(\frac{1}{E_{g1}^2 - \omega_1^2} - \frac{1}{E_{g2}^2 - \omega_1^2} \right)$$

Here the components of $\vec{\sigma}$ are the Pauli matrices, E_{gi} is the energy difference between the conduction band and the ith valence band evaluated at momentum p (1 = heavy hole band, 2 = light hole band, 3 = split-off band), $P = -i < s|p_z|z >$ is the interband matrix element of the momentum, and

$$\overleftrightarrow{R} = \hat{p}\hat{p} - \frac{\overleftrightarrow{I}}{3} \tag{11}$$

where $\overleftrightarrow{I}$ is the unit dyadic. The form of the angular anisotropy of $\overleftrightarrow{A}_p$ and $\overleftrightarrow{B}_p$ is unaltered by the use of the p-dependent wave functions of Kane.

For simplicity we shall in this paper drop the anisotropic terms in $\gamma_{\alpha\beta}(p)$ involving $\overleftrightarrow{R}$. This omits the type of nonparabolic enhancement found by Wolff [3], but which is known to be too small to explain Mooradian's results [1]. It is important to keep the exact form of the energy denominators, however.

In order to evaluate the cross section, we still need to know $g_{\alpha\beta}$. We use the one-band Hamiltonian

$$\mathcal{H} = \sum_{p\sigma} \epsilon_p a^\dagger_{p\sigma} a_{p\sigma} + \frac{1}{2} \sum_q \frac{4\pi e^2}{q^2 \epsilon_\infty} \sum_{p\sigma p'\sigma'} a^\dagger_{p+q\sigma} a^\dagger_{p'-q\sigma'} a_{p'\sigma'} a_{p\sigma}$$

$$+ \omega_\ell \sum_q b^\dagger_q b_q - ie\,(2\pi\omega_\ell)^{1/2} (\epsilon_\infty^{-1} - \epsilon_0^{-1})^{1/2} \sum_{pq\sigma} q^{-1} (b_q - b^\dagger_{-q}) a^\dagger_{p+q\sigma} a_{p\sigma} \tag{12}$$

Here ω_ℓ is the longitudinal optical phonon frequency when no electrons are present, and $b_q, b^\dagger_q$ are phonon operators. After writing equations of motion for $a^\dagger_{p+q} a_p$ and $b^\dagger_q$, we take Fourier transforms and assume $q \ll p_F$. We then find in the RPA

$$(\omega - q\cdot v) g_{\alpha\beta}(p, \omega) = \gamma_{\alpha\beta}\, v\cdot q \left(-\frac{\partial n}{\partial \epsilon}\right)$$

$$+ \delta_{\alpha\beta} \frac{4\pi e^2}{q^2 \epsilon(\omega)}\, v\cdot q \left(-\frac{\partial n}{\partial \epsilon}\right) \sum_{k\sigma} g_{\sigma\sigma}(k\omega)$$

$$- \frac{i}{\tau} \left[g_{\alpha\beta}(p, \omega) - \int \frac{d\Omega_k}{4\pi}\, g_{\alpha\beta}(k, \omega) \right] \tag{13}$$

where the phonons have been eliminated by introducing the longitudinal dielectric constant of the lattice

$$\epsilon(\omega) = \epsilon_\infty \left[1 + \frac{\omega_\ell^2 - \omega_t^2}{\omega_t^2 - \omega^2} \right] \tag{14}$$

We have also added a phenomenological scattering term that conserves energy, spin and particle number.

If we use the fact that $\gamma_{\alpha\beta}$ has the form $\hat{\epsilon}_1 \cdot \hat{\epsilon}_2\, \delta_{\alpha\beta} A_p + i(\hat{\epsilon}_1 \times \hat{\epsilon}_2) \cdot \vec{\sigma}_{\alpha\beta} B_p$ after the anisotropic terms in $\overleftrightarrow{R}$ have been dropped, we find that

$$\sum_{p\alpha\beta} \gamma^\dagger_{\alpha\beta} g_{\alpha\beta} = - \left[L_2 - \frac{4\pi e^2}{q^2} \frac{L_1^2}{\epsilon_T(q, \omega)} \right] (\hat{\epsilon}_1 \cdot \hat{\epsilon}_2)^2 - K_2 (\hat{\epsilon}_1 \times \hat{\epsilon}_2)^2 \tag{15}$$

Here

$$(L_n, K_n) = \frac{2}{(2\pi)^3} \int d^3p\, (A_p^n, B_p^n) \frac{v\cdot q\, (\partial n/\partial\epsilon)}{\omega - v\cdot q + i/\tau} \left[1 - \frac{i}{\tau} \left\langle \frac{1}{\omega - v\cdot q + i/\tau} \right\rangle_\Omega \right]^{-1} \tag{16}$$

where the angular brackets $\langle\,\rangle_\Omega$ denote an average over Ω_p, and

$$\epsilon_T(q, \omega) = \epsilon(\omega) + \frac{4\pi e^2}{q^2} L_0 \tag{17}$$

is the total dielectric constant including the electronic contribution. The scattering cross section is then

$$\frac{d^2\sigma}{d\omega d\Omega} = \left(\frac{e^2}{mc^2}\right)^2 \frac{1}{1-e^{-\omega/T}} \frac{1}{\pi}$$

$$\mathrm{Im}\left\{\left[L_2 - \frac{4\pi e^2}{q^2} \frac{L_1^2}{\epsilon_T(q, \omega)}\right](\hat{\epsilon}_1 \cdot \hat{\epsilon}_2)^2 + K_2 (\hat{\epsilon}_1 \times \hat{\epsilon}_2)^2\right\} \tag{18}$$

Note that the spin-fluctuation contribution $K_2 (\hat{\epsilon}_1 \times \hat{\epsilon}_2)^2$ has no screening denominator. The function L_n is a generalization of that introduced by Wolff [3].

RESULTS

We compare the cross section in Eq. (18) with Mooradian's data, using the GaAs parameters in Table 1 and setting $\omega_1 = 1.17$ eV. Consider first the low-density limit in which $\omega_p << qv_{th}$, where $v_{th} = (2k_B T/m^*)^{1/2}$. Then the screening term involving $4\pi e^2/q^2$ can be neglected. The ratio of charge-density scattering to spin-density scattering should be $\sigma(\hat{\epsilon}_1 \,||\, \hat{\epsilon}_2)/\sigma(\hat{\epsilon}_1 \perp \hat{\epsilon}_2) = \mathrm{Im}\, L_2/\mathrm{Im}\, K_2 = 32.4$ for GaAs at room temperature. As shown in Fig. 3 of [12], Mooradian's room temperature data for $\sigma(\hat{\epsilon}_1 \perp \hat{\epsilon}_2)$ vs. n can be fitted within experimental error over the whole concentration range studied if a straight line of unity slope is drawn with the above ratio of $\sigma(\hat{\epsilon}_1 \,||\, \hat{\epsilon}_2)$ to $\sigma(\hat{\epsilon}_1 \perp \hat{\epsilon}_2)$ in the low-concentration limit.

Next consider the high-concentration limit at low temperatures where the electrons have a degenerate Fermi distribution and the charge-density fluctuations are heavily screened. Eq. (18) predicts in agreement with experiment [12] that for the single-electron excitations the spin-density scattering $(\hat{\epsilon}_1 \perp \hat{\epsilon}_2)$ should be much greater than the charge-density scattering $(\hat{\epsilon}_1 \,||\, \hat{\epsilon}_2)$. A quantitative comparison between theory and experiment for the ratio of $\sigma(\hat{\epsilon}_1 \perp \hat{\epsilon}_2)$ to $\sigma(\hat{\epsilon}_1 \,||\, \hat{\epsilon}_2)$ will not be attempted here since the experimental value for $\sigma(\hat{\epsilon}_1 \,||\, \hat{\epsilon}_2)$ is too small for accurate determination from Mooradian's data and the theoretical value for $\sigma(\hat{\epsilon}_1 \,||\, \hat{\epsilon}_2)$ must be computed with the $\overleftrightarrow{R}$ term retained in the expression for $\overleftrightarrow{A}_p$. Instead we compare the integrated cross section for single-electron spin-density scattering with the sum of the integrated cross sections for charge-density scattering for both longitudinal modes of the coupled plasmon-phonon system. Taking for simplicity $\tau \rightarrow \infty$ and ignoring common factors, we have

TABLE I

GaAs Parameters Used in Scattering Calculations

$E_g(300°) = 1.43$ eV, $E_g(0°) = 1.51$ eV, $\Delta = 0.33$ eV
$\frac{2}{3}\frac{P^2}{mE_g} = 4.94$, $m^* = 0.067$ m (band bottom)
$\frac{\omega_t}{2\pi c} = 273\ \mathrm{cm}^{-1}$, $\frac{\omega_\ell}{2\pi c} = 296\ \mathrm{cm}^{-1}$, $\epsilon_0 = 13.1$, $\epsilon_\infty = 11.1$
For $n = 1.35 \times 10^{18}\ \mathrm{cm}^{-3}$, and with nonparabolicity included,
$\frac{qv_F}{2\pi c} = 73.5\ \mathrm{cm}^{-1}$, $\frac{\omega_p}{2\pi c} = 386\ \mathrm{cm}^{-1}$
$\frac{\omega_+}{2\pi c} = 415\ \mathrm{cm}^{-1}$ (calc.), $\frac{\omega_-}{2\pi c} = 254\ \mathrm{cm}^{-1}$ (calc.)

$$\sigma_{s.e.}(\hat{\epsilon}_1 \perp \hat{\epsilon}_2) \sim \frac{1}{2} B^2 \tag{19}$$

and

$$\sigma_{L_-}(\hat{\epsilon}_1 \| \hat{\epsilon}_2) + \sigma_{L_+}(\hat{\epsilon}_1 \| \hat{\epsilon}_2) \sim \frac{1}{6} A^2 \left(\frac{qv_F}{\omega_+ + \omega_-}\right)\left(1 + \frac{\omega_\ell^2}{\omega_p \omega_t}\right) \tag{20}$$

where ω_+ and ω_- are the frequencies of the coupled longitudinal modes L_+ and L_- (given by the roots of $\epsilon_T = 0$), $\omega_p = (4\pi ne^2/\epsilon_\infty m^*)^{1/2}$ is the plasma frequency, and A and B are to be evaluated at the Fermi surface. Using the parameter values in Table I, we find that the ratio of (19) to (20) is about 0.22 for $n = 1.35 \times 10^{18}\ \mathrm{cm}^{-3}$, whereas the experimental ratio for the sample of this concentration is about 0.20 ± 20% at T = 4.2°K.

As a further check on the theory, we have also compared the observed line shape of the single-electron excitations with the predictions of Eq. (18). For the sample with $n = 1.35 \times 10^{18}$, we get a good fit to the experimentally-observed spectrum at T = 4.2°K if we take $qv_F\tau = 1.4$, corresponding to $\tau = 1.0 \times 10^{-13}$ sec. This is only slightly larger than the value of $\tau = 7.9 \times 10^{-14}$ sec which is found to fit the line shape of both the L_+ and L_- collective modes in the same sample [13].

Finally, we comment on the large cross section observed for single-electron scattering with $\hat{\epsilon}_1 \| \hat{\epsilon}_2$ at higher temperatures in the high-concentration GaAs samples. Here, spin-density fluctuations are not involved. However, we find that there is a large, temperature-dependent contribution from $\mathrm{Im}[L_2 - (4\pi e^2/q^2)L_1^2/\epsilon_T]$, which arises because the momentum dependence of the energy denominators in A_p makes $L_2L_0 \neq L_1^2$. As a result of the strong cancellation involved, it has been found necessary to retain the $\overleftrightarrow{R}$ term in $\overleftrightarrow{A}_p$ for accurate evaluation of the scattering cross section. When this is done, satisfactory agreement between theory and experiment is obtained [14].

REFERENCES

1. A. Mooradian, Phys. Rev. Letters 20, 1102 (1968).
2. See, for example, D. Pines, "Elementary Excitations in Solids," Ch. 3, W.A. Benjamin Inc., New York, 1963.
3. P.A. Wolff, Phys. Rev. 171, 436 (1968).
4. R. Loudon, Adv. Physics 13, 423 (1964); P.A. Fleury and R. Loudon, Phys. Rev. 166, 514 (1968).
5. Y. Yafet, Phys. Rev. 152, 858 (1966); P.L. Kelley and G.B. Wright, Bull. Am. Phys. Soc. 11, 812 (1966).
6. E.O. Kane, J. Phys. Chem. Solids 1, 249 (1957).
7. S. Jha (to be published).
8. R. Loudon, Proc. Roy. Soc. A275, 218 (1963).
9. A.L. McWhorter and P.N. Argyres, paper D-6 this conference.
10. P.A. Wolff, Phys. Rev. Letters 16, 225 (1966).
11. L. Van Hove, Phys. Rev. 95, 249 (1954).
12. A. Mooradian, paper D-2 this conference.
13. A. Mooradian and A.L. McWhorter, Phys. Rev. Letters 19, 849 (1967).
14. A.L. McWhorter, A. Mooradian, and D.C. Hamilton (to be published).

D-5: LIGHT SCATTERING FROM A PLASMA IN A MAGNETIC FIELD

N. Tzoar
The City College of The City University of New York
and
P. M. Platzman, P. A. Wolff
Bell Telephone Laboratories, Incorporated
Murray Hill, New Jersey

INTRODUCTION

In this talk we consider the incoherent (Raman) scattering[1] from plasma in a homogeneous static magnetic field B_o. The Raman scattering cross-section is completely characterized by the frequency and wave number transfer (k, ω) in the scattering event. The wave number and frequency transfers are the difference between the wave numbers and frequencies of the incoming and outgoing photons $(k_1 - k_2, \omega_1 - \omega_2)$. The wave number transfer to the system, k, determines the spatial resolution which we looked at the system. If k is small relative to the screening wave number (k_D, k_{FT}) then the scattering takes place from many electrons coherently, and the spectrum is directly related to that of collective excitations in the plasma. In the other limit, where k is large relative to $k_D (k_{FT})$, the scattering takes place from individual electrons. In general the cross-section is given by

$$\frac{d\sigma}{d\omega d\Omega} = V \left(\frac{d\sigma}{d\Omega}\right)_{Th} \frac{1}{2\pi} \int_{-\infty}^{+\infty} dt\, e^{i\omega t} < n_k(t) n_{-k} >$$

$$\left[\left(\frac{d\sigma}{d\Omega}\right)_{Th} \equiv \frac{e^4}{m^{*2} c^4} (\epsilon_1 \cdot \epsilon_2)^2\right]$$

where n_k is the Fourier transform of the electron density operator[2].

The magnetic field changes the spectrum of the electron fluctuations when, for example, the cyclotron frequency, $\omega_c = eB_o/m^*c$, is of the order of the plasma frequency. Since, the magnetic field is easily tunable experiments which measure the fluctuation spectrum for a range of magnetic fields should yield useful information.

Although density fluctuations can be excited at any angle relative to the field B_0 we will consider the particular geometry where $\underset{\sim}{k} \perp \underset{\sim}{B}_0$. In this geometry the single particle spectrum is discrete and is given by the cyclotron frequency and its harmonics. Density waves perpendicular to B_0 are not Landau damped even when $k \sim k_D$. We can therefore study the fluctuations over a wide range of k. In fact, in this geometry the boundary in k space between the collective and single particle regime is not well defined. It is possible to follow what is clearly a collective resonance for small k, out to large k, where it merges continuously into a single particle resonance spectrum at the cyclotron frequency and its harmonics.

The density fluctuation spectrum for single component plasma in the quasi static approximation is given by $I_m(1/\epsilon(k\omega))$. Here all the effects of the magnetic field are buried in ϵ.

$$\frac{d\sigma}{d\omega d\Omega} = V\left(\frac{d\sigma}{d\Omega}\right)_{Th} \frac{1}{\pi} \frac{\hbar}{1-e^{-\beta\hbar\omega}} \frac{k^2}{4\pi e^2} \times \operatorname{Im}\left[\frac{1}{\epsilon(k,\omega)}\right] .$$

To obtain some understanding of this formula, we first consider the case $k << k_D$ and approximate ϵ by

$$\epsilon(k,\omega) \cong 1 - \tilde{\omega}\frac{\omega_p^2}{\omega}\left[\frac{(1-\lambda)}{\tilde{\omega}^2-\omega_c^2} + \frac{\lambda}{\tilde{\omega}^2-4\omega_c^2}\right] ,$$

$$\lambda = \left(\frac{k}{k_D}\frac{\omega_p}{\omega_c}\right)^2 \qquad \tilde{\omega} = \omega + i/\tau$$

From the expression of the cross-section it is clear that the scattered light will have resonance lines at the zero's of ϵ. The solution of the equation $\epsilon = 0$ gives, in this approximation, two resonance frequencies.

$$\omega_{1,2}^2 = \frac{1}{2}\left[5\omega_c^2 + \omega_p^2 \mp \sqrt{(3\omega_c^2 - \omega_p^2)^2 + 12k^2 V_{th}^2 \omega_p^2}\right]$$

For the case of $\dfrac{\omega_p - \sqrt{3}\,\omega_c}{\omega_c} >> \lambda$ the two resonance lines appear approximately at the hybrid frequency $\omega_1 \approx \omega_p^2 + \omega_c^2$ and at the cyclotron harmonic $\omega_2 \approx 2\omega_c$. The line intensity at the cyclotron harmonic is smaller by a factor of $(k/k_D)^2$ than the line intensity at the hybrid frequency. As a consequence, for small k it is difficult to observe the resonance line at $2\omega_c$. However, for the case $\omega_p = \sqrt{3}\,\omega_c$, i.e., $\omega_1 \approx \omega_2 \approx 2\omega_c$ the two modes interact strongly with one another. The splitting is given by $\omega_1 - \omega_2 = \frac{3}{2}\sqrt{\lambda}\,\omega_c$ and their intensity is more or less equally.

SUMMARY

for $\dfrac{\omega_p - \sqrt{3}\,\omega_c}{\omega_c} >> \lambda$

$$\begin{cases} \omega_1 \approx \sqrt{\omega_p^2 + \omega_c^2} \\ \omega_2 \approx 2\omega_c \\ \dfrac{I_2}{I_1} \propto \left(\dfrac{k}{k_D}\right)^2 \end{cases}$$

and

for $\omega_p = \sqrt{3}\,\omega_c$

$(\omega_1 \approx \omega_2 \approx 2\omega_c)$

$$\begin{cases} \omega_1 - \omega_2 = \dfrac{3}{2}\sqrt{\lambda}\;\omega_c \\ \dfrac{I_2}{I_1} \approx 1 \end{cases}$$

An interesting feature of our result is that the splitting depends on nonlocal effects, i.e., finite k and it is linearly proportional to k/k_D. Finite k effects have only been qualitatively observed in light scattering from semiconducting plasmas because of the small values of $k/k_D (\approx .1)$ which are reached in such experiments; in the absence of a magnetic field one is forced to look for the dispersion of the plasmon to get information about finite k values. This dispersion is quadratic in (k/k_D) and thus exceedingly difficult to observe. The effects are of the order of one percent. The splitting of the Bernstein and upper hybrid will be, in this case, a ten percent effect, and should be observable. To illustrate these points numerical computations of the cross section have been performed. The intensities at the resonance modes as a function of ω_p/ω_c, k/k_D and $\omega_c \tau$ are presented in Fig. 1 - Fig. 4.

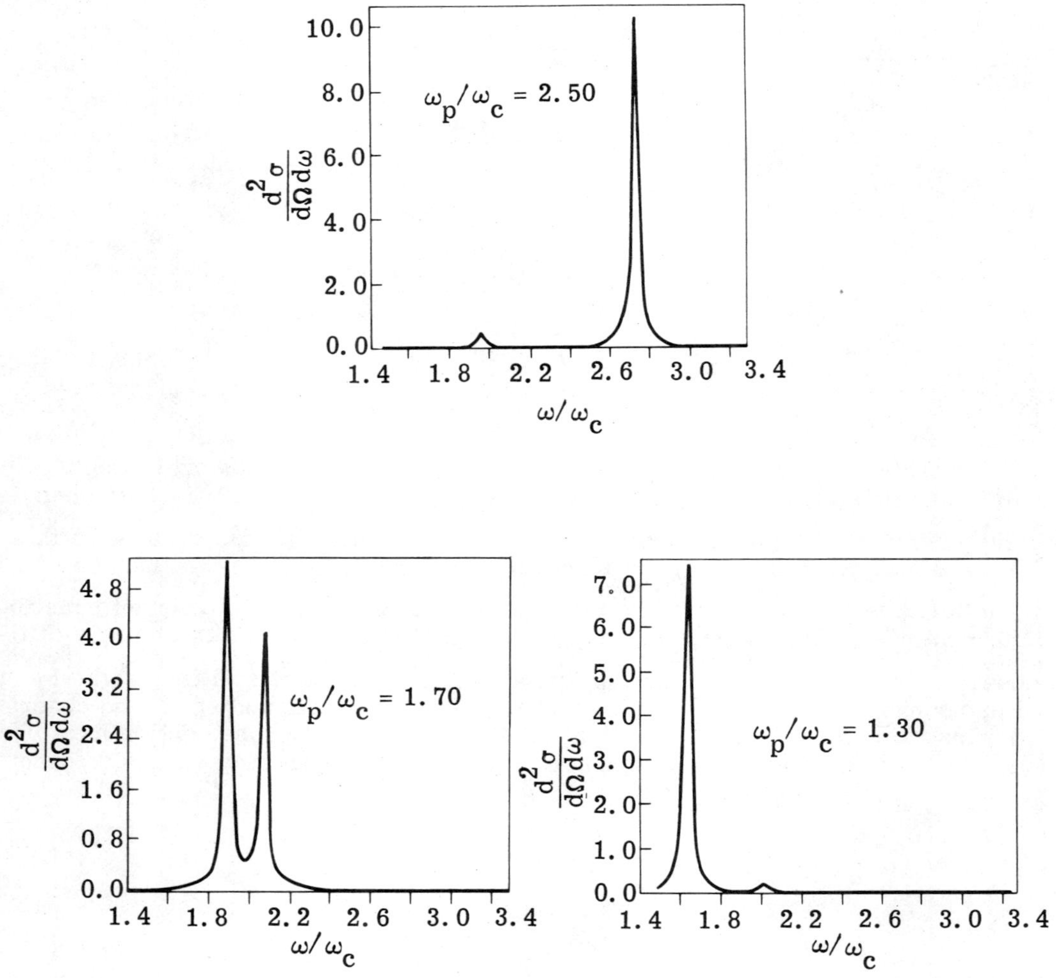

Fig. 1. The scattering cross-section per particle per unit solid angle in units of one tenth the Thompson cross-section is plotted versus the frequency shift of the scattered light. Here $(k/k_D)^2 = 0.005$ and $\omega_c\tau = 50$. When $\omega_p/\omega_c = 1.3$ (top picture) most of the scattered intensity is in the upperhybrid mode at $\omega_p/\omega_c = 1.65$ with small intensity at the cyclotron harmonica mode. For $\omega_p/\omega_c = 1.7$ (middle picture) the two modes have similar intensities. For $\omega_p/\omega_c = 2.5$ (lower picture) the intensity is in the upperhybrid mode at $\frac{\omega}{\omega_c} = 2.7$.

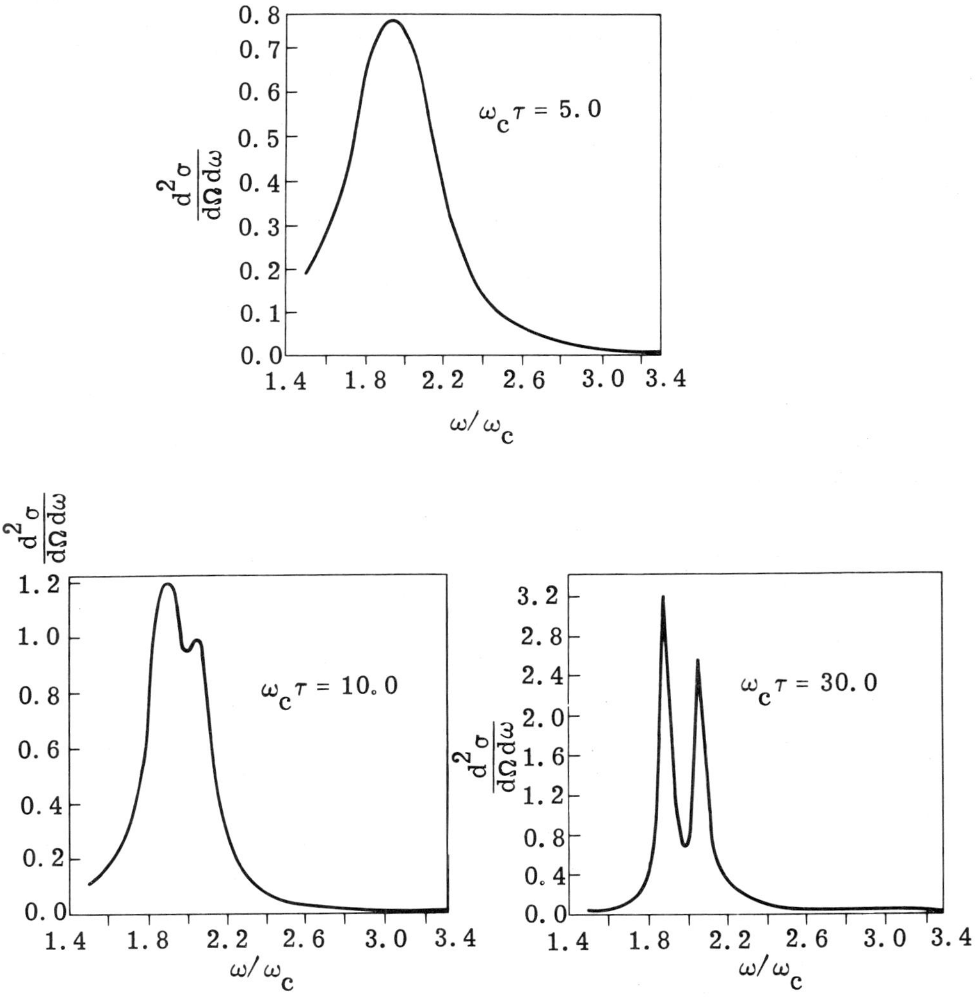

Fig. 2. The spectrum of scattered light is plotted as a function of ω with $(k/k_D)^2 = 0.005$ and $\omega_p/\omega_c = \sqrt{3}$ for $\omega_c\tau = 50$, 10 and 5.

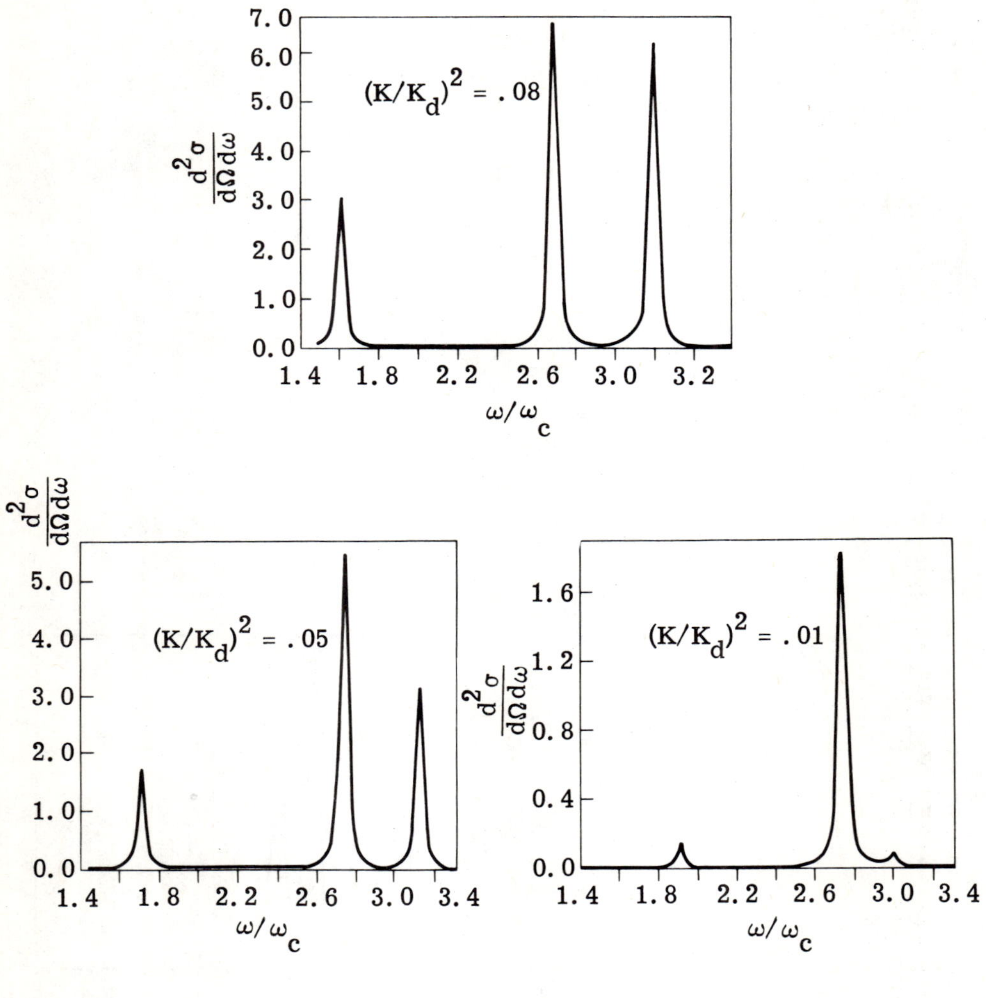

Fig. 3. The scattered spectrum in the neighborhood of the first two cyclotron harmonics for $\omega_p/\omega_c = 2.5$ and $\omega_c\tau = 50$, as a function of k/k_D. When k/k_D increases the mode at the first cyclotron harmonic moves toward ω_c, the upperhybrid mode at $\omega/\omega_c = 2.9$ (for $k \approx 0$) moves toward $2\,\omega_c$.

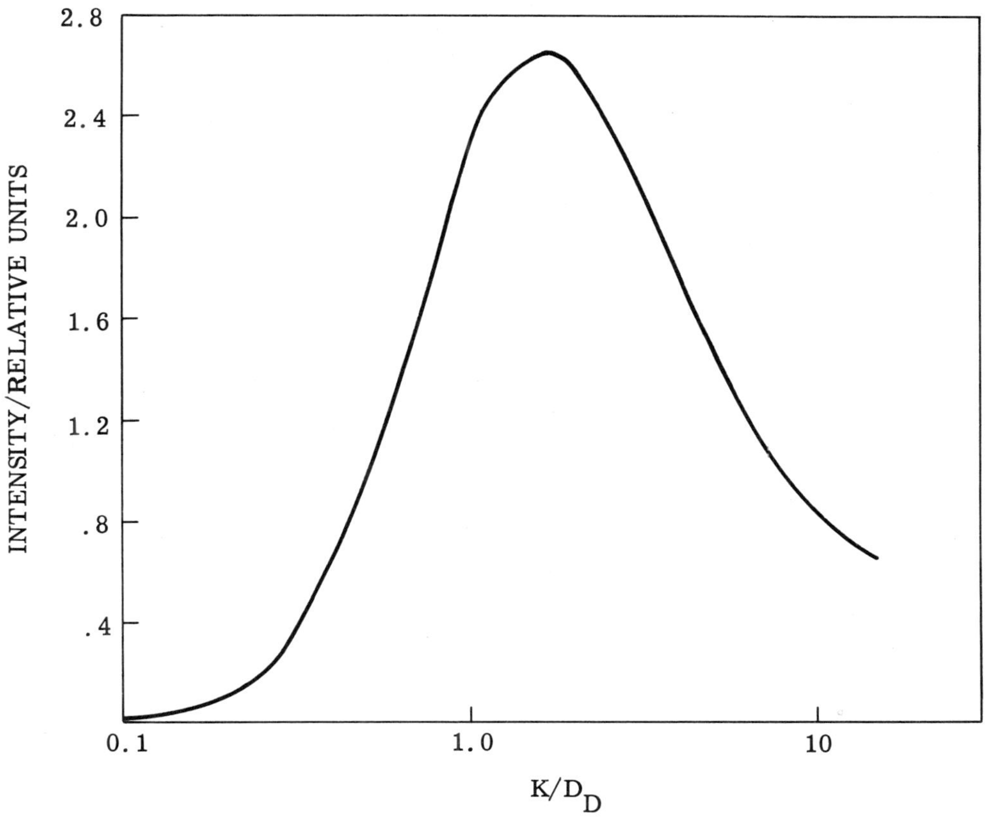

Fig. 4. The intensity of scattered light in the first Bernstein mode, ($\omega = 2\omega_c$ at $k/k_D \approx 0$) is plotted against k/k_D for $\frac{\omega_p}{\omega_c} = 2.5$. As we increase k/k_D the intensity increases until $k/k_D \sim 2$ i.e. $kRc \sim 5$ and then decreases.

CONCLUSIONS

We have shown that it is possible to study the Bernstein modes in plasmas even when $k/k_D < 1$. This is accomplished by suitably picking the magnetic field so that the upper hybrid and Bernstein modes will couple. The dispersion of the Bernstein mode and its coupling to the hybrid mode is interesting because it depends on nonlocal effects which to date have not been quantitatively studied in "semiconducting" plasmas. The perpendicular geometry discussed here allows one to follow the "sharp" resonances in the spectrum out to arbitarily large k values. The behavior of these resonances depends on nonlocal effects in the plasma. The detailed observation, in semiconductors, of the phenomenon described here and comparison of it with the simple theory will be useful in determining the importance of band structure effects beyond the simple effective mass approximation. These effects have been completely neglected in the present treatment.

REFERENCES

1. E.E. Salpeter, Phys. Rev. 120, (1960); M.N. Rosenbluth and N. Rostoker, Phys. Fluids 5, 776 (1962); P.M. Platzman, Phys. Rev. 139, A379 (1965); A.L. McWhorter, "Physics of Quantum Electronisc", 111, P.L. Kelley, B. Lax and P.E. Tannenwald (eds.) McGraw-Hill, New York, 1966; K.W. Bowles, Phys. Rev. Letters 1, 454 (1958); S.A. Ramsden and W.E.R. Davies, Phys. Rev. Letters 16, 303 (1966); H.J. Kunze, E. Füner, B. Kronast, and W.H. Kegel, Phys. Rev. Letters 11, 42 (1964).
A. Mooradian and G.B. Wright, Phys. Rev. Letters 16, 999 (1966); R.E. Slusher, C.K.N. Patel, and P.A. Fleury, Phys. Rev. Letters 18, 530 (1967); A. Mooradian and A.L. McWhorter, Phys. Rev. Letters 19, 850 (1967).
E.E. Salpeter, Phys. Rev. 122, 1663 (1961).
D.T. Farley, J.P. Dougherty, and D.W. Barron, Proc. Roy. Soc. 263 (A) 238 (1961).
2. M.N. Rosenbluth and N. Rostoker, Phys. Fluids 5, 776 (1962); P.M. Platzman, Phys. Rev. 139, A379 (1965).

D-6: RAMAN SCATTERING FROM MAGNETOPLASMA WAVES IN SEMICONDUCTORS

A. L. McWhorter and P. N. Argyres*
Lincoln Laboratory†, Massachusetts Institute of Technology
Lexington, Massachusetts

ABSTRACT

A general theory has been developed for Raman scattering from magnetoplasma waves in semiconductors with arbitrary band structure. The method used is an extension of that previously developed to explain the mixed plasmon-phonon scattering in GaAs. We assume that (1) the excitation of the magnetoplasma mode takes place via single-particle excitations of the electron system by the incident light wave and (2) the magnetoplasma wave is coupled to the single-particle excitations by the vector potential of its electromagnetic field, which in thermal equilibrium is determined by the fluctuation-dissipation theorem for the system under consideration. In the limit where the Raman frequency shift is small compared to the incident frequency, it is possible to make contact with phenomenological theory. For wave vector and dc magnetic field sufficiently small for quantum effects to be neglected, the theory predicts three principal phenomenological scattering mechanisms: free carrier fluctuations, electro-optic effect, and magneto-optic effect.

INTRODUCTION

In this paper we present a general theory for Raman scattering from magnetoplasma waves in semiconductors with arbitrary band structure. It has been necessary to go beyond calculations based on the effective mass approximation [1-3], since not only is this approximation restricted to the case where the incident frequency is small compared to the band gap, but it also completely omits some very important coupling mechanisms.

The method used is an extension of that previously developed to explain the mixed plasmon-phonon scattering in GaAs [4]. We assume that (1) the excitation of the magnetoplasma mode takes place via single-particle excitations of the electron system by the incident light wave and (2) the magnetoplasma wave is coupled to the single-particle excitations by the vector potential of its electromagnetic field, which in thermal

*Present address: Physics Department, Northeastern University, Boston, Massachusetts.
†Operated with support from the U. S. Air Force.

equilibrium is determined by the fluctuation-dissipation theorem. Assumption (1) omits coupling via excitons, which can be important for Raman scattering in insulators [5], but which should not play a role in semiconductors with large free carrier concentrations since the excitons become screened out.

In the limit where the Raman frequency shift is small compared to the incident frequency, it is possible to make contact with phenomenological theory. For wave vector and dc magnetic field sufficiently small for quantum effects to be neglected, the theory predicts three principal phenomenological scattering mechanisms: free carrier fluctuations, electro-optic effect, and magneto-optic effect. Scattering due to lattice deformation would also be present if coupled magnetoplasma-phonon waves were considered.

GENERAL FORMULATION

The electron system is described to lowest order by the Hamiltonian

$$H_0 = \sum_{\nu} \epsilon_{\nu} c^{+}_{\nu} c_{\nu} \tag{1}$$

where c^{+}_{ν}, c_{ν} are the creation and destruction operators for the electron in state $|\nu>$ of energy ϵ_{ν}. This system is taken to interact with a photon field of vector potential

$$A(r) = \sum_{k} \left(2\pi\omega_k/\Omega k^2\right)^{1/2} e_k \left(a_k e^{ik\cdot r} + a^{+}_{k} e^{-ik\cdot r}\right) \qquad (\hbar = 1) \tag{2}$$

where the subscript k stands for both the wave vector k and a polarization index, e_k is a unit polarization vector, Ω is the volume of the sample, and a^{+}_{k}, a_k are the creation and destruction operators for the photon. For simplicity we assume that the solid is lossless and nondispersive at the incident and scattered photon frequencies. The electron-photon interaction energy is given by

$$V = \sum_{\nu\nu'} \left[-\left(e/2c\right) \left(A(r)\cdot v + v\cdot A(r)\right)_{\nu\nu'} + \left(e^2/2mc^2\right) \left(A^2(r)\right)_{\nu\nu'} \right] c^{+}_{\nu} c_{\nu'} \tag{3}$$

in which

$$v = m^{-1}(\pi - eA_0/c) \tag{4}$$

is the velocity operator for the electron in the presence of a dc magnetic field of vector potential A_0, and π is the canonical momentum operator including spin-orbit interaction.

Each electron also interacts with a fluctuating electromagnetic field arising from all the other electrons. We shall assume that the one-electron wave functions are sufficiently nonlocalized for each electron to see statistically the same fluctuating electromagnetic field, which we take to be the total thermal equilibrium fluctuating field of the system. That is, we neglect local correlations and do not subtract out self-field interactions (which approach zero in the limit of large wave-function volume). Then if we denote by a(rt) the vector potential of the fluctuating electromagnetic field, the Fourier transform of the electron interaction energy can be written to first order in $a(r\omega)$ as

$$U(\omega) = \sum_{\nu\nu'} \Big[- \left(e/2c\right) \left(a(r\omega)\cdot v + v\cdot a(r\omega)\right)_{\nu\nu'} + \left(e^2/mc^2\right) \left(A(r)\cdot a(r\omega)\right)_{\nu\nu'}\Big] c^+_\nu c_{\nu'} \tag{5}$$

Here, as in (3), we omit the direct magnetic dipole interaction with the electron spins as this is usually negligible in semiconductors.

Since the different frequency components of a(rt) are statistically uncorrelated, it is convenient to define $\tilde{a}(r\omega)$ such that

$$< a_\alpha (r\omega) a^+_\beta(r'\omega')> = <\tilde{a}_\alpha (r\omega)\tilde{a}^+_\beta(r'\omega)> \delta(\omega - \omega') \tag{6}$$

where the angular brackets denote a thermal average. After expanding $\tilde{a}(r\omega)$ in a Fourier series in k, we may use the fluctuation-dissipation theorem[6] to obtain

$$< \tilde{a}_\alpha (k\omega)\, \tilde{a}_\beta(k'\omega)> = \frac{i}{2\pi\Omega} (n_\omega + 1) \left[D^*_{\beta\alpha} (k\omega) - D_{\alpha\beta}(k\omega)\right] \delta_{kk'} \tag{7}$$

where $n_\omega = [\exp(\omega/k_B T) - 1]^{-1}$ is the Bose-Einstein occupation number and

$$D(k\omega) = 4\pi \left[k^2 I - kk - \frac{\omega^2}{c^2} \epsilon(k\omega)\right]^{-1} \tag{8}$$

is the Green's function for Maxwell's equations, expressed in terms of the dielectric tensor $\epsilon(k\omega)$ of the system.

The Raman cross section for the excitation of a collective wave depends on the probability rate that the photon in state $i(k_i, e_i)$ is incoherently scattered to state $s(k_s, e_s)$ while the electron system returns to its original state of single-particle excitations. If we take the scattering system to be in thermal equilibrium and denote the many-electron state of H_0 by $|m>$, we may write for the differential cross section

$$d^2\sigma/d\Omega_s\, d\omega_s = r_o^2(\omega_s/\omega_i) << M^+_{sm,im} M_{sm,im}>> \tag{9}$$

where $r_o = e^2/mc^2$ is the classical electron radius, and the double angular brackets denote a thermal average over the initial many-electron states $|m>$ as well as over the fluctuating potentials. Explicitly, we find by straightforward perturbation theory

$$M_{xm,im} = (M^{(0)}_{sm,im} + M^{(1)}_{sm,im} + M^{(2)}_{sm,im} \tag{10}$$

where

$$M^{(0)}_{sm,im} = \frac{e}{c} \tilde{a}^+_\alpha (q\omega) \sum_{\nu\nu'} \Big[e^\alpha_i \frac{n_{\nu'} - n_\nu}{\epsilon_{\nu'\nu} - \omega_s} v^s_{\nu\nu'} \left(e^{ik_s\cdot r}\right)_{\nu'\nu} + e^\alpha_s \frac{n_{\nu'} - n_\nu}{\epsilon_{\nu'\nu} - \omega_i} \left(e^{-ik_i\cdot r}\right)_{\nu\nu'} v^i_{\nu'\nu} \Big] \tag{11}$$

$$M^{(1)}_{sm,im} = \frac{em}{c}\, \tilde{a}^{+}_{\alpha}(q\omega) \sum_{\nu\nu'\nu''} n_{\nu} \left[\frac{v^{s}_{\nu\nu''} v^{i}_{\nu''\nu'} v^{\alpha}(-q)_{\nu'\nu}}{(\epsilon_{\nu''\nu}-\omega_s)(\epsilon_{\nu'\nu}+\omega)} + \frac{v^{i}_{\nu\nu''} v^{s}_{\nu''\nu'} v^{\alpha}(-q)_{\nu'\nu}}{(\epsilon_{\nu''\nu}+\omega_i)(\epsilon_{\nu'\nu}+\omega)} \right.$$
$$+ \frac{v^{s}_{\nu\nu''} v^{\alpha}(-q)_{\nu''\nu'} v^{i}_{\nu'\nu}}{(\epsilon_{\nu''\nu}-\omega_s)(\epsilon_{\nu'\nu}-\omega_i)} + \frac{v^{i}_{\nu\nu''} v^{\alpha}(-q)_{\nu''\nu'} v^{s}_{\nu'\nu}}{(\epsilon_{\nu''\nu}+\omega_i)(\epsilon_{\nu'\nu}+\omega_s)}$$
$$\left. + \frac{v^{\alpha}(-q)_{\nu\nu''} v^{s}_{\nu''\nu'} v^{i}_{\nu'\nu}}{(\epsilon_{\nu''\nu}-\omega)(\epsilon_{\nu'\nu}-\omega_i)} + \frac{v^{\alpha}(-q)_{\nu\nu''} v^{i}_{\nu''\nu'} v^{s}_{\nu'\nu}}{(\epsilon_{\nu''\nu}-\omega)(\epsilon_{\nu'\nu}+\omega_s)} \right] \quad (12)$$

$$M^{(2)}_{sm,im} = \frac{e}{c}(e_s \cdot e_i)\, \tilde{a}^{+}_{\alpha}(q\omega) \sum_{\nu\nu'} \frac{n_{\nu'} - n_{\nu}}{(\epsilon_{\nu'\nu}+\omega)} (e^{iq\cdot r})_{\nu\nu'} v^{\alpha}(-q)_{\nu'\nu} \quad (13)$$

Here we have denoted by n_{ν} the occupation number of the one-electron state $|\nu\rangle$ in the many-electron state $|m\rangle$, and have introduced for brevity the operators

$$v^{\alpha}(q) = \frac{1}{2}\left(v^{\alpha} e^{iq\cdot r} + e^{iq\cdot r} v^{\alpha}\right)$$

$$v^{i} = e^{\alpha}_{i}\, v^{\alpha}(k_i) \quad , \quad v^{s} = e^{\alpha}_{s}\, v^{\alpha}(-k_s) \quad (14)$$

Also $\omega = \omega_i - \omega_s$, $q = k_i - k_s$, and $\epsilon_{\nu'\nu} = \epsilon_{\nu'} - \epsilon_{\nu}$. The Einstein summation convention is used throughout for the Greek indices labeling the vector components.

After rearranging and relabelling some terms, we can also write (12) in the form

$$M^{(1)}_{sm,im} = -\frac{em}{c}\, \tilde{a}^{+}_{\alpha}(q\omega) \sum_{\nu\nu'} (n_{\nu'} - n_{\nu}) \left\{ \frac{v^{s}_{\nu\nu'}[v^{i}, w^{\alpha}]_{\nu'\nu}}{\epsilon_{\nu'\nu} - \omega_s} + \frac{[v^{s}, w^{\alpha}]_{\nu\nu'} v^{i}_{\nu'\nu}}{\epsilon_{\nu'\nu} - \omega_i} \right\} \quad (15)$$

where

$$w^{\alpha}_{\nu'\nu} = v^{\alpha}(-q)_{\nu'\nu}/(\epsilon_{\nu'\nu} + \omega) \quad (16)$$

The thermal average over $|m\rangle$ is easily carried out with the use of the expression

$$\langle n_{\nu} n_{\nu'} \rangle = f_{\nu} f_{\nu'} + \delta_{\nu\nu'} f_{\nu} (1 - f_{\nu'}) \quad (17)$$

where f_{ν} is the Fermi-Dirac distribution function for an electron in state $|\nu\rangle$. However, in the limit of large sample volume the contributions arising from the second term of (17) are negligible. Hence, the cross section after averaging over $|m\rangle$ is

$$d^2\sigma/d\Omega_s d\omega_s = r_0^2 (\omega_s/\omega_i) \langle M^{+}_{si} M_{si} \rangle \quad (18)$$

where

$$M_{si} = M^{(0)}_{si} + M^{(1)}_{si} + M^{(2)}_{si} \quad (19)$$

with the $M^{(0,1,2)}_{si}$ given by (11), (12), and (13) with n_{ν} replaced by f_{ν}.

Note that the cross section will automatically have a peak at the frequency of the collective modes, since the fluctuation of $a(q\omega)$ given by (7) peaks at these frequencies. Expression (18) is to be used only for the collective modes; it does not give the cross section correctly for the single-particle excitations because of the neglect of coupling not involving $a(q\omega)$.

CONTACT WITH PHENOMENOLOGICAL THEORY

The expressions for the $M_{si}^{(n)}$ can be given simple physical interpretations by introducing the random-phase approximation for the electron susceptibility $\chi(q\omega)$:

$$\chi_{\alpha\beta}(q\omega) = -\frac{e^2 N}{m\omega^2\Omega}\,\delta_{\alpha\beta} - \frac{e^2}{\Omega\omega^2}\sum_{\nu\nu'}\frac{f_{\nu'} - f_\nu}{\epsilon_{\nu'\nu} - \omega}\,v^\alpha(-q)_{\nu\nu'}\,v^\beta(q)_{\nu'\nu} \tag{20}$$

where N is the total number of electrons. With the use of (20) and the continuity equation, $M_{si}^{(0)}$ and $M_{si}^{(2)}$ may be written

$$M_{si}^{(0)} = -(\Omega/ec)\,\tilde{a}_\alpha^+(q\omega)[e_i^\alpha\,\omega_s\,e_s^\beta\,\chi_{\beta\gamma}(k_s\omega_s)k_s^\gamma + e_s^\alpha\,\omega_i k_i^\beta\,\chi_{\beta\gamma}(k_i\omega_i)e_i^\gamma] \tag{21}$$

$$M_{si}^{(2)} = -(\Omega/ec)\,e_s^\alpha e_i^\alpha\,\omega\,\tilde{a}_\beta^+(q\omega)\,\chi_{\beta\gamma}(q\omega)q^\gamma \tag{22}$$

These expressions involve the electron charge density at ω_s, ω_i and ω. If the incident and scattered waves are transverse, $M_{si}^{(0)}$ vanishes.

Let us consider the long wavelength region where $\chi(k_i\omega_i) \approx \chi(0, \omega_i)$ and also take $\omega \ll \omega_i$. Then by comparing (15) with (20) and using (21) and (22), one can show that

$$M_{si} \approx (m\,\Omega^2\omega_i^2/e^2)\,\delta\chi_{\alpha\beta}^+\,e_s^\alpha\,e_i^\beta \tag{23}$$

where $\delta\chi_{\alpha\beta}^+$ is the fluctuation in the susceptibility at ω_i induced quasistatically by the collective mode through the fluctuation of $\tilde{a}^+(q\omega)$. Hence, from (18) the cross section is

$$d^2\sigma/d\Omega_s\,d\omega_s \approx (\Omega^2\omega_i^4/c^4)\,\langle\delta\chi_{\alpha\beta}\,e_s^\alpha e_i^\beta\,\delta\chi_{\gamma\delta}^+\,e_s^\gamma\,e_i^\delta\rangle \tag{24}$$

in accordance with the usual phenomenological theory. Note that $\langle\delta\chi\,\delta\chi^+\rangle \sim \Omega^{-1}$.

In order to provide a more detailed phenomenological interpretation to the theory, we rewrite the part of $M_{si}^{(1)}$ arising from intraband v^α transitions as

$$M^{(1)}_{si}\,(\mathrm{intra}) = \frac{em}{c}\,\tilde{a}^{+}_{\alpha}(q\omega) \sum_{\nu\nu'\nu''}\left(\frac{1}{2}f_{\nu} + \frac{1}{2}f_{\nu'} - f_{\nu''}\right)\left[\frac{v^{s}_{\nu\nu''}v^{i}_{\nu''\nu'}v^{\alpha}(-q)_{\nu'\nu}}{(\epsilon_{\nu''\nu}-\omega_{s})(\epsilon_{\nu''\nu'}-\omega_{i})}\right.$$

$$\left. + \frac{v^{i}_{\nu\nu''}v^{s}_{\nu''\nu'}v^{\alpha}(-q)_{\nu'\nu}}{(\epsilon_{\nu''\nu}+\omega_{i})(\epsilon_{\nu''\nu'}+\omega_{s})}\right] + \sum_{\nu\nu'} F_{\nu\nu'}\,\delta\rho^{+}_{\nu'\nu} \tag{25}$$

Here

$$\delta\rho^{+}_{\nu'\nu} = -\,(e/c)\,\tilde{a}^{+}_{\alpha}(q\omega)\,v^{\alpha}(-q)_{\nu'\nu}(f_{\nu'} - f_{\nu})/(\epsilon_{\nu'\nu} + \omega) \tag{26}$$

which can be identified as the perturbation in the one-electron density matrix due to $\tilde{a}^{+}_{\alpha}(q\omega)$, and

$$F_{\nu\nu'} = -m\sum_{\nu''}\left[\frac{1}{2}\left(\frac{1}{\epsilon_{\nu''\nu}-\omega_{s}} + \frac{1}{\epsilon_{\nu''\nu'}-\omega_{i}}\right)v^{s}_{\nu\nu''}v^{i}_{\nu''\nu'}\right.$$

$$\left. + \frac{1}{2}\left(\frac{1}{\epsilon_{\nu''\nu}+\omega_{i}} + \frac{1}{\epsilon_{\nu''\nu'}+\omega_{s}}\right)v^{i}_{\nu\nu''}v^{s}_{\nu''\nu'}\right] \tag{27}$$

Next we introduce Stinchcombe's representation [7] for an operator O,

$$\{n_1k_1\,|O|\,n_2k_2\} = \iint dr_1dr_2\,e^{-i(e/c)A_0(R)\cdot r}\,\langle r_1|O|r_2\rangle\,\psi^{*}_{n_1k_1}(r_1)\,\psi_{n_2k_2}(r_2) \tag{28}$$

where $\psi_{nk}(r)$ are Bloch functions, $R = (r_1 + r_2)/2$ and $r = r_1 - r_2$ (spin indices have been suppressed). Then by slightly generalizing Stinchcombe's results [7], we can show for $q \ll k_F$ and $\omega_c \ll kT$ that

$$\sum_{\nu\nu'} F_{\nu\nu'}\,\delta\rho^{+}_{\nu'\nu} = \sum_{nkk'}\{nk|F|nk'\}\{nk'|\,\delta\rho^{+}\,|nk\}$$

$$\approx \sum_{nk}\{n, k + \tfrac{1}{2}q|F|n, k - \tfrac{1}{2}q\}\,[\,\delta f^{+}_{nk}(q, \omega, H_0)$$

$$-(e/c)\tilde{a}^{+}_{\alpha}(q\omega)\,\nabla^{\alpha}_{k}f_{nk}] \tag{29}$$

for the intraband v^{α} transitions, where δf^{+}_{nk} is the solution to the semiclassical Boltzmann equation in the presence of the dc magnetic field H_0, with the fluctuating electric field

$$\delta E_{\alpha}^{+} = -(i\omega/c)\,\tilde{a}_{\alpha}^{+}(q\omega) \tag{30}$$

acting as the driving force. Actually, there is a second, but generally negligible driving force that results from the q-dependence of the factor $v^{\alpha}(-q)_{\nu'\nu}$ in (26); to first order in q this extra driving force is just the gradient of the Zeeman energy arising from the electron orbital angular momentum and the fluctuating magnetic field

$$\delta H^{+} = -iq \times \tilde{a}^{+}(q\omega) \tag{31}$$

The term in $\tilde{a}_{\alpha}^{+}(q\omega)$ in (29) arises from the fact that the Boltzmann equation is in terms of the kinetic rather than the canonical momentum.

The first term of (29) can be combined with $M_{si}^{(2)}$ to yield

$$\frac{m\Omega\,\omega_i^2}{e^2}\, e_s^{\alpha}\, e_i^{\beta} \sum_{nk} G_{nkq}^{\alpha\beta}\, \delta f_{nk}^{+}(q, \omega, H_0) \tag{32}$$

where

$$G_{nkq}^{\alpha\beta} = -\frac{e^2}{m\omega_i^2}\,\delta_{\alpha\beta} + \frac{e^2}{m\Omega\,\omega_i^2}\,\{n, k + \tfrac{1}{2}q \,|\, F_{\alpha\beta} | n, k - \tfrac{1}{2}q\} \tag{33}$$

and $F_{\alpha\beta}$ is defined by $F = e_s^{\alpha} e_i^{\beta} F_{\alpha\beta}$. By rewriting (20) in Stinchcombe's representation, we see that for small ω, q, k_i and H_0

$$G_{nkq}^{\alpha\beta} \approx \frac{\partial \chi_{\alpha\beta}(\omega_0, H_0)}{\partial f_{nk}} \tag{34}$$

where $\omega_0 = (\omega_i + \omega_s)/2$ and $\chi_{\alpha\beta}(\omega_0, H_0) = \chi_{\alpha\beta}(k = 0, \omega_0, H_0)$

In the interband v^{α} terms in $M_{si}^{(1)}$ we put

$$\frac{1}{\epsilon_{\nu'\nu} + \omega} = \frac{1}{\epsilon_{\nu'\nu}} - \frac{\omega}{\epsilon_{\nu'\nu}(\epsilon_{\nu'\nu} + \omega)} \tag{35}$$

The second part of (35) gives a contribution to $M_{si}^{(1)}$ of

$$-\frac{ie\Omega}{m^2}\, e_s^{\alpha}\, e_i^{\beta}\, P_{\beta\alpha}^{\gamma}\, \delta E_{\gamma}^{+} \approx \frac{m\Omega\,\omega_i^2}{e^2}\, e_s^{\alpha}\, e_i^{\beta}\, \frac{\partial \chi_{\alpha\beta}(\omega_0, H_0)}{\partial E_{\gamma}}\, \delta E_{\gamma}^{+} \tag{36}$$

where $P_{\beta\alpha}^{\gamma}$ is the tensor defined by Loudon, [8] generalized to finite temperature, wave vector and magnetic field. All other terms are independent of ω to first order, but involve q. To first order in q and k_i these remaining terms combine to yield the magneto-optic coefficient; for $H_o = 0$ the resulting expression can be reduced after considerable manipulation to that derived by Roth [9].

Hence we finally conclude that the quantity $\delta\chi^+_{\alpha\beta}$ in [24] can be written for sufficiently small ω, q, k_i and H_o in the form

$$\delta\chi^+_{\alpha\beta} = \sum_{nk} G^{\alpha\beta}_{nkq}\, \delta f^+_{nk} + \frac{\partial\chi_{\alpha\beta}(\omega_o, H_o)}{\partial E_\gamma}\, \delta E^+_\gamma + \frac{\partial\chi_{\alpha\beta}(\omega_o, H_o)}{\partial H_\gamma}\, \delta E^+_\gamma \tag{37}$$

with δf^+_{nk} given by the Boltzmann equation, δE^+ by (30) and δH^+ by (31). In general it is necessary to use the q-dependent expression (33) for $G^{\alpha\beta}_{nkq}$ rather than the approximate form (34) since if $G^{\alpha\beta}_{nkq}$ is expanded in powers of q, there is a partial cancellation between the terms linear in q and the free carrier contribution to the magneto-optic term. Eq. (37) can also be used for externally excited waves if the system is not perturbed strongly from thermal equilibrium; δf_{nk}, δE and δH are then to be determined directly from the properties of the wave rather than from the fluctuation-dissipation theorem.

Helicons in semiconductors should show all three types of coupling indicated in (37). The inherent weakness of the magneto-optic coupling is offset by the relatively large magnetic field associated with slow waves like the helicon. In many-valley semiconductors such as PbTe the free carrier coupling can be quite sizable despite the nearly overall charge neutrality. Density fluctuations do occur in the individual valleys [10], and these are generally weighted differently in (37).

Finally, we remark that (37) can be used for single-particle as well as collective excitations if δf_{nk} is reinterpreted as the total thermal fluctuation in occupancy, the independent-particle or Hartree-Fock part as well as the part induced by the fluctuating electric field. For example, in the case where $H_o = 0$ and $q \ll k_F$, one must take

$$\delta f_{nk} = \delta g_{nk} + \frac{e\,\delta E \cdot v_{nk}\, \partial f_{nk}/\partial\epsilon}{i(\omega - q \cdot v_{nk})} \tag{38}$$

with

$$\langle \delta g_{nk} \delta g^+_{nk'} \rangle = -(\omega/\Omega)(n_\omega + 1)(\partial f_{nk}/\partial\epsilon)\, \delta(\omega - q \cdot v_{nk})\, \delta_{kk'} \tag{39}$$

In the quantum calculation, the first term of (38) arises from scattering not involving $\tilde{a}(q\omega)$, i.e., from scattering due just to the electron-photon interaction term (3). If proper account is taken of the correlation between the two parts of δf_{nk}, one can obtain, for instance, Wolff's effective mass expression [3] for the scattering cross section in nonparabolic bands from the revised first term of (37). Also, if spin indices are included, the more general expressions in [11] can be obtained as well.

REFERENCES

1. P.M. Platzman, Phys. Rev. 139, A379 (1965).
2. A.L. McWhorter, "Physics of Quantum Electronics," p. 111, P.L. Kelley, B. Lax, and P.E. Tannenwald (ed.), McGraw-Hill Book Co. Inc., New York, 1966.
3. P.A. Wolff, Phys. Rev. 171, 436 (1968).
4. A.L. McWhorter and P.N. Argyres, Bull. Am. Phys. Soc. 12, 102 (1967); A. Mooradian and A.L. McWhorter, Phys. Rev. Letters 19, 849 (1967).
5. A.K. Ganguly and J.L. Birman, Phys. Rev. 162, 806 (1967).
6. See, for example, A.A. Rukhadze, and V.P. Silin, Usp. Fiz. Nauk 74, 223 (1961), (English translation: Soviet Phys. - Usp. 4, 459 (1961)).
7. R.B. Stinchcombe, Proc. Phys. Soc. (London) 78, 275 (1961).
8. R. Loudon, Advan. Phys. 13, 423 (1964).
9. L.M. Roth, Phys. Rev. 133, A542 (1964).
10. J.N. Walpole and A.L. McWhorter, Phys. Rev. 158, 708 (1967).
11. D.C. Hamilton and A.L. McWhorter, see paper D-4 this conference.

D-7: LANDAU LEVEL RAMAN SCATTERING

G. B. Wright, P. L. Kelley and S. H. Groves
Lincoln Laboratory,* Massachusetts Institute of Technology
Lexington, Massachusets

INTRODUCTION

When a solid containing mobile charge carriers is placed in a magnetic field, the energy of the carriers is quantized in Landau levels. The inelastic light scattering from this system which results in the excitation of electrons from one Landau level to another is termed Landau level Raman scattering. A theoretical discussion of this scattering was first given by Wolff [1], who considered the process in which the Landau level quantum number of the charge carrier increases by two ($\Delta n = 2$), with no change in the spin state ($\Delta s = 0$). Kelley and Wright [2], and Yafet [3], subsequently included the effect of the actual degenerate valence band structure of InSb, and found that processes for $\Delta n = 0$ and $\Delta n = 2$ in which the spin changed ($\Delta s = 1$), were also important. Subsequent experimental investigations by Slusher et al. [4, 5, 6] confirmed these predictions, and also demonstrated that there was an important $\Delta n = 1$, $\Delta s = 0$ process which had not been treated. In this paper, we present the results of a numerical calculation of the single particle Raman cross section, based on the model of Pidgeon and Brown [7] for InSb. Our model assumes that only the lowest Landau level is occupied.

CALCULATION OF THE CROSS SECTION

We have calculated the Landau level Raman scattering cross section on the dipole approximation, $qr_c << 1$, where r_c is the cyclotron radius of the carriers. We obtain

$$\frac{d\sigma}{d\Omega} = \left(\frac{e^2}{m^2c^2}\right)^2 \frac{\omega_s}{\omega_i} \left| \sum_m \frac{\hat{\epsilon}_s \cdot p_{fm} \hat{\epsilon}_i \cdot p_{md}}{\omega_i - \omega_{md}} - \frac{\hat{\epsilon}_i \cdot p_{fm} \hat{\epsilon}_s \cdot p_{md}}{\omega_i - \omega_{fm}} \right|^2 \qquad (1)$$

where d, m, and f refer to initial, intermediate, and final electronic states. We may rewrite (1) in terms of the conduction band effective mass, $m^* \simeq \frac{3\mathcal{E}_g}{4p^2}$ and a Raman weight function, W, which will be of order unity or less for fields up to 100 kilogauss.

*Operated with support from the U. S. Air Force.

$$\frac{d\sigma}{d\Omega} = \left(\frac{e^2}{mc^2}\right)^2 \frac{\omega_s}{\omega_i} \left(\frac{p^2}{\mathcal{E}_g m}\right)^2 W \qquad (2)$$

$$\text{For InSb,} \quad \frac{d\sigma}{d\Omega} = 2.4 \times 10^{-23} \left(\frac{\omega_s}{\omega_i}\right) W_{cm^2}$$

To compute the weight function and the polarization selection rules, we have taken the basis functions appropriate to InSb [7].

$$\begin{array}{ll} a_1(n)\Phi_n \, |S\uparrow> & b_1(n)\Phi_{n+1}|\, iS\downarrow> \\ a_2(n)\Phi_{n-1}|\frac{1}{\sqrt{2}}(X+iY)\uparrow> & b_2(n)\Phi_{n+2}|\frac{i}{\sqrt{2}}(X-iY)\downarrow> \\ a_3(n)\Phi_{n+1}|\frac{1}{\sqrt{6}}[(X-iY)\uparrow +2Z\downarrow]> & b_3(n)\Phi_n|\frac{i}{\sqrt{6}}[(X+iY)\downarrow -2Z\uparrow]> \\ a_4(n)\Phi_{n+1}|\frac{i}{\sqrt{3}}[-(X-iY)\uparrow + Z\downarrow] & b_4(n)\Phi_n|\frac{1}{\sqrt{3}}[X+iY)\downarrow +Z\uparrow]> \end{array} \qquad (3)$$

In InSb, the conduction band will consist primarily of functions subscripted 1, the light and heavy hole bands of a mixture of 2 and 3, and the spin-orbit split off band primarily of functions 4. For $k_z = 0$, the a-set and the b-set will, to a good approximation, not interact with each other. Using this approximation, and the model Hamiltonian of Pidgeon and Brown [7], we have calculated the Raman cross section numerically for various polarizations of input and scattered photons. For low fields, the results will be well described by the analytical expressions of Yafet [3]. For the purpose of describing the Raman processes, we find it convenient to make one further approximation, that of uncoupled bands.

THE UNCOUPLED BAND APPROXIMATION

If we assume that the light and heavy hole bands are formed solely from basis functions of subscript type 2 and 3 in (3), we obtain a 2 x 2 interaction Hamiltonian for the a-set and for the b-set. Then it is easy to obtain from the orthogonality conditions on these functions the following relations between the coefficients; $a_2^+ a_3^+ = - a_2^- a_3^-$, $b_2^+ b_3^+ = -b_2^- b_3^-$, and $a_3^{+2} + a_3^{-2} = b_2^{+2} + b_2^{-2} = 1$, where the plus and minus superscripts denote the light and heavy holes. These relations will be useful in demonstrating interference effects between light and heavy holes.

In Fig. 1 we show the electric dipole transitions possible between the conduction band and the light and heavy hole bands in the uncoupled band approximation. The n-label on the valence band states in the figure is one higher than the n-label of functions 2 and 3 in Eq. (3). Thus the n = 0 level in the figure consists only of a_3 or b_2, while the n = 2 level contains both a_2 and a_3, or b_2 and b_3. The circular polarization symbols are for photon absorption for a transition in the direction of the arrow. Inspection of the basis functions (3) reveals that the $\Delta n = 2$ transitions proceeding through the valence band are made

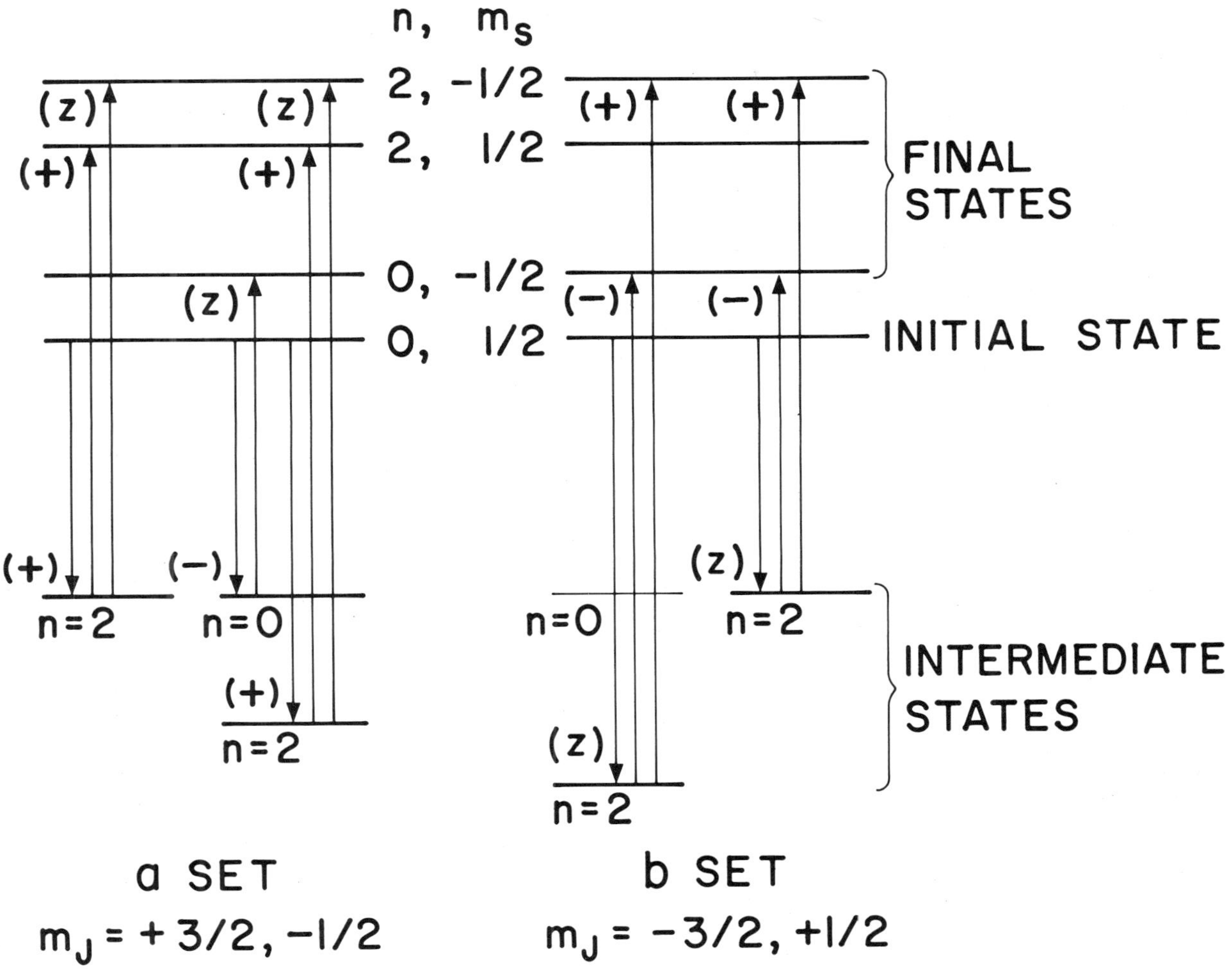

Fig. 1. Electric dipole transitions allowed between conduction band and decoupled valence band. The circular polarization symbols denote an absorptive transition in the direction of the arrow.

possible by the mixture of n = 0 and n = 2 harmonic oscillator functions in the light and heavy hole bands. Similarly, the spin-flip transitions from the a_1 conduction band to the b_1 conduction band are made possible by the mixed spin state of the hole bands. Finally, at this level of approximation, the $\Delta n = 1$, $\Delta s = 0$ transitions do not occur. Below we consider two single particle processes by which these transitions may occur. We now look at the structure of W for the different Raman scattering processes.

THE $\Delta n = 2$, $\Delta s = 0$ TRANSITION

In the uncoupled band system, the conduction band is parabolic, and the Landau levels are evenly spaced. Thus in Equation (1), $\omega_{mi} = \omega_{fm} = \omega_c$, and interference completely cancels any contribution from conduction band intermediate states. Summing over valence band intermediate states, we obtain for $W_{2,\ 1/2}$

$$W_{2,+1/2} = \frac{\omega_g^4}{9}\left[\frac{a_2^+(2)a_3^+(2)}{(\omega_g + \omega_{\ell ha})^2 - \omega_1^2} + \frac{a_2^-(2)a_3^-(2)}{(\omega_g + \omega_{hha})^2 - \omega_i^2}\right]^2$$

$$= \frac{1}{9}\left[\frac{\Delta\omega_a \omega_g(\omega_g^2 + \omega_i^2)}{(\omega_g^2 - \omega_i^2)^2} 2a_2^+(2)\, a_3^+(2)\right]^2 \qquad (4)$$

The subscripts on W denote the change of n, and the final conduction band spin state, and $\Delta\omega_a = \omega_{\ell ha} - \omega_{hha}$ is the difference in energy between the light and heavy holes in a magnetic field. Since $\Delta\omega_a$ increases linearly with field, the cross section increases quadratically. When the interaction with the conduction band is turned on, the conduction band becomes non parabolic. The results of the full numerical calculation are shown in Fig. 2a as the (+, -) polarization. The additional (z, z) polarization process arises from wave function mixing. In addition we notice a strong interference cancellation near 80 kG for both polarizations.

THE $\Delta n = 2$, $\Delta s = 1$ TRANSITION

The $\Delta s = 0$ transitions involved only the a-set intermediate states, while the spin flip transitions use both a- and b-set states. For the spin-flip transitions the relative order of photon absorption and emission is important. We have exhibited this condition by a $\pm$ superscript on W, and show its relation to the polarization selection rules in Table I.

TABLE I

Selection Rules and Weight Notation For Landau Level Raman Scattering[a]

OUT / IN	Z	−	+
Z	0	$W^+_{2,-1/2}$	$W^+_{0,-1/2}$
+	$W^-_{2,-1/2}$	$W_{2,\ 1/2}$	0
−	$W^-_{0,-1/2}$	0	0

[a] The subscripts on the symbols indicate the terminal n, m_s state of the electron. The selection rules apply only to the uncoupled band scheme, and are relaxed for interacting bands.

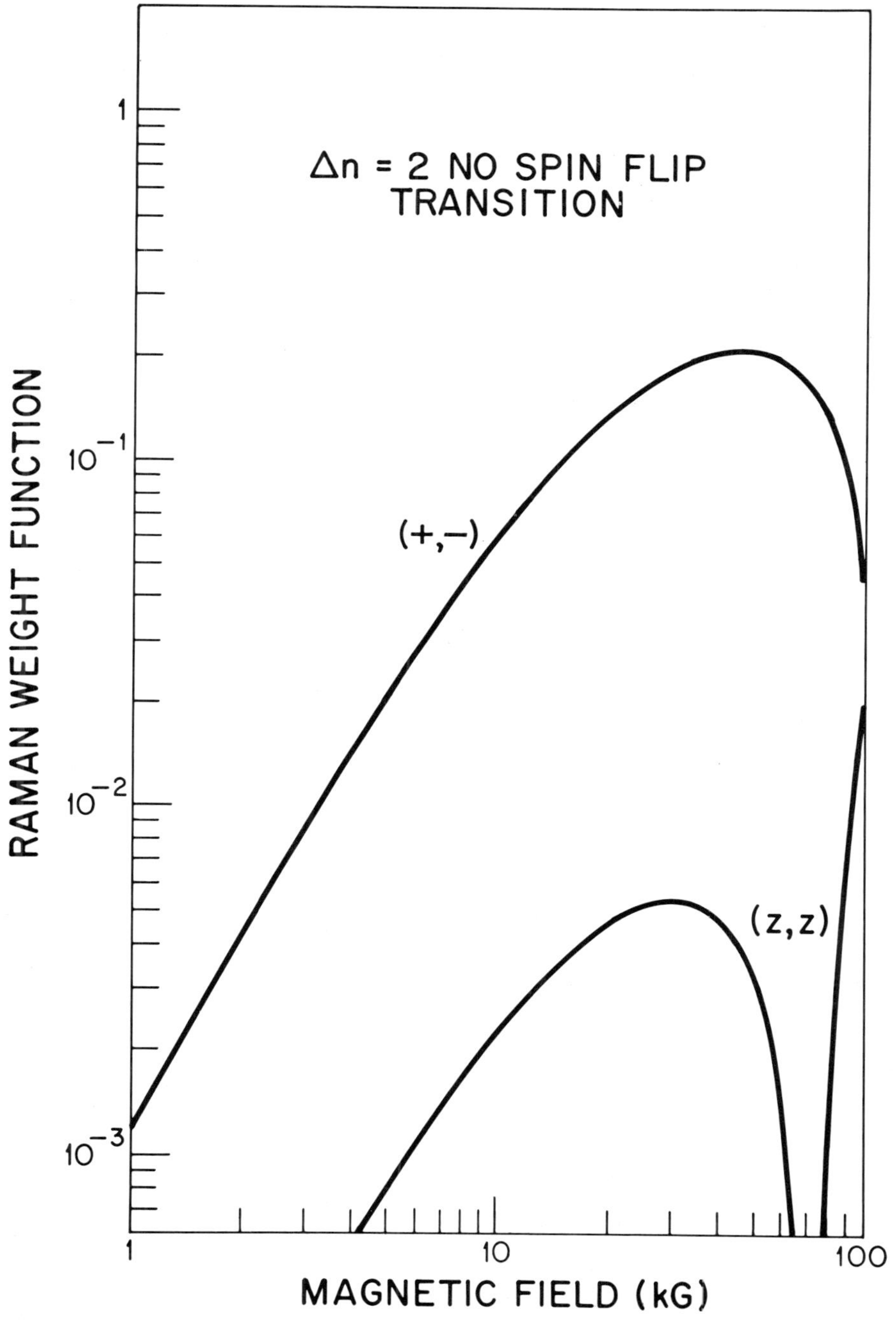

Fig. 2. The $\Delta n = 2$ no spin flip transition. The first and second symbols in parentheses denote polarization of input and scattered photon, respectively. $\hbar\omega = 0.12$ eV.

Evaluation of the sums for $W^{\pm}_{2,-1/2}$ gives

$$W^{\pm}_{2,-1/2} = \frac{2\omega_g^2}{3}\left[\frac{b_1^+(2)b_2^+(2)}{\omega_g \pm \omega_i + \omega_{\ell hb}} + \frac{b_1^-(2)\, b_2^-(2)}{\omega_g \pm \omega_i + \omega_{hhb}} \right. \tag{5}$$

$$\left. + \frac{a_1^+(2)\, a_2^+(2)}{\omega_g \mp \omega_i + \omega_{\ell ha}} + \frac{a_1^-(2)\, a_2^-(2)}{\omega_g \mp \omega_i + \omega_{hha}}\right]^2$$

$$= \frac{1}{6}\left[\frac{\Delta\omega_a(2)\omega_g}{(\omega_g \mp \omega_i)^2} + \frac{0.6\Delta\omega_b(2)\omega_g}{(\omega_g \pm \omega_i)^2}\right]^2$$

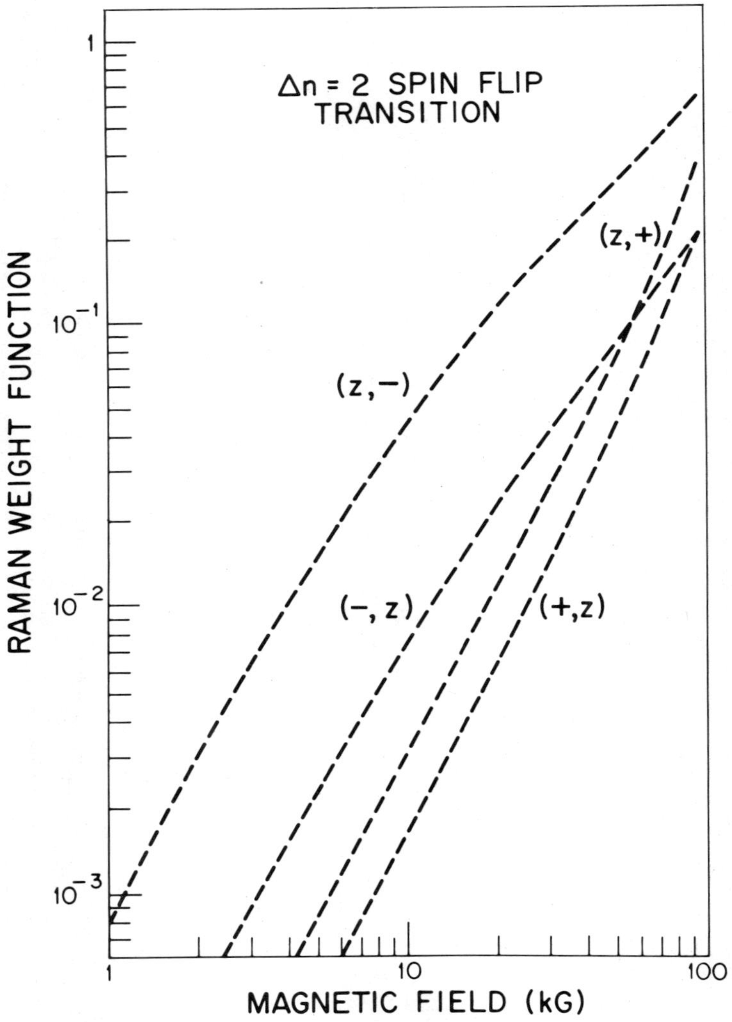

Fig. 3. The $\Delta n = 2$ spin flip transition. $\hbar\omega = 0.12$ eV.

where we have used the results that for the n = 2 level in InSb, $a_1^+ a_2^+ \approx .5$ and $b_1^+ b_2^+ \approx .3$. Cancellation has again made the cross section quadratically dependent upon magnetic field. The numerical results of the full calculation are shown in Fig. 3.

THE $\Delta n = 0$, $\Delta s = 1$ TRANSITION

This transition alone does not suffer the strong light hole-heavy hole cancellation which made the others quadratically dependent upon field. Evaluation of this process in the uncoupled band scheme gives

$$W^{\pm}_{0,-1/2} = \frac{2\omega_g^2}{9}\left[\frac{a_3^{+2}(0)}{\omega_g \mp \omega_i} - \frac{b_3^{+2}(2) + b_3^{-2}(2)}{\omega_g \pm \omega_1}\right]^2 \tag{6}$$

$$= \frac{8}{9}\left[\frac{\omega_g \omega_i}{\omega_g^2 - \omega_i^2}\right]^2$$

The numerical results for this process are shown in Fig. 4. This process does not disappear at zero field, and is the basic mechanism for the spin-density fluctuation scattering discussed by Hamilton and McWhorter [8].

THE $\Delta n = 1$, $\Delta s = 0$ TRANSITION

We consider two single particle processes which might cause the $\Delta n = 1$, $\Delta s = 0$ transitions observed experimentally. The first is made possible by the admixing of the a-set functions with the b-set functions of Eq. (3) via the linear k, zincblende splitting of the valence band. Although this is a small term in the Hamiltonian it does cause an interaction between pairs of a- and b-set heavy hole levels which are nearly degenerate in energy [9]. However, it can be shown quite generally that a process which uses one of these interacting levels as an intermediate state interferes with the process which goes via the other interacting level. Thus, for this process the terms in W cancel to order $\Delta\varepsilon/(\varepsilon_g - \omega_i)$, where $\Delta\varepsilon$ is the separation between the interacting heavy hole states. At 40 kG this factor is ~3×10^{-3}, which occurs squared in W or ~1×10^{-5} in W. Because of this cancellation we conclude that this process is not important.

The second process we consider is made possible by the coupling of the a-set with the b-set function of Eq. (3) which occurs for $k_z \neq 0$. Patel [10] has suggested this as the mechanism for the observed $\Delta n = 1$, $\Delta s = 0$ process and has given an approximate expression for the differential cross section. Since the coupling between the valence and conduction bands is important for this process, and since the intermediate states are the uppermost valence band levels which exhibit the "quantum effects" in energy spacing and wave function composition, we have thought it advisable to use the method of Ref. [7] to calculate this cross section. Since this involves the complexity of a summation over those values of k_z which are occupied by conduction band electrons for each magnetic subband, we have made only the computation which is appropriate to the experimental conditions of Ref. [4].

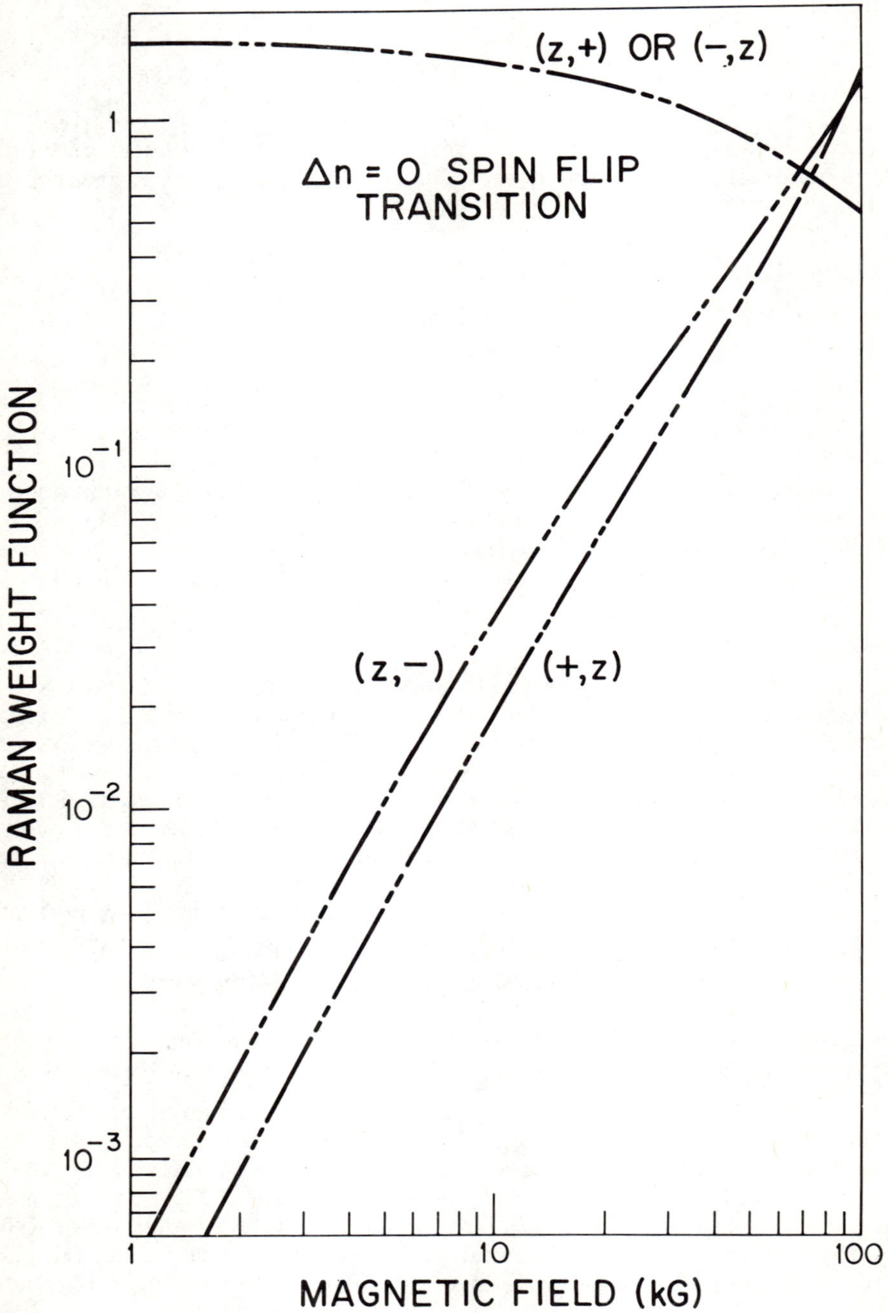

Fig. 4. The $\Delta n = 0$, spin flip transition. $\hbar\omega = 0.12$ eV.

An immediate result, which can be seen from the form of the Landau level wave functions for $k_z \neq 0$, is that only one component (p^+) of the circularly polarized radiation can cause this scattering (with a p_z transition as the other leg of the process). This polarization dependence could provide a necessary, but not sufficient, experimental check on the k_z mixing as the source of the $\Delta n = 1$, $\Delta s = 0$ scattering.

With the assumption of complete statistical degeneracy, we find that only the two $n = 0$ conduction band Landau levels are occupied at 50 kG for the sample concentration of Ref. [4], 5×10^{16} cm^{-3}. From these initial states, 14 valence subbands (split-off band included) and one conduction subband can serve as intermediate states. Our computed summation of these processes gives a weight function $W = 5 \times 10^{-3}$ at 50 kG and about half as large at 100 kG. This is roughly 1000 times weaker than the calculated $\Delta n = 0$, $\Delta s = 1$ cross section of Fig. 4, while the integrated strengths of these transitions in the experiment of Ref. [4] are of the same order of magnitude. This suggests that other mechanisms, such as coulomb interactions, need be invoked to explain the observed strength of the $\Delta n = 1$, $\Delta s = 0$ process.

REFERENCES

1. P.A. Wolff, Phys. Rev. Letters 16, 225 (1966).
2. P.L. Kelley and G.B. Wright, Bull. Am. Phys. Soc. 11, 812 (1966).
3. Y. Yafet, Phys. Rev. 152, 858 (1966).
4. R.E. Slusher, C.K.N. Patel and P.A. Fleury, Phys. Rev. Letters 18, 77 (1967).
5. C.K.N. Patel and R.E. Slusher, Phys. Rev. 167, 413 (1968).
6. C.K.N. Patel and R.E. Slusher, Bull. Am. Phys. Soc. 13, 480 (1968).
7. C.R. Pidgeon and R.N. Brown, Phys. Rev. 146, 575 (1966).
8. D.C. Hamilton and A.L. McWhorter, paper D-4 this conference.
9. C.R. Pidgeon and S.H. Groves, Phys. Rev. Letters 20, 1003 (1968).
10. C.K.N. Patel, Modern Optics 17, 19 (1967).

D-8: LANDAU LEVEL RAMAN SCATTERING IN SEMICONDUCTORS

V. P. Makarov
Lebedev Physical Institute
Moscow, USSR

ABSTRACT

Starting from general theory of Raman scattering (see, e.g. [1]) we have investigated theoretically Raman scattering of light by the electron gas in the conduction band of n-semiconductors subjected to magnetic field. We use the one-electron approximation and a two-band model of the crystal (see, e.g. [2]), which is known to be an excellent approximation of the energy spectrum in a narraw band gap semiconductor such as InSb. We confined ourselves to the case where concentration of electrons in the conduction band n_e and the value of magnetic field $\vec{H}$ are such that electron energy levels with Landau quantum number $n \geq$ I lie above Fermi level.

The differential scattering cross sections for the processes with various possible changes $\Delta n = 0, 1, 2$ and $\Delta S = 0, \div 1, -1$, ($n = 0, 1, 2, 3, \ldots$ is Landau quantum number, $S = +1/2, -1/2$ is spin quantum number) are calculated in the dipole approximation. Polarization selection rules for the various processes are given. The dependences of the scattering cross section on the electron concentration n_e and on the magnetic field H are obtained for the various Δn and ΔS.

Our results differ from those obtained by Wolff[3], Yafet[4] and Kelley and Wright[5] in the number of predicting Raman lines and in the magnetic field dependence of scattering cross sections.

The differential scattering cross sections for the various processes $\Delta n = 0, 1, 2$ and $\Delta S = 0, +1, -1$ are estimated for crystal InSb (incident photon energy is equal 0,12eV). In contrast to results of the works[3, 4, 5] the results, obtained in this work, agree for the most part with the experimental data for InSb[6]. But in order to explain the same experimental results[6] (the line with $\Delta n = 2$, $\Delta S = 0$, e.g.) it seems necessary to take into account the electron-electron interaction.

Ed. note: This paper has been published as V.P. MAKAROV, J.E.T.P. 55, 704 (1968).

REFERENCES

1. W. Heitler, "The Quantum Theory of Radiation," 2nd ed., Oxford University Press, New York, 1954.
2. R. Bowers and Y. Yafet, Phys. Rev., 115, 1165 (1959).
3. P.A. Wolff, Phys. Rev. Letters, 16, 225 (1966); J. Quant. Electr., 2, 659 (1966).
4. Y. Yafet, Phys. Rev., 152, 858 (1966).
5. P.L. Kelley and G.B. Wright, Bull. Am. Phys. Soc., 11, 812 (1966).
6. R.E. Slusher, C.K.N. Patel, and P.A. Fleury, Phys. Rev. Letters, 18, 77, 227 (1967).

E-1: PHONONS AND POLARITONS IN ZINCBLENDE

S. Ushioda* and A. Pinczuk*
Physics Department and Laboratory for Research on the Structure of Matter,
University of Pennsylvania
Philadelphia, Pennsylvania
and
E. Burstein† and D. L. Mills‡
Physics Department, University of California
Irvine, California

ABSTRACT

The data for the Raman (R) scattering intensity ratio of LO and TO phonons I_{LO}/I_{TO} and its frequency dispersion are analyzed in terms of the atomic displacement contribution and the electro-optic contribution to the R scattering tensor $\chi^{(1)}(\omega_o)$. $\chi^{(1)}(\omega_o)$ is separated into a part $\chi^{(1)ex}(\omega_o)$ due to excitons and a part $\chi^{(1)cont}(\omega_o)$ due to continuum excitations, which have different frequency dispersion. $\chi^{(1)cont}(\omega_o)$ will, in general, dominate the scattering process at $\omega_o << \omega_{ex}$ because of the low oscillator strength of the exciton bands. On the other hand, $\chi^{(1)ex}(\omega_o)$ will dominate the scattering process at $\omega_o \approx \omega_{ex}$, providing that the exciton lifetime is long and the exciton strength of the polariton is larger than the continuum exciton strength. The theory of exciton-enhanced R scattering is formulated in terms of the scattering of polaritons by optical phonons via the exciton part of the coupled modes. The expression for the exciton contribution $\chi^{(1)ex}(\omega_o)$ to the scattering tensor is given in terms of the same parameters that determine the exciton contribution to the

*Research supported by the U. S. Army Research Office - Durham.
†On sabbatical leave from the University of Pennsylvania during the 1967-68 academic year.
‡Research supported by the Air Force Office of Scientific Research, Office of Aerospace Research, U.S. Air Force under AFOSR Grant Number 68-1448.

frequency dependent dielectric constant. The effect of temperature on the exciton lifetime and thereby on the scattering efficiency is also discussed.

INTRODUCTION

The scattering of EM radiation by optical phonons in zinc-blende and wurtzite type crystals depends on the atomic displacement contribution and the corresponding electro-optic contribution to the Raman (R) scattering tensor $\chi^{(1)} = \chi^{(1)}(u) + \chi^{(1)}(E)$[1]. The electro-optic contribution $\chi^{(1)}(E)$ is zero for TO phonons, and the scattering intensity ratio I_{LO}/I_{TO} for LO and TO phonons is determined by the relative magnitudes and signs of $\chi^{(1)}(u)$ and $\chi^{(1)}(E)$.

The I_{LO}/I_{TO} ratio for a given material is found to vary with the incident exciting frequency ω_o. The frequency variation of I_{LO}/I_{TO} implies that $\chi^{(1)}(u)$ and $\chi^{(1)}(E)$ have different frequency dispersion. The dispersion of $\chi^{(1)}$ can be separated into a part $\chi^{(1)ex}$ associated with the exciton (bound electron-hole pair excitation) strengths of the incident and scattered polaritons, and a part $\chi^{(1)cont}$ associated with the electron-hole pair continuum excitation strengths, which in general may have different signs. Since photons couple only to discrete exciton bands, but to a continuum of free electron-hole excitations, the two parts have a different character. As pointed out by Birman and Ganguly[2] $\chi^{(1)ex}$ exhibits a resonance enhancement and dominates the scattering process in the vicinity of the exciton absorption bands, i.e. $\omega_o \approx \omega_{ex}$. On the other hand, at $\omega_o << \omega_{ex}$ the exciton strength is generally much smaller than the electron-hole pair continuum excitation strength, (i. e. the oscillator strengths of the exciton absorption bands are very small), so that $\chi^{(1)ex}$ will generally be smaller than $\chi^{(1)cont}$. Since excitons interact with phonons more strongly via the Fröhlich interaction than via the deformation potential interaction, the electro-optic contribution to $\chi^{(1)ex}$ will in general be greater than the atomic displacement contribution.

By formulating the theory of exciton-enchanced R scattering in terms of scattering of polaritions by optical phonons via the exciton parts of the coupled modes, the frequency dispersion of $\chi^{(1)ex}$ can be related to the frequency dependence of the exciton strengths and the group velocity of the polaritons[3]. The theory predicts an appreciable temperature dependence of $\chi^{(1)ex}$, arising from the effect of temperature on the exciton lifetimes which determine the magnitudes of the exciton strengths and the group velocity of the polaritons.

In this paper we review the available data on the I_{LO}/I_{TO} in zincblende and wurtzite type crystals, and discuss the frequency dispersion of I_{LO}/I_{TO} in terms of the dispersion of the exciton and continuum excitation contributions to the R scattering efficiency.

THE RAMAN SCATTERING INTENSITIES OF TO AND LO PHONONS

In zincblende and wurtzite type crystals, there is only one infrared and R active vibration mode of given symmetry, which, depending on the direction of the phonon $\underset{\sim}{q}$ vector, may be either transverse or longitudinal. In zincblende type crystals the

R active mode has F_2 symmetry. Wurtzite type crystals have one R active mode of A_1 and one of E_1 symmetry. By choosing appropriate polarization geometries, each symmetry mode can be investigated separately. For a given symmetry mode the appropriate R scattering tensor has the form[4]:

$$\chi^{(1)} = \left[\frac{\partial\chi(\omega_o)}{\partial u}\right]_E u + \left[\frac{\partial\chi(\omega_o)}{\partial E}\right]_u E = a(\omega_o)\, u + b(\omega_o)\, E = \chi^{(1)}(u) + \chi^{(1)}(E) \qquad (2\text{-}1)$$

where u and E are the atomic displacement and the macroscopic electric field of the optical phonon.

The expression for the I_{LO}/I_{TO} ratio in these crystals (for right angle scattering with the incident and scattered light polarized along the principal axes of the crystal) is given by[4]

$$I_{LO}/I_{TO} = A \left\{1 - \frac{4\pi N}{\epsilon_o}\,\frac{e_T^* b(\omega_o)}{a(\omega_o)}\right\}^2 = A \left\{1 - \frac{\chi^{(1)}(E)}{\chi^{(1)}(u)}\right\} \qquad (2\text{-}2)$$

where

$$A = \left(\frac{\omega_o - \omega_{LO}}{\omega_o - \omega_{TO}}\right)^4 \left(\frac{\omega_{TO}}{\omega_{LO}}\right) \frac{\bar{n}(\omega_{LO}) + 1}{\bar{n}(\omega_{TO}) + 1} \approx 1$$

ω_o is the incident frequency; ω_{TO} and ω_{LO} are the TO and LO phonon frequencies; $\bar{n}(\omega_{TO})$ and $\bar{n}(\omega_{LO})$ are the TO and LO phonon occupation numbers; ϵ_o is the optical dielectric constant; N is the number of unit cells per unit volume; and e_T^* is the ionic effective charge defined by $e_T^* = \frac{\partial M}{\partial u}$, where M is the phonon induced dipole moment. From Eq. (2-2) one sees that $(I_{LO}/I_{TO})/A$ is equal to or greater than unity, either when $\frac{4\pi N}{\epsilon_o}\frac{e_T^* b}{a} \leq 0$ or when $\frac{4\pi N}{\epsilon_o}\frac{e_T^* b}{a} > 2$ and that (I_{LO}/I_{TO}) is less than unity when $0 < \frac{4\pi N}{\epsilon_o}\frac{e_T^* b}{a} < 2$

The R scattering intensity ratio of phonon-polaritons (coupled TO phonon-photon) and TO phonons $I_\pi(\omega_\pi)/I_{TO}$, for the near-forward scattering configuration with the incident and scattered light polarized along the principal axes of the crystal, is given by[4]:

$$I_\pi(\omega_\pi)/I_{TO} = A \left\{1 + \frac{4\pi N}{\epsilon_o}\,\frac{e_T^* b(\omega_o)}{a(\omega_o)} \left(\frac{\omega_{TO}^2 - \omega_\pi^2}{\omega_{LO}^2 - \omega_{TO}^2}\right)\right\}^2 \qquad (2\text{-}3)$$

where ω_π is the phonon-polarition frequency. Since ω_π can be varied by changing the scattering angle θ, one can measure $I_\pi(\omega_\pi)/I_{TO}$ as a function of ω_π and determine the sign of $e_T^* b/a$ from Eq. (2-3).

If $e_T^* b/a$ is positive, the atomic displacement, $\chi^{(1)}(u)$, and the electro-optic, $\chi^{(1)}(E)$, contributions have opposite signs at ω_{LO} and the ratio I_{LO}/I_{TO} depends on the difference in the magnitudes of the two contributions. If $e_T^* b/a$ is negative the two contributions have the same sign at ω_{LO} and they add to produce a greater LO phonon scattering intensity than that of TO phonons.

When Eq. (2-2) is solved for $e_T^* b/a$ in terms of an experimental value of I_{LO}/I_{TO}, one obtains two roots for $e_T^* b/a$. When $(I_{LO}/I_{TO})/A$ is greater than unity, the two roots have opposite signs. The choice between the two roots can be made simply by seeing whether $I_\pi(\omega_\pi)$ increases or decreases with decrease in ω_π. On the other hand, when $(I_{LO}/I_{TO})/A$ is less than unity, the two roots fall between zero and two, and qualitative measurements of the frequency dependence of $I_\pi(\omega_\pi)$ are necessary, in order to choose one of the roots for $e_T^* b/a$.

Thus by combining the data on I_{LO}/I_{TO} and the variation of $I_\pi(\omega_\pi)$ with ω_π, one can determine the relative signs as well as the magnitudes of $e_T^* b$ and a. The magnitude of e_T^* can be found from infrared lattice vibration data, although its sign cannot be determined experimentally. When the electro-optic coefficient b, which is related to the second harmonic generation (SHG) coefficient $d^{2\omega}$ (=b/2), and e_T^* are known, one can determine the magnitude of a.

Little attention has generally been paid to the exact ratios of the LO and TO phonon scattering intensities and data on I_{LO}/I_{TO} taken for well defined geometric and polarization factors are as yet fairly limited. Table I shows some available data on I_{LO}/I_{TO} ratios and related parameters measured at different exciting frequencies ω_0 for various zincblende and wurtzite type crystals. One sees that I_{LO}/I_{TO} ratios assume a wide range of values.

In ZnSe[4] and GaP[5], the intensity ratio is much greater than unity and the phonon-polariton scattering intensity $I_\pi(\omega_\pi)$ was found to decrease rapidly with decrease in ω_π. Thus one concludes that $e_T^* b/a$ is negative, and since the two possible values for $e_T^* b/a$ have opposite signs, the negative root is the appropriate one.

We measured the I_{LO}/I_{TO} ratios of the A_1 and the E_1 modes of ZnO (single crystal provided by 3M Company) at room temperature, using a double grating spectrometer and a 35 mW He-Ne laser as a light source.

For the A_1 mode of ZnO (with the incident and scattering light polarized along z), the value of $(I_{LO}/I_{TO})/A$ is greater than unity and the two roots for $\frac{4\pi N}{\epsilon_0}\frac{e_T^* b}{a}$ are -0.2 and 2.2. We find that the phonon-polariton scattering intensity $I_\pi(\omega_\pi)$ increases with decrease in ω_π. Therefore, in the case of the A_1 mode of ZnO $e_T^* b/a$ is positive and the appropriate choice for $\frac{4\pi N}{\epsilon_0}\frac{e_T^* b}{a}$ is 2.2. For the E_1 mode of ZnO, $(I_{LO}/I_{TO})/A$ is less than unity and the two roots for $\frac{4\pi N}{\epsilon_0}\frac{e_T^* b}{a}$ are 1.5 and 0.45. $I_\pi(\omega_\pi)$ for the E_1 mode is also found to increase with decrease in ω_π, and our preliminary data on $I_\pi(\omega_\pi)$ vs. ω_π indicate that the appropriate choice for $\frac{4\pi N}{\epsilon_0}\frac{e_T^* b}{a}$ is 1.5 in the case of the E_1 mode.

TABLE I

		I_{LO}/I_{TO}	$\hbar\omega_o$	$\hbar\omega_g$(eV)	$\frac{4\pi n}{\epsilon_o}\frac{e_T^*b}{a} = \frac{\chi^{(1)}(E)}{\chi^{(1)}(u)}$	Source
Zincblende						
ZnS		>100	He-Ne	3.54		a
		15	Ar^+		-3.5 or 5.5	b
ZnSe		8.5	He-Ne	2.58	-2.4	a
GaP		1.7	He-Ne	2.24	-0.39	c
Wurtzite						
ZnO	A_1(z, z)	0.83	He-Ne	3.24	2.2	Present work
	E_1	0.18	He-Ne		1.5	Present work
	E_1	~0.5	Ar^+		1.9	d
ZnS	E_1	50	He-Ne	3.65	-7.3 or 9.3	b
	E_1	6	Ar^+		-1.8 or 3.8	b
CdS	E_1	~4	Ar^+	2.41	-1.4 or 3.4	e
AlN	E_1	<1	He-Ne	3.8		f

He-Ne laser line is at 1.96 eV and Ar^+ laser line is at 2.41 eV.

Source

a. S. Ushioda, A. Pinczuk, W. Taylor and E. Burstein, Proceedings of Int. Conf. on II-VI Compound Semi-conductors, Providence, 1967, 1185 (W. A. Benjamin).
b. O. Brafman and S. S. Mitra, Phys. Rev. 171, 931 (1968).
c. W. L. Faust and C. H. Henry, Phys. Rev. Letters 17, 1265 (1966).
d. T. C. Daman, S. P. S. Porto and B. Tell, Phys. Rev. 142, 570 (1966).
e. B. Tell, T. C. Damen, and S. P. S. Porto, Phys. Rev. 144, 771 (1966).
f. O. Brafman, G. Lengyel, and S. S. Mitra, P. J. Gielisse, J. N. Plendl and L. C. Mansur, Solid State Comm. 6, 523 (1968).

It is interesting to note that in all cases except for GaP the electro-optic contribution $\chi^{(1)}(E)$ has a larger magnitude than the atomic displacement contribution, $\chi^{(1)}(u)$, i.e. $\left|\frac{4\pi N}{\epsilon_o}\frac{e_T^*b}{a}\right| = \left|\frac{\chi^{(1)}(E)}{\chi^{(1)}(u)}\right| > 1$. In the case of the A_1 mode of ZnO, $\chi^{(1)}(u)$ and $\chi^{(1)}(E)$ have opposite signs at the LO phonon frequency and $|\chi^{(1)}(E)|$ is slightly larger than $|\chi^{(1)}(u)|$x2. Thus $|\chi^{(1)}(E)| - |\chi^{(1)}(u)| > |\chi^{(1)}(u)|$ and consequently $(I_{LO}/I_{TO})/A$ becomes greater than unity. For the E_1 mode of ZnO, $\chi^{(1)}(u)$ and $\chi^{(1)}(E)$ have opposite signs at the LO phonon frequency and $|\chi^{(1)}(E)| > |\chi^{(1)}(u)|$, but $|\chi^{(1)}(E)|$ is not quite large enough to get $(I_{LO}/I_{TO})/A > 1$.

When the data for different exciting frequencies are available, it is clearly seen that I_{LO}/I_{TO} ratios vary appreciably with ω_o. In ZnS[4, 6] (zincblende), for example, the I_{LO}/I_{TO} ratio is more than 100 at $\omega_o = 15,800\ cm^{-1}$ (6328 Å) and it decreases with increase in ω_o to approximately 15 at $\omega_o = 19,450\ cm^{-1}$ (5145 Å), and a similar dispersion of I_{LO}/I_{TO} is observed for the E_1 mode of ZnS (wurtzite). On the other hand the I_{LO}/I_{TO} ratio for the E_1 mode of ZnO appears to increase as the incident frequency ω_o increases by the same amount. A large enhancement in the I_{LO}/I_{TO} ratio of CdS has been reported in measurements made at 77°K for ω_o near the intrinsic absorption edge ω_g[7]. (See also paper E-2 by R. C. C. Leite, T. C. Damen and J. F. Scott, "Resonant Raman Effect in CdS and ZnSe" this conference.)

THE FREQUENCY DISPERSION OF THE RAMAN SCATTERING EFFICIENCY

In this section we present a discussion of the frequency dispersion of R scattering efficiency by LO and TO phonons in terms of the scattering of polaritons by optical phonons via the exciton and via the continuum (electron-hole pair) excitation parts of the incident and scattered coupled modes.

We separate the R scattering tensor $\chi^{(1)}$ into two parts, a part due to excitons, and a part due to continuum excitations.

$$\chi^{(1)}(\omega_o) = \chi^{(1)ex}(\omega_o) + \chi^{(1)cont}(\omega_o)$$

$$= [\chi^{(1)ex}(u) + \chi^{(1)ex}(E)] + [\chi^{(1)cont}(u) + \chi^{(1)cont}(E)] \qquad (3\text{-}1)$$

where $\chi^{(1)ex}$ is the contribution from scattering via the exciton parts of the incident and scattered polaritons, and $\chi^{(1)cont}$ is the contribution from scattering via the exciton parts of the incident and scattered polaritons, and $\chi^{(1)cont}$ is the contribution from scattering via the continuum excitation parts of the polaritons. Since photons couple only to discrete exciton states, but to a continuum of electron-hole pair excitations, $\chi^{(1)ex}$ and $\chi^{(1)cont}$ have a different character. This follows in part from the fact that the exciton contribution to the dielectric constant $\epsilon^{ex}(\omega)$ exhibits anomalous dispersion at frequencies in the region of the exciton absorption bands, whereas the continuum excitation contribution $\epsilon^{cont}(\omega)$ changes gradually as frequency passes through the interband absorption edge.

The frequencies of polariton modes $\Omega(k)$ in a crystal containing several exciton bands (Fig. 1) are given by the dispersion relation:[8]

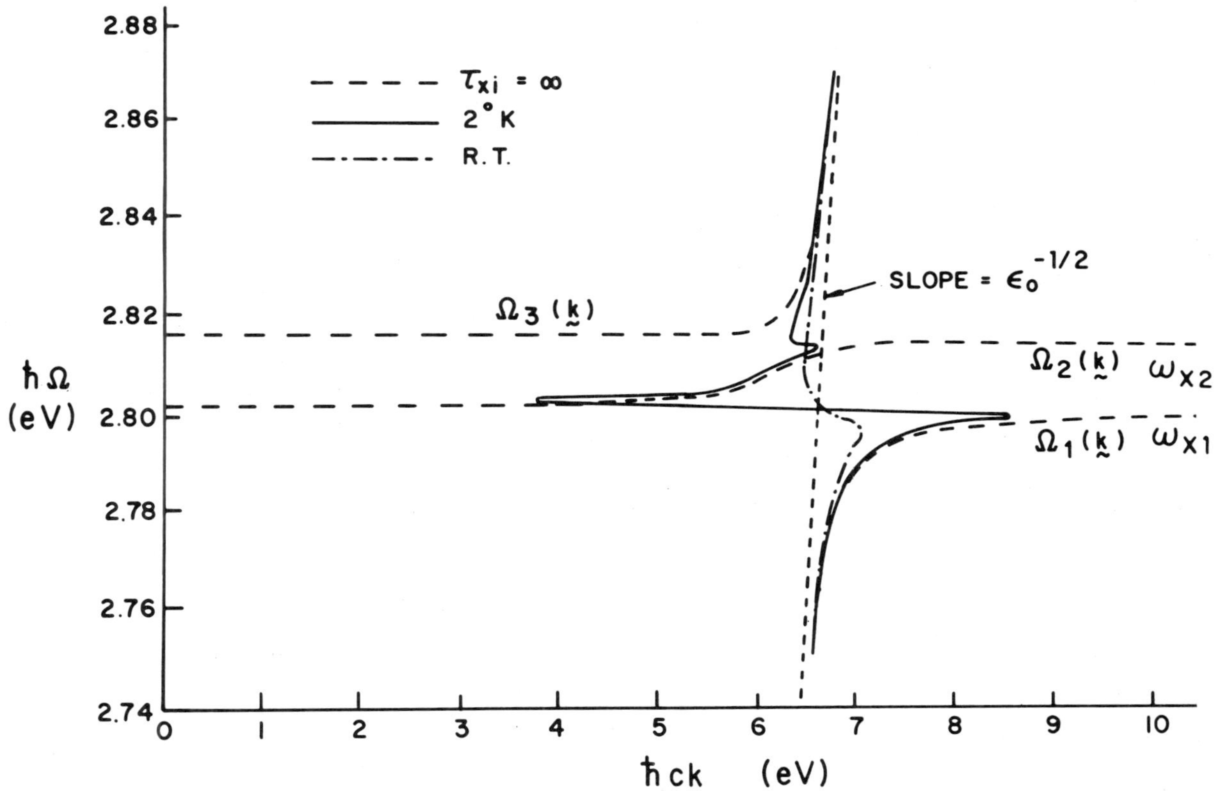

Fig. 1. Calculated polariton dispersion curves for a two exciton band model, based on the parameters for ZnSe: ω_{x1} = 2,800 eV; ω_{x2} = 2.815 eV; $\beta_{x1} = 5 \times 10^{-3}$; $\beta_{x2} = 5 \times 10^{-4}$. For the 2°K curve: $\Gamma_{x1} = 1/\tau_{xi} = 2.5 \times 10^{-3}$ eV and $\Gamma_{x2} = 1/\tau_{x2} = 1 \times 10^{-2}$ eV; and for the R. T. curve: $\Gamma_{x1} = 2.5 \times 10^{-2}$ eV and $\Gamma_{x2} = 1 \times 10^{-1}$ eV.

$$c^2k^2 = \epsilon(\Omega)\,\Omega^2 = \Omega^2\,[1 + \epsilon(\Omega)^{cont} + \epsilon(\Omega)^{ex}]$$

$$= \Omega^2 \left[1 + \epsilon(\Omega)^{cont} + \sum_j \frac{4\pi\,\beta_j\,\omega_{xj}^2}{\omega_{xj}^2 - \Omega^2 - i\,\Omega/\tau_{xj}} \right] \tag{3-2}$$

where ω_{xj} and τ_{xj} are the frequency and lifetime of exciton of type j; $4\pi\beta_{xi}$ is the zero-frequency contribution to the dielectric constant, and we have neglected the wave vector dependence of ω_{xj} and β_j.

At frequencies far removed from the exciton absorption bands, i.e. $\omega_0 << \omega_{xj}$, the exciton strengths of the incident and scattered polaritons are generally much smaller than the continuum excitation strengths, and $\chi^{(1)ex}$ will generally be much smaller than $\chi^{(1)cont}$

As the frequency of the polaritons approaches the exciton bands, the exciton strengths increase sharply and go to unity at $\omega_0 = \omega_{xj}$, i.e. the polaritons become essentially pure excitons in character, (providing the exciton lifetimes τ_{xj} are long), while the continuum excitation strengths remain relatively unchanged, initially, and then decrease essentially to small values when the polaritons approach a pure exciton character. The exciton and continuum excitation strengths determine in part the relative magnitudes of $\chi^{(1)ex}$ and $\chi^{(1)cont}$.

At some ω_0, $\chi^{(1)ex}$ and $\chi^{(1)cont}$ will become comparable in magnitude and, depending on their relative signs, their superposition may have a large effect of the LO and TO phonon scattering intensities. The scattering intensity may, in fact, exhibit minima if $\chi^{(1)ex}$ and $\chi^{(1)cont}$ have opposite signs. Moreover, even in frequency regions where one or the other dominates, the difference in the frequency dispersion of the atomic displacement and electro-optic contributions to $\chi^{(1)ex}$ and to $\chi^{(1)cont}$ may lead to minima in the scattering intensity of LO phonons. Consequently, the frequency dispersion of I_{LO}/I_{TO} may exhibit maxima and minima.

Birman and Ganguly[2] have pointed out that the creation of virtual excitons dominate the enhanced R scattering by TO and LO phonons, and have interpreted the data of Leite and Porto[7] on R scattering by LO phonons in CdS in the region of the intrinsic absorption edge on this basis. We attribute the large $\chi^{(1)ex}(E)/\chi^{(1)ex}(u)$ ratio in CdS at 77°K (inferred from the large I_{LO}/I_{TO} ratio in the resonance region) and the large enhancement of the field induced scattering by LO phonons in InSb[9] to the fact that the exciton-optical phonon interaction via the Fröhlich interaction (macroscopic electric field) is much stronger than the corresponding interaction via the deformation potential. We have formulated the theory of exciton-enhanced R scattering in terms of scattering of polaritons by optical phonons via interaction with the exciton parts of the coupled modes[3]. Ovander[10] has also discussed R scattering from this point of view.

Our theory leads to a frequency dispersion of the scattering efficiency of TO and LO phonons which is qualitatively different from that of Birman and Ganguly[2] who use a perturbation theoretic approach to treat the photon-exciton coupling. The theory also takes into account the effect of temperature on the exciton enhanced R scattering efficiency through its effect on the exciton lifetimes and, thereby, on the exciton strengths of the incident and scattered polaritons.

The derivation of the scattering efficiency/unit solid angle in terms of the exciton-optical phonon scattering matrix element is straight forward. For a model involving two exciton bands, the Stokes scattering efficiency for $\Omega_o < \omega_x$ (which corresponds to the case where the initial and final polariton states are in the same lowest polariton branch) has the form:

$$S_x = \frac{\bar{n}_{op} + 1}{4\pi^2 \hbar^2} \frac{|k_s|^2}{v_g(\Omega_o) v_g(\Omega_s)} \left| \sum_{i,j} M_{ij}(\underset{\sim}{q}) F \mathfrak{s}^{1/2}_{xi}(\Omega_o) \mathfrak{s}^{1/2}_{xj}(\Omega_s) \right|^2$$

$$F = 1/2 \left[1 + \frac{\omega_{xi}\omega_{xj}}{\Omega_o \Omega_s} \right] \approx 1 \qquad (3\text{-}3)$$

where $\underset{\sim}{k}_s = \underset{\sim}{k}_o - \underset{\sim}{q}$ and $\Omega_s = \Omega_o - \omega_{op}$; $v_g(\Omega_o)$ is the group velocity of the incident polariton; $M_{ij}(\underset{\sim}{q})$ is the exciton-optical phonon scattering matrix element connecting the i th and j th exciton states[11]; and $\mathfrak{s}_{xj}(\Omega)$ is the exciton strength which has the form:

$$\mathfrak{s}_{xj}(\Omega) = \frac{4\pi \beta_j \omega^3_{xj} \Omega}{(\omega^2_{xj} - \Omega^2 - i\Omega/\tau_{xj})^2} \frac{v_g(\Omega) \epsilon_o(\Omega)}{c\, \epsilon^{1/2}(\Omega)} \qquad (3\text{-}4)$$

We note that as the incident frequency approaches ω_{x1} from below, $\mathfrak{s}_{x1}(\Omega) \rightarrow 1$ while $\mathfrak{s}_{x2}(\Omega) \rightarrow 0$, since, when $\Omega_o = \omega_{x1}$, the incident polariton is pure band 1 exciton in character.

In the case of TO phonons $M_{ij}(\underset{\sim}{q})$ involves the deformation potential interaction. In the case of LO phonons it involves the deformation potential and the macroscopic electric field. Since the Fröhlich (macroscopic electric field) interaction is much stronger than the deformation potential interaction, we may expect S_{xLO} to be dominated by the electro-optic contribution to $\chi^{(1)ex}$ and to be much larger than S_{xTO}. Further details of the character of $M_{ij}(\underset{\sim}{q})$ are discussed in Ref. [3].

It should be noted that the matrix element of Eq. (3-3) is never infinite, since the exciton strengths only vary from 0 to 1. Thus the behavior of the matrix element differs qualitatively from that of Birman and Ganguly[2] which diverges as $(\omega_{x1} - \omega_o)^{-1}$. Also the expression for S_x has $v_g(\Omega_o)v_g(\Omega_s)$ in the denominator, whereas Girman and Ganguly have c^2/ϵ_o. Since $v_g(\Omega_o)$ decreases as the resonance region is approached (it vanishes at $\Omega_o = \omega_{x1}$ only in the case where spatial dispersion of $\epsilon^{ex}(\Omega)$ is ignored) one obtains an additional enhancement from this source. Since $M_{ij}(\underset{\sim}{q})$ is insensitive to wave vector, the frequency dispersion of $\chi^{(1)ex}$ arises from the frequency dispersion of the exciton strengths of the incident and scattered polaritons and from the frequency dispersion of the group velocity of the incident polaritons.

The effect of the exciton lifetime, τ_{xj}, on the scattering must also be taken into consideration. The lifetimes of the exciton states are known to decrease rapidly with temperature due to phonon-exciton scattering processes and to the increasing rate of thermal ionization. A decrease in τ_{xj} decreases $\epsilon^{ex}(\Omega)$ and thereby decreases g_{xj} and increases $v_g(\Omega_0)$ at $\Omega_0 \approx \omega_{xj}$. (Fig. 1) We may therefore expect exciton enhancement of R scattering to decrease with increasing temperature and to be small when $[4\pi\beta_j/\epsilon^{cont}(\omega)]\,\omega_{xj}\,\tau_{xj} < 1$. The scattering via continuum excitation part, on the other hand, may be expected to be less sensitive to temperature. In fact the effect of temperature on the scattering efficiency can be used to establish whether excitons or continuum pair excitations are involved in the scattering processes.

It should be noted also that similar considerations to those discussed here apply to R scattering by optical phonons in the vicinity of impurity bound exciton absorption bands. A pronounced enhancement of the R scattering may be expected particularly at impurity densities for which $\epsilon^{bx}(\omega)$, the bound exciton contribution to $\epsilon(\omega)$ is larger than $\epsilon^{cont}(\omega)$. Also the effect of temperature on the lifetime of the bound exciton and therefore the effect on the bound exciton strengths and on $\chi^{(1)ex}$ should be qualitatively similar to that encountered in exciton enhanced R scattering.

CONCLUDING REMARKS

When the parameters e_T^* and $b(\omega_0)$ are known from infrared lattice vibration data and from the second harmonic generation (SHG) data, respectively, the atomic displacement susceptibility tensor $a(\omega_0)$ can be evaluated by relative intensity measurements of R scattering as we have shown. There have been efforts to calculate electro-optic coefficient $b(\omega_0)$, but to our knowledge there has been no theoretical calculation of the atomic displacement susceptibility tensor $a(\omega_0)$. Since they are rather fundamental parameters and experimental data are now becoming available, it appears timely for theoreticians to carry out ab initio calculations for these parameters.

As we have pointed out, the frequency dispersion of $\chi^{(1)ex}$ and $\chi^{(1)\,cont}$ may lead to maxima and minima in I_{LO} and I_{TO}. It will be interesting to look for such extrema in the R scattering intensities, when more laser frequency lines become available.

ACKNOWLEDGEMENTS

We wish to acknowledge the assistance of Dr. A. S. Filler in obtaining the Raman data and the assistance of R. Klaffky in making the calculations.

REFERENCES

1. R. Loudon, Advan. in Phys. 13, 423 (1964).
2. J. L. Birman and A. K. Ganguly, Phys. Rev. Letters 17, 647 (1966); A. K. Ganguly and J. L. Birman, Phys. Rev. 162, 806 (1967).
3. E. Burstein, D. L. Mills, A. Pinczuk, and S. Ushioda (to be published).
4. S. Ushioda, A. Pinczuk, W. Taylor, and E. Burstein, "Proceedings of Int. Conf. on II-VI Compound Semi-conductors," Providence, p. 1185 (1967), W. A. Benjamin, 1968.
5. W. L. Faust and C. H. Henry, Phys. Rev. Letters 17, 1265 (1966).
6. O. Brafman and S. S. Mitra, Phys. Rev. 171, 931 (1968).
7. R. C. C. Leite and S. P. S. Porto, Phys. Rev. Letters 17, 10 (1966).
8. J. J. Hopfield, Phys. Rev. 112, 1555 (1958).
9. A. Pinczuk and E. Burstein, paper E-9 this conference.
10. L. N. Ovander, Soviet-Phys. Solid State 3, 8, 1737 (1962).
11. Y. Toyozawa, Prog. Theor. Phys. 20, 53 (1958).

E-2: RESONANT RAMAN EFFECT IN CdS AND ZnSe

R. C. C. Leite, T. C. Damen and J. F. Scott
Bell Telephone Laboratories, Incorporated
Holmdel, New Jersey

INTRODUCTION

Preliminary studies of lattice Raman scattering from CdS[1] disclosed a large enhancement of the Raman cross-section as the excitation photon energy approached that of the band gap. Predictions of this resonance effect were common in the literature, but different theories[2-4] yielded rather different quantitative results. In an effort to discriminate among the several theories we have measured the absolute percentage increase of Raman cross section of transverse and longitudinal optical phonons, as well as polaritons, in CdS and ZnSe. While we have been unable to explain in detail the empirical results of our experiments, we are motivated in our present communication by the belief that their novelty and complexity will prompt others to hypothesize explanations.

EXPERIMENTAL NOTES

The experimental arrangement for 90° and near forward scattering was identical to that used previously for ZnO[5, 6]. The laser lines used in the present experiments were the following: 4880, 4965, 5017, 5145 and 5208 (Ar^+) and 5682 (Kr^+).

Earlier work on resonant Raman scattering[1] was performed by a reflection technique that had many inherent disadvantages; the most important one was the difficulty controlling the effective excitation intensity because of change of scattering volume due to frequency-dependent absorption. This approach allowed a comparison between different Raman lines from the same material but did not furnish reliable information on percentage change of Raman cross sections. We have since devised a simple technique that overcomes these difficulties. A sample slice is sandwiched between two plates, one of quartz and one of calcite. The laser beam is normal to the three plates, and the Raman scattered light is simultaneously collected from the three materials. The ratio of intensities of the Raman lines of the two reference materials, i. e., calcite and quartz, varies with excitation frequency and is a direct measure of the absorption in our sample. Consequently, consideration of the three spectra allows compensation for the highly frequency dependent absorption near the band gap. It is therefore possible to calculate accurate percentage enhancement of the Raman scattering cross sections. All data presented in this communication have been corrected in this way.

CdS was studied at ~10°K in order to permit easier penetration of the incident laser beam and ZnSe was studied at room temperature in order to bring the gap energy as close as possible to the available laser frequencies.

Fig. 1 illustrates the present technique. Notice in this CdS spectrum the Raman lines from the three materials. C and Q stand for calcite and quartz and LO and TO for the longitudinal and transverse optical phonons.

FIRST ORDER RAMAN SCATTERING

Different theoretical approaches to Raman scattering have as a common feature in their expressions for the scattering cross-section terms that either diverge or become extremely large when the frequency of excitation radiation approaches the allowed optical transition frequencies of the material. This effect was observed by Tsenter and

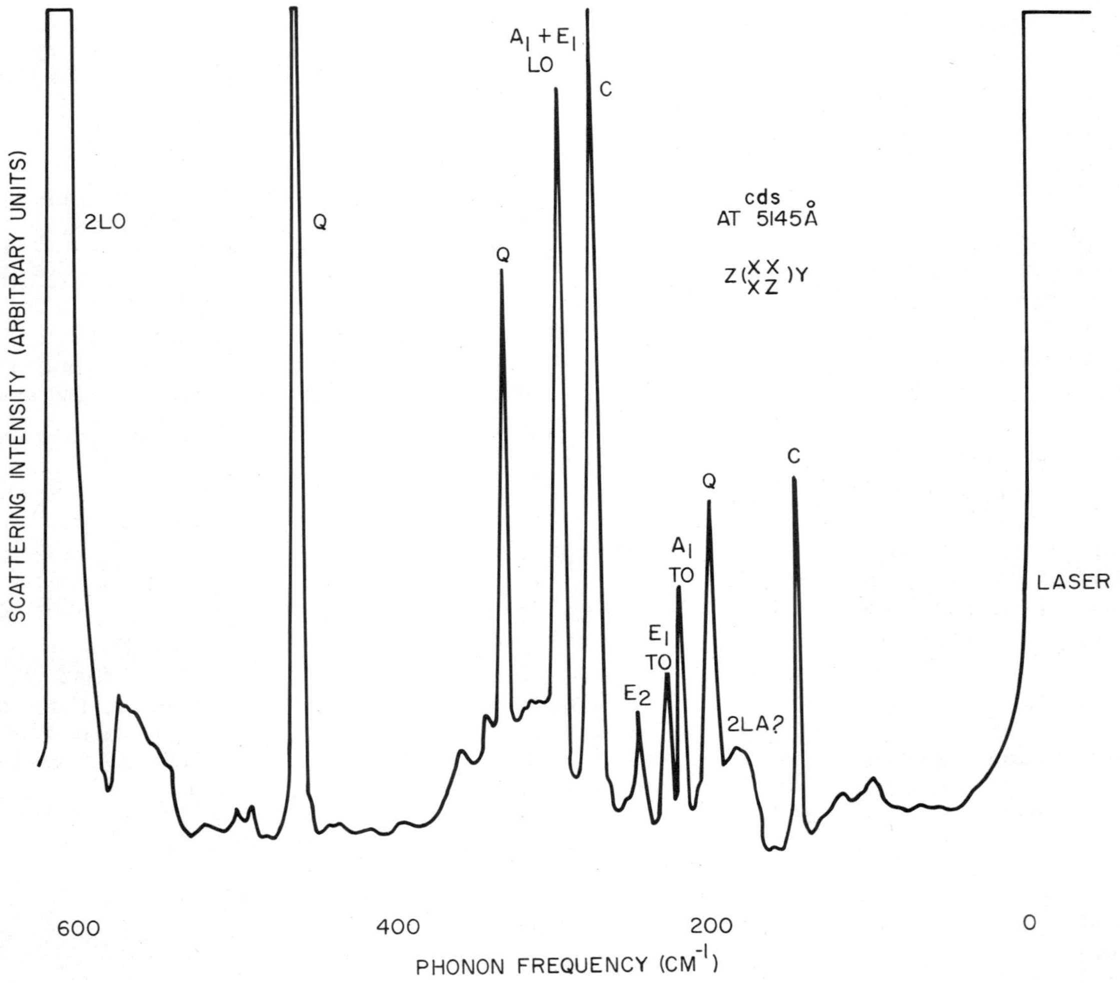

Fig. 1. Spectrum of calcite-CdS-quartz sandwich, as explained in the text.

Bobovich[7] in liquids and shown to yield rather good agreement with theory[2]. Resonant Raman scattering has also been observed recently from solids[1] but no reasonable agreement with theoretical predictions by Loudon[3] was obtained. More recently a different theoretical approach was given by Ganguly and Birman[4]. Loudon's theory deals only with the atomic displacement susceptibility tensor whereas Ganguly and Birman consider also the electrooptic tensor. The latter authors point out the importance near resonance of the electrooptic component upon the longitudinal optical phonons. Indeed, the different frequency dependences of LO and TO phonon intensities shown in Fig. 2 supports their contention. However, no available theory accounts for the TO phonon intensity saturation shown in Fig. 2.

In Fig. 3 we compare our experimental results with computer calculations of Loudon's theory for resonance. The curve labeled 1 is computed by assuming the width of the allowed bands in the semiconductor to be 10 electron volts, whereas curve 2 corresponds to 1 electron volt. Curves 3, 4 and 5 correspond to dominance of terms of type $(E_g - R - \hbar\omega_i)^{-1}$, $(E_g - \hbar\omega_i)^{-2}$ and $(E_g - R - \hbar\omega_i)^{-2}$, respectively. Here R is the exciton dissociation energy and $\hbar\omega_i$ is the excitation photon energy.

Our experimental results indicate that the poles in the terms dominant near resonance are of second order, which further supports the theory of Ganguly and Birman.

SECOND ORDER RAMAN SCATTERING

Loudon pointed out that there are two types of Raman scattering processes and that they give rise respectively to line and continuous spectra. The line spectrum is due to iteration of successive first order Raman scatterings. This process is illustrated in Fig. 4(a), where H_{EL} and H_{ER} represent the first order interactions between the electron and lattice system and electron and radiation system, respectively. The intermediate state ω'_s is so defined that wave vector is conserved between ω_i and ω'_s but not energy. Overtones due to process a are therefore exact replicas in shape of their generating first order processes and should be displaced by twice the energy offset of the first order line from the excitation photon energy; that is, they are due to two $k \approx 0$ phonons. This is not the case of the 2 LO phonon band in Fig. 1 which is broader than the LO phonon and shifted by less than twice the energy of LO phonon. Process a also predicts that Raman cross section varies as the square of that of the one phonon process as a function of the excitation phonon energy. This is observed experimentally as shown in Fig. 2.

Process b in Fig. 4 involves the absorption of the incident photon ω_i leaving the system in a state "a". A phonon is then emitted during a transition to state "b". After the emission of the second phonon the system reaches a state "c" before the emission of the scattered photon ω_s. A third possibility was suggested by Leite and Porto[1] and is included in Fig. 4 as process c. The only formal difference between processes b and c is that the intermediate state is eliminated in the latter, but as pointed out before, the two processes predict rather different frequency dependences of their scattering cross-sections. In both b and c processes energy and wave-vector need to be conserved only during the entire scattering process. The results in Fig. 2 definitely discriminate against process "c" but give reasonable agreement with process "b" when the formalism of Birman and Ganguly is used. However, better agreement is obtained with process a if one assumes that wave vector need not be conserved between the ω_i and ω'_s states. We leave this puzzling point to theoreticians.

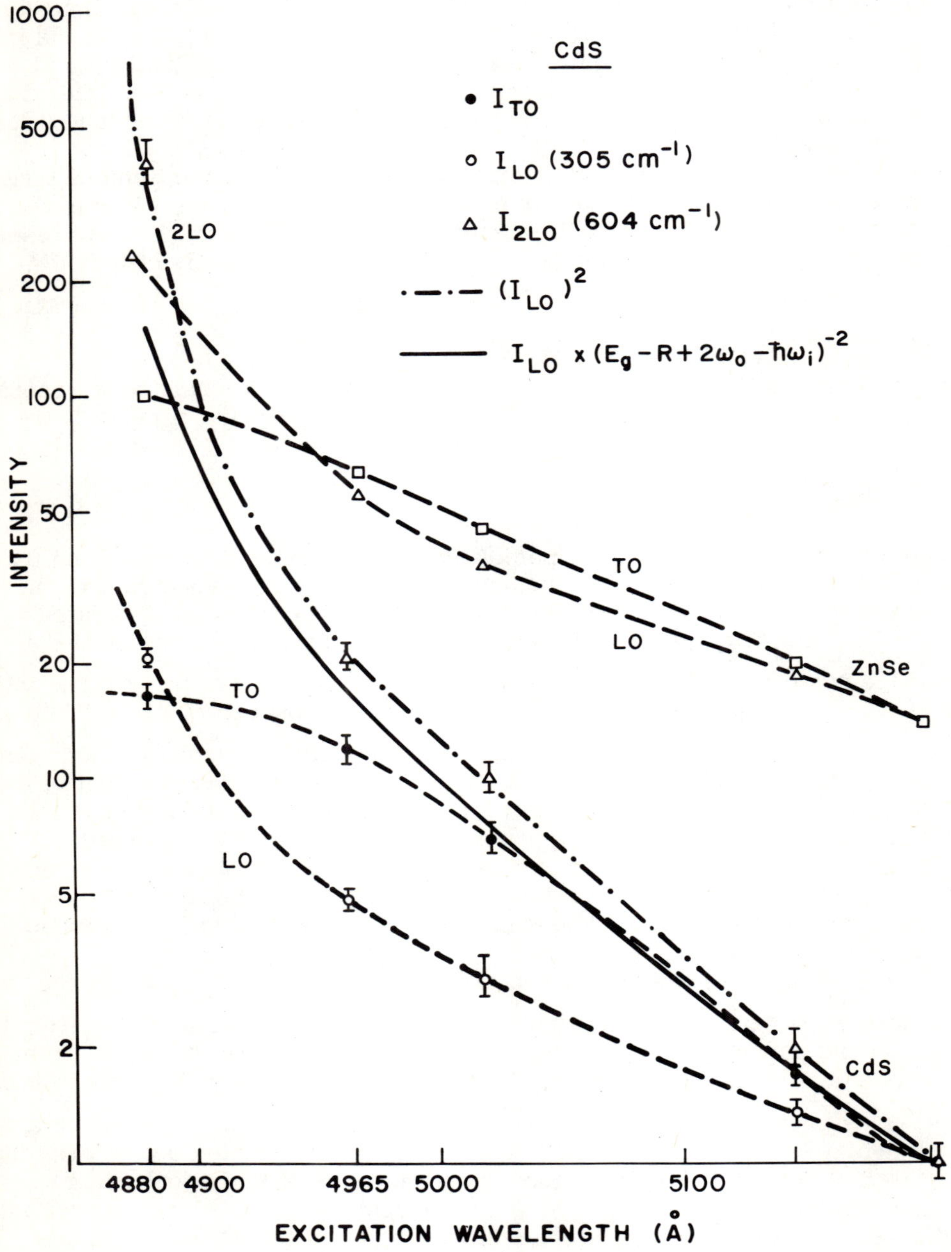

Fig. 2. Percentage enhancement of phonon cross-sections in CdS and ZnSe as a function of photon energy.

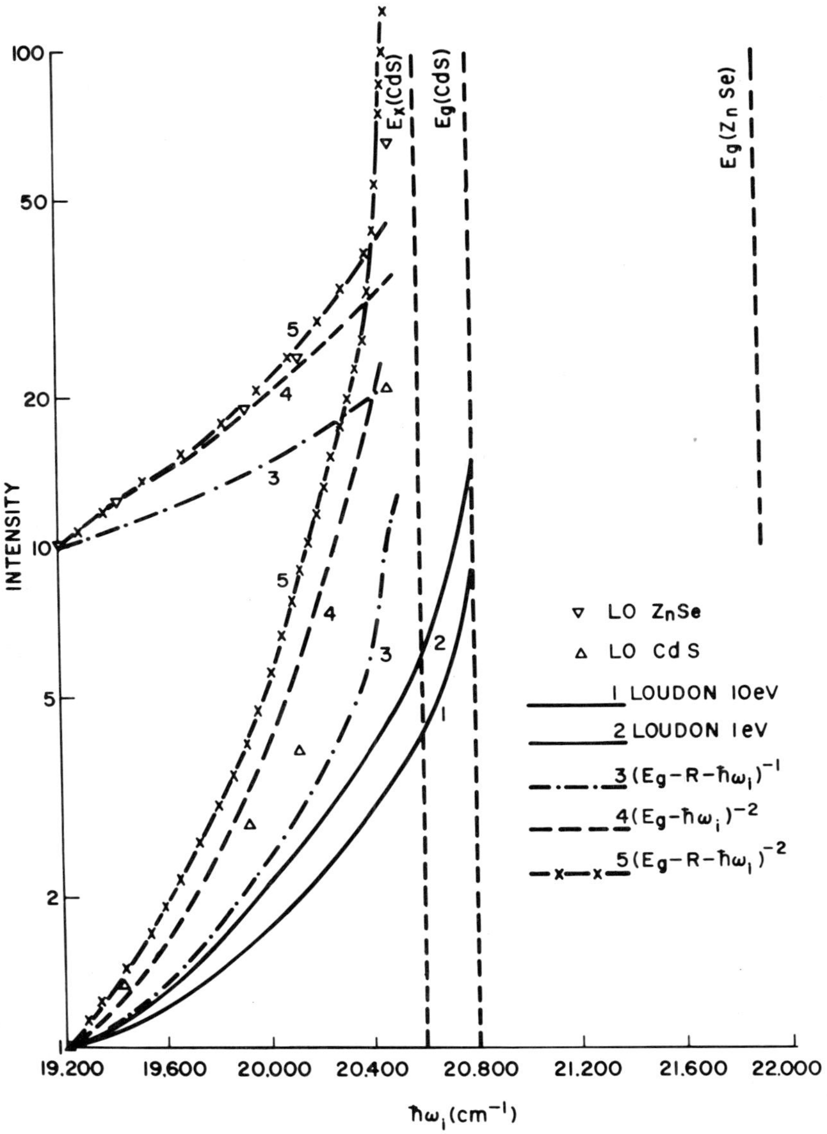

Fig. 3. Comparison of observed cross-section enhancement and that calculable from Loudon's theory.[3]

POLARITONS

The kinematics of light scattering from polaritons have been discussed by several writers[8-10]. For crystals of the zincblende variety the polariton frequency ω is given as a function of internal scattering angle θ by the equation below:

$$C^2 \left(\frac{\partial K}{\partial \omega}\right)^2_{\omega=\omega_L} + C^2 \omega_L (\omega_L - \omega)\omega^{-2} n_L^{\,2} \theta^2 = \frac{\epsilon_0 \omega_0^{\,2} - \epsilon_\infty \omega^2}{\omega_0^{\,2} - \omega^2} \qquad (1)$$

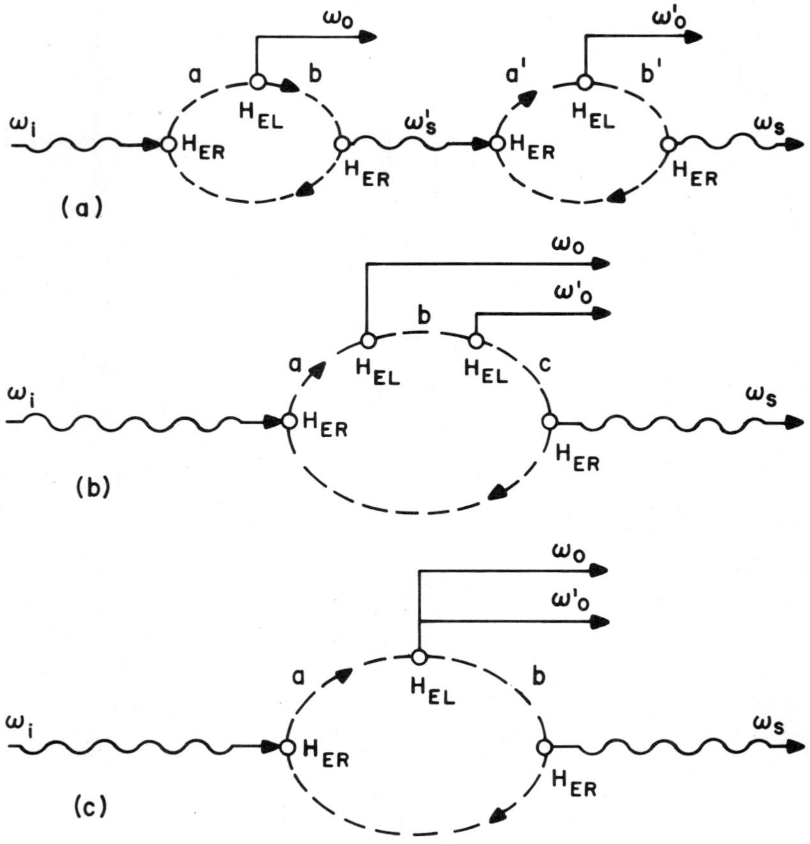

Fig. 4. Schematic diagrams of Raman scattering processes.

Here K is the wave vector of the polariton; c, the speed of light; ω_L, the laser frequency; n_L, the index of refraction at ω_L; ϵ_0 and ϵ_∞, the d.c. and high-frequency indices of refraction of the crystal. Note that even for forward scattering (θ=0) the values of ω are explicitly dependent upon $\left(\frac{\partial K}{\partial \omega}\right)$ at ω_L. $\frac{\partial K}{\partial \omega}$ becomes large near the band gap; hence, if we increase ω_L toward $E_{gap}/\hbar$, the range of polariton frequencies over which the phase-matched three-wave Raman scattering process can occur will be sharply reduced. This is shown in Fig. 5, where $\omega(\theta)$ is calculated for 6328, 5680, 5145 and 4880Å laser excitation using values of $\frac{\partial K}{\partial \omega}$ interpolated from Rambauske's data[11]. Excellent agreement between calculated and experimental values is obtained for 4880Å excitation as shown in Fig. 5. Typical data for 4880Å excitation are shown in Fig. 6. For 5145 or 5680Å lasers the predictions are not as easily compared. For 5145Å excitation phase matching should be obtained down to about 160 cm^{-1}; however, the cross-section for Raman scattering becomes vanishly small[12] near 180 cm^{-1}, and consequently the dispersion $\omega(\theta)$ is difficult to verify below ~ 190 cm^{-1}. Unless polariton spectra are recorded with small slit width and small solid angle of light acceptance, it is extremely difficult to distinguish effects of phase mismatch from intrinsic zeroes in the nonlinear susceptibility. Fig. 5 demonstrates the latter effect. Here the polariton scattering intensities in ZnSe are calculated for several choices of parameters and are compared with experimental values. The parameter which is varied in Fig. 5 is the ratio of electrooptic coefficient $\frac{\partial \chi^1}{\partial E}$ to a mechanical susceptibility coefficient $\frac{\partial \chi^1}{\partial U}$ characterizing response to ionic displacements in a deformation potential[8, 12]. Some data points have been plotted on Fig. 5 for 4880 and 5680Å. Note that while all the data are consistent with the choice of negative sign for the $\frac{\partial \chi}{\partial E}/\frac{\partial \chi}{\partial U}$ ratio, the numerical value of that ratio appears to be frequency dependent. In particular, the electrooptic coefficient appears to be increasing in magnitude near the band gap. This is consistent with our phonon scattering data, which show I(LO)/I(TO) to be greater at 4880Å than at 5680Å. Direct measurement of the electrooptic coefficient of ZnSe near resonance will be reported on at a later date[13].

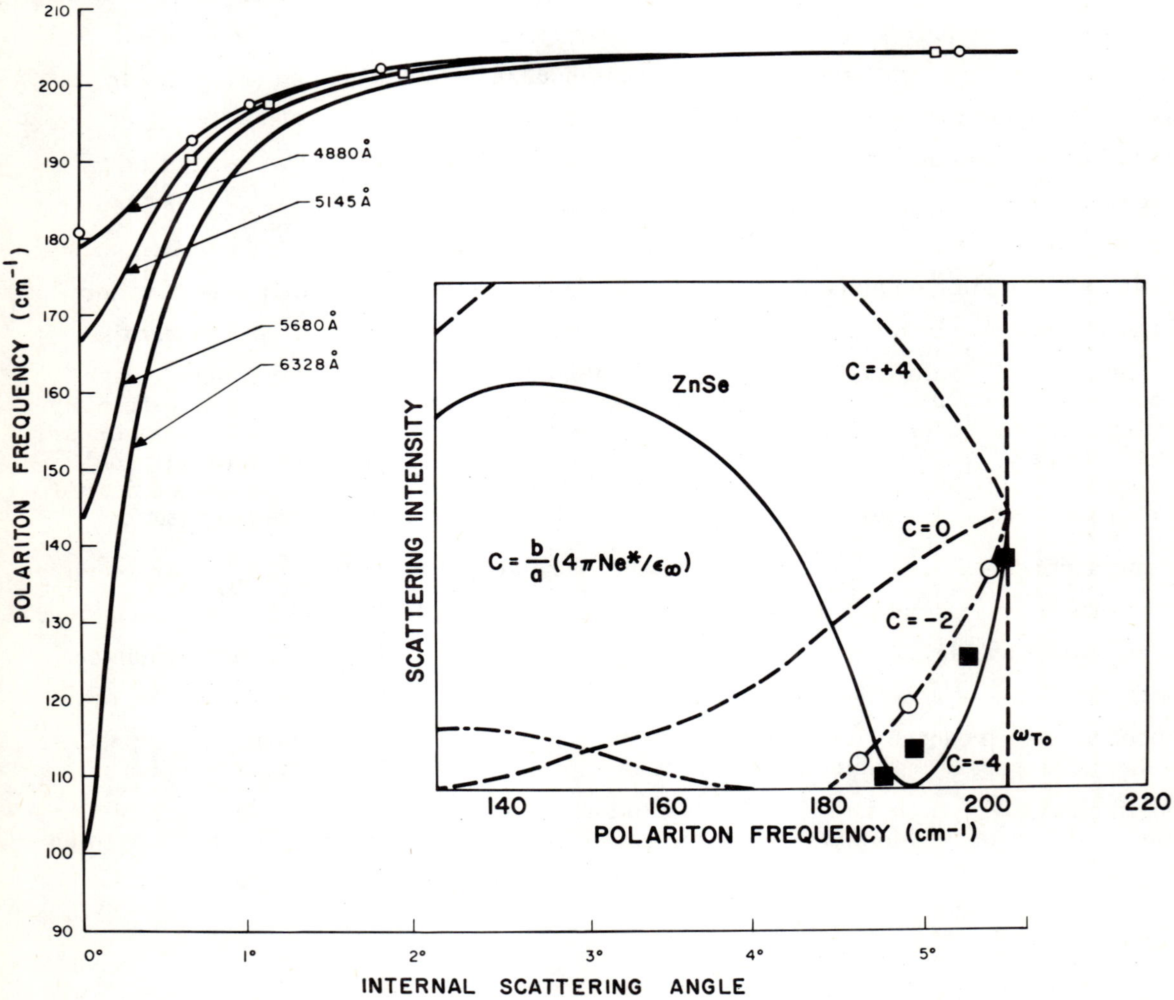

Fig. 5. Polariton dispersion $\omega(\theta)$ in ZnSe for several laser wavelengths: 4880 and 5145Å (argon), 5680Å (krypton), and 6328Å (helium-neon) - Dielectric constants from M. Aven, D.T.F. Marple, and B. Segall, J. Appl. Phys. 32, 2261 (1961). Circles represent experimental values. Inset: polariton intensities, calculated[12] and observed (circles - 5682Å, squares - 4880Å).

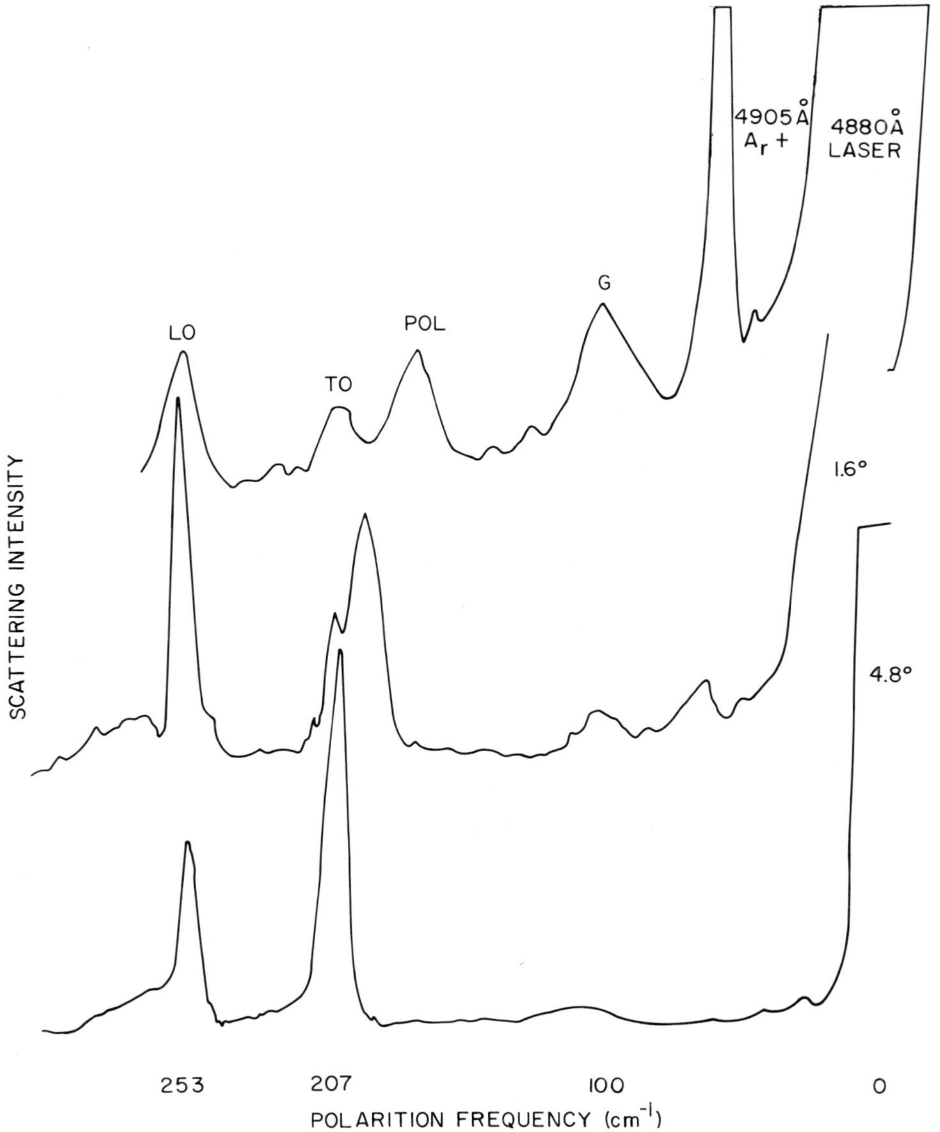

Fig. 6. Polariton spectra for 4880Å excitation and scattering angles of 4.8°, 1.6° and 0.0°. TO phonon peaks are due to reflection following large angle (especially ~ 180°) scattering.

CONCLUSIONS

In summary, we have shown that phonon scattering near resonance exhibits the same characteristics in CdS as in ZnSe. In each material the TO phonon cross-sections saturate as the incident photon energy approaches E_{gap}, while LO phonon cross-sections exhibit poles. On this basis we believe that scattering theories which ignore electrooptic contributions to LO phonon (and polariton) cross-sections will not be able to accurately describe the physical situation near resonance. 2LO and 2LA features in CdS exhibit an enhancement approximately proportional to the square of that for 1LO features, although the 2LO and 2LA features are known to be composed of two large wave-vector phonons (since $\omega_{2LO} \neq 2\omega_{LO}$); this suggests a second-order Raman process in which the intermediate state conserves neither energy nor wave-vector. The polariton scattering is consistent with LO phonon scattering and shows that electrooptic effects contribute to each through the macroscopic electric field. For ZnSe the implication of both LO phonon and polariton cross-sections is that the electrooptic coefficient is noticeably frequency dependent near the band gap, that it becomes larger near E_{gap} and remains opposite in sign from that of the deformation-potential displacement susceptibility tensor.

REFERENCES

1. R.C.C. Leite and S.P.S. Porto, Phys. Rev. Letters 17, 10 (1966).
2. L.N. Ovander, Fiz. Tverd. Tela. 3, 2394 (1961), (Translation: Soviet Physics - Solid State 3, 1737 (1961)).
3. R. Loudon, Advan, Phys. 13, 423 (1964); J. Phys. 26, 677 (1965).
4. A.K. Ganguly and J.L. Birman, Phys. Rev. 162, 806 (1967).
5. B. Tell, T.C. Damen, and S.P.S. Porto, Phys. Rev. 144, 771 (1966).
6. S.P.S. Porto, B. Tell, and T.C. Damen, Phys. Rev. Letters 16, 450 (1966).
7. M. YaTsenter and YaS. Bobovich, Opt. i Spectroskopiya 16, 246, 417 (1964). (Translation: Opt. Spectry. (USSR) 16, 134, 228 (1964).
8. R. Loudon, Proc. Phys. Soc. (London) 82, 393 (1963).
9. J.J. Hopfield and C.M. Henry, Phys. Rev. Letters 15, 964 (1965).
10. J.F. Scott, L.E. Cheesman, and S.P.S. Porto, Phys. Rev. 162, 834 (1967).
11. W.R. Rambauske, J. Appl. Phys. 35, 2958 (1964).
12. S. Ushioda, A. Pinczuk, W. Taylor and E. Burstein, "Proceedings of the II-VI Semiconducting Compounds 1967 Conf.," p. 1185, D.G. Thomas (ed.).
13. E.H. Turner et al. (to be published).

E-3: TEMPERATURE DEPENDENCE OF RAMAN LINEWIDTH AND INTENSITY OF SEMICONDUCTORS*

R. K. Chang†, J. M. Ralston and D. E. Keating
Dunham Laboratory, Yale University
New Haven, Connecticut

ABSTRACT

Pronounced decrease in the silicon Raman intensity as the temperature was increased has been measured with a Nd: YAG laser. A brief extension of resonance Raman effect is made for semiconductors with indirect energy band gap. The progression of the LO and TO Raman active modes of CdSe is presented as the S concentration was increased for various alloys of CdS_xSe_{1-x}. The effect of anharmonic forces in shifting the LO and TO modes of GaAs and in broadening the linewidths of these modes and the triply degenerate mode of silicon has been measured from 10°K to 475°K.

INTRODUCTION

The knowledge of Raman scattering cross-section, linewidth, and vibrational frequency of crystals from normal Raman scattering is pertinent to the investigation of stimulated Raman scattering (SRS). Enhancement of the Raman cross-section by resonance Raman effect and narrowing the linewidth by cooling should significantly lower the threshold of SRS. Furthermore, the use of mixed crystals to change the phonon frequency offers the interesting possibility of tunable Raman lasers.

In this paper, the enhancement of the spontaneous Raman cross-section in silicon is presented in section 2. Here the photon energy of the CW Nd: YAG laser is above and near the indirect electronic transition energy (IETE). The change of the Raman modes in mixed crystals of CdS_xSe_{1-x} is presented in section 3. In particular, the progression of the LO and TO phonon frequencies in CdSe towards the resonant gap mode of Se in CdS is presented as the concentration of S is increased. Similarly, the progression of the LO and TO modes of CdS towards the local mode of S in CdSe is also shown. The temperature dependence of the Raman linewidth and shift of the lattice frequency in GaAs and Si are presented in section 4. The present data would be a good

*Work supported in part by U. S. Air Force Cambridge Research Laboratories, Office of Aerospace Research and the Office of Naval Research.

†Alfred P. Sloan Foundation Research Fellow.

test for detailed theories of resonance Raman effect, of lattice dynamics of disordered crystals, and of anharmonic interactions in solids.

SILICON

The enhancement of the Raman scattering cross-section, when the frequency of the exciting radiation is near the medium's allowed optical transition frequency is known as Resonance Raman Effect (RRE). The theory of RRE in crystals has been treated by Loudon[1] in semiconductors with direct allowed transitions, by a two-band parabolic model. The effect of excitons in RRE of crystals has been included in the papers of Birman and Ganguly[2]. Extensive work on the theory and experiment of RRE in liquids can be found in the review paper by Behringer[3]. However, experimental evidence of RRE in crystals has been limited to the work of Leite and Porto[4] on CdS and that of Leite et al. reported in this conference on CdS and ZnSe.

This paper presents RRE data in silicon measured by a Nd:YAG laser, the photon energy of which (1.165ev) is close to the indirect energy gap of silicon[5]. The Raman intensity $I_{observed}$, collected in 90° geometry, is shown in Fig. 1 as a function of temperature. The characteristic decrease with temperature as shown in Fig. 1, has been observed from samples of different purities, ranging from 20 Ω-cm to 8,000 Ω-cm. Above 220°K, accurate measurement of the Raman intensity was made difficult by the presence of recombination radiation.

A complete theoretical treatment of the RRE in the presence of phonon assisted electronic transitions and excitons [2, 6] is rather involved. A qualitative explanation of the observed decrease in the Raman intensity can be given by extending Loudon's[1, 7] calculation for direct band gap semiconductors. Using a two-band parabolic model, Eq. (24) of Ref. [7] shows the frequency dependence of the resonant term in the Raman tensor for direct gap semiconductors to be

$$R_{(-\omega_1,\ \omega_2,\ \omega_0)} \ \alpha \int_0^{K_{max}} d\underset{\rightarrow}{K}\ \{(\omega_g + \omega_0 - \omega_1 + K^2/2\mu)(\omega_g - \omega_1 + K^2/2\mu)\}^{-1}$$

$$\alpha\ \{(\omega_g + \omega_0 - \omega_1)^{1/2} - (\omega_g - \omega_1)^{1/2}\} \qquad (1)$$

where ω_1, ω_2, and ω_0 are the incident, Stokes, and lattice vibration frequencies respectively. The direct energy gap is $\hbar\omega_g$, the reduced mass μ, and the electronic wave vector is $\underset{\rightarrow}{K}$. The Raman scattered intensity is proportional to the square of the Raman tensor $R_{(-\omega_1,\ \omega_2,\ \omega_0)}$. It can be seen from (1) that the Raman intensity increases rapidly when ω_1 is increased to approach ω_g to within several times the phonon frequency ω_0. The Raman intensity is maximum when $\omega_1 = \omega_g$, and rapidly decreases when $\omega_1 > \omega_g$, partly due to the decrease in the Raman tensor and partly due to the increase in optical absorption in the medium.

For silicon, the electronic transition from the maximum of the valence band at K = O, Γ'_{25}, to the six conduction band valleys along the <100> directions at $\underset{\rightarrow}{K} = Q$, requires the aid of a phonon ω^i_Q with momentum Q. Superscript i stands for TA, LA, TO and LO modes. For temperatures below 77°K, indirect transitions take place mainly by phonon emission, while above this temperature both phonon absorption and emission can

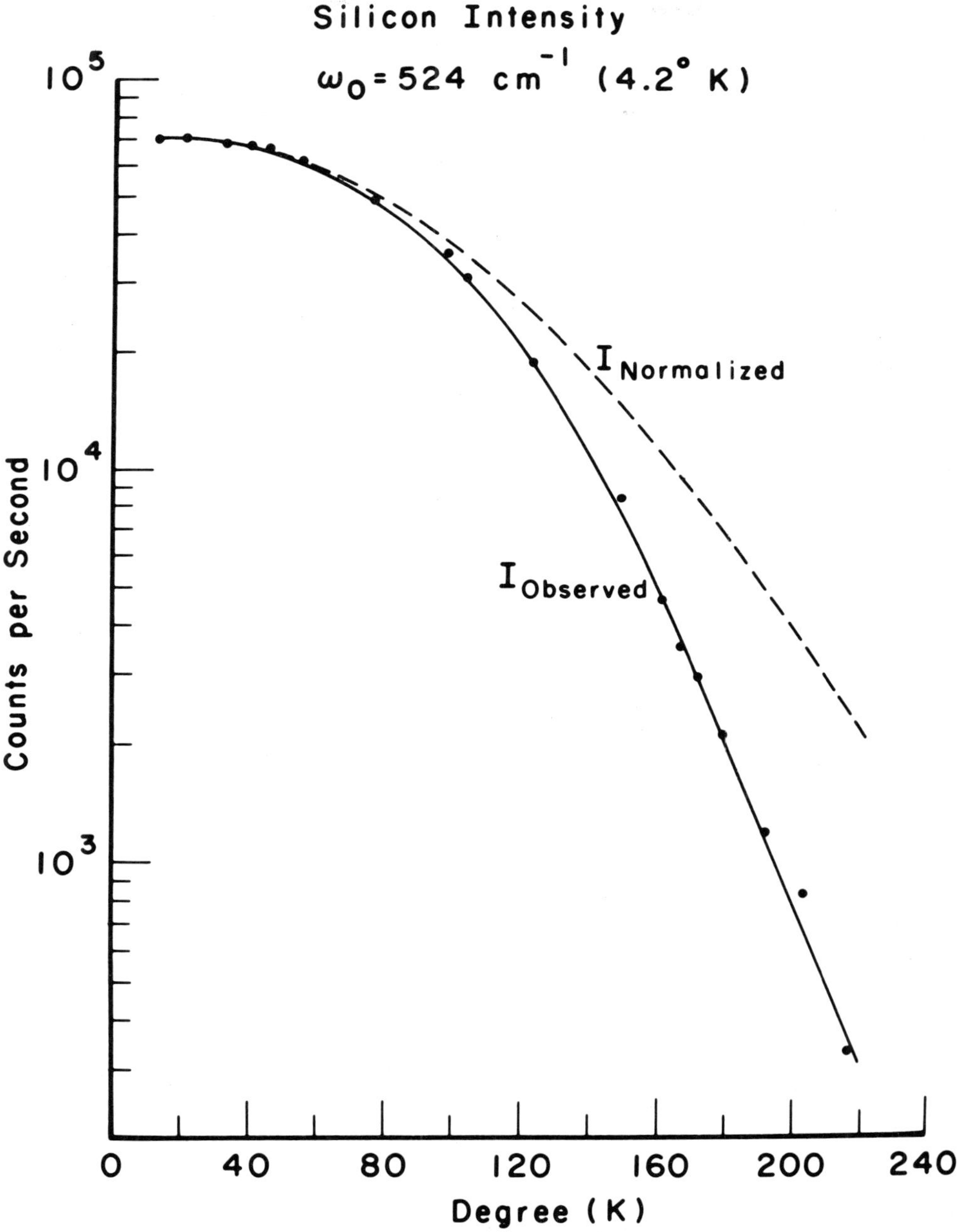

Fig. 1. Spontaneous Raman scattering intensity $I_{observed}$, from silicon as a function of temperature. The dotted curve $I_{normalized}$, shows the Raman intensity normalized to the case of no linear absorption at ω_1 and ω_2.

assist the electronic transitions. Due to electronic-phonon interaction, the indirect band gap of silicon $\hbar\omega'_g$ decreases as the temperature is increased. Such decrease (1.166ev at 0°K to 1.145ev at 220°K) has been measured by Long[8] and Macfarlane et al.[5].

The observed decrease in the silicon Raman intensity can be qualitatively explained by modifying Loudon's arguments for the direct energy gap semiconductors. The important parameter for the indirect semiconductor is not $\hbar\omega'_g$, but is the photon energy necessary for an electron to make an indirect transition (IETE). At the lower temperature ranges, photon energy of $(\omega^i_g + \omega^i_Q)$ would be needed, while at the higher temperature ranges, photon energy of only $(\omega^i_g - \omega^i_Q)$ would be needed. If the incident frequency is fixed as here, the Raman tensor should increase as the IETE decreases and approaches the incident frequency. The Raman tensor would then decrease once the incident photon energy is larger than the IETE. Furthermore, the presence of optical absorption would make the Raman intensity drop even more steeply. Our experimental situation was such that below 20°K the incident laser photon energy was approximately equal to the IETE, $\omega'_g + \omega^{TA}_Q$, where ω^{TA}_Q was the lowest phonon frequency ($\simeq$0.018ev)[5, 9] that must be emitted in order to acquire the necessary momentum Q. That is, below 20°K, $\{\omega_1 - (\omega'_g + \omega^{TA}_Q)\}/\omega_0 \simeq 0$, and thus the Raman tensor was near its maximum. As the temperature was increased, the IETE became less than $\hbar\omega_1$, partly because ω'_g was decreasing and partly because transitions could occur by absorbing a phonon of energy $\hbar\omega^i_Q$. Consequently, as the temperature was increased, the Raman intensity, $I_{normalized}$, would be expected to drop due to the decrease in the Raman tensor. The fact that as the temperature was increased, $\omega_1 > (\omega'_g - \omega^i_Q)$, was supported by the presence of photoluminescence. In fact at 220°K, $\{\omega_1 - (\omega'_g - \omega^i_Q)\}/\omega_0 > 1$. The RFE results on an indirect band gap are in accord with Loudon's conclusion for direct band gap semiconductors that Raman efficiency falls sharply when the exciting frequency is increased beyond the absorption edge.

The measured Raman linewidth of silicon will be presented in section 4 along with that of GaAs.

CdS_xSe_{1-x}

This paper reports the Raman spectra of both the LO and TO modes of CdS_xSe_{1-x} for various concentrations (x = 0, 1/4, 1/2, 3/4, & 1). Starting from the bottom, Fig. 2 shows the change in the Raman spectra as x was varied, while scanning speed, integration time, crystal orientation, incident and scattered polarization, crystal temperature, and laser wavelength were unchanged. The bottom trace of Fig. 2 shows the TO and LO modes of CdSe. As the S concentration was increased, the frequencies of these two modes monotonically converged to the resonant gap mode of Se in Cds. Note that as S was increased, the intensity of the LO mode of CdSe diminished faster than that of the TO mode. Similarly, the top trace of Fig. 2 shows the TO (232cm^{-1}) and LO (305cm^{-1}) modes of CdS. Now as Se was increased, the frequency of these two modes again converged, this time towards the local mode of S in CdSe. Contrary to the previous case (S increasing), the TO intensity of CdS diminished faster than that of the LO mode when the concentration of Se was increased.

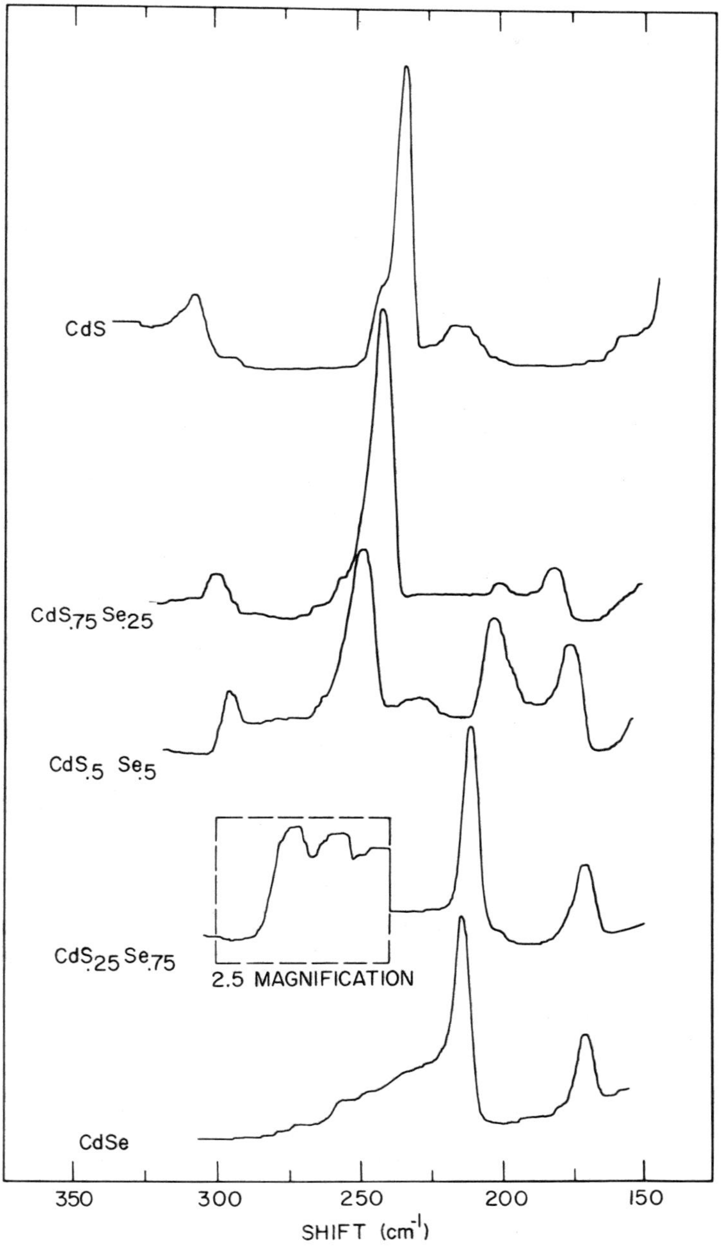

Fig. 2. The Raman spectra of CdS_xSe_{1-x} at 80°K. The bottom trace shows the LO and TO of CdSe. As the S concentration is increased, these two modes converge and diminish in intensity. The top trace shows the LO and TO of CdS. As the Se concentration is increased, the two modes converge and diminish in intensity.

The measured Raman shift of these modes for the various alloys are plotted in Fig. 3. This data for CdS_xSe_{1-x} agrees with that presented by Parrish et al. [10], where the TO modes were obtained by Kramers-Kronig analysis of IR reflectivity data, while the LO modes were measured by Raman scattering, using a He-Ne laser.

The spectra shown in Fig. 2 were measured at 80°K with a CW Nd:YAG laser polarized along the crystal's z-axis and entering along the x-axis. The unanalyzed scattered light was observed along the y-axis. In order to determine the symmetry of the LO and TO modes as presented in Fig. 3, the scattered light was analyzed along the z-axis (for A_1 symmetry, x(zz)y) and along the x-axis (for E_1 symmetry, x(zx)y).

From Raman spectra, the vibrational frequencies of the CdS_xSe_{1-x} and ZnS_xSe_{1-x}[11] alloys were found to have characteristics which were somewhere in between those of ionic crystals $Ca_xSr_{1-x}F_2$[12] and those of the covalent crystals Ge:Si[13]. Disorder in the mixed crystals can add to the anharmonic Raman linewidth by allowing phonons other than $K \simeq 0$ to contribute to 1st-order Raman scattering. This additional Raman linewidth in CdS_xSe_{1-x} alloys, the relative intensity variations of the LO and TO modes in these and other (Cf. Table I) semiconductors, and the RRE in this system are presently under investigation.

LINEWIDTH & LINESHIFT

Both the threshold and gain of SRS are linewidth dependent. However in recent years, the Raman linewidths of only one solid (calcite)[14] and a limited number of liquids have been reported[15]. This paper presents Raman linewidth data of GaAs and Si, and lineshift of GaAs as a function of temperature.

An optical mode can interact with other lattice modes through the lattice force anharmonicities. The principal interactions are the cubic anharmonicities, resulting in the splitting of $K \simeq 0$ optical phonon into two acoustic modes of opposite momentum[16]. Consequently, the linewidth should increase linearly with temperature, and the lineshift should increase or decrease depending on whether ω_0 is greater or less than the sum of the frequencies of the two acoustic phonons[7]. For semiconductors, only that latter inequality has been encountered.

Figs. 4 and 5 show the GaAs lineshift and the Si and GaAs linewidth as a function of temperature. Both GaAs modes decreased linearly at $0.016 cm^{-1}/°K$. Before 77°K, all the linewidths are approximately constant, and the widths of the various modes were in this order, (silicon)> (LO, GaAs)> (TO, GaAs). Above 300°K, both GaAs linewidths increased linearly. Note that these crystal linewidths at low temperature are comparable to that of CS_2 and are less than that of benzene and toluene [15].

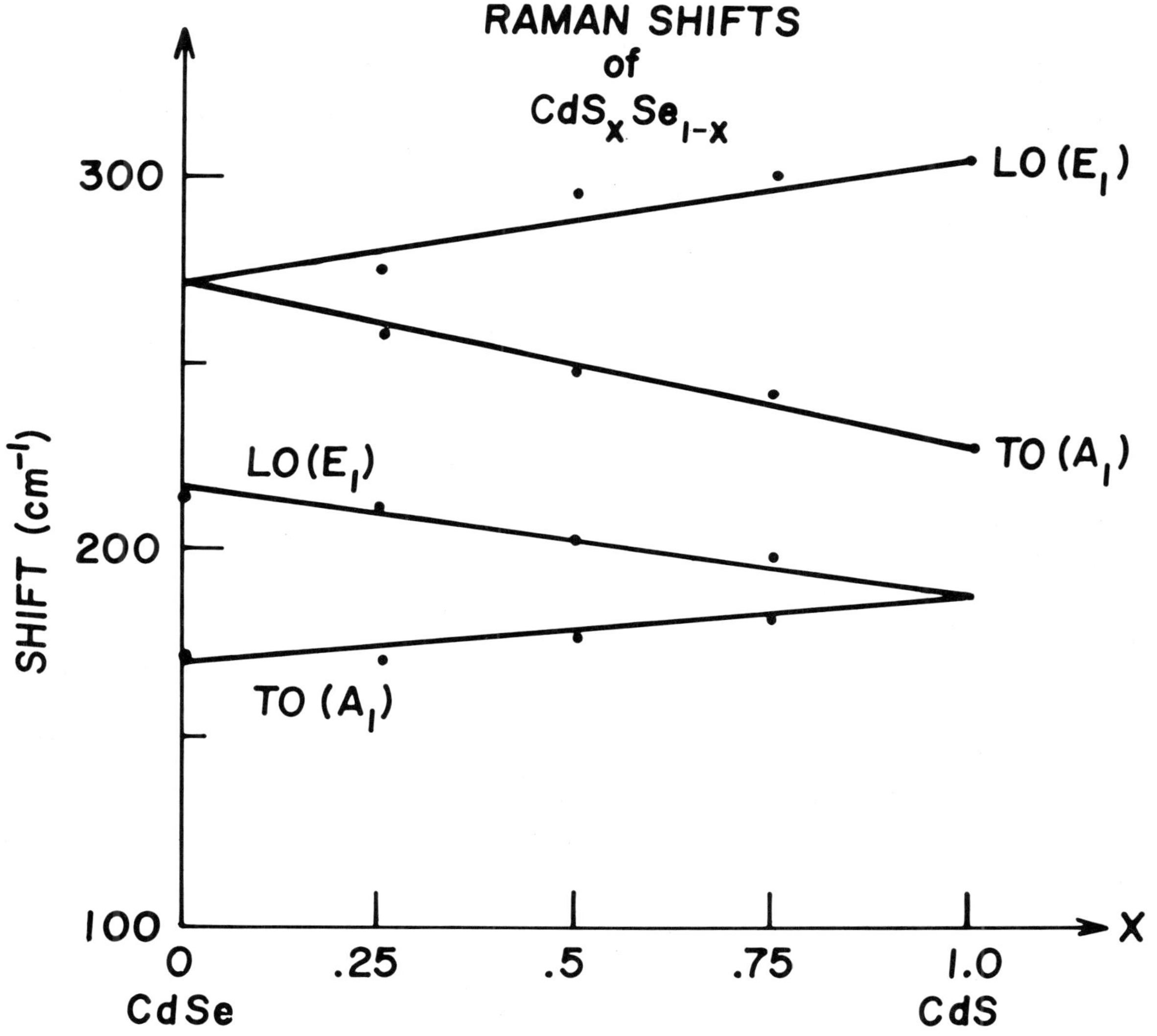

Fig. 3. The frequency of the LO and TO modes of CdS_xSe_{1-x} measured from Fig. 2. As the S concentration is increased, the LO and TO modes of CdSe converge to form the resonant gap mode of Se in CdS, while the local mode of S in CdSe diverges to form the LO and TO modes of CdS.

TABLE I

1st-Order Raman Shifts for Semiconductors.
Present Measurements With Nd:YAG Laser Shown With Asterisks.

COMPOUNDS	(optical gap)	FREQUENCY(cm^{-1})	(TEMP.)	[REF.]	
Ge	0.74ev	300.7±0.5	RT	a	IV
*Si	1.17	523	RT	b	
AlP	1.7	LO=501 TO=440	RT	c	III-V
AlSb	1.6	LO=339.9±0.5 TO=318.9±0.5	RT	d	
*GaAs	1.52	LO=291.9±0.3 TO=268.6±0.3	RT	d	
GaP	2.34	LO=403.0±0.5 TO=367.3±1	RT	b, d	
InP	1.42	LO=345.0±0.3 TO=303.7±0.3	RT	d	
InSb	0.24	LO=200±2 TO=179±1	RT	e	
*CdS	2.58	LO (E_1, A_1)=305 TO (E_1, A_1)=235, 228	LNT	f	II-VI
*CdSe	1.84	LO (E_1, A_1)=214 TO (A_1) =171	LNT	present data	
*CdTe	1.61	LO=171 TO=140	RT	g	
ZnO	3.44	LO (E_1, A_1)=583, 574 TO (E_1, A_1)=407, 381	RT	h	
ZnS	3.9	LO=349 TO=274	RT	i	
*ZnSe	2.9	LO=251±1 TO=204±1	RT	e	
ZnTe	2.39	LO=206 TO=179	RT	e	

a) D. W. Feldman, M. Ashkin, and J. H. Parker, Jr., Phys. Rev. Letters 17, 1209 (1966).
b) J. P. Russell, J. Phys. Radium 26, 620 (1965).
c) S. Z. Beer, J. F. Jackovitz, D. W. Feldman, and J. H. Parker, Jr., Phys. Letters 26A, 331 (1968).
d) A. Mooradian and G. B. Wright, Solid State Comm. 4, 431 (1966).
e) M. Krauzman, C. R. Acad. Sc. Paris 264, 1117 (1967).
f) B. Tell, T. C. Damen, and S. P. S. Porto, Phys. Rev. 144, 771 (1966).
g) A. Mooradian and G. B. Wright, Solid State Res. Report, Lincoln Laboratory, M.I.T., 1, 47 (1968).
h) T. C. Damen, S. P. S. Porto, and B. Tell, Phys. Rev. 142, 570 (1966).
i) O. Brafman, I. F. Chang, G. Lengyel, S. S. Mitra, and E. Carnall, Jr., Phys. Rev. Letters 19, 1120 (1967).

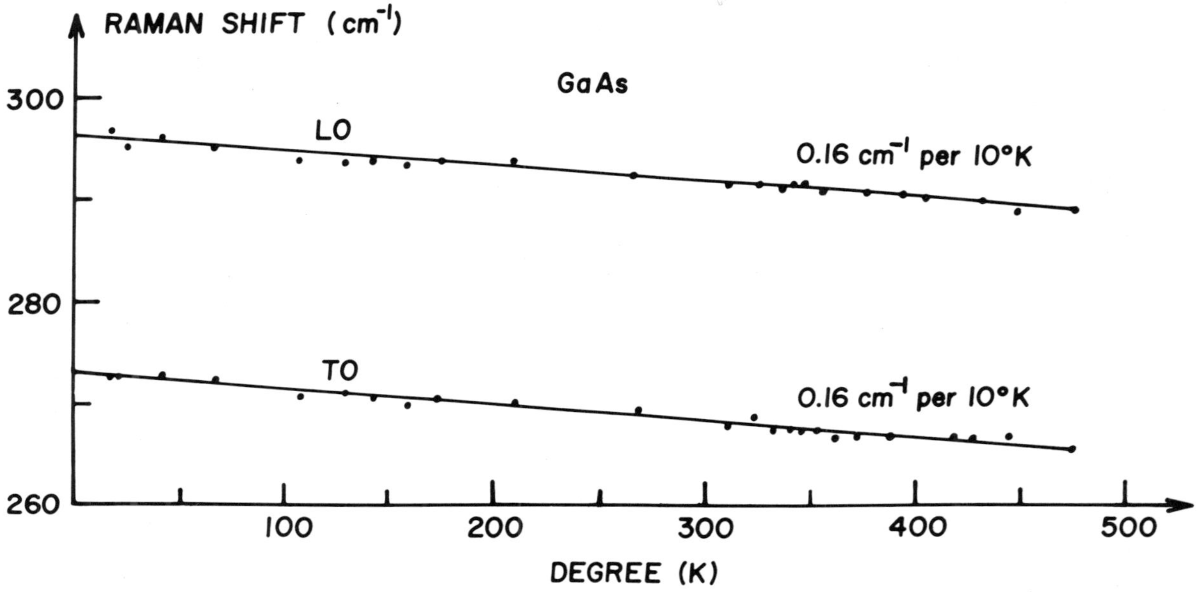

Fig. 4. The Raman shift of the LO and TO modes of GaAs as a function of temperature. An approximate $0.16 cm^{-1}$ per $10^{o}K$ has been observed for both the LO and TO modes.

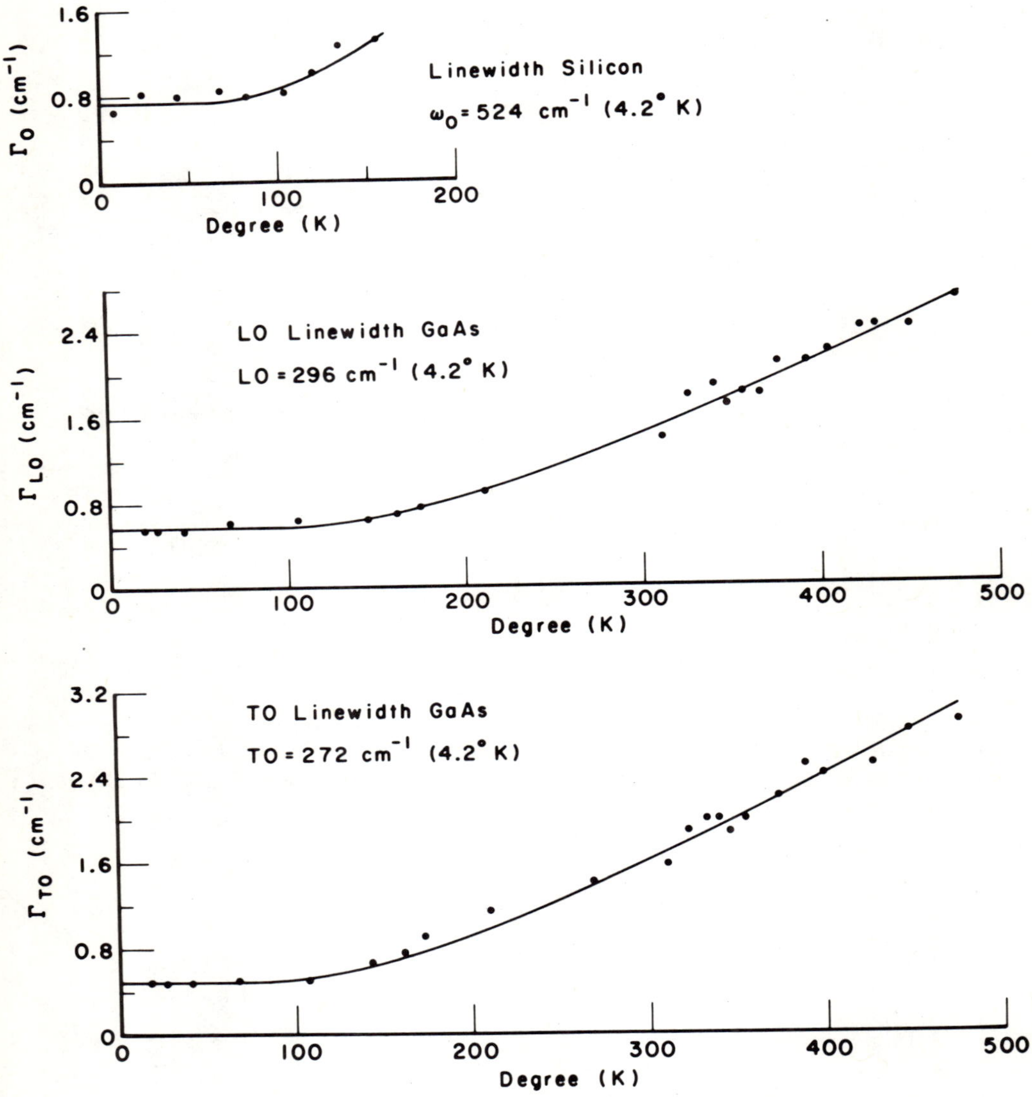

Fig. 5. Raman linewidth measured at various temperatures for silicon and the LO and TO modes of GaAs. No unfolding of the spectrometer slit broadening and the linewidth of the Nd:YAG laser (≈0.5cm^{-1}) has yet been performed. The plotted points represent the difference of the measured Raman linewidth FWHH and that of the laser.

CONCLUSION

CW Nd:YAG laser is particularly suitable for spontaneous Raman studies of semiconductors. The presented results on the resonance Raman effect of indirect band gap silicon, Raman shifts of CdS_xSe_{1-x} alloys, and the linewidths of GaAs and Si, offer tests for detailed theory of RRE, lattice vibrations of disordered crystals, and anharmonic lattice forces. Furthermore, the present work offers pertinent information for SRS by Q-switched lasers, which seems eminently achievable, in ordered and disordered semiconductors.

REFERENCES

1. R. Loudon, J. Phys. Radium 26, 677 (1965).
2. J. L. Birman and A. K. Ganguly, Phys. Rev. Letters 17, 647 (1966); A. K. Ganguly and J. L. Birman, Phys. Rev. 162, 806 (1967).
3. J. Behringer, "Raman Spectroscopy," p. 168, H. A. Szymanski (ed.), Plenum Press. New York, 1967.
4. R. C. Leite and S. P. S. Porto, Phys. Rev. Letters 17, 10 (1966).
5. G. G. MacFarlane, T. P. McLean, J. E. Quarrington, and V. Roberts, Phys. Rev. 111, 1245 (1958).
6. R. J. Elliott, Phys. Rev. 108, 1384 (1957).
7. R. Loudon, Proc. Roy. Soc. A275, 218 (1963).
8. D. Long, J. Appl. Phys. 33, 1682 (1962).
9. F. A. Johnson, Prog. in Semiconductors 5, 179 (1965).
10. J. F. Parrish, C. H. Perry, O. Brafman, I. F. Chang, and S. S. Mitra, "II-VI Semiconducting Compounds 1967 International Conference," p. 1164, D.G. Thomas (ed.). W. A. Benjamin, New York, 1967.
11. O. Brafman, I. F. Chang, G. Lengyel, S. S. Mitra, and E. Carnall, Jr., Phys. Rev. Letters 19, 1120 (1967).
12. R. K. Chang, B. Lacina, and P. S. Pershan, Phys. Rev. Letters 17, 755 (1966).
13. D. W. Feldman, M. Ashkin, and J. H. Parker, Jr., Phys. Rev. Letters 17, 1209 (1966).
14. K. Park, Phys. Letters 22, 139 (1966).
15. W. R. L. Clements and B. P. Stoicheff, App. Phys. Letters 12, 246 (1968).
16. P. C. Klemens, Phys. Rev. 148, 845 (1966).
17. M. Born and M. Blackman, Zeit. f. Physik 82, 551 (1933).

E-4: THEORY OF INTERACTION OF LIGHT WITH INSULATING CRYSTALS*†

Bernard Bendow and Joseph L. Birman
Physics Department, New York University
New York

INTRODUCTION

The present work develops a formal theory of the interaction of radiation in insulators at zero temperature, from which one obtains expressions for the transition probabilities for elementary processes such as Raman scattering and direct absorption. Presentation of the theory is followed by various applications.

Three factors are given special attention in the formalism:

(a) The electromagnetic field must be taken into account at all times, even when no external source of radiation is present;

(b) Both absoption and scattering ought to be considered within a unified viewpoint;

(c) Asymptotic scattering states should be defined unambiguously, and their precise relation to experiment specified.

HAMILTONIAN AND SCATTERING THEORY

The hamiltonian H consists of the free electromagnetic and crystalline fields H_0, and all their mutual interactions V, $H = H_0 + V$. The crystalline fields are assumed to extend throughout a very large but finite crystal, while the electromagnetic field extends throughout all space. In the applications to follow, a model hamiltonian will be employed, with $V = V' + V''$, where

$$H_0 = \sum_{\vec{q}\lambda} qc\, a^{+}_{\vec{q}\lambda} a^{-}_{\vec{q}\lambda} + \sum_{\vec{q}\beta} W(\vec{q}\beta)\, b^{+}_{\vec{q}\beta} b^{-}_{\vec{q}\beta} + \sum_{\vec{q}\gamma\alpha} E(\vec{q}\gamma\alpha)\, B^{+}_{\vec{q}\gamma\alpha} B^{-}_{\vec{q}\gamma\alpha} \quad (1)$$

*This paper is based in part on a thesis to be submitted in partial fulfillment of the requirements for the degree of Ph. D. in physics at New York University by Bernard Bendow.

† Supported in part by the U. S. Army Research Office, Durham, and the Aerospace Research Laboratories, Wright-Patterson Air Force Base, Dayton, Ohio.

$$V' = \sum_{\pm\vec{q}\lambda\gamma\alpha} g \begin{pmatrix} \pm & \pm \\ \vec{q}\lambda\gamma\alpha \end{pmatrix} a^{\pm}_{\vec{q}\lambda} B^{\pm}_{\pm\vec{q}\gamma\alpha} + \begin{pmatrix} \text{similar bilinears} \\ \text{in a-b, b-B, a-a} \end{pmatrix} \tag{2}$$

$$V'' = \sum_{\pm\gamma\gamma'\beta\vec{q}\vec{k}} G \begin{pmatrix} \pm\vec{q}\ \vec{k} \\ \beta\gamma\gamma' \end{pmatrix} b^{\pm}_{\pm\vec{q}\beta} B^{+}_{\vec{q}+\vec{k},\gamma} B^{-}_{\vec{k}\gamma'} \tag{3}$$

The a's are photon, B's exciton, and b's phonon, operators; ($\gamma\alpha$) are the exciton band and interband indices, λ is the photon polarization, and β is the phonon branch; q is the wave vector. Examples of choices of the dispersions and coupling constants may be found in[1] or [2].

In setting up scattering theory, a formalism employed by Wick[3] in meson theory will be adapted and extended. The asymptotic states are chosen as $\Psi_{\pm\infty} = a^{+}_{p} | \alpha \rangle$, where α refers to an exact eigenstate of H. Now, $\Psi_{\pm\infty}$ can be shown to be an eigenstate of H of energy $E_\alpha + pc$, when the photon is far from the crystal. This state, then, represents a free "external" photon impinging on the crystal, as is the case in an experiment. The full wave function satisfying in or outgoing boundary conditions, $| p_\alpha \pm \rangle$, is $\Psi_{\pm\infty}$ plus a scattered wave $\Psi_{\pm \text{scat.}}$, $| p_\alpha \pm \rangle = \Psi_{\pm\infty} + \Psi_{\pm \text{scat.}}$, where

$$\Psi_{\pm \text{scat.}} = \frac{[V, a^{+}_{p}]}{E_\alpha + pc - H \pm i\epsilon} \, | \alpha \rangle \tag{4}$$

Expansion of the scattered wave in states $| \beta \rangle$, $| q\beta \rangle$, $| qq'\beta \rangle$, ..., representing none, one, two, ..., scattered free photons leads to the identification of transition matrix elements:

(a) Absorption:

$$T(p_\alpha \to \beta) = \langle \beta | [V, a^{+}_{p}] | \alpha \rangle \tag{5}$$

(b) One photon scattering:

$$T(p_\alpha \to q\beta) = \langle q\beta - | [V, a^{+}_{p}] | \alpha \rangle \tag{6}$$

Similar expressions represent two outgoing photon processes, etc. These expressions are actually a hierarchy of integral equations. For example, using (4) to substitute for $\langle q\beta - |$ in (6), then expanding the energy denominators in states $| \beta \rangle$, $| q\beta - \rangle$, ..., and finally using (5) and (6) to identify matrix elements, (6) becomes:

$$\begin{aligned} T(p_\alpha \to q\beta) = & \sum_\gamma \left(\frac{T^*(q\beta \to \gamma)\, T(p\alpha \to \gamma)}{E_\alpha - E\gamma + pc + i\epsilon} + \frac{T^*(p\beta \to \gamma)\, T(q\alpha \to \alpha)}{qc - E_\beta + E_\gamma} \right) \\ & + \sum_{k\gamma} \left(\frac{T^*(q\beta \to k\gamma)\, T(p\alpha \to k\gamma)}{E_\alpha + pc - E_\gamma - kc + i\epsilon} + \frac{T^*(p\beta \to k\gamma)\, T(q\alpha \to k\gamma)}{qc + kc - E_\beta - E_\gamma} \right) + \cdots \end{aligned} \tag{7}$$

This result, in fact, points out the interdependence of the absorptive and scattering transition matrix elements. Such chains of coupled equations for the transition matrix elements are well known from scattering theory[4].

THE OHRON APPROXIMATION AND RAMAN SCATTERING

For applications of the equation, the approximate hamiltonian, as given by Eqs. (1) to (3), will be employed. One assumes that the identical wave vector continuum may be employed for both the photons and the crystal. The basic crystalline states are taken as those resulting from an exact diagonalization of $\tilde{H} \equiv H_0 + V'$, and these states will be herein* referred to as "ohrons". These states include the electromagnetic interaction, independent of the existence of an external source, as required. With ohron operators $A^{\pm}$ and eigenenergies $\tilde{\omega}$, one has

$$\tilde{H} = \sum_{q\gamma} \tilde{\omega}(q\gamma) A^{+}_{q\gamma} A^{-}_{q\gamma} \tag{8}$$

where γ is a compound index for all the previous polarizations. Details of such transformations are given in [5-7], and will be discussed at length by one of us elsewhere[8]. With V'' considered as a perturbation, one may expand

$$| \alpha > = | \tilde{\alpha} > + \frac{1}{E - \tilde{H}} V'' | \tilde{\alpha} > + \cdots \tag{9}$$

$$\frac{1}{E - H} = \frac{1}{E - \tilde{H}} + \frac{1}{E - \tilde{H}} V'' \frac{1}{E - \tilde{H}} + \cdots \tag{10}$$

where the tilde indicates ohron quantities. Now, we consider Raman scattering in this framework. This process is here to be viewed as a single photon scattering, where the crystal is originally in its ground state, and ends up with one ohron excited. In lowest order in the trilinear coupling constant, one obtains six terms, analogous to the six terms obtained in[2]. The term corresponding to the resonant term in[2] is

$$\begin{aligned} T_{RES.}(p\,q\,\gamma) = \sum_{\gamma', \gamma''} \Big[& C(\pm q \gamma')\, C(\pm p \gamma'') \\ & \times < p\text{-}\tilde{q}, \gamma \,|\, [V'(A^{\pm}), A^{\pm}_{q\gamma}] \,|\, q\tilde{\gamma}' ; p\text{-}q, \gamma > \\ & \times < q\tilde{\gamma}' ; p\text{-}q, \gamma \,|\, V''(A^{\pm}) \,|\, p, \tilde{\gamma}'' > < p\tilde{\gamma}'' \,|\, [A^{\pm}_{p\gamma''}, V'(A^{\pm})] \,|\, \tilde{O} > \\ & \times \frac{1}{qc - \tilde{\omega}(q\gamma')} \frac{1}{pc - \tilde{\omega}(p\gamma'')} \end{aligned} \tag{11}$$

The C's are the transformation coefficients linking a^{+} to $A^{\pm}$; V' and V'' are to be expressed in terms of the $A^{\pm}$'s. The equation describes the absorption of photon p with excitation of virtual ohron states (p, γ''); these are scattered through V'' to (p-q, γ) and (q, γ'); finally photon q is emitted, leaving the crystal with the single real excitation (p-q, γ).

*Such states have been previously referred to as polaritons[5,7]. They are herein called "ohrons" so as to specialize them to the case at hand.

One notes the complicated wave vector and ohron frequency dependence introduced through the transformation coefficients. As may be inferred from the discussion of the next section, these coefficients are most important in the resonance regions. For Raman scattering, as may be seen from Eq. (11), these occur at crossovers of the free photon and ohron dispersions, i.e., when $\tilde{\omega}(p) = pc$. The singular behavior at resonance may be removed by the introduction of lifetime effects due to V''[11].

Results similar to those obtained in[2], where perturbation theory on H_0 is employed, can be obtained through the following substitutions into the formal theory: $|\alpha> \rightarrow |\alpha>_0 + (E - H_0)^{-1} V |\alpha>_0$; pc, the photon energy $\rightarrow$ pc/n (n $\equiv$ optical index of refraction); and

$$\frac{1}{E - H} \rightarrow \frac{1}{E - H_0} + \frac{1}{E - H_0} V \frac{1}{E - H_0} + \cdots \tag{12}$$

where $|\alpha>_0$ are eigenstates of H_0, and V' does not now include the a-a photon-photon interaction. The result for $T_{RES.}$ becomes

$$T_{RES.} = \sum_{\gamma\gamma'} \Big[{}_0< p - q \,|\, [a_q, V'] \,|\, p - q, q, \gamma' >_0 \tag{13}$$

$$\times {}_0< p - q, q, \gamma' \,|\, V'' \,|\, p\gamma >_0 < p\gamma \,|\, [V', a_p^+] \,|\, 0 >_0$$

$$\times \frac{1}{\frac{qc}{n} - E(q\gamma')} \; \frac{1}{\frac{pc}{n} - E(p\gamma)}$$

where the state $|p\text{-}q>_0$ refers to a phonon. Here the resonances occur at free photon crossovers with $nE(p\gamma)$. This clearly differs from the more exact results of the wave vector dependent crossovers of the ohron levels.

THE OHRON SPECTRUM

From a theoretical point of view, a detailed knowledge of the ohron spectrum is of interest because it represents the exact eigenenergies of a part of a fundamental model hamiltonian in insulator physics. This spectrum may be used, for example, in deriving the thermodynamics of irradiated insulators. The detailed effects of coupling similar to the present ones are of interest in a large variety of solid state problems, such as coupled systems of magnons, photons and phonons[9].

Determination by the experimentalist of the frequencies at which direct absorptive processes[12] proceed serves to verify the theoretically computed crossovers, which in turn permit a check on the correctness of parameters in the theory, such as the various coupling constants, the band gap ΔE, or the effective mass m*. Raman scattering experiments determine the ohron spectrum through the energy conservation condition $pc - qc = \tilde{\omega}(\sqrt{p^2+q^2-2pq\cos\theta}, \gamma)$, as in[10], providing comparison with theoretical spectra.

Fig. 1 shows the typical spectrum arising from coupling of excitons and phonons to light. We have computed the ohron spectrum for the case of two exciton levels and one phonon level, all dispersionless, with coupling constants appropriate to Cds, as listed by[2]. The computed results in the crossover region are shown in Fig. 2 (a) for the exciton region. The phonon crossover, as shown in Fig. 2 (b) is identical to the free crossover because its direct coupling to light was taken to be negligible. The particular forms illustrated follow when simplifications analogous to those employed earlier in[5] are made. These will be described in detail elsewhere[8].

The general behavior indicates a strong influence of the coupling in the crossover regions; one notes the ohron dispersion in the phonon region induced by exciton couplings, even though the direct coupling of the phonon to light has been neglected. The other major effect is the shift, induced by the coupling, of the crossover points. For the ohron in Fig. 2. (a) this shift is to approximately .016 ev above the free exciton crossover (n=1).

SUMMARY AND DISCUSSION

A formal theory of interaction of radiation with insulators has been presented, with special attention to requirements outlined in Section 1. A hierarchy of integral equations results, to which simplifying approximations are applied, using ohron states. The resonant contribution to Raman scattering is obtained, and the relationship with certain other results established. Finally, the characteristics of ohron spectra are considered.

The major differences from perturbation theory on H_O[1, 2, 12] are that we obtain (a) shifts in the resonance positions; and (b) a complicated wave vector dependence of the transition matrix elements. The present theory differs also from the polariton theory of[7], where the asymptotic states were taken as eigenstates of H, and consequently no singularities arose; the present theory would necessitate introduction of lifetimes induced by V''[11] in order to obtain appropriate results in the resonance region. As regards the Raman scattering resonance frequencies, these are identical, within the ohron approximation, to the frequencies at which direct absorption[12] proceeds.

In parallel and future work we intend to: (a) continue to apply the present methods to obtain quantitative results for Raman scattering cross sections and ohron spectra, for various coupling constants in materials such as Cds (computations are presently in progress); (b) discuss in detail other process, such as ohron assisted absorption; and (c) extend the theory to $T > O$.

In order to provide comparison with theoretical computations, it is desirable that experiments be performed which provide measurements of: (a) ohron spectra (as described in Section 4); (b) resonance scattering intensities; and (c) absolute scattering intensities as a function of incident photon frequency. Experimental studies of intensities have been made, for example, as in[13].

ACKNOWLEDGEMENTS

Bernard Bendow would like to thank Dr. L. F. Landowitz for a number of helpful discussions.

Bernard Bendow acknowledges support by NASA trainee ship at New York University, 1964 through 1967.

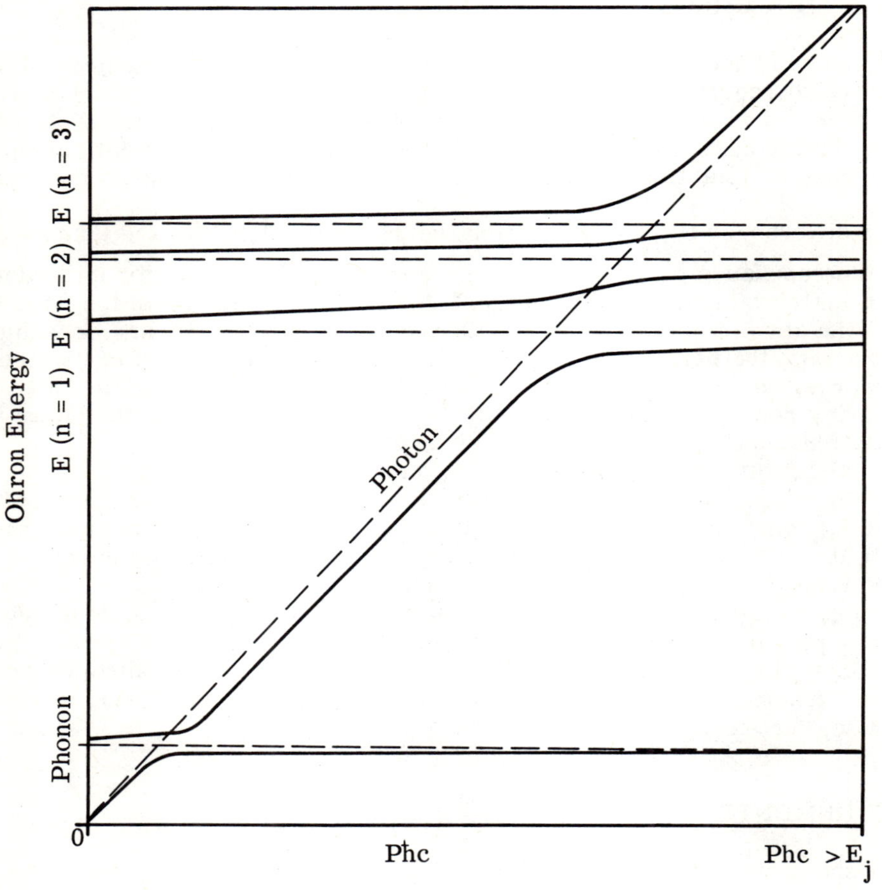

Fig. 1. Typical ohron spectrum.

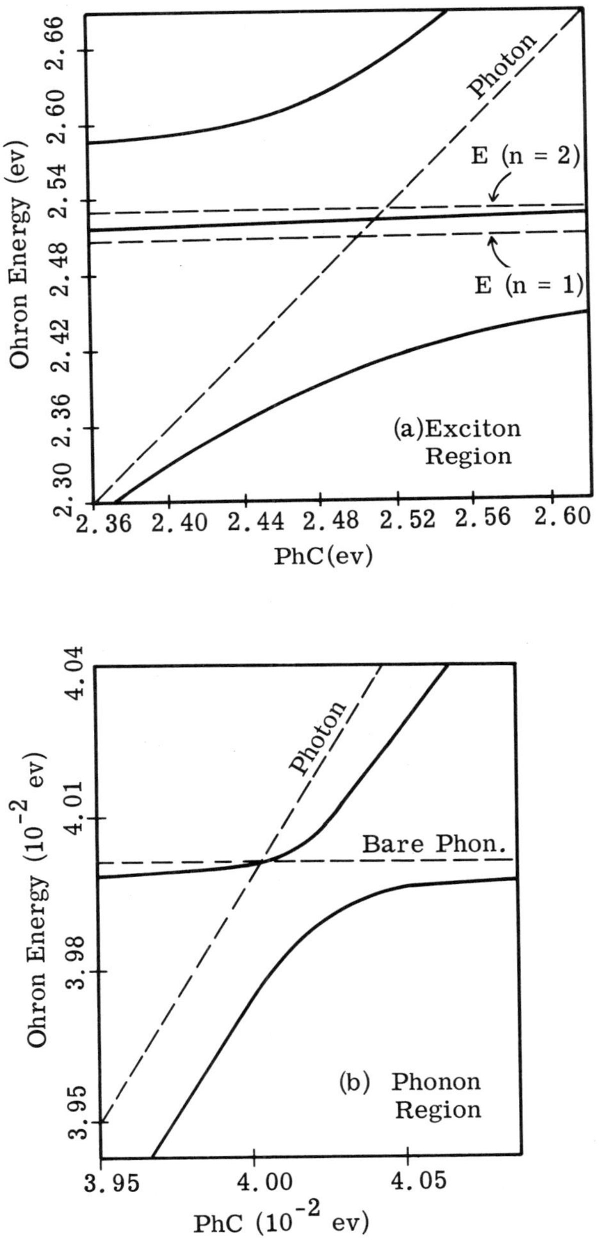

Fig. 2. Crossovers in (a) exciton region and (b) phonon region.

REFERENCES

1. R. Loudon, Proc. Roy. Soc. A275, 218 (1963).
2. A.K. Ganguly and J.L. Birman, Phys. Rev. 162, 806 (1967).
3. G.C. Wick, Revs. Mod. Phys. 27, 339 (1955).
4. A. Messiah, "Quantum Mechanics," II, p. 329, J. Wiley (ed.), New York, 1962.
5. J.J. Hopfield, Phys. Rev. 112, 1555 (1958).
6. V.M. Agranovich, Sov. Phys. JETP 10, 307 (1960).
7. L.N. Ovander, Sov. Phys. Uspekhi 8, 337 (1965).
8. B. Bendow, "Ph.D. Thesis," to be submitted to Physics Dept., New York Univ., N.Y.
9. T. Moriya, Journal of Applied Phys. 39, 1042 (1968).
10. C.H. Henry and J.J. Hopfield, Phys. Re. Letters 15, 964 (1965).
11. A. Messiah, op. cit., chap. XXI, Sec 13.
12. R.S. Knox, "Theory of Excitons," p. 112, Academic Press, N.Y., 1963
13. S. Ushioda, A. Pinczuk; W. Taylor, and E. Burstein, "II-VI Semiconducting Compounds," D.G. Thomas (ed.) W.A. Benjamin, N.Y., 1967.

E-5: RAMAN SCATTERING BY OPTICAL MODES OF METALS*

James H. Parker, Jr., D. W. Feldman and M. Ashkin
Westinghouse Research Laboratories
Pittsburgh, Pennsylvania

ABSTRACT

Raman scattering by the optical modes of the metals Zn, Mg, and Bi has been observed. The measurements were carried out using argon ion laser excitation. One line was observed for each of the hexagonal close packed metals, Zn and Mg, and two lines were observed for Bi. The frequencies that were obtained are: Zn: 70 cm^{-1}, Mg: 120 cm^{-1}, and Bi: 65 cm^{-1} and 90 cm^{-1}. These frequencies agree reasonably well with neutron scattering data. The temperature dependence of the Raman linewidths are discussed for the above metals.

INTRODUCTION

Many metals have structures that allow optical vibrational modes to exist. For example, the common hexagonal close packed structure has two atoms per unit cell and has two optical branches. From symmetry considerations, one or more of the k = 0 optical modes for a given metal should be Raman active. It is the purpose of this paper to describe observations of Raman scattering by the optical modes of the metals Zn, Mg, Bi, Be[1] and $AuAl_2$[1]. The frequencies as well as the linewidths at 300°K are reported along with a brief discussion of the observed temperature dependence of the linewidths.

EXPERIMENTAL

The Raman spectra were observed using an argon ion laser for excitation. The laser was of the graphite capillary type and delivered about 1 watt in the 4880 Å line. The spectra were dispersed by a double tandem grating monochrometer (Spex 1400). Because of the severe problem of laser light being scattered from the metal surface, a 1/4 m grating monochrometer was used, with fixed band pass, as a prefilter to the double monochrometer to further discriminate against the effect of scattered laser light. The

*Work partially supported by Materials Laboratory, Wright-Patterson Air Force Base, Ohio.

spectra were photoelectrically detected using a cooled S-11 photomultiplier (EMI 9502S) having a dark count of about 1 sec^{-1}. The laser beam was chopped within the optical cavity at 90 Hz and the signal from the photomultiplier was amplified and synchronously detected. The integration time constant varied from 60 to 600 sec with the monochrometer scan rate ranging from 0.4 to 4 cm^{-1}/min. While chopping within the optical cavity gave some discrimination against the effect due to argon discharge lines being scattered from the sample, further discrimination was attained by using a long path length between the laser and the sample position as well as interposing a narrow interference filter, centered at the laser line, along this path. Fig. 1 shows the experimental arrangement as described above. The laser light, after passing through the interference filter, was focussed to a small spot on the sample with an angle of incidence to the sample face of ~70-80°. The scattered light was viewed normal to the sample face. Measurements were carried out at both 300°K and 77°K.

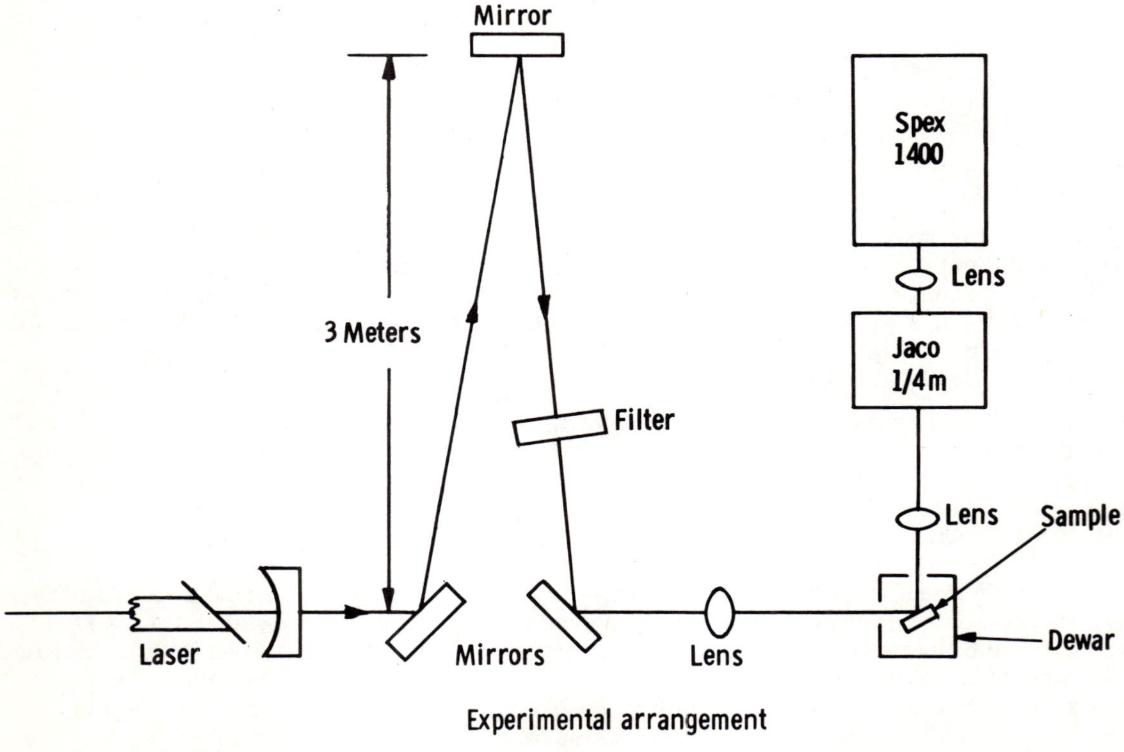

Fig. 1. Schematic diagram of the experimental arrangement.

All of the metal samples were single crystals with the exception of Be which was a polycrystalline ingot. Measurements were made with mechanically polished faces for all of the metals and, in addition, with "as grown" faces for Zn crystals. No special crystallographic orientation was used for the face of the samples.

RESULTS AND DISCUSSION

Figs. 2, 3, 4, 5 and 6 show the Raman lines observed for Zn, Mg, Bi, Be[2] and $AuAl_2$, respectively. The Raman lines shown in the figures are stokes components with the exception of Bi (Fig. 4) which is an anti-stokes component. The Raman character of the observed lines was verified in all cases, either by observing the stokes and anti-stokes for 4880 Å excitation or the stokes component for both 4880 Å and 5145 Å excitation.

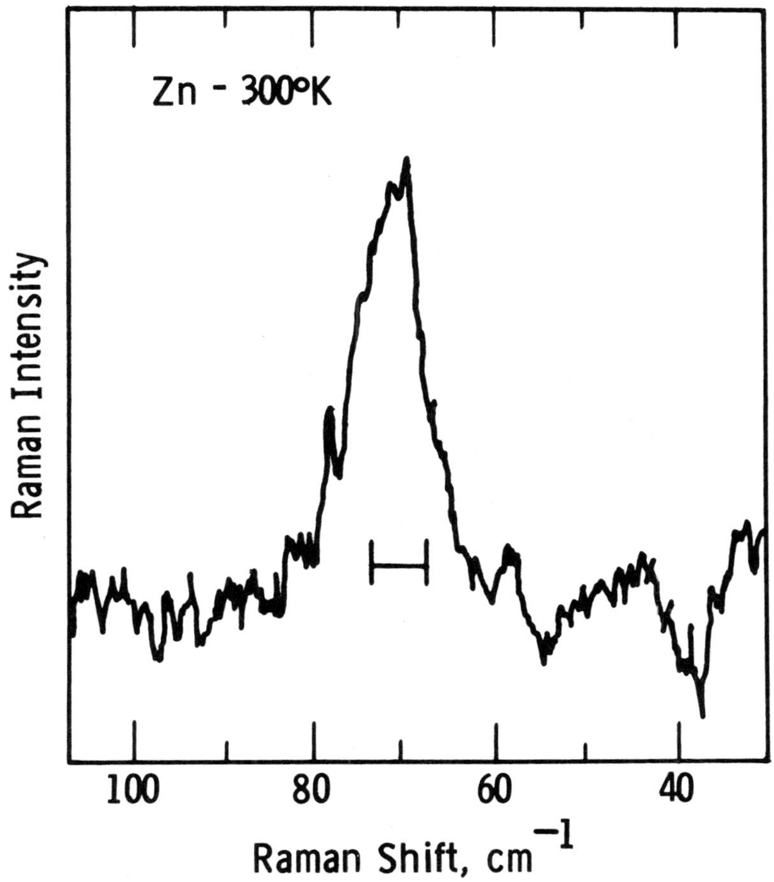

Fig. 2. Observed Raman line (Stokes) for Zn single crystal ("as-grown" face), 300°K. The horizontal line segment indicates the instrumental resolution.

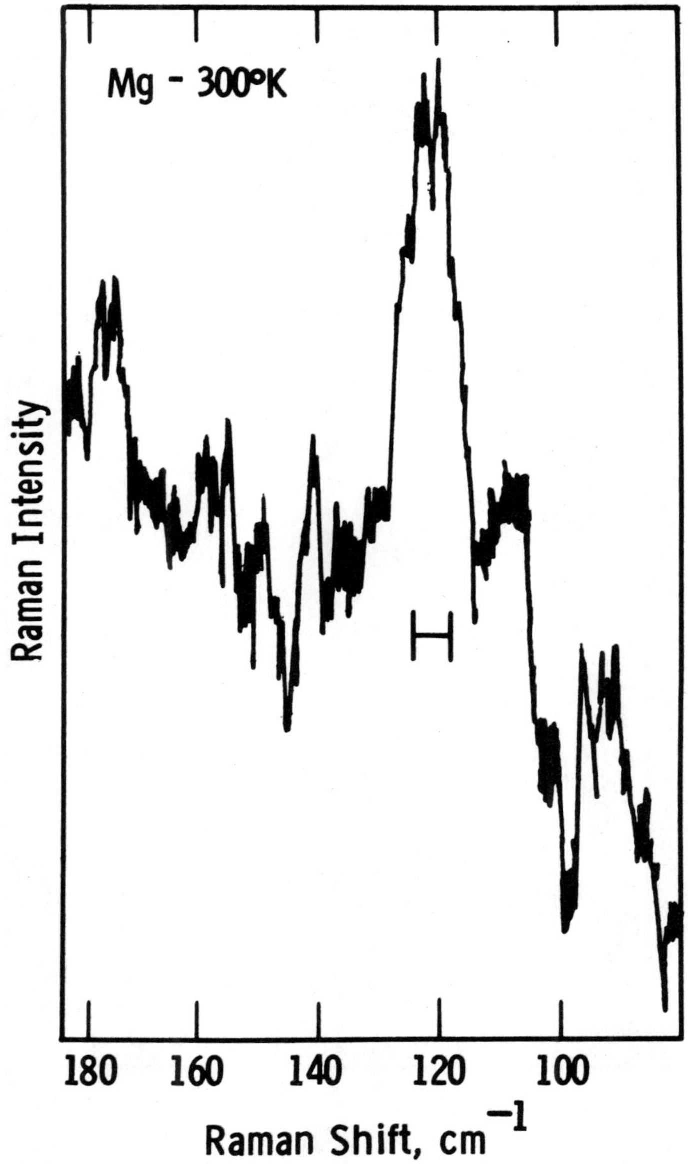

Fig. 3. Observed Raman line (Stokes) for Mg single crystal, 300°K. The horizontal line segment indicates the instrumental resolution.

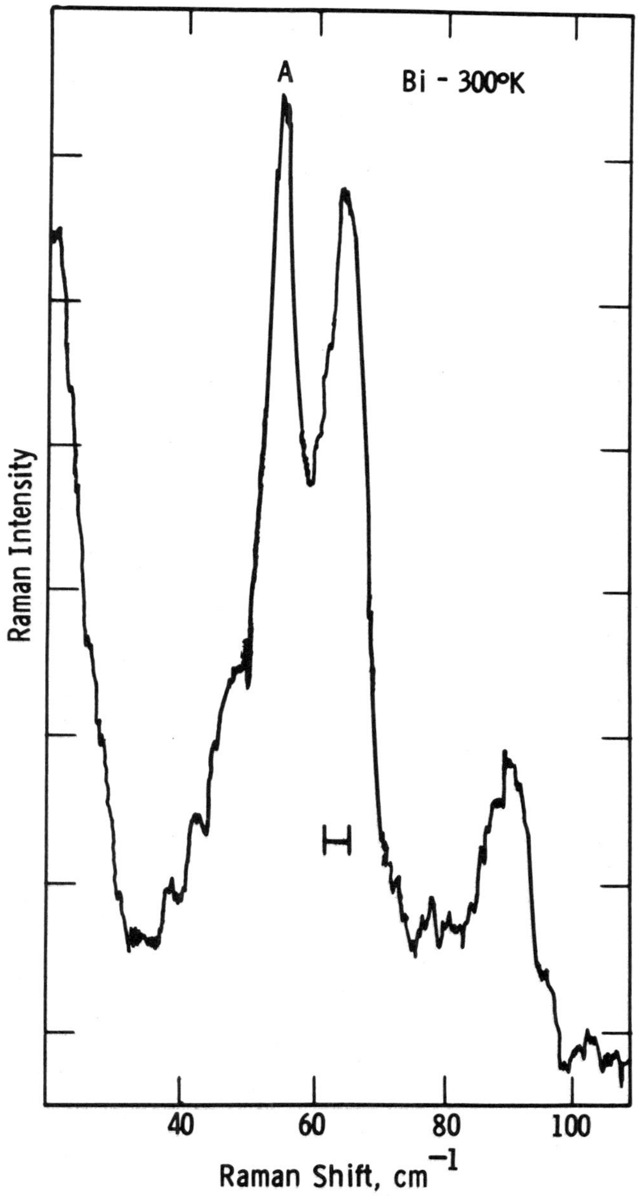

Fig. 4. Observed Raman lines (anti-Stokes) for Bi single crystal, 300°K. The line marked "A" is an argon discharge line. The horizontal line segment indicates the instrumental resolution.

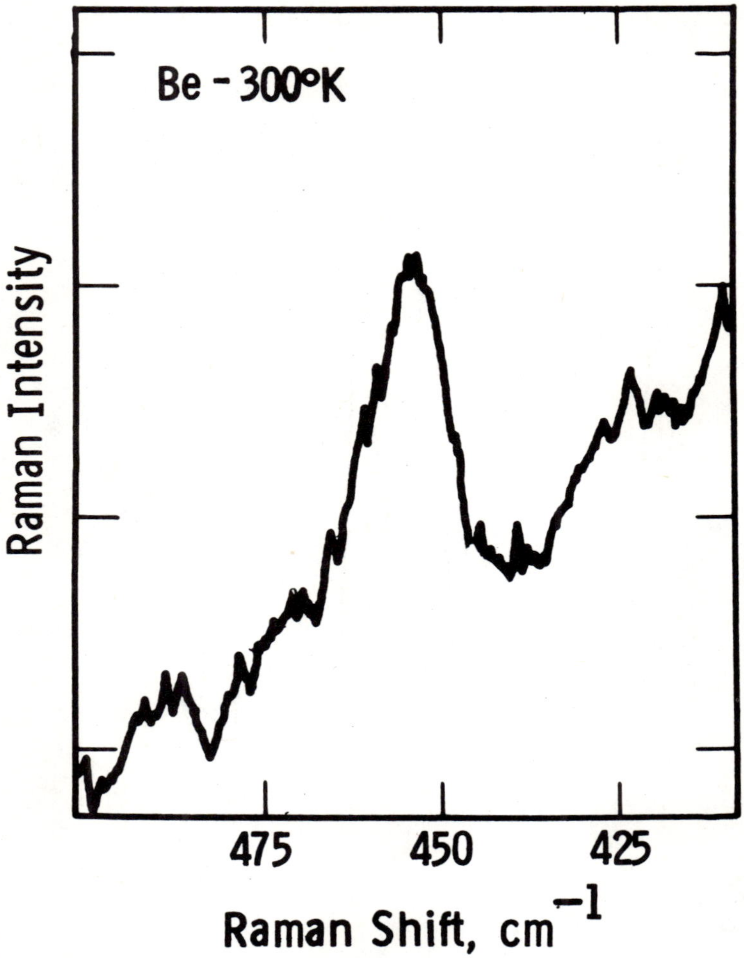

Fig. 5. Observed Raman line (Stokes) for Be, 300°K.

The metals Zn, Mg and Be are all of the hexagonal close packed structure with two atoms in the unit cell. The space group for this structure is D_{6h}^4 and there is one E_{2g} Raman active mode. This is consistent with the results for these metals. The metal Bi has a rhombohedral structure with two atoms in the unit cell and the space group is D_{3d}^5. There are two Raman active modes, A_{1g} and E_g. Fig. 4 shows the two observed lines for Bi, the line marked "A" is an argon discharge line. The alloy $AuAl_2$ is of the CaF_2 structure and has one Raman active mode, which is consistent with our results. In all of these figures, the instrumental resolution is given in each case by the horizontal line segment. Table I lists the observed frequencies and linewidths at 300°K. Also listed in Table I are frequencies that have been obtained by inelastic neutron scattering for Zn, [3] Mg, [4] Bi[5] and Be[6]. It is evident that the agreement is quite good.

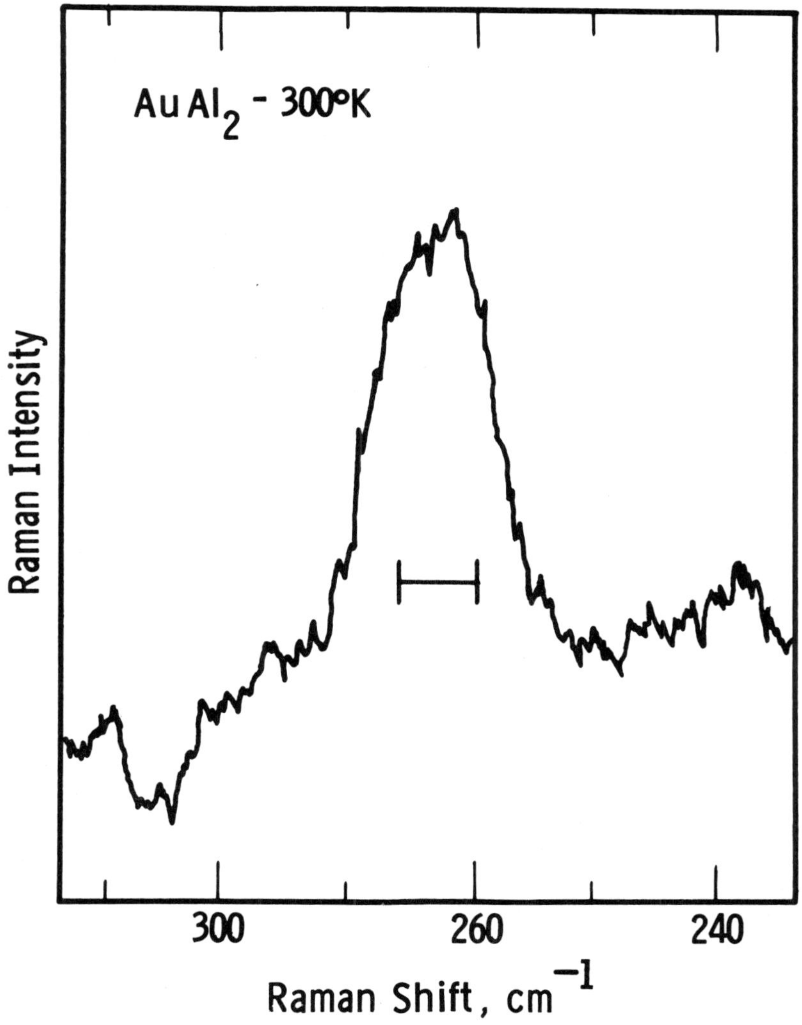

Fig. 6. Observed Raman line (Stokes) for $AuAl_2$, 300°K. The horizontal line segment indicates the instrumental resolution.

TABLE I

Raman Frequencies and Linewidths (300°K)

	ω_{Raman}	$\omega_{Neutron}$	$\Delta\omega$
Zn	70 cm^{-1}	73[(a)] cm^{-1}	11 cm^{-1}
Mg	120	120[(b)]	12
Bi	65, 90	74, 100[(c)]	8, 7
Be	455	460[(d)]	16
$AuAl_2$	266	--	28

(a) Borgonovi et al. (1963)
(b) P. K. Iyengar (1965)
(c) J. L. Yarnell et al. (1964)
(d) R. E. Schmunk et al. (1962)

The scattering intensity from Be was compared to that of Ge. The penetration depth and reflectivity of the laser light is almost identical for these two materials. The ratio of the scattering intensity for Be to that of Ge was found to be ~0. 1. The scattering intensity of the other metals did not differ by more than a factor of two to three from that of Be.

For Zn, a comparison was made between a mechanically polished face and an "as grown" face. Within the accuracy of our measurements, the results were identical for frequency and linewidth. It is therefore reasonable to assume that the linewidth is due predominantly to a finite phonon lifetime and not to inhomogeneous broadening due to strains near the surface. It is interesting to point out that $\Delta\omega/\omega$ for the metals is about a factor of ten larger than that for the elemental semiconductors, Ge and Si[7] or the insulator Al_2O_3. [8]

While frequency shifts of several cm^{-1} were observed for the metals in going from 300°K to 77°K, quantitative results for this aspect of the problem have not, as yet, been obtained.

Within the accuracy of the linewidth measurements (± 10%), the width was found to be unchanged in going from 300°K to 77°K for Zn, Bi and $AuAl_2$. There are indications that this behavior is also true for Be and Mg, but the results for these metals are still somewhat uncertain. If the phonon lifetime were due only to lattice anharmonicity, the linewidth would be expected to be strongly temperature dependent over this temperature range. One possible process that would lead to a temperature independent lifetime is for the optical phonon to decay into an electron-hole pair. Preliminary perturbation calculations for this process give results for hcp material two to three orders of magnitude too small for the linewidth. Another possible process is for the optical phonon to decay into an acoustical phonon and an electron-hole pair. A more complete study of these and related processes is in progress.

ACKNOWLEDGMENT

The authors wish to thank Dr. M. Rubenstein for providing single crystals of Zn.

REFERENCES

1. The frequencies for Be and $AuAl_2$ have been reported previously, see D. W. Feldman, James H. Parker, Jr., and M. Ashkin, Phys. Rev. Letters 21, 607 (1968).
2. The Be spectrum was obtained with the sample inside the optical cavity of the laser, see Ref. [1].
3. G. Borgonoui, G. Caglioti, and J. J. Antal, Phys. Rev. 132, 683 (1963).
4. P. K. Iyengar, G. Venkataraman, P. R. Vijayaraghavan, and A. P. Roy, Inelastic Scattering of Neutrons, IAEA Vienna, I, 153 (1965).
5. J. L. Yarnell, J. L. Warren, R. G. Wenzel, and S. H. Koening, IBM J. of Res. & Dev. 8, 234 (1964).
6. R. E. Schmunk, R. M. Brugger, P.O. Randolph, and K. A. Strong, Phys. Rev. 128, 562 (1962).
7. J. H. Parker, Jr., D. W. Feldman, and M. Ashkin, Phys. Rev. 155, 712 (1967).
8. M. Ashkin, J. H. Parker, Jr., and D. W. Feldman, Solid State Comm. 6, 343 (1968).

E-6: THEORY OF THE RAMAN EFFECT IN METALS

D. L. Mills*, A. A. Maradudin* and E. Burstein†
Department of Physics, University of California
Irvine, California

A theory of the Raman scattering of light by the optical vibration modes of polyatomic metals is presented. A microscopic theory has been developed which describes the modulation of the electronic susceptibility tensor $\chi_{\mu\nu}(\omega)$ by long wavelength optical phonons. We will discuss the various mechanisms which contribute to the change in $\chi_{\mu\nu}$ with excitation of an optical mode. We point out that in a metal, $\chi_{\mu\nu}$ may be modulated by odd parity LO phonons, since carrier density fluctuations are associated with them.

The theory of inelastic scattering of light by a semi-infinite metallic sample has been formulated. We apply the theory to describe the spectrum of the light scattered by both even and odd parity optical modes of long wavelength. An estimate of the strength of the scattering has been made for the two cases, and is compared with recent experimental results of Feldman, Parker and Ashkin.

INTRODUCTION

In recent experiments, Feldman, Parker, and Ashkin[1] have observed the first order Raman scattering of light by the optical vibrations of the elemental metal Be, and the ordered alloy $AuAl_2$. Beryllium is a close packed hexagonal crystal, with two atoms per unit cell. In this structure only the doubly degenerate E_{2_g} optical mode is Raman active. The alloy $AuAl_2$ has the CaF_2 structure, with one Raman active mode of T_{2_g} symmetry. The spectra were obtained with an argon ion laser, and only a single line was observed in each case. The intensities of the Raman lines were quite small, as would be expected,

*Supported in part by the Air Force Office of Scientific Research, Office of Aerospace Research, U.S.A.F. under AFOSR Grant No. 68-1448.

†On Sabbatical leave from the University of Pennsylvania, Philadelphia, Pa. during the 1967-68 academic year. Research supported by the U.S. Army Research Office-Durham.

since scattering takes place only within the skin depth. In fact, the intensities are apparently comparable to those observed in back-scattering from opaque semiconductors, where the skin depth is essentially the same as the skin depth in metals. This implies, of course, that the Raman scattering matrix elements in these metals are comparable to those in the semiconductors.

The mechanism responsible for the scattering of light is the modulation of the electronic susceptibility $\chi_{\mu\nu}(\omega)$ of the metal within the skin depth by the optical vibration modes. Both even and odd parity phonons can scatter light, but the scattering mechanisms in the two cases are different. In the case of Raman active even parity phonons, i.e., those whose symmetries would allow them to be Raman active in non-metallic crystals, there are two possible scattering mechanisms: (1) the change in $\chi_{\mu\nu}(\omega)$ to first order in the displacements of the atoms in a primitive unit cell; and (2) the change in $\chi_{\mu\nu}(\omega)$ due to the change in the electronic occupation numbers associated with the relative displacements of the sublattices. A microscopic theory of these mechanisms is described elsewhere[2]. Both mechanisms allow scattering of light by modes of zero wave vector. In the case of infrared active odd parity phonons, i.e. those whose symmetries allow them to exhibit a linear dipole moment, the scattering mechanism is the change in the electronic susceptibility resulting from the change in the charge density accompanying the displacements of the sublattices. In contrast with the matrix element for the scattering by even parity modes, which is independent of the magnitude of the phonon wave vector, the matrix element for scattering by the charge density fluctuations accompanying odd parity modes is proportional to the magnitude of the phonon wave vector. The susceptibilities which are being modulated contain both intra and interband contributions. In crystals lacking an inversion center, the optical phonons will have mixed parity, and all three mechanisms will contribute to the scattering. In the case of Be, there are no odd parity optical modes, so one expects scattering only by the Raman active, even parity mode mentioned above. On the other hand, the CaF_2 structure also has a triply degenerate, odd parity mode of T_{1u} symmetry which, in principle can scatter light.

In this paper, we present the theory of the inelastic scattering of light in opaque media, where the skin depth must be considered. The theory is applied to metals in the frequency range where the dielectric constant is negative by introducing phenomenological expressions for the modulation of the dielectric tensor by the even and odd parity modes. Expressions for the form of the spectrum of the scattered light are exhibited in each case. The odd parity modes give rise to an asymmetric line, while the even parity modes give a symmetric line. The fact that the line observed in $AuAl_2$ is apparently symmetric indicates that the line is associated with the even parity mode.

SCATTERING OF LIGHT BY OPAQUE SURFACES

We consider the scattering of light from the surface of a semi-infinite, opaque medium. Let the surface lie in the x-y plane, with the material filling the lower half space $z < 0$. We assume the dielectric constant of the material has the form for $z < 0$

$$\epsilon_{\mu\nu}(\underset{\sim}{x}\, t) = \epsilon_0 \, \delta_{\mu\nu} + \delta\, \epsilon_{\mu\nu}(\underset{\sim}{x}\, t) \tag{1}$$

The quantity $\delta\,\epsilon_{\mu\nu}(\underset{\sim}{x}\, t)$ is the change in dielectric constant induced by the presence of a phonon. The constant ϵ_0 is the electronic contribution to the dielectric constant of the undeformed crystal. For a metal, when the frequency of the incident light Ω_0 is below the plasma frequency ω_p, the real part of ϵ_0 is negative. For the incident frequency in the visible range, one has $\Omega_0 \tau >> 1$ in the usual case (i.e. for frequencies not in the range of interband transitions), where τ is the electron relaxation time. In this limit, the imaginary part of ϵ_0 is small compared to the real part. For an opaque semiconductor, the imaginary part of ϵ_0 will generally be large, and comparable in magnitude to the real part. In both cases, the wave vector of light $k = k_1 + ik_2$ propagating in the medium will have a large imaginary part k_2 resulting in a skin depth $\delta = 2\pi/k_2$. In Eq. (1), we have assumed that the crystal is isotropic. This is a convenient, but unessential assumption.

We expand the part $\delta\epsilon_{\mu\nu}(\underset{\sim}{x}\, t)$ in powers of the phonon amplitude. To discuss the first order Raman scattering, we only need to retain the lowest term in the expansion. If $Q_{n\underset{\sim}{q}}$ is the normal coordinate of the normal mode $n\underset{\sim}{q}$ of the crystal, then we write

$$\delta\epsilon_{\mu\nu}(\underset{\sim}{x}\, t) = \sum_{n\underset{\sim}{q}} \frac{\partial\epsilon_{\mu\nu}}{\partial Q_{n\underset{\sim}{q}}} Q_{n\underset{\sim}{q}}(t) \exp(i\, \underset{\sim}{x}\cdot\underset{\sim}{q}) + \text{c.c.} \tag{2}$$

The scattering cross section may be expressed in terms of the vector potentials of the incoming and outgoing scattered radiation. The vector potential $\underset{\sim}{A}(\underset{\sim}{x}, t)$ may be determined from Maxwell's equations. One may write Maxwell's equations in integral form by introducing a suitable Green's matrix $G_{\alpha\beta}(\underset{\sim}{x}\underset{\sim}{x}';t-t')$ as follows:

$$A_\alpha(\underset{\sim}{x}\, t) = A_\alpha^{(0)}(\underset{\sim}{x}\, t) + \left(\frac{\Omega_0}{c}\right)^2 \sum_{\beta\gamma} \int d^3x'dt' G_{\alpha\beta}(\underset{\sim}{x}\underset{\sim}{x}';t-t')\delta\epsilon_{\beta\gamma}(\underset{\sim}{x}'t')A_\gamma(\underset{\sim}{x}'t') \tag{3}$$

where the Green's matrix satisfies

$$\sum_\gamma \left\{ [\underset{\sim}{\nabla} \times (\underset{\sim}{\nabla} \times]_{\alpha\gamma} + \delta_{\alpha\gamma} \frac{\epsilon_0(z)}{c^2} \frac{\partial^2}{\partial t^2} \right\} G_{\gamma\beta}(\underset{\sim}{x}\underset{\sim}{x}', t-t') = \delta_{\alpha\beta}\, \delta(\underset{\sim}{x} - \underset{\sim}{x}')\, \delta(t-t') \tag{4}$$

The function $A_\alpha^{(0)}$ is the solution to the Maxwell equations when $\delta\epsilon_{\mu\nu} = 0$. It describes the simple specular reflection of light from the surface. The Green's matrix $G_{\alpha\beta}(\underset{\sim}{x}\underset{\sim}{x}';t-t')$ has the outgoing boundary condition of the scattering problem built into it. In Eq. (3) and Eq. (4), we have assumed the light frequency Ω_0 large compared to the phonon frequencies.

In Eq. (3), if $\underset{\sim}{A}(\underset{\sim}{x}', t')$ is the vector potential at point $\underset{\sim}{x}'$ and time t' inside the medium, then $\delta\epsilon_{\beta\gamma} A_\gamma$ is the amplitude of the modulated wave at this point. The Green's matrix $G_{\alpha\beta}(\underset{\sim}{x}\underset{\sim}{x}';t-t')$ is a transfer function, which gives one the amplitude of the contribution to the scattered radiation at point $(\underset{\sim}{x}, t)$ from the source $\delta\epsilon_{\beta\gamma}(\underset{\sim}{x}'t')A_\gamma(\underset{\sim}{x}'t')$.

The amplitude of the scattered wave to first order in the phonon amplitude may be found from Eq. (3) by inserting the function $A_\gamma^{(0)}$ into the integral on the right hand side. This is identical to the first Born approximation of quantum mechanical scattering theory.

It is convenient to utilize the translational invariance of the problem in two directions parallel to the surface by introducing the partial Fourier transform for the Green's matrix

$$G_{\alpha\beta}(\underset{\sim}{x}\underset{\sim}{x}';t-t') = \int \frac{d^2k_{||}d\Omega}{(2\pi)^3}\, g_{\alpha\beta}(\underset{\sim}{k}_{||}\,\Omega, zz')\, e^{i\underset{\sim}{k}_{||}\cdot(\underset{\sim}{x}-\underset{\sim}{x}')} e^{i\Omega(t-t')} \tag{5}$$

with a similar transformation applied to the other quantities. Here $\underset{\sim}{k}_{||}$ is the two dimensional wave vector $\underset{\sim}{k}_{||} = (k_x, k_y, 0)$, where k_x and k_y are real.

It will be useful to discuss briefly the kinematics of the scattering process. Consider the interaction of the light with a phonon of wave vector $\underset{\sim}{q}$. In the presence of the opaque surface, the z component of wave vector (normal to the surface) is not conserved. However, the component of wave vector parallel to the surface is conserved. Let $\underset{\sim}{k}_S$ and $\underset{\sim}{k}_I$ be the wave vectors of the scattered and incident light outside the medium, which are real vectors. By consideration of the equations that describe conservation of energy and wave vector, one finds that light scattered in a given direction $\hat{k}_s$ contains contributions from the interaction of the incident light with phonons with a range of values of q_z, but with wave vector component $\underset{\sim}{q}_{||} = \underset{\sim}{k}_{||}^{(s)} - \underset{\sim}{k}_{||}^{(I)}$ parallel to the surface. This is true in the limit that the photon frequency is large compared to phonon frequencies. This remark simplifies construction of the cross section.

To find the amplitude of the scattered wave, one requires $g_{\alpha\beta}(\underset{\sim}{k}_{||}\,\Omega\,;zz')$ for z outside the medium, and z' inside. For this case $g_{\alpha\beta}$ may be written

$$g_{\alpha\beta}(\underset{\sim}{k}_{||}\,\Omega\,, zz') = \left(\frac{\Omega_0}{c}\right) g_{\alpha\beta}\,(\underset{\sim}{k}_{||}\,\Omega)\,\exp\,(i\,k_z\,z)\,\exp\,(i\,k_{z'}z')$$

$$k_{z'}^2 = \epsilon_0\,(\Omega/c)^2 - k_{||}^2 \tag{6a}$$

$$k_z^2 = (\Omega/c)^2 - k_{||}^2 \tag{6b}$$

We now discuss the form of the scattered flux $\phi^{(s)}$. The expression for the flux scattered per unit solid angle per unit frequency is obtained by computing the outgoing flux/unit solid angle from the vector potential given by Eq. (3), and then extracting the intensity of the radiation that has undergone a frequency shift between ω and $\omega + d\omega$. We find[2]

$$\frac{d^2\phi^{(S)}}{d\Omega d\omega} = \frac{cV}{2(2\pi)^4}\left(\frac{\Omega_0}{c}\right)^4 \cos\theta_S \sum_n \int \frac{dq_z}{2\pi}$$

$$\left[\left(k^S_{z'2} + k^I_{z'2}\right)^2 + \left(k^S_{z'1} - k^I_{z'1} - q_z\right)^2\right]^{-1} \frac{\Gamma_n}{\Gamma_n^2 + (\omega - \omega_n(q))^2}$$

$$\times \sum_\alpha \Big| \sum_{\beta\gamma} \sum_\lambda g_{\alpha\beta}(\underset{\sim}{k}^S_{||}\,\Omega_0) \frac{\partial \epsilon_{\beta\gamma}}{\partial Q_{n\underset{\sim}{q}}} \Gamma^\lambda_\gamma(\underset{\sim}{k}_I)\, E^{(0)}_\lambda \Big|^2 < Q^*_{n\underset{\sim}{q}} Q_{n\underset{\sim}{q}} > \tag{7}$$

In this result V is the crystal volume, θ_S the angle between $\underset{\sim}{k}^{(S)}$ and the normal to the surface, $k^I_{z'1} + i\,k^I_{z'2}$ and $k^S_{z'1} + i\,k^S_{z'2}$ are the wave vectors of the incident and scattered radiation in the medium, and Γ_n the intrinsic width of the phonon of frequency $\omega_n(\underset{\sim}{q})$. $E^{(0)}_\lambda$ is the amplitude of the component of the incident electric field of polarization λ, and $\Gamma^\lambda_\gamma(\underset{\sim}{k}_I)$ is a "transfer function" which gives the amplitude of the incident field in the metal, when $E^{(0)}_\lambda$ assumes the value of unity.

In Eq. (9), the factor $\Gamma_n / \{\Gamma_n^2 + (\omega - \omega_n(\underset{\sim}{q}))^2\}$ is the strength of the spectral weight function of the mode $(n\underset{\sim}{q})$ at the frequency ω. The integral

$$\int \frac{dq_z}{2\pi}\left[\left(k^S_{z'2} + k^I_{z'1}\right)^2 + \left(k^S_{z'1} - k^I_{z'1} - q_z\right)^2\right]^{-1}$$

is the scattering coherence length. In the limit as the imaginary parts $k^S_{z'2}$ and $k^I_{z'2}$ approach zero, one recovers the Bragg condition for the components of wave vector normal to the surface, since the integrand peaks strongly when $q_z = k^S_{z'1} - k^I_{z'1}$.

When the skin depth δ is << the light wavelength λ, the incident field produces an electric field inside the metal nearly parallel to the surface, regardless of the incident polarization. Similarly, only the component parallel to the surface of the scattered radiation is transmitted through with efficiency. Thus, when $\delta << \lambda$, one obtains information about those components of $(\partial \chi_{\mu\nu}/\partial Q_n)$ for which both μ and ν refer to directions parallel to the surface. For example, if the incident field is plane polarized in the plane of incidence (the x-z plane), one finds to lowest order in (δ/λ),

$$\underset{\sim}{E}^{(S)} \alpha \left[(\partial \chi_{xx}/\partial Q_0)(\hat{y} \times \hat{k}^{(S)}) - (\partial \chi_{yx}/\partial Q_0)(\hat{x} \times \hat{k}^{(S)})\right]$$

RAMAN SCATTERING BY EVEN PARITY PHONONS IN METALS

We apply the result of Eq. (7) to the discussion of Raman scattering by even parity modes in metals.

In insulators, $(\partial \epsilon_{\mu\nu}/\partial Q_{nq})$ arises from deformation potential effects, i.e. a shift in position of the ions changes the shape and relative positions of the energy bands, as well as the Bloch wave functions. In metals, in addition to these contributions, one has contributions from intra-band processes. Also, changes in electronic occupation numbers corresponding to redistribution of the electrons near the Fermi surface may occur. The dielectric tensor $\epsilon_{\mu\nu}$ will be therefore modulated by these occupation number changes. Thus a number of distinct processes make contributions to $(\partial \epsilon_{\mu\nu}/\partial Q_{nq})$. Of course, each contribution to the Raman tensor has the same symmetry properties.

For an even parity mode of long wavelength, the quantity $(\partial \epsilon_{\mu\nu}/\partial Q_{nq})$ that appears in the integral of Eq. (7) is insensitive to the wave vector of the phonon. Thus this quantity, as well as $g_{\alpha\beta}$ and $\Gamma_{\gamma}^{\lambda}$ may be removed from the integration.

Since the coherence length factor decreases rapidly with q_z for q_z larger than the inverse skin depth, the spectral weight factor, which varies slowly in this region, may be replaced by its value at $\underset{\sim}{q} = 0$ and removed from the integral, i.e. we replace Γ_n and $\omega_n(\underset{\sim}{q})$ by their values Γ_0 and ω_{op} at $\underset{\sim}{q} = 0$.

Furthermore, for frequencies below the plasma frequency, as we mentioned above, the dielectric constant ϵ_0 may be taken to be real and negative. Then $k^S_{z'1} = k^I_{z'1} = 0$, within this approximation. The integral over q_z is then easily evaluated. We obtain

$$\frac{d^2\phi_S}{d\Omega d\omega} = \frac{cV}{4(2\pi)^3}\left(\frac{\omega_0}{c}\right)^4 \frac{\cos\theta_S}{\left(k^S_{z'2} + k^I_{z'2}\right)} \frac{\Gamma_o}{\Gamma_o^2 + (\omega-\omega_{op})^2}$$

$$\times \sum_{\alpha} \Big| \sum_{\beta\gamma} \sum_{\lambda} g_{\alpha\beta} \frac{\partial \epsilon_{\beta\gamma}}{\partial Q_0} \Gamma_{\gamma}^{(\lambda)} E_{\lambda}^{(0)} \Big|^2 < Q_o^* Q_o > \tag{8}$$

The Raman line is a symmetric line, centered about ω_{op} with half width Γ_0. Also, as Ω_0 approaches ω_p, and the skin depth increases, the factor $(k^S_{z'2} + k^I_{z'2})$ approaches zero, so the cross section is enhanced. Thus, one can greatly increase the cross section by working with frequencies near the reflection minimum, i.e. near ω_p.

We now consider the specific form of the Raman tensor $(\partial \chi_{\mu\nu}/\partial Q_0) = (4\pi)^{-1} (\partial \epsilon_{\mu\nu}/\partial Q_0)$ for the even modes in the hexagonal close packed structure. Upon displacing the sublattice κ by the amount $u_\alpha(\kappa)$, the change in the electronic polarizability $\chi_{\mu\nu}$ is

$$\delta \chi_{\mu\nu} = V_c \sum_{\alpha k} \chi_{\mu\nu,\alpha}(\kappa)\, u_\alpha(\kappa)$$

where V_c is the volume of the unit cell. Since the tensor $\chi_{\mu\nu}$ is left unchanged if all the sublattices are translated by the same amount, it follows that $\chi_{\mu\nu,\alpha}(1) = -\chi_{\mu\nu,\alpha}(2)$ for a crystal with two atoms per unit cell. When symmetry considerations are applied to the hcp lattice, one finds that only a single independent element of $\chi_{\mu\nu,\alpha}(\kappa)$ is non-vanishing. Taking the c-axis of the crystal parallel to z, we find that in Loudon's notation[3], the tensor $\chi_{\mu\nu,\alpha}$ is given by

$$\begin{pmatrix} a & 0 & 0 \\ 0 & -a & 0 \\ 0 & 0 & 0 \end{pmatrix} \quad \begin{pmatrix} 0 & 0 & -a \\ -a & 0 & 0 \\ 0 & 0 & 0 \end{pmatrix} \quad \begin{pmatrix} 0 & 0 & 0 \\ 0 & 0 & 0 \\ 0 & 0 & 0 \end{pmatrix}$$

i.e. $\chi_{xx,x} = -\chi_{yy,x} = -\chi_{xy,y} = -\chi_{yx,y} \equiv a,$

with all other $\chi_{\mu\nu,\alpha} \equiv 0$. This result indicates that the mode with displacements parallel to the c-axis does not contribute to the first order spectrum. This result is to be expected from physical considerations. There exists a reflection plane for this structure normal to the c-axis. Displacement of the sublattices parallel to the c-axis thus can not modulate $\chi_{\mu\nu}$, to first order in the displacements, i.e. a positive and a negative displacement by symmetry must produce the same effect on $\chi_{\mu\nu}$.

When this form of $\chi_{\mu\nu,\alpha}$ is combined with the observation above that the field in the metal is nearly parallel to the surface when $\delta \ll (c/\Omega_0)$, polarization studies would be of interest. If the c-axis is parallel to the surface, then no scattering occurs when either the incident or the scattered radiation is polarized parallel to the c-axis. The scattering intensity is a maximum when the incident and scattered light is polarized at right angles to the c-axis. One can observe the variation of the intensity of the scattering by rotating the crystal about the normal to the surface, keeping the incident and scattered beam directions and polarizations fixed. When the c-axis is normal to the surface, the intensity is independent of the polarizations, and orientation of the crystal relative to the plane of scattering.

The magnitude of the scattering efficiency may be estimated from the approximate relation

$$\frac{1}{\phi_I} \frac{d\phi_s}{d\Omega} \cong 2\pi a^2 \delta \left(\frac{\Omega_0}{c}\right)^4 \left(\frac{\hbar}{V_c \mu \omega_{op} |\epsilon_0|^2}\right)$$

where μ is the reduced mass of the unit cell. As in the case of semiconductors, it is difficult to estimate the numerical value of the quantity a, since this involves knowledge of the wave functions and energy band structure. Since the observed scattering efficiency in Be is similar to that in the opaque semiconductors, the polarizability derivative a is comparable to that in semiconductors. (The quantity a is frequency dependent, so it may be larger or smaller for different excitation frequencies, in a given material.)

RAMAN SCATTERING BY ODD PARITY PHONONS IN METALS

We next consider the scattering of light by odd parity phonons by again employing the general result of Eq. (7).

The ionic motion associated with an odd parity LO phonon generates a macroscopic electric field. This field is screened out by the conduction electrons. Thus one has macroscopic fluctuations in conduction electron density associated with excitation of a long wavelength LO phonon of odd parity. Since the dielectric function depends on the electron density, the presence of such an LO phonon changes the dielectric tensor $\epsilon_{\mu\nu}$. The scattering of light by carrier density fluctuations in semiconductors has been investigated theoretically by McWhorter[4], and observed experimentally by Mooradian and Wright[5].

There are two contributions to the change in $\epsilon_{\mu\nu}$ induced by an odd parity optical mode. One arises from intra-band transitions, and one from inter-band transitions. The intra-band contribution may be treated in a simple manner. The $\underset{\sim}{k} = 0$ dielectric function of the electron gas contains a term $-\delta_{\mu\nu}(\omega_p/\Omega)^2$, where $\omega_p^2 = (4\pi ne^2/m^*)$ is the plasma frequency of the electrons. Since this term is proportional to the electron density n, this term is changed upon excitation of an odd parity LO phonon. If δn_q is the amplitude of the density fluctuation, then $\delta\epsilon_{\mu\nu}$ is given by

$$\delta\epsilon_{\mu\nu} = -\delta_{\mu\nu}(\omega_p/\Omega)^2(\delta n_q/n) \tag{8}$$

The contribution to $\delta\epsilon_{\mu\nu}$ from inter-band transitions arises because a change in electron density alters the Fermi occupation factors, and hence changes that portion of the inter-band contribution to $\epsilon_{\mu\nu}$ in which either the initial or final state is in the conduction band. The magnitude of this contribution depends on the details of the band structure of the material, and also on frequency. For example, as in the case of semiconductors, the frequency dependence of $\delta\epsilon_{\mu\nu}$ may exhibit a resonant enhancement when Ω_o is near an inter-band transition.

Thus the total change in $\epsilon_{\mu\nu}$ upon excitation of an odd parity LO phonon has the form

$$\frac{\partial\epsilon_{\mu\nu}}{\partial Q_{LO,\underset{\sim}{q}}} = \left(\frac{\partial\epsilon_{\mu\nu}^{\text{intra}}}{\partial n_q} + \frac{\partial\epsilon_{\mu\nu}^{\text{inter}}}{\partial n_q}\right)\left(\partial n_q/\partial Q_{LO,q}\right)$$

These two contributions will have different frequency dependences, and may have different signs.

Let $u(\underset{\sim}{q}\varkappa)$ be the amplitude of the displacement of sublattice $\varkappa$. Then in the long wavelength limit δn_q is given by

$$\delta n_q = \frac{i}{(VV_c)^{1/2}}\sum_{\varkappa} z_{\varkappa}\,\underset{\sim}{q}\cdot\underset{\sim}{u}(\underset{\sim}{q}\varkappa) \tag{9}$$

where $z_{\varkappa}$ is the effective ionic charge of sublattice $\varkappa$, measured in units of the electronic charge e.

We now proceed to obtain the form of the spectrum of scattered light from Eq. (7). The main difference between the present case, and the discussion of Raman scattering by even parity modes is that δn_q, and consequently $(\partial\epsilon_{\mu\nu}/\partial Q_{LO,q})$ are proportional to the wave vector q of the (odd parity) scattering phonon. This means that $(\partial\epsilon_{\mu\nu}/\partial Q_{LO,\underset{\sim}{q}})$ may no longer be extracted from the integral over q_z. From the structure of the integral in Eq. (7), it is clear that the important values of q_z are much greater than (Ω_0/c), since $k^I_{z'2}$ and $k^S_{z'2}$ are much greater than this quantity. Thus the factor of q^2 that comes from the square of $(\partial\epsilon_{\mu\nu}/\partial Q_{LO,q})$ may be replaced by q_z^2, to a good approximation. The integrand then becomes proportional to $\{ q_z^2/[(k^I_{z'2} + k^S_{z'2})^2 + q_z^2] \}$ $\{\Gamma_n/[\Gamma_n^2 + (\omega - \omega_n(\tilde{q}))^2]\}$ in the important region of the integration. Let us suppose we consider frequency shifts ω near the q = 0 longitudinal phonon frequency ω_{LO}. For large q_z, the first factor (the coherence factor multiplied by q_z^2) approaches unity. However, because of dispersion in the phonon branch, as q_z increases, $\omega_n(\tilde{q})$ changes, until $|\omega - \omega_n(\tilde{q})|$ becomes large compared to typical values of Γ_n. The second spectral weight factor thus falls off with increasing q_z. If one assumes $\omega_n(\tilde{q})$ varies quadratically with q near q = 0, the resulting integral converges, since the integrand falls off sufficiently rapidly with increasing q_z. We have evaluated the integral for an isotropic phonon branch, with $\omega_n(\tilde{q}) = \omega_{LO}\,[1 - \beta\,(\pi q/a_0)^2]$. Here a_0 is the lattice constant, and β is a dimensionless parameter that gives a measure of the curvature of the phonon branch near q = 0. The parameter β may be either positive or negative, and one expects its value will be near unity for most metals. For a two sublattice crystal, when the incident light is normally incident of the surface and the scattered light outside the metal is observed in the direction $\hat{k}_S$, we find for the Stokes component an expression of the following form for the Raman efficiency per unit solid angle, per unit frequency:

$$\frac{1}{\phi_I}\,\frac{d^2\phi_S}{d\Omega\, d\omega} = \frac{\sqrt{2}}{8\pi^2}\left(\frac{\Omega_0}{c}\right)^4\left(\frac{\partial\epsilon}{\partial n_q}\right)^2\left(\frac{\hbar}{V_c\mu\omega_{LO}^2}\right)\left(z_A\frac{\mu}{M_A} - z_B\frac{\mu}{M_B}\right)^2\frac{(1+\bar{n}_{LO})\,f(\hat{k}_s)}{a_0|\beta|(1+|\epsilon_0|)^2}\,g(\omega) \qquad (10)$$

In this expression ϕ_I is the incident flux, $f(\hat{k}_s)$ is of order unity in magnitude, and the function $g(\omega)$ gives the shape of the spectrum of the scattered light. A plot of $g(\omega)$ is given in Fig. 1. Also, in Eq. (10), Γ_0 is the intrinsic width of the q = 0 LO phonon. We have replaced Γ_n by Γ_0 in evaluating the integral.

From Fig. 1, one can see that $g(\omega)$ is strongly asymmetric. The sense of the asymmetry of g provides a measure of the sign of the curvature of the phonon branch near q = 0. If the curvature is positive (β< O), the steep side of the scattered band will be on the low frequency side. The converse is true if the curvature is negative (β> O). The line reported[1] in $AuAl_2$ appears nearly symmetric. This suggests the scattering is from an even parity mode. However, it is difficult to extract the line from the frequency dependent background in an unambiguous manner, so that this assignment is somewhat uncertain.

We can estimate the strength of the scattering from the intra-band contribution to $\partial\epsilon/\partial n_q$ by employing Eqs. (8) and (10). If the intra-band and inter-band contributions to $(\partial\epsilon/\partial n_q)$ have the same sign, this provides an estimate of the lower limit of the scattering efficiency. To estimate the integrated strength of the peak in Fig. 1, we note that the maximum value of g is $\sim (\omega_{LO}/\Gamma_0)^{1/2}$, while the half width of the peak is the order of 2-3 Γ_0. Then for the intraband contribution to $\partial\epsilon/\partial n_q$, the scattering efficiency per unit solid angle is estimated to be the order of

$$\frac{1}{\phi_I}\frac{d\phi_S}{d\Omega} \sim 15\,\frac{(r^*)^2}{a_0}\left(\frac{\hbar}{V_c\omega_{LO}\mu}\right)\left(\frac{\Gamma_0}{\omega_{LO}}\right)^{1/2}\frac{1}{(1+|\epsilon_0|)^2}$$

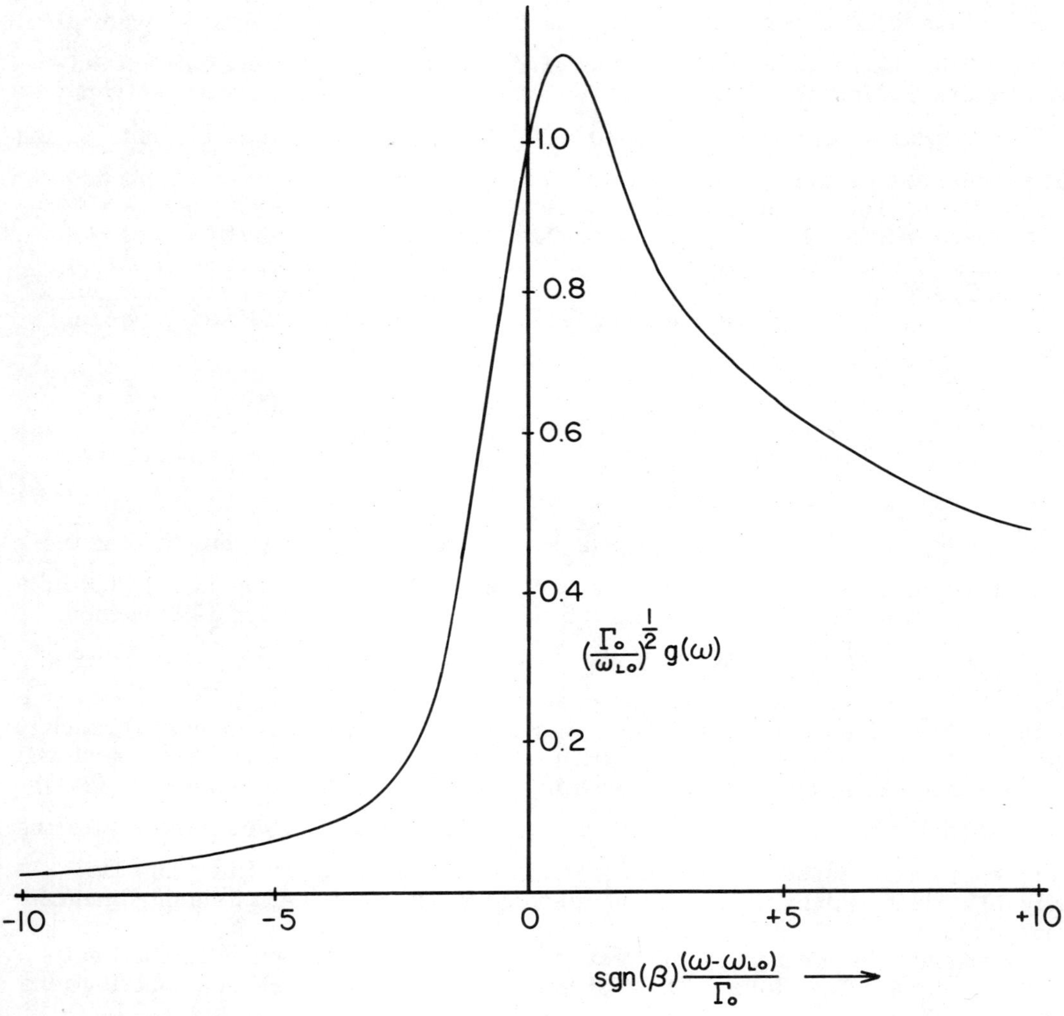

Fig. 1. The shape function g (w) that describes the spectrum of light scattered by odd parity optical phonons.

where $r_* = (e^2/m^*c^2)$ is the classical radius of an electron with effective mass m*. If we take $r_* = 2.5 \times 10^{-13}$ cm., $a_o = 3$A, $V_c = 10^{-23}$ cm.^{3}, $\mu = 40 \times 10^{-24}$ gm., $\omega_{LO} = 200$ cm^{-1}., $(\Gamma_o/\omega_{LO}) = 0.05$, and $(\delta/\lambda) = 0.1$, we find

$$\frac{1}{\phi_I}\frac{d\phi_S}{d\Omega} \cong 10^{-13}$$

This estimate indicates scattering from phonon induced density fluctuations in metals is sufficiently strong to be observed.

CONCLUSIONS

We have presented a theory of the Raman effect in metals which can serve as a basis for the interpretation of the experimental data. At the moment, there are relatively few observations of Raman scattering in metals - only one study has been reported in the literature[1]. Work is now underway in several laboratories, so we can expect a rapid expansion of this area of Raman spectroscopy in the future. The availability of a variety of laser sources should make it possible to obtain spectra in frequency regions where the scattering efficiencies are considerably larger, i.e. near the plasma frequency ω_p. It would be interesting to carry out experiments near the onset of interband transitions, especially when these occur close to the plasma frequency. The theory developed in the text may be applied in a straight-forward manner to a description of Brillouin scattering in metals.

ACKNOWLEDGEMENTS

We are grateful to Dr. J.H. Parker, Jr. for providing us his data prior to publication.

REFERENCES

1. D.W. Feldman, J.H. Parker, Jr., and M. Ashkin, Phys. Rev. Letters.
2. D.L. Mills, E. Burstein, and A.A. Maradudin (to be published).
3. R. Loudon, Advances in Physics 13, 423 (1964).
4. A. McWhorter, "Proc. Int. Conf. on Physics of Quantum Electronics," p. 111, McGraw-Hill Book Co., New York, 1966.
5. A. Mooradian and G.B. Wright, Phys. Rev. Letters 16, 999 (1966); A. Mooradian and A. McWhorter, Phys. Rev. Letters 19, 849 (1967).

E-7: ELECTRIC FIELD INDUCED RAMAN SCATTERING IN CRYSTALS

J. M. Worlock
Bell Telephone Laboratories, Incorporated
Holmdel, New Jersey

INTRODUCTION

In this paper, we wish to investigate the consequences, for Raman scattering phenomena, of imposing uniform external electric fields on samples of crystalline matter. In general, when a field is applied, space group operations are suppressed, and the crystalline symmetry is reduced. When this happens, energy states are mixed and shifted, and Raman selection rules are relaxed. This subject can be conveniently divided into two parts; the qualitative discussion of which states are mixed and which selection rules are altered by the external field, being mostly group theoretical; and the quantitative discussion of the magnitudes of the induced scattering cross sections. We shall discuss these two aspects of the problem, as applied to the particular crystals we have studied, $KTaO_3$ and $SrTiO_3$.

SYMMETRY CONSIDERATIONS

Both $KTaO_3$ and $SrTiO_3$ crystallize in the cubic perovskite structure* for which a unit cell is shown in Fig. 1. These crystals are well known to possess large temperature dependent dielectric constants, which makes them especially responsive to electric fields, and suitable for these experiments.

Also shown in Fig. 1 is a cube with an arrow through it. The cube represents the factor group, O_h, which applies to the cubic perovskites. The arrow, pointing along (001) or z, represents the uniform electric field.

The union of the group of the cube, O_h, and the group of the arrow, $C_{\infty v}$, is the group applicable to the crystal in the electric field. With the field oriented along 001, the union is the group C_{4v}, as shown in the accompanying table.

*We can ignore the 110°K phase transition in $SrTiO_3$, which has no discernible effect on the infrared active phonons.

OBJECT	FACTOR GROUP	OPERATIONS									
Cube	O_h	E	$6C_4$	$3C_2$	$6C_2'$	$8C_3$	i	$6S_4$	3σ	$3\sigma'$	$8S_3$
Cube with Arrow	C_{4v}	E	$2C_4$ (001)	C_2 (001)	-	-	-	-	2σ (100) (010)	$2\sigma'$ (110) $(1\bar{1}0)$	-

It is obviously possible to extend this kind of analysis to other structures and other field directions.

The uniform modes, or long wavelength phonons, in the cubic perovskites, have symmetry F_{1u} or F_{2u}, and these symmetries are changed in the presence of the (001)-directed field as shown in Fig. 2. The infrared active F_{1u} modes, of which there are three, split into A_1 and E components, while the silent mode F_{2u} splits into B_1 and E components. In addition, of course, the infrared active modes are split into longitudinal and transverse components by the influence of the internal fields they carry with them.

The experimental arrangement corresponding to this analysis is shown in the same figure. It is important to specify the phonon propagation direction, for in this arrangement, the A_1 modes are transverse, while in another configuration they could be made longitudinal.

The Raman tensor elements shown in the last column are, in the low field limit, proportional to the electric field E. This gives us a scheme for obtaining the field induced scattering, while discriminating against the second order background. For example $\alpha_{zz} = \lambda E$. Hence the induced electric moment which radiates scattered light is

$$M_z \propto \alpha_{zz} E_z(\text{optical}) = \lambda E E_z(\text{optical}) \tag{1}$$

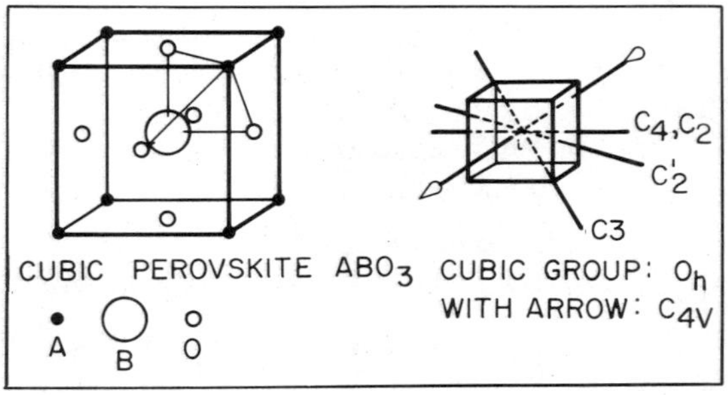

Fig. 1. Unit cell of cubic perovskite crystal O_h^1 (Pm3m) and cube representing factor group O_h with arrow representing electric field in (001) direction.

Each of the quantities in this expression has its characteristic frequency. λ oscillates at the phonon frequency ω, E_z(optical) at the incident light frequency ω_ℓ, and E varies at some low frequency, ω_o, assuming a sinusoidal field $E = E_o \sin \omega_o t$. The dipole moment therefore has components at four frequencies $\omega_\ell \pm \omega \pm \omega_o$. The doublets at $(\omega_\ell \pm \omega)$ split by $2\omega_o$ are too close to be resolved optically, but by their beat give rise to an amplitude modulated current in a photomultiplier: $I \propto E_o^2 \sin 2\omega_o t$. In our experiments we have used a phase sensitive detector tuned to $2\omega_o$, the harmonic of the applied field frequency. Results are shown in Fig. 3, for $KTaO_3$ at 80°K. The intrinsic second order scattering in Fig. 3(a) is quite completely removed in Fig. 3(b) since it is not modulated by the electric field. The phonon made visible in 3(b) is the low frequency infrared active TO phonon whose properties we have studied extensively[1].

ELECTRIC FIELD INDUCED RAMAN SCATTERING

MODE TYPE	POLARIZATION	REPRESENTATION O_h	REPRESENTATION C_{4v}	SPLITTING	RAMAN TENSOR
IR TO	001		A_1	EXTERNAL FIELD	$\alpha_{XX} = \alpha_{YY}, \alpha_{ZZ}$
IR TO	110	F_{1u}	E		α_{XZ}, α_{YZ}
LO	$\bar{1}10$		E	INTERNAL FIELD	α_{XZ}, α_{YZ}
SILENT	001		B_1	EXTERNAL FIELD	$\alpha_{XX} = -\alpha_{YY}$
SILENT	110	F_{2u}	E		α_{XZ}, α_{YZ}
SILENT	$\bar{1}10$				

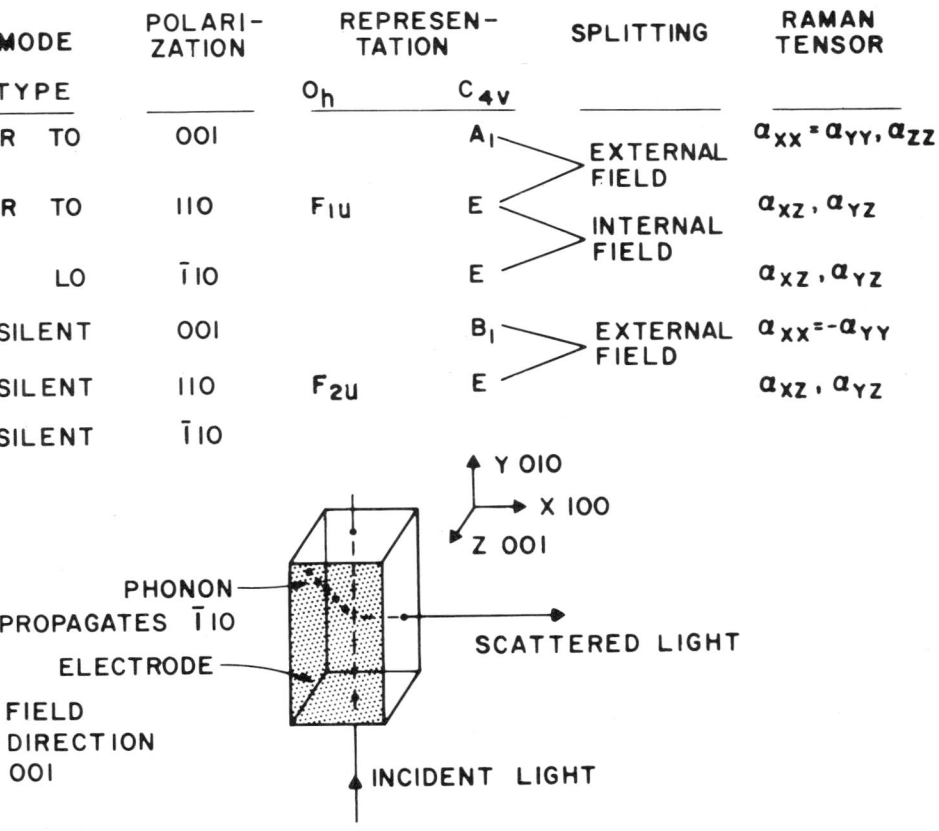

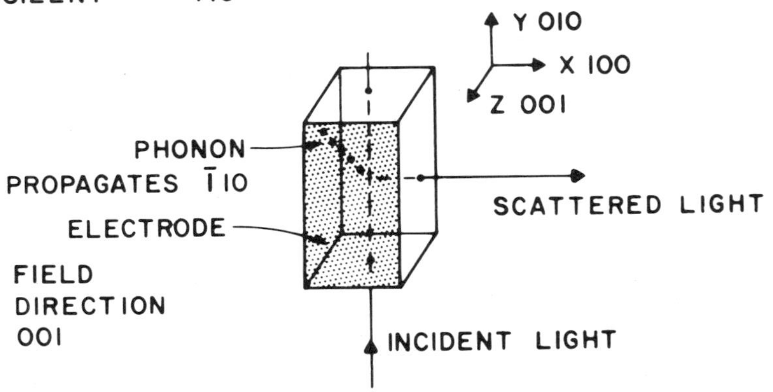

Fig. 2. Symmetries and splittings of various phonon modes in the cubic perovskites. Experimental arrangement shown forces the A symmetry modes to be transverse.

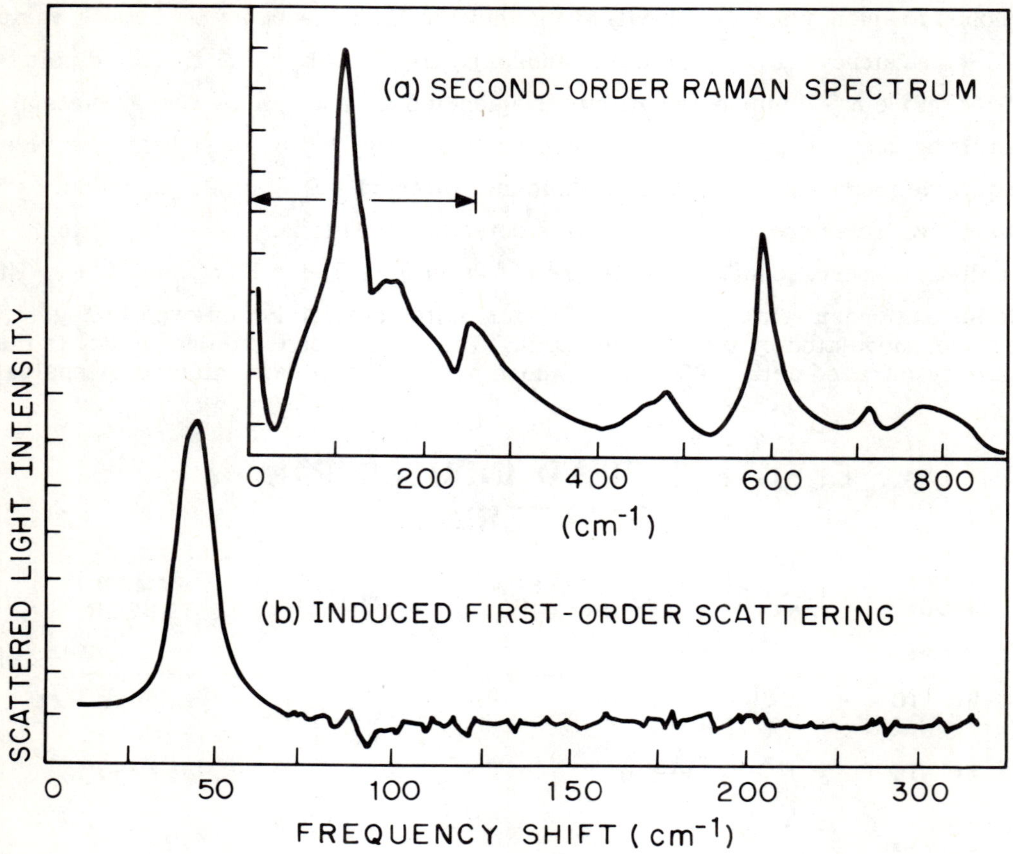

Fig. 3. Raman spectra of $KTaO_3$ at 80°K. (a) Intrinsic second order spectrum taken with no applied electric field. The horizontal arrow indicates the frequency range of Fig. 3b. (b) Electric field induced scattering for the A symmetry low frequency TO mode. E_{ac} was 10,000 v/cm, at 210 Hz. Detection at 420 Hz. Intensity units on scale b are 1/10 as large as those on scale a.

For experiments involving fields for which the crystal response was nonlinear, it was much more convenient to use pulsed fields and gated detection. This technique had the additional advantage of a variable duty cycle.

Our treatment would lead us to expect that the induced Raman cross section would vary as E_o^2 and this behavior is shown in Fig. 4, for the same phonon in $KTaO_3$ at 80°K.

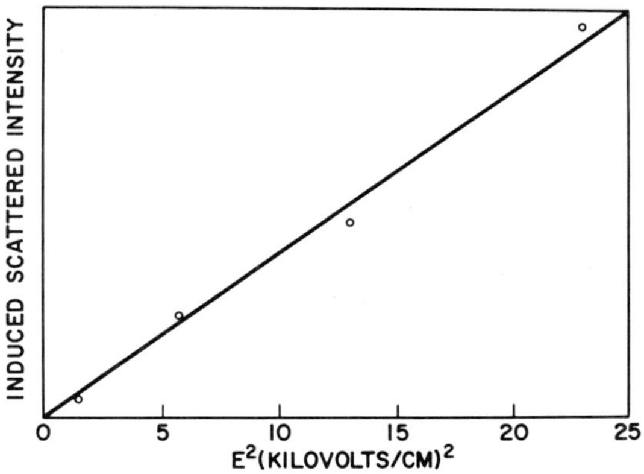

Fig. 4. Electric field dependence of induced scattered intensity for the A symmetry low frequency TO phonon in $KTaO_3$ at $80^{\circ}K$. The solid line represents $\sigma_R \propto E^2$.

CALCULATIONS OF RAMAN SCATTERING EFFICIENCIES

Now let us turn to a discussion of how the electric field modifies the Raman tensor. A completely general theory of electric field induced scattering would include the hyper-raman effect, in which the extra electric field which breaks selection rules is an optical frequency field. However, we shall simplify the treatment a great deal by considering only very low frequency, or d-c fields.

This subject has been analyzed previously by E. Burstein[2] and V. Dvořák[3]. For the purpose of his argument, Burstein noticed that the induced polarization is equivalent to a q = 0 phonon of large amplitude, frozen in. He considered the induced scattering to be analogous to second order scattering, where one of the participating phonon modes is the frozen one, having an equivalent population factor which is very large and proportional to E^2, and having zero frequency. His calculation is similar to one we shall present shortly, but emphasizes different factors, largely because he was interested in the alkali halides as prototypes.

Dvořák gave his attention to the cubic perovskites and derived induced Raman scattering efficiencies from a consideration of the second order electrooptic effect. His conclusions, though expressed in different form, are equivalent to those presented here.

For the remainder of the discussion we shall be guided by the treatment developed by Wemple and DiDomenico[4]. They have presented some of their results, as applied to induced scattering, at this conference (paper A-5).

We define a polarization potential β analogous to the deformation potential well known in semiconductor studies. The effective band gap ϵ is modified when the crystal is polarized: $\epsilon = \epsilon_O + \beta P^2$, where P is the total low frequency polarization of the crystal. If we further imagine that the optical polarizability arises completely from virtual transitions to states near ϵ, the polarizability at optical frequencies ω_ℓ can be written as:

$$\alpha = \frac{\text{constant}}{(\epsilon_O+\beta P^2)^2-\omega_\ell^2} \cong \alpha(P=0) \cdot \left[1 - \frac{2\epsilon_O\beta P^2}{\epsilon_O^2-\omega_\ell^2} + \dots\right] \tag{2}$$

Only terms linear in β are retained in the approximation. We are here primarily interested in the term in polarizability which is sensitive to polarization. This is closely related to the second order electrooptic coefficient

$$\frac{\partial^2\alpha}{\partial E^2} = 4\alpha(E=0)\,\frac{\epsilon_O\beta}{\epsilon_O^2 - \omega_e^2}\left(\frac{\varkappa-1}{4\pi}\right)^2 \tag{3}$$

$\varkappa$ is the dielectric constant. The advantage of the present formulation is that the constant β is largely independent of temperature, and is the same for all oxygen octahedron perovskites[4].

The polarization P has two important components:

$$P_s = \left(\frac{\varkappa-1}{4\pi}\right) E$$

a static or very low frequency component; and P_f, a component of thermal fluctuations at the infrared phonon frequency. The optical polarizability we are interested in, the Raman polarizability, contains the cross-term P_fP_s:

$$\alpha_R \propto \beta\, P_fP_s \tag{4}$$

The radiating dipole moment density is the product of α_R and the optical field amplitude E(optical).

$$M \propto \beta\, P_fP_s\, E(\text{optical}) = \beta\, P_f\,\frac{\varkappa-1}{4\pi} \cdot E \cdot E(\text{optical}) \tag{5}$$

This is obviously identical to (1) where $\beta\, P_f\,\frac{\varkappa-1}{4\pi} \sim \lambda$. The power radiated by M is proportional to its mean square amplitude, so the Raman efficiency varies as:

$$S \propto \beta^2\,\overline{P_f^{\,2}}\,P_s^{\,2} \tag{6}$$

In the perovskites $\varkappa-1 \cong \varkappa$ at all temperatures, so $P_s^{\,2} \propto \varkappa^2K^2$.

We now have to deal with the polarization fluctuations $\overline{P_f^{\,2}}$. For an infrared active mode with effective charge e*, the atomic displacement u_f and dipole moment density P_f are related by $P_f = Ne^*u_f/\epsilon_\infty$ when N is the molecular density, and ϵ_∞ the high frequency dielectric constant. The mean square displacement can be evaluated statistically:

$$\overline{u_f^2} = \frac{(\bar{n}+1)h\omega}{m\omega^2} \tag{7}$$

where ω is the frequency of oscillation, m is some effective mass for the mode, and $\bar{n}$ is the thermal occupation factor. $(\bar{n}+1)$ occurs since we will concentrate on the Stokes component of Raman scattering. $(\bar{n}+1)$ becomes $\bar{n}$ for the anti-Stokes component. Putting these things together,

$$\overline{P_f^2} \propto \frac{e^{*2}}{m\omega^2}(\bar{n}+1)h\omega, \text{ and} \tag{8}$$

$$S \propto \beta^2 \chi^2 E^2 \left(\frac{e^{*2}}{m\omega^2}\right)(\bar{n}+1)\omega \tag{9}$$

We have consistently dropped factors which are independent of temperature and frequency. β remains because of its tensor properties. $(e^{*2}/m\omega^2)$ is the oscillator strength of the mode in question, which can be obtained from infrared studies[5], or from polariton dispersion analysis[6].

In the following, we shall use Eq. (9) to make estimates of relative scattering efficiencies for modes we have observed. The absolute scattering efficiency has been calculated for the A-symmetry soft mode in $KTaO_3$ at room temperature, using a formula from Wemple and DiDomenico[4]. The observed Raman attenuation, for a field of 14,000 v/cm, is 5×10^{-8} cm^{-1}, while the calculated attenuation is 6×10^{-8} cm^{-1}. This agreement is almost embarrassing, since it is better than either the experimental uncertainty (about a factor of 2) or the theoretical precision.

Table I shows a comparison of the measured relative scattering efficiencies[1] with those calculated from (9), at three temperatures for $KTaO_3$. The mode in question is the A-symmetry component of the low frequency infrared active phonon, the soft mode.

TABLE I

T	W	$\bar{n}+1$	$S/S_{300°K}$(est.)	$S/S_{300°K}$(obs.)
300°K	85 cm^{-1}	4.5	1	1
80°K	47 cm^{-1}	3.2	13	10
8°K	25 cm^{-1}	1	130	100

Both the dielectric constant[1] and the oscillator strength[5] vary as ω^{-2}. The agreement between columns 4 and 5 is well within experimental uncertainty.

The relative scattering efficiencies of the A and E transverse soft modes can be estimated by taking account of the tensor nature of β. We notice that for the transverse A mode the static and fluctuating polarization components are parallel, while for the E mode they are perpendicular. The optical polarizations bear the same relationships. The appropriate β coefficient for the perpendicular case is characteristically 0.37 as large as for the parallel case[4]. Hence the scattering efficiency ratio S_E/S_A should be on the order of 14 percent. This compares favorably with the measured ratio of 13 per cent for $SrTiO_3$, but rather poorly for the measured ratio of 4 per cent for $KTaO_3$.

Next, we can estimate the scattering efficiency expected from the highest frequency LO modes. We expect that the displacement eigenvector of this LO mode resembles closely that of the low-frequency TO mode, so that it will have the same e^{*2}/m and the same β coefficient as the transverse E mode. The scattering efficiency, from (9), should be reduced simply by the ratio of frequencies, ω_{TO}/ω_{LO} which is 25/840 for $KTaO_3$, and 10/800 for $SrTiO_3$ at low temperatures. Thus the longitudinal E mode scattering should be a few percent of the transverse E mode scattering, or a few tenths of a percent of the transverse A mode. We have not been able to observe this scattering experimentally.

Finally, we extend the treatment to estimate scattering from the other two transverse A modes, relative to the soft mode, in $KTaO_3$ and $SrTiO_3$, at low temperatures. This amounts to the unsupported assumption that these modes possess the same polarization potential, β, as the soft mode. Equivalently it amounts to the assumption that the electrooptic coefficients are frequency independent.

From (9) we see that the relevant parameters are $e^{*2}/m\omega^2$, the oscillator strength; ω; and $\bar{n}+1$. Table II shows, for $KTaO_3$ and $SrTiO_3$, the calculated and measured ratios $S/S_{soft\ mode}$.

TABLE II

	Transverse A Mode Frequency	Oscillator Strength	$\bar{n}+1$	S/S_{soft}(est.)	S/S_{soft}(obs.)
$KTaO_3$	25 cm^{-1}	5230	1	1	1
	200 cm^{-1}	4.4	1	.006	.06
	556 cm^{-1}	2.5	1	.01	.05
$SrTiO_3$	10 cm^{-1}	6140	1.5	1	1
	175 cm^{-1}	1.8	1	.003	.017
	560 cm^{-1}	1.6	1	.01	.013

The experimental ratios are anywhere from 30 percent to a factor of 10 higher than the estimated ratios, which indicates that the polarization potentials are actually larger for the higher frequency modes than for the soft mode.

The rather general good agreement between our measurements and the estimates presented here is taken as an indication of the general validity of the point of view. Further progress requires the performance of more careful measurements of scattering efficiencies. We should like to point out specifically that more measurements of electric field dependence would be interesting, to explore the $\varkappa^2$ dependence of scattering in the nonlinear dielectric response range. The technique should be extended to other perovskites, notably $BaTiO_3$, for which field induced scattering can be compared to that in the ferroelectric phase. Let us emphasize also in closing that the electric field induced scattering has been interesting not only in studying the scattering processes as discussed here, but also in studying the behavior of the modes themselves, or the lattice dynamics of the crystals.

REFERENCES

1. P.A. Fleury and J.M. Worlock, Phys. Rev. (to be published); J.M. Worlock and P.A. Fleury, Phys. Rev. Letters 19, 1176 (1967); P.A. Fleury and J.M. Worlock, Phys. Rev. Letters 18, 665 (1967).
2. E. Burstein, Raman Scattering Phenomena, in: "Dynamical Processes in Solid State Optics," 1966 Tokyo Summer Lectures in Theoretical Physics, Part 1, R. Kubo and H. Kamimura, (ed.), W.A. Benjamin Inc., New York, 1967.
3. V. Dvořák, Phys. Rev. 159, 652, (1967).
4. M. DiDomenico, Jr., and S.H. Wemple, J. Appl. Phys. (to be published); S.H. Wemple and M. DiDomenico, Jr., paper A-5 this conference.
5. A.S. Barker, Jr., "Ferroelectricity," E.F. Weller (ed.), Elsevier, New York, 1967.
6. J.F. Scott, P.A. Fleury, and J.M. Worlock, Phys. Rev. (to be published).

E-8: THE EFFECT OF ELECTRIC FIELDS ON RAMAN SCATTERING IN DIAMOND*

E. Anastassakis and A. Filler,
University of Pennsylvania
Philadelphia, Pennsylvania
and
E. Burstein†
University of California
Irvine, California

ABSTRACT

The triply degenerate $q \simeq 0$ optical phonons in diamond type crystals have even-parity and are Raman active and infrared inactive. The application of an electric field modifies the symmetry of the optical phonons and, thereby modifies the polarization selection rules for first order Raman scattering. It is therefore possible to observe an electric field induced Raman scattering by the optical phonons for directions and polarizations of the incident and scattered radiation which normally do not lead to first order Raman scattering. Furthermore, the Raman scattering tensor will exhibit a quadratic dependence on the field. (In the case of crystals lacking a center of inversion, such as zinc-blende type crystals, the Raman scattering tensor will exhibit a linear dependence on the applied field.) We report here the observation of electric field dependent Raman scattering by optical phonons in diamond, together with a discussion of the different types of contributions to the electric field dependent Raman tensor in diamond type crystals. The electric field dependence of the Raman tensor also manifests itself as a high order effect in the electric field induced infrared absorption.

INTRODUCTION

In the presence of an external electric field, the center of symmetry of the diamond-type crystals is removed by changing the electron charge distribution of the atoms and by inducing a relative displacement ($\underline{d}_{E}^{(2)}$) of the two sublattices. The subsequent lowering of the symmetry results in new optical effects, known as morphic effects. In lattice dynamics, morphic effects manifest themselves in two major ways: i) as breaking of the

*Research supported in part by the U. S. Army Research Office-Durham and the U. S. Office of Naval Research.

†On sabbatical leave from the University of Pennsylvania, Philadelphia, Pennsylvania during the 1967-68 Academic year.

selection rules which normally forbid certain optical processes such as first order infrared (IR) absorption and Raman (R) scattering of the light. The effects appear as electric field induced IR absorption by an IR inactive mode[1, 11], and electric field induced R scattering by a R inactive mode[2-5]. ii) as modification of the polarization selection rules for vibration modes which are normally IR and R active. For example, the new field-dependent components of the Raman tensor may allow the scattering, for directions and polarizations of the incident and scattered light, which in the absence of the field do not lead to first order R scattering.

We present here a discussion of the nature of the field dependent Raman tensor in diamond-type crystals, together with our preliminary experimental results in diamond.

DISCUSSION

The strength of any first order Raman process depends on the square of the Raman tensor components, which are given by $a(j)_{\mu\lambda\sigma} = \frac{\partial \alpha_{\mu\lambda}}{\partial u_\sigma(j)}$, where $\alpha_{\mu\lambda}$ is the $\mu\lambda$-component of the static electronic polarizability per unit cell and $u_\sigma(j)$ is the σ-component of the relative displacement of the atoms in the unit cell, for the j-type optical phonon. In what follows we ignore the indices $\mu\lambda\sigma$ for simplicity. $\underset{\approx}{a}(j)$ is identically zero for centrosymmetric crystals with odd parity phonons (e. g. NaCl type crystals). $\underset{\approx}{a}(j)$ can be expanded as a function of an applied electric field $\underset{\sim}{E}_A$ as follows:

$$\begin{aligned} \underset{\approx}{a}(E_A, j) &= \underset{\approx}{a}(j)^{(o)} + \frac{da(j)}{dE_A} \underset{\sim}{E}_A + \frac{d^2a(j)}{dE_A\, dE_A} \underset{\sim}{E}_A \underset{\sim}{E}_A + \cdots \qquad (1) \\ &= \underset{\approx}{a}(j)^{(o)} + A(j)^{(1)} \underset{\sim}{E}_A + A(j)^{(2)} \underset{\sim}{E}_A \underset{\sim}{E}_A + \cdots \\ &= \underset{\approx}{a}(j)^{(o)} + \underset{\approx}{a}_E(j)^{(1)} + \underset{\approx}{a}_E(j)^{(2)} + \cdots \end{aligned}$$

The coefficient $A(j)^{(1)} = \frac{da(j)}{dE_A} = \frac{d^2\alpha}{du_j\, dE_A}$ is zero for diamond type crystals. The reason for this is that $\underset{\approx}{a}(j)$ is a macroscopic property of the centrosymmetric crystal and can therefore vary only quadratically with the field. $A(j)^{(1)}$ is one of the coefficients which determine the strength and symmetry of the electric field induced first order R scattering in ZnS and NaCl type crystals[2,]. The quadratic coefficient $A(j)^{(2)} = \frac{da^2(j)}{dE_A dE_A}$ $= \frac{d^3\alpha}{du_j\, dE_A\, dE_A}$ is non-zero for diamond[7]. It is the symmetry of this coefficient, which modifies the zero-field selection rules for R scattering in diamond.

Due to the electrostriction (electric field induced piezoelectricity), the electric field induces a strain $\eta_E^{(2)}$ in addition to the relative sublattice displacement $d_E^{(2)}$. Both parameters $\eta_E^{(2)}$ and $d_E^{(2)}$ are quadratic with the field, and therefore one can write for $a_E^{(2)}(j)$,

$$a_E^{(2)}(j) = \frac{d^2 a(j)}{dE_A\, dE_A} E_A E_A \tag{2}$$

$$= \frac{\partial a(j)}{\partial E_A\, \partial E_A} E_A E_A + \frac{\partial a(j)}{\partial \eta_E^{(2)}} \eta_E^{(2)} + \frac{\partial a(j)}{\partial d_E^{(2)}} d_E^{(2)}$$

for terms leading to one phonon (first order) R scattering. [8]

Any first order R experiment carried out with a static external field would involve contributions from all three terms of Eq. (2). On the other hand, if the applied field is at optical frequencies, such as the EM field provided by a powerful source at ω_ℓ, the second and third term of Eq. (2) do not occur since $d_E^{(2)}$ and $\eta_E^{(2)}$ are negligible at optical frequencies. The first term then, leads to non-linear scattering processes, such as "Three-photon R scattering" and "field induced two-photon R scattering". The stokes and antistokes lines for these effects will appear at $(3\omega_\ell \pm \omega_j)$ and $(2\omega_\ell \pm \omega_j)$ respectively.

SYMMETRY CONSIDERATIONS

We now consider the summetry properties of the coefficient $A^{(2)}(j) = \frac{d^2 a(j)}{dE_A\, dE_A}$. Of the five indices labeling $A^{(2)}(j)$, the first two designate the polarization of the incident and scattered light (interchangeable), the third refers to the polarization of the jth phonon, and the last two indicate any two of the electric field components.

For a given direction of the applied field, only certain components of $A^{(2)}(j)$ enter the scattering process. At this point it is more convenient to refer to the specific experimental configuration used in the laboratory. Our diamond plate was oriented along the [1$\bar{1}$0], [110], and [001] directions, referred to as u, v, and w respectively from now on. The zero-field Raman tensor has the form $a\,|e_{\mu\lambda\sigma}|$ when referred to the crystallographic axes [100], [010] and [001]. a is the single independent component for diamond, and $e_{\mu\lambda\sigma}$ is the Levi-Civita function. The same tensor, when expressed in the uvw system, takes the form

$$\underset{\sim}{a}^{(o)}(j) = \begin{pmatrix} \cdot & \cdot & -a \\ \cdot & \cdot & a \\ \cdot & \cdot & \cdot \\ \cdot & a & \cdot \\ -a & \cdot & \cdot \\ \cdot & \cdot & \cdot \end{pmatrix} \tag{3}$$

where the column index corresponds to the polarization of the j-phonon, and the row index is a supressed index[9] for the polarizations of the incident and scattered light.

When an electric field is applied, the cubic symmetry of the crystal is lowered, the new symmetry class being determined by the direction of the field. If we take $\underline{E}_A || w$ the summetry changes from Q_h to C_{4v} and the triply degenerate R active mode F_{2g} splits into a nondegenerate mode B_2 polarized parallel to the field, and a doubly degenerate mode E polarized in the plane perpendicular to the field. The R tensor now has two independent components which correspond to the sum of the terms to all orders of the expansion given by Eq. (1). If we only consider terms quadratic in the field, in Eq. (1), the representation F_{2g} is reduced only to second order with the field. The new R components for a general direction of $\underline{E}_A$ correspond to the direct representation product

$$F_{2g} \times (F_{1u} \times F_{1u}) = 4F_{2g} + 2E_g + A_{1g} + (\text{R inactive representations}) \tag{4}$$

within the 0_h Group. $(F_{iu} \times F_{iu})_s$ is the O_h representation for the symmetric part of $\underline{E}_A\underline{E}_A$. Expressed in the uvw axes, the new R tensor is given by

$$\underset{\approx}{a}(\underset{\sim}{E}) = \begin{pmatrix} b & b & -a+c-d \\ b & b & a+c+d \\ -2b & -2b & c \\ . & a+d & . \\ -a-d & . & . \\ \sqrt{12}b & -\sqrt{12}b & . \end{pmatrix} \tag{5}$$

where (d, f), b, and c are the field induced tensor components, corresponding to F_{2g}, E_g, and A_{1g} respectively. For $\underline{E}_A || w$ it can be shown that b = c = o.

From (3) and (5) it becomes clear what the effect of the field is on the selection rules. For each of the zero-field components there is a new contribution quadratic with the field. Thus for example, we can write for the a_{232} component

$$a_{232}^{(0)}(j) = a \xrightarrow{\underset{\sim}{E}} \sum_j a_{232}(\underset{\approx}{E}, j) = a + d \tag{6}$$

where

$$d = A_{23233}^{(2)} \ E_3 \ E_3 \tag{7}$$

with all indices referred to the uvw system of axes. The observed change in the R intensity is proportional to

$$(a + d)^2 - a^2 = 2ad + d^2 \approx 2ad \tag{8}$$

for $d \ll a$. Therefore, from the observed magnitude $2ad/a^2$ and Eq. (7) one can deduce a numerical value for the coefficient $A_{23233}^{(2)}$. In addition, new components appear which lead to a field dependent R scattering for configurations which lead to no R scattering for zero field (for example $a^{(0)}_{333} = 0$, whereas $a_{333}(E, A_{1g}) = c \neq 0$).

EXPERIMENTAL RESULTS

Measurements were made using the University of Pennsylvania Raman spectrometer, and a He-Ne laser excitation frequency at 15, 800 cm^{-1}. The diamond plate described in the previous section was kept in a glass vacuum cell, at 80°K. In scanning through the Raman line, two measurements were made at each grating position, with and without field. The transistorized DC voltage supply was turned on and off electronically through a relay for equal time intervals. A typical run is shown in Fig. 1 for a field of 1.33×10^5 V/cm.

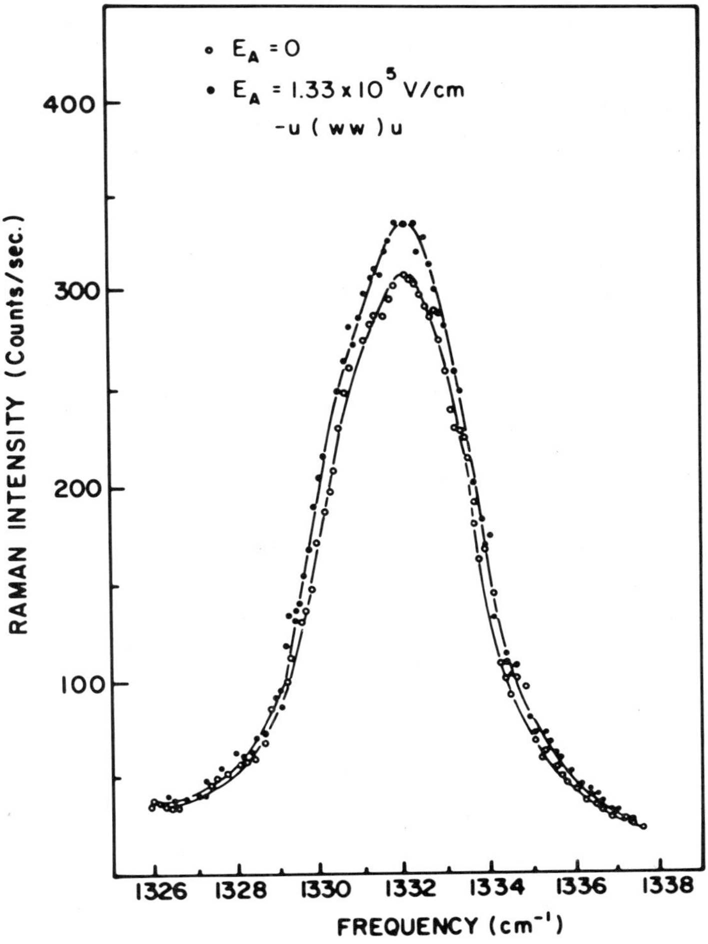

Fig. 1. A typical run of the first order Raman line of diamond in the -u(ww)u configuration with and without an electric field. The slit width was 1mm.

A circular polarizer was placed before the linear polarizer in the incident beam, to assure an equal incident intensity for any polarization.

With the field parallel to the w axis, the symmetry analysis presented in the previous section applies. Back scattering measurements were carried out on the type IIA diamond plate, first in the configuration -u(ww)u, which according to the tensor form (3), should not lead to R scattering in the absence of an applied field. A weak peak was observed instead, which was initially attributed to misalignment and residual strain birefringence. Careful alignment in different configurations, however, did not improve the situation (whereas separate birefringence experiments indicated strong residual strain birefringence). When an electric field was applied in the -u(ww)u configuration, an increase in the intensity was observed, which was quadratic with the field (Fig. 2) in agreement with Eq. (8). From this fractional change, it would be easy to calculate a value for the ration $A^{(2)}_{ij\sigma 33}/a$ [10], if the origin of the residual intensity were well defined, i. e. if it were due to a known misalignment. This is however impossible because

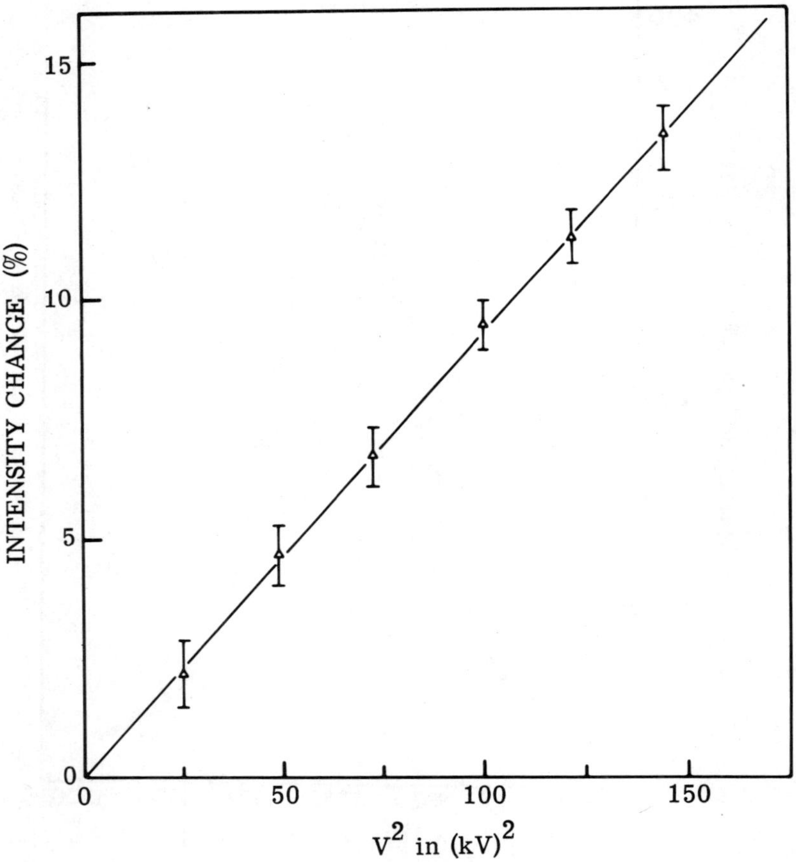

Fig. 2. The electric field dependence of the intensity change of the first order Raman line, in the -u(ww)u configuration.

of the inhomogenious, i.e. random, nature of the residual birefringence. Our experiments only allowed us to obtain a rough estimate of $A^{(2)}_{ij\sigma 33}/a$. The results yield a value of 10^{-7} in cgs units. Such a small value of the coefficient $A^{(2)}_{ij\sigma 33}$ would lead to a field induced intensity $(A^{(2)}_{ij\sigma 33} E_3 E_3)^2$ which for the available fields is far below the noise level and which would therefore be undetectable. In other words the residual R intensity acts as a 'bias' to amplify the effect of the field.

The present observation introduces in R spectroscopy a procedure for detecting a small coefficient through its product with a larger known parameter. By rotation either the crystal or the polarizers through a small angle, one can introduce an arbitrary amount of "bias" and thereby obtain a numerical value for the coefficient.

Measurements in the -u(vw)u configuration gave an approximate value of -10^{-8} cgs for the ratio $A^{(2)}_{23233}/a$. The absolute value of a has been established experimentally by electric field-induced IR absorption data as equal to 4×10^{-16} cm^2 [1, 11]. The sign of a has been established theoretically as (+)[12].

Finally, it is interesting to mention, that the effect of the electric field on the R scattering tensor manifests itself as a higher order effect, in the "Electric field induced IR absorption." The effect was observed by using modulated fields and by detecting the signal at the harmonics of the modulation frequency[11]. Although we examined only the coefficient $A^{(2)}_{23233}$ it is significant that the same sign and value were obtained for the ratio $A^{(2)}_{23233}/a$ as in the R experiments.

ACKNOWLEDGMENT

We thank Mr. A. Pinczuk for helpful discussions and the authorities of Diamond Research Laboratory, South Africa, for kindly supplying the diamond samples.

REFERENCES

1. E. Anastassakis, S. Iwasa, and E. Burstein, Phys. Rev. Letters 17, 1051 (1966).
2. E. Burstein, A.A. Maradudin, E. Anastassakis, and A. Pinczuk, Helvetica Physica Acta 41, 730 (1968).
3. P.A. Fleury and J.M. Worlock, Phys. Rev. Letters 18, 665 (1967).
4. R.F. Schaufele, M.J. Weber, and B.D. Siverman, Phys. Rev. Letters 25A, 47 (1967).
5. J.M. Worlock and P.A. Fleury, Phys. Rev. Letters 19, 1176 (1967).
6. A. Pinczuk and E. Burstein, paper E-9 this conference.
7. All the coefficients entering the even-power terms of Eq. (1) are non-zero for diamond, but zero for NaCl type crystals. The argument is reversed for the odd-power terms. All these coefficients are nonzero for ZnS type crystals.
8. If E(i) is the linear macroscopic electric field of an i-type IR active phonon which has a longitudinal component, a term of the form $\frac{d^2a(j)}{dE_A\, dE(i)} E_A E(i)$, would also occur in (2). A higher order polarizability term due to this term, is $\frac{d^2a(j)}{dE_A\, dE(i)} E_A E(i) u(\omega_j)$. Since E(i) varies at the frequency ω_i, it becomes clear that in the presence of a laser field at the frequency ω_ℓ,

this term leads to an "electric field induced two-phonon (second order) R scattering, at $\omega_\ell \pm 2\omega_j$ for diamond ($\omega_j = \omega_i$), and at $\omega_\ell \pm 2\omega_j$ or $\omega_\ell \pm (\omega_j \pm \omega_j)$ in ZnS type crystals, where the IR active mode at ω_i may or may not be the same as the normally R active mode at ω_j.

9. J.F. Nye, "Physical Properties of Crystals", Oxford, 1964.

10. Because of the residual birefringence, it is resonable to expect a mixing of the configurations -u(vv)u or -u(vw)u even when the experimental geometry has been set up for the configuration -u(ww)u. Thus the observed quadratic change can be attributed to the presence of terms of d and f character, and described by an effective coefficient $A^{(2)}_{ij\sigma 33}$.

11. E. Anastassakis and E. Burstein (to be published).

12. A.A. Maradudin and E. Burstein, Phys. Rev. 164, 1081 (1968). A (-) sign appears in this reference. However, according to these authors, the sign should be changed to a (+), after a calculational mistake was discovered.

E-9: RESONANCE ENHANCED ELECTRIC FIELD INDUCED RAMAN SCATTERING BY LO PHONONS IN InSb

A. Pinczuk and E. Burstein
Department of Physics
and
Laboratory for Research on the Structure of Matter,
University of Pennsylvania, Philadelphia, Pa.

ABSTRACT

We have observed large $q \approx 0$ TO and LO phonon scattering intensities in the backward scattering Raman spectra obtained from n-type and p-type InSb surfaces with 6328 Å (1.96 ev) excitation. These are attributed to resonance enhancement near the E_1 energy gap at 1.89 ev. Further, the $q \approx 0$ LO phonon peak is observed for scattering geometries in which, according to the polarization selection rules for the Raman scattering tensor, it should be forbidden. The occurrence of the LO phonon band is explained by an electric field induced Raman scattering originating in the electric field which is present in the surface depletion layers of n-type and p-type InSb. The effects of temperature changes indicate that excitons take part in the resonance enhancement mechanism of the scattering by LO phonons. The effects of externally applied electric fields support our interpretation of the data.

INTRODUCTION

Recently, we have reported the observation of a surface electric field induced Raman scattering by LO phonons in n-type InSb[1]. The effect was studied by obtaining the Raman scattering spectra of $q \approx 0$ TO and LO phonons for InSb samples with different carrier concentrations at room temperature and at 140°K. Electric field induced Raman scattering was seen previously in paraelectric crystals[2] where fields as low as 400 volts/cm produced an observable effect. In less polarizable crystals, like the group III-V semiconducting compounds, it is expected that larger fields are necessary in order to produce a sizable effect. The surface electric fields encountered in group III-V semiconductors which are of the order of 10^5 volts/cm were found capable of producing field induced bands at room temperature with about the same strength as the first order bands.

*Supported by the U. S. Army Research Office, Durham

The surface field induced scattering was found to show a resonant enhancement due to the proximity of the energy of the laser photons (1.96ev) to the E_1 energy gap of InSb (1.89ev at 300°K[3]). This resonant enhancement increased dramatically with decreasing temperature and at 140°K the intensity of the field induced band was about five times the intensity of the first order band. We also found[1] that the resonant enhancement has a different behavior for the TO phonons than for the LO phonons. It was suggested that such a difference is expected if excitons play an important role in the resonant Raman scattering[4].

In this communication we present new results concerning the behavior of p-type InSb crystals and the effects of externally applied electric fields. A discussion of surface electric fields is given in terms of the present knowledge about semiconductor surfaces which explains several of our results. A phenomenological treatment is given of the electric field induced scattering, and the possible mechanisms which produce the resonance enhancement are considered.

SURFACE FIELDS

It has been shown[5] that the Fermi level at the surface of a large number of group IV and group III-V semiconductors (including InSb) is pinned within the forbidden energy gap at a fixed position (relative to the edge of the conduction band) which is a constant fraction of the fundamental energy gap E_g ($\phi_B \simeq \frac{2}{3} E_g$). ϕ_B was found to be independent of the doping and of temperature in the range between 77°K and 300°K. This pinning effect very likely originates in a large density of surface states at the energy ϕ_B. A depletion layer exists at the surface which can be described as a Schottky barrier with a barrier height at the surface given by:

$$\phi_{on} = \phi_B + \phi_{Fn} \simeq \frac{2}{3} E_g + \phi_{Fn} \qquad \text{n-type} \qquad (1.a)$$

$$\phi_{op} = \phi_B - E_g + \phi_{Fp} \simeq -\frac{1}{3} E_g + \phi_{Fp} \qquad \text{p-type} \qquad (1.b)$$

where ϕ_{Fn} is the position of the Fermi level in the bulk of n-type crystals measured from the conduction band edge and ϕ_{Fp} is the position of the Fermi level in p-type samples measured from the top of the valence band. The electric field in the Schottky barrier will increase with increasing carrier concentration in the bulk. Due to the increasing number of ionized impurities in the space-charge region, the width of the surface layer will decrease with increasing carrier concentration in the volume. In InSb, the Schottky barrier, for the highest concentrations available is about the same as the skin depth (~500 Å) at 6328 Å.

RESONANCE ENHANCEMENT

It was shown in reference [1] for a crystal with $n = 1.4 \times 10^{18}$ cm^{-3} that when the temperature of the sample was lowered to 140°K the intensity of the surface field induced LO phonon band increases by a factor of ~6 while the intensity of the first order TO phonon band remains unchanged. Since no significant change is expected in the surface field in the temperature range between 77°K and 300°K, the observed effect can only result from an increase in the resonant enhancement originating in the changes in the electronic structure of the sample produced by the lowering of its temperature. Resonance enhancements occur with free electron-hole pairs and excitons as intermediate states[4]. Similar enhancements are expected for the TO and LO phonon bands if the resonance mechanism involves free electron-hole pairs. We believe that the experimental result that no changes were observed in the TO phonon band with decreasing temperature is an indication that excitons take part in the resonance enhancement mechanism of the surface field induced LO phonon band. Based on the evidence that the interaction of optical phonons with excitons is large for LO phonons and relatively small for TO phonons, Burstein et al. [6] have suggested that the exciton enhancement will be appreciable only for LO phonons. Evidence for exciton behavior at the E_1 energy gap of InSb has been given by Cardona and Harbeke[3]. On decreasing the temperature of the sample, two effects occur: The E_1 energy gap moves closer to the energy of the laser photons and the lifetime of the exciton corresponding to the E_1 energy gap increases. Both effects result in an increase of the exciton strength at the energy of the laser photons which in turn produces an increase in the exciton-enhanced scattering efficiency of the LO phonons[6].

PHENOMENOLOGICAL THEORY OF FIELD INDUCED SCATTERING

The scattering efficiency is known to be proportional to $|\hat{e}_j \cdot \tilde{\tilde{\alpha}} \cdot \hat{e}_i|^2$ where e_i and e_j are unit vectors along the polarization of the incident and scattered radiations and $\tilde{\tilde{\alpha}}$ is the optical phonon contribution to the electronic polarizability tensor. The components of α_{ij} induced by the electric field in first order are[7].

$$\alpha_{ij} = a_{ijk\ell} u_k E_\ell + b_{ijk\ell} E_k E_\ell \tag{2}$$

(the interference between the field independent and the field induced terms is not considered here because it was not observable in our experiments); u_k and E_k are the optical phonon displacement and macroscopic field; $a_{ijk\ell} E_\ell$ and $b_{ijk\ell} E_\ell$ can be considered as effective field induced deformation potential and electro-optic contributions. $b_{ijk\ell}$ is also an effective quadratic electro-optic tensor.

The components of the Raman scattering tensor (RST) induced by the field can be obtained from Eq. (2). The non-zero components of a general fourth rank tensor are known[8] to be determined by only three parameters:

$$\begin{aligned} b_{xx,xx} &= b_{yy,yy} = b_{zz,zz} = A \\ b_{xy,xy} &= b_{xz,xz} = b_{yz,yz} = B \\ b_{xx,yy} &= b_{xx,zz} = b_{yy,zz} = C \end{aligned} \tag{3}$$

where x, y, z are the three [100] directions.

Our experiments are described in the following coordinate system: $x' \equiv [\bar{1}11]$; $y' \equiv [110]$; $z' \equiv [1\bar{1}2]$. From (3) we obtain the following tensor components of interest:

$$b_{z'z',y'y'} = \frac{1}{2} C + \frac{1}{6} A - \frac{2}{3} B \quad (4.a)$$

$$b_{x'x',y'y'} = \frac{1}{2} C + \frac{1}{6} A - \frac{2}{3} B \quad (4.b)$$

$$b_{x'z',y'y'} = \frac{1}{3\sqrt{2}} A - \frac{\sqrt{2}}{3} B \quad (4.c)$$

$$b_{z'z',z'y'} = b_{x'x',x'y'} = b_{z'z',x'y'} = b_{x'x',x'y'} = b_{x'z',x'y'} = b_{x'z',z'y'} = 0 \quad (4.d)$$

In agreement with our results[1] (the surface field was along [110] ≡ y'), equations (4.d) indicate the TO phonon is not expected to show a field induced back-scattering from the [110] surface. Further, according to equations (4.a), (4.b), and (4.c), the surface field induced LO phonon band is expected in the spectra of the (x'x'), (z'z'), (x'z'), and (z'x') components of the RST. It was found[1] that the LO phonon band was not present in the (x'z') and (z'x') spectra. This is an indication that, unless an accidential cancellation occurs, the parameter C dominates the scattering. This implies that for photons, phonons, and surface fields polarized along [100] directions, only electric fields parallel to the phonon displacements will give a field induced RST which will have only diagonal non-zero components. The reason for this is not yet clear to us although it may be related to the mechanism of resonance enhancement.

EXPERIMENTAL

InSb samples, Te doped (n-type), and Zn doped (p-type), with carrier concentrations in the range from 1.5×10^{15} cm^{-3} to 4.8×10^{18} cm^{-3} were obtained from the Monsanto Company. (110) surfaces were obtained by cleavage in air and no further treatment was given to the surface.

The spectra were excited with a 50 mW He-Ne laser operating at 6328 Å (1.96 ev). The scattered light was analyzed with a double grating spectrometer (designed by A. Filler) in which the gratings were turned in steps equivalent to 0.88 cm^{-1}. The intensity of scattered light was recorded using a S-20 photomultiplier and photon counting electronics.

The backward scattering geometry for opaque crystals used by Russel in Si[9] and by Krauzman[10] in InSb was used to obtain the spectra. The incident light was directed at an angle of about 20° to the normal of a [110] surface and the back-scattered light was collected into the spectrometer. Due to the large refractive index of InSb at 6328 Å, the scattering wave vector $\vec{q} = \vec{k}_{inc} - \vec{k}_{scatt}$ is normal to the surface along [110] ≡ y'.

An external d.c. field along the y'direction was applied to the crystal surface with a 3 volts battery and 0.5 N solution of Kcl in water as transparent electrode using a technique that has been described in detail by Cardona et al[11]. In all cases the samples were biased negatively and the currents through the crystal were of the order of 5μA. The external field induced spectra were obtained by measuring at each grating position the intensity of the scattered light with and without the field.

RESULTS

Fig. 1 shows the spectra for (x'x') component of the RST for two p-type InSb samples with 1×10^{17} cm^{-3} and 4.8×10^{18} cm^{-3} carriers in the bulk. Two bands are seen at

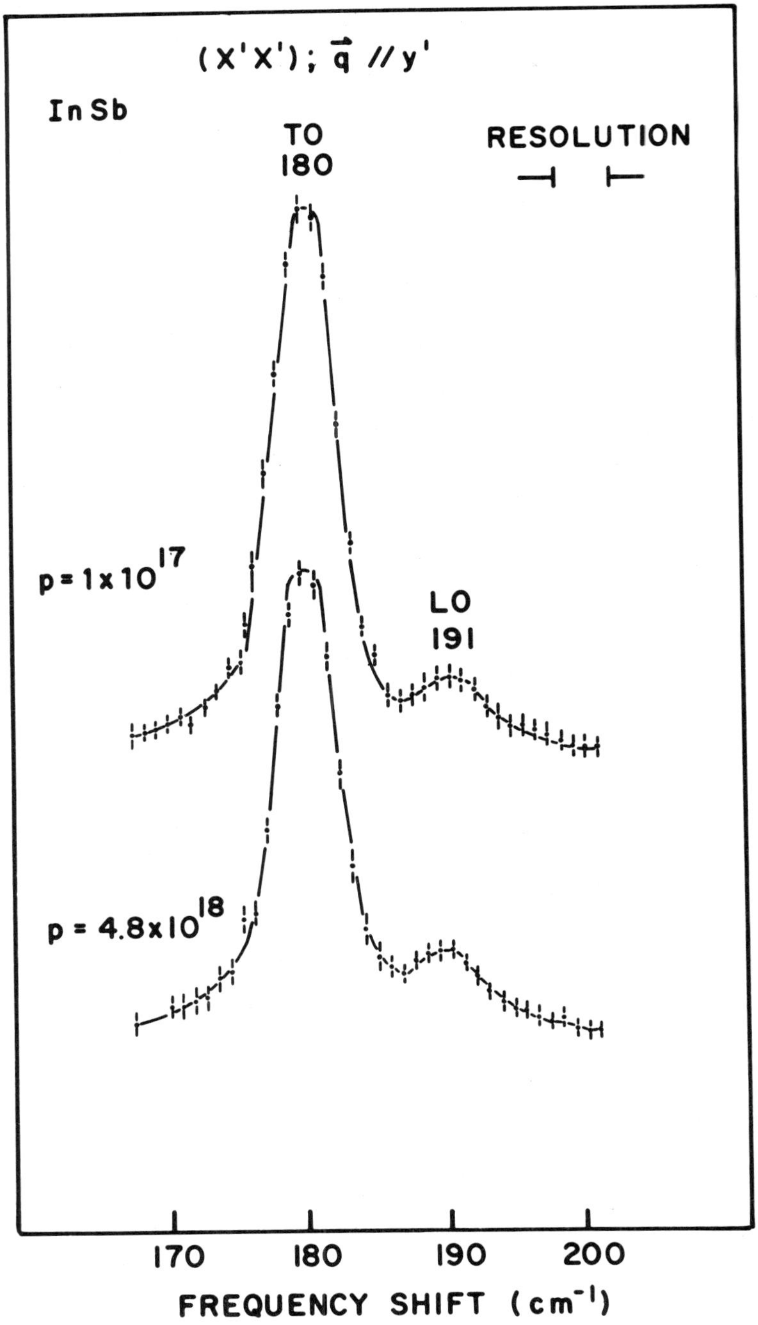

Fig. 1. The (x'x') back scattering spectrum of p-type InSb samples having hole concentrations of 1.0×10^{17} cm^{-3} and 4.8×10^{18} cm^{-3}.

180 ± 3 cm^{-1} and 191 ± 3 cm^{-1} which are assigned to the $q \approx 0$ TO and LO phonons. These frequencies are in good aggrement with the values obtained in IR absorption and reflection[12] measurements but our LO phonon frequency is considerably lower than the value of $200 \pm$ cm^{-1} reported by Krauzman[11]. Contrary to what was found in n-type crystals[1], the LO phonon band has a small intensity which appears to be independent of the concentration of carriers in the bulk of the sample.

Fig. 2 shows the effects of temperature changes on the spectra for (x'x') component of the RST of a sample with $p = 4.8 \times 10^{18}$ cm^{-3}. The strength of the LO phonon band at $T \simeq 160$°K increased to several times its strength at room temperature while the intensity of the TO phonon band remained unchanged. The observed frequency shifts are in agreement with those observed in the IR reflectivity spectra.

Figs. 3 and 4 illustrate the effects of externally applied d.c. electric fields. About the same battery voltages (1.8 - 2.2 volts) are necessary to produce changes in the spectra of n-type and p-type samples. It can be seen that while only minor changes are found in the intensity of the TO phonon band, large changes are produced in the intensities of the LO phonon bands. In n-type samples, a decrease in the intensity of the band is obtained while in p-type samples an increase is observed.

DISCUSSION

It is our purpose now to show that the data presented here on p-type InSb and on the effect of external electric fields can be explained on the same basis as was done for the n-type samples.

The fact that, as shown by Fig. 1, the surface field induced LO phonon band in p-type samples is weak and does not seem to depend on the carrier concentration in the bulk can easily be understood if we realize that: (a) As shown by Eq. (1.a) the modulus of the barrier height is much larger in n-type than in p-type material. (b) Due to the large difference between the electron and heavy hole effective masses ($m_h^* \sim 30\, m_e^*$), changes in the concentration of carriers which produce a large change in ϕ_{on} will produce a much smaller change in ϕ_{op}.

Further, the results of Fig. 2 indicate that the observed resonance enhancement of the LO phonon band in n-type material is also found in p-type samples. This shows that this effect is, as expected, independent of the nature of the dopant and of the properties of the carriers in the bulk of the crystal.

The results of Figs. 3 and 4 concerning the effects of external voltages can be explained in the following way: voltages up to about 1.8 volts are screened by the charge which is induced in the surface states; voltages larger than 1.8 volts can no longer be screened because of complete filling of the surface states. The direction of changes of the band intensities are those expected for the changes induced by the external field in the surface depletion layer.

In summary, we have shown the existence of a resonance enhanced surface field induced Raman scattering by LO phonons in both p-type and n-type InSb. Similar effects are expected in other semiconductors in which the Fermi level at the surface is pinned. Our results also indicate that the mechanism of Raman Scattering can be elucidated by studying the temperature dependence of the band intensities. These phenomena provide a new tool to study surface electronic states by Raman spectroscopy.

ACKNOWLEDGMENT

We would like to acknowledge discussions with D. L. Mills.

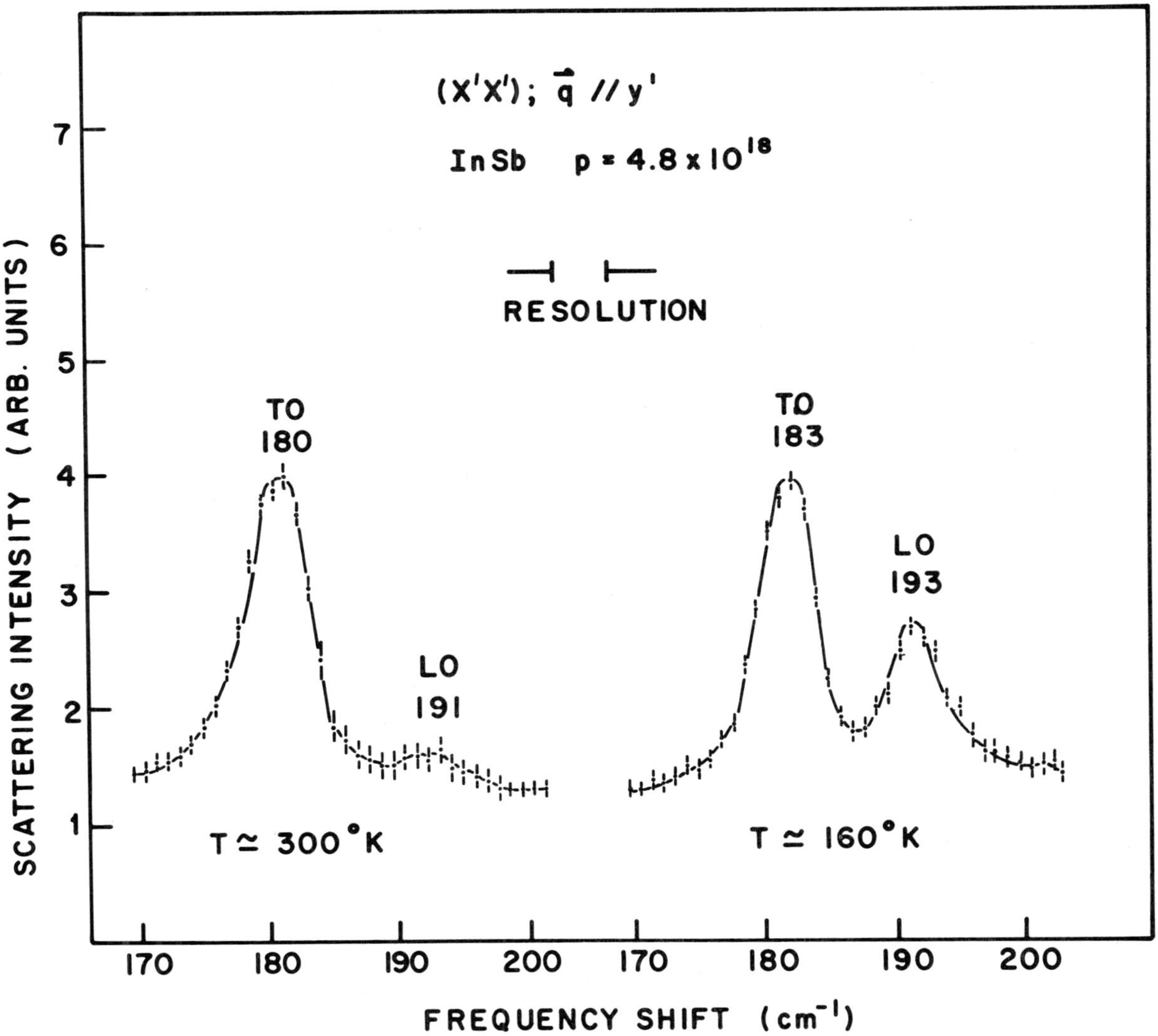

Fig. 2. Temperature dependence of the (x'x') backscattering spectrum of a p-type InSb sample with a hole concentration of 4.8×10^{18} cm^{3}.

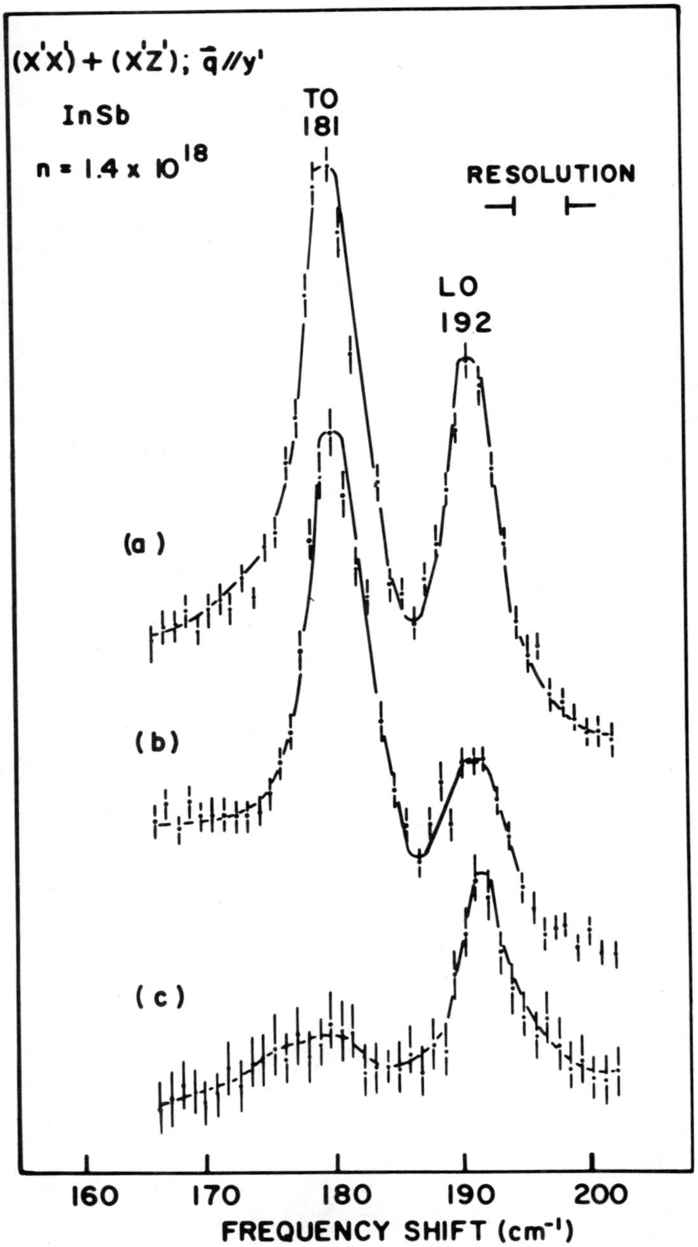

Fig. 3. Effect of an externally applied voltage on the spectrum of a n-type sample with 1.4×10^{18} cm^{-3} electrons. (a) V = 0. (b) V = 2 volts. (c) Difference between (a) and (b) spectra.

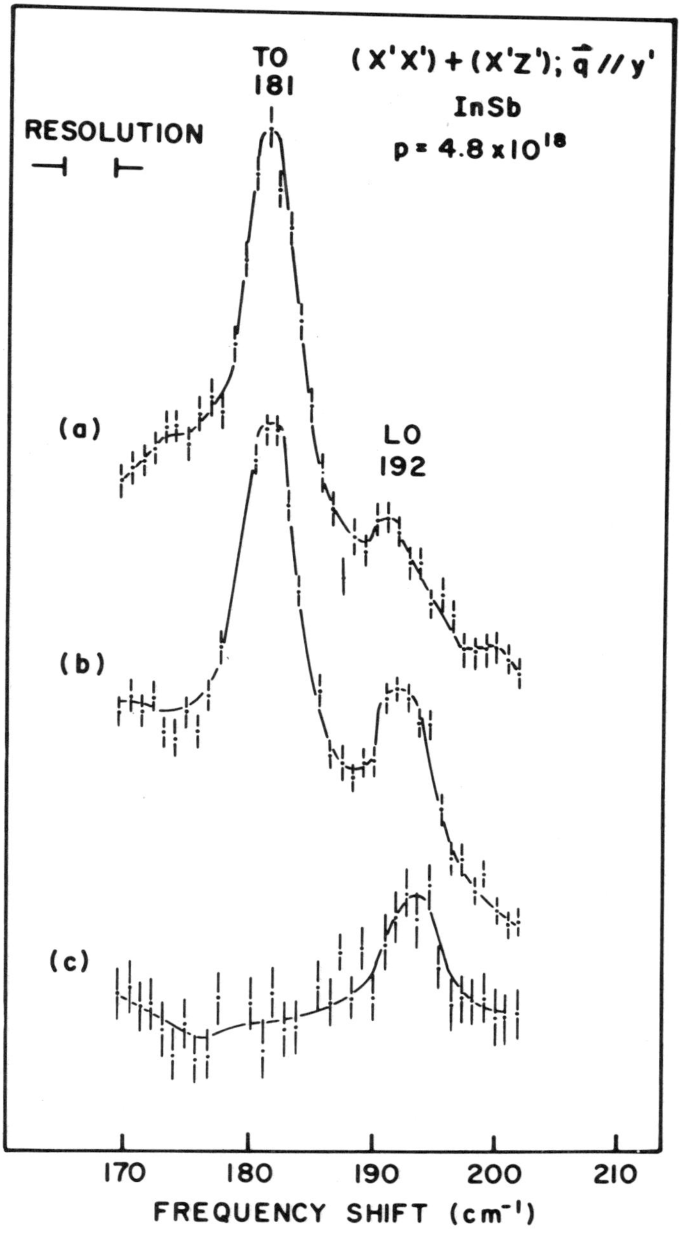

Fig. 4. Effect of an externally applied voltage on the spectrum of a p-type sample with 4.8×10^{18} cm^{-3} holes. (a) V = 0. (b) V = 1.9 volts. (c) Difference between (b) and (a) spectra.

REFERENCES

1. A. Pinczuk and E. Burstein, Phys. Rev. Letters, 21, 1073, (1968).
2. P. A. Fleury and J. M. Worlock, Phys. Rev. Letters 18, 665 (1967).
3. M. Cardona and G. Harbeke, Phys. Rev. Letters 8, 90 (1962).
4. A. K. Ganguly and J. L. Birman, Phys. Rev. 162, 906 (1967), and references therein.
5. C. A. Mead and W. G. Spitzer, Phys. Rev. 134, A713 (1964).
6. E. Burstein, D. L. Mills, A. Pinczuk, and S. Ushioda (submitted for publication).
7. E. Burstein, A. A. Maradudin, and A. Pinczuk, Bull. Am. Phys. Soc. Ser. II 13, 480 (1968); and E. Burstein, A. A. Maradudin, E. Anastassakis, and A. Pinczuk, Helvetica Phys. Acta 41, 730 (1968).
8. J. F. Nye, "Physical Properties of Crystals," Clarendon Press, Oxford, 1964.
9. J. P. Russel, Appl. Phys. Letters 6, 223 (1965).
10. M. Krauzman, C. R. Acad. Sc. Paris 264, B117 (1967).
11. M. Cardona, K. L. Shaklee, and F. M. Pollak, Phys. Rev. 154, 696 (1967).
12. M. Haas, "Physics of III-V Compounds," 3, R. K. Willardson and A. C. Beer (ed.), Academic Press, New York, 1967.

F-1: RAMAN SCATTERING FROM MIXED CRYSTALS*

P. S. Pershan† and W. B. Lacina
Division of Engineering and Applied Physics,
Harvard University
Cambridge, Massachusetts

ABSTRACT

A theory for the optical properties (Raman scattering or I. R. absorption) of crystals containing a small but finite concentration of impurities will be described. One result of this theory will be a qualitative understanding of the various experimental results that have been obtained for different mixed crystals at much larger concentrations. For example, depending on the difference between the impurity atom and the host atom, a single formula describes either the CaF_2-SrF_2 Raman spectra, in which the Raman frequency shifts linearly with concentration, or the Ge:Si Raman spectra in which new lines appear when Si is added to Ge. The intensity of the new lines increase with increasing Si concentration. The theory further demonstrates that it is possible for the Raman modes and I. R. modes, within the same crystal to vary differently with impurity concentration.

Detailed numerical results will be presented for the CaF_2-SrF_2 system. The linear shift of Raman frequency and the increased width of the Raman line with concentration are in quantitative agreement with the experimental observations.

INTRODUCTION

Recent experiments on the optical properties of mixed crystals, either Raman scattering (RS) or infrared absorption (IR), have yielded two characteristic types of spectra. Consider a mixed crystal of the type A_cB_{1-c} where c is the concentration of

*This work was supported in part by the Advanced Research Projects Agency, by the Joint Services Electronics Program (U.S. Army, U.S. Navy, and U.S. Air Force) under Contract N00014-67-A-0298-0006, and by the Division of Engineering and Applied Physics, Harvard University.

†Alfred P. Sloan Foundation Fellow.

crystal A mixed into crystal B. In one type the only change in a given spectral line, either RS or IR, is a linear shift in frequency as c varies from zero to one. The line-width first increases with c, peaks near c = 1/2, and then decreases as c approaches unity. For example, this behavior occurs for the Raman mode in CaF_2-SrF_2 [1], the infrared mode in CoO-NiO [2], and for some of the modes in ZnS-ZnSe [3]. We will designate this type of dependence on concentration as type I.

For the second type of change, designated II, one finds that as A is added to B new spectral lines appear, while the lines observed in the pure B crystal shift only slightly. When c increases further, the new lines grow in intensity, and shift frequency, while the original lines shift further, decrease, and eventually disappear. For example, Fig. 1 shows the behavior of the Raman mode observed for mixed crystals of Si-Ge [4]. A typical infrared spectrum of the same type is shown in Fig. 2 for CdS-CdSe [5].

Some crystals exhibit spectral changes with impurity concentration that are intermediate between types I and II. The purpose of this article is to present a theory for the RS and IR properties of mixed crystals that can describe both of these effects as well as the intermediate cases. We will also demonstrate some numerical results for this theory as applied to CaF_2-SrF_2 mixed crystals which exhibit type I Raman spectral changes, and intermediate IR changes. There is numerous other work on mixed crystal systems in the literature[6-12, 20-22].

BASIC THEORY

The study of Raman scattering and infrared absorption from phonons in crystals can be formulated in terms of the displacement-displacement Green's function. For example, the response of any classical linear system to a driving force F(t) can be written in terms of a retarded Green's function G as

$$x(t) = \int_{-\infty}^{\infty} dt' G(t-t') F(t') \tag{1}$$

For a statistical ensemble of such systems, the average power is obtained from

$$P \sim \langle \dot{x}(t) F(t) \rangle = \int_{-\infty}^{\infty} dt' \dot{G}(t-t') \langle F(t) F(t') \rangle \tag{2}$$

For a stationary random process, the correlation function $R(\tau) = \langle F(t)F(t') \rangle$ depends only upon (t-t'). Then

$$P = \int_{-\infty}^{\infty} d\tau G(\tau) R(\tau) \tag{3}$$

which gives, using Parseval's theorem, an average spectral power density

$$P(\omega) \sim \omega \, \mathrm{Im} G(\omega) R(\omega) \tag{4}$$

where $R(\omega)$ is the Fourier transform of $\langle F(0)F(\tau) \rangle$.

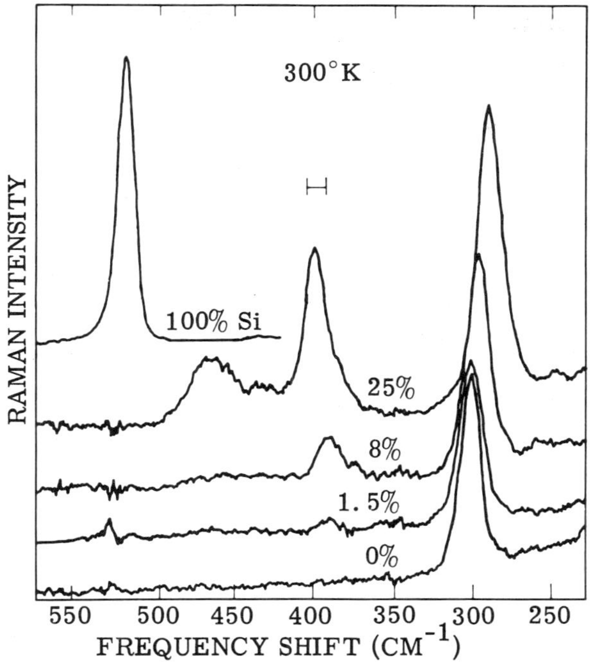

Fig. 1. Raman spectra of germanium-silicon alloys for several compositions. The instrumental resolution is indicated by the vertical lines. This figure is taken from the work of Feldman, Ashkin, and Parker[4].

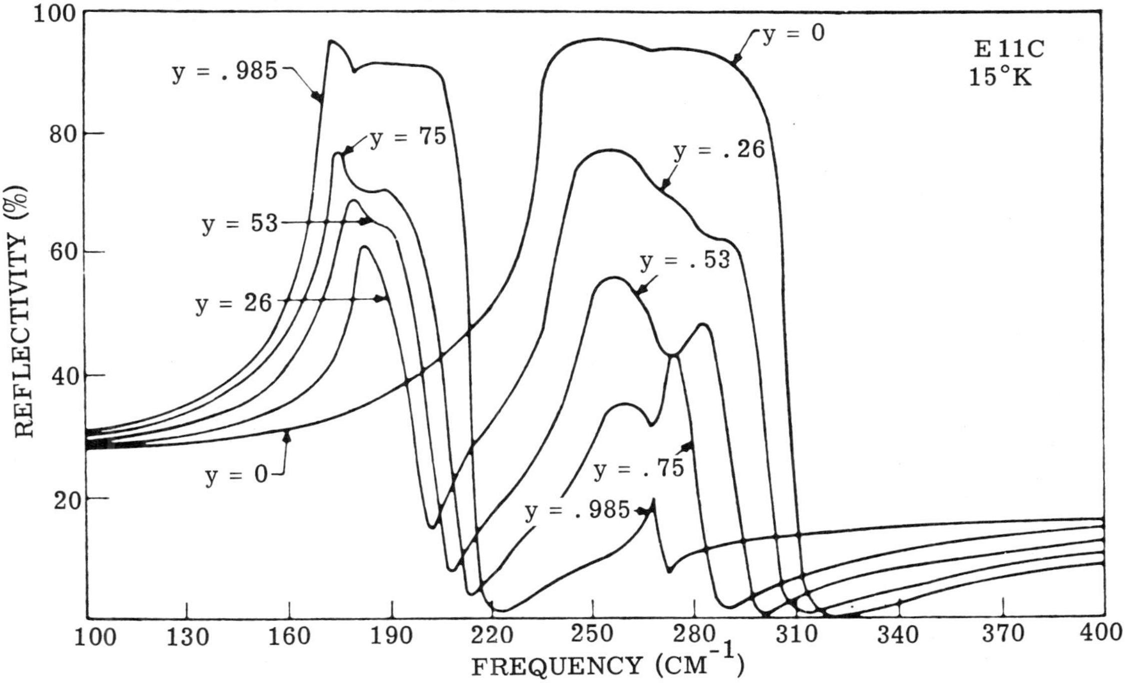

Fig. 2. Experimentally determined far-infrared reflectivity spectra of CdS and four mixed crystals of $CdSe_yS_{1-y}$ at 15°K with $\vec{E}\,||$ (c axis). A smooth line was drawn through experimental points. This figure is reproduced from the work of Verleur and Barker[5].

This represents one of the simplest examples of how one relates Green's functions for particular systems to observed power spectra.

For harmonic oscillator systems, the equations of motion for the Green's functions are the same in either a classical or quantum-mechanical formulation. For an arbitrary crystal lattice, we can define a set of Green's functions quantum mechanically by

$$G^{o}_{\alpha\beta}(\ell\kappa, \ell'\kappa';t) = \frac{1}{i}\theta(t)\left\langle[u_{\alpha}(\ell\kappa, t), u_{\beta}(\ell'\kappa', 0)]\right\rangle_T \tag{5}$$

where $u_{\alpha}(\ell\kappa, t)$ is the nuclear displacement in the direction $\alpha(\alpha = x, y, z)$ at site (ℓ, κ) with primitive cell $\vec{R}_{\ell}$ and basis index κ. The unit step function, $\theta(t)$, is defined to be 0 for $t < 0$ and 1 for $t > 0$. There are several excellent and readable review articles[12-19] on the application of Green's function techniques to defect problems in solids.

In general one deals with the Fourier transform of the Green's function, which satisfies $\underline{G}^{-1}(\omega) = \underline{M}\omega^2 - \underline{\Phi}$, where $\underline{\Phi}$ is the matrix of force constants in the harmonic approximation. For a perfect lattice, with translational invariance, the force constant matrix $\underline{\Phi}_o(\ell\kappa, \ell'\kappa')$ depends only upon $(\ell-\ell')$, and the mass matrix $\underline{M}_o$ does not depend on the cell index ℓ. The unperturbed Green's function matrix, $\underline{G}_o^{-1}(\omega) = \underline{M}_o\omega^2 - \underline{\Phi}_o$, can be expressed in a spectral representation in terms of the phonon eigenfrequencies $\omega_{\vec{\kappa}\sigma}$ and vectors $\underline{w}(\kappa|\vec{\kappa}\sigma)$ of the perfect lattice:

$$G^{o}_{\alpha\beta}(\ell\kappa, \ell'\kappa';\omega) = \tag{6}$$

$$\frac{1}{NM_{\kappa}^{1/2}} \sum_{\vec{k}\sigma} e^{i\vec{k}\cdot(\vec{R}_{\ell\kappa}-\vec{R}_{\ell'\kappa'})} \frac{w_{\alpha}(\kappa|k\sigma)w^{*}_{\beta}(\kappa'|\vec{k}\sigma)}{\omega^2 - \omega^2_{\vec{\kappa}\sigma}} \frac{1}{M_{\kappa'}^{1/2}}$$

If we define a defect matrix $\underline{V} = (\underline{M} - \underline{M}_o)\omega^2 - (\underline{\Phi} - \underline{\Phi}_o)$, then the Green's function $\underline{G}(\omega)$ of the perturbed lattice can be expressed in terms of $\underline{G}_o$ and $\underline{V}$ by $\underline{G} = (\underline{1} + \underline{G}_o\underline{V})^{-1}\underline{G}_o$. For a single isolated impurity, the defect matrix $\underline{V}$ affects only a small subspace of sites, and the techniques of matrix partition and group theory are useful for simplifying the calculations. For the cases of interest here, only Green's functions in the impurity subspace are required, and the matrices involved are reduced from $10^{23} \times 10^{23}$ to a much smaller dimension for which we can obtain explicit solutions.

Generalizing on the simple example discussed in the introduction, the formulation of physical properties in terms of Green's functions involves the evaluation of quantities as a function of $\omega+i\epsilon$ where ϵ is an infinitely small positive number. If we denote matrices defined only over the defect space by small letters, then

$$\underline{g}(\omega+i\epsilon) = [\underline{1} + \underline{g}_o(\omega+i\epsilon)\underline{v}]^{-1}\underline{g}_o(\omega+i\epsilon)$$

This equation leads to several interesting phenomena. At those frequencies ω, for which the pure crystal has no phonon modes, in the limit of $\epsilon \to 0$ $\underline{g}_o(\omega + i\epsilon)$ is a real number and if it should happen that $\lim_{\epsilon \to 0} \det(\underline{1} + \underline{g}_o(\omega + i\epsilon)\underline{v}) = 0$ for some such frequency one obtains "local modes" [13-15]. Because modes with a frequency outside of the band are not propagated by a perfect crystal, the so-called local modes are actually characterized by a high degree of spatial localization and the vibrations of the atoms in such a mode fall off rapidly away from the defect site.

If the real part of $\lim_{\epsilon \to 0} \det(\underline{1} + \underline{g}_o(\omega + i\epsilon)\underline{v})$ vanishes at frequencies for which the pure crystal does have phonon modes, $\lim_{\epsilon \to 0} \underline{g}_o(\omega + i\epsilon)$ is complex and one observes the so-called "resonance modes". They have a linewidth that can be related to the imaginary part of $\underline{g}_o(\omega + i\epsilon)\underline{v}$. There is extensive discussion in the literature of local modes and resonance modes, and no further attempt will be made to review the results here.

RAMAN SCATTERING

By arguments similar to the one given above for the simple harmonic oscillator, one can show that the intensity per unit solid angle for RS radiation is given by [18, 19],

$$I(\omega) \sim \sum_{\alpha\beta\gamma\delta} n_\alpha n_\beta i_{\alpha\gamma,\beta\delta}(\omega) E_\gamma E_\delta^* \tag{7}$$

where $\hat{n}$ is a unit polarization vector of the scattered radiation, and $\underline{E}$ is the (complex) amplitude of the electric field for the incident radiation. The function $i_{\alpha\gamma,\beta\delta}(\omega)$ is a fourth-order tensor which, for most practical cases, possesses the same symmetry as the elastic constants of the crystal; it can be expressed as the Fourier transform of a correlation function of the electronic polarizability [18, 19],

$$i_{\alpha\gamma,\beta\delta}(t) = \frac{1}{2\pi} \int_{-\infty}^{\infty} dt\, e^{i\omega t} \left\langle P_{\beta\delta}(t) P_{\alpha\gamma}(0) \right\rangle_T \tag{8}$$

where $P_{\beta\delta}(t)$ is an operator in the Heisenberg picture. The angular brackets here represent an average over a thermodynamic ensemble for the lattice vibrational states. Eq. (8) can be reduced to an expression containing a displacement correlation function by expanding the electronic polarizability $\underline{P}(t)$ in terms of the nuclear displacements:

$$P_{\alpha\beta}(t) = P^o_{\alpha\beta} + \sum_{\ell\varkappa\mu} P_{\alpha\beta,\mu}(\ell\varkappa) u_\mu(\ell\varkappa, t) + \dots \; . \tag{9}$$

The first term in (9) contributes to Rayleigh scattering (elastic), the second term to one-phonon Raman scattering, the next to second-order Raman scattering, and so on. Since we are not interested here in Rayleigh scattering, we shall neglect the first term in (9). For first-order Raman scattering, we obtain

$$i_{\alpha\gamma,\beta\delta}(\omega) = \sum_{\substack{\ell\varkappa\mu \\ \ell'\varkappa'\nu}} P_{\alpha\gamma,\mu}(\ell\varkappa) I_{\mu\nu}(\ell\varkappa,\ell'\varkappa';\omega) P_{\beta\delta,\nu}(\ell'\varkappa') \tag{10}$$

where

$$I_{\mu\nu}(\ell\varkappa,\ell'\varkappa';\omega) = \frac{1}{2\pi}\int_{-\infty}^{\infty} dt\, e^{i\omega t} \left\langle u_\mu(\ell\varkappa,t) u_\nu(\ell'\varkappa',0)\right\rangle_T \tag{11}$$

This thermodynamic correlation function is related to the displacement-displacement Green's function, and for positive ω, Eq. (7) becomes

$$I(\omega) \sim [I+n(\omega)] \ \mathrm{Im} \ \Sigma P_{\alpha\gamma,\mu}(\ell\varkappa) G_{\mu\nu}(\ell\varkappa,\ell'\varkappa';\omega+i\epsilon) \times P_{\beta\delta,\nu}(\ell'\varkappa') n_\alpha n_\beta E_\gamma E_\delta^* \tag{12}$$

where $G_{\mu\nu}(\ell\varkappa,\ell'\varkappa';\omega+i\epsilon)$ is defined above, and the quantity $n(\omega) = [\exp(h\omega/kT) - 1]^{-1}$ is just the thermal average of the phonon occupation number. For a perfect crystal the coupling coefficients are independent of the unit cell index ℓ and the sum over ℓ, ℓ' serves simply to project out the phonons at the center of the Brillouin zone (i. e. at $\underline{k} = 0$). The space-group operations of the host lattice determine specific relations between $P_{\alpha\gamma,\mu}(\ell,\varkappa)$ that then serve to select which particular zone center ($\underline{k} = 0$) phonons are Raman active. Examination of this symmetry provides one point of view for discussing the selection rules for RS.

For an imperfect lattice containing defects, translational symmetry is destroyed and scattering becomes possible from other phonons besides those at $\vec{k} = 0$. In general one has now to consider two separate effects which modify the observed RS. Firstly the presence of a defect at some site, can modify the coupling coefficients $P_{\alpha\gamma,\mu}(\ell\varkappa)$ in the vicinity of the defect. Thus even if the defect is of such a nature that it does not seriously alter the lattice dynamics, this effect can induce significant changes in the RS spectra. This effect will be most important when the defect introduces new strong electronic absorption bands at optical wavelengths. For example, the RS from U-centers[18] is most likely dominated by this effect. It would be rather hopeless to try and predict how the $P_{\alpha\gamma,\mu}(\ell\varkappa)$ vary near the defect and the only practical approach is to treat the variations in $P_{\alpha\gamma,\mu}(\ell\varkappa)$ as phenomenological constants that reflect the point-group symmetry of the defect site. In this case it becomes rather impossible to discuss anything except the low concentration limit of non-interacting defects. The problem reduces to calculations of the Green's functions in the vicinity of the defect, and subsequently using the new $P_{\alpha\gamma,\mu}(\ell\varkappa)$ to project out the Raman active part of the phonon spectra by means of Eq. (12). Interactions between defects become just too complicated to treat.

Schematically the spectrum of a crystal of N unit cells, with a concentration c of defects, distributed in some random configuration γ can be written

$$I(\omega, c, \gamma) \sim \operatorname{Im} \operatorname{Tr}\{\underline{P}(c, \gamma)\underline{G}(c, \gamma)\underline{P}(c, \gamma)\} \tag{13}$$

The observed spectra is obtained by taking an average over all configurations γ

$$\left\langle I(\omega, c)\right\rangle_\gamma \sim \operatorname{Im} \operatorname{Tr}\left\langle \underline{P}(c, \gamma)\underline{G}(c, \gamma)\underline{P}(c, \gamma)\right\rangle_\gamma \tag{14}$$

Defining $\underline{P}(c, \gamma) = \left\langle \underline{P}(c)\right\rangle_\gamma + \delta\underline{P}(c, \gamma)$,

$$\begin{aligned} I(\omega, c)_\gamma \sim \operatorname{Im} \operatorname{Tr}\{ & \left\langle \underline{P}(c)\right\rangle_\gamma \left\langle \underline{G}(c)\right\rangle_\gamma \left\langle \underline{P}(c)\right\rangle_\gamma \\ & + \left\langle \underline{P}(c)\right\rangle_\gamma \left\langle \underline{G}(c, \gamma)\delta\underline{P}(c, \gamma)\right\rangle_\gamma \\ & + \left\langle \delta\underline{P}(c, \gamma)\underline{G}(c, \gamma)\right\rangle_\gamma \left\langle \underline{P}(c)\right\rangle_\gamma + \left\langle \delta\underline{P}(c, \gamma)\underline{G}(c, \gamma)\delta\underline{P}(c, \gamma)\right\rangle_\gamma \} \end{aligned} \tag{15}$$

In the most general cases the $\delta\underline{P}(c, \gamma)$ would not be neglected and more complex averages would have to be taken. In the low concentration limit one might take $\left\langle \underline{P}(c)\right\rangle_\gamma \approx \underline{P}(0)$, $\left\langle \delta\underline{P}(c, \gamma)G(c, \gamma)\right\rangle \approx c\,\delta\underline{P}(1/N, i)\,\underline{G}(1/N, i)$, and $\left\langle \delta\underline{P}(c, \gamma)\underline{G}(c, \gamma)\delta\underline{P}(c, \gamma)\right\rangle \approx c\,\delta\underline{P}(1/N, i)\underline{G}(1/N, i)\delta\underline{P}(1/N, i)$, where the calculations, carried out for a single defect at some arbitrary site "i", yield a result independent of i.

INFRARED SPECTRA

A general discussion of the IR properties of mixed crystals is available elsewhere[12, 13]. In the approximation that all charges are replaced by some sort of average (i. e. $\delta\underline{P}(c, \gamma) = 0$) for crystals with only one infrared-active mode,

$$\epsilon(\omega) - \epsilon(\infty) \sim \left\langle \vec{k} = 0,\ TO \middle| \underline{M}_o^{1/2}\underline{G}(c, \omega+i\epsilon)\underline{M}_o^{1/2} \middle| \vec{k} = 0, TO \right\rangle \tag{16}$$

where the TO subscript indicates the projection of the IR active transverse optical phonon. Arguments similar to those previously given can be done for cases in which $\delta\underline{P}(c, \gamma) \neq 0$.

THE AVERAGE GREEN'S FUNCTION

There are several discussions of average Green's function techniques and numerous applications of these to the random crystal problem[18, 23-27]. Many of these involve complicated diagrammatic techniques common to many body theory. Some of these same results can be obtained more easily by a differential technique[28]. We will assume familiarity with[28] and thus restrict ourselves to some brief comments.

In many cases, it may be reasonable to assume that a mixed-crystal system formed from two similar isomorphs (such as the $Ca_xSr_{1-x}F_2$ and $Sr_xBa_{1-x}F_2$ systems) does not involve any appreciable changes in the $P_{\alpha\gamma,\mu}(\ell\varkappa)$ coefficients that characterize the pure host. A less stringent assumption, which accomplishes the same final result, is to assume merely that the polarizability coefficients for the mixed crystal have the same structure, on the average, as the pure (fluorite) crystal has. One would then take the $\delta\underline{P}(c,\gamma) = 0$.

Assuming the $\underline{P}$'s can be taken as average values, $\langle \underline{P}(c,\gamma)\rangle$, independent of ℓ, and with the same symmetries as for the pure crystal, the lattice sum in Eq. (12) will project out only $\vec{k} = 0$ part of the Raman active phonon modes. For the mixed crystal, in contrast to the pure crystal, $\vec{k}$ is not a good eigenvalue and instead of projecting out a number of discrete Raman active lines, Eq. (15) will now yield a spectral distribution. Written schematically, the averaged Eq. (14) becomes

$$I(\omega) \sim \mathrm{Im}(\underline{nE}:\langle \underline{P}\rangle)\cdot \langle \underline{G}(c,\omega+i\epsilon)\rangle\cdot (\underline{nE}:\langle \underline{P}\rangle) \tag{17}$$

One can show that if a crystal contains a small concentration of defects ($c << 1$) the averaged Green's function, from which one obtains the spectral information (e. g., Eqs. (16) and (17)) is given to lowest order in c as [28]

$$\underline{G}(c;\vec{k},\omega)_{\sigma\sigma'} = \sigma_{\vec{k}\vec{k}'}[\underline{1} + \underline{G}_0(\vec{k},\omega)\underline{F}(c;\vec{k},\omega)]^{-1}_{\sigma,\sigma'}\underline{G}_0(\vec{k},\omega)_{\sigma'\sigma'}, \tag{18}$$

where σ, σ' indicate the particular phonon branches. To lowest order in c, $\underline{F}$ can be expressed in terms of the isolated defect problem:

$$\underline{F}(c;\vec{k},\omega)_{\sigma\sigma'} \approx c \sum_{\substack{\ell_1\varkappa_1\alpha \\ \ell_2\varkappa_2\beta}} \exp[-i\vec{k}\cdot(\vec{R}_{\ell_1\varkappa_1} - \vec{R}_{\ell_2\varkappa_2})]w_\alpha(\varkappa|\vec{k}\sigma) \tag{19}$$

$$\times\, w^*_\beta(\varkappa'|\vec{k}\sigma')\langle \ell_1\varkappa_1\alpha|\underline{v}_0(\underline{1} + \underline{G}_0\underline{v}_0)^{-1}|\ell_2\varkappa_2\beta\rangle$$

where $\underline{v}_0$ is the defect matrix for a single impurity placed at the origin.

GENERAL DISCUSSION

Physical insight into the optical properties of mixed crystals can be obtained by considering the special case that a small concentration of defects are added to a pure crystal that has only one Raman active phonon mode. Substitution of Eq. (18) into Eq. (17), assuming the $P_{\alpha\beta,\mu}(\ell\varkappa)$ have average values, results in a spectral density of the form

$$I(\omega) \sim \mathrm{Im}[\omega^2 + i\epsilon - \omega_R^2 + \underline{F}(c;\vec{k} = 0, \omega + i\epsilon)_{\sigma_R\sigma_R}]^{-1} \tag{20}$$

From Eq. (19) $\underline{F}(c, \omega+i\epsilon)_{\sigma_R\sigma_R}$ is the projection of the average of $\underline{v}_o(\underline{1} + \underline{g}_o\underline{v}_o)^{-1}$ on the Raman active mode. For those defects which exhibit either local or resonance modes $\underline{g}_o\underline{v}_o$ is of order -1 and $\underline{F}$ has considerable structure. Fig. 3 shows a schematic example. This structure has the effect of producing a local (resonance) in $I(\omega)$ at ω_1 (near the local (resonance) mode at ω_o) as well as shifting the Raman mode by $-F(c, \omega_R)$. This type of spectral change has been observed numerous times[3-5, 9-11]. On the other hand, for weak defects, in which $g_o v_o << 1$, one can see from Eq. (19) that $\underline{F}(c, 0, \omega)_{\sigma_R, \sigma_R}$ is just equal to what one would expect from a virtual crystal approximation[1, 2, 8]. Because of the relative complexity of the computational aspects of

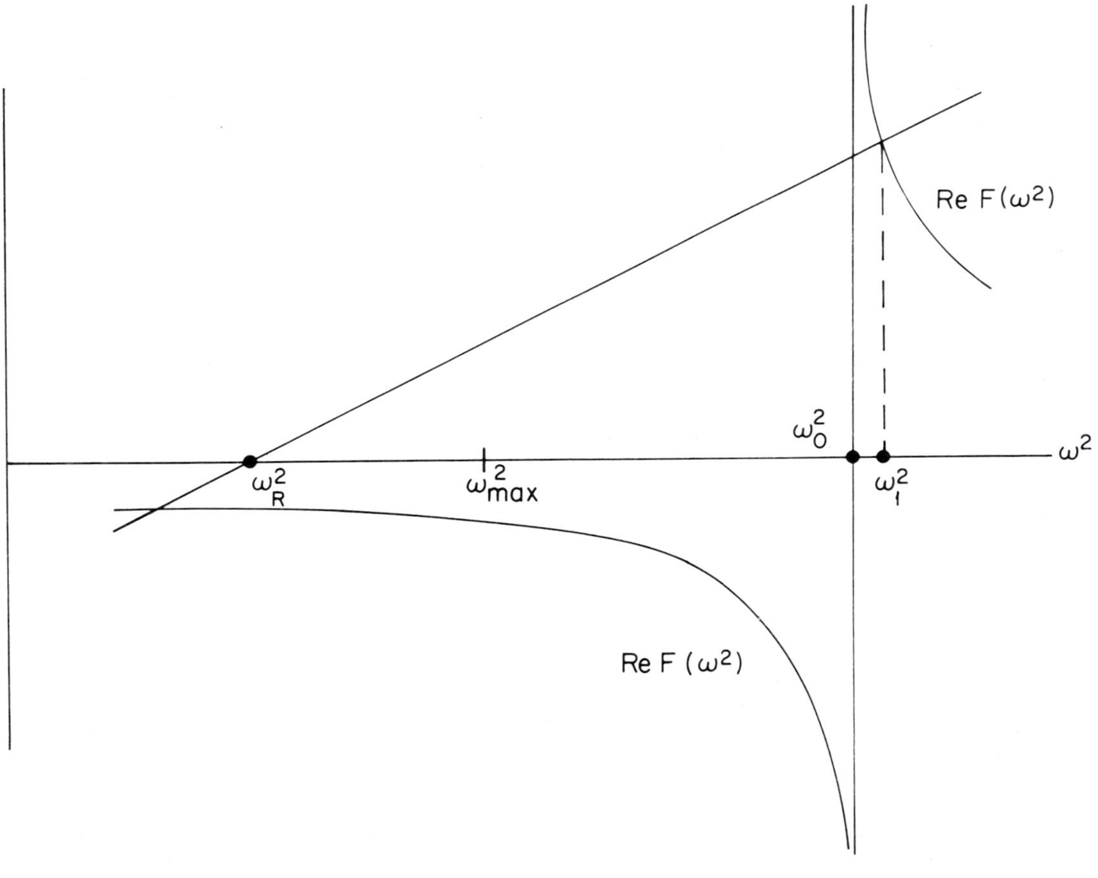

Fig. 3. Schematic example of behavior of proper self-energy near local mode.

these problems, Green's functions have not been calculated for many of the systems which have been studied experimentally. However, Professor Balkanski[29] has recently applied these techniques with success, in order to interpret most of the experimental data on mixed crystals of the II-VI type. In many cases, the assumptions leading to the approximation $\langle P\,G\,P\rangle = \langle P\rangle \langle G\rangle \langle P\rangle$ may not be justified, and for those problems, the average Green's function formulation is clearly inadequate to describe the situation completely.

NUMERICAL RESULTS FOR CaF_2-SrF_2

The first step in explaining the spectral changes that result from adding SrF_2 to CaF_2 is to specify a model from which one can calculate the Green's functions for pure CaF_2. As a practical matter, this first step requires the calculation of the eigenfrequencies $\omega^2(\vec{k}, \sigma)$ and the eigenvectors $\hat{W}(\varkappa | \vec{k}, \sigma)$ throughout the first Brillouin zone. In order to keep the calculation of manageable magnitude, the model we have used consists of a lattice of rigid (nonpolarizable) ions which interact through long-range Coulomb forces and short range repulsive forces. Although a "shell model" would be desirable for the pure CaF_2, it would expand the dimensions of the defect problem beyond our capacity. The rigid ion model corrected for electronic polarizabilities is a reasonably good approximation that has successfully predicted a number of experimentally observed quantities[30].

The details of our model for CaF_2 will be published elsewhere. Basically we assume five parameters to specify the short range force constant matrices between nearest neighbor Ca-F, F-F, Ca-Ca ions. The long range Coulomb forces are specified by the lattice constant and the effective charges on the Ca^{++} and F^- ions.

The $\omega(\vec{k}, \sigma)$ and the density of states curves obtained are shown in Fig. 4. The Raman (ω_R) and the transverse and longitudinal (ω_{TO} and ω_{LO}) frequencies are shown. The five short range constants and the effective charge are determined from experimental knowledge of these three frequencies and the three elastic constants.

In order to make the computational aspect of the mixed fluorite problem manageable, we shall assume that long-range Coulomb forces are not affected by the impurity, and that there are no force constant changes associated with the short-range interaction between the divalent metal ions.

We assume that the defect matrix $\underline{v}_o$ connects only the 9 atoms involved in the (SrF_8) subspace. It describes the change in the mass of the central cation (Sr^{++}), the change in the short range forces between the Sr^{++} and its eight nearest neighboring F^- ions, as well as the F^--F^- forces for the eight F^- ions in the complex.

Nine atoms, each with three degrees of freedom, lead to 27 x 27 dimensional matrices in the subspace of the impurity. Using the point group symmetry of the defect site one can reduce the number of independent Green's functions that must be calculated to 13. The details of this computation will be presented elsewhere.

Figs. 5 and 6 show the calculated proper self-energy functions (real and imaginary parts) that are relevant for the Raman and infrared modes. Considering the differences between the experimental and theoretical reflectivity spectra for the pure CaF_2 crystal the calculations for the (25% Sr, 75% Ca)F_2 crystal are in remarkably good agreement with the data (see Fig. 7). The disagreement for the pure crystal is one indication of the fact

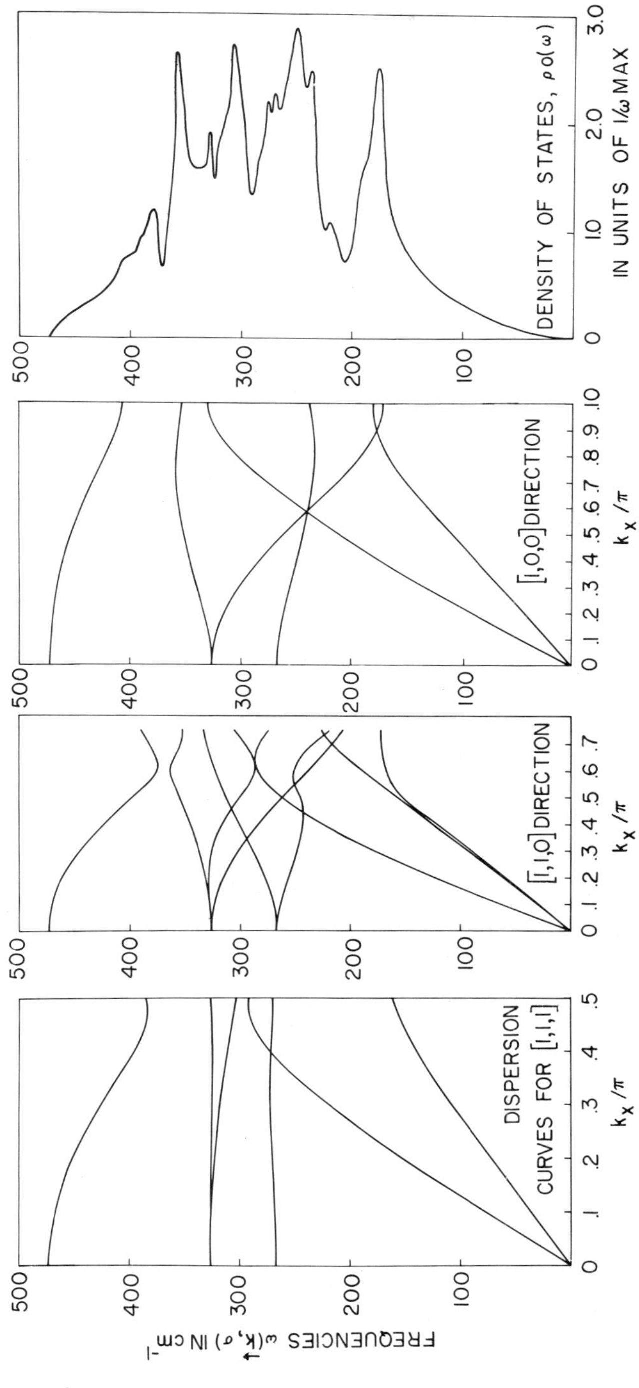

Fig. 4. Calculated phonon dispersion curves and density of states for CaF_2.

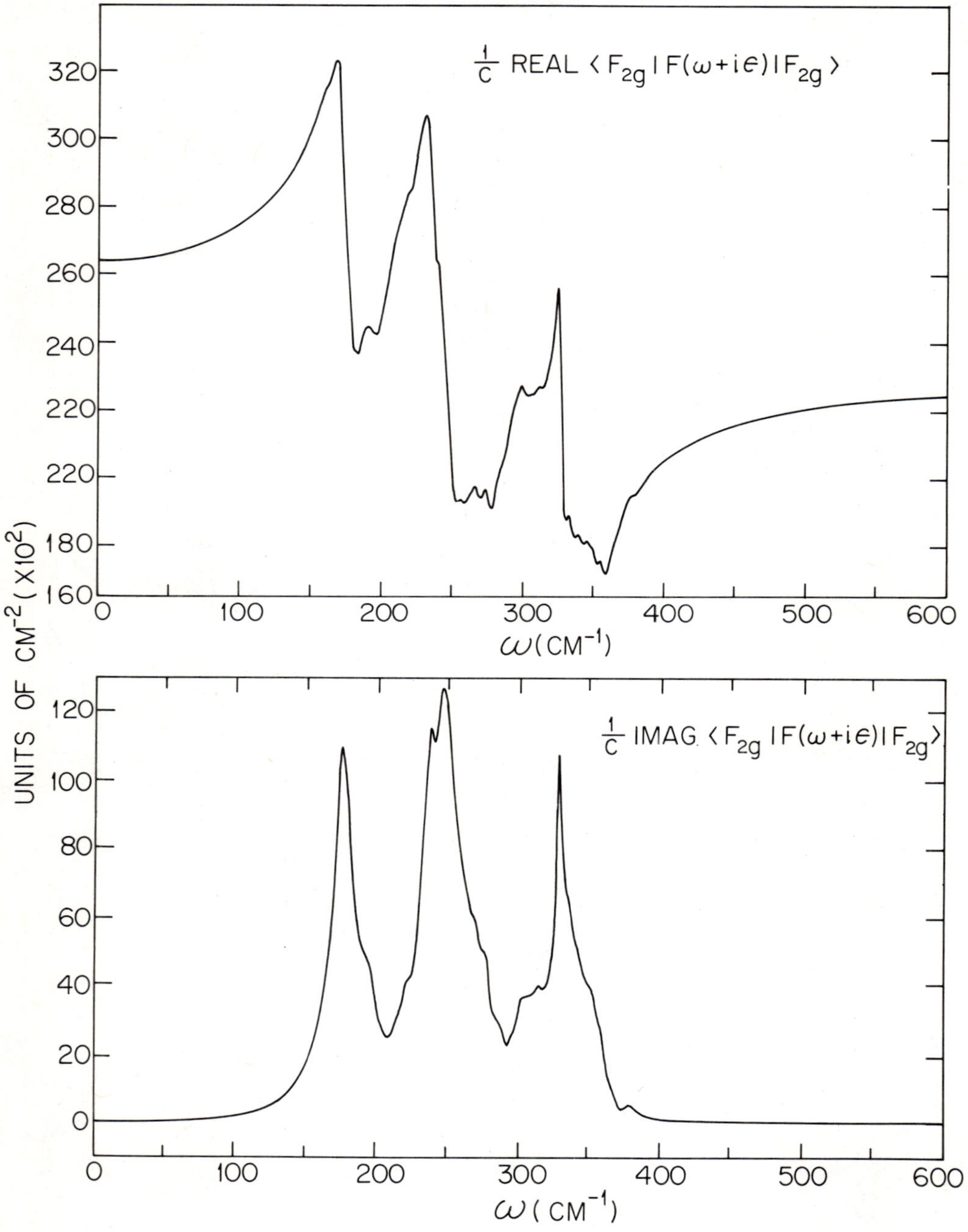

Fig. 5. Real and imaginary parts of the proper self-energy at $\vec{k} = 0$ that is relevant for the Raman scattering: $\frac{1}{c}\left\langle F_{2g}, \vec{k} = 0 \middle| \underline{F}(\omega+i\epsilon) \middle| F_{2g}, \vec{k} = 0 \right\rangle$.

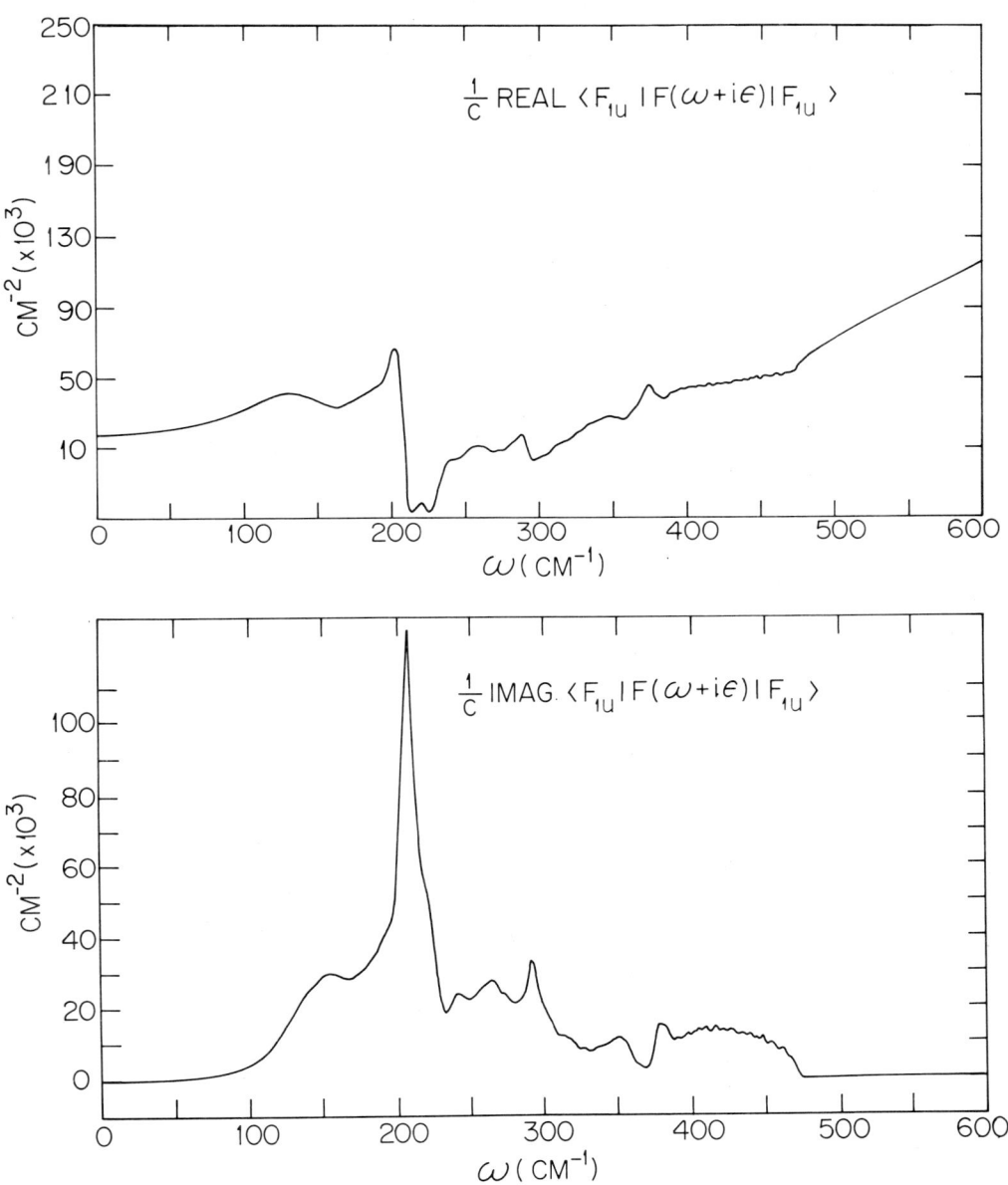

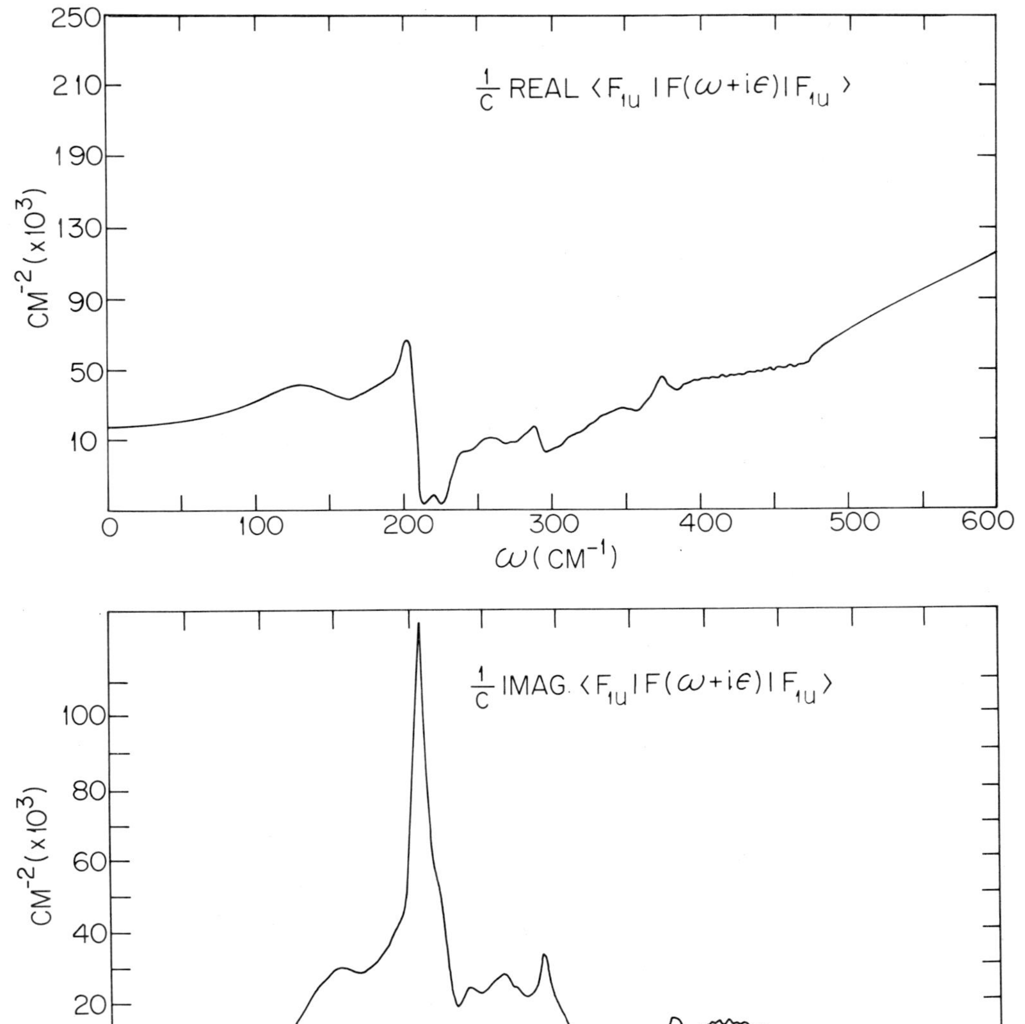

Fig. 6. Real and imaginary parts of the proper self-energy at $\vec{k} = 0$ that is relevant for infrared properties: $\frac{1}{c}\left\langle F_{1u}, \vec{k} = 0 \middle| \underline{F}(\omega+i\epsilon) \middle| F_{1u}, \vec{k} = 0 \right\rangle$.

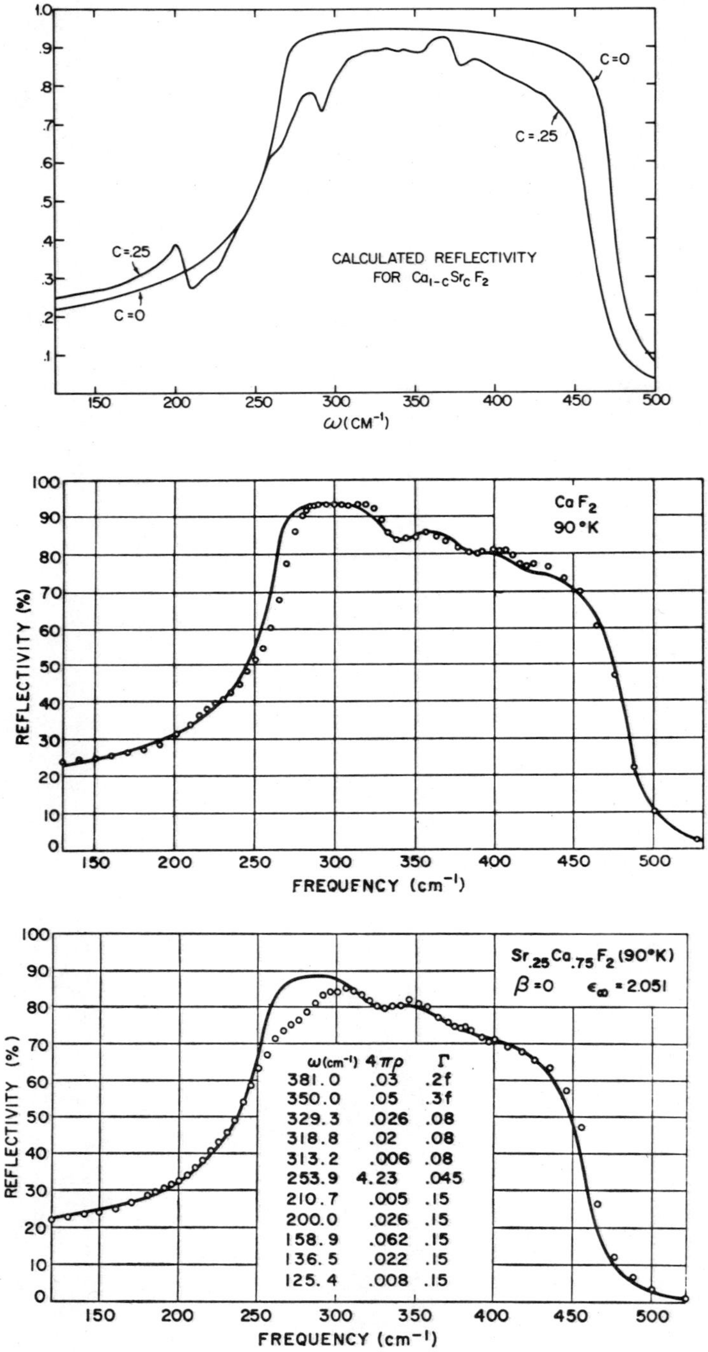

Fig. 7. Experimental reflectivity curves from Ref. [6], and calculated reflectivity curves using present theory.

that the crystal is not really well approximated by a purely harmonic solid. For such an ideal system the reflectivity would be 100% in the reststrahlen band. The theoretical curve for the mixed crystal does describe the general decrease of reflectivity, as compared with the pure crystal, and the rounding off of the edge at 280 cm^{-1}. The anomolous structure at 210 cm^{-1} is due to the sharp peak in the self energy function that is shown in Fig. 6. This coincides with the minimum in the density of states function calculated for the pure crystal (see Fig. 4). A more sophisticated model for either the pure crystal or the impurity might eliminate this structure, however it would be interesting to see whether or not one can observe any structure in the reflectivity spectra when one cools the sample to 4°K.

The calculations for the Raman mode are in perfect agreement with experiment[1]. When the results displayed in Fig. 5 are substituted into Eq. (20) one can calculate a theoretical Raman line shape for any concentration of impurity. The center of the calculated lines shift linearly with concentration in perfect agreement with experiment. The predicted line width variations are shown, in Fig. 8, to agree with the observed line widths.

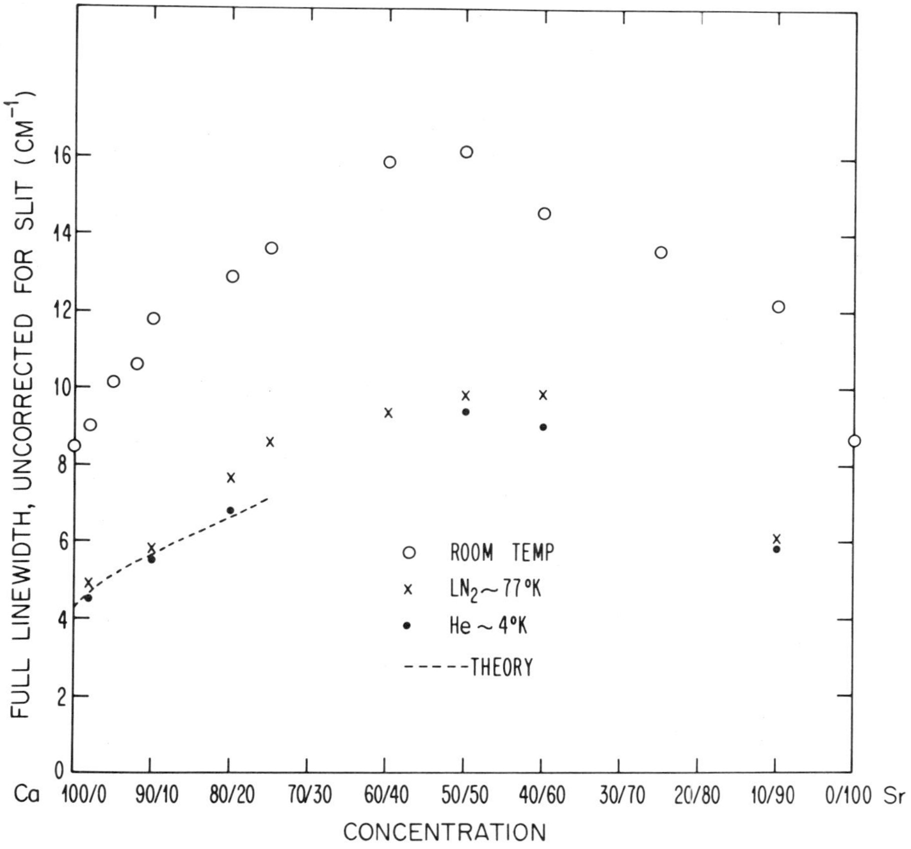

Fig. 8. Linewidth data for the Raman-mode in mixed crystals of $Ca_xSr_{1-x}F_2$.

REFERENCES

1. R.K. Chang, B. Lacina, and P.S. Pershan, Phys. Rev. Letters 17, 755 (1966).
2. P.J. Giellise, J.M. Plendl, L.C. Mansur, R. Marshall, S.S. Mitra, R. Mykolajewyca, and A. Smakula, J. Appl. Phys. 36, 2447 (1965).
3. O. Brafman, I.F. Chang, G. Lengyel, S.S. Mitra, and E. Carnall, Phys. Rev. Letters 19, 1120 (1967).
4. D.W. Feldman, M. Ashkin, and S.H. Parker, Jr., Phys. Rev. Letters 17, 1209 (1966).
5. H.W. Verleur and A.S. Barker, Jr., Phys. Rev. 155, 750 (1967).
6. H.W. Verleur and A.S. Barker, Jr., Phys. Rev. 164, 1169 (1967).
7. F. Von Oswald, Z. Naturforsch 14a, 374 (1959).
8. R.F. Potter and D.L. Stierwalt, "Proc. International Conf. on Phys. of Semiconductors," p. 1111, Donod, Paris, 1964.
9. M. Balkanski, R. Beserman, and J.M. Besson, Solid State Commun. 4, 201 (1966)
10. Y.S. Chen, W. Shockley, and G.L. Pearson, Phys. Rev. 151, 648 (1966).
11. H.W. Verleur and A.S. Barker, Jr., Phys. Rev. 149, 715 (1966).
12. Y.A. Izyumov, Adv. in Phys. 14, 569 (1965).
13. A.A. Maradudin, Report Progr. Phys. 28, 331 (1965).
14. A.A. Maradudin, Solid State Phys. 18, 273 (1966); Solid State Phys. 19, 1 (1967).
15. R.J. Elliott, "Phonons in Perfect Lattices and in Lattices with Point Imperfections," R.W.H. Stevenson (ed.), Plenum Press, Inc., N.Y., 1966.
16. R.J. Elliott, Argonne National Laboratory, Report No. ANL-7237 (1966).
17. R.J. Elliott, "Proc. International Conf. on Lattice Dynamics," 459, Copenhagen, 1963.
18. N.X. Xinh, J. Phys. (France) 28, Supp. No. 2, C1-103 (1967); Westinghouse Research Report No. 65-9F5-442-P8 (1965).
19. R. Loudon, Adv. in Phys. 13, 423 (1964).
20. M. Balkanski, J. Phys. (France) 28, Supp. No. 2 Cl-14 (1967).
21. T.P. Martin, Phys. Rev. 160, 686 (1968); 170, 779 (1968).
22. M.V. Klein and H.F. MacDonald, Phys. Rev. Letters 20, 1031 (1968).
23. D.W. Taylor, Phys. Rev. 156, 1017 (1967).
24. R.J. Elliott and D.W. Taylor, Proc. Roy. Soc. (London) A296, 161 (1967).
25. J.S. Langer, J. Math. Phys. 2, 584 (1961).
26. Hin-Chiu Poon and Arthur Bienenstock, Phys. Rev. 141, 7105 (1966); 142, 466 (1966).
27. R.W. Davies and J.S. Langer, Phys. Rev. 131, 163 (1963).
28. P.S. Pershan and W.B. Lacina, Phys. Rev. 168, 725 (1968).
29. M. Balkanski (private communication).
30. S. Ganesan and R. Srinivasan, Can J. Phys. 40, 74 (1962); 40, 91 (1962).

F-2: RAMAN SCATTERING FROM LATTICE VIBRATIONS OF $GaAs_xP_{1-x}$

N. D. Strahm* and A. L. McWhorter
Lincoln Laboratory, † Massachusetts Institute of Technology
Lexington, Massachusetts
and
Electrical Engineering Department, Massachusetts Institute of Technology
Cambridge, Massachusetts

ABSTRACT

First-order Raman scattering from the lattice vibrations of the mixed semiconductor $GaAs_xP_{1-x}$ has been studied with both He:Ne and YAG:Nd lasers. Polarization studies have allowed separation of the Raman spectra into two parts, which have the scattering symmetries expected of transverse and longitudinal phonon polarizations in the zincblende structure. The spectra show lattice oscillations in two distinct frequency bands, one of which (345-405 cm^{-1}) is identified with GaP, and the other of which (250-295 cm^{-1}) with GaAs. The prominent feature of each band is a pair of peaks, the lower peak having the symmetry of a transverse phonon, the upper having the symmetry of a longitudinal phonon. In the arsenic band of the phosphorus-rich samples there is a weaker resonance whose strength relative to the "main" transverse resonance increases with decreasing arsenic concentration and is in fact stronger than the "main" resonance in an $x \approx 0.07$ sample. Two weak spectral peaks occur in the phosphorus band; in the $x \approx 0.07$ sample they correlate with peaks in the phonon density of states of GaP. The nature of this fine structure indicates an inadequacy of the Verleur-Barker mixed crystal model in the interpretation of Raman and infrared spectra. A theory is also developed for the Raman scattering spectrum of an isotropic, multiply-resonant, infrared-active lattice.

INTRODUCTION

The mixed semiconductor $GaAs_xP_{1-x}$ has the zincblende structure and hence optical phonons that are both infrared and Raman active. Infrared reflectivity measurements by Chen, Shockley and Pearson (CSP)[1] and by Verleur and Barker (VB)[2] have shown

*Supported by the National Science Foundation during a portion of this work.
†Operated with support from the U. S. Air Force.

lattice vibrations in two distinct frequency bands, the higher band (the "phosphorus" band) in the range of frequencies of lattice vibrations of GaP and the lower band (the "arsenic" band) in the range of those of GaAs. In addition, each band contains some fine structure, which VB used to fit the parameters of their dynamical model of mixed crystals.

We report here Raman spectra of $GaAs_xP_{1-x}$ in the frequency range 240-420 cm^{-1}. Temperature dependence of the spectra shows this frequency region to be dominated by first-order processes. By using polarization to separate the spectra into parts characteristic of transverse and longitudinal phonons, much more detailed information has been obtained than is possible by infrared reflectivity. The results are in general agreement with the infrared work, but the position and nature of the fine structure indicate that the VB model is inadequate.

The experimental techniques involved in the polarization studies are described in Sec. 2 and the results are presented and discussed in Sec. 3. A general theory for Raman scattering from a multiply-resonant, infrared active, isotropic crystal is derived in Sec. 4 in terms of quantities which can be obtained from models of the lattice dynamics and of the Raman photon-phonon coupling. Adjustment of the parameters of the models to fit the experimental spectra should eventually enable one to disentangle the effects of lattice dynamics from those of the deformation and electro-optic coupling in the Raman scattering from mixed crystals.

EXPERIMENTAL

The absorption edge of $GaAs_xP_{1-x}$ at 300°K varies from 1.43 eV for GaAs to 2.20 eV for GaP. One consequence of the shifting edge is that crystals with greater than ~ 50% arsenic content have a band gap at room temperature less in energy than the photon energy of the helium-neon laser (1.96 eV). With increasing arsenic content the laser beam becomes heavily absorbed in the crystal, resulting in a reduced scattering length, an absorption of the scattered light, and luminescence from laser-excited electrons. Consequently, the neodymium doped YAG laser (1.17 eV) was used for the study of arsenic-rich samples. Even so, luminescence from n-type crystals of carrier concentrations 10^{17} - 10^{18} cm^{-3} masked the Raman scattering.

The samples were epitaxially grown {111} platelets,* typically 4 mm on a side and 0.2 mm thick, with carrier densities of about 10^{15} cm^{-3}. The compositions studied in percent arsenic were ≈ 7%, 15%, 32.5%, 52%, 71.5%, 94.6%. All concentrations except the ≈ 7% one were determined for the infrared studies by x-ray analysis of lattice parameters. The ≈ 7% concentration was determined simply by fitting the observed frequency of the Raman lines onto smooth curves extrapolated from the other concentrations.

The laser beam was propagated in the plane of the platelets along both the ±[$11\bar{2}$] and ±[$\bar{1}10$] directions, and observation was made at right angles along [111] direction perpendicular to the platelet face. The polarization of the laser beam and analyzed scattered light were taken either perpendicular or parallel to the scattering plane formed by the incident beam and scattered light directions. Part of the scattered light entering the spectrometer originates as Raman light which is scattered in the direction opposite to that of observation, but which is reflected into the spectrometer from the back surface of the platelet. With this effect taken into account, the polarization

*These samples were kindly loaned by Professor G.L. Pearson of Stanford. Some were crystals used in the CSP infrared work.

dependences of the scattering tensors given by Loudon[3] for these propagation axes were verified within experimental uncertainty for the main spectral peaks of the phosphorus band in the 15% and 32.5% arsenic samples.

RESULTS AND DISCUSSION

Room temperature spectra of the 32.5% arsenic sample with the laser beam in the [$11\bar{2}$] direction are shown in Fig. 1 for two polarization configurations, and similar spectra for the 15% arsenic sample are shown in Fig. 2. In the absence of reflection of the laser beam and of the scattered light inside the sample, we expect in the ($\perp\perp$) configuration to obtain light scattered only from longitudinal phonons and in the ($\perp\parallel$) configuration to see scattering only from transverse phonons. The internal reflection of the scattered light adds a transverse phonon spectrum to that of the ($\perp\perp$) configuration of about one third the intensity obtained in the ($\perp\parallel$) configuration.

The 32.5% arsenic spectra are representative of the mixed crystals in that the prominent feature of each band is a pair of peaks, the lower peak originating from transverse (TO) phonons, the upper from longitudinal (LO) phonons. Fine structure is observed at frequencies below the main arsenic band and between the TO and LO peaks of the phsophorus band. These general features are in agreement with the infrared observations. All spectra can be consistently interpreted as being the superposition of these two (in general overlapping) TO and LO spectra.

The TO-LO pair of the arsenic band is resolved in the 32.5% arsenic sample only by the use of polarization. In the spectra of the 15% sample, although the ratio of LO to TO strength varies between the two polarization configurations by an order of magnitude in the phosphorus band, the ratio in the arsenic band hardly changes. This indicates that the two peaks of the arsenic band are not the TO-LO pair but represent two distinct resonances of arsenic in GaP. From the composite of unpolarized spectra shown in Fig. 3 we see that in the arsenic band it is the upper frequency peak which corresponds to the main TO-LO pair, the splitting being unresolved. The lower frequency peak is the stronger of the two in the 7% arsenic sample but it loses strength relative to the upper frequency peak with increasing arsenic concentration. At 32.5% and 52% arsenic concentration it is seen as a shoulder and as a tail respectively.

With increasing arsenic concentration the transverse spectral strength of the phosphorus band shifts from the peak into the frequency region between the TO-LO pair. The fine structure observed in the low arsenic concentration samples, however, arises from longitudinal modes. The fine structure comes out more clearly at low temperatures as illustrated in the 7% arsenic spectrum in Fig. 4.

Frequencies of the room temperature resonances are plotted in Fig. 5 along with the frequencies of the VB model. The model parameters used here were very slightly adjusted from those of VB in order to fit more closely the Raman data of GaP and GaAs. Each band of the model has four transverse and four longitudinal modes alternating in frequency. The four transverse modes and the high frequency, strong longitudinal mode of each band are indicated in the figure. In the arsenic band the three lower frequency longitudinal modes are almost degenerate with the three lower transverse modes. In the phosphorus band they lie close to the three upper frequency transverse modes at low arsenic concentration and drop to the three lower transverse frequencies with increasing arsenic. We will not go into details of the model, but one expects the relative strength in the transverse part of each band to shift from the lowest transverse mode to the upper modes with increasing arsenic concentration such as is observed in the phosphorus band. The clustering of like cations around gallium sites determines the manner in which the shift occurs. The fact that the "main" mode of the arsenic band retains so much of the band strength at low arsenic concentration requires, in terms of the model, that considerable clustering occurs.

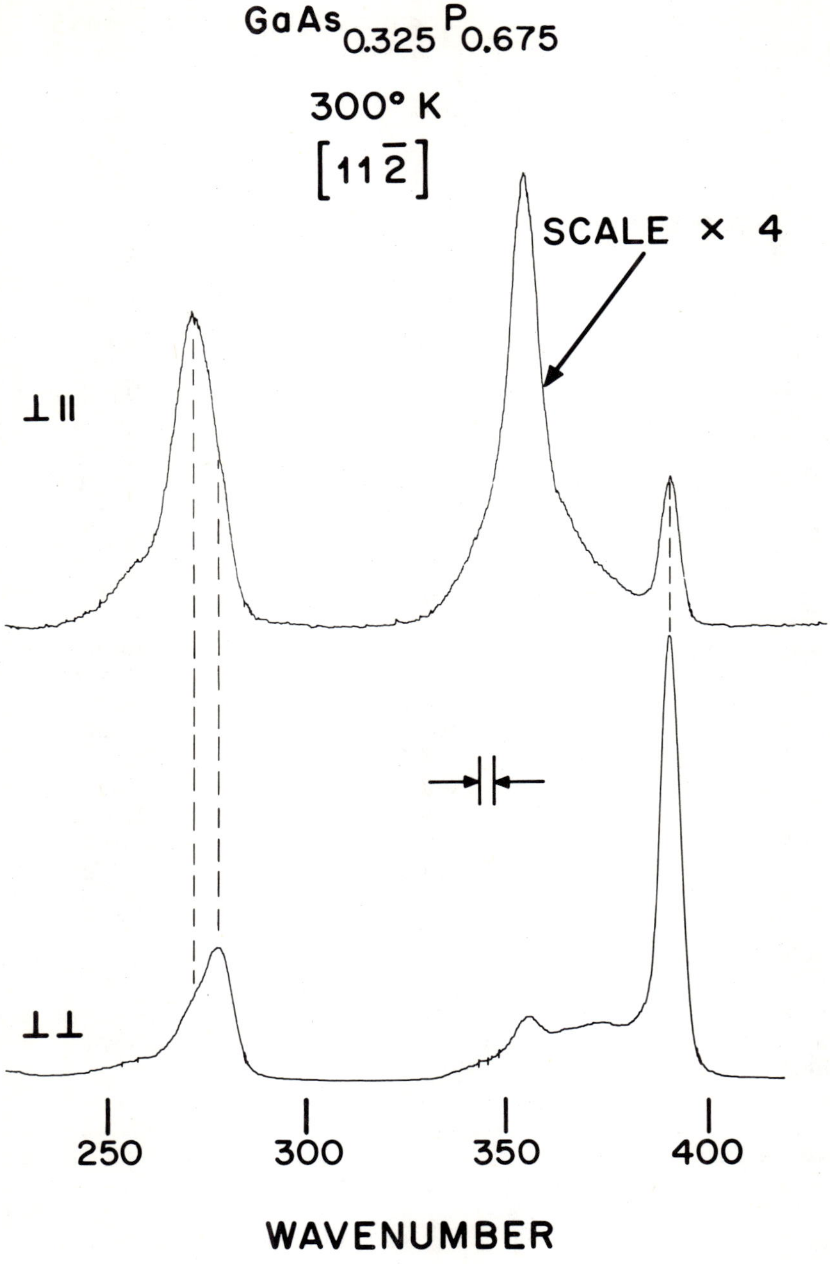

Fig. 1. Polarized Raman spectra of room temperature $GaAs_{0.325}P_{0.675}$. The first symbol of the notation (⊥ ⊥) and (⊥ ∥) refers to the laser beam polarization, and the second to the polarization of the analyzed scattered light with respect to the scattering plane. The spectrum of longitudinal phonons is expected in the (⊥ ⊥) configuration and of transverse phonons in the (⊥ ∥) configuration. Internal reflections of laser beam and scattered light prohibits complete separation.

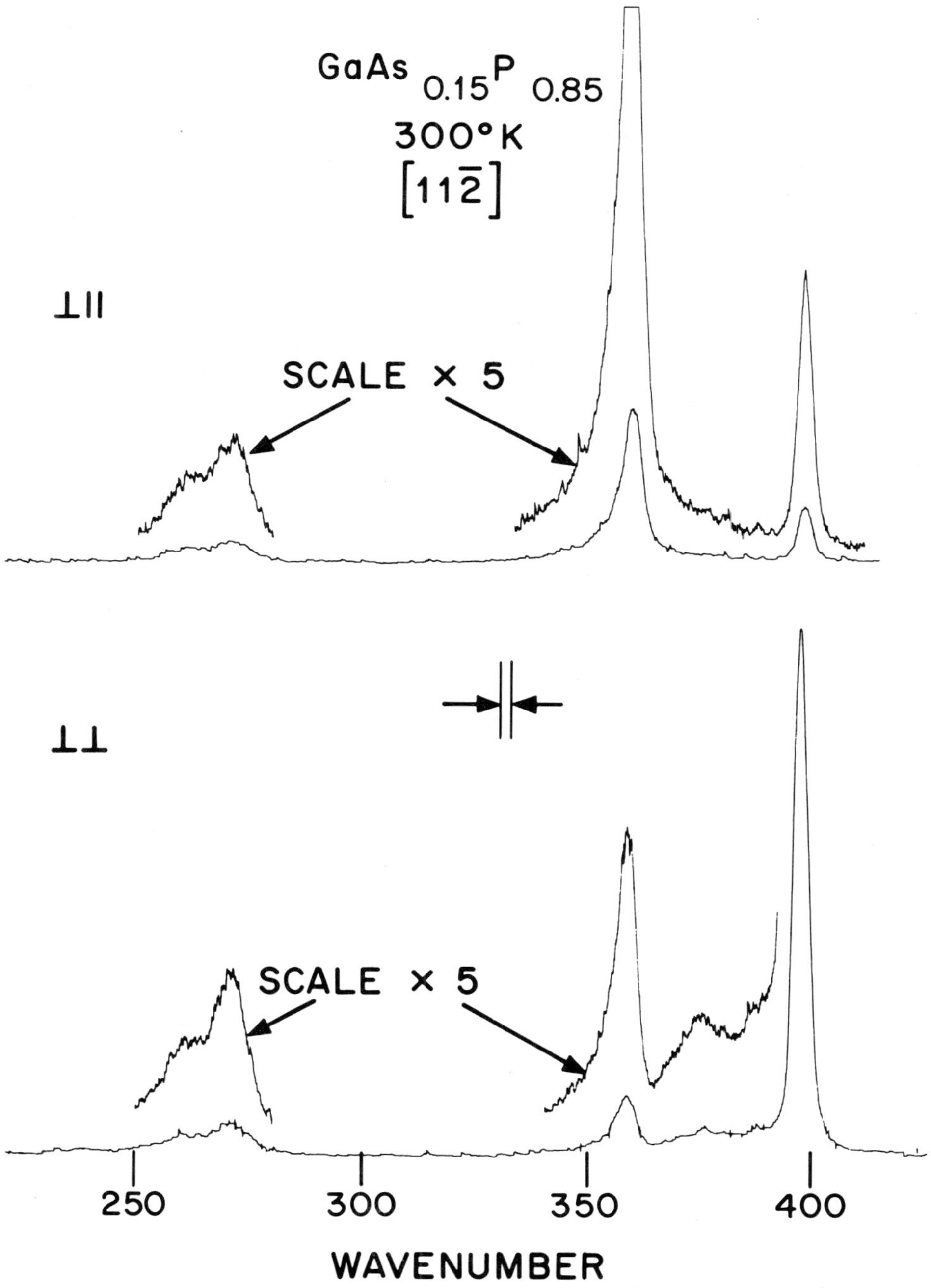

Fig. 2. Polarized Raman spectra of room temperature $GaAs_{0.15}P_{0.85}$, taken with the same polarization configurations as in Fig. 1. The arsenic band shows only a slight dependence on polarization.

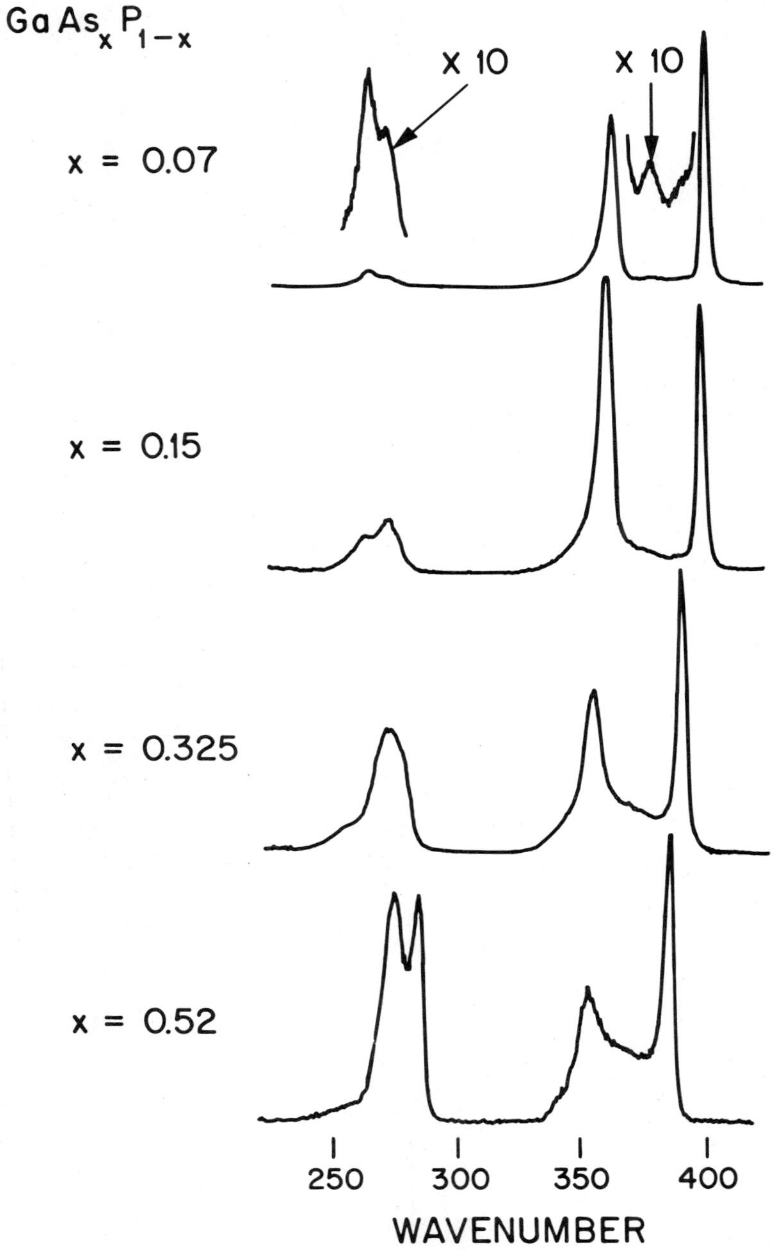

Fig. 3. Composite of unpolarized spectra of the phosphorus-rich samples. The spectra of the three lowest arsenic concentration samples were taken at room temperature. The 52% arsenic spectrum was taken at liquid nitrogen temperature and is shifted $\approx 2\ cm^{-1}$ toward lower frequencies to match the frequency peaks of room temperature spectra. Details of room temperature spectra in the latter sample are masked by luminescence.

We note that the model predicts too low a frequency for the fine structure of each band. Also the arsenic band of the model does not support the longitudinal mode in mid-concentration samples at high enough frequency. The latter discrepancy is consistent with the former in that, if all the infrared strength of the arsenic transverse band were placed at the frequency of the main mode (such as in the simpler model of CSP) the TO-LO frequency difference agrees with experiment.

We also note that although the transverse peak of the phosphorus band becomes quite broad in frequency as arsenic is added, the width of the longitudinal peak does not appreciably increase. This phenomenon should be quite general in infrared-active mixed crystals in which the longitudinal peak is well removed from the transverse. The frequency spread of the transverse peak can result from a distribution of energy eigenvalues rather than from a decay in time of the eigenfunctions. The highest frequency longitudinal peak, however, results from a macroscopically-determined "in-phase" oscillation of all the lesser-frequency transverse eigenfunctions. Its frequency width is determined by a decay in time rather than a statistical distribution of eigenvalues. If the combined infrared strength of the transverse modes is not sufficient to displace the longitudinal mode, the latter will also appear as a broad peak, overlapping the transverse.

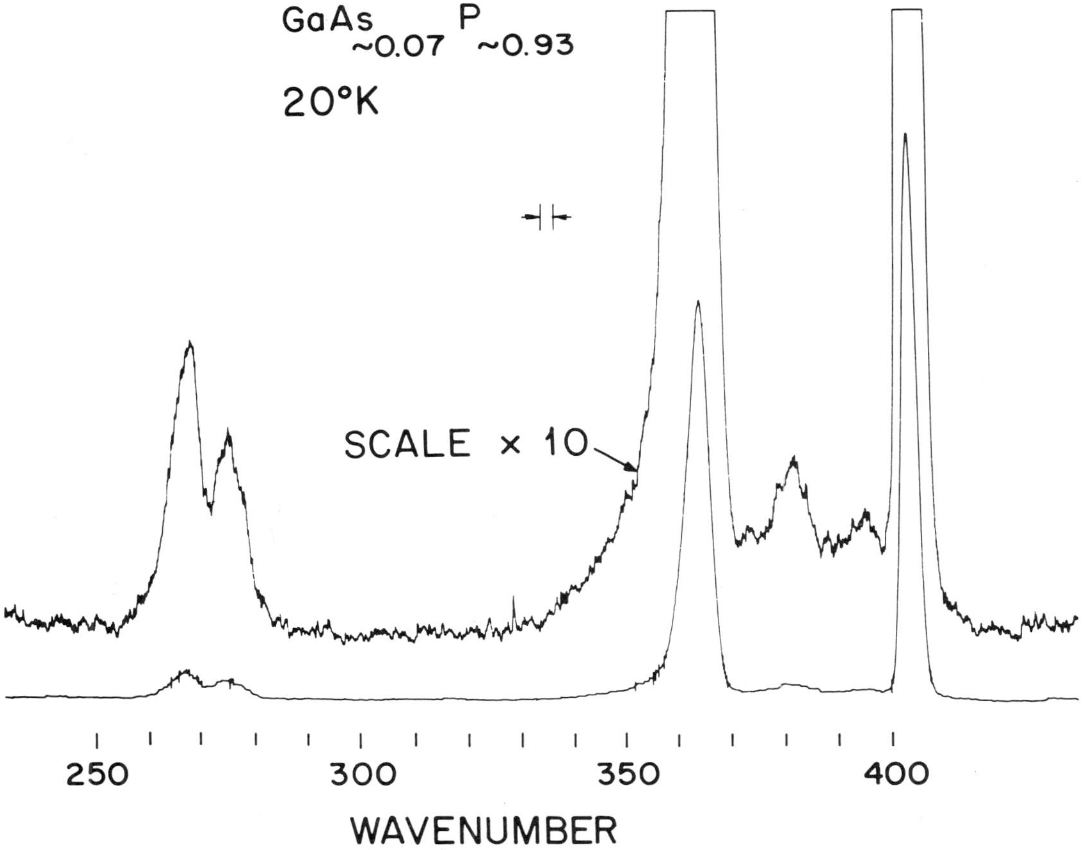

Fig. 4. Low temperature unpolarized spectrum of $GaAs_{0.07}P_{0.93}$. Line widths are less than at room temperature and the fine structure in the phosphorus band is more clearly defined.

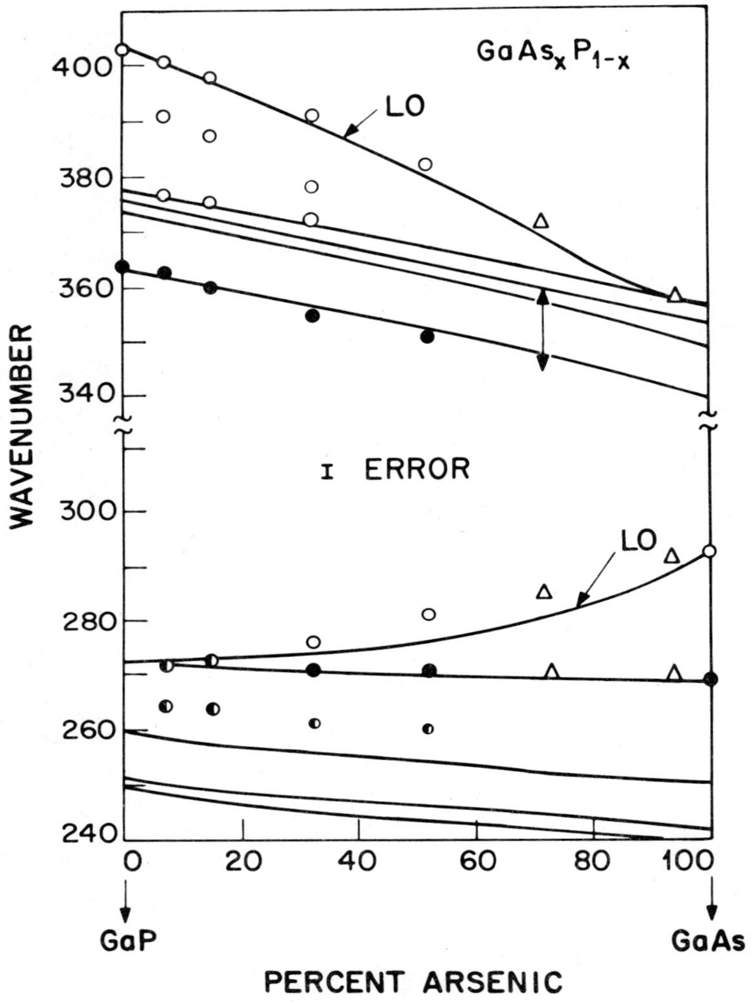

Fig. 5. Comparison of experimental room temperature Raman resonances with frequencies of the Verleur-Barker model. The open circles indicate experimental peaks (or shoulders that become peaks at low temperature) which have the polarization symmetry of longitudinal vibrations. The closed circles indicate vibrations of transverse phonon symmetry. The half-closed circles represent peaks that showed both symmetries, and the triangles indicate peaks whose symmetry was not checked. The lowest frequency points in the arsenic band of the 32.5% and 52% samples represent a broad shoulder and tail respectively. Structure in the phosphorus band becomes difficult to identify with increasing arsenic concentration; the main transverse peak loses strength relative to the continuum which forms in the middle of the band. The double arrow of the 72% sample indicates the broad spectrum. The solid curves correspond to the transverse modes and to the strongest of the four longitudinal modes; the other three longitudinal modes, which alternate in frequency with the transverse modes, are not shown.

A light mass, such as phosphorus, isolated in the GaAs lattice should create a localized mode, which (following VB) we assume to have a frequency obtained by extrapolation of the mixed crystal phosphorus band toward GaAs. An interpretation of the modes of arsenic observed in GaP as obtained by extrapolation in the other direction has been made possible by the recent GaP phonon dispersion curves obtained at 300°K by neutron spectroscopy and the fitting of a shell-model to the data[4]. In the 7% arsenic concentration sample the two modes at $264.5 \pm 1.0\ cm^{-1}$ and $271.4 \pm 1.0\ cm^{-1}$ are between the acoustic and optical phonon bands; consequently these resonances would extrapolate to "gap" modes, which are localized in space. One of the modes at least must correspond to a cluster of greater than one arsenic atom. The fine structure in the phosphorus band seems to occur at frequencies corresponding to high density of one-phonon states in the optical band of GaP. For instance the peak at $377.0 \pm 1.0\ cm^{-1}$ corresponds to the frequency of an LO phonon at the zone-edge L point and to minima in the optical branches along $<110>$ directions. The $390.5 \pm 1.0\ cm^{-1}$ peak is just below a peak in the density of states near the K-point of the zone.

Vibrations at frequencies within the bands of the host lattice are the "in-band" modes of isolated defect theory. By arguments similar to those used by Dawber and Elliott[5] in describing infrared absorption by isolated defects, one can show that the "in-band" modes can contribute as much to Raman scattering as the localized and gap modes, and that the "in-band" spectrum is heavily weighted or even controlled by the one-phonon density of states. These conclusions provide some theoretical basis for the identification of observed structure with critical points in the phonon dispersion curves of the host crystal.

Two relatively strong low-frequency Raman bands ($\sim 100\ cm^{-1}$ and $210\ cm^{-1}$) are observed in the phosphorus-rich samples. These bands are contributed in part by two-phonon processes, but probably also contain one-phonon "in-band" modes of the acoustic branches. The frequencies correspond well with acoustic zone-edge phonons in the neutron data. No "in-band" structure can be identified in the GaAs end of the mixed crystals. A major reason is the lower sensitivity of the YAG laser Raman system.

THEORETICAL SPECTRA

The first-order Raman spectrum can be obtained directly from the fluctuation spectrum of the dielectric constant $< \delta\epsilon\,\delta\epsilon^{\dagger} >_{q\omega}$ when the Raman frequency shift is small compared to the incident frequency. We outline a derivation of the spectrum in the small wave vector approximation for a multiply-resonant, infrared-active, isotropic crystal. This result is valid for regular crystals of many atoms per unit cell and also applicable to models of mixed crystals in which a relatively small number of lattice coordinates are assumed adequate to describe the system. It is assumed that the equations of motion for the lattice can be put into the form

$$(\omega^2 - \omega_n^2 + i\Gamma_n\omega)\,\delta u_n + \beta_n \delta E = -\,\delta f_n \qquad (1)$$

where δu_n are normalized displacements of the relevant lattice "oscillators," ω_n their resonant frequencies (frequencies of transverse vibration at wave vector large compared to that of the polariton region), δE the macroscopic electric field and δf_n, in the simplest interpretation, Langévin noise generators which induce the thermal fluctuation in the medium. Any lattice model linear in its parameters should reduce to this form.

The fluctuation spectrum of the noise generators is related to the phenomenologically introduced loss coefficients Γ_n through the fluctuation-dissipation theorem, such that

$$\langle \delta f_m \, \delta f_n^{\dagger} \rangle_{q\omega} = \frac{\hbar}{\pi} \delta_{mn} \Gamma_n \omega \, (n_\omega + 1) \tag{2}$$

where n_ω is the Bose-Einstein occupation factor. We have given the fluctuations appropriate to Stokes scattering. The fluctuations in the dielectric constant can be expressed in terms of the lattice oscillators and electric field as

$$\delta\epsilon = \sum_n \frac{\partial\epsilon}{\partial u_n} \delta u_n + \frac{\partial\epsilon}{\partial E} \delta E \tag{3}$$

However, we may eliminate the electric field from (3) by the constraint imposed by Maxwell's equations:

$$\delta D = \epsilon_\infty \delta E + 4\pi\delta P = \epsilon_\infty \delta E + 4\pi \sum_n \beta_n \delta u_n \tag{4}$$

$$= \begin{cases} 0 & \text{longitudinal} \\ (qc/\omega)^2 \, \delta E & \text{transverse} \end{cases}$$

One can show through a general Hamiltonian formulation that the β_n in (4), the "charge" associated with oscillator n, is the same as the β_n introduced in (1).

We can now solve for the coordinate fluctuations from (1) with the use of (2) and (4). More formally, we could also use the fluctuation-dissipation theorem for systems of many variables[6] to get the lattice fluctuations from (1) and (4) alone and derive (2). In either case the resulting spectrum for the longitudinal case is given by

$$\langle \delta\epsilon\delta\epsilon^{\dagger} \rangle_{q\omega} = \frac{\hbar}{\pi} (n_\omega + 1) \, \mathrm{Im} \left[\sum_n \frac{\varphi_n^2}{\xi_n} - \frac{4\pi}{\epsilon(\omega)} \left(\sum_n \frac{\varphi_n \beta_n}{\xi_n} \right)^2 \right] \tag{5}$$

where

$$\varphi_n = \frac{\partial\epsilon}{\partial u_n} - \frac{4\pi\beta_n}{\epsilon_\infty} \frac{\partial\epsilon}{\partial E}$$

$$\xi_n = \omega_n^2 - \omega^2 - i\Gamma_n \omega$$

and

$$\epsilon(\omega) = \epsilon_\infty + \sum_n 4\pi\beta_n^2/\xi_n$$

is just the dielectric function. The spectrum for scattering from transverse phonons and polaritons is obtained by replacing ϵ_∞ by $\epsilon_\infty - (qc/\omega)^2$ in the expressions for φ_n and $\epsilon(\omega)$. In analogy to the designation of β_n (or β_n^2/ω_n^2) as the "infrared strength," the quantity φ_n (or φ_n^2/ω_n^2) can be called the "Raman scattering strength." It consists of two parts, the deformation contribution, $\partial\epsilon/\partial u_n$, and the electro-optic contribution, proportional to $\beta_n \partial\epsilon/\partial E$.

For right-angle scattering from transverse phonons, the electro-optic contribution is negligible because of the $(qc/\omega)^2$ term in the denominator. Also in this case the second term in (5) is negligible.

The application of (5) to a mixed crystal requires: (a) a model of the lattice dynamics (such as those of CSP or VB), from which one obtains equations of motion of the form of (1), and hence ω_n and β_n; and (b) a model or calculation of the optical-frequency dielectric constant as a function of the lattice parameters, from which the coefficients of (3) are obtained. Modeling of the deformation and electro-optic coefficients in mixed crystals is complicated by the fact that the electron (and exciton) band structure changes with crystal composition. Resonant enhancement of the Raman photon-phonon coupling by virtue of the small (and varying) difference between laser energy and electron band gap can affect the deformation and electro-optic coefficients differently. We are presently using various approximations in conjunction with the VB model to compute spectra from (5) for comparison with the experimental data. The results will be reported elsewhere.

ACKNOWLEDGMENT

We wish to thank Dr. G. B. Wright for suggesting this problem and for considerable help throughout the course of the experimental work. We gratefully acknowledge the loan of the samples by Professor G. L. Pearson of Stanford. The aid of Dr. Aram Mooradian and the use of his YAG laser system is greatly appreciated.

REFERENCES

1. Y.S. Chen, W. Shockley, and G.L. Pearson, Phys. Rev. 151, 648 (1966).
2. H.W. Verleur and A.S. Barker, Jr., Phys. Rev. 149, 715 (1966).
3. R. Loudon, Advan. Phys. 13, 423 (1964).
4. J.L. Yarnell, J.L. Warren, R.G. Wenzel, and P.J. Dean, "International Conference on Inelastic Neutron Scattering," Copenhagen, May 1968 (to be published).
5. P.G. Dawber and R.J. Elliott, Proc. Phys. Soc. 81, 453 (1963).
6. L.D. Landau and E.M. Lifshitz, "Statistical Physics," p. 403, Pergamon Press, London, 1958.

F-3: THE RAMAN SPECTRA OF $PbTiO_3$ AND SOLID SOLUTIONS OF $NaTaO_3$ - $KTaO_3$ AND $KTaO_3$ - $KNbO_3$.

C. H. Perry* and N. E. Tornberg
Spectroscopy Laboratory† and Research Laboratory of Electronics‡,
Massachusetts Institute of Technology
Cambridge, Massachusetts

INTRODUCTION

The Raman spectra of $PbTi0_3$ and the mixed crystal systems $(Na_x : K_{1-x}) Ta0_3$ and $K(Ta_y :Nb_{1-y})0_3$ have been studied within the temperature range 10 - 900°K for X = 0, 0.12, 0.40 and 0.85 and for Y = 0, 0.25, 0.65, 0.89, and 1.0.

Complementary far infrared reflectance measurements over a similar temperature range have been made on $PbTi0_3$, $KTa0_3$, $KNb0_3$, and some of the mixed crystals. The frequencies of the zone center transverse and longitudinal modes were obtained from a Kramers-Kronig analysis of the reflection spectra and provide a starting point for the interpretation of the Raman data.

As most of the samples investigated were multi-domain single crystals or ceramics. no detailed study of the spectra for different directions of phonon propagation and polarization could be obtained. The symmetry assignments to the observed mode frequencies have been made but the vibrations observed are not necessarily purely longitudinal or transverse. In the majority of the cases treated here, the anisotropy in the interatomic forces predominates over the long range electrostatic force. This leads to a larger splitting between the frequencies of phonons polarized parallel and perpendicular to the C axis than between the frequencies of corresponding longitudinal and transverse phonons. This effect is lessened when the wave vector is not strictly parallel or perpendicular to the C axis. Thus the frequencies presented here are likely to be less widely separated and not as distinct as they would be for ideal materials.

The Raman spectra were obtained using a Cary model 81 spectrophotometer and a 50 mW. Spectra-Physics laser operated at 6328 Å. High and low temperature cells were incorporated in the sample illuminator and the temperature was measured using calibrated thermocouples and a Germanium resistance thermometer attached to the

*Present address: Department of Physics, Northeastern University, Boston, Massachusetts

†This laboratory is supported in part by the National Science Foundation Grant No. GP-4923.
‡This work is supported in part by the Joint Services Electronics Program (Contract DA 28-043-AMC 02536(E)) and by NASA Grant NGR 22-009-(237).

samples. Three scattering geometries were used: conventional right angle scattering, back scattering using a wide angle collection lens, and oblique angle scattering.

$PbTiO_3$

$PbTiO_3$ is ferroelectric at room temperature and has a Curie temperature at about 760°K. The structure in the ferroelectric phase involves a large tetragonal distortion from the perovskite lattice and is isomorphic with tetragonal $BaTiO_3$ (space group C_{4v} -P 4mm). The lattice parameters are[1] a = 3.904 Å, c = 4.152 Å, c/a = 1.063 at 300°K.

The dielectric constant as a function of temperature has been measured on ceramic samples[2]. Above the Curie point (T_c), Curie-Weiss law behavior is observed. The spontaneous polarization is considerably larger than in $BaTiO_3$ but the ferroelectric domains are substantially the same. A small "c" domain face (approximately 3 mm X 2mm) was observed in a multidomain crystal using a polarizing microscope and showed reasonable extinctions indicating that the "c" axis lay approximately in the polished surface. Nevertheless, Raman spectra observed in it still showed considerable mixing of the mode symmetries, and a ceramic sample was found to give better data due to its larger surface area, so the "oriented" sample was used primarily in the polarized infrared reflectance measurements.

Above 760°K, $PbTiO_3$ exhibits a very weak, broad second order Raman spectrum. Below this temperature - in the tetragonal phase - a first order spectrum is superimposed and increases in strength as the temperature is lowered. The spectrum at several representative temperatures is shown in Fig. 1. The bands in $PbTiO_3$ observed just below T_c at about 65, 120, 185, 280, and 495 cm^{-1} show a shift of approximately 4, 5, 8, 2, and 3.5 cm^{-1} per 100°K respectively to higher frequency as the temperature is lowered. The shift decreases with decreasing temperature in a manner similar to the variation in cell volume.

Barker[3] has written the Lyddane-Sachs-Teller relation for a multimode crystal with zero damping in differential form:

$$\frac{d\epsilon_o}{\epsilon_o} = \frac{2d\omega_{L_1}}{\omega_{L_1}} + \frac{2\,d\omega_{L_2}}{\omega_{L_2}} + - - - - \frac{2d\omega_1}{\omega_1} - \frac{d\omega_2}{\omega_2} - - - -$$

where ω_j are the transverse phonon force parameters (related to the transverse optical phonon frequencies), ω_{L_j} are the longitudinal phonon frequencies and ϵ_o is the static dielectric constant at the temperature under consideration. However, this equation is strictly valid only for a particular vibration direction, and since ϵ_o was measured on a ceramic and our Raman data covers both polarizations, only a rough check on the relative contributions of the modes can be made. At room temperature $\frac{d\epsilon_o}{\epsilon_o}$ is approximately 0.002. The 65 cm^{-1} mode contributes $\approx$ 0.0007 ($\approx$35%), the 120 cm^{-1} mode $\approx$0.0003 ($\approx$15%), the 185 cm^{-1} mode $\approx$0.0005 ($\approx$25%) and the 495 cm^{-1} mode $\approx$0.0001 ($\approx$5%).

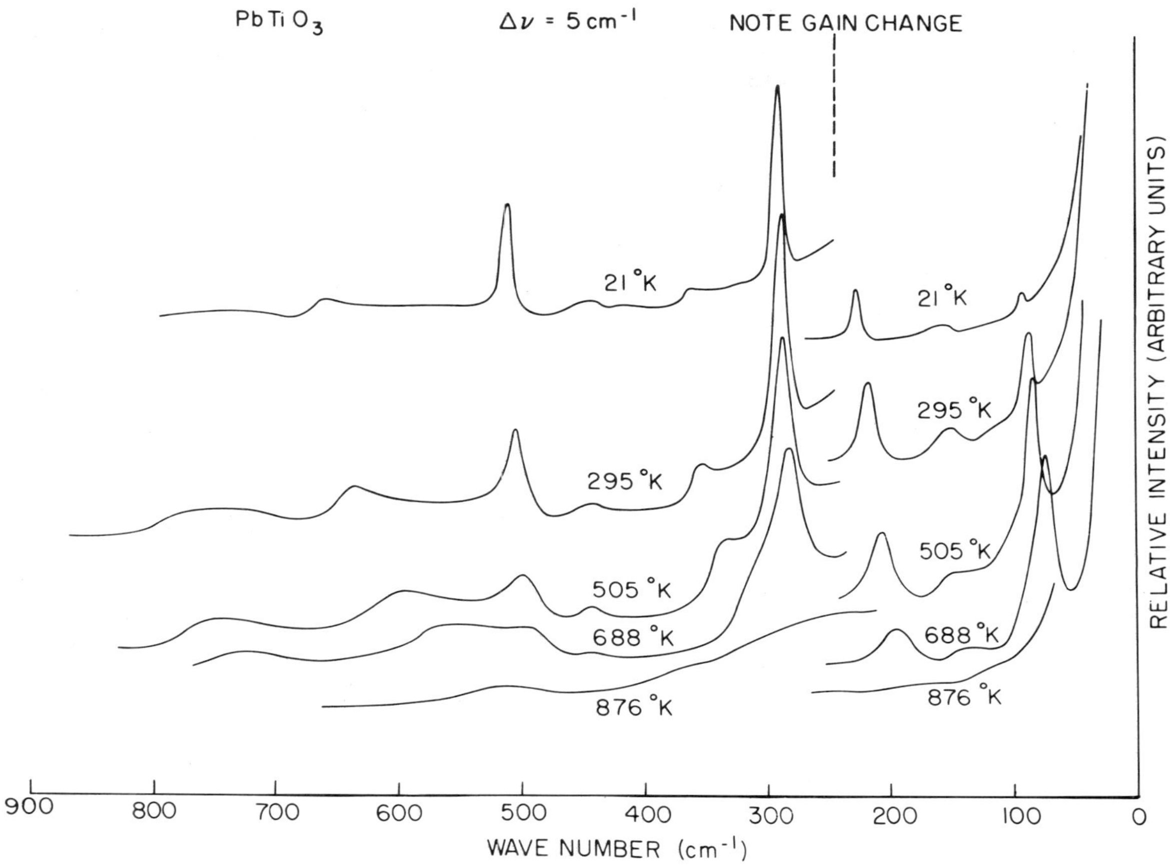

Fig. 1. Raman spectrum of $PbTi0_3$ as a function of temperature.

As can be seen the two E modes account for ≈60% of the observed value of $\frac{d\epsilon_0}{\epsilon_0}$. However, the contribution may be as high as 80% as the value of ϵ_0 could be too low due to the porousness of the ceramic samples. The longitudinal modes hardly shift at all with temperature and do not contribute significantly to $\frac{d\epsilon_0}{\epsilon_0}$. The lowest E mode is relatively the "softest" but the other E mode also contributes significantly to the temperature dependence of ϵ_0 and both may approach some type of instability in the paraelectric phase.

Fig. 2 shows the temperature dependence of the phonon frequencies. The symmetries of the modes are identified on the figure and the labeling of the modes is the same as that used by Cowley[4] for $SrTi0_3$.

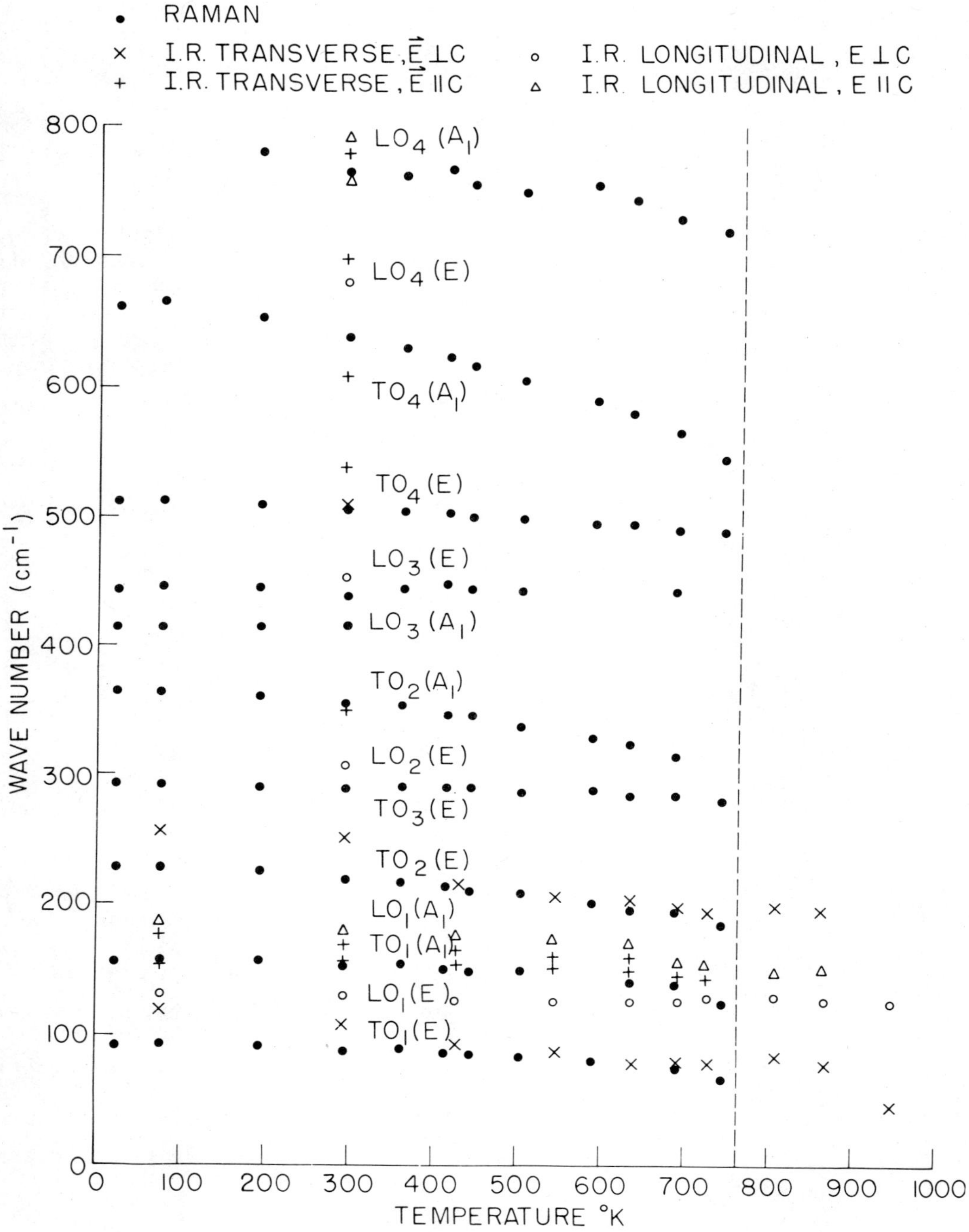

Fig. 2. Temperature dependence of the phonon frequencies taken from the infrared and Raman studies.

$NaTaO_3$ - $KTaO_3$

The dielectric data as a function of temperature on this mixed crystal system have been outlined by Davis[5]. In the mixed system the Curie point rises to a maximum of 65°K for a sample containing 48% $NaTaO_3$ and falls for higher concentrations. At approximately 72% $NaTaO_3$ the ferroelectric transition disappears. The departure from Curie-Weiss law behavior as T_c is approached is more significant as the concentration of $NaTaO_3$ is increased. For materials in the range 35-50% $NaTaO_3$, an anomalous "plateau" in the dielectric behavior is observed. This could arise from a second order ferroelectric transition at 50°K and a transition to a structurally different ferroelectric phase (e.g. tetragonal) at 37°K as observed in the 40% sample. As primarily second order Raman spectra were observed, the infrared results helped to positively identify the presence of any first order bands. Fig. 3 shows the Raman spectrum of $(Na_{0.4}:K_{0.6})TaO_3$ as a function of temperature from 30-583°K. The interpretation of the second order Raman spectrum of $KTaO_3$ has been discussed by Perry, Fertel and McNelly[6] and by Neilson and Skinner[7]. A similar analysis has been applied to the

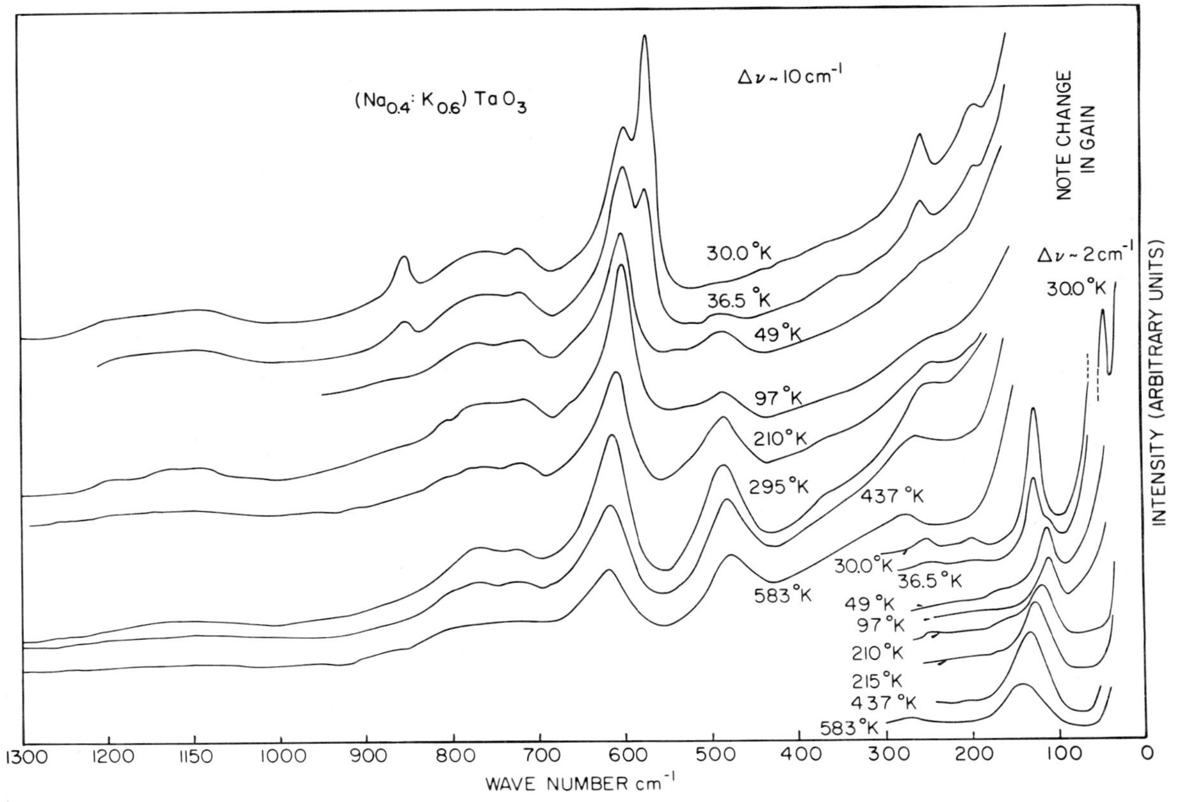

Fig. 3. Temperature-dependent Raman spectrum of $(Na_{0.4}:K_{0.6})TaO_3$.

mixed crystals. The temperature dependence of the intensities of the bands has been used to differentiate between possible combination and difference processes which contribute to the multi-phonon spectrum. Although a classical Brillouin zone is not well defined for these mixed crystals, the multi-phonon spectrum behaves as if it were due only to critical points at the edge of a pseudo-Brillouin zone. Fig. 4 shows the frequency vs. concentration plot at room temperature and a possible interpretation of some of the second order bands observed. At higher frequencies the various combinations become somewhat speculative.

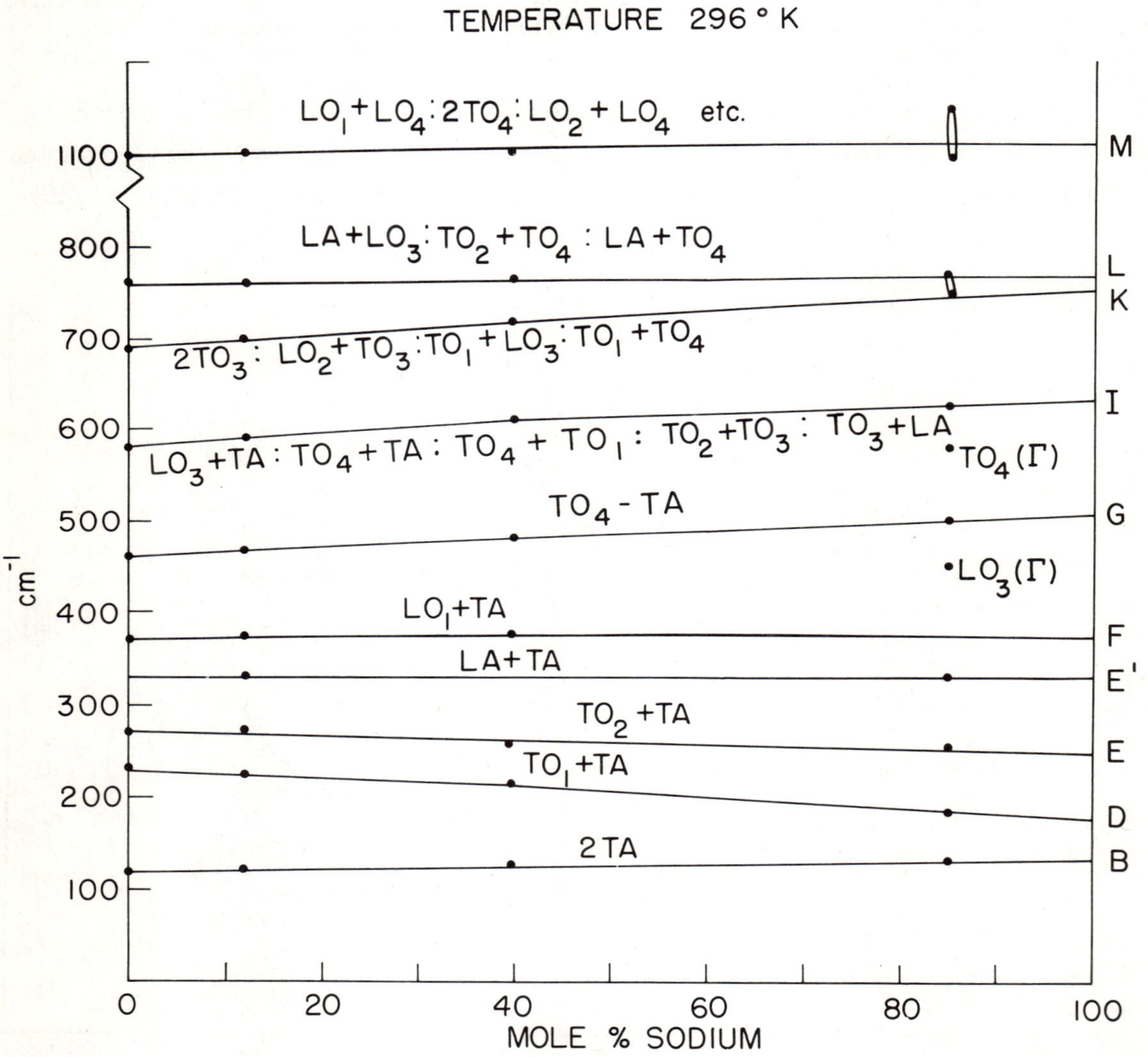

Fig. 4. Phonon combination assignment and their frequency dependence on composition.

In Fig. 3, once again, it can be noted that at 49°K the $(Na_{0.4}:K_{0.6})$ spectrum displays a very small first order contribution. At the lowest two temperatures, however, additional bands appear at 42, 128, 200, 225, 572, and 850 cm^{-1}. These are indicative of a structural change, but at a temperature below that established for the ferroelectric transition. Similar bands are present in the 85% $NaTa0_3$ sample's spectra, the distorted perovskite structure allowing all zone center modes to be Raman active.

The band around 255 cm^{-1} in the two crystals with highest Na content could be the normally inactive "F_{2u}" mode in the cubic phase now being Raman active. Table I shows a comparison between the infrared and Raman vibrations having their origin as first order modes. The generally good agreement helps confirm the supposition that for all sodium concentrations in the paraelectric phase, the Raman spectrum is essentially a profile of the combined density of states.

TABLE I

Comparison of Infrared - Raman Frequencies at K ≈ O at ≈ 10°K

	0% Na	12% Na	40%Na;R @ 30°K	85% Na ;	R @ 24°K
TO_1	25 / -	48 / -	39, 129 / 42, 128	57, 132 /	65, 130
LO_1	183 / -	183 / -	186 / -	147 /	150
TO_2	196 / 2nd	198 / 2nd	198 / 200	196 /	205
LO_2/TO_3	- / order	- / order	- / 255	246, 258, 270/	246, 261, 270
LO_3	421 / only	420 / only	414 / -	414 /	450
TO_4	547 / -	540 / -	570 / 572	584 /	580
LO_4	837 / -	819 / -	849 / 850	864 /	

$KTa0_3$ - $KNb0_3$

Except for the highest Ta concentrations, the mixed crystals $K(Ta:Nb)0_3$ exist in cubic, tetragonal, orthorhombic and rhombohedral phases at successively lower temperatures. In the cubic phase, the Raman spectra for all compositions are entirely second order and resemble those of $(Na:K)Ta0_3$ in its cubic phase. In the other phases, a 1st order spectrum is superimposed. At each phase transition this first order spectrum changes markedly but in no cases are all Raman-allowed transitions visible as discreet peaks. This could arise either from an insufficient distortion from cubic symmetry or from the ambiguity of phonon wave vector orientation with respect to the major axes of the crystal, as previously discussed. Representative spectra from the four phases of pure $KNb0_3$ are shown in Fig. 5. The temperature dependence of the frequencies associated with lines identifiable exclusively as first order is very small, generally being no greater than 1% per 100°K. This is usually less than the error in determining the frequency and indicates that the associated phonons play a negligible part in the variation in dielectric constant and the phase transition. The behavior of the

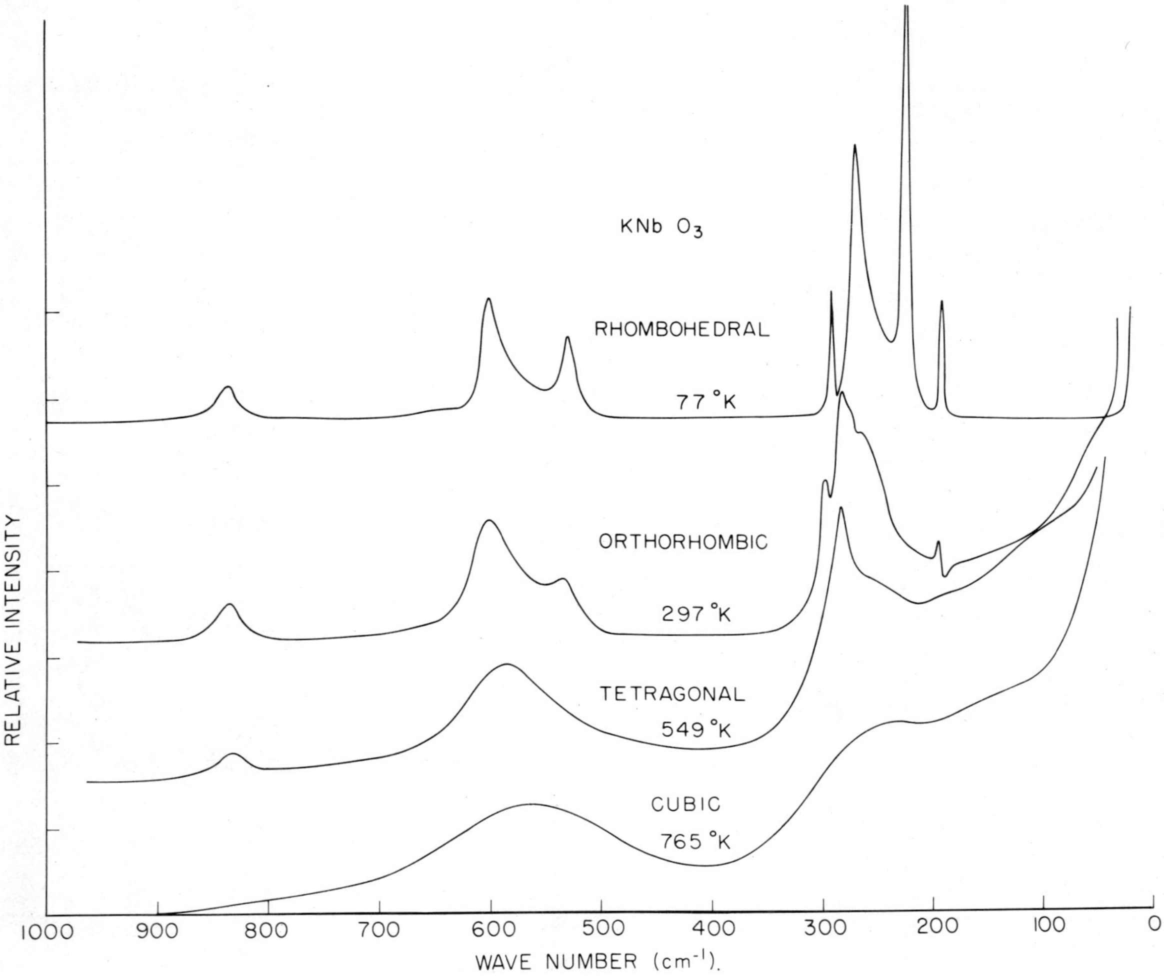

Fig. 5. Raman spectra of $KNb0_3$ in four phases.

scattering near the exciting line, however, would suggest that in the majority of cases an overdamped phonon of low frequency exists and plays the major role in the phase transition. Infrared studies tend to substantiate this.

The changes in the spectra at the phase transitions have allowed a careful study of the corresponding transition temperatures. These are found to agree well with Triebwasser [8] through the range of composition to within the error with which it is known, and considerable hysteresis is noted for those samples with 100% Nb and 75% Nb. The technique employed here, to obtain these temperatures, involved the repeated scanning of a narrow portion of the Raman spectrum while the sample temperature was slowly varied. The region of the spectrum was chosen to contain a feature that changed dramatically at the transition in question. A representative series of measurements of the rhombohedral/orthorhombic transition in pure $KNb0_3$ is shown in Fig. 6.

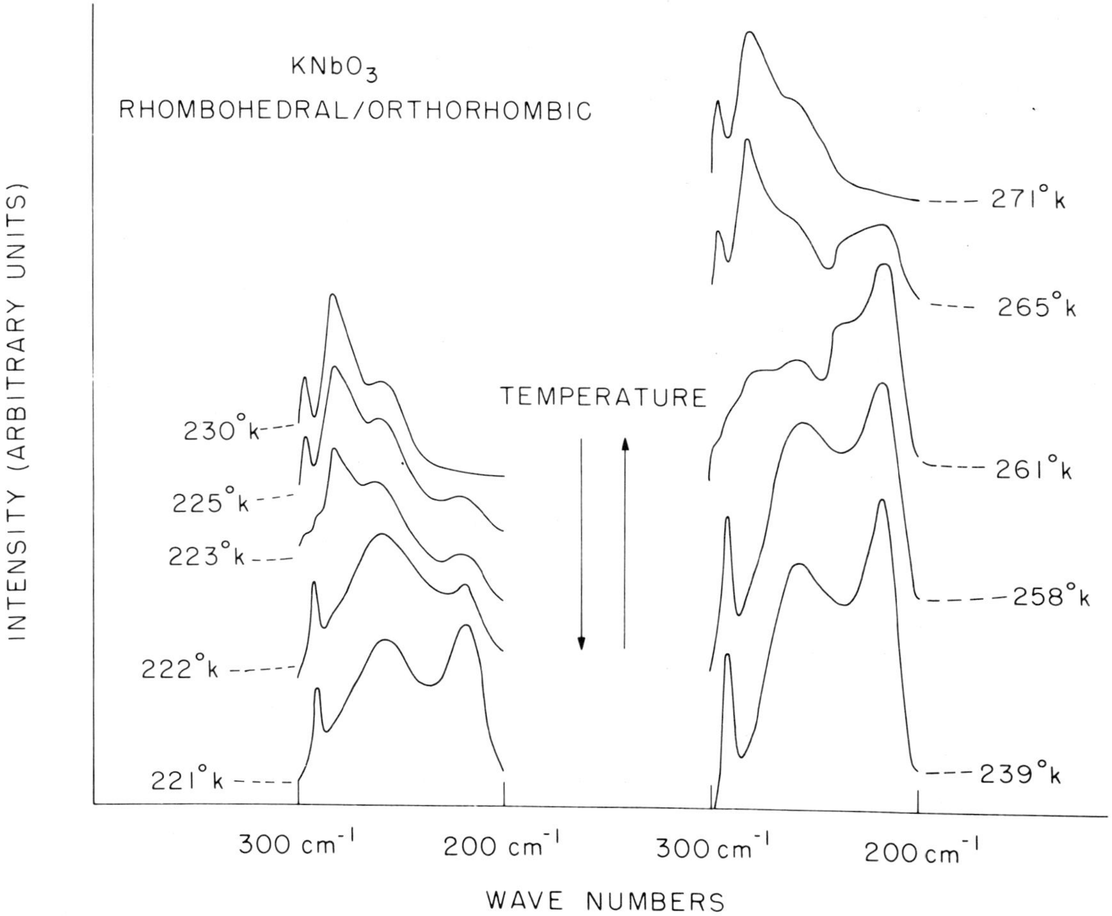

Fig. 6. Location of rhombohedral/orthorhombic phase transition temperature in $KNbO_3$ and associated hysteresis by means of changes in Raman spectra.

Some of the frequencies may be unambiguously assigned phonons on the basis of the $(Na{:}K)TaO_3$ work and the infrared measurements of $KNbO_3$. These are, referring to Fig. 5, LO_4 (838 cm^{-1}), TO'_4 (604 cm^{-1}), and TO_4 (530 cm^{-1}), where TO'_4 and TO_4 are the highest transverse optic mode split by anisotropy in the crystal. In addition, work on other perovskites indicates that the frequency at 292 cm^{-1} (rhombohedral phase) or 298 cm^{-1} (orthorhombic phase) may be identified with the TO_3 mode which in the Cubic phase has "F_{2u}" symmetry and is therefore infrared and Raman inactive.

REFERENCES

1. G. Shirane, R. Pepinsky, and B. C. Frazer, Acta Cryst. 9, 131 (1956).
2. G. Shirane and S. Hoshino, J. Phys. Soc. (Japan) 6, 265 (1951).
3. A.S. Barker, Jr., "Ferroelectricity," p. 238, E. F. Weller (ed.), Elsevier Publishing Co., Amsterdam, 1967.
4. R. A. Cowley, Phys. Rev. 134, A981 (1964).
5. T. G. Davis, "S. M. Thesis," Dept. Elect. Eng., Mass. Inst. of Tech., 1965.
6. C. H. Perry, J. H. Fertel and T. F. McNelly, J. Chem. Phys. 47, 1619 (1967).
7. W. G. Nilsen and J. G. Skinner, J. Chem. Phys. 47, 1413 (1967).
8. S. Triebwasser, Phys. Rev. 114, 63 (1959).

F-4: IMPURITY INDUCED RAMAN SCATTERING IN SOLIDS

R. S. Leigh and B. Szigeti
University of Reading, England

ABSTRACT

An impurity atom in a solid can affect the Raman activity of the atoms in its neighbourhood in two ways: (a) by short-range interactions; (b) if the charges of the impurity and of the host atoms are different, then there is also an effect of longer range due to the electrostatic field of the impurity. Using the methods recently developed for infra-red absorption in a similar situation, expressions are derived for the first-order Raman scattering arising from the electrostatic field of the impurity. NaCl and CsCl lattices are considered. The result is expressed in terms of constants of the pure material. Peaks are expected in the Raman scattering both at ω_ℓ and at ω_t.

INTRODUCTION

In three recent papers we discussed the infra-red absorption by charged impurity centres in solids[1-3]. We pointed out that the presence of charged impurity atoms makes the vibrations of the neighbouring atoms infra-red active, even if they would not be active in the pure crystal. This infra-red activity is due partly to short-range interactions with the impurity, and partly to the effect of its electrostatic field. This second effect extends to quite a few atomic distances and contributes a substantial amount to the impurity absorption. It was possible to derive formulae for this effect which, by comparison with the experimentally measured absorption, allowed interesting conclusions also with regard to the electronic properties of the pure material[3].

It is therefore of interest to develop a similar theoretical method for the treatment of Raman scattering by atoms near a charged impurity. For experimental reasons probably the main interest centres on those materials which are not Raman active in the pure state. In this paper we shall therefore consider crystals with the NaCl or CsCl structures and containing an impurity whose charge differs from the host atoms. As in the infra-red work, we shall assume that the impurity concentration is sufficiently low so that the impurities do not interact, and we can therefore consider a crystal containing a single impurity atom. It is true that a charged impurity is usually compensated by an oppositely charged impurity or defect in its neighbourhood, but in many cases the compensating defect is not too near to the impurity, and this is the situation we have in mind in this paper.

Let E denote the macroscopic electric field, P the polarization and q the long-wave optical lattice displacement in a volume element. Considering E and q as independent coordinates, for a pure ionic crystal we can write

$$\frac{dP}{dE} = \frac{\partial P}{\partial_q E} + \frac{\partial P}{\partial_E q}\frac{dq}{dE} \tag{1}$$

where $(\partial/\partial_q E)$ indicates that q is kept constant. E, P and q are of course vectors but this need not be indicated at this stage. Let ϵ_s denote the static and ϵ_o the "optical" dielectric constant, i. e. the dielectric constant at frequencies where the lattice cannot follow at all while the electronic polarization does not lag behind the field. Clearly

$$\begin{aligned} \epsilon_s - 1 &= 4\pi\, dP/dE \\ \epsilon_o - 1 &= 4\pi\, \partial P/\partial_q E \end{aligned} \tag{2}$$

We would stress that these equations as well as all the equations which follow are written in terms of the macroscopic field E and are therefore independent of any assumption concerning the effective field.

We shall consider the Raman scattering at 'optical' frequencies, i. e. at frequencies below all the electronic resonances. In a pure or impure material, the Raman scattering by a normal mode from a volume element whose size is small compared with the wave length of the incoming radiation is given by $(d/dQ)\int \epsilon_o dv$ where Q is the normal coordinate of the vibration and the integral extends over the volume element considered. If the volume element were large compared with the wavelength then the integral would have to be replaced by a more elaborate expression. In a similar way it can be shown that the total first order Raman scattering of the nth atom is determined by $(d/du_n)\int \epsilon_o dv$ where u_n is the displacement coordinate of the nth atom (cf. Xinh, Maradudin and Horsefall[4]). Written in component form, we can thus define R_n the Raman tensor of the nth atom by

$$R_{n\xi}^{\xi'\xi''} = \frac{1}{4\pi}\frac{d}{du_{n\xi}}\int \epsilon_o^{\xi'\xi''}\, dv = \frac{d}{du_{n\xi}}\int \frac{\partial P_{\xi'}}{\partial E_{\xi''}}\, dv \tag{3}$$

where ξ, ξ' and ξ'' denote x, y or z components and the integral extends over the volume affected by the displacement $u_{n\xi}$. We shall see that for the case considered that volume extends over quite a few interatomic distances but is nevertheless small compared with the wavelength of an incoming radiation. There are cases, however, when the latter condition is not fulfilled; in view of what has been said earlier, (3) has then to be replaced by a more complicated expression.

In Eq. (3), if the displacement $u_{n\xi}$ creates an electrostatic field then this effect is included, i. e.

$$\frac{d}{du_{n\xi}} = \frac{\partial}{\partial_E u_{n\xi}} + \frac{dE}{du_{n\xi}}\frac{\partial}{\partial_u E}$$

but the displacements of all the other atoms are to be kept constant during the differentiation.

As is well known, the vibrations of an atom are Raman active if the atom moves in an asymmetric potential. Due to short-range interactions, the neighbours of an impurity are in an asymmetric position and hence are Raman active. This effect is well established. In addition, however, if the impurity is differently charged from the host atoms then due to its electrostatic field an appreciable asymmetry is created also on further neighbours. The main purpose of this paper is to calculate R_n for such further neighbours using macroscopic electrostatics.

A part of this electrostatic effect has been discussed by Gurevich, Ipatova and Klochikhin[5] by microscopic methods. Our treatment is entirely different as it is completely analogous to the methods developed for infra-red absorption[1] and the results are obtained in terms of measurable macroscopic constants of the pure material.

CALCULATION OF R_n

For lack of space, we can only outline here the main steps in the calculation of R_n. The calculation proceeds in a way which is exactly analogous to the way the impurity-induced apparent charge was calculated in connection with the infra-red absorption[1]. We calculate R_n for an atom whose distance from the impurity is large compared with atomic dimensions but small compared with the wave length of radiation. For such an atom, R_n is due only to the electrostatic field exerted by the impurity and, using macroscopic methods, it can be calculated exactly as a function of its position relative to the impurity. Using the same arguments as in the infra-red paper[1] we then assume that the result represents a reasonable approximation even for near neighbours of the impurity, from second neighbours outwards, but it is, of course, quite invalid for nearest neighbours for which non-electrostatic interactions are very important.

We note that R_n is by definition independent of the displacement considered. Therefore if we have calculated R_n from any particular displacement the result is generally valid. Since we want to use macroscopic methods we build up a displacement from long lattice waves. Further, since the infra-red calculations showed[1] that photoelastic effects are negligible beyond first neighbours, we shall consider a displacement consisting only of long optical waves. This optical displacement is restricted to a region whose size is large compared with atomic size but small compared with the crystal. Its distance from the impurity is also large compared with atomic dimensions. The nth atom is situated inside this region. The optical displacement q varies very slowly from one unit cell to the next inside the region and is zero outside the region. For brevity, we shall represent the whole displacement by the collective coordinate U. Further, we shall denote by the vector $\vec{r}$ the position of a point relative to the impurity and the suffix r will denote the value of a quantity at the point $\vec{r}$. We shall need the integral $\int \epsilon_{or}(U)dv$.

Since the optical displacement may create an electrostatic field, the effect of the collective coordinate U on ϵ_{or} can be written, in first order,

$$\epsilon_{or}(U) - \epsilon_o(0) = \frac{d\epsilon_{or}}{dU}U = \left[\frac{\partial\epsilon_{or}}{\partial_E q_r}\frac{dq_r}{dU} + \frac{\partial\epsilon_{or}}{\partial_U E_r}\frac{dE_r}{dU}\right]U \tag{4}$$

Let e' denote the difference between the charge of the impurity and of the host atom which it replaces. Since in the region considered the effect of the impurity is purely electrostatic, in the required order Eq. (4) can be written

$$\epsilon_{or}(U) - \epsilon_o(0) = \left[\left(\frac{d}{de'}\frac{\partial\epsilon_{or}}{\partial_E q_r}\right)\frac{dq_r}{dU} + \left(\frac{d}{de'}\frac{\partial\epsilon_{or}}{\partial_U E_r}\right)\frac{dE_r}{dU} + \frac{\partial\epsilon_{or}}{\partial_U E_r}\frac{d}{de'}\frac{dE_r}{dU}\right] e'U \qquad (5)$$

where it is understood that all the derivatives represent values for e' = 0, U = 0. $\partial\epsilon_o/\partial_U E$ represents the linear change of ϵ_o with an electric field in the pure material, which for NaCl and CsCl is zero. Hence the last term in the square brackets vanishes. In the other two terms the effect of e' is uniquely expressed by F_r^0, the electrostatic field it exerts at $\vec{r}$ in the undisplaced configuration. To be exact, $F_r^0 = e'\vec{r}/\epsilon_s r^3$. Thus e'(d/de') may be replaced by F^0(d/dE). We thus have

$$\epsilon_{or}(U) - \epsilon_o(0) = \left[\left(\frac{d}{dE}\frac{\partial\epsilon_o}{\partial_E q}\right)\frac{dq_r}{dU} + \left(\frac{d}{dE}\frac{\partial\epsilon_o}{\partial_q E}\right)\frac{dE_r}{dU}\right] F_r^0 U \qquad (6)$$

Since the various derivatives represent values for $F^0 = 0$ and U = 0, the derivatives of ϵ_o are those for the pure material and do not depend on position.

As $\epsilon_o(0)$ does not depend on u_n, the integrand in Eq. (3) is to be replaced by the right-hand-side of Eq. (6). Since $q_r = (dq_r/dU)\,U$, the integral of the first term in (6) takes the form

$$\frac{d}{dE}\frac{\partial\epsilon_o}{\partial_E q}\int F_r^0\, q\, dv$$

This integral is clearly restricted to the region where $q_r \neq 0$, i.e. to a region small compared with the wave length of radiation. The second term is more complicated as $(dE_r/dU)\,U$ is the dipolar field created by the displacement and this field extends over the whole crystal. However, this field is multiplied by F_r^0, i.e. we have the product of a dipole field and a monopole field, and this product is of course effectively restricted to a region not much larger than the region where $q_r \neq 0$. This proves our statement after Eq. (3), namely that the displacement of an atom only affects ϵ_o within a radius small compared with the radiation wave length.

For F_r^0 we insert the expression given in the paragraph after Eq. (5). For evaluating $(dE_r/dU)\,U$, the displacements were expanded in a Fourier series. If we put $I^{\xi'\xi''} = \int \epsilon_o^{\xi'\xi''} dv$, the expression obtained can be written in the form

$$I^{\xi'\xi''} = \int f_1^{\xi'\xi''}(\vec{r})\, q_x\, dv + \int f_2^{\xi'\xi''}(\vec{r})\, q_y\, dv + \int f_3^{\xi'\xi''}(\vec{r})\, q_z\, dv$$

where the f are functions of position but not of displacement. Since the nth atom is in the region where the displacements are not zero, the argument presented on top of page 217 of the paper first quoted[1] applies exactly to the displacement we have considered here and we have

$$\frac{d}{du_{nx}} I^{\xi'\xi''} = \pm \frac{1}{N} f_1^{\xi'\xi''}(\vec{r}_n); \quad \frac{d}{du_{ny}} I^{\xi'\xi''} = \pm \frac{1}{N} f_2^{\xi'\xi''}(\vec{r}_n);$$

$$\frac{d}{du_{nz}} I^{\xi'\xi''} = \pm \frac{1}{N} f_3^{\xi'\xi''}(\vec{r}_n)$$

where N is the number of ion pairs per unit volume and the sign is + or - according to whether the nth atom is positive or negative. By this method the various elements of R_n are obtained.

THE CONSTANTS WHICH ENTER INTO R_n

In Eq. (6) $U(dE_r/dU)$ represents a dipolar field and it was convenient to express this field in terms of the apparent charge of the ions as defined in a recent paper[1]. If η denotes the apparent charge of an ion in the pure material, then it follows from the definition of the apparent charge[1] and from the expression for the dielectric constant[6] that

$$\eta = \pm \left(\frac{\epsilon_s - \epsilon_o}{4\pi N} m_{red} \omega_t^2 \right)^{\frac{1}{2}} \tag{7}$$

where m_{red} denotes the reduced mass of an ion pair and the sign in front is to be chosen according to whether the ion is positive or negative.

Since $\frac{d}{dE} \frac{\partial \epsilon_o}{\partial_E q}$, which occurs in Eq. (6), is probably very difficult to measure, we eliminate it by the relation

$$\frac{d^2 \epsilon_o}{dE^2} = \frac{d}{dE} \frac{\partial \epsilon_o}{\partial_q E} + \left(\frac{d}{dE} \frac{\partial \epsilon_o}{\partial_E q} \right) \frac{dq}{dE} \tag{8}$$

For the two second derivatives of ϵ_o which we then need we introduce the abbreviations

$$\rho = \frac{d^2 \epsilon_o}{dE^2} \tag{9}$$

$$\lambda = \frac{d}{dE} \frac{\partial \epsilon_o}{\partial_q E} \tag{10}$$

In cubic crystals such fourth-rank tensors have at most three independent elements, which in the contracted notation may be written as ρ_{11}, ρ_{12}, ρ_{44} and λ_{11}, λ_{12} and λ_{44}. For instance,

$$\rho_{11} = d^2 \epsilon_o^{xx}/dE_x^2$$

$$\rho_{12} = d^2 \epsilon_o^{xx}/dE_y^2$$

$$\rho_{44} = d^2 \epsilon_o^{xy}/dE_x dE_y$$

and similarly for λ. In addition, with the use of Eq. (2) it is readily verified that

$$\left.\begin{array}{ll} & \lambda_{12} = \lambda_{44} \\ \text{but} & \rho_{12} \neq \rho_{44} \end{array}\right\} \tag{11}$$

Concerning the definition of ρ in Eq. (9) we remark that in a static electric field there is also an elastic strain proportional to E^2 which in general would contribute to $d^2\epsilon_o/dE^2$ through the photo-elastic effect. But it follows from our procedure that this effect is not included in our $d^2\epsilon_o/dE^2$, i.e. our $d^2\epsilon_o/dE^2$ includes the change of q due to E, but not the strain induced by E^2. Thus if for $d^2\epsilon_o/dE^2$ a measured value were to be used which includes this strain effect then a correction would have to be applied. Regarding this correction, cf. for instance Zheludev[7].

RESULTS FOR R_n

Using the preceeding notation, the results for the elements of R_n can be written

$$R_{n\xi}^{\xi'\xi''} = R_{n\xi}^{\xi''\xi'} = \frac{e'\eta_n}{\epsilon_s} \frac{\lambda_{12}}{\epsilon_o} \frac{3\xi_n \xi'_n \xi''_n}{r_n^5} \tag{12a}$$

$$R_{n\xi}^{\xi'\xi'} = \frac{e'\eta_n}{\epsilon_s} \frac{\xi_n}{r_n^3} \left\{ \frac{\rho_{12} - \lambda_{12}}{\epsilon_s - \epsilon_o} + \frac{1}{2\epsilon_o} \left[\lambda_{11} \left(\frac{3\xi'^2_n}{r_n^2} - 1 \right) - \lambda_{12} \left(\frac{3\xi'^2_n}{r_n^2} + 1 \right) \right] \right\} \tag{12b}$$

$$R_{n\xi}^{\xi\xi'} = R_{n\xi}^{\xi'\xi} = \frac{e'\eta_n}{\epsilon_s} \frac{\xi'_n}{r_n^3} \left\{ \frac{\rho_{44} - \lambda_{12}}{\epsilon_s - \epsilon_o} + \frac{\lambda_{12}}{\epsilon_o} \left(\frac{3\xi_n^2}{r_n^2} - 1 \right) \right\} \tag{12c}$$

$$R^{\xi\xi}_{n\xi} = \frac{e'\eta_n}{\epsilon_s}\frac{\xi_n}{r_n^3}\left\{\frac{\rho_{11}-\lambda_{11}}{\epsilon_s-\epsilon_o}+\frac{1}{2\epsilon_o}\left[3\lambda_{11}\left(\frac{\xi_n^2}{r_n^2}-1\right)-\lambda_{12}\left(\frac{3\xi_n^2}{r_n^2}-1\right)\right]\right\} \qquad (12d)$$

Eqs. (12) determine all the 27 elements of R_n. The labels ξ, ξ' and ξ'' denote x, y or z components, and in these four equations it is understood that $\xi'' \neq \xi' \neq \xi$. As before, $\vec{r}_n$ denotes the position of the nth atom relative to the impurity, and ξ_n, ξ'_n and ξ''_n are its three components. The apparent charge η_n is given by Eq. (7) where the sign has to be chosen according to whether the nth atom is positive or negative. For materials with NaCl or CsCl structure we have not found any measured values for ρ and λ, but the measurement of these quantities would probably not be too difficult. Eqs. (9) and (10) show that ρ is obtained by measuring the change of ϵ_o due to a strong static field, and λ can be obtained by a rather similar measurement except that in that case a strong high frequency field is also required.

THE PEAKS AT ω_ℓ AND ω_t

As ρ and λ are not known, we cannot calculate the magnitudes of the R_n. From analogy with the infra-red results[1, 3] we may expect, however, that the total Raman scattering due to the electrostatic field of the impurity is probably comparable in magnitude to the Raman scattering produced by short-range interactions between the impurity and its nearest neighbours.

The Raman scattering due to the electrostatic field of the impurity extends over almost the entire region of lattice frequencies, but it can be shown that it exhibits peaks at both ω_ℓ and ω_t, i. e. at the frequencies of both long longitudinal and long transverse optical waves. In this paper we can only give a very rough outline of the proof of this statement.

In analogy to R_n, let R_j denote the Raman tensor of the jth lattice vibration. Considering the longitudinal waves first, it can be shown that the effect of the impurity extends over a range which is proportional to the wavelength of the mode. Therefore if $R_{j\alpha}$ is one of the elements of R_j, we can write

$$R_{j\alpha} \propto \frac{1}{k_j}$$

where $\propto$ means "proportional to", k_j is the wave number of the mode and the factor of proportionality depends on directions but not on the magnitude of k. For small wave numbers we can write

$$\omega_k = \omega_\ell - ak^2$$

where ω_ℓ denotes the longitudinal frequency for zero wave number and ak^2 is a small correction. We also note that the Raman scattering intensity is proportional to the square of the elements of R_j and that the density of lattice modes in k space is proportional to k^2dk. Therefore, the total intensity scattered in a particular direction by all the longitudinal modes with wave numbers between 0 and k' is proportional to

$$\int_0^{k'} \frac{1}{k^2} k^2 dk \propto k' \propto \sqrt{\omega_\ell - \omega_{k'}}$$

Hence, if ω is a frequency slightly less than ω_ℓ the total scattering into the direction considered by all the longitudinal modes in the frequency range between ω_ℓ and ω is proportional to $\sqrt{\omega_\ell - \omega}$. Therefore, if $I(\omega)$ denotes the scattering intensity at frequency ω, we have

$$I(\omega) \propto -\frac{d}{d\omega}\sqrt{\omega_\ell - \omega} \propto \frac{1}{\sqrt{\omega_\ell - \omega}} \tag{13}$$

which has a peak at ω_ℓ.

In the neighbourhood of ω_ℓ, i.e. of $k = 0$, we can expand all the quantities in a power series in k and the calculation presented is of course based on the leading term. The next term would have an additional factor of k^2 which in the region considered is negligible, and (13) should therefore represent a reasonable first approximation. However, due to the usual secondary effects we expect the peak to be very much less sharp than given by Eq. (13).

In the case of very long transverse waves there is the polariton effect. This however is restricted to a very narrow region of k space and therefore in the present context we can neglect it. The argument presented above for longitudinal waves is then also valid for transverse waves, except that the various factors of proportionality are different and ω_ℓ is replaced by ω_t. In analogy to Eq. (13), for the scattering intensity by long transverse waves we thus get

$$I(\omega) \propto \frac{1}{\sqrt{\omega_t - \omega}} \tag{14}$$

Again, we expect that the actual peak is much less sharp than given by this expression. We intend to publish the calculations in more detail elsewhere.

REFERENCES

1. R.S. Leigh and B. Szigeti, Proc. Roy. Soc. A. 301, 211 (1967).
2. R.S. Leigh and B. Szigeti, Phys. Rev. Letters 19, 566 (1967).
3. R.S. Leigh and B. Szigeti, "Proc. of 1967 Irvine Conference on Localized Excitations in Solids," p. 159 Plenum Press, New York, 1968.
4. N.X. Xinh, A.A. Maradudin, and R.A. Coldwell-Horsfall, Jour. de Phys. 26, 717 (1965).
5. L.E. Gurevich, I.P. Ipatova, and A.A. Klochikhin, Soviet Physics, Solid State, 8, 2608 (1967).
6. B. Szigeti, Trans. Faraday Soc. 45, 155 (1949).
7. I.S. Zheludev, Soviet Physics, Uspekhi 9, 97 (1966).

F-5: MICROSCOPIC THEORY OF LATTICE RAMAN SCATTERING IN CRYSTALS CONTAINING IMPURITIES

Achintya K. Ganguly
The Bayside Laboratory, Research Center of General Telephone & Electronics Laboratories Incorporated
Bayside, New York
and
Joseph L. Birman*
Physics Department, New York University
University Heights, New York

ABSTRACT

In this paper we develop a general theory of lattice Raman scattering by an insulator containing isolated mass defects which have no net charge with respect to the perfect lattice. The theory given here is specialized to the case where the lattice force constants and the electronic wave functions of the crystal are assumed unperturbed by the defect. The Raman scattering tensor has been obtained as a function of the incident photon frequency and all lattice frequencies. The Raman tensor has poles at incident photon energies equal to creation of virtual excitons. This divergence appears in all types of lattice vibrations: local, resonance and band modes.

INTRODUCTION

In a recent paper[1] (which will be referred to as paper I) we presented a microscopic theory of lattice Raman scattering from perfect, insulating crystals. A novel result of that theory was the existence of a pole in the Raman scattering tensor as function of incident photon frequency, at energy equal to that of a crystal quasi-particle: the free exciton. The presence of that pole, as well as certain other quantitative predictions of the theory such as intensities, and intensity ratios of Raman scattering by lattice phonons, appears to be confirmed by presently available experimental results.

*Work supported in part by the U.S. Army Research Office (Durham) under Grant No. DA-ARO-(D)-31-124-G424, and the Aerospace Research Laboratories, Office of Aerospace Research, Wright-Patterson AFB, Dayton, Ohio, under Contract No. Af(33)(615)-1746.

In the present paper we give the results of an extension of that theory to the case of a crystal containing isolated substitutional mass defects with no net charge compared to the perfect lattice. Some examples of this type of system are CdS:Se mixed crystals, and crystals with isoelectronic defects such as GaP:N. In any case, the introduction of the impurity will alter the normal modes of vibration of the crystal from their values in perfect crystal. Also the impurity may produce changes in the electronic states of the crystals. For example, certain isoelectronic defects appear to bind an exciton, while others do not. If the defect does not produce a bound exciton state, it seems to be a good approximation to use the unperturbed (perfect) crystal electronic eigenfunctions in the theory. The case of isoelectronic defects which do bind the excitons will not be treated here, but some discussion of this case will be given later. The work given here is based on the assumption that the lattice force constants and the electronic states of the crystal are unchanged. As before, it is assumed that the electronic system, which is "virtually excited," is described by Wannier exciton eigenfunctions.

In section II we first set up the formal Hamiltonian for the interacting crystal electron, photon, and phonon fields. Next the coupling parameters are determined for a prototype diatomic crystal such as CdS wurtzite. The parameters are given in terms of integrals over the electronic basis and in principle include the effect of frequency and wave vector dispersion. We then apply a canonical transformation to separate out the terms which cause one phonon Raman scattering. In section III the spectral decomposition of the first order Raman tensor is obtained. The scattered radiation forms a continuous spectrum. We then discuss the new results of our theory.

THEORY

We consider an insulator having N unit cells and r atoms in each cell. We suppose that the defect atom is located at atomic site 1 in the Oth unit cell. The Hamiltonian of the system of electrons, phonons and radiation may be written as

$$H = H^{(0)} + H^{(1)} + H^{(2)} \tag{1}$$

The unperturbed Hamiltonian $H^{(0)}$ is the sum of the following three terms:

$$H_{exciton} = \sum_{cv\lambda\underset{\sim}{K}} E_{\lambda\underset{\sim}{K}}(cv)a^{+}_{\lambda\underset{\sim}{K}}(cv)a_{\lambda\underset{\sim}{K}}(cv) \tag{2}$$

$$H_{Radiation} = \sum_{\underset{\sim}{\eta}\underset{\sim}{\epsilon}} \hbar\omega_{\underset{\sim}{\eta}\underset{\sim}{\epsilon}}(d^{+}_{\underset{\sim}{\eta}\underset{\sim}{\epsilon}} d_{\underset{\sim}{\eta}\underset{\sim}{\epsilon}} + \frac{1}{2}) \tag{3}$$

$$H_{Lattice} = \sum_{s} \hbar\omega_{s}(b^{+}_{s}b_{s} + \frac{1}{2}) \tag{4}$$

Here $a^{+}_{\lambda\underset{\sim}{K}}(cv)$ and $a_{\lambda\underset{\sim}{K}}(cv)$ are the creation and the annihilation operators of the excitons having quantum number λ and wave vector $\underset{\sim}{K}$ formed from conduction band c and valence band v. The energy spectrum $E_{\lambda\underset{\sim}{k}}(cv)$ of the excitons is "hydrogenic"[2]:

$$E_{n\underset{\sim}{K}}(cv) - E_{g} = -R/n^{2} + \hbar^{2}|\underset{\sim}{K}|^{2}/2(m^{*}_{e} + m^{*}_{h}) \tag{5}$$

where $R = \mu e^4/2\hbar^2 \varkappa^2$ is the exciton Rydberg, $\varkappa$ is the dielectric constant, m_e^* and m_h^* are the effective masses of the electron and the hole, respectively, and μ is their reduced mass. E_g is the band gap. The integer n is the principal quantum number for the discrete states. For the continuum states, the internal energy is $\hbar^2 |\underset{\sim}{k}|^2/2\mu$, where $\hbar\underset{\sim}{k}$ is the relative momentum. ω_s represents the frequencies of the normal modes (labelled by s) of the imperfect crystal. s takes all values from 1 to 3rN. $\underset{\sim}{\eta}$, $\underset{\sim}{\epsilon}$, $\omega_{\underset{\sim}{\eta}\underset{\sim}{\epsilon}}$ are the wave vector, unit polarization vector and frequency of the photons. $\omega_{\underset{\sim}{\eta}} = (c/\varkappa_\infty^{1/2})|\underset{\sim}{\eta}|$, where $\varkappa_\infty$ is the optical dielectric constant and c is the velocity of light in vacuum. $d^+_{\underset{\sim}{\eta}\underset{\sim}{\epsilon}}$, $d_{\underset{\sim}{\eta}\underset{\sim}{\epsilon}}$ and b_s^+, b_s are the creation and the annihilation operators of the photons and the phonons, respectively. The operators d^+, d and b^+, b obey Bose commutation relation. a^+ and a satisfy Bose commutation relation approximately.

The perturbation terms arise from the exciton-phonon (H_{eL}) and the exciton-photon (H_{eR}) interactions. $H^{(1)}$ may be written as the sum of $H_{eL}^{(1)}$ and $H_{eR}^{(1)}$ given by

$$H_{eL}^{(1)} = \sum_{\substack{cv\lambda\underset{\sim}{K} \\ s}} \left\{ g_s(cv\lambda\underset{\sim}{K}) a^+_{\lambda\underset{\sim}{K}}(cv) b_s^+ + g_s^*(cv\lambda\underset{\sim}{K}) a_{\lambda\underset{\sim}{K}}(cv) b_s^+ \right\} + \text{h.c.} \tag{6}$$

$$H_{eR}^{(1)} = \sum_{\substack{cv\lambda\underset{\sim}{K} \\ \underset{\sim}{\eta}\underset{\sim}{\epsilon}}} \left\{ f_{\underset{\sim}{\eta}\underset{\sim}{\epsilon}}(cv\lambda\underset{\sim}{K}) a^+_{\lambda\underset{\sim}{K}}(cv) d_{\underset{\sim}{\eta}\underset{\sim}{\epsilon}} \delta_{\underset{\sim}{K},\underset{\sim}{\eta}} + f^*_{\underset{\sim}{\eta}\underset{\sim}{\epsilon}}(cv\lambda\underset{\sim}{K}) a_{\lambda\underset{\sim}{K}}(cv) d_{\underset{\sim}{\eta}\underset{\sim}{\epsilon}} \delta_{\underset{\sim}{K},-\underset{\sim}{\eta}} \right\} + \text{h.c.} \tag{7}$$

$H^{(2)}$ is the sum of $H_{eL}^{(2)}$ and $H_{eR}^{(2)}$, where

$$H_{eL}^{(2)} = \sum_{\substack{cv\lambda\underset{\sim}{K} \\ c'v'\lambda'\underset{\sim}{K}' \\ s}} G_s(cv\lambda\underset{\sim}{K}, c'v'\lambda'\underset{\sim}{K}') a^+_{\lambda\underset{\sim}{K}}(cv) a_{\lambda'\underset{\sim}{K}'}(c'v') b_s^+ + \text{h, c.} \tag{8}$$

$$H_{eR}^{(2)} = \sum_{\substack{cv\lambda\underset{\sim}{K} \\ c'v'\lambda'\underset{\sim}{K}' \\ \underset{\sim}{\eta}\underset{\sim}{\epsilon}}} \left\{ F_{\underset{\sim}{\eta}\underset{\sim}{\epsilon}}(cv\lambda\underset{\sim}{K}, c'v'\lambda'\underset{\sim}{K}') a^+_{\lambda\underset{\sim}{K}}(cv) a_{\lambda'\underset{\sim}{K}'}(c'v') d_{\underset{\sim}{\eta}\underset{\sim}{\epsilon}} \delta_{\underset{\sim}{K}-\underset{\sim}{K}',\underset{\sim}{\eta}} \right\} + \text{h.c.} \tag{9}$$

$H_{eL}^{(1)}$ is the term which creates or annihilates an exciton with the emission of a phonon. $H_{eL}^{(2)}$ scatters an exciton emitting a phonon. Similar processes with photons are described by the terms $H_{eR}^{(1)}$ and $H_{eR}^{(2)}$. Note that owing to the loss of the translational symmetry in the phonon field in the imperfect crystal, there is no wave-vector selection rule on the exciton-phonon interaction although the exciton-photon term is the same as in the perfect crystal.

The coupling parameters in Eqs. (6-9) can be obtained by the procedure used in the case of the perfect lattice[1]. In the following, the equations of our previous paper I will be referred to by I: (28), etc. Thus $g_s(cv\lambda \underset{\sim}{K})$ and $G_s(cv\lambda K, c'v'\lambda'\underset{\sim}{K}')$ are given in I: (10) and I: (11) but now $\delta\varphi$ in these equations refers to the electron-phonon interaction in the perturbed lattice. The relevant matrix elements between Bloch conduction and valance band states are denoted here as $\langle c|\delta\varphi|v\rangle$. We now show the specific forms of these matrix elements.

In our model, the electronic states remain unperturbed by the introduction of the impurity. Hence the coupling parameters f and F are the same as in perfect lattice. So for electron-photon interaction we have

$$\langle c|\delta\varphi|v\rangle = -(e/m)(2\pi\hbar/V\varkappa_\infty \omega_{\underset{\sim}{\eta}\underset{\sim}{\epsilon}})^{1/2} \langle c|\underset{\sim}{\epsilon}\cdot \underset{\sim}{p}|v\rangle \tag{10}$$

where $\varkappa_\infty$ is the optical dielectric constant, V is the volume of the crystal and $\underset{\sim}{p}$ is the momentum of the electron. We have assumed the wave vector independence of the matrix elements of p. Eq. (10) is to be used for the various matrix elements in the expressions for $f_{\eta\epsilon}(cv\lambda\underset{\sim}{K})$ and $F_{\eta\epsilon}(cv\lambda\underset{\sim}{K}, c'v'\lambda'\underset{\sim}{K}')$.

The exciton-phonon coupling parameters g and G are, however, different from those in the perfect lattice due to the change in the normal modes of the lattice. In the deformation potential approximation, the electron-phonon interaction energy $\delta\varphi$ is given by[3]:

$$\delta\varphi = \frac{1}{a} \sum_{\ell, b} \underset{\sim}{D}_{\ell, b}(\underset{\sim}{r}) \cdot \underset{\sim}{u}\binom{\ell}{b} \tag{11}$$

where $\underset{\sim}{D}_{\ell, b} = \partial\varphi_0/[\partial \underset{\sim}{u}\binom{\ell}{b}/a]$ is the deformation potential, φ_0 is the equilibrium lattice potential and a is the lattice constant. The displacement $\underset{\sim}{u}\binom{\ell}{b}$ of the atom b in the ℓth unit cell is

$$\underset{\sim}{u}\binom{\ell}{b} = \sum_s \underset{\sim}{B}\left(\begin{smallmatrix}\ell\\ b\end{smallmatrix}\middle| s\right) \left(\frac{\hbar}{2\omega_s}\right)^{1/2} b_s^+ + \text{h.c.} \tag{12}$$

where $\underset{\sim}{B}\left(\begin{smallmatrix}\ell\\ b\end{smallmatrix}\middle| s\right)$ is the eigenvector of the atom b in the ℓth unit cell in normal mode s. For a single mass defect[4]

$$\underset{\sim}{B}\left(\begin{smallmatrix}\ell\\ b\end{smallmatrix}\middle| s\right) = \frac{\gamma}{N} \sum_{\underset{\sim}{q}j} \underset{\sim}{e}^*\left(b\middle|\begin{smallmatrix}\underset{\sim}{q}\\ j\end{smallmatrix}\right) \frac{\underset{\sim}{e}\left(1\middle|\begin{smallmatrix}\underset{\sim}{q}\\ j\end{smallmatrix}\right)\cdot \underset{\sim}{B}\left(\begin{smallmatrix}0\\ 1\end{smallmatrix}\middle| s\right)}{\omega_{qj}^2 - \omega_s^2} \left(\frac{M_1}{M_b}\right)^{1/2} \omega_s^2 \, e^{-i\underset{\sim}{q}\cdot \underset{\sim}{R}_\ell} \tag{13}$$

in which/$B\left(\begin{smallmatrix}0\\ 1\end{smallmatrix}\middle| s\right)$ is the eigenvector of the defect atom, $\underset{\sim}{e}\left(b\middle|\begin{smallmatrix}\underset{\sim}{q}\\ j\end{smallmatrix}\right)$ and ω_{qj} are the eigenvectors and eigenfrequencies of the perfect lattice normal modes labelled by wave vector $\underset{\sim}{q}$ and

and branch index j, M_b denotes mass of atom b, and $\gamma = (M_1 - M_1')/M_1$, M_1 is the original mass at lattice site 1 and M_1' is the mass of the substituted atom. Using Eq. (12)-(13) it can be shown for the deformation potential interaction that

$$\langle c|\delta\varphi|v\rangle = \left(\frac{\hbar}{2M_1 N\omega_s}\right)^{1/2} \frac{1}{a} \sum_{\underset{\sim}{q}j} A_s(\underset{\sim}{q}j) \left\{ \sum_b \left(\frac{M_1}{M_b}\right)^{1/2} \underset{\sim}{e}(b|{}^{q}_{j}) \cdot \langle c|\underset{\sim}{D}_b|v\rangle \right\} \delta_{\underset{\sim}{K},-q} \tag{14}$$

where

$$A_s(\underset{\sim}{q}j) = \frac{M_1^{1/2}\gamma\omega_s^2}{N^{1/2}} \; \frac{\underset{\sim}{e}(1|{}^{q}_{j}) \cdot \underset{\sim}{B}({}^{0}_{1}|s)}{\omega_s^2 - \omega_{\underset{\sim}{q}j}^2} \tag{15}$$

$A_s(\underset{\sim}{q}j)$ reflects the change in the electron-phonon interaction due to the introduction of the mass defect. We have again neglected the wave vector dependence of the matrix elements of $\underset{\sim}{D}_b$.

In polar crystals, there is an additional electron-phonon interaction arising from the polarization associated with the longitudinal modes[5]. The polarization arising from the ℓth unit cell (correct to first order in displacement) is

$$P_\alpha(\ell) = \frac{1}{v_0} \sum_{b\beta} P_{\alpha\beta}({}^{\ell}_{b})\, u_\beta({}^{\ell}_{b}) \tag{16}$$

v_0 is the volume of unit cell, $P_{\alpha\beta}({}^{\ell}_{b}) = \partial P_\alpha(\ell)/\partial u_\beta({}^{\ell}_{b})$. α, β refer to the cartesian components. We assume that $P_{\alpha\beta}({}^{\ell}_{b})$ in the defect lattice is the same as in a perfect lattice. It is then independent of ℓ. In wurtzite crystals a point defect corresponds to the point group c_{3v}. In this case $P_{\alpha\beta}(b) = 0$ if $\alpha \neq \beta$ and $P_{xx} = P_{yy} \neq P_{zz}$ where z is along the c-axis. In cubic crystals $P_{\alpha\alpha}$ is isotropic. We will neglect the anisotropy in $P_{\alpha\alpha}$ in wurtzite crystals. $P_{\alpha\alpha}$ can then be related to the static dielectric constant $\varkappa_0$, the optical dielectric constant $\varkappa_\infty$ and the frequency ω_ℓ of the longitudinal modes in the perfect lattice. Using the polarization given by Eq. (12), (13) and (16) and the values of $\underset{\sim}{e}(b|{}^{q}_{j})$ of the polar modes in CdS, the interaction energy between the electrons and the polar modes is

$$\delta\varphi = ie \left(\frac{1}{\varkappa_\infty} - \frac{1}{\varkappa_0}\right)^{1/2} \left(\frac{2\pi\hbar\omega_\ell}{V}\right)^{1/2} \left(\frac{\omega_\ell}{\omega_s}\right)^{1/2} \sum_{\underset{\sim}{q}j} A_s(\underset{\sim}{q}j)(\underset{\sim}{\xi}_j \cdot \underset{\sim}{q}) \frac{e^{-i\underset{\sim}{q}\cdot\underset{\sim}{r}}}{|q|^2} \tag{17}$$

$\underset{\sim}{\xi}_j$ is the unit polarization vector in the branch j. Eq. (17) is true also for cubic crystals. Eq. (17) then gives

$$<c|\delta\varphi_0|v> = ie\left(\frac{1}{\kappa_\infty} - \frac{1}{\kappa_0}\right)^{1/2}\left(\frac{2\pi\hbar\omega_\ell}{V}\right)^{1/2}\left(\frac{\omega_\ell}{\omega_s}\right)^{1/2}\sum_{\underset{\sim}{q}j} A_s(\underset{\sim}{q}j)\frac{<e|\hat{q}\cdot \underset{\sim}{p}|v>}{m(\mathcal{E}_c - \mathcal{E}_v)}\delta_{\underset{\sim}{K}, -\underset{\sim}{q}} \tag{18}$$

$\mathcal{E}_c$ is the band edge. $\hat{q}$ is a unit vector in the direction $\underset{\sim}{q}$. In obtaining Eq. (18), exp $(-i\underset{\sim}{q}\cdot r)$ in Eq. (17) was expanded and the lowest order term in q was kept.

A canonical transformation of the total Hamiltonian is now performed such that the terms linear in the exciton operators are eliminated as in I:(24) - I:(28). The transformed Hamiltonian $\hat{H}$ is then as in I:(26) with the generator S chosen such that $i\,[S, H^{(0)}] = H^{(1)}$. In this case S is given by

$$S = \sum_{\substack{cv\lambda\underset{\sim}{K}\\ \underset{\sim}{\eta}\underset{\sim}{\epsilon}}}\left\{\frac{f^*_{\underset{\sim}{\eta}\underset{\sim}{\epsilon}}(cv\lambda\underset{\sim}{K})a_{\lambda\underset{\sim}{K}}(cv)d_{-\underset{\sim}{\eta}\epsilon}}{E_{\lambda\underset{\sim}{K}}(cv) + \hbar\omega_{-\underset{\sim}{\eta}\underset{\sim}{\epsilon}}} - \frac{f_{\underset{\sim}{\eta}\underset{\sim}{\epsilon}}(cv\lambda\underset{\sim}{K})a^+_{\lambda\underset{\sim}{K}}d_{\underset{\sim}{\eta}\underset{\sim}{\epsilon}}}{E_{\lambda\underset{\sim}{K}}(cv) - \hbar\omega_{\underset{\sim}{\eta}\underset{\sim}{\epsilon}}}\right\}\delta_{\underset{\sim}{K}, \underset{\sim}{\eta}} \tag{19}$$

$$+ \sum_{cv\lambda\underset{\sim}{K}s}\left\{\frac{g^*_s(cv\lambda\underset{\sim}{K})a_{\lambda\underset{\sim}{K}}(cv)b^+_s}{E_{\lambda\underset{\sim}{K}}(cv) - \hbar\omega_s} - \frac{g_s(cv\lambda\underset{\sim}{K})a^+_{\lambda\underset{\sim}{K}}(cv)b^+_s}{E_{\lambda\underset{\sim}{K}}(cv) + \hbar\omega_s}\right\} + \text{h.c.}$$

$\hat{H}$ contains the commutators of all orders of S with $H^{(1)}$ and $H^{(2)}$ [see I:(26)]. One phonon Raman scattering arises from the terms in the commutator $[S, [S, H^{(2)}]]$. These terms involve three successive interactions: (1) exciton-photon interaction absorbing an incident photon, (2) exciton-phonon interaction creating or destroying a phonon, and (3) exciton-photon interaction emitting a scattered photon. These real transitions are accompanied by three virtual exciton transitions. The commutator contributes six terms corresponding to the various time ordering of the three interactions.

ONE PHONON RAMAN SCATTERING

Suppose n_1, n_2 and n_s are the numbers of the incident photons, the scattered photons and the phonons respectively. In one-phonon scattering let the initial state of the system be specified by $|i) = |n_1, n_2, n_s, 0>$ and the final state with the emission of a phonon by $|f> = |n_1 - 1, n_2 + 1, n_s + 1, 0>$. The zero in the ket refers to the electronic ground state. We treat spontaneous emission only and set $n_2 = 0$. ω_1 and ω_2 will denote the frequencies of the incident and the scattered photons. By first order time-dependent perturbation theory, the transition probability W per unit time in the deformation potential approximation is

$$W = \frac{4\pi^3 e^4}{h^3 m^4 a^2 \kappa_\infty^2 M_1 N}\sum_{s, \eta_2}\frac{n_1(n_s+1)}{\omega_1\omega_2\omega_s}\left|\sum_i R^{(i)}_{12}(-\omega_1, \omega_2, \omega_s)\right|^2 \delta(\omega_1 - \omega_2 - \omega_s) \tag{20}$$

The Raman tensor $R_{12}^{(i)}$ is given by

$$R_{12}^{(i)} = \frac{N\hbar^2}{V} \sum_{\substack{cv\lambda \\ c'v'\lambda' \\ j}} \Big\{ \langle v|\underset{\sim}{\epsilon}_2 \cdot \underset{\sim}{p}|c\rangle \langle cv|d_j^{(i)}|c'v'\rangle \langle c'|\underset{\sim}{\epsilon}_1 \cdot \underset{\sim}{p}|v'\rangle U_{cv\lambda\underset{\sim}{\eta}_2}(0) U_{c'v'\lambda'\underset{\sim}{\eta}_1}(0)^*$$

$$\times \frac{(M_1/N)^{1/2}\, \gamma\, \omega_s^2\, \underset{\sim}{e}(1|\underset{\sim}{\eta}_1 \underset{j}{-} \underset{\sim}{\eta}_2) \cdot \underset{\sim}{B}(\begin{smallmatrix} O \\ 1 \end{smallmatrix}|s)}{(\omega_s^2 - \omega^2_{\underset{\sim}{\eta}_1 - \underset{\sim}{\eta}_2, j})[E_{\lambda\underset{\sim}{\eta}_2}(cv) - \hbar\omega_1 + \hbar\omega_s][E_{\lambda'\underset{\sim}{\eta}_1}(c'v') - \hbar\omega_1]} + 5 \text{ other terms} \tag{21}$$

where

$$\langle cv|d_j^{(i)}|c'v'\rangle = \sum_b \left(\frac{M_1}{M_b}\right)^{1/2} e^{(i)}\left(b|\underset{\sim}{\eta}_1 \underset{j}{-} \underset{\sim}{\eta}_2\right) \Big\{ q_e \langle c|D_b^{(i)}|c'\rangle \delta_{vv'} - q_h \langle v'|D_b^{(i)}|v\rangle \delta_{cc'} \Big\} \tag{22}$$

The subscripts 1 and 2 on $R_{12}^{(i)}$ refer to the polarization directions of the incident and the scattered photons, while the superscript refers to the phonon polarization direction. η_1 and η_2 will be assumed to be zero. The transition probability for absorbing a phonon is obtained by replacing ω_s by $-\omega_s$ everywhere in Eq. (20) except in the denominator.

The energy $I(\omega_2)$ at frequency ω_2 scattered into solid angle $d\Omega$ in the unit volume of the crystal per unit time is then

$$I(\omega_1 - \omega_s) = \frac{v_0 e^4 J_0 d\Omega}{2\hbar^3 m^4 a^2 M_1 c^4} \frac{(n_s + 1)(\omega_1 - \omega_s)^2}{\omega_s\, \omega_1^2} \left| \sum_i R_{12}^{(i)} \right|^2 g(\omega_s) d\omega_s \tag{23}$$

where J_0 is the energy flux of the incident radiation and $g(\omega_s)$ is the density of states of the phonons in the perturbed lattice. $g(\omega_s)$ has different values for the band modes and the localized modes.

Eq. (20) and (23) show that the first order Raman scattering gives a continuous spectrum in the defect lattice instead of the line spectrum found in a perfect lattice. $M_1|B(\begin{smallmatrix} 0 \\ 1 \end{smallmatrix}|S)|^2$ is of order unity for a localized mode and of order $1/rN$ for a band mode. The whole of the scattering in the continuum is thus of comparable magnitude with that in the localized modes if the latter exist.

$R_{12}^{(i)}$ can be approximately calculated for a two band model. Using the energy spectrum (5) and the standard hydrogenic wave functions $|U_\lambda(0)|^2$ for discrete and continuum states, one obtains for $R_{12}^{(i)}$

$$R_{12}^{(i)} \sim \sum_{jb} \langle v|\underset{\sim}{\epsilon}_2 \cdot \underset{\sim}{p}|c\rangle \left(\frac{M_1}{M_b}\right)^{1/2} e^{(i)}(b|{}^0_j)\,[\langle c|D_b^{(i)}|c\rangle - \langle v|D_b^{(i)}|v\rangle] \langle c|\underset{\sim}{\epsilon}_1 \cdot \underset{\sim}{p}|v\rangle \tag{24}$$

$$\times \frac{M_1^{1/2}\gamma\,\omega_s^2 \underset{\sim}{e}(1|{}^0_j)\cdot \underset{\sim}{B}({}^0_1|S)}{N^{1/2}(\omega_s^2 - \omega_{0j}^2)} \left\{ \frac{1}{\pi a_0^3} \sum_n \frac{\hbar^2}{n^3[E_g - R/n^2 - \hbar\omega_1 + \hbar\omega_s][E_g - R/n^2 - \hbar\omega_1]} \right.$$

$$\left. + \frac{1}{2\pi^2} \int_0^{k_{max}} \frac{2\pi(2\mu R/\hbar^2 k^2)^{1/2}}{1-\exp\{-2\pi(2\mu R/\hbar^2 k^2)^{1/2}\}} \cdot \frac{\hbar^2 k^2\, dk}{(E_g - \hbar\omega_1 + \frac{\hbar^2 k^2}{2\mu})(E_g - \hbar\omega_1 + \hbar\omega_o + \frac{\hbar^2 k^2}{2\mu})} \right\}$$

where a_0 is the exciton Bohr radius. The first term inside the curly bracket is the contribution from the discrete states and the second term arises from the continuum.

The energy denominator in Eq. (24) shows the resonant behavior of $R_{12}^{(i)}$. When $\hbar\omega_1$ approaches the exciton energy $(E_g - R)$, the term with $n = 1$ dominates and $R_{12}^{(i)}$ diverges as $1/(E_g - R - \hbar\omega_1)$. This divergence is common to all types of phonon modes: band and localized. The ω_1 dependence of the intensity of the scattered light in our theory is quite different from that of Maradudin et al. [6].

In the case of polar interaction, the scattered energy $I(\omega_2)$ is obtained in a similar way.

$$I(\omega_1 - \omega_s) = \frac{\pi e^6 \omega_\ell^2 J_0 d\Omega}{\hbar^3 m^6 c^4} \left(\frac{1}{\varkappa_\infty} - \frac{1}{\varkappa_0}\right) \frac{(n_s + 1)(\omega_1 - \omega_s)^2}{\omega_1^2\,\omega_s} \left| P_{12}(-\omega_1, \omega_1 - \omega_s, \omega_s)\right|^2 g(\omega_s)d\omega_s \tag{25}$$

where the Raman tensor P_{12} is

$$P_{12} = \frac{N\hbar^3}{V} \sum_{\substack{cv\lambda \\ c'v'\lambda' \\ j}} \langle v|\underset{\sim}{\epsilon}_2 \cdot \underset{\sim}{p}|c\rangle \left[\frac{q_e \langle c|\hat{\eta}\cdot \underset{\sim}{p}|c'\rangle \delta_{vv'}}{\mathcal{E}_c - \mathcal{E}_{c'}} - \frac{q_h \langle v'|\hat{\eta}\cdot \underset{\sim}{p}|v\rangle \delta_{cc'}}{\mathcal{E}_{v'} - \mathcal{E}_v}\right] \langle c'|\underset{\sim}{\epsilon}_1 \cdot \underset{\sim}{p}|v'\rangle$$

$$\times \frac{M_1^{1/2}\gamma\omega_s^2 \underset{\sim}{e}(1|{}^0_j).\,\underset{\sim}{B}({}^0_1|S)U_{cv\lambda 0}(0)U_{c'v'\lambda'0}(0)^*}{N^{1/2}(\omega_s^2 - \omega_{0j}^2)[E_{\lambda 0}(cv) - \hbar\omega_1 + \hbar\omega_s][E_{\lambda' 0}(c'v') - \hbar\omega_1]} + 5 \text{ other terms} \tag{26}$$

$\hat{\eta}$ is a unit vector in the direction $\underset{\sim}{\eta}_1 - \underset{\sim}{\eta}_2$. We assume that $\underset{\sim}{\eta}_1, \underset{\sim}{\eta}_2 \cong 0$. Σ' means that c = c', v = v' is excluded from the sum. For this reason two-band calculation cannot be done in this case. However, we can infer from the presence of the term $(E_{\lambda'0}(c'v') - \hbar\omega_1)$

in the denominator of Eq. (26) that P_{12} will diverge at photon energies which correspond to the creation of virtual excitons.

DISCUSSION

Eq. (23-26) are the main results of this paper. We have shown that the Raman tensor in an imperfect crystal also diverges when the incident photon energy approaches the exciton energy. This resonant behavior is shown by all types of lattice vibrations--band, resonance and local modes. In our theory we have used the perfect crystal electronic states, i.e., free excitons. Even for the isoelectronic case, one might expect bound exciton states. Then the virtual intermediate states will be different and the pole of the Raman tensor will be shifted. The matrix elements of the various interaction terms will also be different resulting in a change in the intensity of the scattered radiation. Thus the resonance phenomena may be used to probe whether or not the isoelectronic defect binds the exciton. In CdS the pole in Raman tensor occurs at energy 2.544 eV. A local mode of frequency 339 cm^{-1} appears in CdS when S is replaced by 0[7]. According to our preliminary calculation the intensity of light scattered by this mode would increase by a factor of 100 when $\hbar\omega_1$ is varied from 2.4 eV to 2.53 eV. No experiments have yet been reported to test this prediction.

REFERENCES

1. A.K. Ganguly and J.L. Birman, Phys. Rev. 162, 806 (1967).
2. R.S. Knox, Solid State Physics, Supplement 5, 37 (1963).
3. M.A. Ivanov, M.A. Krivoglaz, and V.F. Los, Soviet Physics-Solid State 8, 2294 (1967).
4. P.G. Dawber and R.J. Elliot, Proc. Roy. Soc. A273, 222 (1963).
5. H. Frohlich, Advan. Phys. 3, 325 (1954).
6. A.A. Maradudin, Solid State Physics 19, 1 (1966).
7. P. Pfeuty, J.L. Birman, M.A. Nusimovici, M. Balkanski, "Localized Excitations in Solid," p. 210, Plenum Press, 1968.

F-6: RAMAN SCATTERING BY THE HYDROXYL ION IN ALKALI HALIDES*

Wayne R. Fenner and Miles V. Klein,
Department of Physics and Materials Research Laboratory,
University of Illinois, Urbana, Illinois

ABSTRACT

Raman scattering measurements made on hydroxyl doped alkali halide crystals show three types of excitation. (1) A narrow line with polarization parallel to the polarization of the exciting light [(xx) geometry] is observed in the region of the infrared stretching frequency. The half-width is nearly independent of temperature in contrast to the strong temperature broadening of the infrared absorption. The Raman lines, measured at 300°K, 77°K, and 5°K show a shift to higher frequencies with decreasing temperature. NaCl:OD^- shows a Raman stretching line, in contrast to the infrared case where the stretching oscillator strength is essentially zero. (2) A perpendicularly polarized Raman line [(xy) geometry] corresponding to the direct librational absorption line in the infrared is observed in KBr:OH^- and KCl:OH^- at helium temperature. Our failure to see this line in other systems is probably due to insufficient sensitivity. (3) A broader emission with parallel polarization is seen in the region of 50 cm^{-1} in KBr:OH^- and NaCl:OH^-. This may be related to part of the far infrared spectrum reported on KBr:OH^-. The infrared absorption seen as a 30 cm^{-1} sideband to the main stretching band has not been observed by Raman scattering.

INTRODUCTION

We report on Raman scattering of laser light by hydroxyl ions in some alkali halide crystals. One motivation for this work was our desire to correlate the Raman lines with near infrared absorption measurements made by Wedding and Klein[1] in the hope of providing insight into the mechanisms responsible for both the infrared absorption and the Raman scattering.

*Supported in part by the National Science Foundation under contract GP 6581 and also by the Advanced Research Projects Agency under contract SD-131.

Since this work is largely based on the infrared measurements of Wedding and Klein, it is worthwhile to review briefly their results. Fig. 1 shows a typical absorption spectrum for the hydroxide ion in an alkali halide host crystal. The three main features are common to almost all such spectra in other host crystals. The prominent peak at 3641 cm^{-1} is due to the stretching of the O-H bond. The two curves shown are for liquid helium and liquid nitrogen temperatures. Data taken at room temperature reveal only two very diffuse peaks, one at the stretching frequency and the other about 300 cm^{-1} higher. At helium temperature this latter peak occurs at 3938 cm^{-1} and has been assigned to a combination band due to the stretching plus librational motion of the OH^- dipole in the lattice. The librational frequency is then found by subtraction to be about 297 cm^{-1}. This librational or rocking motion is thought to be best understood at the present time[2]. We shall have more to say about it later.

The third common feature of the infrared spectra is represented by the broad band at 3673 cm^{-1}. This has also been assigned to a sideband of the stretching motion; however its detailed origin is not understood at this time. For want of a better name we call it band X. Its separate existence as a 32 cm^{-1} energy level has been confirmed by direct absorption measurements in the far infrared[3] and by depressions in the low temperature thermal conductivity[4]. The sharp peak at 3666 cm^{-1} has not been observed in the far infrared, and its origin is not yet clear.

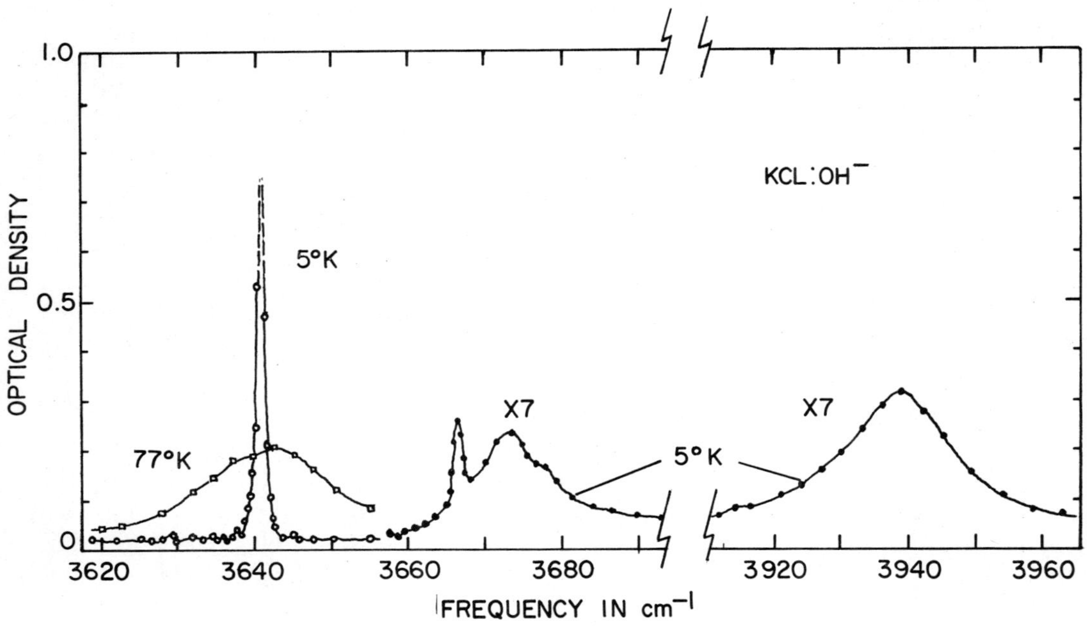

Fig. 1. The Infrared absorption of $KCl:OH^-$. Crystal thickness, 12mm. Hydroxyl concentration, 250 ppm. The "X7" portions of the curve are expanded vertically sevenfold. [Refs. 1 and 4.]

EXPERIMENTAL DETAILS

The Raman measurements, which at this point must be called preliminary, were taken using a home made Argon ion laser. The 4880 line was employed at powers of about 100 mw. The spectra were analyzed with a tandem double monochromator utilizing 50 mm high curved slits. A fiber optics device was utilized to match the image of the laser beam in the crystal with the curved entrance slits. The detector was an ITT FW 130 photomultiplier tube operated at -25°C. Photon counting techniques were used. A polaroid analyzer was used in front of the entrance slits. The scattering angle was 90°.

RAMAN SCATTERING BY THE LIBRATIONAL ENERGY LEVEL

In Fig. 2 we compare the Raman data and the infrared absorption data[2] (measured directly in the 300 cm^{-1} region) for the hydroxyl librator in KBr at helium temperature. The Raman peak position agrees well with the infrared peak, and the Raman linewidth, although instrument limited, is not inconsistent with the infrared linewidth.

Studies of the dichroism induced in the ultraviolet "OH" band by an electric field at low temperatures have shown that the O-H axis takes up an equilibrium position along one of the six equivalent <100> directions[5]. Field-induced dichroism measurements on the 3900 cm^{-1} librational sideband in KCl and RbCl have shown that the transition dipole moment for the librator is perpendicular to the O-H axis[2]. The <100> orientation of the molecular ion seems to imply that the symmetry at the defect site is C_{4v}. The infrared dichroism results then say that the librational transition belongs to the doubly degenerate representation E of C_{4v}. First order Raman scattering by this E mode would yield a completely depolarized spectrum of the (xy) type with no diagonal or polarized scattering of the (xx) type. This is exactly what the Raman data of Fig. 2 reveal, apart from a small amount of parallel or (xx) polarization scattering, that can be attributed to the large collection angle of about f/1.6.

It is shown in Ref. [2] that the infrared data on the librational energy level can be explained by a simple model in which the OH^- ion undergoes nearly harmonic torsional or angular oscillations away from its <100> equilibrium direction about an axis that nearly coincides with the molecular center of mass. Suppose a given ion has its equilibrium O-H axis along the +x direction and is undergoing librational oscillation in the xy plane so that the instantaneous small angle between the O-H axis and the +x axis is θ. Then the first order change in the polarizability tensor has the components

$$\delta\alpha_{xy} = \delta\alpha_{yx} = (\alpha_{xx} - \alpha_{yy})\ \theta$$

All the other components are zero. Here α_{xx} and $\alpha_{yy} = \alpha_{zz}$ are the diagonal elements of the static polarizability tensor associated with the OH^- ion. The resulting Raman scattering will be completely depolarized and of the (xy) symmetry type.

We have observed the same Raman scattering by the librator at 305 cm^{-1} in KCl at helium temperature. Observations were not made at higher temperatures in either KBr or KCl because of interference with the two phonon Raman spectrum of the host lattice.

RAMAN SCATTERING BY OH^- AND OD^- STRETCHING MODES

The remainder of this paper is devoted to the main stretching band, where the Raman line is different from the infrared line. Fig. 3 shows the Raman and infrared data for the O-H stretching mode in NaCl. At room temperature there is a strong, well-defined Raman line that is only slightly wider than the helium temperature line. The Raman line shifts to lower frequency with increasing temperature in an essentially linear fashion. This behavior is in striking contrast with that of the infrared stretching line, which broadens very rapidly with increasing temperature, so much so that the line is nearly unobservable at room temperature. In addition the center of the infrared line shifts to higher frequency with increasing temperature. This lack of correlation between thermal behavior of Raman and infrared lines seems to be unique with OH^---Callender and Pershan have found good correlations in the cases of CN^- and NO_2^- in alkali halides and are in agreement with us concerning OH^-[6].

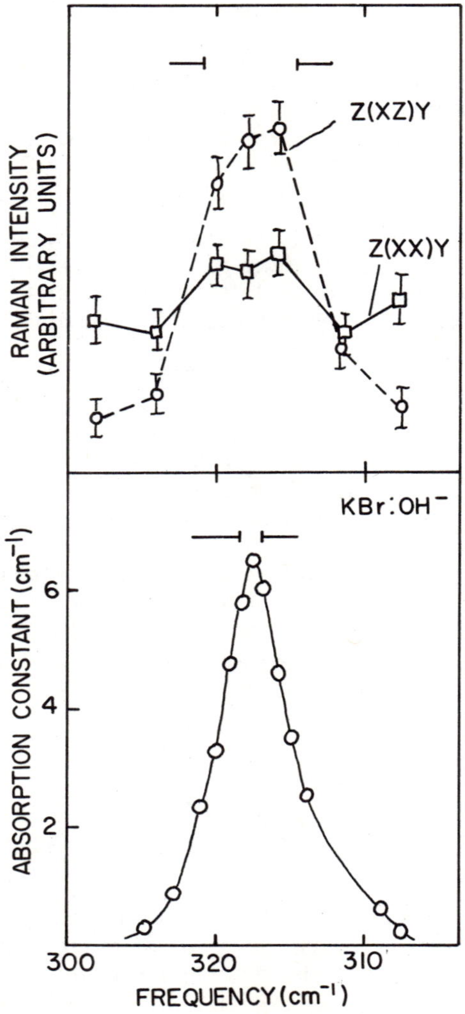

Fig. 2. The hydroxyl librator in KBr. Top curves, Raman spectra on a 1500 ppm crystal at 5°K. Bottom curve, direct infrared spectrum on a 240 ppm crystal at 8°K [Ref. 2].

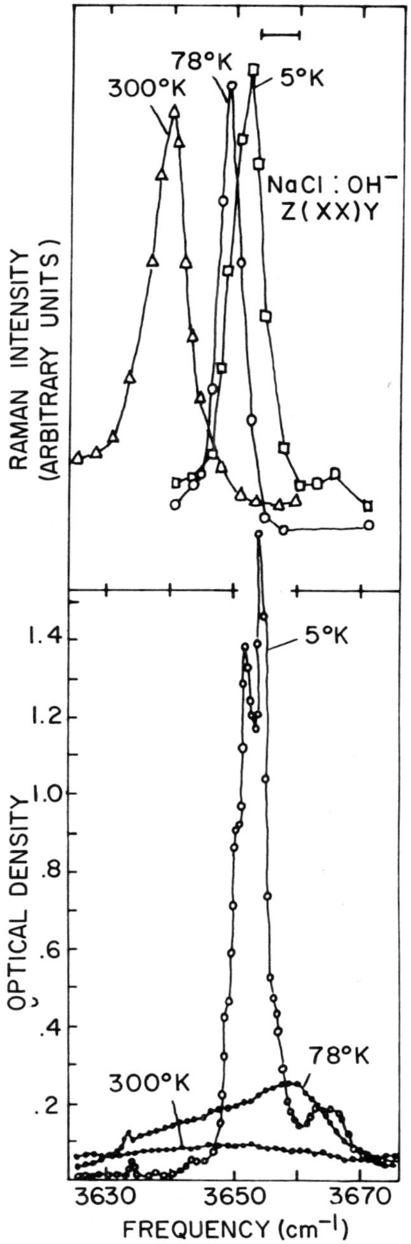

Fig. 3. The hydroxyl stretching mode in NaCl. Upper curves, Raman spectra on a 1900 ppm crystal at three temperatures. Lower curves, infrared spectrum on a 20mm thick crystal containing 860 ppm OH^- [Ref. 1].

The measured (xx) polarization of the Raman stretching band is consistent with the picture of a stretching motion that would induce first order changes in the diagonal components of the polarizability tensor.

The Raman and infrared spectra of the OD^- stretching band in NaCl are compared in Fig. 4. The OD^- stretching band has never been seen in NaCl and must have an oscillator strength less than 1/100 that of OH^-. This essentially zero oscillator strength has been attributed to a cancellation of two contributions to the first derivitive of the dipole moment with respect to the stretching coordinate, namely, the positive contribution of the outward-moving positive deuteron and the negative contribution of assumed outward moving non-bonding electrons. Quite a different coupling to the electronic structure is of course involved in Raman scattering, and this sort of cancellation does not occur. The OH^-/OD^- frequency ratio is precisely that observed for the stretching modes in KCl and KBr[1].

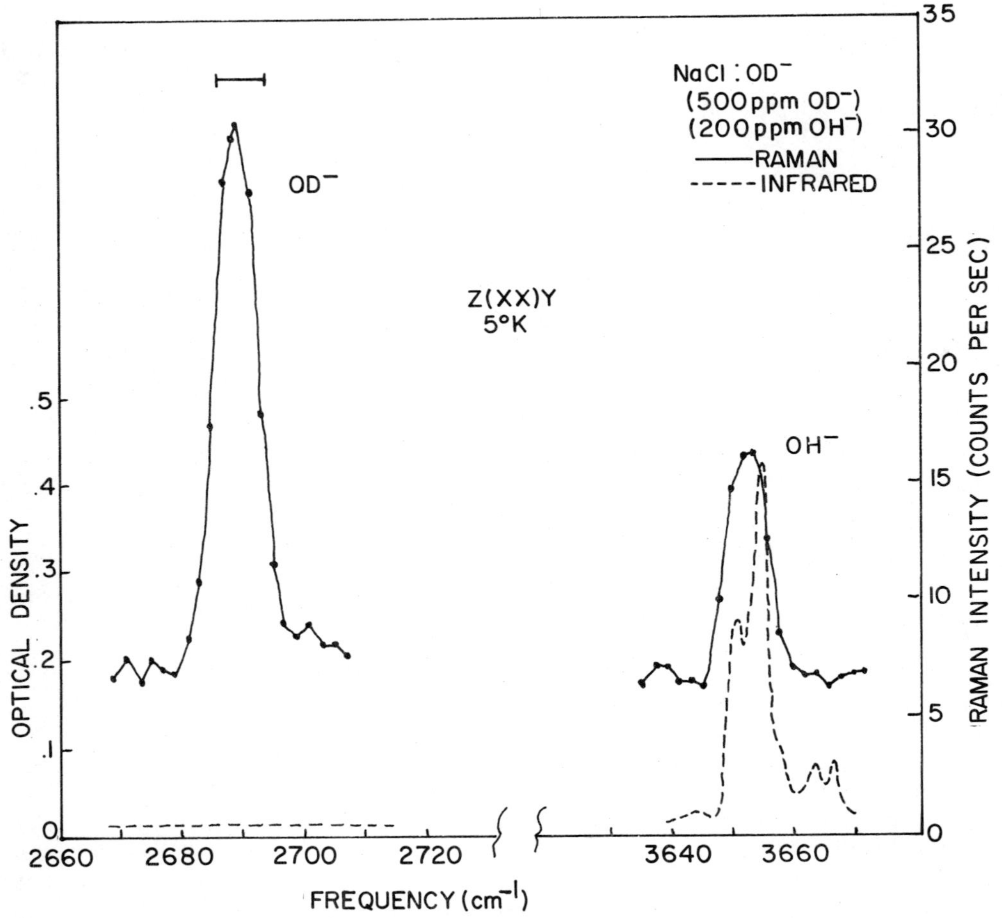

Fig. 4. The OH^- and OD^- stretching modes in NaCl. Solid curves, Raman spectra on a crystal having the indicated concentrations. Dashed lines, infrared spectra. No infrared OD^- spectrum was found for this crystal, as indicated schematically by the horizontal dashed line. The OH^- infrared spectrum is that for a 13 mm thick 470 ppm crystal [Ref. 1].

We have also taken Raman data on OH^- stretching modes in KBr and KCl. They are very similar to the OH^- and OD^- data for NaCl, and will not be shown here. The measured peak positions for all the Raman and infrared lines are collected in Table I.

TABLE I

Peak Positions (cm^{-1})

	Raman (this work)	Infrared (Refs. 1 and 2)	Temperature
KBr (librator)	313	312.7 (309.7 as sideband)	He
KBr (stretch)	3618	3618	He
KBr (stretch)	3618	3618	N_2
KBr (stretch)	3613	3611	Room
KCl (librator)	305	297.5 (as sideband)	He
KCl (stretch)	3643	3641	He
KCl (stretch)	3640	3641	N_2
KCl (stretch)	3631	3639	Room
NaCl (stretch)	3652	3654.5	He
NaCl (stretch)	3648	3659	N_2
NaCl (stretch)	3638	3650-3660 (very broad)	Room
NaCl (OD^- stretch)	2689	not observed	He
	2688	not observed	N_2
	2683	not observed	Room

CONCLUDING REMARKS

The Raman and infrared stretching bands seem to be nearly identical at helium temperatures. The mechanism operable in broadening the infrared line seems to be absent in the Raman case. In Reference 1 it is suggested that the low-lying band X energy level is responsible in some way for broadening the infrared stretching band. It would therefore be valuable if band X could be seen in Raman spectra, either directly, or in combination with the stretching mode. The stretching plus X infrared combination band has the same E-type symmetry as the librational band[1]; we would then predict that the resulting Raman lines would be depolarized, of the (xy) type.

We have not yet seen band X in Raman spectra, although we have seen a broad Raman line with parallel or (xx) polarization at 50 to 60 cm^{-1} in hydroxyl-doped NaCl, KBr, and KCl. This may be related to part of the far infrared spectrum, as seen, for example in the work of Bosomworth on KBr:OH^-[3].

REFERENCES

1. B. Wedding and M. V. Klein, Phys. Rev. 177, (No. 3) (1969).
2. M. V. Klein, B. Wedding, and M. A. Levine, "Libration of the Hydroxyl Ion in Alkali Halide Crystals," (submitted to Physical Review).
3. D. R. Bosomworth, Solid State Commun. 5, 681 (1967).
4. C. K. Chau, M. V. Klein, and B. Wedding, Phys. Rev. Letters 17, 521 (1966).
5. U. Kuhn and F. Lüty, Solid State Commun. 2, 281 (1964); H. Paus and F. Lüty, Phys. Status Sol. 12, 341 (1965).
6. R. H. Callender and P. S. Pershan, Raman Spectra of Molecular Impurities in Alkali Halides, paper F-7 this conference.

F-7: RAMAN SPECTRA OF MOLECULAR IMPURITIES IN ALKALI HALIDES*

R. H. Callender and P. S. Pershan
Division of Engineering and Applied Physics, Harvard University
Cambridge, Massachusetts

ABSTRACT

Spontaneous Raman measurements have been made on various alkali halide crystals doped with CN^-, OH^-, or NO_2^-. Two effects have been observed: spectra due to the internal or stretching vibrations of the molecule (in the range of 1000-4000 cm^{-1}) and lower frequency modes (0 ~ 1000 cm^{-1}) due to either the rotational or translational degrees of freedom of the undistorted molecule. The Q(0) mode of the stretching vibration is less than .5 cm^{-1} even at 300°K and is more than two orders of magnitude more intense than any combinational mode. Differences between the observed I. R. and Raman spectra for the same impurity host system can be explained in terms of the rotational motion of the impurity. The low frequency Raman structure is harder to understand since modes of the impurity may resonate with the phonon spectrum of the pure crystal. We have had some success in analyzing signals from CN^- doped systems in the low frequency region as arising from an impurity induced first order Raman effect.

INTRODUCTION

We have made spontaneous Raman measurements on various alkali halides doped with CN^-, OH^-, and NO_2^-. These molecules generally substitute for the anion of the pure crystal; and for the two linear dipole impurities it is believed that the host crystal tends to align them along [100] directions. In view of the fact that these molecules are asymmetric and possess internal degrees of freedom, new rotational and vibronic modes are introduced into the lattice. Many of these systems (CN^- [1], NO_2^- [2], and OH^- [3, 4]) have been studied extensively in those parts of the infrared region where the

*This work was supported in part by the Advanced Research Projects Agency, by the Joint Services Electronics Program under Contract N00014-67-0298-0006, by the Division of Engineering and Applied Physics, Harvard University, and by the National Aeronautics and Space Administration under Grant NGR 22-007-126.

reststrahlen bands are weak. These measurements complement and extend those results since the selection rules for the two measurements are different and since the alkali halides have no first order Raman effect to obscure low energy excitations. We divide this paper into two parts: results from the low frequency (0-1000 cm^{-1}) spectra due to rotational and/or translation motion of the molecule and results from the high frequency (1000-4000 cm^{-1}) internal stretching vibrations of the impurity molecule.

LOW FREQUENCY

We present here low frequency data on CN^- in KCℓ, KBr, and NaCℓ hosts. KCℓ type crystals possess O_h symmetry and have therefore Γ_1^+, Γ_{12}^+, and Γ_{25}^+ Raman active vibrational symmetries. The alkali halides have no k = 0 phonons with these symmetries and thus no first order Raman activity. They do, however, possess a fairly large second order spectrum. By studying these crystals near 10°K, we were able to freeze out most of this background.

The data is presented in Figs. 1, 2, and 3. Spectrum for the symmetry combination $|\alpha_1|^2 + \frac{4}{3}|\alpha_{12}|^2 + |\alpha_{25}|^2$ of the pure hosts is also included. The resolution for all the low frequency data is 8 cm^{-1}. For the NaCℓ and KCℓ cases there is a very small contribution from the intrinsic second order spectrum in the region of the impurity levels. The dominant impurity induced scattering from KBr:CN^- has Γ_{12}^+ symmetry. Fortunately, the Γ_{12}^+ contribution from the second order spectrum of pure KBr in the spectral region of the impurity signals is less than 3% of the impurity signal. In fact for all of the samples studied here, the second order background is not of any significance. For KCℓ:CN^- impurity induced Raman spectra is observed with Γ_1^+, Γ_{12}^+, and Γ_{25}^+ symmetry, while for NaCℓ:CN^- one finds Γ_1^+ and Γ_{25}^+ modes.

This data does not lend itself to easy interpretation. In the first place the rotational degrees of freedom of the impurity would be expected to contribute to the Raman signal in this frequency range. Some success for these levels has been obtained for the infrared results using a Devonshire model. For details see references [5] and [6]. For the KBr:CN^- system we have used the same parameters as the infrared case [1] and Sauer's calculation [6] and have plotted them as lines in Fig. 3. We note that the density of lines is quite high and that without detailed calculations of the transition probabilities and linewidths one could "explain" nearly any impurity Raman scattering.

In the second place we believe it is quite likely that the Raman signals observed from the impurity may also be due to an induced first order Raman signal from the pure crystal. We shall illustrate this point using the KBr:CN system. Using the convention that Raman scattering can be represented by a fourth rank tensor [7] $I_{\alpha\gamma,\beta\lambda}$ of the form

$$I_{\alpha\gamma,\beta\lambda}(\Omega) = \frac{1}{2\pi} \int_{-\infty}^{\infty} e^{-i\Omega t} < P_{\alpha\gamma}(0)\, P_{\beta\lambda}^{+}(t) > dt$$

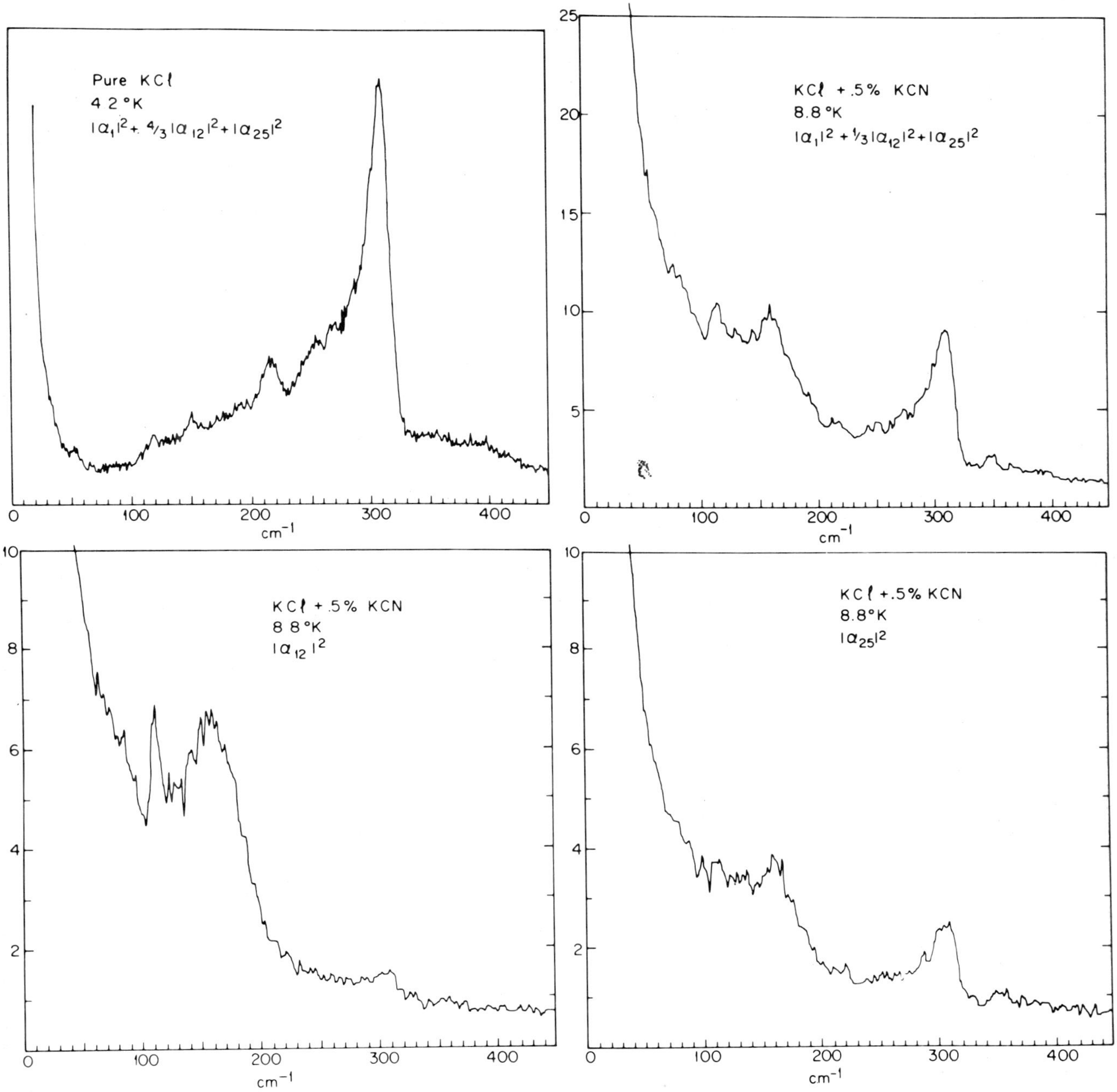

Fig. 1. Right angle low frequency Raman scattering for pure KCℓ and KCℓ:CN^- at helium temperatures with an instrumental resolution of 8 cm^{-1}. The scattering geometries are shown as sums of the three Raman active polarizability tensors. The KCℓ:CN^- data has been normalized to show relative intensities for the three scattering geometries.

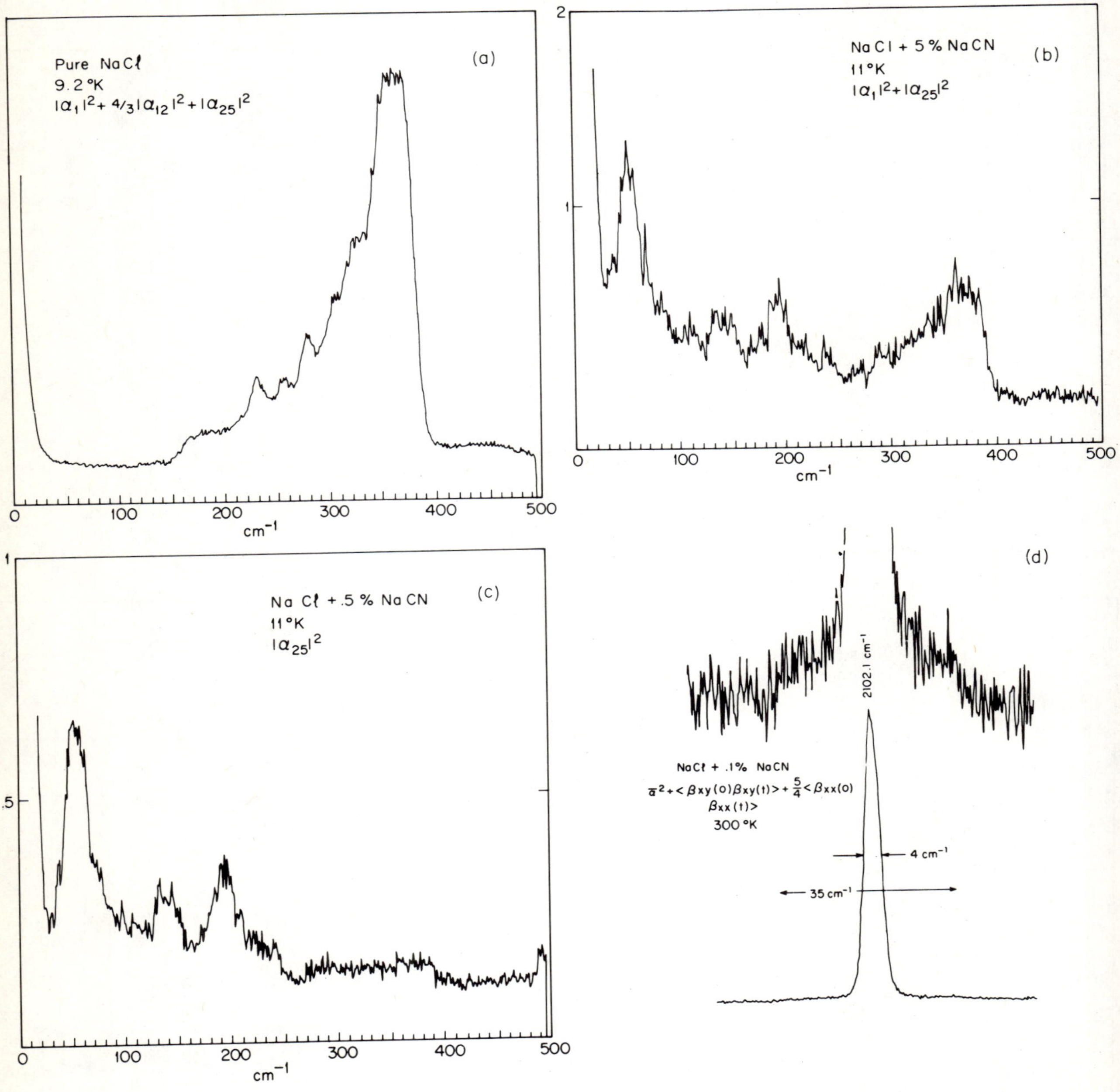

Fig. 2. Parts a, b, and c give right angle low frequency Raman scattering for pure NaCℓ and NaCℓ:CN$^-$ at helium temperatures with an instrumental resolution of 8 cm^{-1} for various scattering geometries. $|\alpha_{12}|^2_{NaC\ell:CN^-} < \frac{1}{20}|\alpha_1|^2_{NaC\ell:CN^-}$. The impurity signals have been normalized to show the same relative vertical scale. Part d shows Raman scattering in the region of Q(0) mode of CN$^-$ in NaCℓ at 300^{0}K with instrumental resolution of 4 cm^{-1}. The scattering geometry is defined by the sum of the two independent correlation function $< \beta_{xx}(0)\ \beta^+_{xx}(t) >$ and $< \beta_{xy}(0)\ \beta^+_{xy}(t) >$ as indicated.

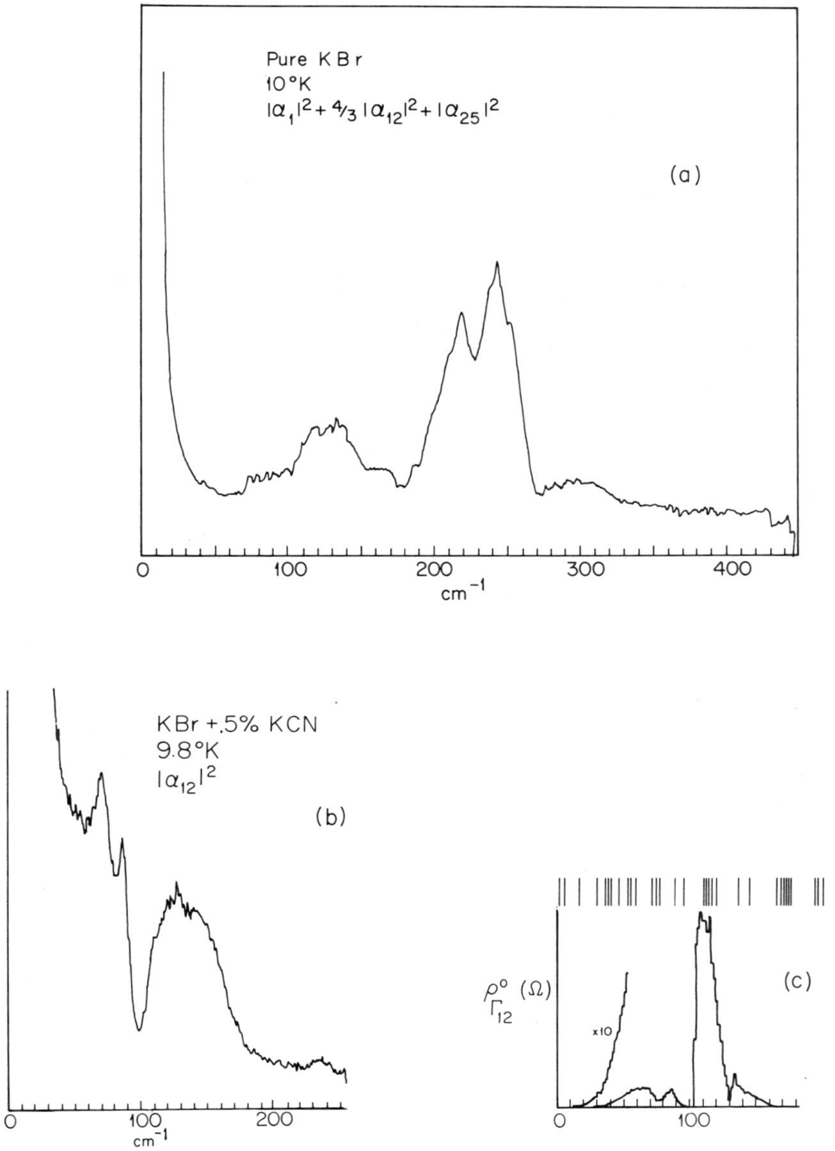

Fig. 3. Parts a and b give right angle low frequency Raman scattering for pure KBr and $KBr:CN^-$ respectively at helium temperatures with instrumental resolution of 8 cm^{-1}. $|\alpha_{12}|^2$ scattering is the only significant impurity signal. In part c the lines indicate $|\alpha_{12}|^2$ scattering from CN^- as predicted by a Devonshire model and the histogram approximates $\rho^o_{\Gamma_{12}}(\Omega)$ as calculated by Timusk and Klein [9].

where Ω is the change in frequency and $P_{\beta\lambda}(t)$ is the polarizability at time t, Benedek and Nardelli[8] show that for a symmetry Γ, the spectral density $S_{\Gamma}(\Omega)$, for light scattered by this effect is of the form

$$S_{\Gamma}(\Omega) \propto \frac{N(\Omega)}{\Omega} \rho_{\Gamma}(\Omega)$$

In this case $N(\Omega) = [1 - \exp \frac{-\hbar\Omega}{kT}]^{-1}$ for Stokes radiation and $\rho_{\Gamma}(\Omega)$ represents the density of states for the impure crystal of the Γ symmetry. If we assume that the CN^- molecules are spherically symmetrical and further that the force constants of the CN^- to K^+ are not much different than Br^- to K^+ and also that only nearest neighbor interactions are important, then $\rho_{\Gamma}(\Omega) = \rho^{o}_{\Gamma}(\Omega)$ where $\rho^{o}_{\Gamma}(\Omega)$ is the projected density of states on Γ symmetry of the pure crystal. For the case of KBr, Timusk and Klein have calculated $\rho^{o}_{\Gamma_{12}}(\Omega)$ [9] and in Fig. 3 we compare this to our impure crystal results. Even for the rather strict assumptions we have used, many of the features of the $KBr:CN^-$ scattered light can be explained rather well using this model.

STRETCHING VIBRATION SPECTRA

Table I summarizes the data we have taken as to position and half width of the stretching vibrations for the various impurities and hosts. All measurements have been taken at room temperature except where indicated. The linear molecules have one

TABLE I

Raman Data for the Stretching Vibrations of Different Impurity Host Combinations. Except Where Noted, All Data is at 300°K.

Dopant	Host	Q (0) (cm^{-1})	Line Width (cm^{-1})
CN^-	KCℓ	2085	< .5
	KCℓ (8.5°K)	2087.7	< .3
	KBr	2076	< 2
	KBr (9°K)	2078.1	< .3
	NaCℓ	2103	< 1
	NaCℓ (10°K)	2106.8	< .3
	RbCℓ	2078.7	< .6
	RbBr	2069.0	< .8
OH^-	KCℓ	3629	~ 5
	KCℓ (10°K)	3643	< .8
	KBr	3599	~.1
NO_2^-	KCℓ	1323 (A_1)	3
		806 (A_1)	< 4

stretching vibration, which has a diagonal polarizability tensor in the molecular frame of reference. NO_2^- has C_{2v} symmetry and its stretching vibrations decompose into $2A_1$ + B_1, both of which are nondegenerate. The B_1 mode has an offdiagonal polarizability tensor and has as yet not been observed by us. Sideband data for the stretching vibration from CN^- in NaCℓ at 300°K is shown in Fig. 2. There are several features that should be noted. First, the Q(0) mode, i.e. the stretching vibration itself of the CN^- dopant series, is generally narrower than the resolution of our instrument (about .5 cm^{-1} at best) even at room temperature and, except for a shift of a few wave numbers, exhibits no changes from 300°K to 2°K in the samples we have measured. Second, the position of the Q(0) mode except for room temperature OH^- doped crystals agrees to within experimental with infrared data on the same impurity-host system. Third, there are not in general any strong sidebands for any of the impurity-host systems. Our work in this area is in a preliminary stage at present, but the intensity of any sideband is at least a factor of 10 smaller than the Q(0) mode. In agreement with infrared absorption data, the width of the NaCℓ:CN^- sideband follows a T^2 dependence down to 200°K where we are no longer able to resolve it.

The structure of the CN^- stretching vibration region in NaCℓ and KCℓ can be understood in terms of the rotational like motion of the molecule. Starting from Eq. (1), we assume the rotation or tumbling motion of the molecules are ergodic in the sense that time averages and statistical averages are equivalent. Secondly, we assume the average symmetry of the impurity site for a tumbling molecule reflects the O_h point symmetry of the substituted site. Finally, and only as a first order approximation, we neglect rotational-vibrational coupling. For a tumbling, or rotating, molecule the polarizability tensor can be written as $\underset{\approx}{\alpha}(t) = [\bar{\underset{\approx}{\alpha}} + \underset{\approx}{\beta}(t)]Q(t)$ where Q(t) is the molecular coordinate describing the vibronic motion. For example, $Q(t) \sim \exp^{i\omega_0 t}$ where ω_0 is the stretching frequency. We take $\bar{\underset{\approx}{\alpha}}$ to be a scalar and $Tr\underset{\approx}{\beta} = 0$. The time dependence $\underset{\approx}{\beta}(t)$ follows from the tumbling motion of the molecule. If $\underset{\sim}{\epsilon}^i$ and $\underset{\sim}{\epsilon}^s$ represents the polarizations of the incident and scattered one can show that

$$I(\Omega) \sim (\bar{\alpha})^2 (\epsilon^s \cdot \epsilon^i)^2 \, \delta(\Omega - \omega_0)$$

$$+ (1/2\pi) \int_{-\infty}^{\infty} < (\epsilon^i \cdot \beta(0) \cdot \epsilon^s)(\epsilon^s \cdot \beta^+(t) \cdot \epsilon^i) > x[\exp i(\Omega - \omega_0)t]\, dt$$

Taking the substituted site as having O_h symmetry, one has only three independent quantities: $<\beta_{xx}(0)\,\beta^+_{xx}(t)>$, $<\beta_{xy}(0)\,\beta^+_{xy}(t)>$, and $<\beta_{xx}(0)\,\beta^+_{yy}(t)>$. We further find since $Tr\underset{\approx}{\beta} = 0$ that $<\beta_{xx}(0)\,\beta^+_{xx}(t)> = -2 <\beta_{xx}(0)\,\beta^+_{yy}(t)>$. The scattering has two separate parts: a narrow line, not broadened by rotational motion, due to spherical components of the polarizability tensor and rotational sidebands that can be expressed in terms of the two correlation functions referred to above.

This argument is readily generalized to molecules with more than one internal degree of freedom (like NO_2^- for example). One finds, in general, that for any given mode there is a sharp line whose integrated intensity goes as $<(Tr\underset{\approx}{\alpha}(t))^2>$ and a broad band whose integrated intensity goes as $<\underset{\approx}{\beta}(t)^2>$. For this latter part one might roughly assign a band width of the order of τ_c^{-1}, where τ_c is some type of tumbling time.

The spectrum of NO_2^- is readily interpreted from the above since the one Raman active internal vibration which is not observed at room temperature is the one with B_1 symmetry. Of the three Raman active modes for this molecule, this is the only one for which $Tr\underset{\approx}{\alpha}(t) = 0$.

ACKNOWLEDGMENTS

We would like to acknowledge helpful conversations with Professor R. O. Pohl and Dr. V. Narayanamurti. Most of the crystals studied here were grown for us by Mr. Gerhard Schmidt at Cornell University. We would like to express our appreciation to Mr. Schmidt.

REFERENCES

1. W. D. Seward and V. Narayanamurti, Phys. Rev. 148, 463 (1966).
2. V. Narayanamurti et al., Phys. Rev. 148, 481 (1966).
3. C. K. Chan et al., Phys. Rev. Letters 17, 521 (1966).
4. R. Bosomworth, Solid State Commun. 5, 681 (1967).
5. A. F. Devonshire, Proc. R. Soc. A153, 601 (1936).
6. P. Sauer, Z. Phys. 194, 360 (1966).
7. A. A. Maradudin, "Solid State Physics," 18 and 19, F. Seitz and D. Turnbull (ed.), Academic Press Inc., New York, 1966.
8. G. Benedek and G. F. Nardelli, Phys. Rev. 154, 872 (1967).
9. T. Timusk and M. V. Klein, Phys. Rev. 141, 662 (1965).

F-8 : SOME THEORETICAL ASPECTS OF SECONDARY RADIATION DURING VIBRATIONAL RELAXATION OF LUMINESCENCE CENTERS

V. V. Hizhnyakov, K. K. Rebane, I. J. Tehver
Institute of Physics and Astronomy,
Estonian SSR Academy of Sciences
Tartu, USSR

The secondary radiation emerging at resonance photoexcitation of the impurity center can be described by the following Eqs. [1-3]:

$$W(\omega_0, \Omega) = \frac{B}{2\pi} \int_{-\infty}^{\infty} d\mu \, e^{i(\Omega - \omega_0)\mu}$$

$$\times \iint_{0}^{\infty} d\tau d\tau' \, e^{i\omega_0(\tau - \tau')} A(\tau\tau'\mu) \quad (1)$$

Here only two electronic levels are considered. The following denotations are used: ω_0, Ω - frequencies of the exciting and the secondary radiation resp. :

$$B = \overline{B} \mid M^0_\lambda M^0_{\lambda'} \mid^2 \quad (2)$$

$$A(\tau\tau'\mu) = \Big\langle \overline{M}^*_\lambda \, e^{\frac{i}{\hbar} H_1 \tau' - \frac{\hat{\alpha}}{2}\tau'} \, \overline{M}^*_{\lambda'}$$

$$\times \, e^{\frac{i}{\hbar} H_0 \mu} \, \overline{M}_{\lambda'} e^{-\frac{i}{\hbar} H_1 \tau - \frac{\hat{\alpha}}{2}\tau} \, \overline{M}_\lambda e^{-\frac{i}{\hbar} H_0(\mu + \tau' - \tau)} \Big\rangle_o \quad (3)$$

$$\overline{M}_\lambda \equiv M_\lambda / M^0_\lambda$$

where M_λ is the electronic matrix element, M_λ^o its value if the vibrational coordinates are taken at the minimum of the adiabatic potential of the ground state; H_o, H_1 are the vibrational Hamiltonians of electronic states o and 1; $\hat{\alpha}$ is the Hermitian operator commuting with H_1 the eigenvalues of which in the eigenstates $|1v''\rangle$ of H_1 are equal to the radiative decay constants $\alpha 1v''$ of states 1v''; $\langle \cdots \rangle_o$ signifies the temperature averaging over vibrational substates of the ground electronic state. $\overline{B}$ proves to be a factor having a relatively weak dependence on frequencies ω_o and Ω (more thoroughly v. [4]).

The spectrum of the secondary radiation is determined by the correlation function $A(\tau\tau'\mu)$, which depends on the model of vibrations and the vibronic interaction in the impurity center.

In our previous papers[1-3] it was shown that by choosing an appropriate physical model (i. e. a model taking account of the fact that in the impurity center the duration of the vibrational relaxation is 10^3 - 10^4 times shorter than the optical lifetime of the excited electronic state) the Eqs. (1-3) describe not only the Rayleigh and Raman scattering (RS), which is often assumed, but also the luminescence.

The spectrum of the secondary radiation can be represented as a superposition of two spectra (and interference corrections to them): (1) the zero-phonon (Rayleigh) scattered line with its vibrational recurrences; (2) the zero-phonon (pure-electronic) line of luminescence with its vibrational recurrences.

The spectrum (2) contains not only the whole spectrum of the ordinary luminescence $I_\lambda(\tau)$ but also a small addition to it, the so-called "hot luminescence" (HL). The latter may be interpreted as the luminescence from the high vibrational levels of the excited electronic state whose role in the ordinary luminescence becomes apparent only at high temperatures, much above the crystal temperature.

The interference terms have no influence on the ordinary luminescence but lead to the asymmetry of HL and Raman scattering (RS) lines. Besides, they may lead to the merging of adjacent RS and HL lines. The corrections from interference terms are the more essential the closer the frequency of the excitation is to the maximum of some of the vibronic lines (sub-bands) of the absorption spectrum. Below we consider HL on the example of a simple, physically reasonable model.

In considering the spectrum of the secondary radiation in the resonance case, the Condon approximation may serve as a first approximation; i. e. one may assume the independence of the electronic matrix element of the vibrational coordinates.

In the Condon approximation $\hat{\alpha}$ proves to be a c number (constant of radiative decay $\alpha_{\hat{1v}''}$ is equal to α, independent of vibrational states $|1v''\rangle$), and $\overline{M}_\lambda = 1$. Therefore $A(\tau\tau'\mu)$ has the form

$$A(\tau\tau'\mu) = \Big\langle e^{\frac{i}{\hbar} H_1 \tau'} e^{\frac{i}{\hbar} H_0 \mu} e^{-\frac{i}{\hbar} H_1 \tau} \tag{4}$$

$$\times e^{-\frac{i}{\hbar} H_0(\mu + \tau' - \tau)} \Big\rangle_o e^{-\frac{\alpha}{2}(\tau + \tau')}$$

Now it is necessary to assume a concrete model - to determine the form of vibrational Hamiltonians and the character of vibrational relaxation.

We assume that the vibrational Hamiltonians H_o and H_1 differ only in the value of the minimum points of potential energy curves.

Since we have to consider the vibrational relaxation as an essential property in the process, the anharmonic coupling between vibrations is, of course, taken into account. Mathematically:

$$H_1 = e^{\sum_i x_{oi} \frac{\partial}{\partial x_i}} H_o \, e^{-\sum_i x_{oi} \frac{\partial}{\partial x_i}} + \hbar\omega_{10} \qquad (5)$$

where x_i and x_{oi} are the coordinates of the i-th normal vibration and the change of its equilibrium position resp., $\hbar\omega_{10}$ the energy of the zero-phonon transition.

If the condition (5) is satisfied the correlation function $A(\tau\tau'\mu)$ is easily expressed through the function $g(x) = \langle \nabla(x)\nabla(0)\rangle_o - \langle\nabla\rangle^2_o$, $(\nabla \equiv \sum_i x_{oi} \frac{\partial}{\partial x_i})$:

$$A(\tau\tau'\mu) \simeq \exp\left[-\frac{\alpha}{2}(\tau' + \tau) + i\omega_{10}(\tau'-\tau) + g(\mu) \right. \qquad (6)$$

$$\left. + g(\mu + \tau' - \tau) + g(-\tau) + g(\tau') - g(\mu - \tau) - g(\mu + \tau') \right]$$

($g(x)$ determines also the ordinary absorption and luminescence spectra[5]).

Further we assume that in the luminescence center there exists only one local vibration which interacts with crystal modes. This interaction leads to the decay of the local vibration which may be described by the exponential law with the decay constant Γ. We also set $T = 0$.

In such a case there is a simple expression for $g(x)$ allowing the integrations over μ, τ, τ' to be carried out exactly:

$$g(x) = \xi^2 \left(e^{i\omega x - \Gamma |x|} - 1 \right) \qquad (7)$$

where ω is the frequency of the oscillator, ξ^2 the dimensionless Stokes losses parameter, i.e. the energy of the Stokes displacement expressed by the number of local vibrational quanta.

By presenting all functions $\exp[g(x)]$ in the formula (6) in the form of series

$$\exp[g(x)] = \sum_{\rho=0}^{\infty} \frac{\xi^{2\rho}}{\rho!} e^{i\omega\rho x - \Gamma_\rho |x|} e^{-\xi^2} \qquad (8)$$

and by subdividing the integration domain into suitable parts the integrations may be carried through as a result of which the following expression for the intensity distribution in the secondary radiation spectrum is obtained:

$$W(\omega_o, \Omega) = \frac{B}{2\pi} e^{-2\xi^2} \sum_{m, m', \ell, \ell', \rho, k = 0}^{\infty} (-1)^{\ell+\ell'} \frac{\xi^{2(m+m'+\ell+\ell'+\rho+k)}}{m!m'!\ell!\ell'!k!\rho!}$$

$$\times \Bigg\{ - \frac{f^*(m' + \ell' + k)\ f(m - \ell - k)}{i[\Omega - \omega_o + \omega(\rho + k + \ell + \ell')] - \Gamma(\rho + k + \ell + \ell')} \tag{9}$$

$$+ \frac{1}{i[\Omega - \omega_{10} + \omega(\rho - m + \ell')] - \alpha/2 - \Gamma(\rho + m + \ell')}$$

$$\times \quad f^*(m + \ell' + k)\ f(m - \ell - k) \ - f(m + \ell + k)\ f^*(m' + \ell' - k)$$

$$- \frac{f^*(m' + \ell' - k) - f^*(m' + \ell' + k)}{i\omega(m - m' + \ell - \ell') + \alpha + \Gamma(m + m' + \ell + \ell')} + Ce \Bigg\}$$

where $f(m \pm \ell \pm k) \equiv \dfrac{1}{-i[\omega_{10} - \omega_o + \omega(m + \ell + k)] - \alpha/2 - \Gamma(m \pm \ell \pm k)}$

Let us consider this result in the case of excitation outside the pure-electronic line ($|\omega_o - \omega_{10}| >> \alpha$) on the assumptions $\Gamma >> \alpha$ and $\omega >> \Gamma$.

In the above described simplified model, which has equidistant levels, the distinguishing of HL of RS according to its spectral characteristics will be possible only when the frequency of the exciting light ω_o does not fall on the maximum of some absorption line.

Thus, in order to distinguish HL from the scattering in the given model it is assumed that the excitation frequency does not coincide with any of the absorption band maxima but falls on the border of one of the absorption lines. That implies we take in the function $f(m \pm \ell \pm k)$, representing the role of absorption in the Eq. (9),

$$\omega_{10} - \omega_o + \omega(m + \ell + k) \equiv x \neq 0. \tag{10}$$

Let us fix now the following combinations of summing subscripts

$$m + \ell + k = m' + \ell' + k \equiv M\ , \quad \rho + k + \ell + \ell' = P \tag{11}$$

and as a consequence of (11):

$$m - \rho - \ell' = M - P = L\ , \quad m' + \ell' = m + \ell = M - k \tag{12}$$

It is evident that M is the number of the vibrational recurrence of the pure-electronic line in the absorption spectrum upon which the excitation frequency falls, P the number of vibrational recurrence Rayleigh line (the number of RS line), and L is the number of the vibrational recurrence of the pure-electronic line in the luminescence spectrum (in the anti-Stokes region of the luminescence spectrum the indexes L are positive). The equidistance of levels and the conservation of vibrational frequencies in the electronic transition assumed in our model lead to the fact that the HL lines in the Stokes region of the spectrum lie exactly on the lines of the ordinary luminescence. Therefore we do not consider here the problem of discriminating HL from the ordinary luminescence for the whole spectrum but confine ourselves to the lines in the anti-Stokes region.

We shall take in the Eq. (9) the terms corresponding to the fixed values of M and P on condition M > P (i.e. the region under study is $\Omega > \omega_{10}$) and neglect in them the terms which are of the order α/Γ and Γ/ω or less. We obtain

$$W(\omega_0, \Omega) = \frac{B}{2\pi} e^{-2\xi^2} \sum_{k=0}^{P} \sum_{\ell=0}^{P-k} \sum_{\ell'=0}^{P-k-\ell} (-1)^{\ell+\ell'}$$

$$\times \frac{\xi^{2(2M+P-2k-\ell-\ell')}}{k!\ell!\ell'!(M-\ell-k)!(M-\ell'-k)!(P-\ell-\ell'-k)!} \times \frac{\Gamma}{x^2+\Gamma^2M^2} \tag{13}$$

$$\frac{\Gamma^2 PM(P+M-k)(P+M-2\ell-2k) + \Delta\Omega_k^2 M(P-2\ell-k) + x(M-k)[xP - 2\Delta\Omega_k(P-\ell-k)]}{(M-k)\left[\Delta\Omega_k^2 + \Gamma^2P^2\right]\left[\Delta\Omega_\lambda^2 + \Gamma^2(P+M-2\ell-2k)\right]}$$

Here the following denotations are used: $\Delta\Omega_k \equiv \Omega - \omega_0 + \omega P$, $\Delta\Omega_\lambda \equiv \Omega - \omega_{10} - \omega L = \Delta\Omega_k - x$ (L = M - P), i.e. $\Delta\Omega_k$ and $\Delta\Omega_\lambda$ represent the frequencies of the secondary radiation spectrum as counted from the "maxima" of RS lines $\omega_0 - \omega P$ and the "maxima" of HL lines $\omega_{10} + \omega L$ resp.

From the result obtained it can be seen that all the terms under sums in the formula (13) include as a factor the product of two Lorentz curves with their maxima at the frequency $\omega_0 - \omega P$ of the "maximum" of the P-th RS line and at the frequency $\omega_{10} + \omega L$ of the "maximum" of the L-th HL line.

It is evident that if $|x|$ is sufficiently great ($|x| \gg \Gamma(M+P)$) then the mentioned product yields a spectral curve with two maxima (one near $\omega_0 - \omega P$, the other near $\omega_{10} + \omega L$) one of which corresponds to RS, the other to HL. The terms in the numerators of the formula (13), depending on the frequency $\Delta\Omega_k$ change their mentioned form to some extent but the two maxima in the resultant spectrum corresponding to RS and to HL, however, always remain in case of sufficiently great $|x|$.

Let us dwell more thoroughly upon the case of small Stokes losses ($\xi^2 \ll 1$). Then the main contribution to the spectrum is made by the term $k = P(\ell = \ell' = 0)$ which is equal to

$$W(\omega_0, \Omega) = \frac{B}{\pi} e^{-2\xi^2} \frac{\xi^{2P}}{P!} \left[\frac{\xi^{2(M-P)}}{(M-P)!} \right]^2 \frac{\Gamma P}{\left[\Delta\Omega_k^2 + \Gamma^2 P^2 \right] \left[\Delta\Omega_\lambda^2 + \Gamma^2 (M-P)^2 \right]} \quad (14)$$

From this formula it follows that in case of sufficiently great M and not too great $|x|$ ($|x| < \Gamma M$) in the vicinity of the pure-electronic line (small $M - P \ll M$) the spectrum is almost of the Lorentz shape with the maximum at the point $\Omega = \omega_{10} - \omega(M - P)$ and with the half-width $\Gamma(M - P)$. Thus, near the pure-electronic line the secondary radiation spectrum represents, in fact, HL lines. Only in case of sufficiently great $|x| \geq \Gamma M$ the spectral curves of the mentioned lines may have one more (shifted by x), much wider (with the half-width of the order ΓP) and much less intensive maximum (with its peak intensity smaller by $[P/(P-M)]^2$ times) due to RS.

Analogously, near the Rayleigh line (small $P \ll M$) the spectrum represents RS. In case of sufficiently great x these lines as well as HL lines are almost of the Lorentz shape: their Lorentz shape is slightly distorted by HL on the wing of the line.

The width of HL lines increases linearly with the line number $L = M - P$ as well as the width of RS lines increases with the increase of their number P.

An analogous general analysis can be carried out also for the limit case of great Stokes losses ($\xi^2 \gg 1$). From (13) it is also not difficult to obtain the results for intermediate values of Stokes losses if one takes concrete (small) values of L.

The improvement of the model (taking account of the change of the local vibration frequency in the electron transition) may essentially change the character of this part of the results which concerns the discrimination of HL from RS.

A correct consideration of the character of the widening of vibronic bands caused by local vibrations must also have a great importance - usually the main contribution to their width is made not by the process of vibrational relaxation but simply by the interaction of the electronic transition with crystal phonons (Stokes losses over crystal vibrations). The interference effects are, as is known, especially sensitive to the interactions which may lead to the change of the phase.

The calculation made above of the model and the simplest additional considerations allow one to formulate the following conditions in case of which the observation of HL is more favorable: (1) distinct vibrational structure of the spectrum, (2) excitation to high vibrational levels, (3) intermediate Stokes losses ($\xi^2 \sim 2 \div 3$), (4) change of the frequency of the local vibration in the electronic transition. The last circumstance results in the fact that a part of HL lines in the anti-Stokes region is out of resonance with absorption lines and the reabsorption decreases; in the Stokes region - out of resonance with the lines of ordinary luminescence.

Attempts to observe HL experimentally and determine the time of the vibrational relaxation of a local mode were made[6] on the example of the molecular NO_2^- impurity ion in the KCℓ crystal.

REFERENCES

1. I. J. Tehver and V. V. Hizhnyakov, Izv. Akad. Nauk Est. SSR, Ser. fiz.-tekh. i mat. Nauk 15, 9 (1966).
2. K. Rebane, V. Hizhnyakov and I. Tehver, Izv. Akad. Nauk Est. SSR 16, 207 (1967).
3. V. Hizhnyakov and I. Tehver, Phys. Stat. Sol. 21, 755 (1967).
4. I. J. Tehver, Dissertation, Tartu 1968 (in Russian).
5. M. Lax, J. Chem. Phys. 20, 1752 (1952); Y. Y. Perlin, Uspekhi fiz. Nauk 80, 553 (1963).
6. K. Rebane, P. Saari, Izv. Akad. Nauk Est. SSR, Ser. fiz.-mat. Nauk 17, 241 (1968).

F-9: RAMAN SPECTRA OF F CENTERS*

C. J. Buchenauer, D. B. Fitchen and J. B. Page, Jr.
Laboratory of Atomic and Solid State Physics,
Cornell University
Ithaca, New York

The Raman spectra of F centers are of particular interest because they provide the most direct information about the phonons responsible for the F center's line-width. In addition, they yield information on the manner in which the vibrational modes of the host lattice are perturbed by the presence of F centers.

The feasibility of Raman studies of F centers in alkali halides using laser sources was first demonstrated by Worlock and Porto[1]. In spite of the relatively low concentrations of F centers which are attainable, a relatively large Raman cross section can be obtained due to a resonant enhancement: when irradiating near the F band the Raman cross section varies approximately as the inverse fourth power of the energy difference between the incident radiation and the F absorption band. Furthermore, since first-order Raman scattering from a pure alkali halide crystal is forbidden, any observed first-order spectra must be due to the addition of F centers.

Henry and Slichter[2] have treated the near-resonance first-order Raman scattering by F centers under the following assumptions:

i) the scattering center has octahedral symmetry,
ii) the scattering occurs from a virtual 1s-2p like transition, and
iii) the electron-phonon interaction is of the form

$$H_{ep}(\underline{r}_e, Q) = \Sigma_i V_i (\underline{r}_e) Q_i$$

where the Q_i's are the lattice normal coordinates and $\underline{r}_e$ denotes the position of the electron. Under these conditions, the only Raman-active modes will be of the symmetry types Γ_1 (A_{1g}), Γ_{12} (E_g), and $\Gamma_{25'}$ (T_{2g}). In addition, three independent measurements can be performed which allow a unique determination of the contribution from each symmetry type; the types of modes contributing to the spectra for various orientations of the incident and scattered electric field polarizations are given in Table I.

*Work supported by the Advanced Research Projects Agency through the Materials Science Center at Cornell University, Report #1038.

TABLE I

First-Order Selection Rules

	$\Gamma_1+\Gamma_{12}$	$\Gamma_{25'}$	$\Gamma_1+\Gamma_{25'}$	Γ_{12}
Incident Field Polarization	(100)	(100)	(110)	(110)
Scattered Field Polarization	(100)	(010)	(110)	$(1\bar{1}0)$

The apparatus used in obtaining the results reported here employed a pulsed argon laser whose rate of firing was electronically adjusted so that its average output intensity remained constant. Since in all cases an appreciable fraction of the incident light was absorbed by the sample, the average incident power was held below 300 μ watts. The laser beam was focused on x-irradiated samples having F-center densities of $1\text{-}4 \times 10^{16}$/cc. Light scattered at 90° was analyzed with a double monochrometer of f/6 entrance aperture. Gated photon counting circuitry was employed with a cooled phototube detector.

Fig. 1 shows the F-center Raman spectrum for KF. The incident electric field polarization is along (100). Three broad peaks are observed which fall on the high-energy side of three extrema in the theoretical KF phonon density of states[3]. The structure does not continue beyond the maximum frequency of the unperturbed lattice phonons.

Qualitatively new features were expected for host lattices having a gap in their phonon densities of states. The possibility of localized or resonant modes occurring in or near the gap was made plausible by the expected large force-constant reduction in the neighborhood of an F center.

The most interesting results have been found for NaBr, which has a frequency gap from 105 cm^{-1} to 126 cm^{-1} between the acoustical and optical phonons as determined by recent inelastic neutron scattering experiments[4]. Fig. 2 shows the unpolarized Raman spectrum for colored NaBr as compared with that for the pure crystal. The most striking feature in the Raman spectrum is a strong resonance peak at 136 cm^{-1}, just above the well resolved gap region. The peak is actually a doublet (separation about 11 cm^{-1}) and is $\geq 97\%$ parallel polarized at 78°K. The low scattering in the gap region indicates that below the resonance peak the spectrum is first order, while above the resonance higher-order scattering is observed which shows peaks at multiples of the resonance frequency up to the fourth order.

Fig. 3 shows the Stokes and anti-Stokes spectra at three different temperatures. Although the mode's peak height in the Stokes spectrum decreases markedly with increasing temperature, its half-width does not increase appreciably in experiments from 6° to 194°K.

Spectra obtained using excitation at the F-band peak energy (5145 Å) and at the F-band half-maximum position (4880 Å) are compared in Fig. 4. Further experimental comparisons of such spectra have shown no discernible differences except for a considerable enhancement of the Raman cross section for 5145 Å excitation.

As a guide to understanding these results, the Γ_1 and Γ_{12} contributions to the first-order spectrum were computed assuming that the electron-phonon interaction involves

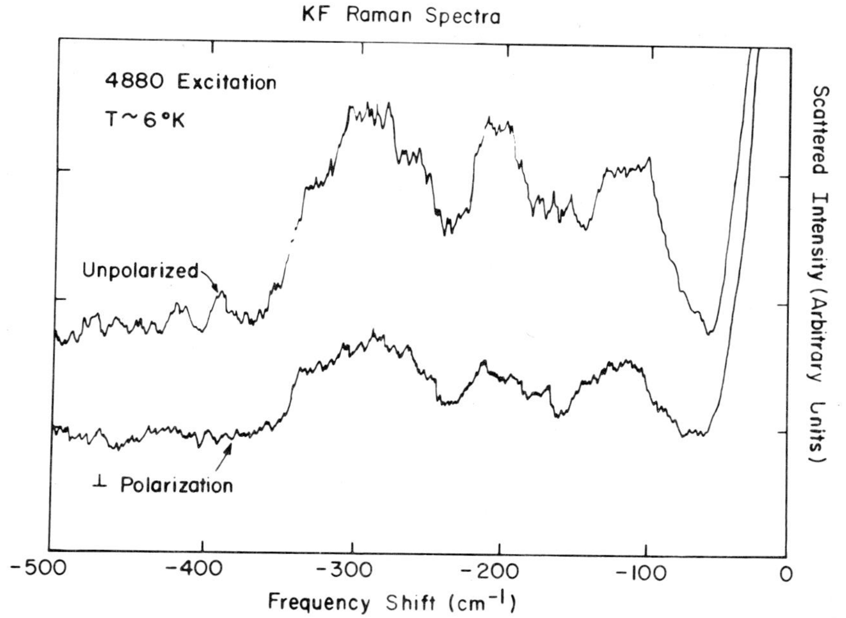

Fig. 1. Raman spectra of F centers in KF. The incident electric field polarization is along (100).

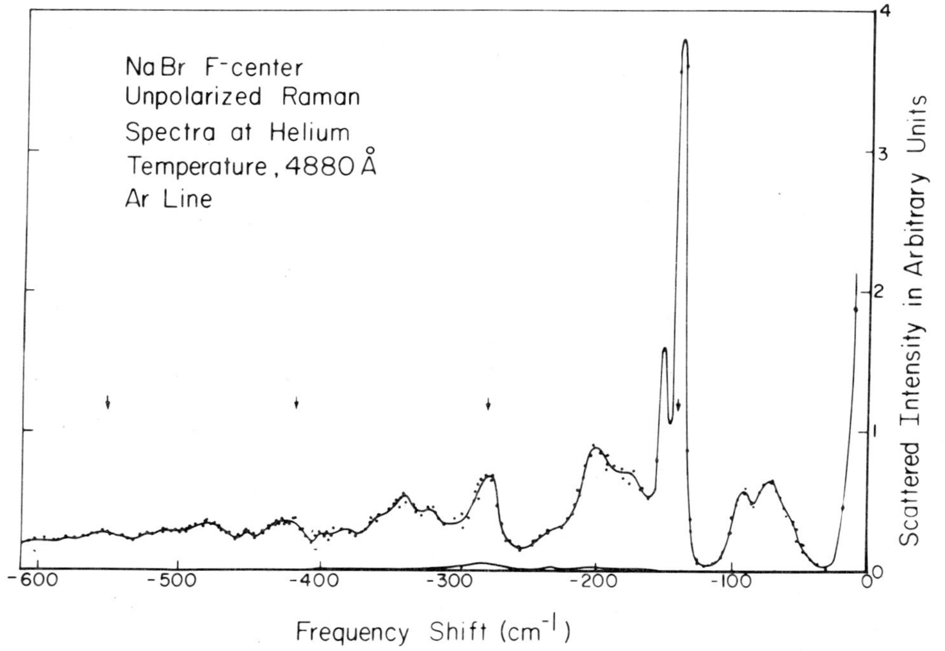

Fig. 2. Unpolarized Raman spectra of NaBr containing F centers (top curve) and of pure NaBr (bottom curve).

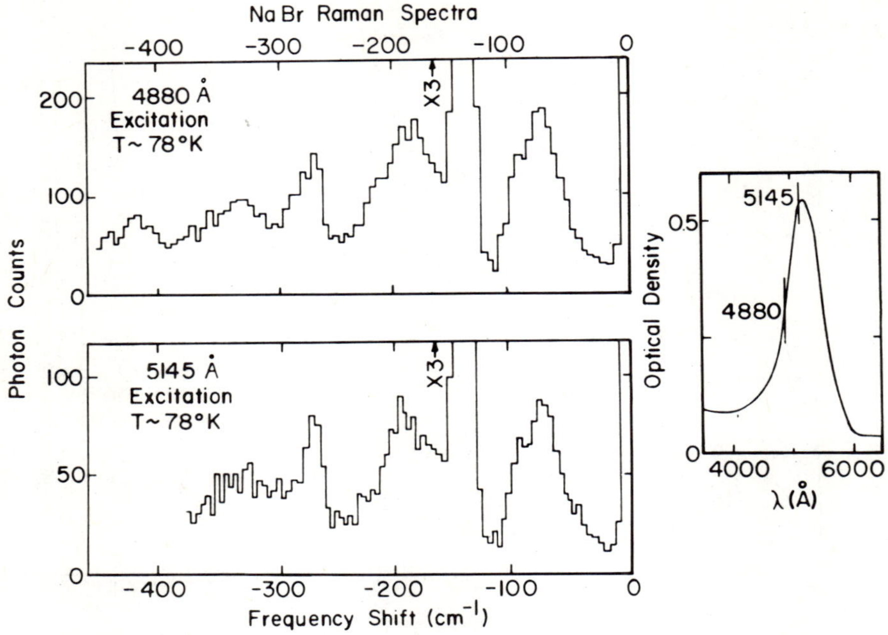

Fig. 3. Stokes and anti-Stokes spectra of F centers in NaBr as a function of temperature.

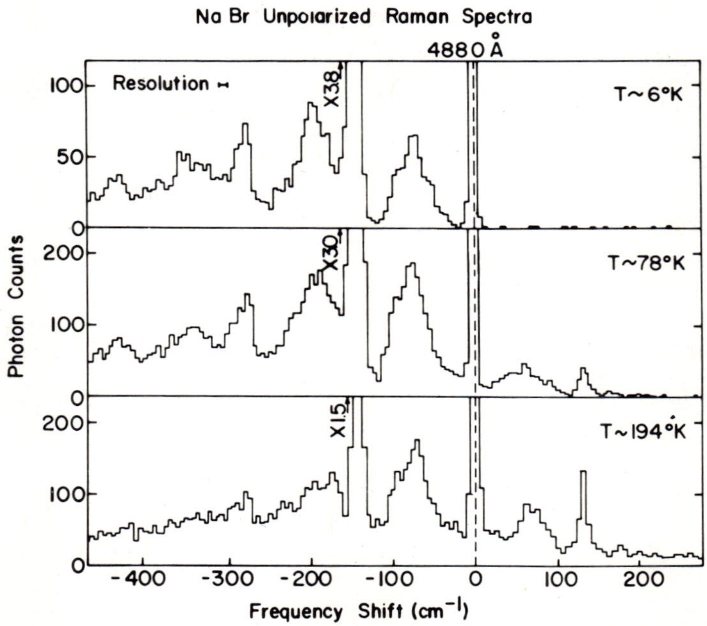

Fig. 4. Comparison of unpolarized Raman spectra of F centers in NaBr obtained using 4880 Å and 5145 Å excitation.

just the defect's six nearest neighbors. In this case, the near-resonance theory[2] yields the result that the contribution to the line shape from modes of a particular symmetry Γ_i is proportional to the nearest-neighbor projected density of states $\rho_{\Gamma_i}(\omega^2)$[5, 6]. The computation of this quantity requires a knowledge of the force-constant perturbations arising from the presence of the defects as well as a knowledge of the unperturbed phonon frequencies and polarization vectors. The shell model as used to fit the recently obtained[4] NaBr dispersion curves was utilized to evaluate the latter quantities at 64, 000 $\underline{q}$ vectors in the Brillouin zone. The force-constant perturbation was assumed to consist of a change Δk in the short-range longitudinal force constants between the defect site and each of the six nearest neighbors; this picture was previously used by Benedek and Nardelli[6] to analyze the NaCl and KCl F-center Raman data of Worlock and Porto[1].

Fig. 5 shows a comparison of the experimental parallel (100) first-order spectrum with the Γ_1 projected density of states computed for a force-constant change $\Delta k/k$ of -0.55, where k is the value of the unperturbed short-range longitudinal force constant between nearest neighbors. This value of $\Delta k/k$ was chosen so as to give the best fit to the resonance frequency, and the scale was adjusted so that one experimental point in the acoustic region was reproduced. The resulting theoretical spectrum gives very good agreement with the measured line shape in the acoustic region. Above the resonance, the agreement is not good, but the fact that the experimental spectrum does not decrease sharply at the maximum lattice frequency (200 cm^{-1}) indicates that two-phonon processes are involved in this region. The Γ_{12} spectrum for $\Delta k/k = -0.55$ is shown in Fig. 6, and it is seen to bear little resemblance to the experimental data. Thus we conclude that the resonance is of Γ_1 symmetry and involves a substantially weakened nearest-neighbor force constant.

Recently a more complete set of experimental data has been obtained. Parallel and perpendicular spectra taken at 78 °K and 6 °K are shown in Figs. 7 and 8 for incident electric field polarizations along (100) and (110). Several interesting features are seen in these data. Both sets of perpendicular spectra are strikingly similar, which is somewhat surprising in view of the fact that according to Table I the (100) and (110) data should give the unique contributions from the $\Gamma_{25'}$ and Γ_{12} modes, respectively. On the other hand, the parallel (100) and (110) data do show noticeable differences in the acoustic region. Furthermore, the small peak occurring in the 78 °K perpendicular spectra near the resonant mode position is seen to undergo a dramatic increase upon going to liquid helium temperature, while no discernible change occurs for the resonance peak in the corresponding parallel spectra in this temperature range. The strong temperature dependence of the 136 cm^{-1} line in the perpendicular spectra is qualitatively similar to that reported for low-lying infrared-active resonances associated with $KBr{:}Li^+$ and $KI{:}Ag^+$[7] (see addendum).

At helium temperature, the 136 cm^{-1} peak is $\approx 90\%$ parallel polarized in both the (100) and (110) spectra. This substantiates the earlier conclusion that the resonance in the parallel spectra is of Γ_1 symmetry.

Recent exploratory Raman spectra have been obtained for F centers in RbF. The possibility of qualitatively new features in these spectra had been anticipated because of the small anion to cation mass ratio for RbF (in contrast to the large ratio for NaBr) and because of the anomalously small F-band transition cubic-strain coupling coefficient in this material [8]. In the resulting spectra, shown in Fig. 9, the positions of the $\underline{q} = 0$

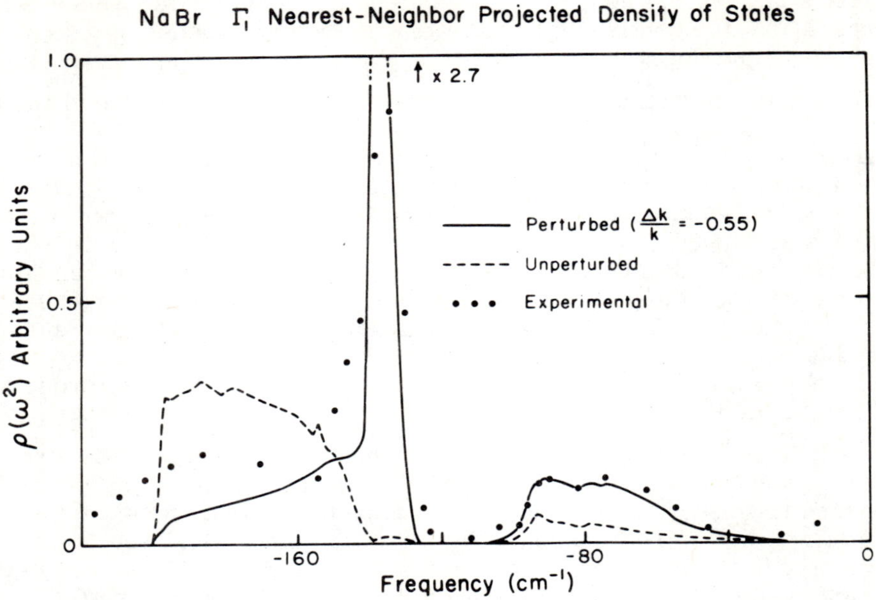

Fig. 5. Theoretical first-order Γ_1 contribution to the NaBr:F-center Raman line shape computed for a nearest-neighbor force-constant weakening of 55% compared with the (100) 6°K data of Fig. 7. The calculation for zero force-constant weakening is also given for reference.

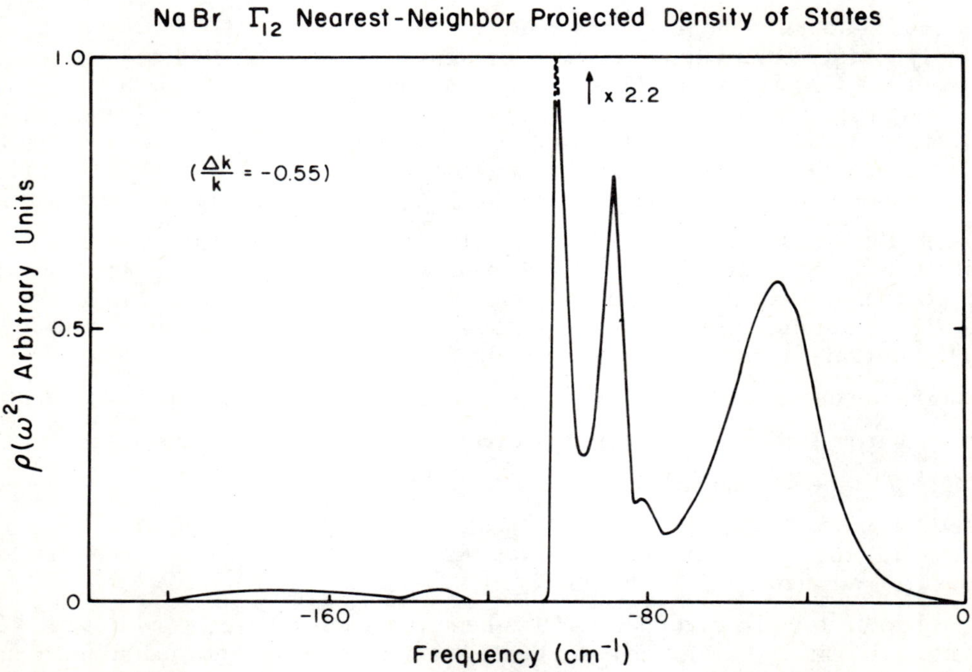

Fig. 6. Theoretical first-order Γ_{12} contribution to the NaBr:F-center Raman line shape computed for a nearest-neighbor force-constant weakening of 55%.

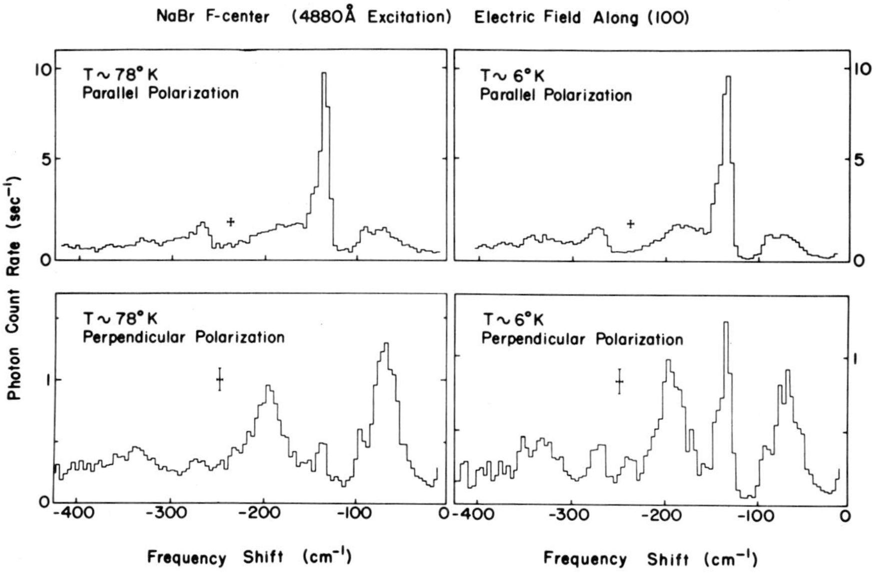

Fig. 7. Polarized Raman spectra of F centers in NaBr taken at 78°K and 6°K with the incident electric field polarized along (100).

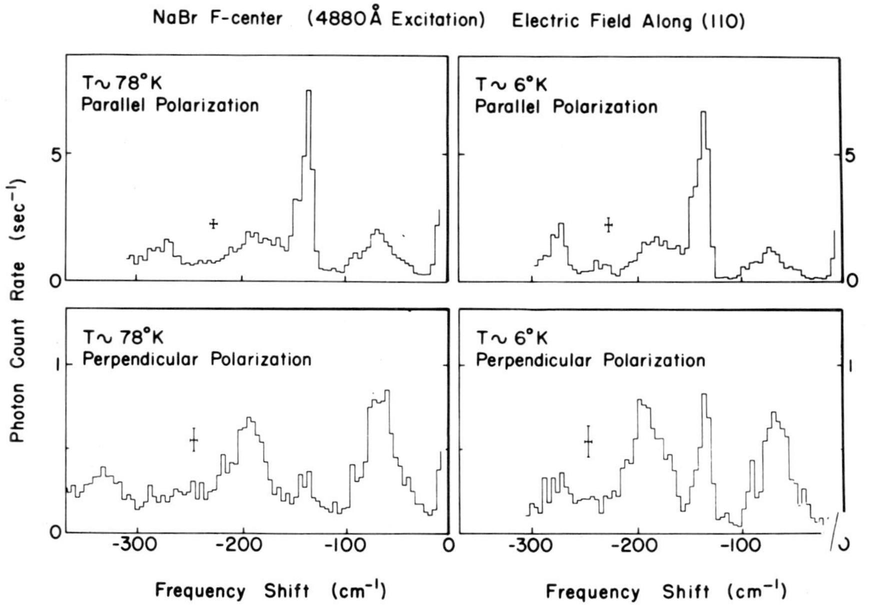

Fig. 8. Same as Fig. 7 but with the incident polarization along (110).

transverse and longitudinal optical phonon frequencies [9] are indicated. It is seen that the main contribution to the unpolarized spectrum arises from phonons in the high-frequency optical region, whereas in NaBr the dominant contribution was seen to arise from phonons in the low-frequency optical region. In addition, two narrow (half-widths $\approx$ 12 cm^{-1}) polarized peaks occur: one near 170 cm^{-1} in the perpendicular spectrum and another near 125 cm^{-1} in the parallel spectrum. Additional work on F centers in RbF and other alkali halides is continuing.

It has been seen that the Raman spectra of F centers vary considerably for different host lattices. Sharp resonances, multi-order scattering, and strong temperature-dependent effects have been observed. Further work in this field should yield valuable information about the dynamical properties of F centers.

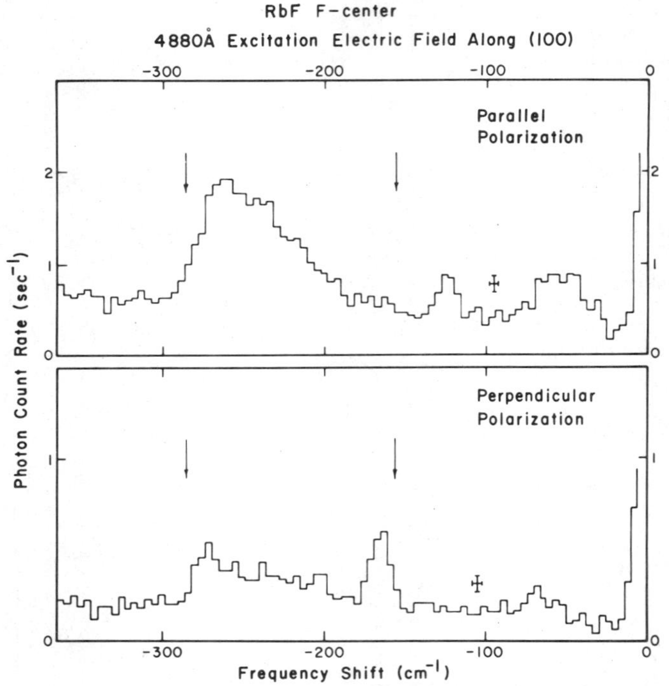

Fig. 9. Polarized Raman spectra of F centers in RbF taken at 78°K with the incident electric field polarized along (100). The arrows indicate the positions of the $\underline{q} = 0$ LO and TO frequencies as given in reference[9].

ACKNOWLEDGEMENT

The authors express their gratitude to Dr. W. J. L. Buyers for sending us the recently obtained NaBr inelastic neutron scattering results and their shell model interpretation.

REFERENCES

1. J. M. Worlock and S. P. S. Porto, Phys. Rev. Letters 15, 697 (1955).
2. C. H. Henry and C. P. Slichter, "Physics of Color Centers," W. B. Fowler (ed.), 351, Academic Press Inc., New York, 1968.
3. A.M. Karo and J. R. Hardy, Phys. Rev. 129, 2024 (1963).
4. W. J. L. Buyers (private communication).
5. T. Timusk and M. V. Klein, Phys. Rev. 141, 664 (1966).
6. G. Benedek and G. F. Nardelli, Phys. Rev. 154, 872 (1967).
7. S. Takeno and A. J. Sievers, Phys. Rev. Letters 15, 1020 (1965).
8. C. J. Buchenauer and D. B. Fitchen, Phys. Rev, 167, 846 (1968).
9. E. Burstein, "Phonons and Phonon Interactions," T. A. Bak (ed.), 296, W. A. Benjamin, Inc., New York. 1964.

ADDENDUM

Recent data obtained from additively colored NaBr show one difference from previous data. The residual peak at 136 cm^{-1} in the perpendicular spectrum of Fig. 8 is approximately one fourth as large as that shown for x-irradiated samples and must, therefore, be considered a preparation-dependent phenomenon.

ERRATUM

Due to an error in experimental conditions, the data presented on the lower half of Fig. 9 is incorrectly identified. This data corresponds to an experiment with incident electric field along (100) and analyzed electric field along (011) which gives the $\Gamma_{25'}$ contribution to the spectrum rather than the Γ_{12} contribution as was indicated. Recent experiments have shown the scattering due to Γ_{12} modes to be several times weaker than that due to $\Gamma_{25'}$ modes. Thus the Γ_1 interpretation of the parallel spectrum resonance at 136 cm^{-1} remains correct.

F-10: THEORETICAL INVESTIGATION OF THE F- CENTER RAMAN SPECTRA IN Na Br : FIRST AND SECOND ORDER PROCESSES

G. Benedek and E. Mulazzi
Istituto di Fisica dell'Universita and Gruppo Nazionale di Struttura della Materia del Consiglio Nazionale delle Ricerche
Milan, Italy

ABSTRACT

A previous theory of the first-order Raman scattering induced by color centers in alkali halides has been extended to include second-order contributions. We discuss under what conditions the second (or higher) order spectra of the imperfect lattice display the peculiarities of the projected densities for the perturbed two (or many) phonon states. The coupling coefficients weighting the above densities, namely the second-order strain derivatives of the electronic polarizability tensor, are qualitatively related to some macroscopic observable quantities.

Our theoretical results are then used in the interpretation of the F-center induced Raman spectra of NaBr, which have been measured recently by Buchenauer and Fitchen. A defect model whose perturbation extends as far as fourth neighbors, including the change in nearest neighbor (nn) force constant and the elastic relaxation effects, is used.

Excellent agreement is found between theoretical and experimental data. From the analysis of the experimental F-center Raman spectra, information on the coupling constants for the phonon hole-electron pair interaction and on the separation between 2p and 2s states is also obtained.

INTRODUCTION

First-order Raman scattering is forbidden in perfect alkali halides, since the first order derivatives of the electron polarizability tensor with respect $\underset{\sim}{q}=0$ displacements vanish identically when each ion of the lattice is an inversion symmetry center. Such condition no longer holds when substitutional defects are introduced into the crystal and first order Raman spectra appear. These spectra reflect the peculiar vibrational structure of the defect quite well, provided that localized electron states of the defect are

*This work is sponsored in part by Consiglio Nazionale delle Richerche and in part by E.O.A.R. under Grant AF-EOAR 67-8.

involved in the polarization processes assisting the Raman transition. The defect is therefore expected to exhibit a strong optical band, well separated from the UV band of the host lattice, and the corresponding electron transition to be coupled strongly with the lattice.

If the frequency ω_0 of the incident light beam falls in the region of the defect optical absorption band (resonance case) the second (and higher) order Raman processes, which are allowed in the host lattice, are also strongly influenced by the presence of defects. Actually it is expected that for transitions between defect states whose wave functions do not extend beyond the defect perturbation region the second (and higher) order Raman spectra can also be reasonably well described in terms of the perturbed densities of two-phonon states as projected onto the defect perturbation subspace. This point is discussed in the next section. The possibility of working on the projected densities for one- and two-phonon states greatly simplifies our theoretical analysis of the experimental Raman spectra. A best fit method reveals a powerful tool for investigating the coupling coefficients of the color-center transition with the lattice vibrations, the separation between 2p and 2s excited states, the elastic properties around the defect, and, finally for testing the dynamical models for our imperfect lattice.

THEORY

The theory of the defect-induced first-order Raman scattering has been developed in detail in some previous papers [1-2]. The coupling coefficients weighting the projected densities of the one-phonon states were explained in the symmetry coordinate representation; they were related to the experimental stress coefficients for the optic absorption band, by the help of suitably defined local elastic constants[3].

The second order contribution to the fourth-order Raman tensor of our imperfect lattice, say $i^{(2)}_{\alpha\beta\gamma\delta}(\Omega)$, can be readily expressed by

$$i^{(2)}_{\alpha\beta\gamma\delta}(\Omega) = \frac{\hbar^2}{2\Omega} \sum_{\ell\ell'mm'} \frac{1}{\sqrt{M_\ell M_m M_{\ell'} M_{m'}}} \frac{\partial^2 P^*_{\alpha\beta}(u)}{\partial u_\ell \partial u_m} \frac{\partial^2 P_{\gamma\delta}(u)}{\partial u_{\ell'} \partial u_{m'}} (\ell, m | g^{(2)}(\Omega^2) | \ell'; m') \qquad (1)$$

where $\Omega = \omega_0 - \omega$ is the Raman frequency shift between incident (ω_0) and scattered (ω) frequencies, u_l and M_l are the displacement and mass of the l-th ion of our imperfect lattice, respectively, $g^{(2)}(\Omega^2)$ is the two-phonon density matrix and has the dimension of Ω^{-2}; in Eq. (1) its elements in the lattice displacement representation are used. In Eq. (1) the dependence of the electron polarizability tensor $P_{\alpha\beta}$ on any lattice strain u is explained. In principle, all virtual dipole transitions τ between ground and excited electron states of the crystal give some contribution to the polarizability for the lattice in the ground state [4]:

$$P_{\alpha\beta} = \frac{Re}{3\hbar} \sum_\tau \frac{2\omega_\tau}{\omega_\tau^2 - \omega_0^2 - i\omega_0 H_\tau} M_\alpha^{(\tau)*} M_\beta^{(\tau)} \qquad (2)$$

where ω_τ is the transition frequency. In order to work in the resonance region, we introduced an "ad hoc" half-width term H_τ for the τ absorption line (or band). $M^{(\tau)}$ is the dipole moment for the dipole transition τ. In (2), the oscillator strength f_τ can be conveniently used instead of $M^{(\tau)}$, since

$$M_\alpha^{(\tau)*} M_\beta^{(\tau)} = \frac{3\hbar e^2 f_\tau}{2m\omega_\tau} m_\alpha^{(\tau)} m_\beta^{(\tau)} \tag{3}$$

where $\underset{\sim}{m}^{(\tau)}$ is a unit vector. As concerns the dependence on the strain u, we assume that both $\omega_\tau \equiv \omega_\tau(u)$ and $f_\tau \equiv f_\tau(u)$ are functions of u; however $f_\tau(u)$ has no linear term, i.e.

$$\frac{\partial f_\tau(u)}{\partial u} = 0 \tag{4}$$

The second-order strain derivative of the polarizability is then

$$\frac{\partial^2 P_{\alpha\beta}}{\partial u_\ell \partial u_m} = \frac{e^2}{m} \Sigma_\tau m_\alpha^{(\tau)} \operatorname{Re} \left\{ \frac{2 f_\tau}{(\omega_\tau^2 - z)^3} \left[(3\omega_\tau^2 + z) \frac{\partial \omega_\tau}{\partial u_\ell} \frac{\partial \omega_\tau}{\partial u_m} - \omega_\tau(\omega_\tau^2 - z) \frac{\partial^2 \omega_\tau}{\partial u_\ell \partial u_m} \right] + \frac{1}{\omega_\tau^2 - z} \frac{\partial^2 f_\tau}{\partial u_\ell \partial u_m} \right\} \tag{5}$$

with $z = \omega_o^2 + i\omega_o H_\tau$.

Hereafter we shall work in the resonance case of the F-center, namely $|\omega_o - \omega_F| \lesssim H_F$, where ω_F is the peak frequency of the F band. This condition is not inconsistent with the restriction used in Kleiman [4] derivation $|\omega_o - \omega_F| >> \Omega$, where Ω is of the order of the vibrational frequencies, because for the F-center it is usually $H_F \sim 30$ times Ω. In resonance conditions the summation over the transitions τ of Eq. (5) can be restricted to the three independent 1s → 2p transitions of the F-center, say τ = F; which is to say that the contribution to the second-order Raman spectra coming from exciton transitions of the host lattice is neglected.

Such approximation turns out from the following considerations:

i) the oscillator strengths of the F-center and exciton transitions are both close to unity;

ii) the excited state wave function of the exciton transition is probably more extended in space than that of the F center; thus temperature and strain dependence should be larger for ω_F than for $\omega_{exciton}$;

iii) $(\omega_{exciton} - \omega_F)^3$ is nearly 10^5 times larger than H_F^3; such a large separation between F and UV bands enables one to neglect all terms with τ other than F in Eq. (5), because of the factor $(\omega_\tau^2 - z)^{-3}$. Nevertheless, exciton contributions may remain appreciable or comparable with F-center contributions, since the fractional concentration of F-centers is seldom larger than 10^{-4}.

According to the above approximations, in Eq. (1) the summations run over the lattice sites around the defects only, say the lattice sites belonging to defect perturbation subspace. Then, we can work in the symmetry coordinates for our subspace, which transform according to the irreducible representations Γ of the O_h point group.

Now, we observe that f_F depends on u only through the mixing of the F-center wave functions which is produced when the region around the defect is strained. Such mixing is appreciable for those couples of states whose frequencies are close enough. For the 1s → 2p transition only the mixing between 2p and 2s states may be of some interest. It follows that f_F could depend appreciably on polar (i.e. Γ_{15}) strains, as produced by an electric field (Stark effect) while dependence of f_F on elastic (i.e. Γ_1, Γ_{12} and Γ'_{25}) strains is probably negligible. Vice versa, the first and second order derivatives of ω_F with respect to elastic strains give the most important contribution; remember that

$$\frac{\partial \omega_F}{\partial u(\Gamma_1)} = \frac{A}{\sqrt{3a}} \; ; \; \frac{\partial \omega_{F,x}}{\partial u(\Gamma_{12}t)} = -2 \frac{\partial \omega_{F,y}}{\partial u(\Gamma_{12}t)} = \frac{4B}{\sqrt{6a}} \tag{6}$$

$$\frac{\partial \omega_{F,xy}}{\partial u(\Gamma'_{25}z)} = - \frac{\partial \omega_{F,x\bar{y}}}{\partial u(\Gamma'_{25}z)} = \frac{C}{3a}$$

where indexing of ω_F denotes the orientation of the 2p state involved in the transition; the meaning of the stress coefficients A, B, C and their relation to the experimental stress shifts through the local elastic constants are supposed well known: the reader is referred to previous works [5, 6]. The magnitude of the second order derivatives of ω_F can be inferred from the temperature dependence of A, B and C [7]. However, the first term in the right-hand member of Eq. (5), if does not vanish identically, is usually much larger than the second term, proportional to $\partial^2\omega_F/\partial u \partial u$. For example, consider $\partial^2 p_{xx}/\partial u(\Gamma_1)^2$ in resonance conditions for a NaCl host lattice ($H_F \simeq 0.25$ eV). The comparison between the first and second term inside the square brackets in Eq. (5) reduces to that between

$$A\eta \quad \text{and} \quad \frac{3\sqrt{3}}{2} \left| \omega_F - \omega_o \right| \frac{\partial \ln A}{\partial T} \leq \frac{3\sqrt{3}}{4} H_F \frac{\partial \ln A}{\partial T} \tag{7}$$

where $\eta \simeq 1.2 \times 10^{-4}$ $(°K)^{-1}$ is the thermal expansion and T absolute temperature. From a previous paper [7] we take $A \simeq 4$ eV and $\partial \ln A/\partial T \simeq 2.6 \times 10^{-4}$ $(°K)^{-1}$; the last value is obtained taking into account that at T = 0°K the elastic constants are temperature independent. Thus, $A\eta$ is found to be at least 20 times larger than the second expression (7). The above statement does not hold for terms containing tetragonal (Γ_{12}) strain derivatives, since actually for NaBr, as well as NaCl and other alkali halides, B << AW, while $\partial \ln A/\partial u(\Gamma_1)$ and $\partial \ln B/\partial u(\Gamma_1)$ have the same magnitudes [7]. However, for the same reason, Γ_{12} terms will give only minor contributions to the whole Raman spectrum. Therefore, the main contributions for parallel and perpendicular polarized Raman spectra, related to the derivatives of P_{xx} and P_{xy} respectively, are (denote $\partial^2 P_{\alpha\beta}/\partial u(\Gamma)\,\partial u(\Gamma') \equiv P_{\alpha\beta/\Gamma\Gamma'}$)

$$P_{xx/\Gamma_1\Gamma_1} \simeq D(\omega_0)\,A^2 \;;\; P_{xx/\Gamma'_{25}\Gamma'_{25}} \simeq \frac{1}{3} D(\omega_0)\,C^2 \tag{8}$$

$$P_{xy/\Gamma_1\Gamma'_{25}} = \frac{1}{\sqrt{3}} D(\omega_0)\,AC \qquad \text{with } D(\omega_0) = \frac{2e^2 f_F}{3a^2 m}\,\mathrm{Re}\,\frac{3\omega_F^2 + z}{(\omega_F^2 - z)^3}$$

Minor contributions come from

$$P_{xx/\Gamma_1\Gamma_{12}} \simeq -2\,P_{yy/\Gamma_1\Gamma_{12}} \;;\; P_{xx/\Gamma_1\Gamma'_{25}} \;;\; P_{xy/\Gamma'_{25}} \tag{9}$$

$$P_{xx/\Gamma_{12}\Gamma_{12}} \simeq 4\,P_{yy/\Gamma_{12}\Gamma_{12}} \;;$$

their expressions can be easily derived from Eqs. (5) and (6). An important contribution to parallel-polarized spectra could come from $P_{xx/\Gamma_{15}\Gamma_{15}}$, where Γ_{15} stands for a polar strain, as produced by an external electric field. Assuming that Γ_{15} strains produce mixing between 2p and 2s states and using the standard theory of Stark effect, we get easily

$$P_{xx/\Gamma_{15}\Gamma_{15}} \simeq -\frac{2f_F e^2}{m}\,\mathrm{Re}\,\frac{1}{\omega^2 - z}\left[\frac{e\tilde{f}^* a}{e^*}\,\frac{(2p,x|x|2s)}{E_{2p} - E_{2s}}\right]^2 \tag{10}$$

where $e(2p,x|x|2s)$ is the dipole matrix element between 2p and 2s states, E_{2p} and E_{2s} are the respective energies, e^* the effective charge of the lattice, $\tilde{f}^*$ the local nn effective force constant for our defect. [8] Stark effect measurements would inform us about the element $(2p,x|x|2s)$, but unfortunately we know very little about the energy separation between 2s and 2p states, at present. However, assuming $|E_{2p} - E_{2s}| \sim 0.2$ ev, $(2p|x|2s) \simeq 2\text{x}10^{-8}$ cm [9], $\tilde{f}^* \simeq 1\text{x}10^4$ g sec^{-1} (see below) and a $\sim 3\text{x}10^{-8}$ cm, we found that $P_{xx/\Gamma_{15}\Gamma_{15}}$ is nearly equal to $P_{xx/\Gamma_1\Gamma_1}$ and its contribution must be included.

Now it is interesting to make a comparison between the amplitudes of first and second order Raman spectra. Since both one and two phonon projected densities are normalized to unity, the comparison is between the coefficients

$$[n(\omega_\Gamma) + 1]\,\frac{\hbar}{2\Omega M_+}\,P^2_{\alpha\beta/\Gamma} \quad \text{and} \quad \frac{\hbar^2}{4\Omega^2 M_+^2}\,P^2_{\alpha\beta/\Gamma\Gamma'}\cdot[n(\omega_\Gamma) + n(\omega_{\Gamma'}) + 1] \tag{11}$$

where M_+ is the mass of n.n. ion.

In the above expressions the temperature factors depending on the occupation number $n(\omega)$ have been introduced; only the Stokes contributions, namely $\Omega = \omega_\Gamma > 0$ and $\Omega = \omega_\Gamma + \omega_{\Gamma'}$ with ω_{Γ}, $\omega_{\Gamma'} > 0$, have been considered. Because of the temperature factors, T=0°K is the less favorable case to observe the second-order spectra. However, we see that Γ_1 components

$$\sqrt{\frac{\hbar}{2\Omega M_+}}\,\frac{P_{xx/\Gamma_1\Gamma_1}}{P_{xx/\Gamma_1}} = \frac{1}{\sqrt{3a}}\sqrt{\frac{\hbar}{2\Omega M_+}}\,\frac{A}{H_F} \sim \text{unity for F center} \tag{12}$$

It appears that first and second order Raman amplitudes have the same order of magnitude; the reason is that a very strong coupling occurs between the hole-electron pair and phonons, which is to say that A, B and C are of the same order of magnitude of the transition frequency ω_F. However, we must take into account that in the experimental spectra obtained with low enough concentration of defects or far from resonance conditions the host lattice contributions coming from the exciton transitions can produce a strong second-order background.

NUMERICAL RESULTS

At T=0°K, the two-phonon densities as projected onto the sub-space of our defect perturbation, can be expressed for each couple of irreducible representations (Γ, Γ'), in terms of convolutions of one-phonon densities $\rho_{rs}(1)(\Gamma, \omega^2)$ as follows:

$$(\Gamma_r;\Gamma'r'|g^{(2)}(\Omega^2)|\Gamma s;\Gamma's')\quad \rho^{(2)}_{rr',ss'}(\Gamma\Gamma';\Omega^2) = \tag{13}$$
$$= \int_0^\infty d\omega^2 \rho^{(1)}_{rs}(\Gamma;\omega^2)\,\rho^{(1)}_{r's'}(\Gamma';\Omega-\omega)^2)$$

Indices r, s for irr. rep. Γ (r', s' for Γ' irr. rep.) run from 1 to the number of times $n(\Gamma)$ the irr. rep. Γ is contained in our perturbation model, $n(\Gamma)$ depending on the spatial extension of the defect perturbation. We assume, however, our hole-electron pair to exert non-vanishing forces on the nearest neighbor ions only, so that, regardless to the extension of our defect model, indices r, s, r' and s' will refer only to nearest neighbors (see previous paper [2, 7].

Our aim is to apply the above theory to the F-center in NaBr, in order to give an interpretation of some recent experimental results due to Buchenauer et al. [10, 11], concerning with the Raman spectra of that system.

The eigenvectors and polarization vectors of NaBr are calculated by the deformation dipole model and T = 0°K data [12]. The calculated frequency spectrum of NaBr exhibits a gap in the region $1.95 < \omega < 2.34 \times 10^{13}$ sec^{-1}, which seems to be slightly (few per cent) lower than the experimental gap, as derived from Raman or U-center infrared sideband spectra [13]. However, such small discrepancy does not create any relevant trouble in our discussion. The defect model here used is that due to Gethins et al. [14] and also adopted in our previous work (extended model) [7]. It includes, beyond the local change of mass, the change λ of central nn force constant and the effects due to the elastic relaxation ξ namely to the change γ of nn-fourth neighbor force constant.

The values of λ and γ (or ξ) for the F-center in NaBr should be comparable with those previously reported for other crystals [7]. However, in order to get the best fit of the experimental spectra, the projected perturbed densities $\rho^{(1)}_{rs}(\Gamma, \omega^2)$ and their convolutions were computed for different values of λ and γ. Since the coupling coefficients A, B and C are not well known because of the lack of experimental stress data on the F band in NaBr, it is difficult to attribute the correct weight to the contribution of each projected density. However it seems possible to reproduce quite well the experimental shape of both parallel (| |) and perpendicular (|) Raman spectra by a suitable superimposition of some few projected one and two-phonon densities. The calculated projected one and two-phonon densities of our interest are shown in Fig. 1 in

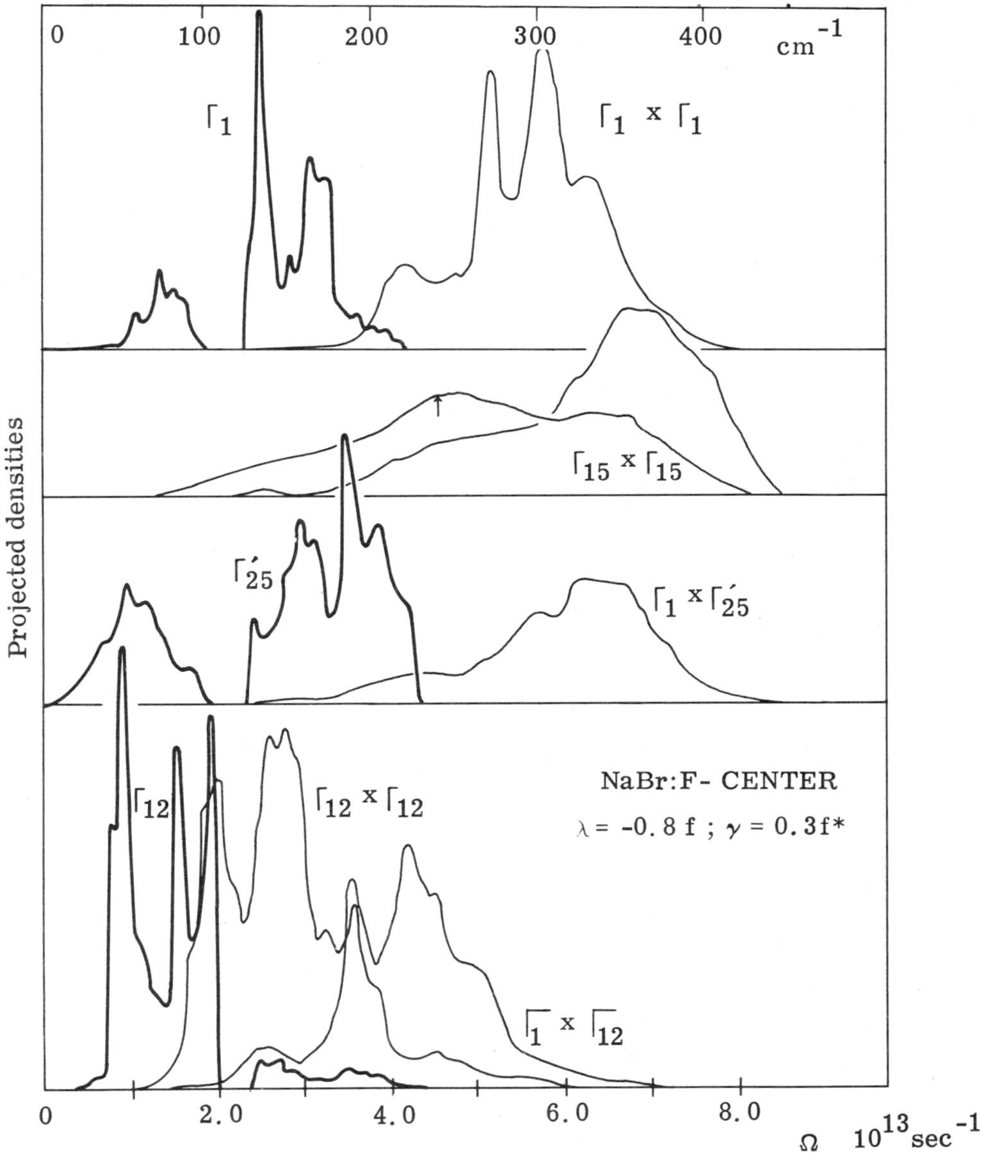

Fig. 1. One-phonon (thick line) and two-phonons (thin line) perturbed projected density of states in NaBr.

arbitrary units (normalization to unity is understood) for the values

$$\lambda = -0.8\,f^{*} \quad ; \quad \gamma = -0.3\,f^{*} \quad \text{or} \quad \xi = -1.5\% \tag{14}$$

which appear to give the best fit. $f^{*} = 2.95 \times 10^{4}$ g sec^{-2} is the NaBr nn effective force constant; also for the F-center in NaBr (ground state) a small inward elastic relaxation ξ seems to be reliable, as for the other crystals [7].

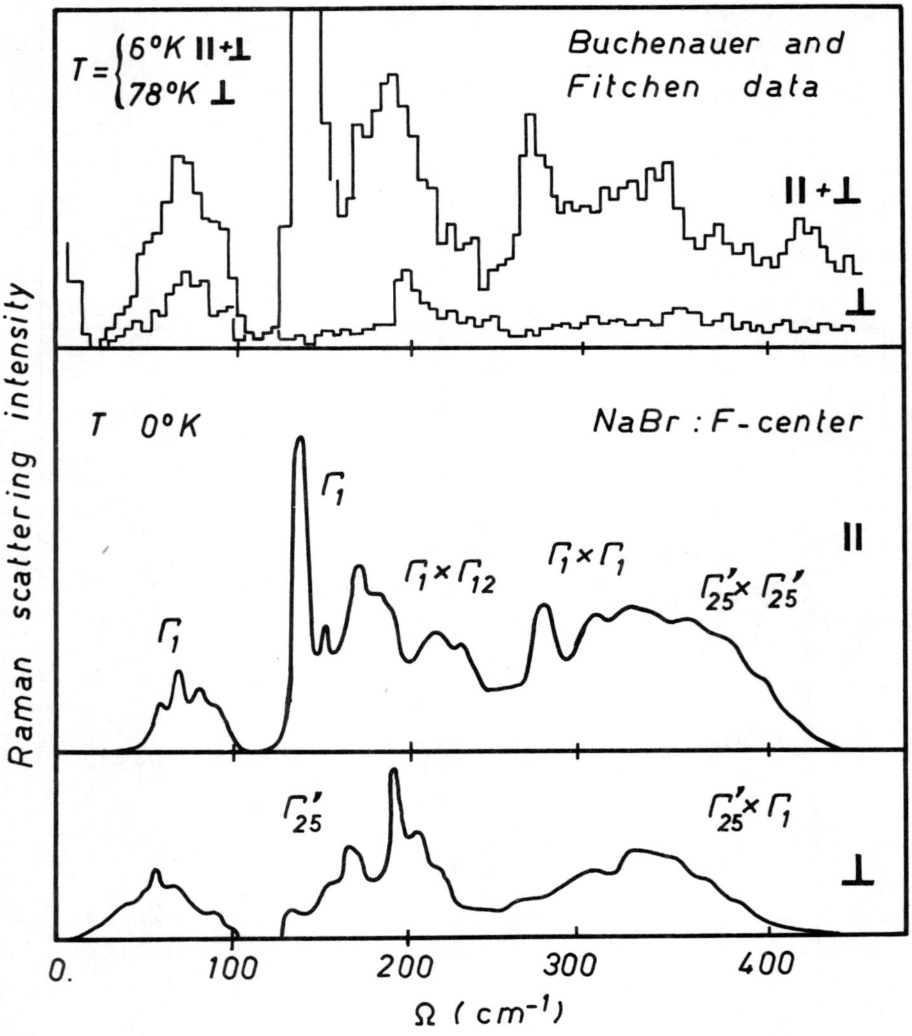

Fig. 2. Raman scattering intensity versus the frequency shift Ω for the F-center in NaBr. The experimental data and the theoretical results found by using the λ and γ fitted parameters are compared.

The theoretical best fit spectra are then shown in Fig. 2: they are in excellent agreement with T=0°K experimental data. Such agreement is much better than that previously obtained for the F-center in KCl and NaCl crystals with the same theoretical model [2, 7] (see for example, the comments by C.H. Henry and C.P. Slichter [15]); in our opinion, the preceding discrepancies were probably connected with the not high degree of reproducibility of the previous experimental data [16]. The irreducible representations and their Kronecher products which enter the final spectra are indicated, and appear to be just those which have been predicted by the formal theory of the previous Section.

Γ_1 and $\Gamma_1 \times \Gamma_1$ give the main contributions to first and second order $||$ spectra, respectively: the sharp peak in the $\Gamma_1 \times \Gamma_1$ term is just the overtone of the sharp peak in Γ_1 term. This fact, as noted first by Buchenauer et al. [10, 11], demonstrates that also second order spectrum is due to the F-center rather than to the host lattice excitations. In fact, the unperturbed $(\Gamma_1 x \Gamma_1)$ density which is associated to the exciton, was calculated and found to exhibit no structure at $\Omega = 5.4 \times 10^{13}\ \mathrm{sec}^{-1}$. $\Gamma'_{25} \times \Gamma'_{25}$ term is comparable with $\Gamma_1 \times \Gamma_1$ term which means that A ~ C; from experiment, however, C is usually quite smaller than A and probably this is true also for NaBr. Such discrepancy could be reduced by introducing $\Gamma_{15} \times \Gamma_{15}$ term. This term is very sensitive to any change of λ and γ and it is difficult to establish its actual shape; however, this term also is not allowed to be very large because it would fill the pseudogap centered at $4.5 \times 10^{13}\ \mathrm{sec}^{-1}$ (overtone of the gap). Therefore the coefficient (10) should neither exceed $P_{xx/\Gamma_1\Gamma_1}$ nor to be too small; say

$$\left| E_{2p} - E_{2s} \right| \sim 1/2\ H_F \simeq 0.14\ \mathrm{eV} \tag{15}$$

This result is consistent with a recent Rhyner and Cameron discussion based on Stark effect measurements [17].

As concerns $\perp$ spectra the predicted Γ'_{25} and $\Gamma_1 \times \Gamma'_{25}$ terms explain quite well the observed structures.

Finally, $\Gamma_1 \times \Gamma_{12}$ term gives an interpretation of the structures in the region $3.5 < \Omega < 4.1 \times 10^{13}\ \mathrm{sec}^{-1}$, while all the other terms derived from Γ_{12} density (tetragonal modes) must be extremely small. The above conclusions could be tested as soon as experimental stress data will be available for deriving A, B and C. However, for a qualitative discussion we could extrapolate the values of A, B and C by means of the empirical plots (A, B, C) versus the ionic radii ratio r_-/r_+ as given by Schnatterly [6], Buchenauer et al. [18] and corrected by the use of local rather than host elastic constants. We get

$$A \cong 4.0\ \mathrm{eV} \quad ; \quad B \cong 0.0\ \mathrm{eV} \quad ; \quad C \cong 1.0\ \mathrm{eV} \tag{16}$$

these values are qualitatively coherent with the above conclusions. They also justify the amplitude of the second-order with respect to the first-order spectrum, according to Eq. (12). Nevertheless, some residual discrepancies of the calculated shape of second-order spectrum and of the amplitude of $\Gamma'_{25} \times \Gamma'_{25}$ terms with respect to the experimental data could be explained by taking into account the existence of the non-negligible second-order spectrum coming from the exciton transitions.

Following the method previously described [7] and using the above values (14) and (16), we calculate also the local elastic constants, the T=0°K half-width $H_F(0)$, Huang-Rhys factor S(0) and band asymmetry μ, and the shift $\Delta\,\Omega$, of the peak frequency with respect to the pure electronic transition frequency for the F-center in NaBr

$$\bar{c}_{11} = 0.207 \times 10^{12} \text{ dyne cm}^{-2} \qquad H_F(0) = 0.289 \text{ eV}$$

$$\bar{c}_{12} = 0.112 \times 10^{12} \text{ dyne cm}^{-2} \qquad S(0) = 53$$

$$\mu = 2.228\ 10^{-4} \text{ eV}^3 \qquad (17)$$

$$\bar{c}_{44} = c_{44} = 0.105 \times 10^{12} \text{ dyne cm}^{-2} \qquad \Delta\,\Omega_1 = 0.73 \text{ eV}$$

The above $H_F(0)$, which includes the spin orbit splitting $H_{s.o.} = - 28.2\ 10^{-3}$ eV [6] well compares with the experimental value $H_F(0) = 0.291$ eV [19]

COMPARISON WITH THE U-CENTER SIDEBANDS

From the point of view of lattice dynamics the F-center and U-center have quite similar properties, i.e. very small defect masses and comparable values of both parameters λ and γ. However, the excitation of a single U-center localized vibrational mode couples to the Γ_1, Γ_{12} and Γ'_{25} in a completely different way with respect to the optical transition of the F-center. In the U-center case the coupling originates through anharmonicity and gives rise to sidebands; it is well known that Γ_{12} (i.e. tetragonal) modes give the main contribution to the sideband, while for the electronic transition the Γ_{12} terms are negligible. On the other hand, the arguments of previous sections could apply to the U-center UV band and suggest that Raman spectra of the U-center would not be dissimilar from those of the F-center. Stress experiments on the U-center band would support or criticize this point of view. We can conclude, however, that the pioneer calculations by Sennett [20] of the U-center Raman spectra are probably not reliable since the weights attributed to the single terms following a criterium similar to that used for sidebands are not realistic.

ACKNOWLEDGEMENTS

We are greatly indebted to Dr. D.B. Fitchen for informing us about his experimental results before publication and for many helpful discussions. We thank also Prof. G.F. Nardelli for several comments.

REFERENCES

1. A.A. Maradudin, "Sol. State Phys.," 19, F. Seitz and D. Turnbull (ed.) Academic Press Inc., N.Y., 1966; N.X. Xinh, "Thesis," Faculté des Sciences de l'Université de Paris, 1966 (unpublished).
2. G. Benedek and G.F. Nardelli, Phys. Rev. 154, 872 (1967).
3. G. Benedek and G.F. Nardelli, Phys. Rev. 167, 837 (1968).
4. D.A. Kleinman, Phys. Rev. 134, A423 (1964).
5. W. Gebhardt and K. Maier, Phys. Status Solidi 8, 303 (1965).
6. S.E. Schnatterly, Phys. Rev. 140, A1364 (1965).
7. G. Benedek and E. Mulazzi (to be published).
8. G. Benedek and G.F. Nardelli, Phys. Rev. 155, 1004 (1967).
9. C.H. Henry, S.E. Schnatterly, and C.P. Slichter, Phys. Rev. 137, A583 (1965).
10. C.J. Buchenauer, D.B. Fitchen, and J.B. Page, "Color Centers in Alkali Halides", p. 52, Int. Symposium in Roma, 1968.
11. D.B. Fitchen (private communication).
12. J.R. Hardy, Phil. Mag. 7, 315 (1961).
13. R. Zeyhez and H. Bilz, "Localized Excitations in Solids", p. 767, R.F. Wallis (ed.), Plenum Press, New York, 1968.
14. T. Gethins, T. Timusk, and E.J. Woll, Phys. Rev. 157 (1967).
15. C.H. Henry and C.P. Slichter, "Physics of Color Centers" p. 351, W. Beall Fowler (ed.), Academic Press Inc., New York, 1968.
16. J.M. Worlock and S.P.S. Porto, Phys. Rev. Letters 15, 697 (1965).
17. C.R. Rhymer and J.R. Cameron, Phys. Rev. 169, 710 (1968).
18. C.J. Buchenauer and D.B. Fitchen, Phys. Rev. 167, 846 (1968).
19. G. Spinolo (private communication).
20. C.T. Sennet, J. Phys. Chem. Solids (GB) 26, 1097 (1965).

F-11: RAMAN SCATTERING BY ADDITIVELY COLORED SrF_2 AND BaF_2 CRYSTALS*

O. Brafman and S. S. Mitra
Department of Electrical Engineering,
University of Rhode Island
Kingston, R. I.

Additive coloration in SrF_2 produces two absorption bands[1]. One of these is the F-band with a peak around 4450 Å and the other has a peak at about 6340 Å. The origin of the latter is not yet clear. It is not produced by x-irradiation, but invariably occurs with the F-band in additively colored crystals with relatively lower intensity. Additively colored BaF_2 also exhibits two absorption bands[2, 3] at approximately 4200 Å and 6500 Å. The latter is known to be the F-band.

It is expected that additional information may be obtained from the Raman spectrum of crystals containing color centers when compared with that of pure crystals. This is so because of the breakdown of the selection rules due to destruction of the translational symmetry in the presence of imperfections. Under certain conditions, it is thus possible for a photon to interact with a single phonon of any k value. This was observed by Worlock and Porto[4] in Raman scattering from NaCℓ and KCℓ containing F-centers. Impurity induced band mode excitation have also been observed in the infrared spectrum[5]. This is true for other kinds of imperfections as well[6].

Crystals from different sources were additively colored with Ca metal at a tempertture between 700-800°C. The concentration of F-centers varied between 10^{17} to $10^{19}/cm^3$. Raman spectra were excited by a He-Ne laser (6328Å) with an output of 50 mW. Perpendicular geometry was used. The spectra were analyzed by means of a Spex 1400 double monochromator in conjuction with a photon counting system. Two different slit width 4 cm^{-1} and 7 cm^{-1} were used in order to make sure that there is no instrumental broadening of the bands. The reported peak positions are reliable within ± 3 cm^{-1}. Stokes and anti-Stokes spectra were recorded at 90°K and 300°K.

*Work supported in part by the U. S. Air Force In-House Laboratory Independent Research Fund under Contract No. AF19 (628-6042).

An equipment grant from the Advanced Research Projects Agency, Grant No. DA-ARO-D-31-124-G754, is gratefully acknowledged.

Both SrF_2 and BaF_2 exhibit[7] cubic symmetry of space group O_h^5. The two non-equivalent fluorine atoms as well as the cation lie on face centered cubic lattices. F_{1u} mode is infrared active and it is split in the vicinity of k = 0 to transverse and longitudinal branches. The triply degenerate $k \simeq 0$ F_{2g} mode is Raman active.

Figs. 1a and 1b show the absorption bands produced by the additive coloration in SrF_2 and BaF_2 respectively, indicating the 6328 Å excitation line. It can be seen that in both cases the exciting line falls well inside an absorption band. The Raman scattering spectrum of additively colored SrF_2 shown in Fig. 2 consists of two parts. A relatively strong band appears at 283 cm^{-1} (at 300°K) which is also observed in pure SrF_2[8]. This band is due to the first order Raman scattering F_{2g} type mode at k = 0. The rest of the spectrum consists of three bands at 115, 171 and 336 cm^{-1}. These bands are not observed in pure SrF_2 and are attributed to the coloration of the crystals. Fig. 3 shows the Raman spectrum of additively colored BaF_2. Again beside the first order long wavelength Raman band of F_{2g} type at 240 cm^{-1}[9] two other bands are observed at about 81 and 300 cm^{-1}. These are not seen in the Raman spectrum of pure BaF_2 and are attributed to the coloration of the crystals.

Occurrence of both the Stokes and anti-Stokes Raman bands in the additively colored crystals and their non-appearance (except for the allowed first order band) in the pure crystal or in slightly colored crystals gives evidence to the fact that these additional bands are, in fact, due to Raman scattering and not due to fluorescence or other spurious effects. The peak positions shift (4-7 cm^{-1}) towards higher frequencies on cooling. The change of the intensity with temperature is comparable with that of the allowed first order Raman band and rules out the possibility that these additional bands are due to multiphonon processes. The half-widths of the bands remain essentially unchanged with temperature. The temperature dependence does not show evidence for localized vibration modes, either. It was qualitatively observed that as the concentration of the color centers increases the intensity of the first order Raman band decreases. At the same time the intensity of the other bands increases gradually. At higher concentrations when the crystals become almost opaque, the intensity of all the bands is reduced due to reabsorption of the scattered light. Because of the high opacity and internal reflections, it was also not possible to make use of the polarization dependence of intensity in assigning the various phonon-modes.

No calculated phonon density of states or neutron scattering data are available for SrF_2. The present results can, however, be compared with the results of Elliott et al. [10] derived from two phonon sidebands involving a local mode phonon of H^- and a critical point phonon in SrF_2:H^- crystals. The fact that the infrared transmission of SrF_2:H^- and SrF_2:D^- exhibit sidebands, separation of which from the main local mode band are same (this is true for BaF_2 as well) serves as a proof that the host lattice plays a dominant role in determining these frequencies. A comparison of the present results with those of Elliott et al. [10] for SrF_2 show that in the present work the frequencies are always higher by 6-13 cm^{-1} (taking into account an approximate temperature dependence). This small difference can be regarded as being within the limits of the experimental error. But there is a tendency of <u>all</u> the frequencies to

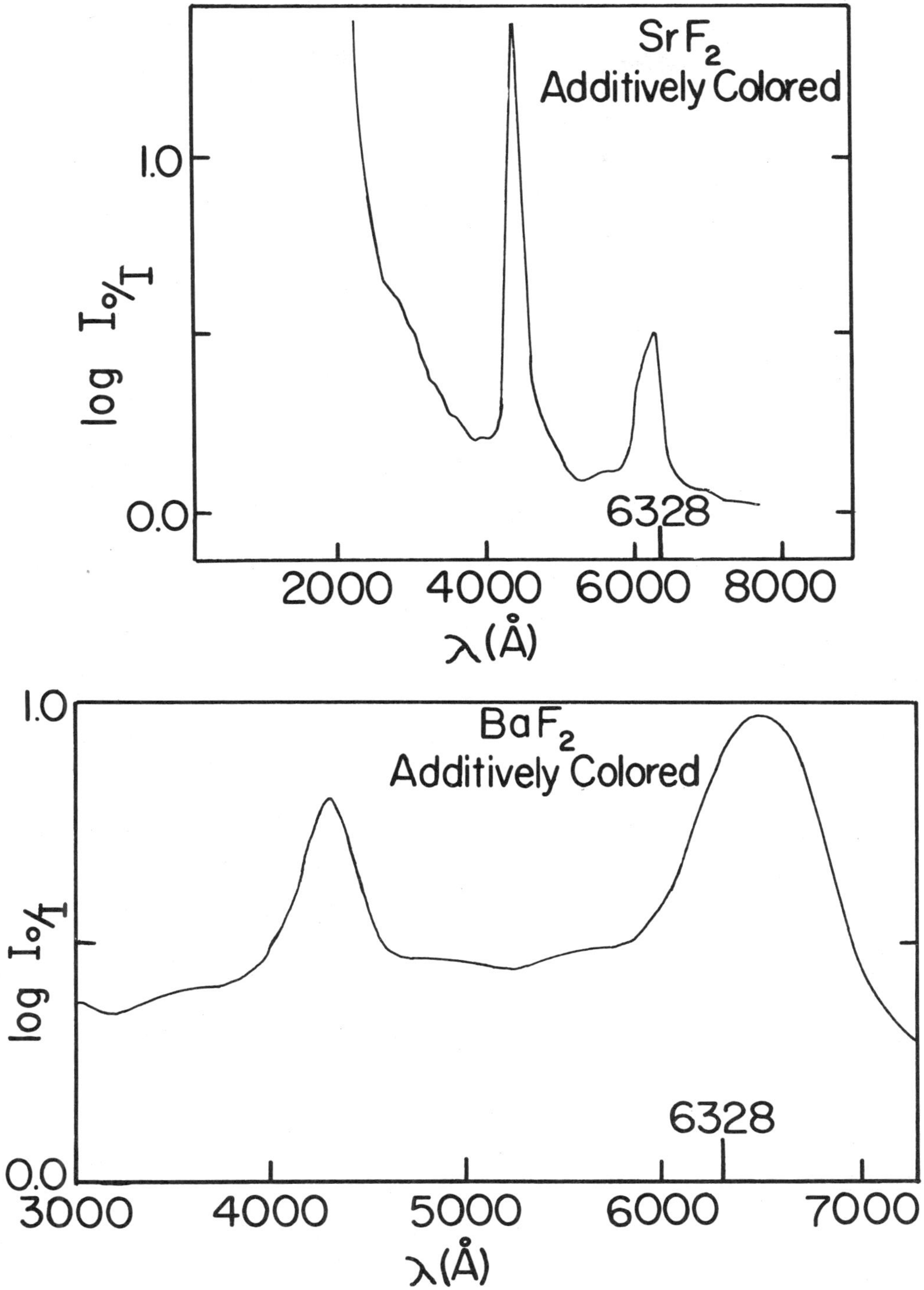

Fig. 1. Optical absorption bands produced by additive coloration in (a) SrF_2, (b) BaF_2. The 6328 Å excitation line is indicated (Ref. [2].

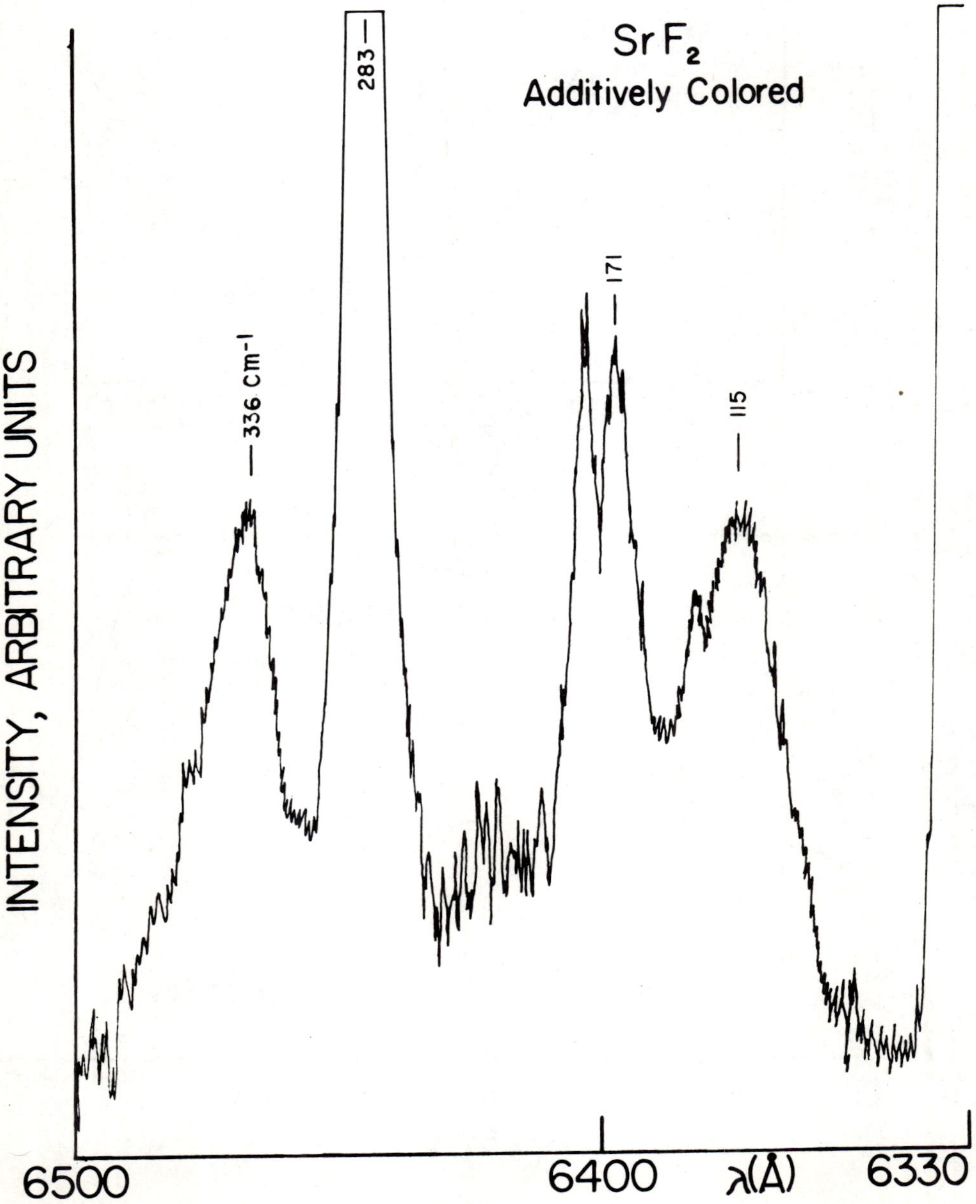

Fig. 2. Raman spectrum of additively colored SrF_2 at room temperature.

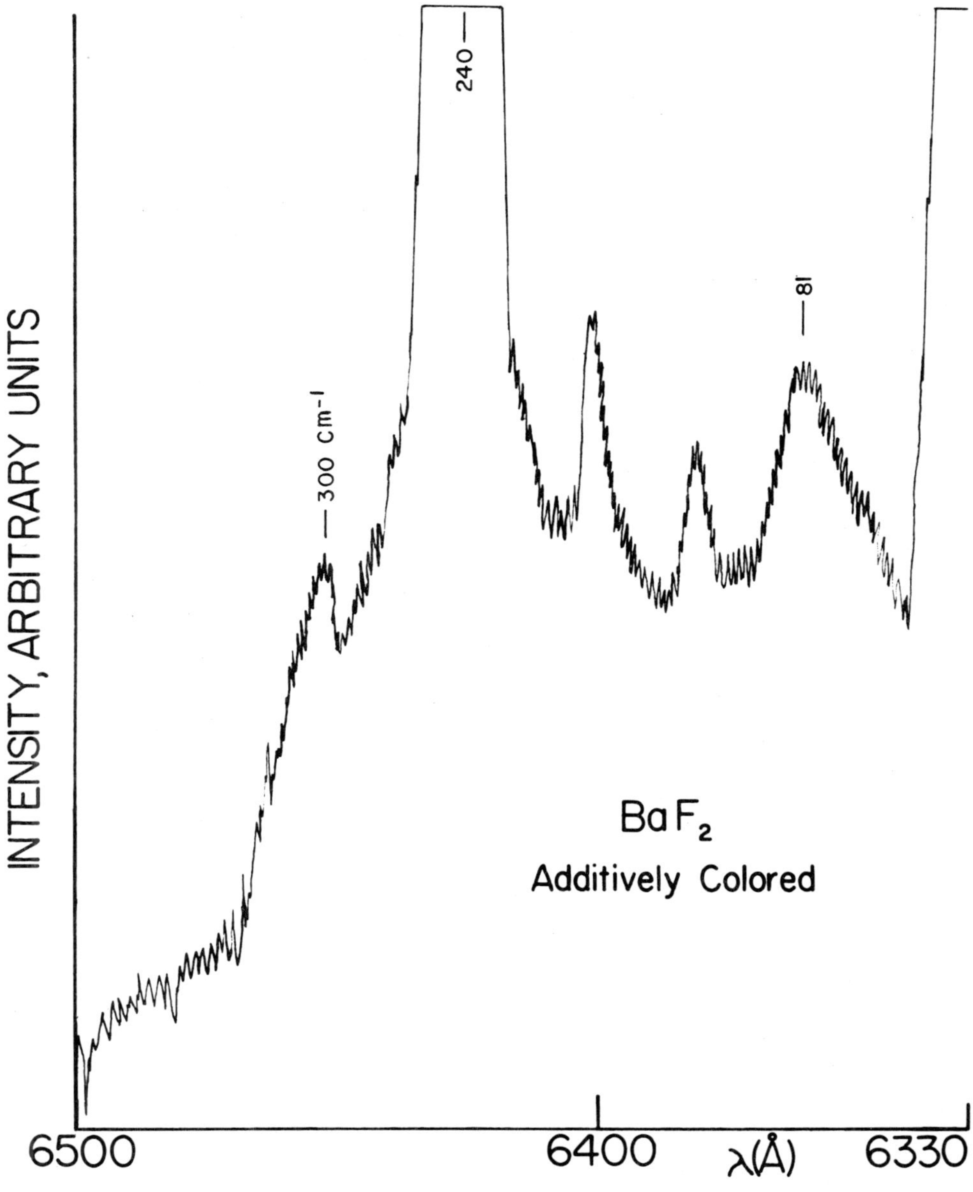

Fig. 3. Raman spectrum of additively colored BaF_2 at room temperature.

differ in the same sense and this may point out to something else. The phonons participating in the two phonon process of $SrF_2:H^-$ sidebands may not necessarily have exactly the same k value and thus also not the same frequencies as those of the phonons in the Raman scattering which involves single phonons only. This assumption gets further support from earlier work. When the results of Worlock and Porto[4] on Raman scattering from F-centers in NaCℓ and KCℓ is being compared to the infrared sidebands frequencies of $NaC\ell:H^-$ and $KC\ell:H^-$ obtained by Fritz[11] it becomes obvious that all frequencies obtained from Raman measurements are higher than the corresponding frequencies obtained from the two phonon infrared spectra of U-centers. It should also be mentioned that comparing his results to a calculation made by Karo[12], Fritz[11] shows that the experimentally obtained sideband frequencies are lower than those corresponding to the frequencies of maxima in density of states.

In BaF_2, the low frequency band compares very well with that obtained by Elliott et al. [10] while the high frequency band is higher by about 20 cm^{-1} than the one derived from a study of U-center sidebands. For BaF_2 new unpublished neutron data[13] were available to us. The low frequency band of 81 cm^{-1} can be correlated to the TA at the point X which is given as 95 cm^{-1} from the neutron data. If another point on the $\Delta_5(A)$ branch which is not at the extreme end of the Brillouin zone is assumed, a much better fit is obtained. The high frequency band at 300 cm^{-1} corresponds to 311 ± 6 cm^{-1} which is the LO at the X point. No such correlation can be made for the corresponding mode derived from the U-center sidebands. Relying on the neutron data it seems then reasonable to assign TA(X) = 81 cm^{-1} and LO(X) = 300 cm^{-1} for BaF_2 and similarly by comparison TA(X) = 115 cm^{-1}, LA(X) = 171 cm^{-1} and LO(X) = 336 cm^{-1} for SrF_2.

It was mentioned previously that both SrF_2 and BaF_2 crystals contained more than one kind of centers and it was not possible to associate the Raman bands with any particular centers. However, the near resonance condition between the optical transition of the color center and the exciting light, although not essential, might be important for the enhancement of the Raman cross section. If this is assumed then different kinds of centers contributed to the Raman scattering in SrF_2 and BaF_2. For a further check of this point, it would be worthwhile to study the Raman spectrum of the same crystals using sources of different wavelengths.

We would like to note that an attempt to repeat a similar experiment for additively colored CaF_2 for which experimental[14] and theoretical[15] data on phonon spectrum are available was unsuccessful. This was due to a very intense luminescence and a large concentration of traps.

ACKNOWLEDGMENT

The authors are indebted to Mr. R. A. Shatas and Mr. G. A. Tanton for additively coloring the crystals and also for the preprint of their paper.

REFERENCES

1. P. Gorlich, H. Karas, and A. Koch, Phys. Stat. Sol. 12, 203 (1965).
2. G.A. Tanton, A. Mukerji, R.A. Shatas, and J.E. Williams, J. Chem. Phys. (to be published).
3. P. Feltham and I. Andrews, Phys. Stat. Sol. 10, 203 (1965).
4. J.M. Worlock and S.P.S. Porto, Phys. Rev. Letters 15, 697 (1965).
5. M.V. Kelin and H.F. MacDonald, Phys. Rev. Letters 20, 1031 (1968).
6. M. Balkanski and W. Nazarewicz, J. Phys. Chem. Solids 23, 573 (1962).
7. R.W.G. Wyckoff, "Crystal Structures," John Wiley and Sons, New York, 1965.
8. I. Richman, Phys. Rev. 133, 1364 (1964).
9. R.S. Krishnan and P.S. Narayanan, Indian J. Pure Appl. Phys. 1, 196 (1963).
10. R.J. Elliott, W. Hayes, G.D. Jones, H.F. MacDonald, and C.T. Sennett, Proc. Roy. Soc. 289, 1 (1965).
11. B. Fritz, "Lattice Dynamics," p. 485, R.F. Wallis (ed.), Pergamon, London, 1965.
12. A.M. Karo, J. Chem. Phys. 31, 1489 (1959).
13. Brookhaven National Laboratory. Private communication with J. Hurrell.
14. D. Cribier, B. Farnoux, and B. Jacrot, Phys. Letters 1, 187 (1962).
15. S. Ganesan and R. Srinivasan, Can. J. Phys. 40, 74 (1962).

F-12: THE MEMORY FUNCTIONS IN MAGNETIC RESONANCE

Andrei N. Weissmann
Babes-bolyai Univ. Cluj-Romania

A quantum-statistical method for studying the correlation mechanism in a "hot reservoir" under interaction with an external field is developed and applied to a general spin system. The method is based on the use of the so called "dynamical characteristic function" technique[9] and is adequate to be applied in other more general thermodynamic systems too.

HISTORY

In recent years there has been a renewed interest in the study of the memory effects which appear in various thermodynamic and quantum systems. Both, from the physical and mathematical point of view this problem is strongly connected with that of the problem of correlation. Ever since the pioneering studies of Wigner and Seitz on the cohesive energy of metals[1,2] the correlation problem has been the object of considerable theoretical interest. They defined at first the correlation energy of the electron liquid as the difference between the Hartree-Fock approximation ground state energy, and any better calculation. However, this definition is not a very consistent one. It gives no answer concerning the microscopic causal mechanism just of that effect which is named: correlation. The mathematical study of these effects leads to a better comprehension of them, and in recent years a series of excellent papers has contributed to this comprehension. We have to mention here, besides a great number of relevant papers from this field, the work of Roy J. Glauber[3] in which he introduced first the concept of time-dependent correlations in connection with neutron scattering by crystals, then the basic work of Leon Van Hove[4], the N. N. Bogoljubov and S. V. Tjablikov's paper[5], the essential contribution of R. Zwanzig[6] and the wonderful work of Philippe Nozieres[7].

THE DENSITY MATRIX

Let us consider a system of N independent spins ($S = \frac{n}{2}$, where $n = \pm 1, \pm 2, \ldots$) in a typical resonance arrangement. The i-th spin is considered immersed in the "hot bath" which consists of the other N-1 spins, and an external magnetic field

$\bar{h} = h_1 (\tau \cos \omega t - \tau \sin \omega t) + h_0 \bar{k}$ is applied on the system. The spin system - before the interaction with the external magnetic field - is considered in thermodynamic equilibrium at temperature $T_B = (k\beta)^{-1}$. Then the radiation field is "switched on" and after a time a new equilibrium is established - the equilibrium which represents a specific balance between the energy absorbtion from the radiation field by some partner of the system, and the energy transfer of them to the "hot reservoir" - to the bath.

Consequently the irradiated spin system gives a "response" to the excitation. This "response" is closely connected with those correlations which exist between different parameters characterizing the bath. If this "response" depends also on the time passed from the beginning of the interaction with external field, then these correlations will be described by the memory functions. At very low temperature the normal modes of a spin system are of a wave-like character - that is there exists also long-range interactions - and the bath correlation time will depend also on the collective oscillation modes - on the magnons - of the system. These modes will produce a damping, the correlation time will suddenly decrease, and will appear the "memory" in the system.

The behaviour of such a system will depend not only on its present but on its past too. Thus the double time-dependence of the memory functions is essential; it reflects the internal causal mechanism responsible for the memory effects.

If $\bar{S}_i$ denotes the i-th spin operator, and taking into account the most general exchange interactions between the spins (symmetric, assymmetric, and anisotropic exchanges), the Hamiltonian of the system will be

$$H = H_1 + H_2 + H_3 + H_4 \tag{1}$$

where

$$\left\{ \begin{aligned} H_1 &= -\gamma \vec{h}_0 \cdot \vec{S}_i - \gamma \vec{h}_0 \cdot \sum_j{}' \vec{S}_j \\ H_2 + H_3 &= \frac{1}{2} \sum_{i<j}{}' J^{\alpha\beta} S_i^{\alpha} S_j^{\beta} \\ H_4 &= -\gamma \vec{h}_1 \cdot \sum_j \vec{S}_j \end{aligned} \right. \tag{2}$$

We can express (1) in the second quantisation formalism by two transformations. One of them is

$$\left\{ \begin{aligned} S_{ix} &= S(C_i^+ + C_i) \\ S_{iy} &= -iS(C_i^+ - C_i) \\ S_{iz} &= 2S(C_i^+ C_i - \tfrac{1}{2}) \end{aligned} \right. \tag{3}$$

where C_i^+ and C_i are the Pauli's exciting and deexciting operators for the spin system

$$\left\{\begin{aligned} &\left\{C_i, C_i^+\right\} = 1 \\ &C_i^2 = C_i^{+2} = 0 \\ &\left[C_i, C_i^+\right] = \left[C_i, C_j\right] = \left[C_i^+, C_j^+\right] = 0 \quad i \neq j \end{aligned}\right. \tag{4}$$

and the other is

$$\left\{\begin{aligned} S_i^+ &= \left(\frac{2S}{N}\right)^{1/2} \sum_q e^{-i\vec{q}\cdot\vec{r}_i} b_q + \cdots \\ S_i^- &= \left(\frac{2S}{N}\right)^{1/2} \sum_q e^{i\vec{q}\cdot\vec{r}_i} b_q^+ + \cdots \\ S_{iz} &= S - \frac{1}{N} \sum_{q,q'} e^{i(\vec{q}-\vec{q}')\cdot\vec{r}_i} b_q^+ b_{q'} \end{aligned}\right. \tag{5}$$

the Hollstein-Primakoff type transformation with the magnon operators b_q and b_q^+.

$$\left\{\begin{aligned} &\left[b_q, b_{q'}^+\right] = \delta_{qq'} \\ &\left[b_q, b_{q'}\right] = \left[b_q^+, b_{q'}^+\right] = 0 \end{aligned}\right. \tag{6}$$

Maintaining only the linear and bilinear terms in spin-wave operators, the Hamiltonian (1) becomes

$$\left\{\begin{aligned} H_1 &= -\omega_0 2S(C_i^+C_i - \tfrac{1}{2}) - \omega_0 2S \sum_j{}' (C_j^+C_j - \tfrac{1}{2}) \\ H_2 &= \frac{S^2}{2} \sum_{i<j} \{J\,[2\,(C_i^+C_j + C_iC_j^+) + 4N_iN_j - 2(N_i + N_j) + 1\,] \\ &\quad + J^{xy}\,[-i(C_i^+C_j^+ - C_i^+C_j - C_iC_j^+ - C_iC_j)] \\ &\quad + J^{yx}\,[-i(C_i^+C_j^+ + C_i^+C_j - C_iC_j^+ - C_iC_j)] \\ &\quad + J^{yz}\,[-2i(C_i^+N_j - C_iN_j - \tfrac{1}{2}(C_i^+ - C_i))] \end{aligned}\right. \tag{7}$$

$$+ J^{zy}\,[-2i(N_iC_j^+ - N_iC_j - \tfrac{1}{2}(C_j^+ - C_j))]$$

$$+ J^{zx}\,[2(N_iC_j^+ + N_iC_j - \tfrac{1}{2}(C_j^+ + C_j))]$$

$$+ J^{xz}\,[2(C_i^+N_j + C_iN_j - \tfrac{1}{2}(C_i^+ + C_i))]\Big\}$$

$$H_3 = J S^2N(N-1) - \omega_0 NS + \sum_q \omega_q b_q^+ b_q$$

$$+ \frac{1}{2}\sum_{i<j}\sum_q \{ [J S^2 \left(\frac{2S}{N}\right)^{1/2} + \frac{iS^2}{2}(J^{xy} - J^{yx})]\, e^{-i\vec{q}\cdot\vec{r}_j}\, C_i^+ b_q$$

$$+ [J S^2\left(\frac{2S}{N}\right)^{1/2} - \frac{iS^2}{2}(J^{xy} - J^{yx})]\, e^{i\vec{q}\cdot\vec{r}_j}\, C_i b_q^+ \}$$

$$H_4 = -\omega_1 S \sum_j{}' (C_j^+ e^{i\omega t} + C_j e^{-i\omega t}) - \frac{1}{2}\omega_1 \sum_j{}'\sum_q \left(\frac{2S}{N}\right)^{1/2}$$

$$\times\ [b_q^+ e^{i\vec{q}\cdot\vec{r}_j}\, e^{i\omega t} + b_q\, e^{-i\vec{q}\cdot\vec{r}_j}\, e^{-i\omega t}]$$

where we used the usual notations and we put $\omega_0 = \gamma h_0$; $\omega_1 = \gamma h_1$ and $N_i = C_i^+C_i$. The Heisenberg equations of motion for the $C = C_i + C_j$ Pauli operators and for the b_q boson (magnon) operators are

$$\begin{cases} i\dfrac{dc}{dt} = (A + B)\,c + 2E\,b_q + D \\[2ex] \dfrac{i}{2}\dfrac{db_q}{dt} = Gc + F\,b_q + Ke^{i\omega t} \end{cases} \tag{8}$$

where

$$\begin{cases} A = 4S^2JN - 2S\omega_0 \\ B = -iS^2\,(J^{xy} - J^{yx}) \\ D = -2iS^2\,(J^{yz} - J^{zy}) - i(J^{zx} - J^{xz})\Big]\ N^2 + 2\omega_1 S N e^{i\omega t} \\ E = 2\sum_j \Big[JS^2\left(\frac{2S}{N}\right)^{1/2} + iS^2\,\frac{1}{2}\,(J^{xy} - J^{yx})\Big]\ e^{-i\vec{q}\cdot\vec{r}_j}\ N \end{cases} \tag{9}$$

$$\begin{cases} F = \sum_q \left[2JS(N-1)(\gamma_q - 1) + \omega_o\right] = \sum_q \omega_q \\ G = 2\sum_j{}' \left[JS^2\left(\frac{2S}{N}\right)^{1/2} - iS^2 \frac{1}{2}(J^{xy} - J^{yx})\right] e^{i\vec{q}\cdot\vec{r}_j} \\ K = -\frac{1}{2}\,\omega_1 \sum_j \left(\frac{2S}{N}\right)^{1/2} e^{i\vec{q}\cdot\vec{r}_j} \end{cases}$$

Solving the system (8) with adequate initial conditions we get the new canonically conjugate operators

$$\begin{cases} Q(t) = \left[Q(o)\cos\frac{at}{2} + P(o)\sin\frac{at}{2}\right]\cos\frac{bt}{2} \\ \quad + \left[\pi_2(o)\cos\frac{at}{2} - K_2(o)\sin\frac{at}{2}\right]\sin\frac{bt}{2} \\ P(t) = \left[P(o)\cos\frac{at}{2} - Q(o)\sin\frac{at}{2}\right]\cos\frac{bt}{2} \\ \quad + \left[K_2(o)\cos\frac{at}{2} + \pi_2(o)\sin\frac{at}{2}\right]\sin\frac{bt}{2} \end{cases} \tag{10}$$

where

$$\begin{cases} a = A + B + 2F \\ b^2 = (A + B - 2F)^2 + 16EG \end{cases} \tag{11}$$

and $Q(o)$, $P(o)$, $K_2(o)$ and $\pi_2(o)$ are constants obtained easily from (9) and from the initial conditions. Thus

$$\begin{cases} Q(o) = \left(\frac{1}{2}\right)^{1/2}\left[C(o) + C^+(o)\right] \\ P(o) = -i\left(\frac{1}{2}\right)^{1/2}\left[C(o) - C^+(o) - 2k_{10}\right] \\ K_2(o) = \left(\frac{1}{2}\right)^{1/2}\left[K_2 + K_2^+\right] \\ \pi_2(o) = -i\left(\frac{1}{2}\right)^{1/2}\left[K_2 - K_2^+\right] \end{cases} \tag{12}$$

where

$$k_{10} = \frac{D_o F - 2EK}{2EG - (A+B)F}$$

$$K_2 = \frac{A+B - 2F}{b} C(o) + \frac{4E}{b} b_2(o) + \frac{2D}{b} + \frac{2\omega}{b} \frac{D_{10}G - (A+B)K}{(A+B)F - 2EG} \tag{13}$$

$$- \frac{(A+B-2F)}{b} k_{10}$$

The dynamical characteristic function of the bath has been obtained by Glassgold's technique[9].

He has introduced a set of unitary operators

$$\Omega(\xi, \eta) = e^{i(\xi Q + \eta P)} \tag{14}$$

defined by the power series expansion of the exponentials; ξ and η are continuous real variables in the range $-\infty$ to $+\infty$. Now the DCF is

$$\rho(\xi, \eta; t) + \mathrm{Tr}\,[\rho(t)\,\Omega^+(\xi, \eta)] \tag{15}$$

and then the expected density matrix is

$$\rho(t) = \frac{1}{2\pi} \int d\xi \int d\eta\ \rho(\xi, \eta; t)\ \Omega(\xi, \eta) \tag{16}$$

From (10), (14) and (15) we have the DCF of our spin system

$$\rho_s(\xi, \eta; t) = \rho_s \left[\xi \left(\frac{at}{2}\right) \cos\frac{bt}{2},\ \eta \left(\frac{at}{2}\right) \cos\frac{bt}{2};\ 0 \right] e^W \tag{17}$$

$$\times \exp -\frac{1}{2}\left(\xi^2 + \eta^2\right) \left\{ \frac{1}{2} |\alpha|^2 \sin^2\frac{bt}{2} + \sum_q |\beta|^2 \left[(e^{\omega_q/kT} - 1)^{-1} + \frac{1}{2} \right] \right.$$

where $\rho_s(\xi, \eta; o)$ is the initial DCF of the system

$$\rho_s(\xi, \eta; o) = \mathrm{Tr}_s \left\{ \rho_s(o)\, e^{-i\left[Q(o)\xi \left(\frac{at}{2}\right) + P(o)\eta \left(\frac{at}{2}\right) \right.} \cos\frac{bt}{2} \right\} \tag{18}$$

and

$$\begin{cases} W = i\sqrt{2}\gamma\eta \\ \alpha = \dfrac{A+B-2F}{b} \\ \beta = \dfrac{4E}{b} \end{cases}$$

are terms expressed by the constant of motion, and exchange integrals in agreement with (9).

THE MEMORY FUNCTIONS

Now we introduce the two-particle temperature dependent Green functions in a q space[8]

$$G(1,2,3,4) = i \left\langle\!\left\langle T\left[C^{+}_{q2}(t_2)\, C_{q1}(t_1)\, C^{+}_{q3}(t_3)\, C_{q4}(t_4) \right]\right\rangle\!\right\rangle \tag{20}$$

where T is the usual time-ordering operator, and $\langle\!\langle o \rangle\!\rangle$ denotes the expectation values of an operator $\hat{o}$ for a system with Hamiltonian H in thermodynamic equilibrium. The defining equation is

$$\langle\!\langle o \rangle\!\rangle = \frac{\mathrm{Tr}\left[e^{-\beta(H - \mu N)}\, o\right]}{\mathrm{Tr}\left[e^{-\beta(H - \mu N)}\right]} \tag{21}$$

where $\beta = \frac{1}{k_B T}$, μ is the chemical potential and N is the total number operator, defined by

$$N = \sum_q N_q = \sum_q C^{+}_q C_q \tag{22}$$

In the time-ordening we shall use the condition:

$$t_1 = t_2 = t^1 \;;\; t_3 = t_4 = t \quad : t^1 > t \tag{23}$$

while the states will be denoted by

$$\left\{ \begin{array}{ll} q_1 = q_i & \\ q_2 = q_i - q & \\ q_3 = q_j & \\ q_4 = q_j - q & (i, j = 1, 2, \ldots, N) \end{array} \right. \tag{24}$$

Then we have

$$G(1,2,3,4) = G_{ij}(q,t) = i \left\langle\!\left\langle T\left[C^{+}_{qi-q}(t')\, C_{qi}(t')\, C^{+}_{qj}(t)\, C_{qj-q}(t)\right]\right\rangle\!\right\rangle \tag{25}$$

Summing (25) over i and j, results in,

$$\sum_{i<j} G_{ij}(q,t) = i\left\langle\!\!\left\langle T\left[\rho^{+}_{q(i)}(t')\,\rho_{q(j)}(t)\right]\right\rangle\!\!\right\rangle \tag{26}$$

while

$$\sum_{i} C^{+}_{qi}\, C_{qi-q} = \rho_q = \sum_{i} C^{+}_{qi+q}\, C_{qi} = \rho^{+}_{-q} \tag{27}$$

Using now the Bogoljubov-Tjablikov spectral theorem[5] which makes the connection between the Green function and correlation functions

$$K_{ij}(t') = \left\langle \rho^{+}_{qi}(t')\,\rho_{qj}(t)\right\rangle = \lim_{\epsilon\to 0} i \int_{-\infty}^{\infty} \frac{e^{-i\omega(t'-t)}}{e^{\beta\omega}-1} \left[G_{ij}(q,\omega+i\epsilon) - G_{ij}(q,\omega-i\epsilon)\right] d\omega \tag{28}$$

where

$$G_{ij}(q,\omega) = \frac{1}{2\pi}\int_{-\infty}^{\infty} G_{ij}(q,t)\, e^{i\omega t}\, dt \tag{29}$$

we have the expected memory functions

$$K_{oo'}(t+\tau) = \left\langle O(t+\tau)O'(t)\right\rangle = \mathrm{Tr}\left[\rho_o O(\tau) O'\right] = \sum_{i<j} K_{ij}(t') = \sum_{i<j}\left\langle \rho^{+}_{qi}(t+\tau)\,\rho_{qj}(t)\right\rangle \tag{30}$$

where we have put $t' = t + \tau$.

In the $K_{oo'}(t')$ are present all the pair correlation functions $K_{ij}(t')$ and so it is able to explain completely the inner correlation mechanism in a system of N particles in interaction.

From (30) with (16) and (17) we have the expected memory functions.

It is interesting to compare the strength of magnon terms which appear in the low-lying energy states due to the long-range correlations with those, in which this term vanishes, because this comparison is able to relate us the inner causal mechanism of correlation.

For this reason we calculate explicitly the memory functions for a spin system with $S = \frac{1}{2}$ in two cases, taking into account only the symmetric exchange interactions $\tau^{\alpha\beta} = \delta_{\alpha\beta}$.

a) with damping in the case when magnons exist and

b) without damping, when magnon terms vanish - the case when only short-range interactions are considered.

a) Applying the same technique as before with the modified (9) where the off-diagonal terms in $\tau^{\alpha\beta}$ vanish, and taking into account that the initial density matrix for the N independent spins is the product

$$\rho_s(o) = \rho_{si}(o) \times \rho_{s2}(0) \times \ldots \times \rho_{sn}(0) \tag{31}$$

we have from (18)

$$\rho_s(\xi, \eta; 0) = e^{\sqrt{2}\, k_{10}\eta}\, e^{-\frac{1}{4}(\xi^2 + \eta^2)} \cos^2 \frac{bt}{2} \tag{32}$$

Now, from (17) and (32) we have the expected DCF

$$\rho_s(\xi, \eta; t) = e^{i\varphi_q \eta}\, e^{-\frac{1}{4}(\xi^2 + \eta^2)}\, \psi_q \tag{33}$$

with

$$\begin{cases} \varphi_q = \sqrt{2}\,(\gamma - ik_{10}) \\ \psi_q = 1 + \sum_q |\beta|^2 \left\{ (e^{\omega_q/kT} - 1)^{-1} + \frac{1}{2} \right\} \end{cases} \tag{34}$$

where γ is a real function of ω_0, ω_1 and ω_2.

From (16) using (33), (14) and (10) we calculate the density matrix in q space

$$\rho_q(t) = \frac{2}{\psi_q} \exp\left(-\frac{|\varphi_q|^2}{\psi_q}\right) \exp\left[-i\sqrt{2}\,\varphi_q(C - C^+ - 2k_1) - 1 - 2k_1(C - C^+ - k_1)\right]$$
$$\times \exp it \left[(i\sqrt{2}\,\varphi_q + 2k_1) \left\{ -\omega_0(C + C^+) - \frac{1}{2} N^{1/2} \sum_q (e^{-iq\cdot r_i} b_q + e^{iq\cdot r_i} b_q^+) \right\} \right] \tag{35}$$

Now from (28) and (30) we have the memory function of the system with N spins of $\frac{1}{2}$ in the case when magnons exist

$$K_{oo'}(t+\tau) = \frac{8}{(2-\beta)\psi_q^2} \exp\left(-\frac{|\varphi_q|^2}{\psi_q}\right) \exp\left(F \exp\left[-2\omega_i ft\right]\right) \exp -F\tau \quad (36)$$

where F and f are complex functions dependent on magnon frequencies, spin number and exchange integrals.

b) In this case only the transformation (3) is applied, that is, we consider no magnon terms. Than the new Heisenberg equations are

$$\begin{cases} i\dfrac{dC_i}{dt} = A'C_i + B'C_j + D' \\ i\dfrac{dC_j}{dt} = A'C_j + B'C_i + D' \end{cases} \quad (37)$$

with adequate A', B', D'. The DCF is then obtained by the same technique as has been shown above

$$\rho_s(\xi,\eta;t) = e^{\sqrt{2}K(\xi+\eta)}\, e^{-\frac{1}{4}(\xi^2+\eta^2)} \quad (38)$$

where $K = \frac{D'}{A'+B'}$ and now the density matrix in q space will be

$$\rho_q(t) = 2e^{-(1+2K^2)}\, e^{-2ik(C^+ + C)}\, e^{-2iK\omega_o t(C^+ - C)} \quad (39)$$

From (30), (39) we have the memory function of the system with N spins of $\frac{1}{2}$ in the case when only short-range interactions are considered and thus magnon terms vanish.

$$K_{oo'}(t+\tau) = \frac{8}{2-\beta} \exp\left(\frac{\omega_1^2}{\omega_o^2} N^2\right) \exp\left[-2i\omega t\right] \exp(-ft) \quad (40)$$

A glance to the (36) and (40) functions relates the following conclusions:

The aspect of the memory functions in the above mentioned two cases is essentially the same, therefore the correlation mechanism evolution of the spin system in both cases is essentially the same.

In the low-lying energy states, the long-range interactions introduce a damping in the correlation mechanism, which appears in concrete manner in the F, f, φ_q and ψ_q functions.

The presence of the magnons frequencies ω_q and the magnon occupation numbers η_q in the power index of the (36) causes a rapid decrease of the correlation time: $t = \tau_c^m$, which in comparison with the correlation time $t = \tau_c^s$ in absence of damping (40) shows that the appearance of magnons favors the memory of the system.

$\tau_c^m \ll \tau_c^s$ means that the absence of damping increases the correlation time, which overtakes the relaxation time of the system and so the system loses his memory and he turns into a Markoffian one.

Our results contain as special case the Van Hove's spin correlation functions, the Argyres and Kelley's[10] correlation functions, and many other results.

Our method is quite general, and therefore is able to be applied to more complex cases, not only to the magnetic resonance but also to all the thermodynamic systems which consist of a "hot bath" expressed by a finite number of degrees of freedom in interaction with a quite arbitrary radiation.

REFERENCES

1. E.P. Wigner and F. Seitz, Phys. Rev. 43, 804 (1933).
2. E.P. Wigner and F. Seitz, Phys. Rev. 46, 509 (1934).
3. R.J. Glauber, Phys. Rev. 87, 189 (1952).
4. L. Van Hove, Phys. Rev. 95, 249 (1954).
5. N.N. Bogoljubov and S.V. Tjablikov, D.A.N. SSSR. 126, 53 (1959).
6. R. Zwanzig, Phys. Rev. 124, 983 (1961).
7. Ph. Nozieres, "The theory of interacting Fermi systems," W.A. Benjamin N.Y., 1964.
8. G.E. Brown, Many-Body Problems, NORDITA, (1967).
9. A.E. Glassgold and D. Holliday, Phys. Rev. 139A, 1717 (1965).
10. P.N. Argyres and P.L. Kelley, Phys. Rev. 134A, 98 (1964).

G-1: THE INVESTIGATION OF SOME NON-LINEAR OPTICAL PHENOMENA IN LIQUIDS, GASES AND CRYSTALS

I. L. Fabelinskii
P. N. Lebedev Physical Institute
Moscow, USSR

The results of some investigations realized this year in the section of non-linear optics and hyper-acoustics of the P. N. Lebedev Physical Institute of the USSR Academy of Sciences are being shortly set forth in this paper. Our investigations are a continuation of a series of the studies of four-photon interaction in liquids. They relate to the investigation of the stimulated Mandelstam-Brillouin scattering (SMBS) in liquids when their viscosity changes from few fractions of a pause to glass state. They refer as well to the SMBS phenomenae in gases when the pressure changes from 100 to 2 atm and to the studies of a new phenomenon of stimulated temperature (entropy) scattering of light (STS) in gaseous hydrogen depending on the pressure. An anti-Stokes broadening of the one-mode pulse of the laser radiation propagating in the non-linear medium has been observed and examined. The stimulated combination scattering (SCS) has been studied, and the first observations of infrared radiation (IR) in quartz crystals at the helium temperatures have been performed. The research workers I. M. Arefiev, S. V. Krivokhizha, D. I. Mash, V. V. Morozov, V. S. Starunov, I. Sh. Zlatin, and the research students Yu. I. Kyzylasov, G. I. Zaitzev took part in the investigations mentioned.

FOUR-PHOTON INTERACTIONS IN LIQUIDS

Minck, Terhune and Rado[1] were the first to observe and rightly interpret the effect of four-photon interaction in SCS of gases.

The four-photon interaction due to the same mechanism as in stimulated wing of the Relay line in liquids[2] was investigated theoretically by Chiao, Kelley and Garmire[3], and Starunov[4]. It was experimentally observed for the degenerate case by Carman, Chiao and Kelley[5] and for the undegenerate case by Arefiev, Zaitzev, Kysylasov, Starunov and the author of the report[6, 7].

One can clearly see from Fig. 1 a scheme of a new experiment made by Starunov and Kyzylasov[8] where the four-photon interaction is effectively revealed. The light from the laser single-mode pulse with the power from 100 to 150 Mw is focused inside the sample made of a fused quartz. The SMBS back scattered light in the fused quartz is amplified in the laser returns back and comes out of the fused quartz having two spectral lines of an exciting light and component of SMBS (both having nearly equal intensity). The light is now focused by a lens inside the cell filled by CS_2 where the four-photon interaction takes place. In Fig. 1 one can see below a result of the four-photon interaction.

Nine Stokes and three anti-Stokes components are observed on the spectrum the distance between which corresponds to the distance between the components of SMBS in the fused quartz.

If one examines thoroughly each of such "quartz" components on the interferometer of Fabry-Perot, one can see that each component produces in its turn Stokes and anti-Stokes components, their shift corresponding to SMBS in CS_2. The effect of four-photon interaction is responsible for this case as well.

The phenomenon described is indicative of the fact that one of the reasons of formation of the anti-Stokes components in SMBS is the result of the four-photon interaction, the nature of which may be different.

The region occupied practically by the equi-distant components of SMBS depends on the laser power and may have several dozens of wave numbers, power of each component being of the order of one Mw. That is why beside other aspects the results of this work are of interest not only because they explain formation of the anti-Stokes components, they are of interest as well because they can be successfully used for solving other problems along with the problems connected with the tunable generator.

MANDELSTAM-BRILLOUIN STIMULATED SCATTERING IN VISCOUS LIQUIDS

The existence of a fine structure of the Rayleigh line in glasses and very viscous media was an object of thorough experimental investigation starting from the thirties. The Mandelstam-Brillouin components (MBS) were not observed before regardless of attempts made by experimentators. Probably, this is the cause that there were opinions that the MBC cannot be observed in glasses because of strong attenuation of hypersound at big viscosities.

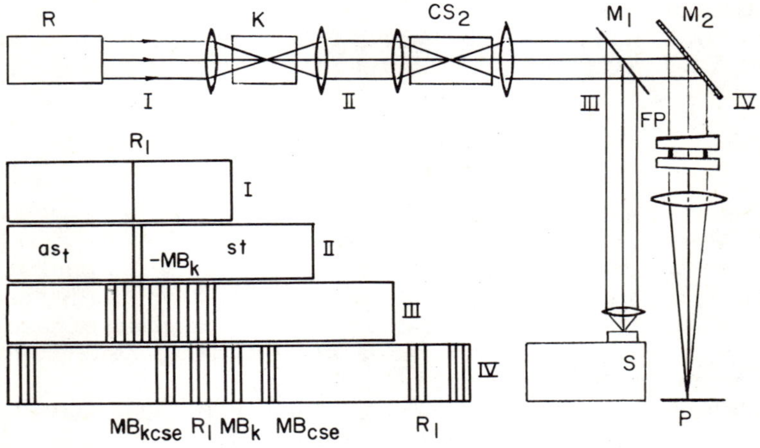

Fig. 1. A set-up diagram and spectrograms of stimulated Mandelstam-Brillouin scattering in a vistrous silica. The scattered light is amplified by a four-photon interaction in carbon disulpide.

However, after the appearance of the works by Pesin and mine[2. 34. 35] where the MBC were detected in glycerol and triacetine in the whole range of viscosity changes up to a vitreous state, and after the work by Flubacher, Leadbetter, Morrison and Stoicheff[11] who observed MBC in a fused quartz which was due to longitudinal and transversal hypersound waves the doubt concerning the existence of MBC disappears. However, a problem remains concerning the dependence of velocity and absorption of hypersound upon the viscosity as well as explanation of various experimental results obtained. Our results of measurements of the hypersound velocity and that of the ultrasound with respect to their temperature dependence are in agreement with the results of the American authors[12] when such comparison is possible. Our hypersound measurements are in contradiction with the experimental data obtained by Rank, Kiess and Fink[13].

A dependence of the hypersound velocity upon the temperature is given in Fig. 2 for triacetine. The solid curves correspond to the theory[15], the points are for the ultrasound data, and the crosses are for the results of hypersound measurements. The spectra of SMBS for different temperatures in triacetine and 1, 2-propilenglycol are given in Fig. 3. One can see from Fig. 2 that the path of the curves for the ultrasound and hypersound is generally alike.

The new ones are the results given in Fig. 4 for 1, 2-propilenglycol in these media for the whole range of temperatures. The results of these investigations cannot be described by the equations of the theory with one or some relaxation time. It is proved

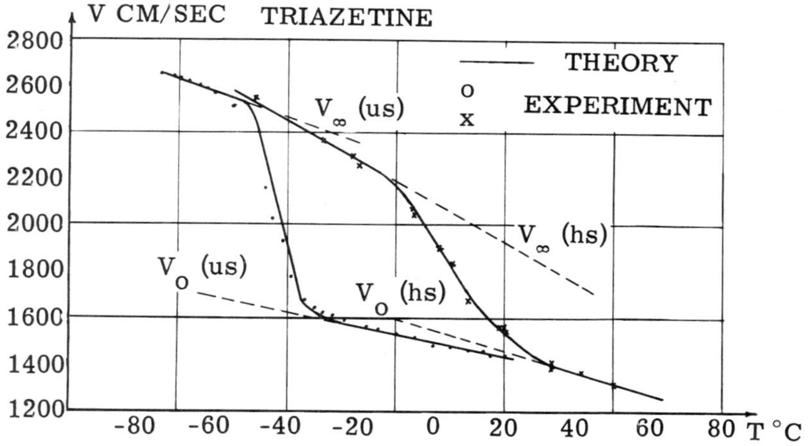

Fig. 2. Dependence of velocity of hypersound and ultrasound upon the temperature in triazetine. The solid curves are theoretical[15]. The crosses and points are the experimental data for hypersound and ultrasound, respectively. 3. Hypersound data obtained from SMBS.

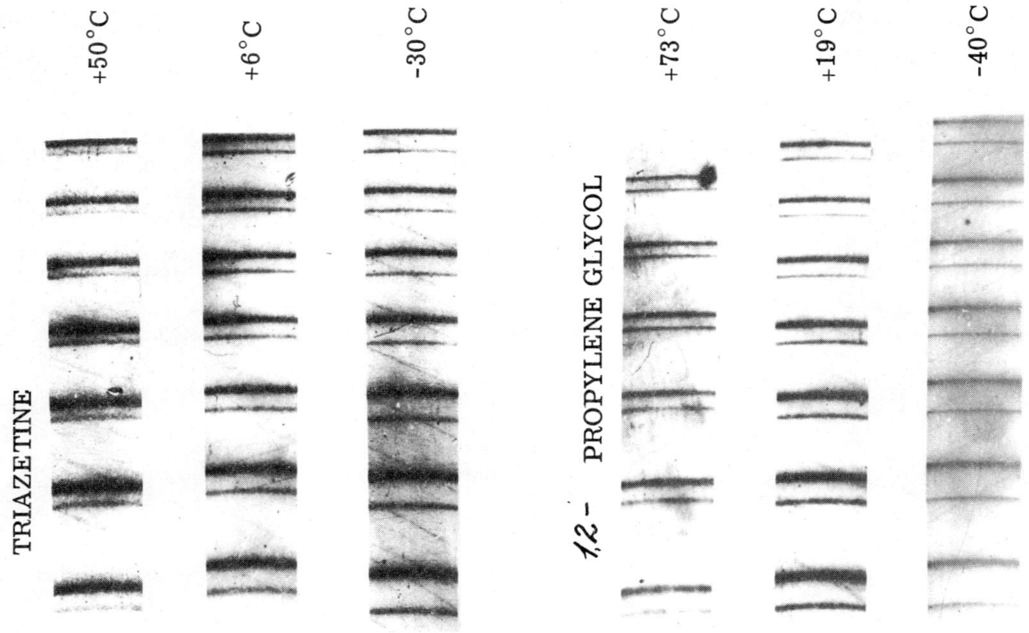

Fig. 3. Spectra of SMBS in triazetine and propilene-glycol for different temperatures.

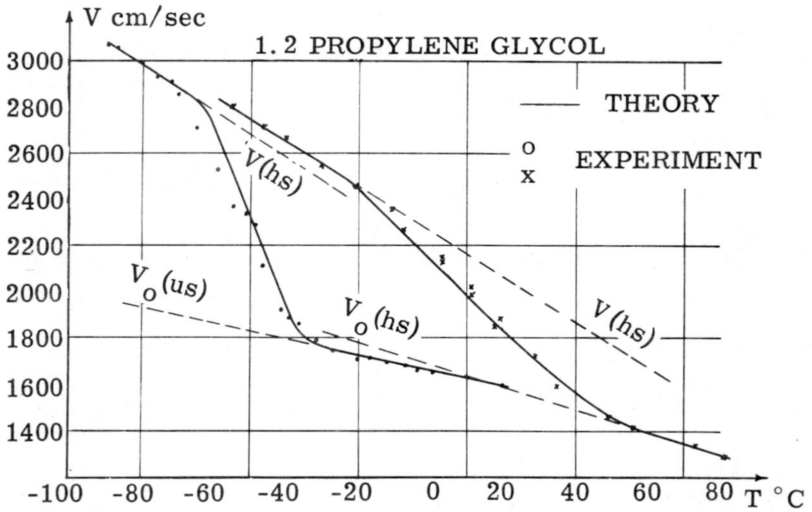

Fig. 4. Dependence of velocity of hypersound and ultrasound upon the temperature in: 1,2 propyleneglycol. The solid curves are theoretical[15]. The crosses and points are the experimental data for hypersound and ultrasound, respectively.

in the work by Krivokhizha and myself[14] that the experimental results cannot be described by the relaxation theory* of Mandelstam-Leontovich or the variations of this theory having a single relaxation time. There were formulated the conditions for the theory claiming to describe the test.

Isakovich and Chaban[15] developed a nonlocal theory of the sound propagation in a viscous medium which well describes all the peculiar conformities of the phenomenon and which has in its basis a hypothesis about the unhomogeneous structure of the liquids. Suspensions with the size of ~ 1000 molecules having rather sharp boundaries are dispersed in a continuous medium. Such a hypothesis is not easy to be understood, and one can hardly be accoustomed to it, though the equations of this theory well describe the experimental results.

MANDELSTAM-BRILLOUIN STIMULATED SCATTERING IN GASES

Earlier in our works and some works of American physicists it was shown that the velocity of hypersound in gases defined from the MBC position in the spectrum is less than the adiabatic value of this magnitude. Goldblatt and Litovitz[16] showed that if one works with a single-mode laser the velocity of the hypersound defined from the SMBS components turns out to be adiabatic. However, in our case the laser worked in a single-mode regime, and the velocity differed from an adiabatic one. That is why we suppose that the matter is not only because of single-mode laser (we predicted it from the very beginning), but the matter is that the light intensity of excited scattering should be as low as possible. In the test that we carried out we took a cell filled with nitrogen at the pressure 125 atm into which the light was focused from the single-mode pulse with the power approximately 100 Mw.

The focusing lenses had a focal distance from 3 to 30 cm. When the light was focused by the lens of 3 cm the velocity of sound found from the position of MBC was 300 m/sec. When the focus reached 30 cm the velocity was 393 m/sec, i. e. it was of adiabatic value. So, the position of the MBC at a single-mode exciting radiation depends upon the intensity. The lower the intensity, the closer is the hypersound velocity obtained from the MBC to its adiabatic value. Now, it is difficult to give a physical explanation to it. It may be that an increased absorption of the nonlinear hypersound in gases influences the position of the components. But this may be discussed only after obtaining a joint solution of nonlinear equations of Maxwell and the equations of hydrodynamics. May be, the interaction of molecules in a strong field of a light wave increases the effective mass[17] and is able to influence the hypersound velocity.

STIMULATED MANDELSTAM-BRILLOUIN SCATTERING IN GASEOUS HYDROGEN AT LOW PRESSURES

There arises a need of an additional amplification of stimulated scattering in order to observe SMBS in gases at low pressures. In this work an amplification was made due to trigger mechanism or super-regenerate amplification of a laser mode.

*The results described by the spectrum of the relaxation times were obtained in the works by Litovitz et al. [12]. Of course, any curve can be described having an infinite number of parameters. It becomes more difficult to acquire a physical sense of such an operation and each parameter used.

The mechanism of the trigger amplification consists in that during an excess of the pumping threshold of the laser and during an additional amplification in the gas cell the single-mode regime of generation can be transferred into a two-mode or biharmonious one. The second mode would appear on the frequency corresponding to the maximum of amplification in the gas cell.

One can control the amplification process as it was shown by Basov, Morozov and Oraevskii[18] by selecting the pumping intensity or by choice of a quality of the modes of vibration for a given region of spectrum within the width of the luminescence line of ruby R_1. In the conditions of our experiment, the single-mode generation regime was held at 50% exceed of pumping of the generation threshold ($\Delta\nu \leq 0.01\ cm^{-1}$).

In the investigations of the SMBS in gaseous hydrogen under the pressure of 100 to 2 atm we used a trigger mechanism of amplification with 25+30% exceed of the generation threshold.

A dependence of the SMBS Stokes component shift is shown in Fig. 5 against the pressure or the value 1/y where $y = \pi^{2/3} \frac{\Lambda}{\bar{l}}$ (here Λ is the wavelength of the hypersound, and $\bar{l}$ is the mean free path of a molecule), is the parameter of the molecular theory[19, 2, 20].

One can see from Fig. 5 that with the decrease of pressure (1/y is increasing) $\Delta\nu$ decreases faster than it can be predicted by the calculation at an adiabatic approach (crosses in Fig. 5) and slower than it can be predicted by the isothermic approach (a cross in the circle). However, the tendency of changing is right. With the decrease of the pressure the hypersound velocity is expected to approach an isothermic value and to become an isothermic one when $\Lambda \sim \bar{l}$.

STIMULATED TEMPERATURE SCATTERING OF LIGHT IN GASES

The stimulated temperature (enthropy) scattering (STS) of light in gases is of the same nature as it is in the case of liquids. The conditions are more favourable for the studies of this new nonlinear optical phenomenon in gaseous hydrogen at a low

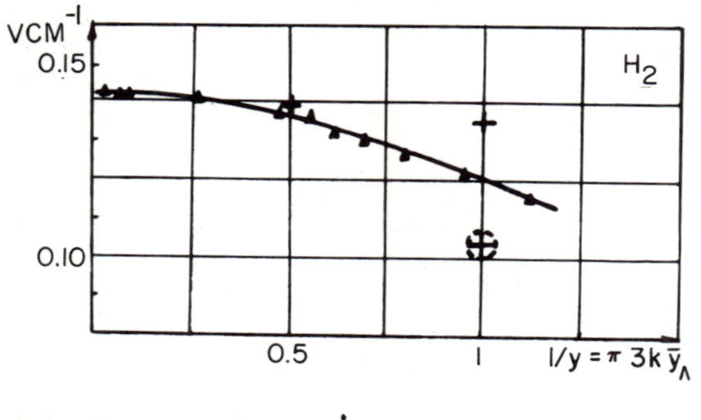

Fig. 5. Dependence of shift of the Stokes SMBS component in hydrogen depending on $\frac{1}{y} = \pi^{-2/3} \cdot \frac{\bar{l}}{\Lambda}$ in an interval of pressures of 2 to 100 atm. Δ - experimental data. + - calculated CMBS position in a thermal scattering having an adiabatic approximation[15]. (+) - calculated position of CMBS in a thermal scattering with an isothermic approximation[15].

pressure for which the halfwidth of the central component is much bigger as compared with other gases.

The STS is believed to be observed in all the cases with an exception of water at the temperature about 4°C.

The phenomenon of the STS is a result of the nonlinear interaction of the intensive exciting light and primary weak light scattered due to enthropy fluctuations when a pumping of energy takes place from an intensive light to weak light waves and the waves of enthropy or temperature.

The theory developed by Starunov[21] suggests that the nonlinear interaction is produced by electrocalorical effect described by Landau and Livshitz[22]. This effect causes the variations of the temperature per a unit of time as follows:

$$\Delta \dot{T} = \frac{T}{4\pi \rho C p} \cdot \left(\frac{\partial \epsilon}{\partial T}\right)_p \vec{E} \cdot \dot{\vec{E}}$$

The equation of the temperature conductivity with an account of this temperature variation gives a possibility of finding an expression of temperature changes ∂T if one knows the field $\vec{E}$.

The nonlinear equation of the temperature conductivity along with the nonlinear equation by Maxwell where the nonlinear addition to the polarization is $P^{NL} = \left(\frac{\partial \epsilon}{\partial T}\right)_p \partial T E$ opens a way toward a complete solution of the problem concerning the STS.

The electrocalorical effect plays an eminent role provided that $\left| 2nT \left(\frac{\partial \epsilon}{\partial T}\right)_p \chi (\vec{k}_o - \vec{k}_1)^2 \Big/ \epsilon'' \omega_o \right| > 1$. If absorbers of light are not put into a transparent medium this effect seems to prevail all the time.*

It follows from the solution of the nonlinear equations of Maxwell and temperature conductivity that the anti-Stokes component will damp, while the Stokes component will increase according to the following law at an amplification coefficient $g_{1T} > 0$:

$$E_1^2(x) = E_1^2(o) \exp (g_{1T} \cdot x)$$

where $E_1(o)$ is the strength of the electric field of the initially scattered light on thermal fluctuations,

$$g_{1T} = -2K_\omega + B|k_1| \frac{\Omega/\Omega_m}{1 + (\Omega/\Omega_m)^2} E_o^2$$

E_o is the strength of the electric field of the exciting light, $2k_\omega$ is the light losses coefficient and $B = T \left(\frac{\partial \epsilon}{\partial T}\right)_p \Big/ 16\pi n^2 \rho C p$. $\Omega_m = \vec{q}^2 \chi$; $(\vec{q}^2 = \vec{k}_o - \vec{k}_1)$; χ is the coefficient of

*In this case another phenomenon which was observed by Rank, Cho, Foltz and Wiggins[23] and stated to be caused by a direct absorption of light is not under consideration.

temperature conductivity, $\vec{k}_0$ and $\vec{k}_1$ is the wave vector of the exciting and scattered light, correspondingly.

The maximum of g_{1T} corresponds to Ω_m, that is the halfwidth of the central line of a fine structure.

The Mandelstam-Brillouin components are always present in liquids along with the central component.

In gases (speaking generally) the MBC are present when $\bar{\ell}/\Lambda \ll 1$, and they are absent when $\frac{\bar{\ell}}{\Lambda} \gg 1$. In the latter case only the central component of a fine structure is present. At high pressures the shape of the central line is Lorentzian and its half-width is small[2]. With the decrease of pressure the width of the central component is increasing, and the limit at very small pressures becomes a Doppler one.

These physical interpretations of quality are confirmed by the calculation[20] which is difficult to be obtained analytically even for a simple case. So, one has to be satisfied with the curves obtained on computer.

To observe the STS a mechanism of trigger amplification was used again. We used a mode separator which damped the MBC. The pictures were obtained on the Fabry-Perot etalon with a region of dispersion of 0.166 cm^{-1}; 0.312 cm^{-1} and 0.5 cm^{-1}.

The spectrograms of STS in gaseous hydrogen are given in Fig. 6; a) laser radiation, b) P = 6.5 atm, c) P = 2.5 atm, d) P = 1.5 atm. One can easily see in this picture an augmentation of the line shift while the pressure decreases.

A

B

C

D

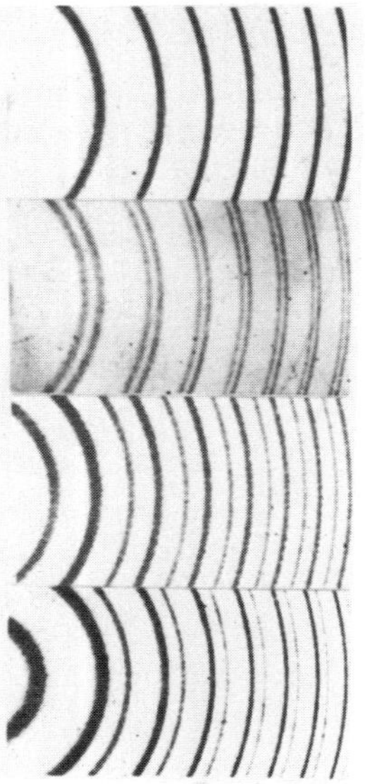

Fig. 6. Spectra of stimulated temperature (enthropy) light scattering in hydrogen at different pressures P. a) Exciting line of a laser, b) P = 6.5 atm, c) P = 2.5 atm, d) P = 1.5 atm. The dispersion region of the Fabry-Perot etalon is 0.166 cm^{-1}.

A dependence of shift of the STS Stokes component is presented in the next Fig. 7 as a function of 1/y or pressure. The pressure is decreasing from 6.5 atm (b) up to 1.5 atm (d). Here the dispersion region of the Fabry-Perot etalon is 0.166 cm^{-1}.

It follows from the figure that at values y >> 2 the experimental values of the half-width are in agreement with the calculated values of this term[20]; they are lying on a straight line corresponding to the hydrodynamic theory[2, 19].

Deviation from the hydrodynamic law starts at y < 2. With the further decrease of y, $\Delta\nu$ tends to a value of 0.11 - 0.12 cm^{-1} corresponding to Doppler halfwidth.

Thus, a new nonlinear optical effect is observed, and it is proved that its basic features are in agreement with the conclusions of the theory developed before.

BROADENING OF THE FREQUENCY SPECTRUM OF A POWERFUL SINGLE-MODE LIGHT PULSE IN A GASEOUS NON-LINEAR MEDIUM.

A significant spectrum broadening of the pulse toward the anti-Stokes region was observed at different pressures during propagation of a light single-mode pulse with a power of 150 Mw focused into the cell (f = 3 cm) filled with a gaseous nitrogen, hydrogen and helium.

In this experiment the feedback between the nonlinear medium and laser was damped by a polarized shutter.

The broadening is registered when the light is under investigation which passes in the direction of propagation of a primary light pulse. During the observations of light scattered or reflected backwards (at an angle of 180° toward the propagation of a primary pulse) there was not observed a broadening of the pulse spectrum.

In the light scattered backwards in nitrogen in an interval of pressures of 50 - 150 atm a Stokes Mandelstam-Brillouin component was detected.

In Fig. 8 there are presented microphotograms of spectrum of light which passed the region of focus in gaseous nitrogen at 15, 35 and 100 atm, respectively.

As one can see from the microphotograms (Fig. 8) a continuous anti-Stokes broadening lessens unsmoothly in the anti-Stokes region. However, it consists of more or less

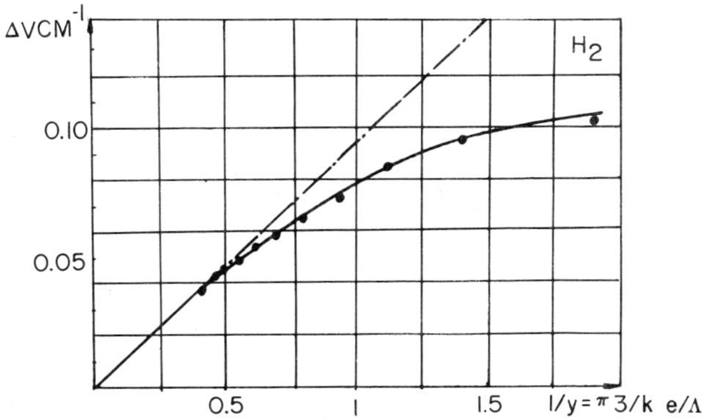

Fig. 7. Dependence of shift of the Stokes component of stimulated temperature (enthropy) light scattering upon the value $\frac{1}{y}$. The straight line presents hydrodynamic approximation.

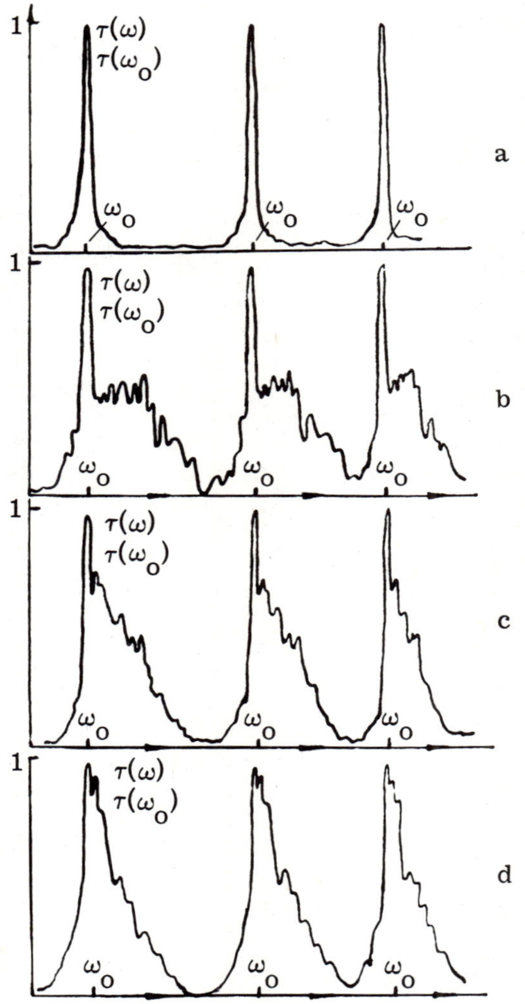

Fig. 8. Microphotograms of the laser radiation spectrum that passed a region of focusing in gaseous nitrogen at different pressures. Microphotograms are reduced to one scale. a) Non-focused laser radiation, b), c), d) light spectrum passed through the focusing volume at 15 atm, 35 atm, and 100 atm, respectively.

clearly seen maxima, or in other words, it has a structure. A total magnitude of broadening is decreasing with the increase of pressure, and the component amplitude of the structure is increasing; though the structure itself does not become more distinct.

The intermediate halfwidth of the anti-Stokes frequency broadening is decreasing with the pressure increase.

A dependence of such halfwidth on the pressure in nitrogen in an interval of 1 to 100 atm is given in Fig. 9. As it follows from the halfwidth measured with respect to increase of pressure in an interval already mentioned the halfwidth becomes one third as great. Perhaps, the phenomenon observed may be explained by a time-dependent change in a nonlinear part of the gas refraction index in the time-dependent field of a laser light pulse.

As a result of such a change of nonlinear part of refraction index an additional phase of light wave $\phi(t)$ dependent on time arises. The frequency pulse spectrum change is related to this additional phase.

Ostrovsky[24, 25, 26], and Jocnek and Landauer[27] proved theoretically a possibility of such a change of an initial pulse spectrum propagating in an nonlinear medium yet before arising of a self-focusing.

Even when the self-focusing does not take place the pulses of a duration of $\tau \sim 10^{-8} - 10^{-9}$ sec are expected to considerably broaden ($\Delta\omega >> \tau^{-1}$) without a marked distortion of an enveloping line of the amplitudes of an initial pulse.

In the cases when a self-focusing arises for single-mode, and especially, for multi-mode initial pulses the spectral changes are even more and may exceed 200 cm^{-1}. The latter case was considered theoretically. The phenomenon of such broadening was discovered experimentally for liquids[28-31]. The latter case does not seem to have any connection with the conditions of our experiment.

We shall make a try to estimate whether the phenomenon observed by us is due to a time variation of the phase as a result of nonlinearity of refraction index when the self-focusing has not arisen yet.

Let a part of the phase of a flat light wave $\phi(t,x)$ dependent on time be expressed through the relation[25, 26]:

$$\phi(t,x) = -\frac{n'(t,x)}{n_o} kx$$

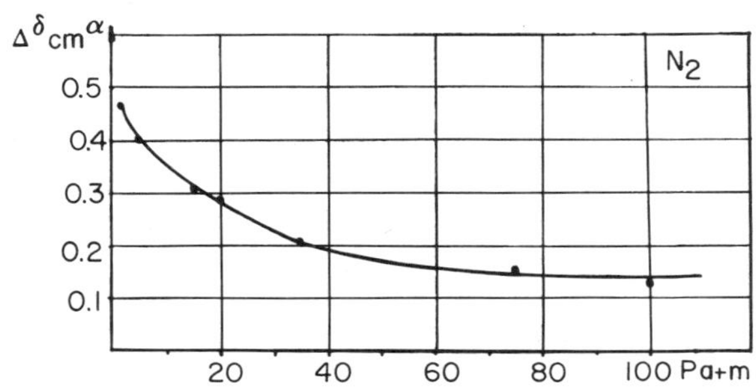

Fig. 9. Dependence of an averaged halfwidth of the light spectrum upon the pressure (2-100 atm).

An initial phase is taken here as equal to 0, and $k = \frac{2\pi n_o}{\lambda}$, where n_o is refraction index in the absence of a strong field; λ is the wavelength of a light in a vacuum; n' is a nonlinear part of the refraction index.

For the case when $\phi \ll \pi$, the width of a pulse spectrum becomes: $\Delta\omega \sim \tau^{-1}$. When ϕ becomes of the order of π or even greater, the frequency pulse changes are the following as it follows from the latter equation.

$$\Delta\omega = \frac{\partial\phi}{\partial t} = -\frac{kx}{n_o}\left(\frac{\partial n'}{\partial t}\right)_{max}$$

As it comes out from the previous equation the distance where ϕ becomes equal to π acquires the form: $x_1 = \frac{\pi n_o}{kn'}$.

Thus, it becomes clear from the equation for $\Delta\omega$ that the value $\Delta\omega$ depends upon the sign $(\frac{\partial n'}{\partial t})$, while its value is dependent upon the absolute magnitude of $(\frac{\partial n'}{\partial t})$ and the value n'.

The further problem is to derive an expression for n' through the values measured which are responsible for nonlinearity of the refraction index.

The nonlinearity of the refraction index can be caused by electrocalorical effect, effects of Kerr and striktia. The evaluation shows that the anti-Stokes shift may be related to the electrocalorical effect which produces a correct sign of shift and the order of the magnitude provided that the field intensity before the plasma production reaches a value of the order of 10^7 v/cm.

The structure of the spectrum observed seems to be related to a spatio-temporal structure of generation of an exciting light pulse.

Here we only try to intrepret this phenomenon, since one cannot say the nature of this phenomenon is completely clear.

STIMULATED COMBINATIONAL LIGHT SCATTERING AND STIMULATED INFRARED RADIATION AT HELIUM TEMPERATURE

Lines of combination scattering 467 cm^{-1} and 130 cm^{-1} as well as their sum and difference frequencies were observed in our previous investigations of stimulated scattering in a quartz monocrystal at helium temperatures[32, 33].

A spectrum of combination scattering is presented in Fig. 10. Several lines of such spectrum were observed by Tannenwald[34, 35]. Besides the lines of combinational scattering and the lines of SMBS we recorded a line which cannot be referred either to the line of SCS or to the line of SMBS. Its physical nature is not established yet. Probably, this line is a result of a specific character of self-focusing at helium temperatures and produced, thus, considerable additional phase shifts of a light wave of a laser pulse. In some cases a shift is believed to be possible which may have a discrete and nonsymmetrical character[26]. It goes without saying that the above said does not imply that there is found a complete explanation of the line discovered.

In our recent experiments with a quartz monocrystal at the temperature about 10°K we observed a longwave infrared radiation. In our first tests we registered a sum flow of an infrared radiation in the direction of propagation of the exciting light. At an angle of 90° the infrared radiation was absent. A beam of a ruby laser with a power of

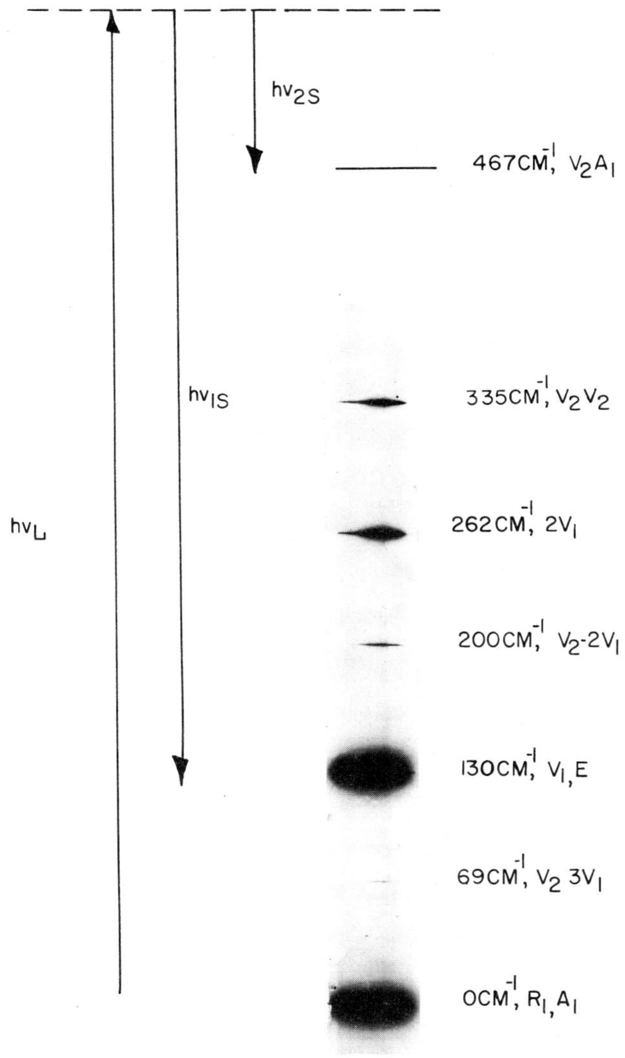

Fig. 10. Spectrum of SCS of light in a quartz monocrystal at a temperature of 9°K.

100-200 Mw and duration of ~ 12 nsec was directed by a lens of f = 5 cm into the quartz crystal along the optical axis. The crystals 2-5 cm long were fitted on a cold conductor of a helium cryostat with windows made of a fused quartz (3 mm thick).

Close to the output window of the cryostat a thermopile with a window of crystal quartz (2 mm thick) was situated. A galvanometer with a sensitivity of 0.02 microvolts per division was switched to the thermopile. The nearest visible infrared radiation and, possibly, ultraviolet one were cut off by filters of a black photographic paper, black polyetilene or teflon covered by a film of turpentine soot. The infrared radiation signal was up to 1 microvolt.

The signal disappeared when a glass plate was added to the filter. The signal became less by 30-40% when the glass plate was substituted by a paraffin plate 2.5 mm thick. This corresponds to paraffin absorption in the range of 40-100 microns. One can possibly explain this IR the following way. If the gain coefficient of the Stokes radiation of SCS is higher for the frequency 467 cm^{-1} (A_1) than for the frequency 130 cm^{-1} (E), then it becomes possible that the infrared radiation generation at transition A_I E has a frequency of 337 cm^{-1} (Fig. 10).

It seems possible to obtain infrared generation at transition from the A_I level to other levels which are not detectable in the SCS spectrum. Such levels can be named the levels of principal oscillations of 266 cm^{-1} and 400 cm^{-1} of the E symmetry (for quartz).

The estimations show that if a spontaneous line of 130 cm^{-1} has a halfwidth of the order of 0.1 cm^{-1} at low temperature[34, 35], it becomes real enough to create an inverse population between a basic state of A_I(O) and E (130 cm^{-1}) and to obtain an infrared generation at a frequency of 130 cm^{-1} having a quartz sample less than 5 cm. The conditions of generation will be more favourable if one bears in mind a possibility of obtaining repeated SCS.

The problem of a possible infrared radiation generation in a quartz crystal was already under consideration in literature[35, 36]. Now this radiation is observed. So, it is thought significant to examine thoroughly its spectral composition and to find out its power and other characteristics.

REFERENCES

1. R.W. Minek, R.W. Terhune, and W.G. Rado, Appl. Phys. Letters 3, 181 (1963).
2. I.L. Fabelinskii, "Molecular scattering of light," Plenum Press, New York, 1968.
3. R.Y. Chiao, P.L. Kelley, and E. Garmire, Phys. Rev. Letters 17, 1158 (1966).
4. V.S. Starunov, DAN USSR 178, 65 (1968).
5. R.L. Carman, R.Y. Chiao, and P.L. Kelley, Phys. Rev. Letters 17, 1281 (1966).
6. G.I. Zaitzev, Yu.I. Kyzylasov, V.S. Starunov, and I.L. Fabelinskii, JETF, Pis'ma, 6, 695 (1967).
7. I.M. Arefev, I.L. Fabelinskii, Yu. I. Kyzylasov, and V.S. Starunov, Phys. Letters.
8. Yu.I. Kyzylasov and V.S. Starunov, JETF, Pis'ma, 7, 160 (1968).
9. M.S. Pesin and I.L. Fabelinskii, DAN USSR 129, 299 (1959).
10. M.S. Pesin and I.L. Fabelinskii, DAN USSR 135, 1114 (1960).
11. P. Flubacher, A.I. Leadbetter, J.A. Morrison, and B.P. Stoicheff, J. Phys. Chem. Solids 12, 53 (1960).
12. K.F. Herzfeld and T.A. Litovitz, "Absorption and Dispersion of Ultrasonic waves," Acad. Press., New York, 1959.

13. D.H. Rank, E.M. Kiess, and U. Fink, J. Opt. Soc. Amer. 56, 163 (1966).
14. S.V. Krivokhizha and I.L. Fabelinskii, JETF 50, 3 (1966).
15. M.A. Isakovich and I.A. Chaban, DAN USSR 65 (2), (1965), JETF, 23, 893 (1966).
16. N.R. Goldblatt and T.A. Litovitz, J. Acoust. Soc. Am. 41, 1301 (1967).
17. G.A. Askarian, JETF, Pis'ma 3, 166 (1966).
18. N.G. Basov, V.N. Morozov, and A.N. Oraevskii, JETF 49, 895 (1965).
19. I.L. Fabelinskii, "Fine structure of the line of Rayleigh scattering in gases." Experimental and theoretical investigations on physics (in memory of G.S. Landsberg). Acad. Press, USSR, Moscow, 1959.
20. S. Yip and M. Nelkin, Phys. Rev. 135A, 1241 (1964).
21. V.S. Starunov, Phys. Letters 26A, 428 (1968).
22. L.D. Landau and E. M. Lifshitz, "Electrodynamics of continuous media." Addison Wesley Reading, Mass., 1960.
23. D.H. Rank, C.W. Cho, N.D. Foltz, and T.A. Wiggins, Phys. Rev. Letters 19, 828 (1967).
24. L.A. Ostrovskii, "Report at the 2d Symposium on non-linear optics," preprint, Novosibirsk, 1966.
25. L.A. Ostrovskii, JETF, Pis'ma 6, 807 (1967).
26. L.A. Ostrovskii, "Report at the 3d Symposium on non-linear optics," preprint, Erevan, 1967.
27. R.J. Jocnek and R. Landauer, Phys. Letters 24A, 228 (1967).
28. R.G. Brewer, Phys. Rev. Letters 19, 8 (1967).
29. R.G. Brewer and C.H. Townes, Phys. Rev. Letters 18, 196 (1967).
30. Y. Ueda and K. Shimoda, Japan J. Appl. Phys. 6, 628 (1967).
31. F. Shimizu, Phys. Rev. Letters 19, 1097 (1967).
32. S.V. Krivokhizha, D.I. Mash, V.V. Morozov, V.S. Starunov, and I.L. Fabelinskii, JETF, Pis'ma 3, 378 (1966).
33. I.L. Fabelinskii and V.S. Starunov, Appl. Optics 6, 1793 (1967).
34. P.E. Tannenwald, J. Appl. Phys. 12, 4788 (1967).
35. P. Tannenwald, "Report at the 3d Symposium on non-linear optics," preprint, Erevan, 1967.
36. F. De Martin, J. Appl. Phys. 37, 4507 (1967).

G-2: EXAMPLES OF CRYSTAL BRILLOUIN SCATTERING POLARIZATION SELECTION RULES

Robert W. Gammon
Catholic University of America
Washington, D. C.

Laser light sources and sensitive photoelectric detection allow the present day experimenter to obtain crystal Brillouin scattering spectra for well defined scattering vector directions and completely polarized incident and scattered radiation. Thus we now have a very powerful tool for exploring the elastic and photoelastic anisotropy in crystals from a combination of measurements of the frequency shifts of the components and the intensities of the components in polarized spectra.

To exhibit these possibilities we have returned to the complete phenomenal theory of the intensity of Brillouin scattering of Born and Huang and extracted from it a simple matrix method for computing the scattering tensor for any given strain wave. This scattering tensor shows the polarization selection rules which will hold for observing the given strain wave. We will show that even in low symmetry crystals it is possible to find orientations that give selection rules for polarized scattering which allow the three strain waves of a direction to be completely distinguished.

From this work it is possible to state some rules-of-thumb which even though non-rigorous are often useful in understanding observed spectra. Assume that the crystal optic anisotropy is small or that we have chosen the scattering plane to lie parallel to a principle plane of the dielectric ellipsoid. Call polarization perpendicular to the scattering plane V and polarizations in the plane H. For given incident and scattered light directions in a principle plane there are four possible polarized spectra; VV, VH, HV, and HH. (Take the first letter to designate the incident light polarization and the second letter the scattered light polarization.) We expect to see: (1) The longitudinal accoustic mode appears in the VV and HH spectra and is the component with the largest frequency shift. (2) The transverse mode with its displacement in the scattering plane appears only in HH scattering and only for scattering angles other than 90°. (3) The transverse mode with its displacement out of the scattering plane will appear only in VH or HV scattering. If the modes are not purely transverse and longitudinal then they will share the polarized scattering characteristics of the modes with which they are mixed.

For isotropic solids and pure modes along symmetry directions in cubic crystals the above rules are rigorous. Lower symmetry materials in principle have more complicated scattering tensors but often the rules still work.

G-3: THERMAL BRILLOUIN SCATTERING STUDY OF THE ATTENUATION OF HYPERSOUND IN QUARTZ

Alan S. Pine
Lincoln Laboratory, * Massachusetts Institute of Technology
Lexington, Massachusetts

INTRODUCTION

The velocity and attenuation of 28 GHz longitudinal phonons along the X-axis of α-quartz have been measured in a temperature range from 100° to 600°K. High resolution Brillouin scattering techniques are employed in this regime where conventional ultrasonics methods are not applicable.

The velocity or Brillouin shift variation with temperature agrees within experimental error with previous low frequency quartz-controlled oscillator data. The hypersonic damping is obtained from the spectral width of the Brillouin lines and is compared to an extrapolation of ultrasonics data by a theory due to Kwok[1]. This rather general theory incorporates the earlier results of Woodruff and Ehrenreich[2], Bömmel and Dransfeld[3], and Maris[4] as special cases. Although not all of the parameters of the theory are adequately known, the magnitude and the trend of the data may be explained.

The study of microwave frequency phonons began about a decade ago when Baranskii[5] piezoelectrically generated and detected 1 GHz sound waves in quartz. Since then this hypersonic regime has been extended to about 100 GHz by refined pulse-echo techniques. The hypersound has been used as a probe for studying phonon interactions with electrons, spins, photons, defects and other phonons. This latter anharmonic process has received the most attention and is the subject of this paper.

Investigation of the frequency and temperature dependence of acoustic attenuation in quartz using microwave transduction was carried out in many laboratories, most notably by Bömmel and Dransfeld[3], Jacobsen[6], Maris[4], Nava et al.[7], and Thaxter and Tannenwald[8]. Practical detection sensitivity limits the pulse-echo method to attenuations of less than ~ 10 db/cm. Thus the higher frequency measurements are restricted to cryogenic temperatures where the attenuation is low.

Light scattering experiments are not subject to the same restrictions. Phonons of wavelength down to half that of the light in the medium may be probed. Here one relies on thermally excited, rather than transduced, hypersound. Of course the temperature needs to be high enough to obtain measurable scattering intensity and the spectrum needs to be broad enough to resolve with available spectrometers. This latter condition limits the minimum attenuation measurable to ~100 db/cm in quartz.

*Operated with support from the U.S. Air Force.

It is seen then that thermal Brillouin scattering complements the conventional transducer methods - but not perfectly since there is no overlap between the accessible frequency-temperature regimes of the two methods. Hybrid techniques such as transducer generation and light scattering detection of the hypersound[9] have served to bridge the 10 to 100 db/cm gap. A novel experiment in this intermediate regime utilizing stimulated Brillouin scattering is reported by Dransfeld at this conference[10].

The theory of sound absorption in solids involves the scattering of thermal phonons by the hypersonic wave. These thermal phonons are characterized by a relaxation time, τ, which is accessible from independent thermal conductivity and heat capacity measurements. If the hypersound has frequency ω, then the region around $\omega\tau = 1$ contains a breakpoint of the theory. This condition is fulfilled outside the region available to transducer experiments for frequencies above ~ 3 GHz, and it occurs near room temperature for the phonons studied here.

EXPERIMENTAL CONSIDERATIONS AND RESULTS

The techniques of high resolution, low level Brillouin spectroscopy in quartz and other solids are described by Durand and Pine[11]. A survey of related light scattering experiments and an expanded discussion of the considerations pertinent to crystals are included there. We present here a brief review of the analysis of scattering from longitudinal hypersonic waves in quartz. The differential scattered light power is

$$\frac{d^2 P_\ell(\omega_s)}{d\Omega d\omega_s} = \frac{\pi^2 kTLP_o(\epsilon-1)^2 \sin^2\psi}{2\lambda_o^4 \rho v_\ell^2(\hat{q})} \times \frac{(\Gamma_q/\pi)}{(\omega_s - \omega_o \pm \omega)^2 + \Gamma_q^2} \tag{1}$$

Here the s and o subscripts refer to the scattered and incident light; L is the length of the scattering volume, $d\Omega$ is the solid angle of collection; ω, Γ_q and $v_\ell(\hat{q})$ are the frequency, damping rate, and velocity of the hypersonic phonon of wavevector, $\underset{\sim}{q}$. ψ is the angle between $\underset{\sim}{E}_o$ and the scattering direction. The intensity is proportional to temperature since $\hbar\omega \ll kT$ and, compared to liquids, scattering from solids is very weak chiefly because of the velocity factor in the denominator.

Correspondingly the Brillouin shift is given by

$$\omega = \omega_o\,(2n_o\, v_\ell(\hat{q})/c)\,\sin(\theta/2) \tag{2}$$

where θ is the scattering angle. Since a range of angles $\delta\theta$ is collected with a finite $d\Omega$, the shift undergoes an angular spread, $\delta\omega_\theta$, due to the explicit $\sin(\theta/2)$ term and the implicit angular dependence of the anisotropic velocity. This contributes a linewidth unrelated to the spectral distribution in (1) which to second order in $\delta\theta$ is

$$\frac{\delta\omega_\theta}{\omega} = \left(\frac{1}{2}\cot\frac{\theta}{2} + \frac{A}{2}\right)\delta\theta - \left(\frac{1}{8} - \frac{B}{4}\right)\delta\theta^2 \tag{3}$$

The coefficients A and B simply represent the expansion of the velocity with respect to angle. It is essential to eliminate the linear term in $\delta\theta$ to obtain the resolution required for quartz. This is accomplished by backscattering from pure longitudinal phonons at the extrema of the high velocity surface; for then $\theta = \pi$ and $A = 0$. In fact the X-axis waves studied here are located at a velocity surface minimum; hence B is positive and the second order coefficients partially compensate. A residual angular broadening of ~ 6 MHz occurs here for the solid angle needed to obtain reasonable intensity.

The high resolution Brillouin spectrometer is similar to that described in DP[11]. A single mode He-Ne laser is the source; a piezoscanned confocal spherical Fabry-Perot is the interferometer; a low dark count photomultiplier feeding a synchronously swept multichannel analyser registers the spectrum. Typical Brillouin spectra of quartz at various temperatures are shown in logarithmic and linear display in Fig. 1. Note that the intense Rayleigh peak triggers the analyser so that slow drift of the spectrometer is automatically compensated in the accumulated spectrum. Such stabilization is necessary because of the long integration times required to enhance the signal-to-noise ratio of the extremely weak scattering from solids.

The natural quartz crystal was obtained from Valpey in the form of a Brewster-angle ended cylinder, 10 cm long by 2.5 cm diameter. Temperature is measured with a thermocouple in mechanical contact with the quartz circumference less than 1 cm from the scattering volume. Thermal environment is provided by an evacuated dewar or oven with a single window to minimize spurious scattering.

The experimental Brillouin shift and linewidth (full width at half maximum) are plotted as a function of temperature in Fig. 2. The predicted thermal variation of the velocity of sound is taken from Atanasoff and Hart's[12] measurements of the frequency of a quartz-controlled oscillator placed in an oven. Then with known expansion coefficients and tabulated refractive index data, the temperature dependence of the Brillouin shift is determined. The agreement of the data with these earlier low frequency measurements is excellent below 600°K. The absolute accuracy of the Brillouin shift is only 0.2% because of calibration errors in the interferometer spacing. However the relative precision of the measurements is 0.02% which compares favorably with the oscillator data.

Above 600°K the data are erratic and deviate from the predicted values. This is due to large thermal gradients induced by radiative transfer to the oven windows and poor material thermal conductivity. Here not only are the temperature measurements in error, but the light scattering volume encompasses severe enough gradients to artificially broaden the spectrum due to the large velocity shift with temperature. For these reasons measurements above 600°K are too unreliable to present here. Below 100°K the resolution of the interferometer becomes inadequate and the scattering intensity decreases beyond an acceptable level with the available laser power of 150 μwatts.

The linewidth data are plotted in Fig. 2B and are compared to several theories which are discussed in the next section. These theories are all normalized to fit the 1 GHz, 120°K data of Bömmel and Dransfeld[3] by adjustment of the "Gruneisen" constant. Linewidths are obtained by convolving Lorentzian curves with the instrumental function and fitting these to the experimental spectra. An averaged angular broadening, $\delta\omega_\theta$, is then subtracted off. This rough deconvolution is believed to result in errors smaller than the data scatter.

THEORETICAL DISCUSSION

We now review the theory of sound absorption in solids. Shortly after the first experiments at microwave frequency, Bömmel and Dransfeld[3] and Woodruff and Ehrenreich[2] developed theories based on Akhieser's[13] mechanism of viscous damping. In this classical description the sound wave modulates the frequencies of the thermal phonons by anharmonicity. The modulated phonons are no longer in thermal equilibrium but relax towards it by collision processes. Such relaxation increases the entropy of the medium which extracts energy from, and damps, the driving sound wave.

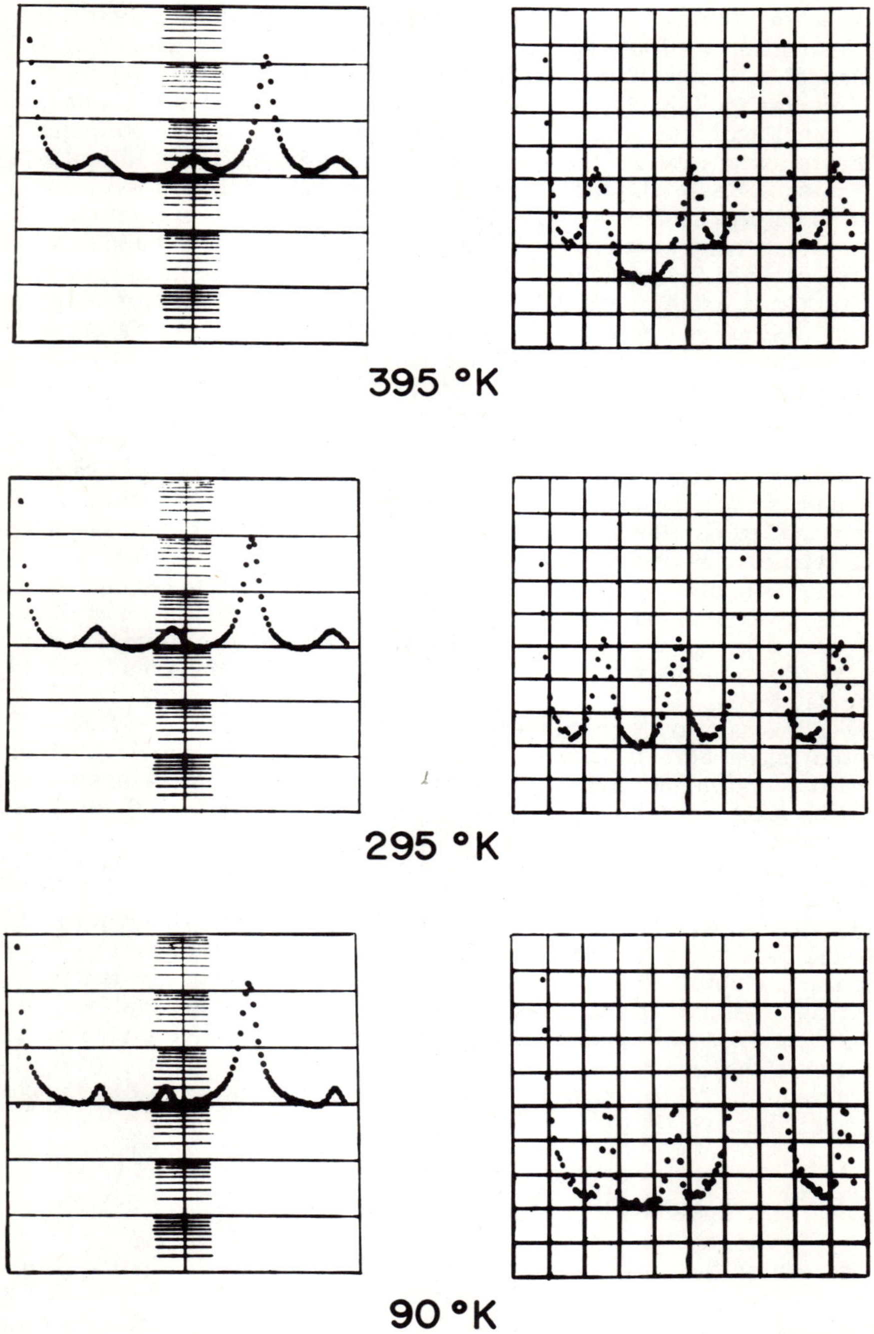

Fig. 1. Logarithmic and linear display of Brillouin spectra of α-quartz at various temperatures. Linear vertical scale 500 counts/div. Interorder spacing 751 MHz.

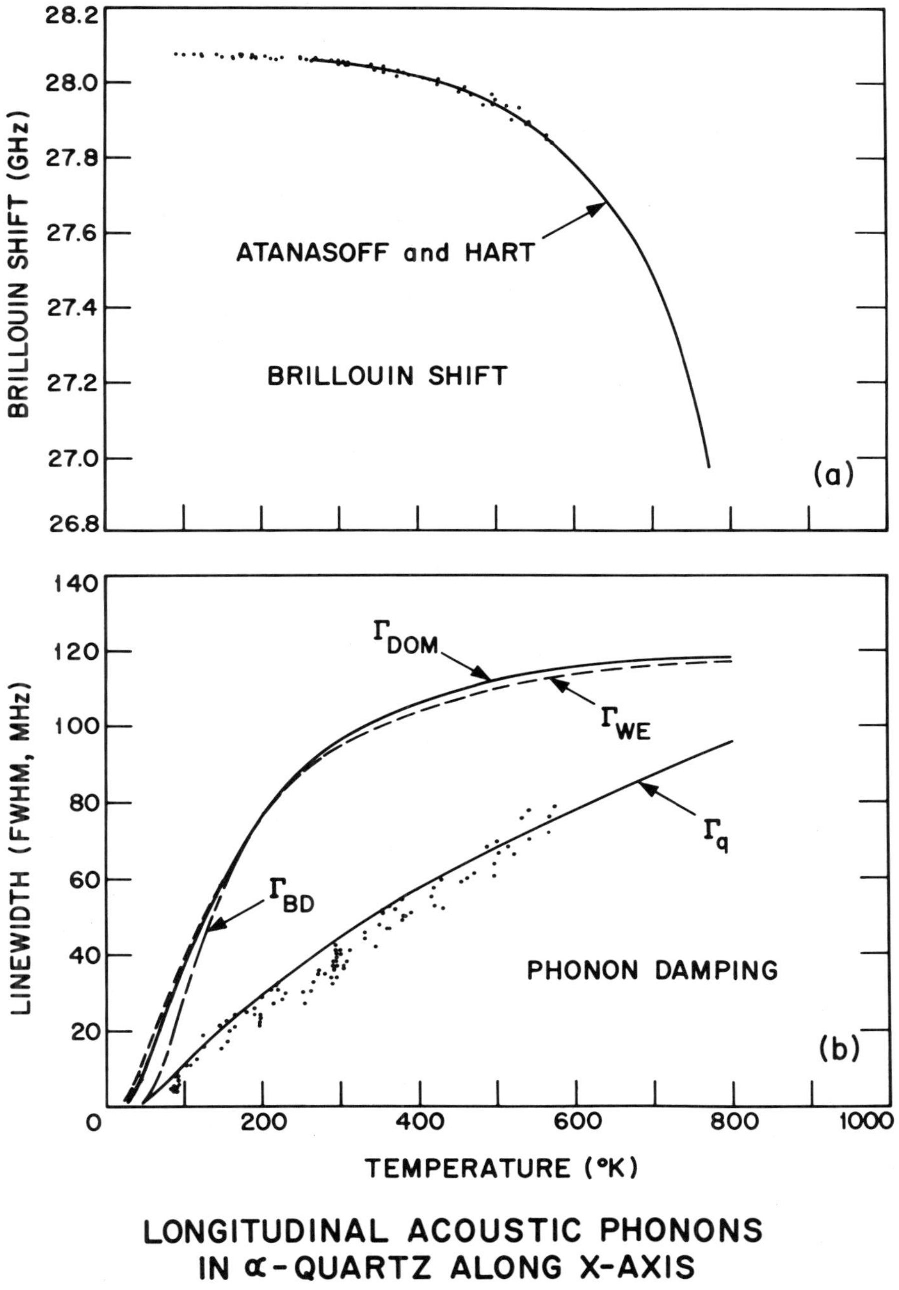

Fig. 2. Brillouin shift (a) and linewidth (b) for X-axis hypersonic waves in α-quartz as a function of temperature. The theoretical curves are discussed in the text.

Nava et al. [7] shed more light on the latter process by demonstrating that the thermal phonons are simultaneously amplitude and frequency modulated. The energy in the upper sideband may then exceed that of the lower, representing a net absorption of the driving wave. These derivations are rigorously correct only in the $\omega\tau < 1$ regime although more widespread validity has been noted. Bömmel and Dransfeld neglected the velocity of the thermal phonons whereas Woodruff and Ehrenreich assumed a dispersionless velocity.

Maris[4] then presented a theory which incorporated phonon dispersion into the $\omega\tau \gg 1$ perturbation limit first studied by Landau and Rumer[14]. He applied his result to his measurements on quartz which were not all taken with $\omega\tau \gg 1$ satisfied. A first order quantum mechanical perturbation theory, valid for all $\omega\tau$ in a dispersionless medium, was derived by Simons[15]. We consider here a formulation due to Kwok[1] which includes the effect of dispersion and the full range of $\omega\tau$, since these conditions are appropriate to the light scattering experiment. All the aforementioned theories are easily derived from this theory as special cases; and, in some instances, the limitations on the derivations are not as restrictive as previously supposed.

Kwok employs the Green's function techniques of many-body theory to obtain a general expression for phonon damping. Similar results had been obtained by others studying various aspects of the anharmonic interaction[16, 17], but they were not evaluated for the specific problem of hypersonic attenuation. The principal advantages of the many-body techniques over standard perturbation theories are that the thermal occupation factors and the relaxation broadening or "dressing" of the thermal phonons are treated more naturally.

Normally, only two processes contribute to phonon damping when cubic terms in the anharmonicity are considered. The first is a decay into lower frequency phonons and the second is a scattering of a higher frequency phonon. Conservation of energy and wavevector allows a longitudinal phonon to decay into two transverse phonons, or one transverse and one longitudinal, or, with little dispersion, into two collinear longitudinal phonons. The damping rate from these processes is on the order of $(\hbar q^5/\rho)[(kT/\hbar v_\ell q) + 1/2]\ \sec^{-1}$. For microwave phonons this corresponds to Γ_q from 10^{-2} to $10\ \sec^{-1}$, so the decay process is completely negligible. However for high energy phonons, the rapid q dependence makes this an important mechanism.

The dominant contribution to the hypersonic attenuation arises from the scattering of thermal phonons. The result given by Kwok for this mechanism may be cast into the form

$$\Gamma_q = \frac{\hbar\gamma^2\omega}{16\rho v_\ell^2}\frac{2}{(2\pi)^2}\int_0^Q dq_1\, q_1^2\, \omega_1^2 \left(-\frac{\partial n}{\partial\omega}\Big|_{\omega_1}\right) \times \left\{\frac{1}{\beta_1}\tan^{-1}\left[\frac{\beta_1\omega\tau_1}{1+\left(\frac{\omega\tau_1}{2}\right)^2(1-\beta_1^2)}\right]\right\} \tag{4}$$

Here Q is the Brillouin zone edge wavevector, n is the boson factor $[\exp(\hbar\omega/kT)-1]^{-1}$, ω_1 and τ_1 are the frequency and life-time of the thermal phonon of wavevector q_1, β_1 is the ratio $(\partial\omega_1/\partial q_1)/v_\ell$. γ^2 is an averaged dimensionless "Gruneisen" constant of order unity which measures the anharmonicity. The approximations of elastic isotropy and $kT \gg \hbar\omega$ have been used. The latter condition is well satisfied in this experiment; the former is not.

Elastic anisotropy, higher order phonon interactions, and defect scattering, although relevant, are ignored because detailed calculations do not exist. For $\omega\tau >> 1$ Herring[18] has considered the effect of anisotropic velocity surfaces and Shiren[19] has calculated an effect of anharmonic anisotropy. In quartz both the velocity and attenuation are sensitive functions of crystal direction, but corrections to Γ_q are difficult to estimate in the regime of this study.

Even Eq. (4) cannot be evaluated exactly since ω_1 and τ_1 are inadequately known functions of wavevector and temperature. The "dominant phonon" approximation is usually made if τ_1 does not vary too rapidly with q_1. Then β_1 and τ_1 are evaluated at $\omega_1 \sim kT/\hbar$, near where the rest of the integrand has a peak and the curly bracketed term may be removed from the integral. Noting that $-(\partial n/\partial\omega) = (T/\omega)(\partial n/\partial T)$ and recalling the definition of the specific heat $C_v = (2/(2\pi)^2) \int_0^Q dq_1 q_1{}^2 \hbar\,\omega_1 (\partial n/\partial T)_{\omega_1}$; we have

$$\Gamma_{DOM} = \frac{\gamma^2 \omega T C_v}{16\rho v_\ell{}^2} \left\{ \frac{1}{\beta_T} \tan^{-1} \left[\frac{\beta_T\,\omega\tau_T}{1+\left(\frac{\omega\tau_T}{2}\right)^2 (1-\beta_T{}^2)} \right] \right\} \tag{5}$$

In the dispersionless case, $\beta_T = 1$, and (5) reduces essentially to the result of Woodruff and Ehrenreich,

$$\Gamma_{WE} = \frac{\gamma^2 \omega T C_v}{16\,\rho\,v_\ell{}^2} \tan^{-1} \omega\tau_T \tag{6}$$

Similarly the theory of Bömmel and Dransfeld is reproduced as $\beta_T \to 0$

$$\Gamma_{BD} = \frac{\gamma^2 \omega T C_v}{16\,\rho\,v_\ell{}^2} \; \frac{\omega\tau_T}{1 + (\omega\tau_T/2)^2} \tag{7}$$

Neither (6) nor (7) is restricted to $\omega\tau < 1$. Γ_{DOM} is equivalent to the result of Simons if β is interpreted as the ratio of the velocities of the scattered thermal phonon branch to the hypersonic branch. A trigonometric identity applied to Γ_{DOM} yields Maris' expression in the limit $\omega\tau >> 1$,

$$\Gamma_M = \frac{\gamma^2\,\omega T\,C_v}{16\,\rho\,v_\ell{}^2\,\beta_T} \left\{ \frac{\pi}{2} - \tan^{-1} \frac{\omega\tau_T}{2} (1-\beta_T) \right\} \tag{8}$$

The ramifications of the various expressions (5) through (8) are thoroughly discussed by the respective authors. It should be noted that the factor τ_T above is 2τ in the other derivations. This difference arises from the definition of the lifetime. Here we use the phonon amplitude relaxation time whereas an energy or occupation number relaxation was used previously[20]. τ_T is deduced from thermal conductivity and heat capacity measurements. Woodruff and Ehrenreich give τ for quartz below 100°K; above 100°K we extrapolate according to T^{-1} in order to compute the theoretical curves in Fig. 2(b).

The major differences among the dominant phonon theories arise at very low temperatures where pulse-echo techniques are exploited. The light scattering experiment is rather insensitive to these differences, but as seen from Fig. 2(b), the data are poorly explained by any of these theories. A better fit to the data is provided by computing Γ_q using an assumed model for the frequency and lifetime of the thermal phonons. This avoids the dominant phonon approximation. For the calculation graphed, we have used a simple dispersion model $\omega_1 = (2v_\ell Q/\pi)\sin(\pi q_1/2Q)$ and the lifetime as given from the damping rate due to decay processes as explained above. The lattice constant $a_o = \pi/Q$ is taken as 5 Å for quartz. The agreement of this model with the data is remarkable considering the total neglect of umklapp processes which are responsible for τ_T. Also it should be mentioned that the derivation of the decay rate is strictly valid only in the extremes $kT >> \hbar v_\ell q$ and $kT << \hbar v_\ell q$. The success of this model should be regarded only as indicative of the failings of the previous calculations. Although the Kwok expression, (4), demonstrates the relationship between many of the previous theories, Mills[20] cautions against its application for all $\omega\tau$ since important "vertex" corrections have been ignored.

ACKNOWLEDGMENTS

The author is greatly indebted to Dr. Georges Durand for the basic experimental design and many useful discussions in the early stages of the study. Also the background and critical judgment of Dr. Peter Tannenwald was drawn upon appreciatively during this work.

REFERENCES

1. P.C. Kwok, "PhD Thesis," Physics Department, Harvard University, 1965.
2. T.O. Woodruff and H. Ehrenreich, Phys. Rev. 123, 1553 (1961).
3. H.E. Bömmel and K. Dransfeld, Phys. Rev. 117, 1245 (1960).
4. H.J. Maris, Phil Mag. 9, 901 (1964).
5. K.N. Baranskii, Soviet Phys.-Doklady 2, 237 (1958).
6. E.H. Jacobsen, "Quantum Electronics," C.H. Townes (ed.), p. 468, Columbia University Press, New York, 1960.
7. R. Nava, R. Azrt, I. Ciccarello, and K. Dransfeld, paper G-4 this conference.
8. J.B. Thaxter and P.E. Tannenwald, Appl. Phys. Letters 5, 67 (1964).
9. C.D.W. Wilkinson and D.E. Caddes, J. Acoust. Soc. Amer. 40, 498 (1966).
10. G. Winterling, W. Heinicke, and K. Dransfeld, paper G-4 this conference.
11. G. Durand and A.S. Pine, IEEE J. Quant. Electronics 4, (1968).
12. J.V. Atanasoff and P.J. Hart, Phys. Rev. 59, 85 (1941).
13. A. Akhieser, J. Phys. I, 277 (1939).
14. L. Landau and G. Rumer, Physik Z. Sowjetunion 11, 18 (1937).
15. S. Simons, Proc. Phys. Soc. 83, 749 (1964).
16. A.A. Maradudin and A.E. Fein, Phys. Rev. 128, 2589 (1962).
17. R.A. Cowley, "Phonons in Perfect Lattices," R.W.H. Stevenson (ed.), p. 170, Plenum Press. New York, 1966.
18. C. Herring, Phys. Rev. 95, 954 (1954).
19. N.S. Shiren, Phys. Letters 20, 10 (1966).
20. D.L. Mills (private communication).

G-4: OPTICAL DETERMINATION OF THE ULTRASONIC ABSORPTION IN QUARTZ AT 29 GHz

G. Winterling, W. Heinicke and K. Dransfeld
Physik-Dept. der Technischen Hochschule
München, Germany

We have measured the ultrasonic absorption in crystalline quartz at a frequency of 29 GHz by a new optical method using two consecutive light pulses [1]. This technique enabled us to extend accurate absorption measurements to 200°K, where no other ultrasonic data are available yet.

Ultrasonic waves were excited in a z-cut quartz rod by Stimulated Brillouin Scattering using a standard giant pulse from a ruby laser having a pulse time of about 20 nanoseconds. A small fraction of the incident light beam was split off and - after travelling a distance of 15 m and a corresponding time delay of 50 nanoseconds - it was directed into the target crystal, traversing it in the same direction as the earlier first pulse.

While the first pulse was used to set up ultrasonic waves at the frequency of 29 GHz, the second time delayed pulse served as a probe to measure the ultrasonic intensity "left over" after the delay time: When the second pulse arrives it will still be partially reflected from the ultrasonic wave depending on the ultrasonic intensity at the time the second pulse arrives. By observing the back-scattered intensity of the time delayed pulse one can directly measure the lifetime of ultrasonic phonons in absolute units. In particular, it is possible to measure also very short lifetimes, i. e. as short as the duration of the primary light pulse = 20 nanoseconds. The corresponding attenuation coefficent is as high as 100 cm^{-1}.

As the first application of this method Fig. 1 shows abosrption measurements in crystalline quartz at 29 GHz as a function of temperature extending up to almost 200°K. The most interesting feature is the absence of a temperature independent plateau and the absence of a quadratic frequency dependence, both of which are so well established at lower frequencies. This behaviour is discussed theoretically.

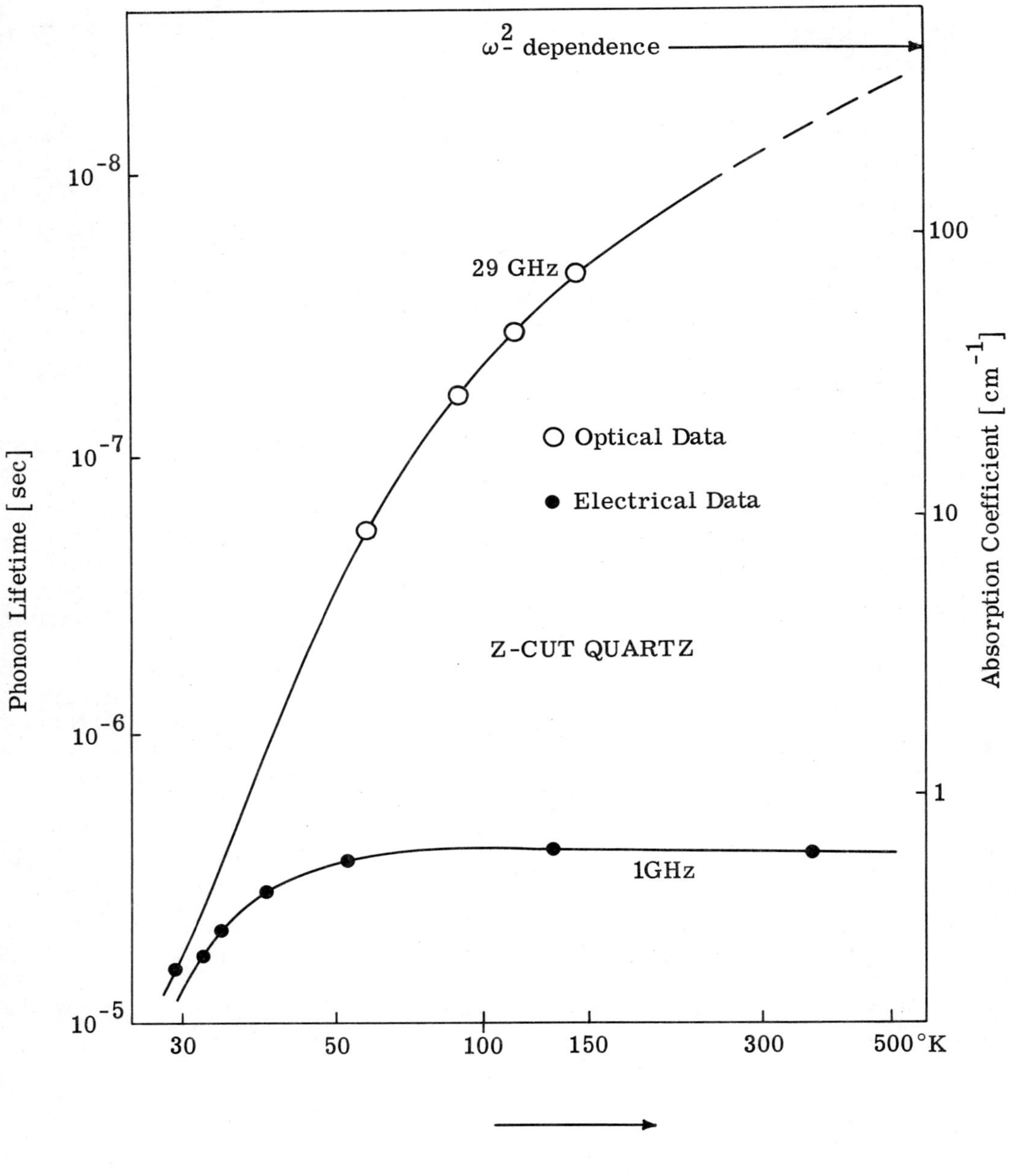

Fig. 1. Absorption measurements in crystalline quartz vs. temperature. The triangles show our optical measurements at 29 GHz and the dotted-dashed line an ω^2-extra polation for 29 GHz from previous 1 GHz data included for comparison.

REFERENCES

1. G. Winterling and W. Heinicke, Phys. Letters 27A, 329 (1968).

G-5: THE RAYLEIGH-BRILLOUIN SPECTRA OF AMMONIUM CHLORIDE*

Paul D. Lazay, Joseph H. Lunacek, Noel A. Clark and George B. Benedek
Physics Department, Center for Materials Science and Engineering,
Massachusetts Institute of Technology
Cambridge, Massachusetts

INTRODUCTION

We have measured the spectrum of light scattered in single crystals of NH_4Cl. This crystal is an interesting system to study as it undergoes a second order lambda transition at T_λ = -30.55 °C. This transition is of the order-disorder type involving the ordering of the NH_4-tetrahedra in the cubic unit cell. We have made measurements on the spectrum of light scattered by thermally excited hypersonic sound waves (the Brillouin spectrum) and on the spectrum of light scattered by non-propagating critical fluctuations in the optical dielectric constant (the Rayleigh spectrum). Specifically, we have used a high power (80 mw) multi-mode He-Ne laser and a 12 m grating spectrograph to measure two purely longitudinal mode velocities and one purely transverse mode velocity as a function of temperature in the interval -50 to +50 °C. We also measured the Brillouin intensities of these modes. We have measured the intensity of the elastically scattered light in the same interval. In addition, we have used a high power (15 mw) stabilized single frequency He-Ne laser, a pressure scanned flat Fabry-Perot interferometer with an instrumental width of 250 MHz, and a light detection system having a dark current of .2 counts per second to measure the natural width of the <110> longitudinal mode Brillouin component in the frequency range 8 to 24 GHz at 24 °C.

THE BRILLOUIN SPECTRUM

We have used the Brillouin light scattering technique to determine the hypersonic sound velocities in NH_4Cl. The sound velocity, V, is related to the observed Brillouin splitting, Δf, by the formula of Brillouin

*This work supported by the Advanced Research Projects Agency (Project Defender) and monitored by the U. S. Army Research Office-Durham under contract DA-31-124-ARO-D-425.

$$\Delta f/f_0 = 2n(V/c)\sin(\Theta/2) \tag{1}$$

where f_0 is the incident light frequency, n the refractive index of the scattering medium, c the velocity of light in vacuo, and Θ is the scattering angle. We have measured the temperature dependence of the velocities of three modes: (a) a purely longitudinal mode in the <100> direction with velocity $V_L(100) = \sqrt{c_{11}/\rho}$, (b) a purely longitudinal mode in the <110> direction with velocity $V_L(110) = \sqrt{(c_{11} + c_{12} + 2c_{44})/2\rho}$, and (c) a purely transverse mode having velocity $V_T = \sqrt{c_{44}/\rho}$ in both the <100> and <110> directions.

In Fig. 1 we give our data on the transverse velocity together with the ultrasonic results of Garland and Renard[1]. Our velocities correspond to a sound frequency of about 9 GHz while the ultrasonic velocities were measured at 10 MHz. Our velocities consistently fall below the ultrasonic values, but the difference is small (.5%) and we may conclude that the transverse velocity shows little or no dispersion. The increase in the velocity in the ordered phase below T_λ can be explained as arising from the rapid reduction in the unit cell volume[2]. The Brillouin intensity I_T of this mode ($I_T \propto p_{44}^2/V_T^2$) decreases almost linearly with decreasing temperature. An unexpected feature of these intensity measurements is that our room temperature photoelastic

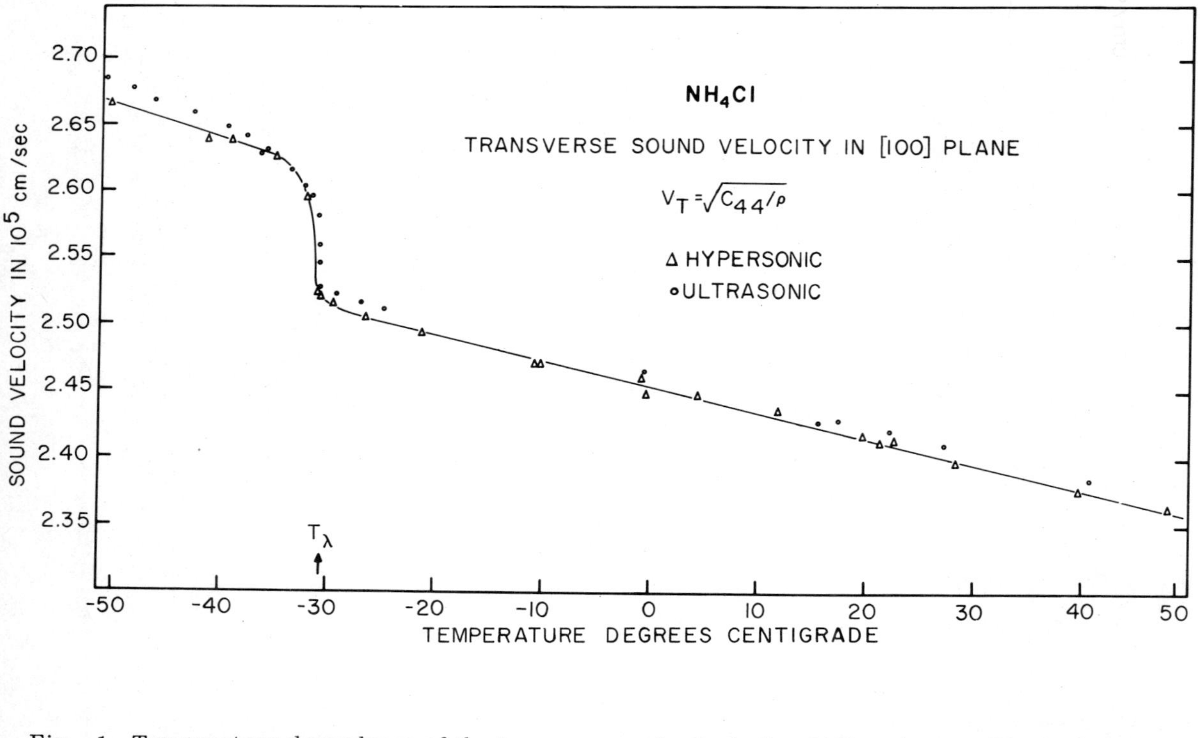

Fig. 1. Temperature dependence of the transverse velocity in the <100> planes. This velocity is independent of propagation direction in the <100> planes.

constant p_{44} is 2.75 times larger than the statically measured value[3]. The p_{ij} are elements of the photoelastic tensor as defined in Nye's book "Physical Properties of Crystals."

In Fig. 2 we give our data on the longitudinal velocity $V_L(110)$ and also ultrasonic measurements[1]. Our velocities correspond to a frequency of about 18 GHz and the ultrasonic velocities to a frequency of 10 MHz. Several features of these data are immediately obvious. Even at room temperature, where we are out of the critical

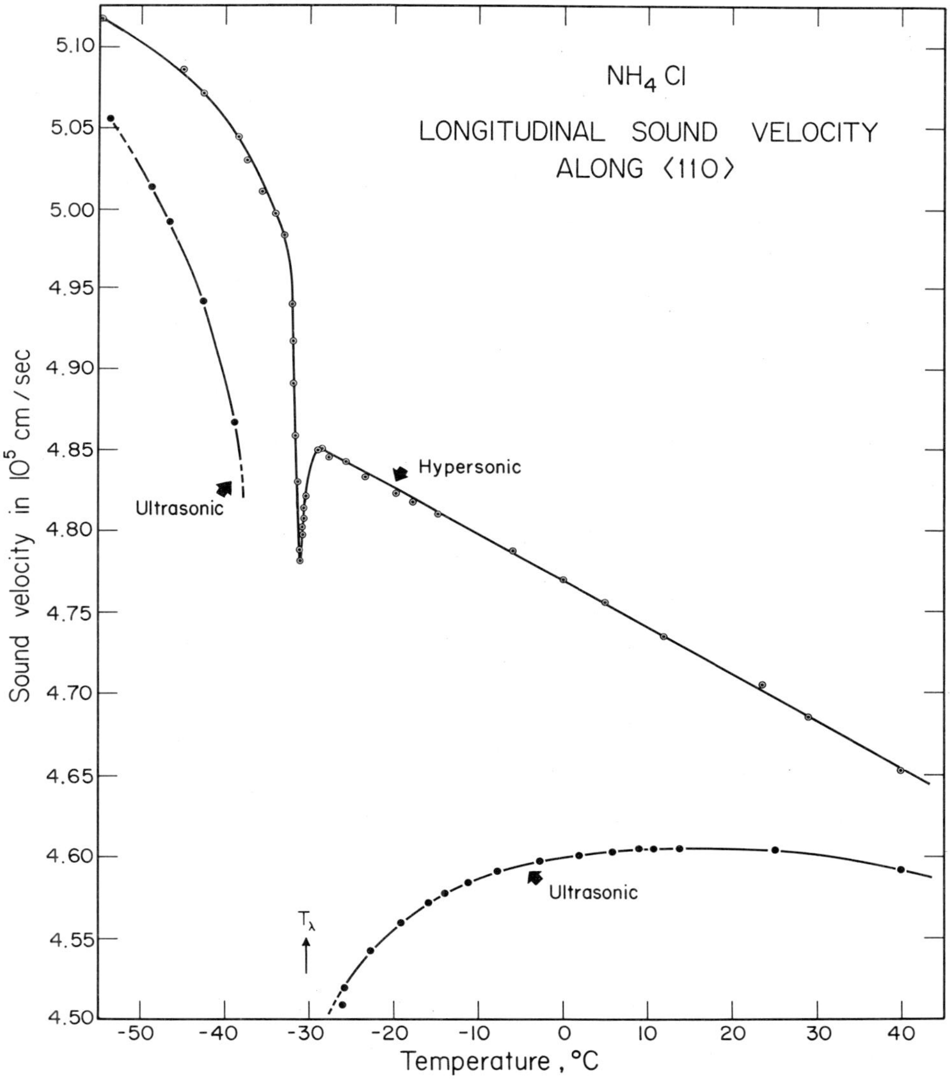

Fig. 2. Temperature dependence of the longitudinal sound velocity in the < 110 > direction.

region ($\epsilon = (T - T_\lambda)/T_\lambda = .22$), there is a sizeable dispersion in the longitudinal sound velocities. This dispersion is a temperature dependent quantity. It is also clear that at a frequency of 18 GHz a non-critical or linear temperature dependence of the velocity with temperature persists much closer to the critical point than at 10 MHz. From the linear region above T_λ we can determine the thermoelastic coefficients $T_{ij} = \frac{d(\ln C_{ij})}{dT}$ For CsCl type crystals there is an empirical law that $T_{44} > T_{11}$ (the opposite is true for NaCl type crystals)[4]. NH_4Cl (a CsCl type crystal) satisfies this law both for the hypersonic and the ultrasonic elastic constants. It is interesting to note that our T_{11} is roughly twice the ultrasonic value, and roughly twice the value for most CsCl crystals[5]. The intensity of the <110> longitudinal mode ($I_L(110) = p_{12}^2/V_L(110)^2$) has a maximum at 35°C, decreases as $T \to T_\lambda$, has a minimum at T_λ, and increases smoothly below T_λ until at -50°C the intensity is 1/2 its 35°C value.

Our results on the elastic and photoelastic constants at 300°K are summarized in Table I. The photoelastic constants were measured by making a direct comparison of the <110> longitudinal mode intensity with the Brillouin intensity of toluene. We use the effective photoelastic constant of toluene given by Fabelinskii[6]. At room temperature $I_L(110)$ is 22.4 times smaller than the Brillouin intensity in toluene, and the transverse mode intensity is 24.2 times smaller than $I_L(110)$.

TABLE I

The Elastic and Photoelastic Constants of NH_4Cl at 300°K

Elastic constants in 10^{10} dynes/ cm^2			
	Ultrasonic[1]	Hypersonic	% Change
c_{11}	38.15	39.18	+ 2.69
c_{12}	8.754	10.23	+ 16.8
c_{44}	8.875	8.791	- .98
Photoelastic constants			
	Static[3]	Hypersonic	% Change
p_{11}	.1449	.121	- 16.3
p_{12}	.2397	.244	+ 1.02
p_{44}	.0247	.068	+ 275
p_{11}	Toluene [4]	1.60	

In addition to these velocity measurements as a function of temperature, we have also measured the natural width of the <110> longitudinal mode Brillouin component as a function of scattering angle at a fixed temperature of 24°C. Fig. 3 shows a typical trace of the spectrum. This spectrum consists of a large central component, a weak transverse mode, and the longitudinal mode of interest. The Brillouin components have a natural width because the phonons responsible for the scattering have a finite lifetime.

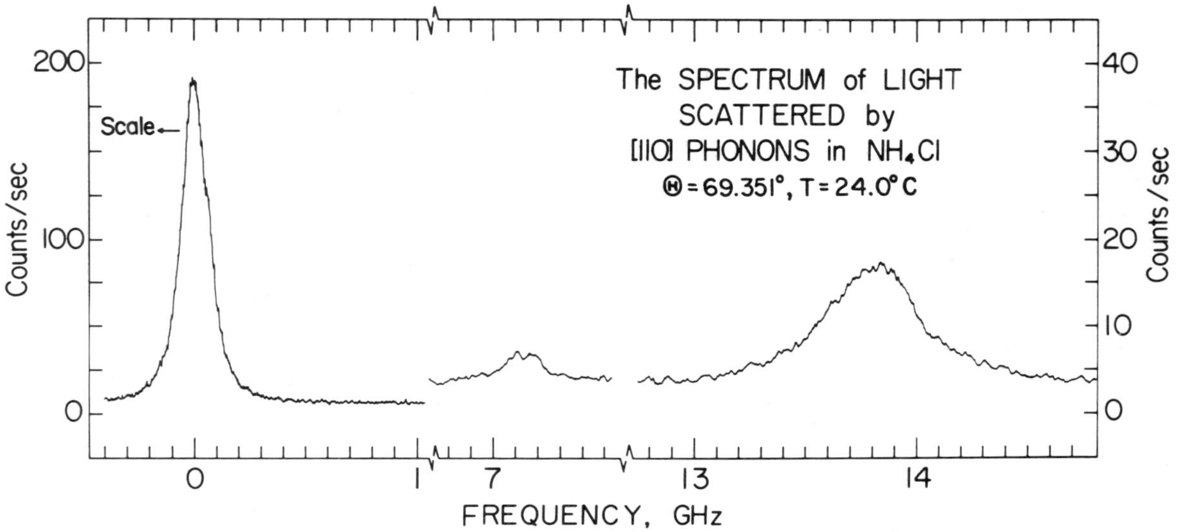

Fig. 3. Typical trace of the spectrum used in the Brillouin linewidth measurements. Scale on the left is for the central component (located on the left), scale on the right for the Brillouin components.

If the phonon amplitude decays exponentially in time with a decay rate Γ, then the spectral line has a Lorentzian shape with a half-width at half height $\Gamma/2\pi$ Hz. We may relate this width to the spatial decay constant or attenuation α, which is the quantity measured in ultrasonic experiments, via the relation

$$\Gamma/2\pi(\mathrm{Hz}) = \alpha\,(\mathrm{cm}^{-1})\ V\ (\mathrm{cm/sec})/2\pi \tag{2}$$

In Table II we present our results on the linewidths $\Gamma/2\pi$ as a function of scattering angle Ⓗ . Γ is the half width at half height of the Lorentzian line.

TABLE II

Brillouin Linewidths of the <110> Longitudinal Phonons in NH_4Cl at 24°C

Ⓗ	Δf(GHz)	V(10^5 cm/sec)	$\Gamma/2\pi$(MHz)	α (cm^{-1})
36.3°	7.641	-----	110 ± 10	1440 ± 10%
69.35	13.776	4.682	175 ± 10	2400
90.36	17.198	4.688	220 ± 10	2900
138.20	22.762	4.712	270 ± 15	3550

This data indicates that α varies much more slowly than f^2. It should be emphasized that the quoted values of $\Gamma/2\pi$ have been corrected for artificial broadening due to the finite acceptance angle of the spectrometer. This correction never amounted to more than 12% of the natural linewidth. As is conventional in the theory of ultrasonic relaxation, we may interpret these data in terms of a relaxation process describing the exchange of energy between the phonons and some as yet unknown internal degree of freedom. We assume that the rate at which this energy exchange takes place can be described by a single temperature dependent relaxation time $\tau(T)$. This same model has been used to interpret ultrasonic attenuation measurements near T_λ [7, 8]. According to such single relaxation time theories the sound velocity and attenuation are given by the equation:

$$(V_o/V)^2 = 1 - \left[1 - (V_o{}^2/V_\infty{}^2)\right] \left(\omega^2\tau^2/(1+\omega^2\tau^2)\right) \tag{3}$$

$$\frac{\Gamma}{\Delta f} = \alpha\lambda = \pi \left(\frac{V}{V_o}\right)^2 \left[1 - \left(\frac{V_o}{V_\infty}\right)^2\right] \frac{\omega\tau}{1+\omega^2\tau^2} \tag{4}$$

where V_o and V_∞ are the zero and in finite frequency sound velocities, $\Gamma/2\pi$ is the Brillouin line width of Eq. (2), Δf is the Brillouin splitting of equation (1), and $\alpha\lambda$ is the attenuation per wavelength. From these two equations we may derive the expression

$$\omega\tau = \frac{\left(\frac{V}{V_o}\right)^2 - 1}{2\Gamma/\omega} \tag{5}$$

where $\omega = 2\pi\,\Delta f$.

A plot of the right hand side of this equation versus ω(which is the same as $2\pi\Delta f$) should yield a straight line of slope τ. Fig. 4 shows such a plot of our data. We have used the ultrasonic (10 MHz) value of 4.605×10^5 cm/sec for V_o. The value of τ is $\tau(24°C) = 1.4 \times 10^{-11}$ sec, corresponding to a relaxation frequency $\nu_R \equiv \frac{1}{2\pi\tau} = 11.6$ GHz. We can also determine from Eq. (4) the maximum attenuation per wavelength (which occurs for $\omega\tau = 1$)

$$\alpha\lambda\,\Big|_{max} = \pi \,\frac{1 - \left(\frac{V_o}{V_\infty}\right)^2}{1 + \left(\frac{V_o}{V_\infty}\right)^2} = .085 \pm .004 \tag{6}$$

and a total dispersion of 2.7% at 24°C.

The ultrasonic velocity and attenuation measurements near T_λ [1, 8] can be combined with our velocity results near T_λ (which are in effect V_∞ since $\tau \to \infty$ as $T \to T_\lambda$ and therefore $\omega\tau >> 1$) to yield a $\tau_{crit}(T)$. This τ_{crit} is presumably the relaxation time of the critical fluctuations in the long range order which are responsible for the anomalously

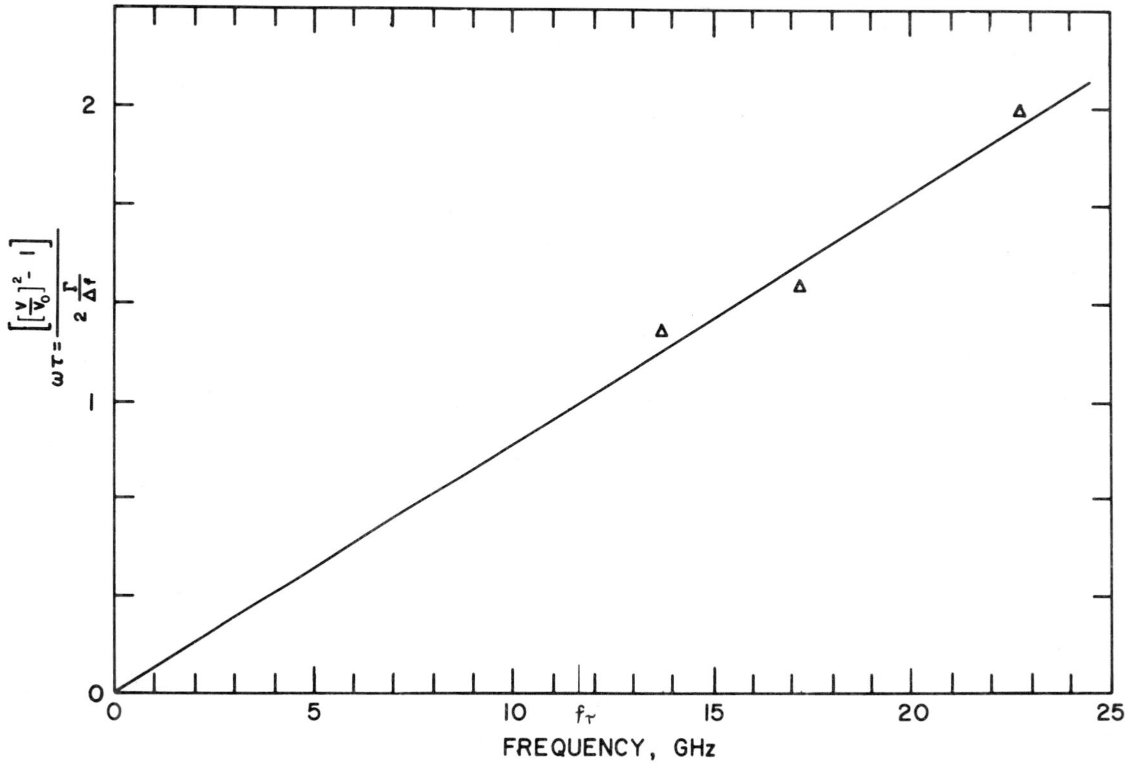

Fig. 4. Analysis of the linewidth data using Eq. 5. Slope of this line gives the relaxation time τ.

large ultrasonic attenuation in the critical region. This curve of τ_{crit} versus T yields a $\tau_{crit} = 8.6 \times 10^{-11}$ seconds at 24°C. This time is nearly identical with the reorientation time of the NH_4 - tetrahedra as determined from proton spin-lattice relaxation times[9]. Both these times are 6.1 times longer than our relaxation time. Because of this factor of 6 it appears that we are observing a velocity dispersion and phonon lifetime that cannot be explained as due to a remnant of the critical fluctuations in the order near T_λ. It is possible that the relaxation we observe could be due to some other process, for example, relaxation associated with the lifetimes of thermal phonons that determine the thermal conductivity[10, 11]. However, estimates of these lifetimes from thermal conductivity and heat capacity data indicate these relaxation times are much shorter than that measured here. It is also of course possible that our simple minded single relaxation time model is an inadequate one, and that more sophisticated analysis of the data along the lines adopted by Pine should be undertaken[12].

We are extending our linewidth measurements into the critical region. In the case where critical fluctuations in the order determine the phonon lifetimes, we expect τ to be proportional to $|T - T_\lambda|^{-1}$, and hence the linewidth Γ to be given by

$$\Gamma = \left[\left(\frac{V_\infty}{V_0}\right)^2 - 1\right]\frac{1}{2\tau} = (\text{constant})\left[\left(\frac{V_\infty}{V_0}\right)^2 - 1\right]\left|T - T_\lambda\right| \tag{7}$$

i.e. the linewidth narrows as $T \to T_\lambda$.

THE RAYLEIGH SPECTRUM

We have measured the intensity of the central component in the spectrum in the temperature interval - 50 to 50°C. The elastically scattered light intensity is, in many cases, due to imperfections in the crystals. The Landau-Placzek ratio indicates that the Rayleigh intensity should be about one fifth the Brillouin intensity. Optical quality crystals can minimize this imperfection scattering, although in our best crystals at room temperature the central component intensity exceeded the Brillouin intensity, generally by a factor of about 3. Theories based on the Landau free energy expansion predict that there should be a critical opalescence[13] in the vicinity of second order phase transition points in solids. One can expect such an opalescence only when the critical fluctuations modulate the optical dielectric constant.

Our measurements are shown in Fig. 5. The central component intensity is depolarized. The polarized intensity (incident and scattered field polarizations perpendicular to the scattering plane, 90° scattering) has a maximum at T_λ, being 10.6 times more intense than at room temperature. The depolarized intensity (scattered field polarization in the scattering plane) has a step-like increase at T_λ, being 4.3 times as intense in the ordered phase as in the disordered phase. This step-like behavior is in marked contrast to the temperature dependence of the <u>total</u> depolarized intensity[14]. Our measurements were taken on crystals that had not previously been through the transition. In each case the laser beam was positioned so as to produce minimum scattered intensity at room temperature. When crystals were repeatedly taken through the transition, the intensity increased with each transit, while the position of the maximum in the polarized intensity remained the same. It was noticed that near T_λ the illuminated region of the crystal appeared to scatter inhomogeneously. The observed increase in the central component intensity may be due to critical fluctuations in the order, especially since this increase extends over a temperature interval above and below T_λ. The polarized intensity exhibits a temperature dependence of the form:

$$I \,\alpha\, |T - T_c|^\mu, \text{ where } \mu \approx 1.0 \text{ for } T < T_c \text{ and } \mu \approx .63 \text{ for } T > T_c.$$

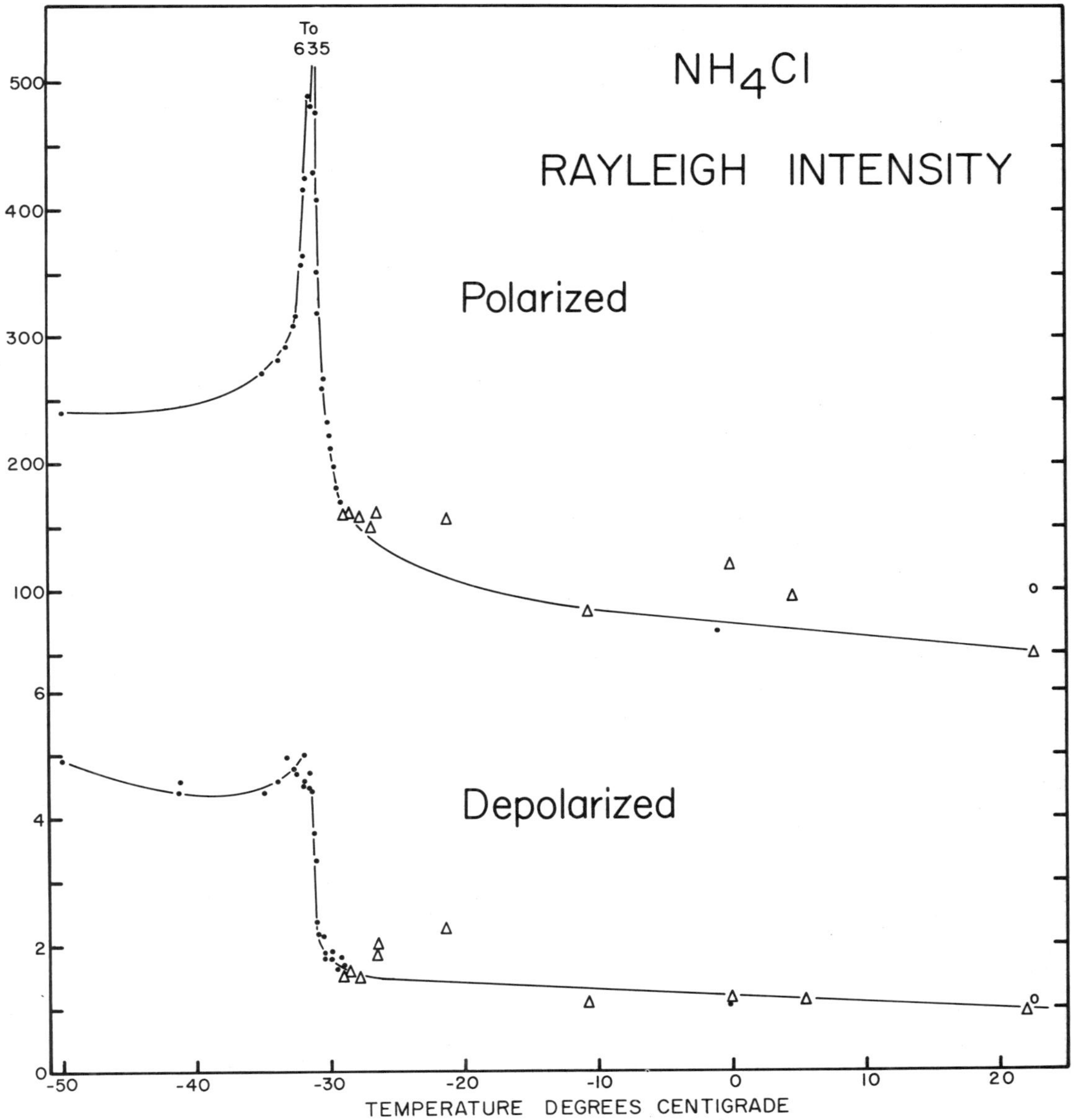

Fig. 5. Temperature dependence of the polarized and depolarized Rayleigh intensity. Note the two intensity scales. The room temperature Rayleigh intensity of toluene is 314 on these intensity scales.

SUMMARY

From our measurements on the Rayleigh-Brillouin spectrum on NH_4Cl we have determined the hypersonic sound velocities in the <100> and <110> directions, three elastic constants and two photoelastic constants, and the central component intensity in the interval -50 to 50°C. The observed increase in the intensity of the elastically scattered light may be due to critical opalescence. From our linewidth measurements we have determined the lifetimes of <110> longitudinal phonons. From a simple model we have extracted the relaxation time at 24°C. We have shown that our lifetimes cannot be explained on the basis of critical fluctuations in the degree of order.

REFERENCES

1. C.W. Garland and R. Renard, J. Chem. Phys 44, 1130 (1966).
2. C.W. Garland and R.A. Young, J. Chem Phys 48, 146 (1968).
3. T. Narasimhamurty, Current Science 23, 149 (1954).
4. G. Liebfried and W. Ludwig, Advances in Solid State Physics 12, 368 (1961).
5. S. Haussühl, Acta Cryst. 13, 685 (1960).
6. I. L. Fabelinskii, "Molecular Scattering of Light," pp. 282 and 563, Plenum
7. Press, New York, 1968.
8 C.W. Garland and J.S. Jones, J. Chem. Phys 42, 4194 (1965).
9. E.M. Purcell, Physica 17, 282 (1951).
10. H.E. Bommel and K. Dransfeld, Phys. Rev. 117, 1245 (1960).
11. T.O. Woodruff and H. Ehrenreich, Phys. Rev. 123, 1553 (1961).
12. A. Pine, paper G-3 of these proceedings.
13. V.L. Ginzburg, Soviet Physics Uspekhi 5, 649 (1963).
14. O.A. Shustin, JETP Letters 3, 320 (1966).

G-6: OPTICAL PROBING OF MAGNETOELASTIC WAVES

Archibald W. Smith
IBM Watson Research Center
Yorktown Heights, New York

ABSTRACT

The use of an optical probe to observe magnetoelastic (ME) waves in yttrium iron garnet gives new information not obtainable by other methods. The interaction of the light with the elastic and magnetic parts of a transverse ME wave gives rise to some unusual polarization properties, which are explained by a simple theory of the diffraction. For the laser polarized parallel or perpendicular to the plane of diffraction, a relation is found between the laser polarization giving maximum diffracted intensity and the sign of the frequency shift of the diffracted light. For a frequency shift of given sign, the optimum laser polarization is reversed for waves on the quasielastic portions of the upper and lower branches of the ME dispersion curves. When the spin-wave component is larger (near the turning points in the bar), the optimum polarization is the same on both branches, contrary to the theory. This anomaly is probably due to the large amplitude of the ME signals used in the experiments.

INTRODUCTION

An unusual optical polarization effect has been observed in the diffraction of light from transverse magnetoelastic (ME) waves [1-3]. With the plane of the input polarization either parallel or perpendicular to the plane of diffraction, the orientation for maximum diffracted signal is found to depend on the sign of the frequency shift of the diffracted light. The optimum polarization is reversed for the two branches of the ME dispersion curve, except that near the turning point the optimum polarization for the upper and lower branches is the same. A simple theory of the effect is presented which is in agreement with the observations, except near the turning point.

THEORY

The geometry of the experiment and the coordinate system are shown in Fig. 1. Since the internal field H_i and the magnetoelastic wavevector $\underline{k}_m$ are coincident with the crystal [001] direction (z-axis), only transverse elastic waves are coupled to the magnetization [4]. Under the perturbation of the transverse ME wave, the optical susceptibility becomes [5, 6]

$$\chi_{ij}=n^2\begin{pmatrix}1 & 0 & n^2p\epsilon_{zx}\\ 0 & 1 & n^2p\epsilon_{yz}\\ n^2p\epsilon_{zx} & n^2p\epsilon_{yz} & 1\end{pmatrix}+iKM_z\begin{pmatrix}0 & 1 & -\alpha_y\\ -1 & 0 & \alpha_x\\ \alpha_y & -\alpha_x & 0\end{pmatrix} \quad (1)$$

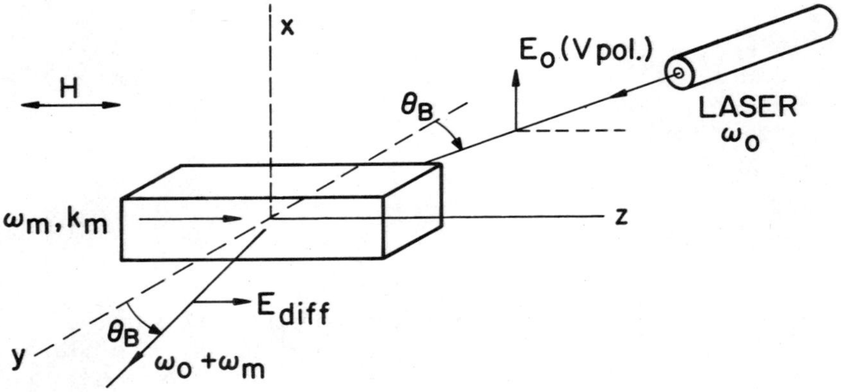

Fig. 1. Experimental arrangement and coordinate system. Transverse ME waves are propagated in the YIG bar, and observed by diffraction of the laser beam.

The first matrix on the right gives the elasto-optic effect due to shear strains ϵ_{zx} and ϵ_{yz}, while the second matrix gives the magneto-optic effect due to the magnetizations $\alpha_x = M_x/M_z$ and $\alpha_y = M_y/M_z$. Here n is the refractive index, p the elasto-optic constant and K the magneto-optic constant ($K = n\,\lambda\,\phi_F/\pi M$ where ϕ_F is the optical Faraday rotation). We assume that $n^2\,p\epsilon \ll 1$ and $\alpha \ll 1$, and thus $M_z = M$ the saturation magnetization. Since the diffraction angle θ_B inside the crystal is always small, we ignore the small Faraday rotation of the laser beam proportional to $\sin\theta_B$. Using the relation $P_i = \chi_{ij}E_j$, the change in optical polarization δP_i due to the ME wave is to first order

$$\begin{pmatrix}\delta P_z\\ \delta P_x\end{pmatrix} = (n^4p\epsilon_{zx} \pm iKM\alpha_y)\begin{pmatrix}E_x\\ E_z\end{pmatrix} \quad (2)$$

where E_i is the incident optical amplitude.

Under the present conditions the eigenmodes for the ME wave are circularly polarized [4]. We take the time variations as follows: $\alpha = \alpha_x \pm i\alpha_y = \alpha_t e^{\mp i\omega_m t}$, $\epsilon = \epsilon_{zx} \pm i\epsilon_{yz} = \epsilon_t e^{\mp i\omega_m t}$ and $E_i = E_i e^{i\omega_o t}$, where ω_m and ω_o are the ME and optical frequencies respectively. We omit the spatial variations $e^{-i\underline{k}\cdot\underline{r}}$ since they lead to the Bragg relation $\sin\theta_B = k_m/2k_o$, and give the relative wave directions for the upshifted and downshifted frequencies of the diffracted wave [6, 7]. In Fig. 1, a wave travelling from left to right is upshifted. Substituting the time variations, we obtain

$$\begin{pmatrix} P_z \\ P_x \end{pmatrix} = \frac{1}{2}\Big[(n^4 p\epsilon_t \mp KM\alpha_t)e^{i(\omega_o + \omega_m)t} + (n^4 p\epsilon_t \pm KM\alpha_t)e^{i(\omega_o - \omega_m)t}\Big] \tag{3}$$

where it is understood that only one term is nonzero as determined by the wavevectors.

The standard small-amplitude analysis of ME waves [4, 8, 9] yields the following relation between α_t and ϵ_t:

$$\alpha_t = \pm\frac{\gamma b}{M}\;\frac{\epsilon_t}{\omega_m \mp \gamma(H_i + Dk_m^2)} \tag{4}$$

where the upper and lower signs refer to the + and – rotating modes, respectively, b is the ME interaction constant, D an exchange constant, and γ the gyromagnetic ratio taken positive. It is clear from the right hand denominator that only the positively rotating mode has an appreciable magnetic contribution. The dispersion relation for this mode splits into upper ($H_i < \omega_m/\gamma$) and lower branches ($H_i > \omega_m/\gamma$), and a sign change in the right-hand denominator gives a phase change of π for α^+ between the two branches. ME waves can be directly generated by an r.f. magnetic field only on the upper branch at $H_i = \omega_m/\gamma$, at which field $k_m \to 0$. This is known as the turning point in a sample with a non-uniform H_i [8-9]. In a bar such as used here, H_i has a maximum at the center and there are two conjugate turning points.

We now define $\chi_P = n^4 p$ and $\chi_F = K\gamma b/[\omega_m - \gamma(H_i + Dk_m^2)]$ for positive rotation, and obtain for the upper branch

$$\begin{pmatrix} P_z \\ P_x \end{pmatrix} = \frac{1}{2}\epsilon_t\Big[(\chi_p \mp \chi_F)e^{i(\omega_o+\omega_m)t} + (\chi_p \pm \chi_F)e^{i(\omega_o-\omega_m)t}\Big]\begin{pmatrix} E_x \\ E_z \end{pmatrix} \tag{5}$$

For the lower branch the signs are reversed. If either E_x or E_z is zero, the diffracted signal is polarized perpendicular to the input polarization, and is sensitive to the relative phase of ϵ and α. If $E_x = E_z$, i.e., input polarization at $\frac{\pi}{4}$ to the plane of diffraction, then the diffracted signals polarized at $\frac{\pi}{4}$ and $3\pi/4$ are given by $P(\frac{\pi}{4}) = \chi_p\epsilon_t E(\frac{\pi}{4})$ and $P(\frac{3\pi}{4}) = \chi_F\epsilon_t E(\frac{\pi}{4})$. Hence the elasto-optic and magneto-optic contributions can be separated by using a polarizer in the diffracted beam. When both rotating modes are present, the diffracted amplitudes must be combined taking the phase difference into account [10] $(P_i)^2 = (P_i^+)^2 + (P_i^-)^2 + 2P_i^+P_i^- \cos 2\phi_A$, where ϕ_A is the acoustic Faraday rotation measured from the x-axis, and the superscripts refer to the rotating modes.

EXPERIMENTAL

The polarization effects have been observed with both acoustic [2] and r.f. magnetic [1] excitation of the ME waves. The results from the two cases are consistent and are in qualitative agreement with the theory except for one systematic deviation: the phase change of π for α^+ does not occur at the predicted field of $H_i = \omega_m/\gamma$, but at a slightly lower value. We will describe only the main features of the polarization effects here, since a quantitative analysis is not feasible at present.

The basic geometry of the experiment is shown in Fig. 1. The ME signals had a frequency of 1100 MHz and were pulsed with a width of 0.5 μ sec. The diffraction angle was $\theta_B = 9.5°$ in the elastic limit for the laser wavelength of 1.15μ. The laser beam was focused to an angular width of approximately 0.01 radians. We discuss results only for the input laser polarization parallel or perpendicular to the plane of diffraction. For convenience these orientations will be referred to as the horizontal or H, and the vertical or V polarizations, respectively, since experimentally the plane of diffraction was horizontal.

Transverse acoustic excitation was provided by an X-cut $LiNbO_3$ transducer bonded to the end of the YIG bar[2]. The acoustic power density entering the YIG was typically 1 to 10 W/cm^2. In this case both the positively (upper branch) and negatively rotating modes are present. Typical magnetic field scans with the H and V polarizations are shown in Fig. 2. For the down shifted case (transducer on right as in Fig. 4) the V polarization gives a systematically larger signal for $H_i < \omega_m/\gamma$. This is consistent with the theoretical production of $P_z \propto (\chi_p + \chi_F) E_x$ from Eq. (5), assuming p and K to be positive. K is known to be positive for YIG, but the sign of p is not known. The intensity oscillations in Fig. 4 result from the acoustic Faraday rotation. For $H_i > \omega_m/\gamma$, the intensity

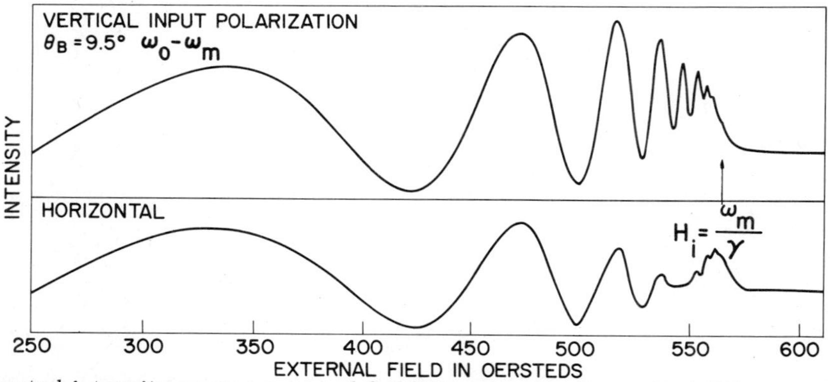

Fig. 2. Diffracted intensity versus external field for acoustically excited ME wave on quasi-elastic portion of upper branch (at center of bar).

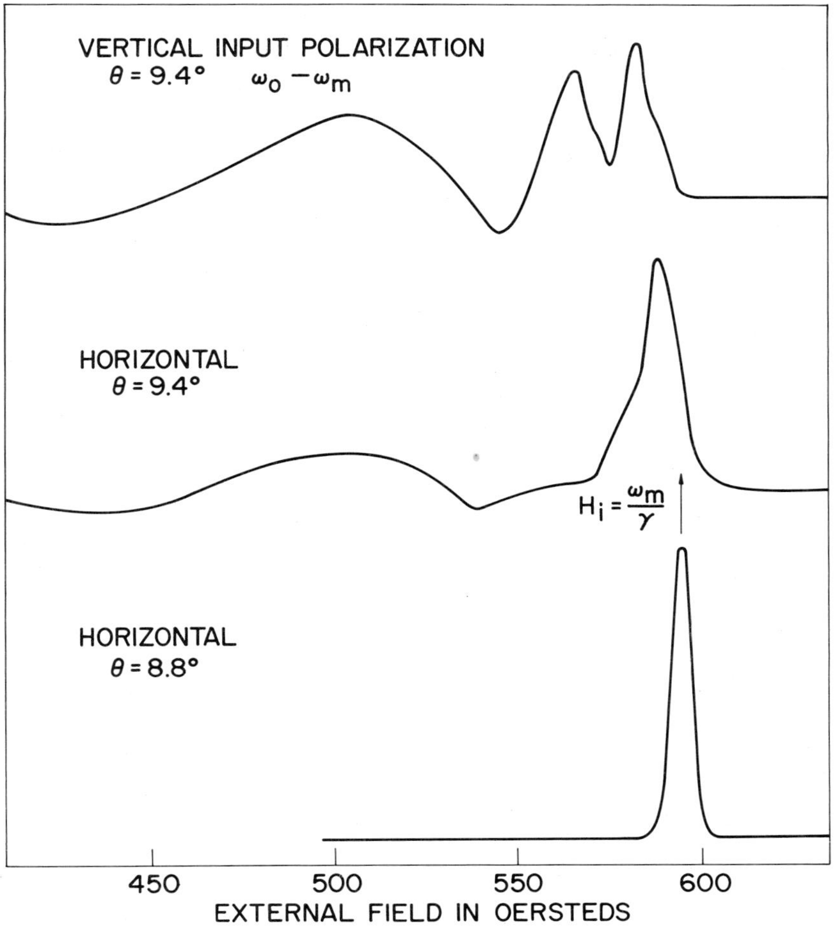

Fig. 3. Diffracted intensity versus external field for acoustically excited ME wave, showing reversal of optimum polarization near turning point (3mm from transducer).

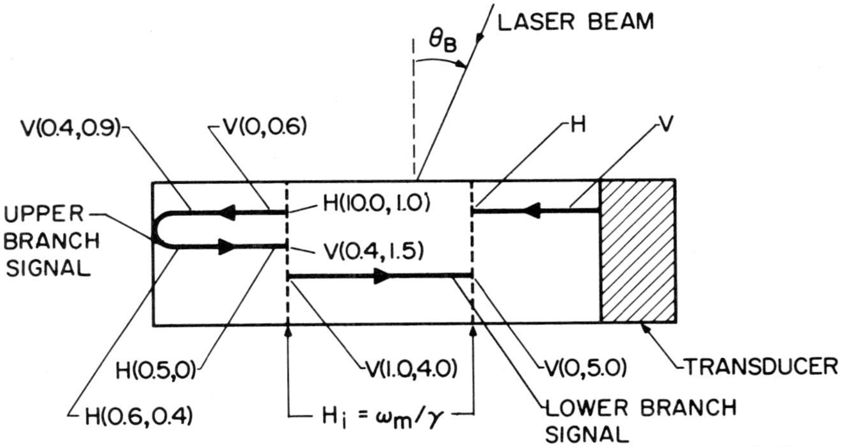

Fig. 4. Schematic view of ME waves observed in bar with r.f. magnetic excitation, with optimum polarizations indicated. The signals are centered on the bar axis, but have been displaced vertically in the figure for clarity. Number pairs are intensities for H and V input polarizations, respectively.

becomes independent of the input polarization, which is expected since only the negatively rotating mode is present here.

We now consider the polarization behavior near the turning point ($H_i \approx \omega_m/\gamma$). When the positively rotating component approaches the turning point in the bar the wave number k_m decreases, and if the laser beam is located at the turning point, the diffraction angle decreases ($\sin\theta_B = k_m/2k_o$). This behavior is illustrated in Fig. 3 for the turning point 3mm from the transducer. It is clear that the diffraction from the ME wave near the turning point is a maximum for the H polarization, i.e., at $\pi/2$ radians to the optimum polarization for the quasi-elastic part of the upper branch (see also Table I of Ref. (2) which refers to the downshifted case). This change of $\pi/2$ radians in the optimum polarization implies a phase change of π radians for α^+ occuring on the upper branch at a field slightly below the predicted value of ω_m/γ.

We now turn to the results for r.f. magnetic excitation of the ME waves, using a ball or hairpin electrode against the end of the YIG bar, and r.f. drive powers in the range 30 to 300 watts [1]. In this case the positively rotating component is generated at the turning point in the bar nearest the electrode. Signals are observed on both the upper and lower branches, as illustrated in Fig. 4. The upper branch signal initially travels toward the end of the bar and the diffracted signal is downshifted. As shown in Fig. 4, the H polarization is optimum at the turning point, changing to V in the quasi-elastic region, in agreement with the acoustic drive results. The number pairs indicate typical intensities for the H and V polarizations, respectively, on an arbitrary scale. After reflection, the diffracted signal is upshifted and the optimum polarizations are reversed as expected. The signal on the lower branch through the center of the bar is upshifted, and the V polarization is optimum, in agreement with the theory (Eq. 5), and with the upper branch results at the turning point.

CONCLUSIONS

It is clear that the π phase change between α^+ and ϵ^+ is consistently occuring for H_i, a few oersteds below the predicted value of ω_m/γ. The reason for this discrepancy is not known at present, but it is probable that the assumptions of uniform field H_i and small amplitude which were made in deriving the relations (4) between α^+ and ϵ^+ are not adequate under the present conditions. With the present high amplitude signals, the behavior at the crossover region may be more complicated than predicted by the simple theory, and the assignment of the observed signals near $H_i = \omega_m/\gamma$ to the upper and lower branches may not be entirely justified. However, the assignment should be valid elsewhere. A related problem is that with r.f. magnetic excitation, the strongest signals near $H_i = \omega_m/\gamma$ are observed near the elastic limit $\theta_B = 9.5°$, whereas we might expect a considerable range in θ_B at the turning point since k_m approaches zero.* As previously noted [1] the generation of a signal on the lower branch is difficult to understand on the basis of small amplitude analysis. It would be desirable to perform similar experiments at lower ME power densities, but the experiments become considerable more difficult due to background scattering from imperfections.

*The range of θ_B from 9.5° to 7.5° previously reported[1]for the upper branch resulted partly from a 2° error in the orientation of the end face relative to the bar axis. In a second more accurately cut bar, the intensity at the lower values of θ_B was considerably less.

Finally, we note that in observing diffraction from magnetostatic modes, which represent the extension of the upper branch to $k_m = 0$ for finite samples, Collins and Wilson [3] found an optimum H or V polarization in agreement with that found here for the quasi-elastic part of the upper branch. Thus the main features of the analysis are borne out under widely varying conditions of magnetoelastic excitation.

REFERENCES

1. A.W. Smith, Phys. Rev. Letters 20, 334 (1968).
2. A.W. Smith, IEEE Transactions on Sonics and Ultrasonics SU-15, 161 (1968). On p. 164 of this paper, second column, the words perpendicular and parallel on lines 7 and 4, respectively, from the bottom should be interchanged.
3. J.H. Collins and D.A. Wilson, Appl. Phys. Letters 12, 331 (1968).
4. C. Kittel, Phys. Rev. 110, 836 (1958).
5. J.F. Nye, "Physical Properties of Crystals", Oxford University Press, London, 1957.
6. B.A. Auld and D.A. Wilson, J. Appl. Phys. 38, 3331 (1967).
7. E.I. Gordon, Proc. IEEE 54, 1391 (1966).
8. W. Strauss, Proc. IEEE 53, 1485 (1965).
9. E. Schlömann, R.I. Joseph, and T. Kohane, Proc. IEEE 53, 1495 (1965).
10. R. W. Dixon, J. Appl. Phys. 38, 3634 (1967).

G-7: FABRY - PEROT ANALYSIS OF THE ACOUSTOELECTRIC INTERACTION IN CdS

R. W. Smith
RCA Laboratories
Princeton, New Jersey

In an acoustoelectric interaction, the attenuation or amplification of an elastic wave in a piezoelectric semiconductor with an applied electric field, is a function of the relative velocity of the wave and the drifting carriers. Through this type of interaction waves with a wide range of $\vec{K}(\omega, \vec{n})$ vector, and which include thermal noise phonons, can be amplified to very high levels of acoustic flux. Under some conditions the flux may increase monotonically along the crystal, and under other conditions a discrete, high energy density disturbance, or acoustoelectric domain, may form and propagate through the crystal. Most probes that can be used to study the acoustoelectric interaction are physically incapable of resolving all of the component waves which superpose to form the complicated acoustic flux pattern in the crystal. These probes essentially observe a collective acoustic Poynting vector and a corresponding group velocity in the crystal. Brillouin scattering may be the unique example of an experimental technique capable of analyzing the plane wave components of the disturbance, since the Bragg condition is equivalent to constructive interference from planes of constant phase. The Doppler frequency shift of the Brillouin components is determined by phase velocity, and the intensity of light scattered into the components is proportional to the Poynting vector of the acoustic plane waves. Acoustic anisotropy is one of the important factors involved in the superposition of component waves, propagating off axis relative to a principal crystallographic direction, in an acoustoelectric domain. The full potential of the Brillouin technique can only be realized by observing the scattered light through a high resolution interferometer, such as a Fabry-Perot.

Although we are interested in acoustoelectric domains, and have observed Brillouin scattering from them, one of our long-term objectives has been to observe the thermal background and then follow the excitation of the Brillouin components as a function of applied voltage. Then with the thermal background as a reference level, we should be able to obtain a better understanding of the origin and the subsequent growth of the acoustoelectric interaction. What we report here then is the first step in a program to obtain the interaction spectrum for the acoustoelectric effect.

The data described in this paper was obtained from a CdS crystal whose V-I curve, Fig. 1, has a long current saturation branch exhibiting no current oscillations. This is assumed to indicate a monotonic distribution of acoustic flux and the absence of a discrete macroscopic domain. As a matter of practical convenience we set the interferometer to observe 2.5×10^9 Hz T_2 mode shear waves propagating in the basal plane and parallel to the applied field. The c axis of the crystal is vertical, the incident 6328 Å laser beam is horizontally polarized, and the Fabry-Perot analyzer is vertical. The sharply focused probe beam passes through the center of the crystal. There are two formidable experimental problems. One is due to the poor optical quality of the CdS, which contains imperfections giving a large Tyndall scattering component. The second problem is the effective light loss (by a factor of 10^{-3} to 10^{-4}) imposed by the duty cycle restriction necessary to limit power dissipation in the crystal. We use two ratemeters operating on a time sharing basis to record the scattered light intensity. Meter No. 1 is on only when the voltage pulse is applied, and meter No. 2 is on for the remaining time in the cycle.

Fig. 2 shows the output of the two ratemeters over three Fabry-Perot orders. The T_2 shear mode thermal Brillouin components are seen at $\Delta\nu = \pm 2.5 \times 10^9$ Hz relative to the central light scattering component, ν_0. The asymetry is incidental and is due

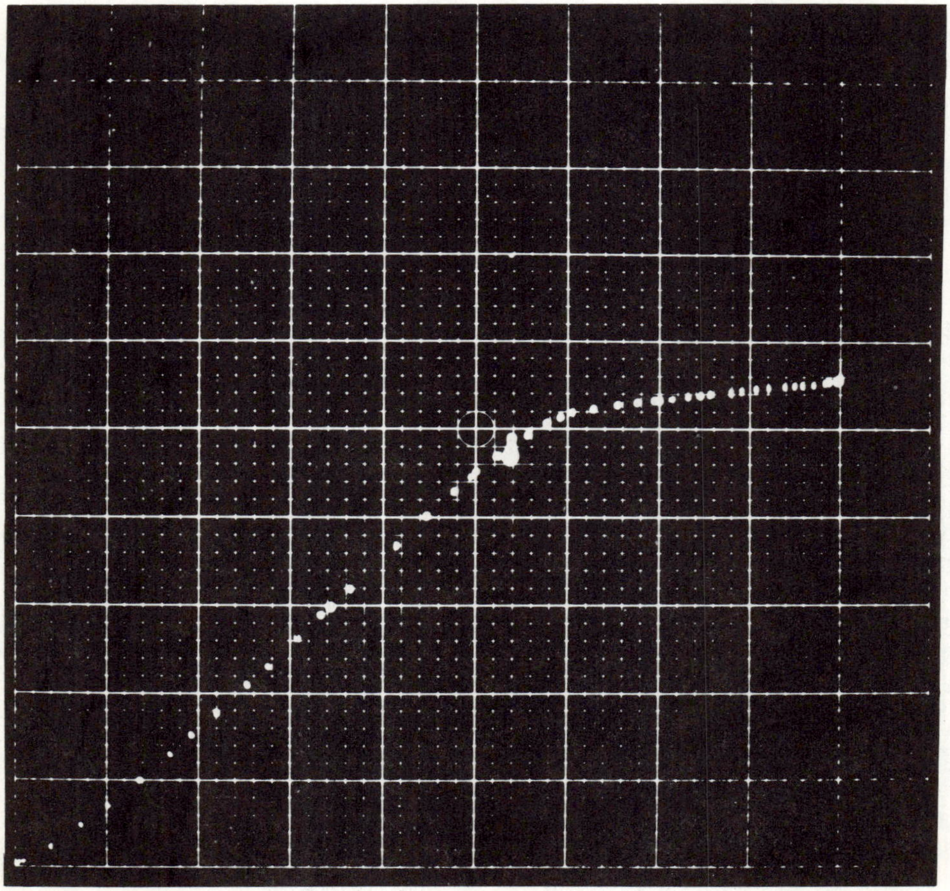

Fig. 1. V-I curve. I = 1/2 A/div, V = 100 V/div. Voltage pulse on 20 μs at 60 pps. Crystal 3.4 x 4.5 x 6.3 mm.

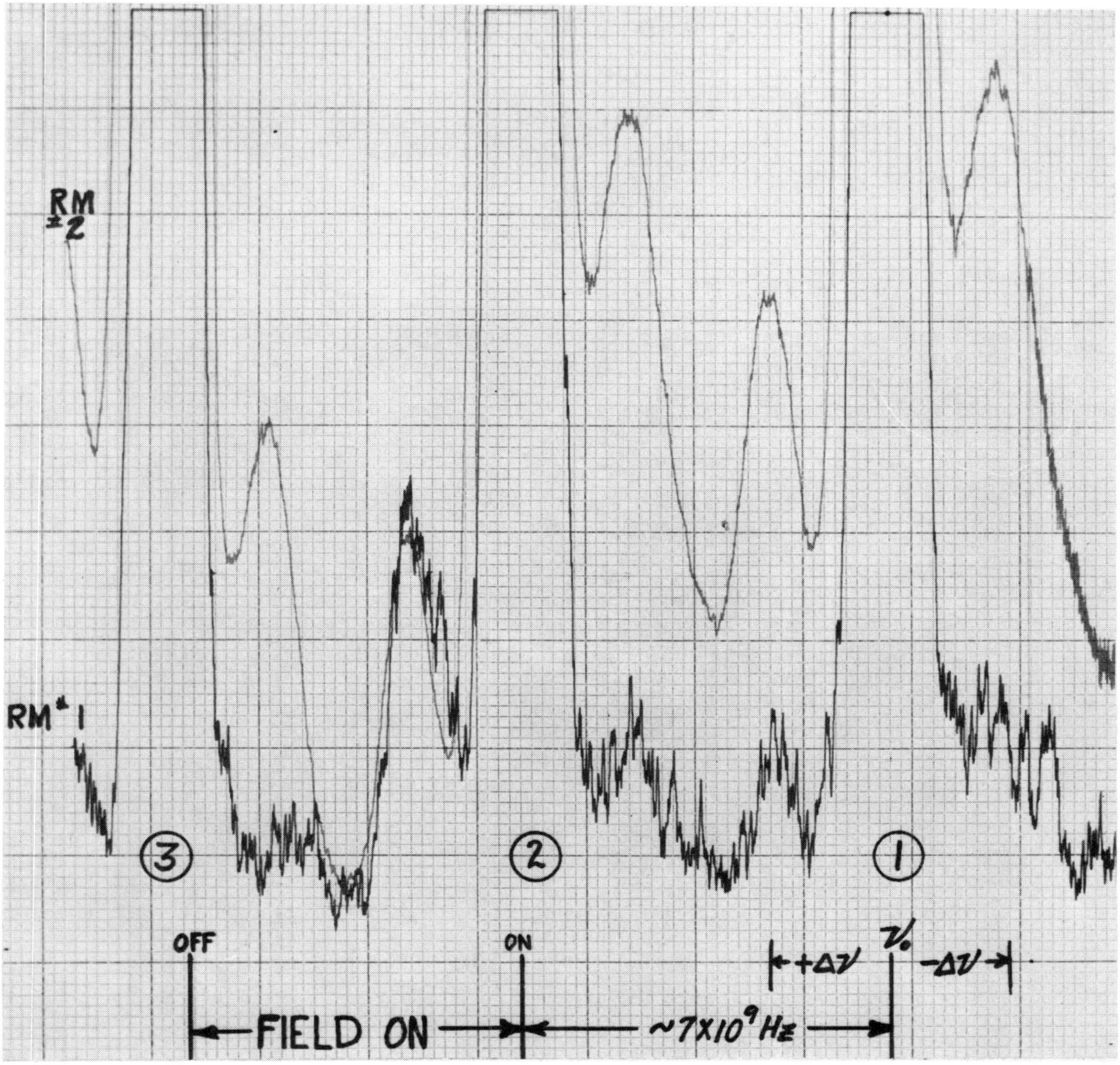

Fig. 2. Scattering intensity vs. frequency. Three Fabry-Perot orders. Free spectral range 7×10^9 Hz. Incident beam 6328 Å, H polarized, $\theta_{IN} = 22.2^o$. Scattered beam V polarized, $\theta_{OUT} = 30^o$. Scattering in basal plane.

to either crystal imperfections or possibly an instrumental error. The voltage applied during the time the field is on was intentionally set to increase the scattering intensity of the 2nd order + $\Delta\nu$ component to a few times the thermal level. Note also that the - $\Delta\nu$ component in the 3rd order, which corresponds to thermal phonons propagating against the field, is reduced in intensity. If the direction of the applied field is reversed, the Brillouin component stimulated is reversed, as expected.

Fig. 1 shows the V-I curves for the crystal and the intensified spot indicates the voltage applied during the scattering curve scan. Fig. 3 shows the initial increase in scattering intensity as a function of applied voltage. The steep increase in acoustic flux has been followed through at least 6 decades.

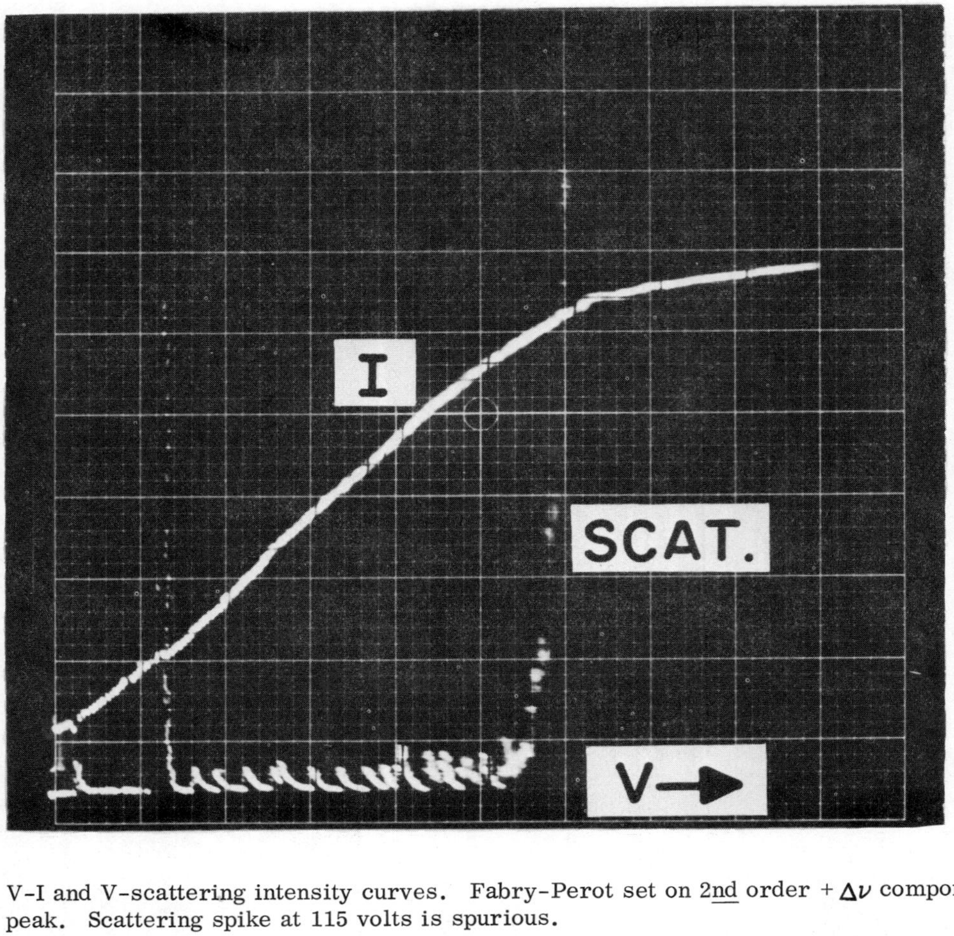

Fig. 3. V-I and V-scattering intensity curves. Fabry-Perot set on 2nd order + $\Delta\nu$ component peak. Scattering spike at 115 volts is spurious.

G-8: BRILLOUIN SCATTERING STUDIES OF ACOUSTOELECTRIC EFFECTS IN PIEZOELECTRIC SEMICONDUCTORS

J. Zucker, S. Zemon, E. M. Conwell and A. K. Ganguly
General Telephone & Electronics Laboratories
Bayside, New York

INTRODUCTION

Brillouin scattering has proved itself a powerful tool for studying acoustoelectric effects in piezoelectric semiconductors. Through its use we have been able to determine the detailed characteristics of the acoustic modes generated in a semiconductor under a variety of experimental conditions. This information should make it possible to construct more accurate models for acoustoelectric effects in the nonlinear regime.

In this paper, we first present a summary of the important characteristics, as determined by Brillouin scattering, of acoustic shear waves generated in CdS and ZnO by electrons with drift velocity greater than the shear wave velocity. We then discuss our recent observations of acoustic parametric amplification in CdS and ZnO, observations made possible only by the use of the Brillouin scattering technique.

EXPERIMENTAL TECHNIQUES

In Fig. 1a is shown the geometry for the experimental data reported in this paper. Light of wavelength λ_o is incident at an angle θ_i to the normal as an ordinary ray (electric vector E_i in the basal plane). After refraction it has wave vector k_i, the free-space wave vector multiplied by n_ω, the ordinary index of refraction. Scattering by a shear wave with wave vector q rotates the plane of polarization by 90° and transforms it into an extraordinary ray (electric vector E_s perpendicular to the basal plane) which, if we neglect the small energy change due to absorption or emission of a phonon, has wave vector k_s equal to the free-space value multiplied by n_e, the extraordinary index of refraction. Since $|k_s|$ may be quite different from $|k_i|$ the Bragg law must be modified for this case[1]. The resulting relationship for obtaining the frequency f of the acoustic wave and its angle α to the current direction from the measured angles θ_i and θ_s (see Fig. 1a) is shown graphically in Fig. 1b. For further details of the measurements the reader should see references[2] through [6].

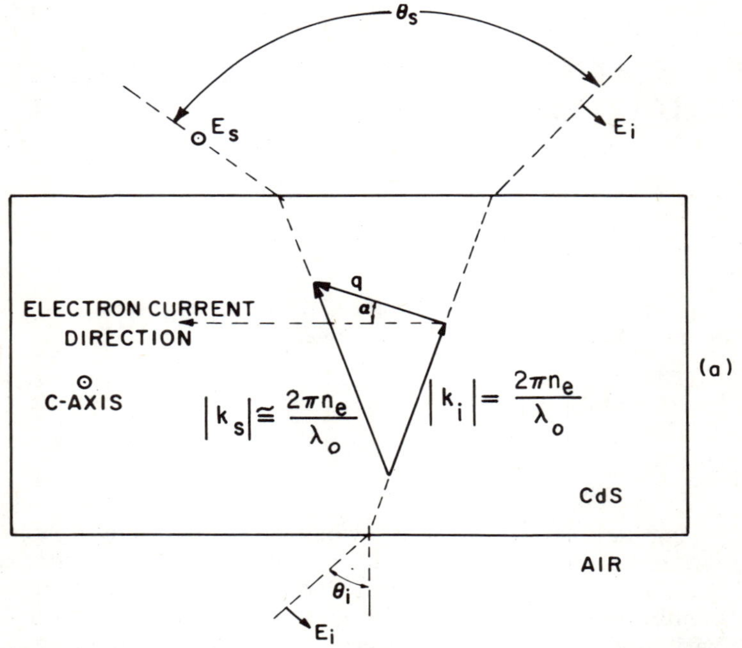

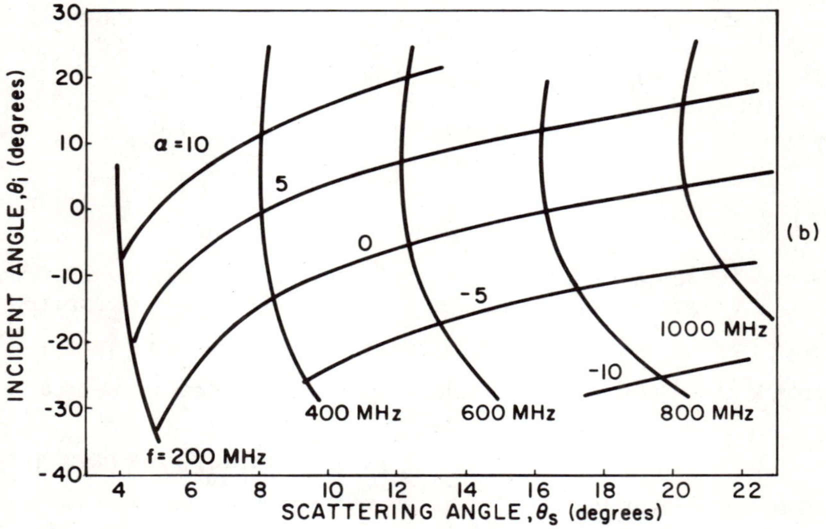

Fig. 1. Brillouin scattering by shear waves in CdS for the case that scattering plane and incident light polarization direction are parallel to the basal plane and scattered light polarization is parallel to the c-axis.

a) Conservation of momentum $\underline{k}_s = \underline{k}_i + \underline{q}$ for the case $|\underline{k}_s| \neq |\underline{k}_i|$ due to birefringence.

b) Transformation from θ_i and θ_s to acoustic frequency and propagation direction for this geometry.

The semiconducting CdS and ZnO samples reported on here had carrier concentrations in the range 1 to 3 x $10^{15}/cm^3$, and dimensions of the order of 1 mm x 1 mm x 5 mm. Mobilities were about 300 $cm^2/V\cdot sec$ for CdS, 180 $cm^2/V\cdot sec$ for ZnO. Current flow was perpendicular to the c axis and along the long dimension of the sample.

RESULTS - INTERNALLY GENERATED WAVES

When drift velocity v_d of the electrons is sufficiently greater than the sound velocity v_s so that rate of gain of acoustic waves due to the electronic interaction exceeds their rate of attenuation due to other processes, amplification of acoustic noise results. Depending on details of contacts, sample homogeneity, time-variation of applied voltage, etc., this amplification may produce either a stationary or a time dependent condition. In the former case, the current drops to a value (saturation) well below ohmic and there is a stationary region of high acoustic flux (domain). In the latter the current oscillates in time between the ohmic and saturation values, due to the periodic formation and propagation from cathode to anode of a region of high acoustic flux. The ZnO data reported here are taken on stationary domains, the CdS data on moving domains. Some data have also been obtained on stationary domains in CdS, and they do not differ essentially from the moving domain data.

The characteristics measured for the acoustic waves were the frequency spectrum, net gain as a function of frequency, net loss (including nonelectronic loss) as a function of frequency, frequency of maximum intensity f_{mi} and frequency of maximum net gain f_{mg} as functions of carrier concentration, temperature and acoustic intensity, and angular distribution. A typical frequency spectrum is shown in curve a of Fig. 2. Such frequency spectra were taken on many samples and studied in detail. The salient findings of these studies (see also [2-4, 8]) were that, in the nonlinear region, in which current has fallen to its saturated value, f_{mi} and f_{mg} have lower values than the frequency f_{mg} predicted by linear theory[7]. Furthermore, f_{mi} and f_{mg} continue to decrease, and their amplitudes continue to grow, as the waves propagate down the sample even though the current maintains its saturation value. In semiconducting CdS the frequency f_{mg} falls typically to one-tenth the value predicted by the linear theory. The frequencies f_{mg} and f_{mi} are not generally the same, the latter being higher because the frequency distribution requires a finite path length to adjust to a new gain characteristic. In photoconducting CdS and in ZnO, f_{mi} was found to vary as $n^{1/2}$, n being the carrier concentration, in agreement with linear theory[7]. Also in agreement with linear theory in ZnO, the ratio in db of the acoustic intensity at the angle α to the acoustic intensity at 0° was found to vary as $-\alpha^2$[4]. In CdS, on the other hand, the maximum intensity was found off axis.

Nonelectronic loss L_o was found to vary from sample to sample. In general, the greater the loss, the smaller the frequency dependence. In one semiconducting CdS sample, L_o was proportional to f, going from 18 dB/cm at 0.38 GHz to 160 dB/cm at 4 GHz[8]. L_o was also measured in an insulating CdS sample by feeding acoustic waves into it from a semiconducting CdS sample bonded to it with Echobond. To within experimental error, the acoustic loss was constant, equal to 210±10dB/cm from 0.24 to 0.38 GHz. These are the highest loss values that were seen.

The acoustic flux characteristics in CdS were also measured simultaneously with the current and the electric field intensity profile[8]. This showed that the electric field across the moving domain is not a direct measure of the acoustic flux intensity, contrary to what has been assumed in the past[9]. For example, the average domain field E_d was found to increase by only about 10%, from 4.5 kV/cm to 5 kV/cm, for a ten-fold increase in acoustic flux intensity φ that took place during current saturation. Only when the total current was set equal to zero by balancing out the acoustoelectric current was it found that $E_d \propto \varphi$ [8]. It was also found that the mobility decreased with increasing

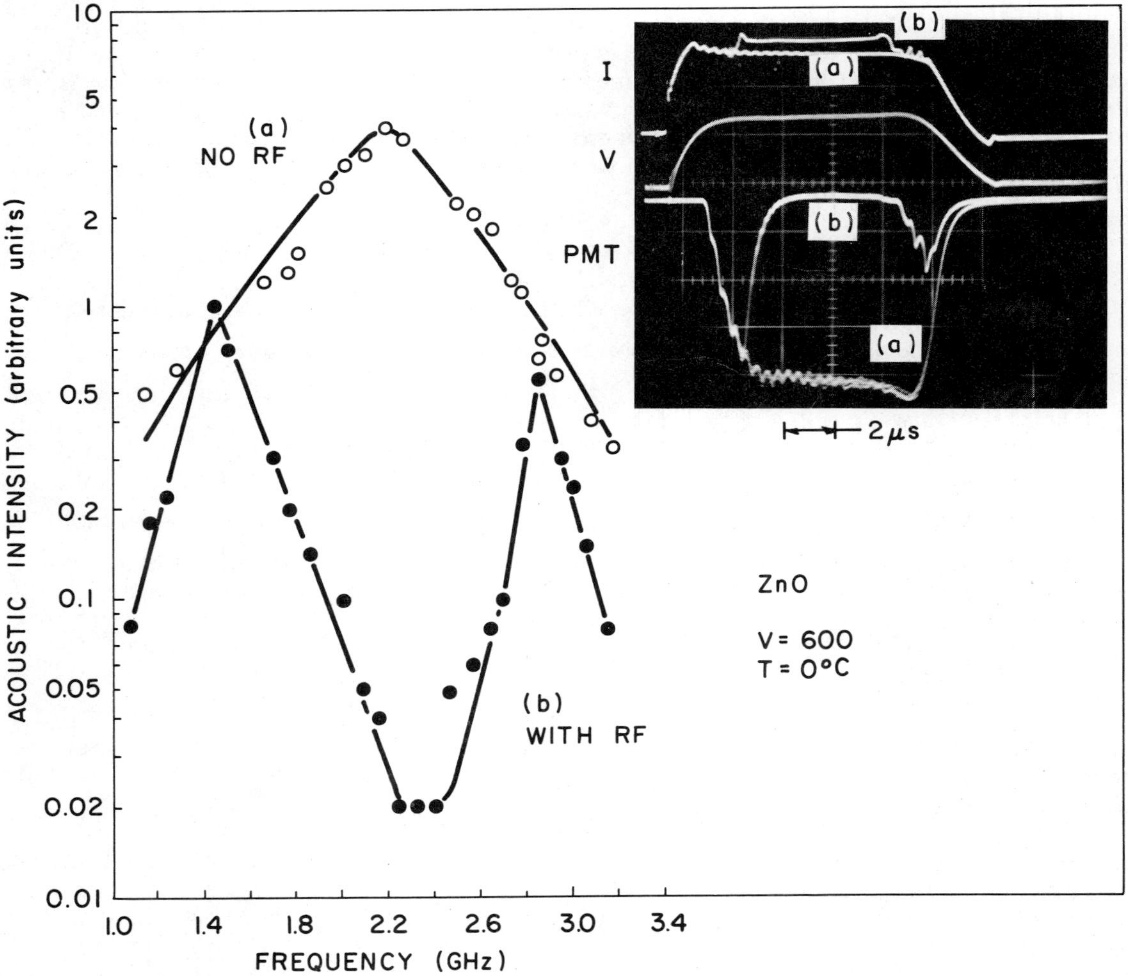

Fig. 2. Acoustic intensity as a function of frequency in ZnO sample at $T = 0^{\circ}C$ for $v_d > v_s$. Curve (a), no microwave input. Curve (b), microwave input of 100 watts at 2.85 GHz. In the oscillogram the scales for the various traces are: I, 0.2A/div; V, 500V/div; scattered light signal, 0.05 mA/div. The angle of the scattered light has been so chosen that the signal is proportional to the acoustic intensity at 2.2 GHz.

flux intensity, even for $v_d < v_s$. For v_d antiparallel to the acoustic propagation direction, however, μ was found not to deviate from its low-field value. Also, for this direction of v_d the electronic loss was found to be consistent with the predictions of linear theory. This set of results could be fitted to an empirical model for acoustoelectric interactions, the details of which are given in Ref. [8].

PARAMETRIC INTERACTIONS

Significant insight into nonlinear acoustic interactions has been afforded by our observations of acoustic parametric interactions in CdS and ZnO. In these experiments, in addition to a dc field, the samples are excited with microwave electromagnetic power which is then partially converted to acoustic power, e.g. shear waves at the microwave frequency, inside the samples. For $v_d < v_s$ it was found that introduction of these shear waves at frequency f resulted in the generation of shear waves with frequency f/2, which in turn generated shear waves at f/4 in some cases[5, 6]. By observing with Brillouin scattering the dependence of the amplitude at f/2 on the power at f, on v_d/v_s and on position, we were able to conclude that f acted as a pump to amplify a signal at f/2 originating in the thermal noise[6]. Since these experiments were done we have also seen parametric amplification of an acoustic signal introduced from outside the sample in the same manner as the pump.

The nonlinearity responsible for these effects is the interaction of the bunched space charge associated with one acoustic frequency with the electric field associated with another acoustic frequency. We have set up coupled wave equations[6] for the three waves involved in the parametric process--signal, pump and idler. The equations are similar in form to these obtained for the case of optical parametric amplification[10], although the details of the interaction are, of course, quite different for that case. We have solved these equations for the case that signal and idler amplitudes are much smaller than the pump amplitude, i.e., for the beginning of the parametric amplification process. Since for $v_d < v_s$ there is significant loss, electronic as well as nonelectronic, the pump amplitude (or the strain due to the pump wave) must have a minimum or threshold value for the signal to have net gain. Our calculations show that, for the parameters of our experiments on photoconducting CdS, the threshold is a minimum for signal and idler frequencies equal to one-half the pump frequency. This accounts for the fact that the half-frequency was the one selected from the noise to be amplified.

We have now extended our studies to $v_d > v_s$, particularly in semiconducting samples, and find that introduction of a coherent acoustic signal causes a radical change in the internally generated acoustic noise. In Fig. 2 it is seen that the introduction of a 2.85 GHz signal into a semiconducting ZnO sample causes the single-peaked noise spectrum to change into a double-peaked spectrum, the peaks being at 2.85 GHz and 1.42 GHz, the half-frequency. The oscillogram in the figure shows directly the depletion of acoustic flux at 2.2 GHz, the noise peak, by the introduction of the coherent signal. As can be guessed from the figure, the total acoustic flux in the presence of the signal is smaller, specifically 0.13 times its value in the absence of the signal. The reduction of total flux is, as expected, accompanied by a rise in current, also shown in the oscillogram. It is noted that there is a small periodic variation in the current in the absence of the coherent signal. This current oscillation is caused by a similar oscillation in the acoustic flux intensity, also seen in the oscillogram, which is due to the presence of a sample inhomogeneity. The large reduction in the acoustic flux intensity upon application of the coherent signal apparently eliminates these current oscillations.

The very substantial transfer of energy from pump to signal and idler suggests that the waves are at least approximately phase-matched[10], i. e. the three wave vectors form a closed triangle. Calculations for the parameters of the ZnO sample used in these experiments indicate that for phase matching (with the dispersion predicted by the linear theory[7]) the k vectors of signal and idler would have to lie at angles of 6° to 9° to the pump wave vector, symmetrically on either side. The uncertainty in angles is due to the uncertainty in the value of the electromechanical coupling constant. Measurement of the angular distribution of the 1.42 GHz signal shows it to be symmetrically peaked at ±6° from the propagation direction of the pump, as shown in Fig. 3, more or less in agreement with the expectation of phase matching. The asymmetry in intensity of the two peaks probably arises from the fact that the peak of the pump signal happened to make an angle of 3° with the current direction.

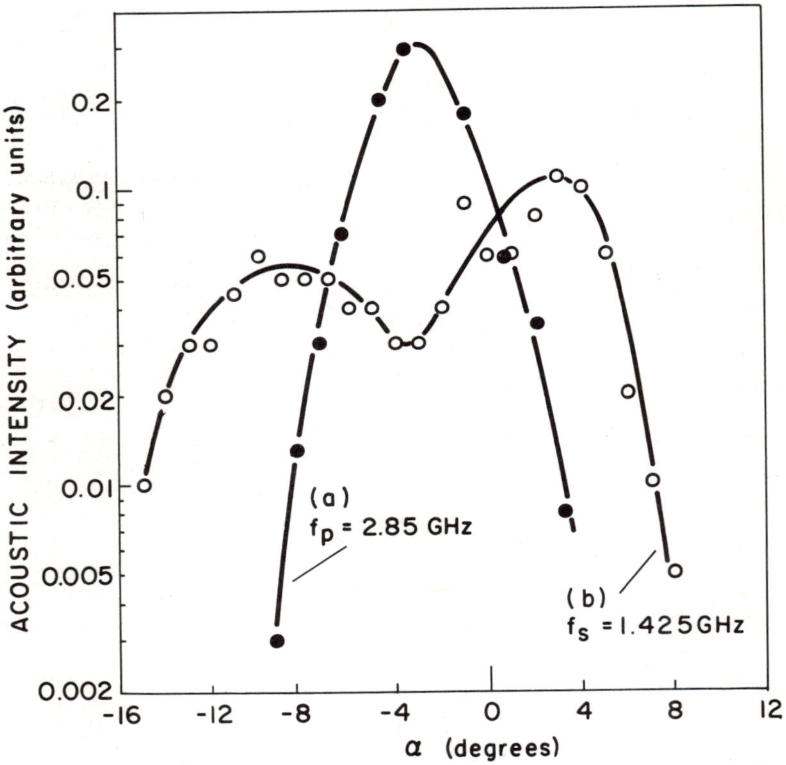

Fig. 3. Acoustic intensity as a function of propagation direction in the ZnO sample of Fig. 2 at $T = 0^{\circ}C$ for $v_d > v_s$. Curve (a), pump frequency, 2.85 GHz. Curve (b), signal frequency, 1.42 GHz.

SUMMARY

By means of Brillouin scattering we have been able to obtain a great deal of information about the phenomena of current saturation and acoustic domains in piezoelectric semiconductors. This has yielded some insight into the nonlinear effects that underlie these phenomena. In particular, we suggest that the progressive decrease in the frequency of maximum gain as the domain moves may be due to a combination of the parametric interactions described in the last section and the decrease in effective conductivity discussed in Ref. [8]. In any case, our data should make it possible to construct better theories of current saturation and domain formation.

ACKNOWLEDGMENTS

The authors acknowledge S. Stone for furnishing the laser and J. Baldovin, L. Johrdan and L. Vivenzio for technical assistance.

REFERENCES

1. V. Chandrasekharen, Proc. Ind. Acad. Sci. A33, 183 (1951). See also L. L. Hope, Phys. Rev. 166, 883 (1968).
2. J. Zucker and S. Zemon, Appl. Phys. Letters 9, 398 (1966); Erratum: J. Zucker and S. Zemon, Appl. Phys. Letters 10, 212 (1967).
3. S. Zemon, J.H. Wasko, L.L. Hope and J. Zucker, Appl. Phys. Letters 11, 40 (1967).
4. J. Zucker, S. Zemon and J.H. Wasko, "II-VI Semiconducting Compounds," D.G. Thomas (ed.), p. 919, W.A. Benjamin, Inc., New York, 1967.
5. S. Zemon, J. Zucker and J.H. Wasko, Proc. IEEE 56, 778 (1968).
6. S. Zemon, J. Zucker, J.H. Wasko, E.M. Conwell and A.K. Ganguly, Appl. Phys. Letters 12, 378 (1968).
7. D.L. White, J. Appl. Phys. 33, 2547 (1962).
8. J. Zucker, S. Zemon and J.H. Wasko, "Proceedings of the 9th International Conference on the Physics of Semiconductors," Moscow, 1968 (to be published).
9. W.H. Haydl and C.F. Quate, Phys. Letters 20, 463 (1966). See also I. Balberg and A. Many, Phys. Letters 24A, 707 (1967); N.I. Meyer and E. Moosekilde, Phys. Letters 24A, 155 (1967).
10. See, for example, J.A. Armstrong, N. Bloembergen, J. Ducuing and P.S. Pershan, Phys. Rev. 127, 1918 (1962).

G-9: BRILLOUIN SCATTERING IN LITHIUM NIOBATE*

R. J. O'Brien, G. J. Rosasco† and A. Weber
Department of Physics, Fordham University
Bronx, N. Y.

INTRODUCTION

With the advent of the laser renewed interest has been stimulated in the phenomenon of the scattering of light by thermal phonons first proposed by Brillouin[1]. In spectroscopy this has become an important method for the determination of the velocity of sound in liquids and solids. Since information is obtained in the hypersonic region this method serves as a complementary one to ultrasonics.

This paper reports the results of a Brillouin study in a single crystal of ferroelectric lithium niobate ($LiNbO_3$, symmetry class C_{3V}). Experiments were performed using a high resolution grating spectrograph. Through the use of derived selection rules, the observed Brillouin shifts for different phonon directions were assigned to specific modes of vibration. From algebraic expressions for the velocities appropriate elastic and piezoelectric constants were computed for the hypersonic region. These results are compared to those measured in the ultrasonic region.

THEORY

Using the conservation laws of energy and momentum one can easily derive the approximate expression for the Brillouin shifts, $\Delta\nu$, for the case of an anisotropic crystal

$$\frac{\Delta\nu}{\nu_A} = \frac{\omega_P}{\omega_A} \simeq \frac{v_P}{c} \sqrt{n_A^2 + n_B^2 - 2n_A n_B \cos\theta} \tag{1}$$

where ν_A is the incident frequency, v_P the phase velocity of the phonon, c the speed of light in vacuum, n_A and n_B are the indices of refraction for the incident and scattered

*Work supported in part by a grant from the National Science Foundation.

†American Can Co. Fellow.

wave respectively, for a given k_A and k_B of specific polarization and θ is the scattering angle. An exact result for the frequency shifts has been derived by Chandrasekharan[2]. Since the acoustic wavelengths experimentally encountered are of the order of several thousand Ångstroms, a classical theory of elastic waves and their interaction with electromagnetic radiation is deemed adequate.

In treating strongly piezoelectric materials, the contribution of internal fields to the elastic stiffness of the medium cannot be neglected. Any solution for wave propagation in a piezoelectric medium must satisfy simultaneously Maxwell's field equations and Newton's second law.

The macroscopic elastic disturbances in the medium are described by Newton's second law, i. e. the second time derivative of the displacement u, for a macroscopically small volume element is related to the force on the element which in turn is given by the divergence of a stress tensor, T_{ij}. Thus

$$\rho \frac{\partial^2 u_i}{\partial t^2} = F_i = \sum_j \frac{\partial T_{ij}}{\partial x_j} \tag{2}$$

where ρ is the mass density. The strain tensor, S_{ij}, defined in the usual manner can be written as

$$S_{ij} = \frac{1}{2}\left(\frac{\partial u_i}{\partial x_j} + \frac{\partial u_j}{\partial x_i}\right) \tag{3}$$

For quasi static fields in non-magnetic, non-conducting, charge free media, Maxwell's equations are given by

$$\nabla \times E = 0 \qquad \text{and} \qquad \nabla \cdot D = 0 \tag{4}$$

where E is the stress induced internal field. The coupled mechanical and electrical equations of state for adiabatic conditions are

$$T_{ij} = \sum_{k,\ell} C^E_{ijk\ell} S_{k\ell} - \sum_m e_{mij} E_m \tag{5}$$

and

$$\frac{D_m}{4\pi} = \sum_{i,j} e_{mij} S_{ij} + \sum_n \frac{\epsilon_{mn} E_n}{4\pi} \tag{6}$$

where $C^E_{ijk\ell}$ is the elastic constant tensor for constant field, e_{mij} the piezoelectric tensor and ϵ_{mn} the dielectric constant tensor. Simultaneous solutions of Eqs. (2),

(4) - (6) by plane harmonic waves yields the displacements u as solutions to the normal mode equation

$$\rho \omega^2 u_i = \sum_{j,k,\ell} C^*_{ijk\ell}\, u_k\, k_j\, k_\ell \tag{7}$$

where

$$C^*_{ijk\ell} = C^E_{ijk\ell} + 4\pi \frac{\sum_{p,q} e_{pij}\, e_{qk\ell}\, \hat{k}_p \hat{k}_q}{\sum_{m,n} \epsilon_{mn}\, \hat{k}_m \hat{k}_n} \tag{8}$$

is the stiffened elastic constant[3] and the $\hat{k}$'s are the direction cosines of the phonon. The electric field[4] as functions of u are expressed in the following form

$$E_n = -4\pi i k_n \frac{\sum_{m,i,j} e_{mij}\, \hat{k}_i \hat{k}_m u_j}{\sum_{m,n} \epsilon_{mn}\, \hat{k}_m \hat{k}_n} \tag{9}$$

The effect of the elastic and electric disturbances in the medium on the optical properties is given in terms of the photoelastic, $p_{ijk\ell}$, and electro-optic coefficients, r_{ijm}. As functions of E and the strains $S_{k\ell}$ this can be written as

$$\Delta B_{ij} = \sum_m r_{ijm} E_m + \sum_{k,\ell} p_{ijk\ell} S_{k\ell} \tag{10}$$

where ΔB_{ij} represents the changes in the coefficients of the index ellipsoid and r_{ijm} are the high frequency zero strain electro-optic coefficients[5]. The relationship between the elastic wave and the electric field given by Eq. (9) leads to an effective photoelastic constant[3].

$$P_{ijk\ell} = p_{ijk\ell} - 4\pi \frac{\sum_n r_{ijn} \hat{k}_n \sum_m e_{mk\ell} \hat{k}_m}{\sum_{m,n} \epsilon_{mn}\, \hat{k}_m \hat{k}_n} \tag{11}$$

The changes in the dielectric tensor ϵ_{ij} are linearly related to the strains $S_{k\ell}$ which can be written in the principal axis system as follows

$$\epsilon_{ij} - \epsilon^o_i \delta_{ij} \equiv \Delta\epsilon_{ij} = -n_i^2 n_j^2 \sum_{k,\ell} P_{ijk\ell} S_{k\ell} \tag{12}$$

where the n's are the principal indices of refraction and $P_{ijk\ell}$ is the effective photo-elastic constant. The results of the calculation for the scattered intensity inside the medium for a particular mode designated by 'a' is [6]

$$I^a \Big|_B^A = \frac{\pi^2 kT}{2} \frac{I_o}{\lambda_A^4} \frac{n_B}{n_A} \left[\frac{\left(\sum_{ij} \hat{B}_i \, \tilde{\Delta\epsilon}^a_{ij} \, \hat{A}_j \right)^2}{\rho (v^a_P)^2} \right]_{\mu = 0} \tag{13}$$

where $\mu = k_A - k_B \pm k_P$, I_o is the incident intensity in vacuum, $\hat{A}$ and $\hat{B}$ are the unit vectors in the direction of the electric field for the incident and scattered light, respectively, $n_A = n\,(k_A, \hat{A})$ and $n_B = n\,(k_B, \hat{B})$ are the indices of refraction. The quantities

$$\tilde{\Delta\epsilon}^a_{ij} = -n_i^2 n_j^2 \sum_{k,\ell} P^a_{ijk\ell} \left(S^a_{k\ell} \right)' \tag{14}$$

and

$$\left(S^a_{k\ell} \right)' = \frac{S^a_{k\ell}}{|u^a_P|\,|k_P|} = \frac{1}{2} \left[(\hat{u}^a_P)_k (\hat{k}_P)_\ell + (\hat{u}^a_P)_\ell (\hat{k}_P)_k \right] \tag{15}$$

are the "normalized polarizability" and "normalized strain" tensor respectively.

TRIGONAL CLASS: SYMMETRY C_{3V} (3m)

For crystal of C_{3V} (3m) symmetry the only non-vanishing elastic (constant field), piezoelectric and principal dielectric constants are

$C_{11} = C_{22}$	and	$e_{15} = e_{24}$
C_{12}		$e_{16} = e_{21} = -e_{22}$
$C_{13} = C_{23}$		$e_{31} = e_{32}$
C_{33}		e_{33}
$C_{14} = -C_{24} = C_{56}$		
$C_{44} = C_{55}$		$\epsilon_{11} = \epsilon_{22}$
$C_{66} = 1/2\,(C_{11} - C_{12})$		ϵ_{33}

where two indices are used according to the conventional scheme. It should be noted that the piezoelectric stiffening of the constants C_{ij}^{E} may lead to constants C_{ij}^{*} having a different scheme than that above.

EXPERIMENTAL APPARATUS AND PROCEDURE

A helium-neon laser light source operating at 6328Å and 20 mw was used to obtain Brillouin spectra which were photographed on the ninth order of a high resolution grating spectrograph. The laser beam was focused in the crystal for different polarizations and crystal orientations. In order to solve for all elastic constants the crystal was of a special cut which permitted phonon propagation along four distinct and soluble directions. Alignment of the crystal was accomplished through auto-collimation and all angles measured to ±0.5° to ±0.7°. The crystal was contained in a temperature control unit at a temperature of 25±1/4°C. All measured shifts, computed velocities and constants are for this temperature. In photographing the Brillouin spectra the diaphragm on the slit was divided, lengthwise, into two halves. The upper half contained an analyzer crossed with respect to one on the lower half. The analyzer was oriented relative to the incident laser polarizations and scattering plane. This procedure permitted observation of the different polarizations of the Brillouin lines and at times served to distinguish between lines of different polarizations otherwise not distinguishable due to the small birefringence of lithium niobate. All Brillouin shifts, accurate to within ± 0.005 cm^{-1}, were measured on a photoelectrically equipped Mann comparator.

DISCUSSION AND RESULTS

The experimentally determined velocities of the scattering phonons were computed from Eq. (1) which for a scattering angle $\theta = 90°$ takes the following form:

$$v_P = \frac{\Delta\nu}{\nu_A} \frac{c}{\sqrt{n_A^2 + n_B^2}}$$

where $c = 2.997925 \times 10^{10}$ cm/sec, $\nu_A = 15798.0010\ cm^{-1}$, n_A and n_B are the appropriate indices of refraction for incident and scattered light. For lithium niobate, a uniaxial crystal, the values of the principal indices n_o and n_e at 6328Å and 25°C are 2.286 and 2.200 respectively[7]. Table I lists the experimentally determined phonon velocities and comparison is made to those measured in ultrasonics[8].

The calculation of the elastic constants from the velocity expression obtained from Eq. (7) was routine with the exception of C_{14}. From the $(0, \frac{1}{\sqrt{2}}, \frac{1}{\sqrt{2}})$ phonon direction C_{14}^{E} could be calculated but consistency did not result when compared to the value of C_{14}^{E} calculated from the (0, 1, 0) direction. For the (0, 1, 0) phonon direction the piezoelectric coupling constants entered into the calculations and C_{14}^{E} proved to be

TABLE I

Comparison of Acoustic Velocities in $LiNbO_3$

VELOCITIES (m/sec)			
Phonon Direction	Mode	Present Work	Bateman and Spencer [8]
(1, 0, 0)	L	6543	6548.73
	S		4759.76
	S	3947	4034.06
(0, 1, 0)	QL	6770	6837.89
	QS	4470	4466.67
	S	3905	3940.43
(0, 0, 1)	L	7308	7330.59
	S	3562	3588.5
	S	3562	3588.5
$(0, \frac{1}{\sqrt{2}}, \frac{1}{\sqrt{2}})$	QL	7315	7300[b]
	S	3997	4000[b]
	QS	3990[a]	4000[b]

L = longitudinal, S = shear, QL = quasi-longitudinal, QS = quasi-shear.

[a] This value is calculated from the reported set of constants.

[b] These values are estimates from the graphs in Reference [8].

very sensitive to the values [9] used for e_{15} and e_{22}. By simultaneously solving for e_{15} and e_{22} from the (0, 1, 0) direction along with the values C^E_{14}, calculated from the $(0, \frac{1}{\sqrt{2}}, \frac{1}{\sqrt{2}})$ direction, a consistent set of values was obtained which is presented in Table II along with a comparison to ultrasonics[3, 9]. Velocity calculations using this reported set of elastic and piezoelectric constants have given consistent agreement for other directions and modes of propagation in this and another single crystal sample.

Recent work appearing in the literature on the variation of properties in lithium niobate as a function of melt composition [10] suggests that this could be a source of disagreement between our results and those determined from ultrasonic measurements.

Our present estimates indicate that P_{44} and P_{14}, the unmodified photoelastic constants, are small. Additional work is in progress to determine the magnitudes of these two constants.

TABLE II

Material Constants of $LiNbO_3$

	Present Work	Bateman and Spencer[9]	Warner[3]
	Elastic Constants (10^{11} N/m^2)		
c^E_{11}	2.0125	2.01564	2.03
c^E_{12}	0.5791	0.55610	0.53
c^E_{13}	0.7108	0.69340	0.75
c^E_{14}	0.0943	0.08120	0.09
c^E_{33}	2.4010	2.41179	2.45
c^E_{44}	0.5963	0.60523	0.60
c^E_{66}	0.7167	0.72977	0.75
	$\rho = 4.7 \times 10^3$ kg/m^3		
	Piezoelectric Constants[3, 9] (coul/m^2)		
e_{15}	3.705	3.67	3.7
e_{16}	-2.264	-2.40	-2.5
e_{31}	0.53	0.53	0.2
e_{33}	1.71	1.71	1.3
	Dielectric Constants[3]		
	$\epsilon_{11}/\epsilon_o = \epsilon_{22}/\epsilon_o = 44$	$\epsilon_{33}/\epsilon_o = 29$	

ACKNOWLEDGMENTS

The authors wish to express their indebtedness to Dr. J. J. Barrett of Perkin-Elmer Co., for the loan of the $LiNbO_3$ samples and to Dr. M. B. Schulz of Raytheon Co. for orienting and cutting one of the samples.

REFERENCES

1. L. Brillouin, Ann. Phys. (Paris) 17, 88 (1922).
2. V. Chandrasekharan, Proc. Ind. Acad. Sci. 33A, 183 (1951); Proc. Natl. Inst. Sci. Ind. 19, 547 (1953).
3. A.W. Warner, M. Onoe, and G.A. Coquin, J. Acoust. Soc. Am. 42, 1224 (1967).
4. J. Chapelle and L. Taurel, Compt. Rend. (Paris) 240, 743 (1955).
5. R. O'B. Carpenter, J. Opt. Soc. Am. 40, 225 (1950).
6. L. Cecchi, "Thesis", University of Montpellier, France (1964).
7. G.D. Boyd, R.C. Miller, K. Nassau, W.L. Bond, and A. Savage, Appl. Phys. Letters 5, 234 (1964).
8. E.G. Spencer, P.V. Lenzo, and A.A. Ballman, Proc. IEEE 55, 2093 (1967). Values from T.B. Bateman and E.G. Spencer measurements are cited in this reference.
9. E.G. Spencer, private communication.
10. J.G. Bergman, A. Ashkin, A.A. Ballman, J. M. Dziedzic, H.J. Levinstein, and R.G. Smith, Appl. Phys. Letters 12, 92 (1968).

G-10: BRILLOUIN SCATTERING IN PARAMAGNETIC CRYSTALS

W. Low and J. Bronstein
Microwave Division, Department of Physics,
The Hebrew University
Jerusalem, Israel

The importance of Brillouin scattering in paramagnetic systems lies in the fact that it is relatively easy to generate microwave phonons of different frequencies, to detect these optically and to study their interaction with the paramagnetic system. Detection by optical techniques of microwave phonons is, apart from the ease of detection, of some advantage since signal to noise that can be obtained by high energy photon counting is superior to that usually obtained by conventional macroscopic microwave detection techniques.

One moreover may obtain information which is difficult and in some cases impossible to obtain by conventional acoustic spin resonance techniques. Consider the incidence of laser light on an insulating crystal with a dilute paramagnetic impurity. In general, the Hamiltonian will consist of

$$\mathcal{H} = \mathcal{H}_L + \mathcal{H}_E + \mathcal{H}_S + \mathcal{H}_{L\text{-}S} + \mathcal{H}_{L\text{-}E} + \mathcal{H}_{S\text{-}E} \tag{1}$$

$\mathcal{H}_L$ represents the acoustical lattice part and can be written with the notation of second quantization as

$$\mathcal{H}_L = \sum_{\underset{\sim}{k}, i} \hbar\,\omega(\underset{\sim}{k}) \left[a^{+}_{\underset{\sim}{k}i}\, a_{\underset{\sim}{k}i} + 1/2 \right] \tag{2}$$

where $\hbar\,\omega(\underset{\sim}{k})$ is the energy of the phonon wave vector $\underset{\sim}{k}$ and polarization i.

$\mathcal{H}_E$ represents the radiation field due to the laser and is given by

$$\mathcal{H}_E = \sum_{\underset{\sim}{q}, \lambda} \hbar\omega(\underset{\sim}{q}) \left[b^{+}_{\underset{\sim}{q}\lambda}\, b_{\underset{\sim}{q}\lambda} + \frac{1}{2} \right] \tag{3}$$

where λ is the polarization and $\hbar\omega(\underset{\sim}{q})$ the energy of the phonon; and the spin field can be represented by

$$\mathcal{H}_S = \sum_{j, m} \epsilon_m\, P^{j}_{mm} \tag{4}$$

where ϵ_m are the energy levels of the different spin levels, and P^j_{mm} are the statistical operators of the paramagnetic ion j.

The three interaction terms, $\mathcal{H}_{L-S}$, $\mathcal{H}_{L-E}$ and $\mathcal{H}_{S-E}$ couple the three fields. Such a system for strong coupling parameters is difficult to solve. If one of the coupling parameters between two fields is stronger than the others, and all of these relatively weak, one may regard the third field as a small perturbation and the problem becomes tractable.

The acoustic phonon wavelength and the light wavelength used at present in Brillouin scattering extend over a macroscopic part of the sample. The solution of the equations can be effected by a semi-macroscopic approach in which the electric field of the scattered light and the spectral density are evaluated. Green's function or many body problem techniques will yield essentially the same results in this approximation.

Experiments can be designed in which one of the interaction parameters is especially strong.

A relatively simple case which we will discuss is when paramagnetic spins are strongly coupled to the acoustic phonons and the light source is the probing field. In this case the dominant interaction Hamiltonian is $\mathcal{H}_{L-S}$, which can be written

$$\mathcal{H}_{L-S} = \sum_{j,m,n} \sum_{\underset{\sim}{k},i} R_{mn} P^j_{mn} \left(a_{\underset{\sim}{k},i}\, e^{i\underset{\sim}{k}\cdot\underset{\sim}{r}_j} - a^+_{\underset{\sim}{k},i}\, e^{-i\underset{\sim}{k}\cdot\underset{\sim}{r}_j}\right) \tag{5}$$

Rather than solving Eq. (5) for different paramagnetic ions in different point symmetries caused by the various crystal fields, it is convenient to use a semi classical description in terms of a spin Hamiltonian.

In the conventional language of paramagnetic resonance, one can describe the behavior of the energy levels in a magnetic field by a spin Hamiltonian. For the iron group the Hamiltonian, conveniently used, is written as

$$\mathcal{H}'_S = \beta \underset{\sim}{H} \cdot \underset{\approx}{g} \cdot \underset{\sim}{S} + \underset{\sim}{S} \cdot \underset{\approx}{D} \cdot \underset{\sim}{S} + \underset{\sim}{I} \cdot \underset{\approx}{T} \cdot \underset{\sim}{S} \tag{6}$$

where the g factors give the first order Zeeman effect, D the initial splitting, T the hyperfine tensor, S the effective electronic spin, and I the nuclear spin. For rare earth ions, a slightly more complicated Hamiltonian is used

$$\mathcal{H}''_S = \beta \underset{\sim}{H} \cdot \underset{\approx}{g} \cdot \underset{\sim}{S} + \sum_{m,n} B^m_n O^m_n \tag{7}$$

O^m_n are spherical harmonic operators. The number of terms which have to be used is restricted by point symmetry considerations and the character of the ground state.

We now can take over the formalism developed for the acoustic spin resonance experiments [1]. The phonon spectrum will be modulated by the strain e and hence change the parameters g, D, T or B^m_n .

Let us write, for example

$\delta D_{ij} = G_{ijkl}\, e_{kl}$ where G_{ijkl} is a coupling

parameter of the fourth rank.

The attenuation of the acoustic wave in a paramagnetic sample is given for the Hamiltonian given by Eq. (6) as

$$\alpha_o = \frac{n\omega}{4\rho h v_o^3} \left[\sum_{\substack{i,j\\k,l}} |< \beta \left(H_i S_j + H_j S_i \right) \overline{F_{ijkl}} + (S_i S_j + S_j S_i) \overline{G_{ijkl}} + + (I_i S_j + I_j S_i) \overline{L_{ijkl}} > |^2 \right] g(\omega) \tag{8}$$

where the F_{ijkl} and L_{ijkl} are the coupling tensors for $\underset{\sim}{G}$ and $\underset{\sim}{T}$, and a similar expression for Eq. (7). The attenuation is to this order independent of the amplitude strain.

In general, this can be written as

$$\alpha_o = \frac{n\omega}{4\rho h v_o^3} G^2 \, |<M>|^2 \, g(\nu) \tag{9}$$

where n is the difference in population between the spin states, ρ the density of the crystal, G the generalized spin-phonon coupling constant (which is a function of the phonon frequency), $<M>$ the generalized matrix element connecting the spin levels, v_o the velocity of sound in the system, and $g(\nu)$ the line shape function. From equation (9) it is seen that the spin-acoustic phonon interaction will change the acoustic attenuation. The effect on the Brillouin scattering is, therefore, to change the intensity of the scattered Brillouin components, and in higher order also the position of the Brillouin component. The Brillouin components, in the absence of paramagnetic impurities, have usually a line width which depends on a number of "order" parameters which affect the sharpness of the elastic moduli in the crystal. If this line width is smaller than $g(\nu)$, the Brillouin component will be broadened because of the shortened lifetime of the phonon in the paramagnetic system. The value of the matrix elements is a function of the magnetic field and these effects can be detected as well.

A particularly interesting case is when the external magnetic field is adjusted so that the separation between the Zeeman levels $\hbar\omega$ is nearly or completely resonant with the acoustic frequency. Assuming fairly strong spin phonon coupling and within the harmonic approximation of ionic displacements in the crystal, one will obtain anomalous dispersion. The waves consist of mixed spin-acoustic waves, the attenuation $\alpha(\omega)$ is given to first order, for $\omega_o \sim \omega$ by

$$\alpha = \alpha_o \frac{\Gamma}{(\omega-\omega_o)^2 + \Gamma^2} \tag{10}$$

where Γ is the approximate inverse of the characteristic spin-lattice relaxation time. Together with the anomalous attenuation is a change in the phase velocity [2],

$$\left(\frac{v}{v_o}\right)^2 = 1 + \sum_{i,j} \frac{\overline{e_{ij}^2}\, B_{ij}}{\omega_{ij}^2 - \omega^2 + \Gamma_{ij}^2 - 2i\omega\Gamma_{ij}} \tag{12}$$

where ε_{ij} is a strain parameter, and B_{ij} is proportional to the ratio of the magnetization and the restoring force.

Taking the data from conventional acoustic spin resonance, one can evaluate this change in the sound velocity used for some non Kramers ion group elements, this change may be of the order of a few percent. Hence, in general, each of the Brillouin components will be split into components [3]. The separation will depend on the strength of the generalized interaction constant G. If the Brillouin components are sufficiently narrow, this fine structure could be resolved. This resonance can be expected whenever the energy level separation coincides nearly with the acoustic frequency which is defined by the direction of the observation. It is in principle possible to evaluate the fine and hyperfine structure tensors using this technique.

We shall briefly summarize what experiments are feasible. Some of these are being actively pursued in our laboratory.

(a) Detection of the attenuation due to spin-acoustic coupling in paramagnetic systems can be made. The interaction constants can be measured as a function of the phonon frequency into the microwave region. Anisotropy in the attenuation can be determined.

(b) The spin Hamiltonian parameters can be obtained particularly for non Kramers doublets. This is not different from conventional spin-resonance acoustic techniques. However, these experiments can be done with ease and as a function of the acoustic frequencies. Since the Brillouin scattering detects both longitudinal and transverse components, the spin-acoustic interaction with these different branches can be found.

(c) The attenuation is dependent on Γ. In the usual resonance techniques, one restricts oneself because of experimental considerations to long spin-lattice relaxation. It is however possible, using Brillouin scattering techniques, to measure short times as well, in the Raman region.

(d) Brillouin scattering techniques, coupled with spin resonance, may elucidate important details regarding the phonon bottleneck and the diffusion of phonons from this bottleneck. Saturation of the Zeeman levels by microwave frequency at low temperatures creates presumably a narrow band of phonon frequencies corresponding to the energy difference between the two Zeeman levels.

(e) As the paramagnetic concentration is increased, the interpretation of the acoustic attenuation is becoming increasingly difficult to evaluate. Very few data are at present available of acoustic attenuation as a function of concentration of paramagnetic impurities. A study of this, using Brillouin scattering, may yield interesting information regarding cooperative phenomena caused by exchange interaction.

(f) More difficult experiments, but not inconceivable experiments, are those on excited optical states, on triplet states in organic compounds, on paramagnetic radiation and F centers and others.

A few remarks regarding the experimental arrangements. One of the great difficulties in Brillouin scattering is the usually large Rayleigh scattering which often masks the Brillouin wings. The Rayleigh scattering is caused by a number of mechanisms such as thermal diffusion, mass diffusion. However, by far the largest contributing factors are different types of crystal defects, which scatter light strongly. It is, therefore, a necessity to choose crystals in which the Rayleigh scattering is held to a minimum. In the case of diamagnetic crystals with paramagnetic impurities, this scattering is enhanced since these very impurities are additional scattering centers. Conventional techniques such as piezoelectric or pressure scanning Fabry - Perot interferometers do not discriminate between the Brillouin and Rayleigh scattering.

However, in paramagnetic systems, one may modulate the magnetic field and use phase sensitive detection. This discriminates against the Rayleigh scattering, but not completely. The magnetic susceptibility will give rise to low frequency fluctuations which will be modulated with the modulation frequency of the magnetic field. The magnetic Rayleigh scattering has, to our knowledge, not been studied and is also of great interest. This modulation is smaller than that of the Brillouin components; the effect of the spin-acoustic coupling can then be relatively easily detected.

In addition, methods of optical detection of E.S.R. can be used for Brillouin scattering. As one passes through resonance, the lines are split as indicated before. The polarization and the angular distribution of the light is slightly altered.

This can be easily seen in the particle picture. We have to conserve energy and moments of the incident photon, the created or absorbed phonon, the spin, and the scattered light. The polarization and the selection rules will be changed, in particular, in the region of anomalous dispersion. One can, therefore, utilize this in detection of the fine and hyperfine structure even in the case when the individual splittings of the additional components are not resolved with the Fabry-Perot scanning technique, by looking for changes in the polarization or light changes in the angular dependence of the scattered radiation.

Finally, in many cases it may be convenient to use an external acoustic source. A transducer may be applied to the crystal and the transducing frequency varied. In this case, one generates a strong band of phonons from which the light will be scattered. Since the number of phonons is large, one may discriminate against the Rayleigh scattering in even not so perfect crystals. The sensitivity can be considerably enhanced, as shown by Cohen and Gordon [4]. There are, however, some differences between this type of experiment and the conventional spontaneous Brillouin scattering type of experiment. First of all, one measures the frequency dispersion $\Delta\omega$ rather than the wavelength dispersion $\Delta \underset{\sim}{k}$. A more subtle difference is that with an external source the boundary conditions are different, and this may lead to slight differences.

We have restricted ourselves in this paper only to the case when one of the interactions $\mathcal{H}_{L-S}$ is the dominant. It is obvious that experiments can be devised in which the other interactions may become of equal importance, or even stronger. The effects of this is at present under consideration. Further, the possibility to extend these to other dilute magnetic systems, as in metals or in alloys, should be explored.

REFERENCES

1. See, for example, the review paper by E.B. Tucker, "Physical Acoustics 4 A," W.P. Mason (ed.), Academic Press, 1966. This paper contains all the major references to acoustic spin resonance data.
2. E.H. Jacobsen and K.W.H. Stevens, Phys. Rev. 129, 2036 (1963).
3. B.I. Kochelaev, Soviet Physics Doklady 11, 130 (1966); S.A. Al'tshuler and B.I. Kochelaev, Soviet Physics, J.E.T.P. 22, 600 (1966).
4. E.I. Gordon and M.G. Cohen, Phys. Rev. 153, 201 (1967).

H-1: SPECTRUM OF LIGHT SCATTERED BY CRITICAL FLUCTUATIONS*

George B. Benedek
Department of Physics and Center for Materials Sciences and Engineering,
Massachusetts Institute of Technology
Cambridge, Massachusetts

ABSTRACT

Near the critical point of second order phase transitions the order parameter describing the system shows very large amplitude fluctuations which relax back to equilibrium ever more slowly as the critical point is approached. This paper will present a brief review of experimental studies of the intensity and spectrum of light scattered near critical points of simple fluids and two component critical mixtures. These provide detailed information on the divergence of the fluctuations and their relaxation times in fluids. As has been pointed out by Ginzburg analogous behavior is to be observed in solids near second order phase transitions. We also propose a measurement of the spectral width of quasielastically scattered light as an indicator of the existence of critical opalescence in a solid. Estimates of the line width for such scattering are given for the case of NH_4Cl.

In early 1962 V. L. Ginzburg presented a paper before the Scientific Council of the U.S.S.R. Academy of Sciences which convened in honor of the memory of G.S. Landsberg. Ginzburg's paper was entitled "The Scattering of Light Near Points of Phase Transition in Solids." The paper appeared later in Uspekhi [1]. Ginzburg's discussion of this topic is so appropriate to the subjects which are about to be discussed in this session of the conference that I feel that I can do no better than take his paper as a model for my own brief talk.

Ginzburg first reminds us that the hallmark of the critical phenomena which accompany second order phase transitions is the occurrence of very large fluctuations in the parameter describing the ordered phase of the system. The order parameter in a pure fluid is the density ρ . In a two component mixture it is the concentration of one constituent. In a ferromagnet it is the magnetization. In a ferroelectric it is the polarization. In a superfluid the order parameter is the number of partioles in the zero momentum state. In a superconductor the "gap parameter" describes the order.

*This research was supported by the advanced Research Projects Agency under contract SD-90 and by the U. S. Army Research Office-Durham under contract DA-31-124-ARO-D-425.

The order parameter takes on very large amplitude fluctuations near the critical point because the work required to produce such fluctuations goes to zero there. The order parameter is a thermodynamic variable whose average value describes the thermodynamic state of the system. Statistical mechanics however shows us that the thermodynamic variables fluctuate around their average value with a probability distribution which is Gaussian. The mean square amplitude of this Gaussian distribution can be related to the second derivative of entropy S as [2].

$$\langle \Delta \eta^2 \rangle = \frac{k}{\frac{\partial^2 S}{\partial \eta^2}} \tag{1}$$

where η in general is the thermodynamic variable whose fluctuations we are determining. Near T_c if η is the order parameter $\partial^2 S/\partial \eta^2 \to 0$ in accordance with the definition of the term "second order" phase transition. If the order parameter is coupled to the dielectric constant ϵ this large amplitude fluctuation in η produces a large fluctuation in ϵ and hence a very large scattering of light. We therefore have the phenomenon of critical opalescence: the striking increase in the scattering of light near a second order phase transition. We can see from this line of reasoning that opalescence is a characteristic concomitant of a critical phase transition. Ginzburg gives examples of such opalescence for the case of a single component fluids near the critical temperature and a mixture near its critical mixing temperature and concentration. Ginzburg also draws our attention to the large increase in the intensity of light scattered near the $\alpha \rightleftarrows \beta$ transition in quartz which occurs near 850°K. He argues that this is an example of critical opalescence in a solid. We shall hear shortly a paper by H. Z. Cummins and S.M. Shapiro in which Ginzburg's interpretation is brought into question.

The burden of my present remarks is to pose a question and suggest a means of answering it. The question is: "Has critical opalescence been observed in a solid?" To answer it I shall suggest that the quasielastic spectrum of the light strongly scattered from NH_4Cl near its second order phase transition at -30°C be measured. To see how the quasielastic spectrum is an unfailing indicator of the onset of critical fluctuations I must depart momentarily from the gospel according to Ginzburg to discuss the spectrum of light scattered from a fluid near its critical point.

Irreversible thermodynamics has as one of its basic postulates the notion that if a thermodynamic variable departs from its equilibrium value by an amount $\Delta \eta$, it will relax back to the value $\Delta \eta = 0$ at a rate determined by the slope of the entropy - η curve evaluated at the off-equilibrium position. From this postulate it follows at once that the rate of relaxation Γ_η is related to $\partial^2 S/\partial \eta^2$ once again by a relation now of the form [2]

$$\Gamma_\eta = \beta \frac{\partial^2 S}{\partial \eta^2} \tag{2}$$

where β is some kinetic or transport coefficient. In particular systems we can identify the precise form of Eq. (2). For example, for a pure fluid Γ_η has the form

$$\Gamma_{S_K} = \frac{\Lambda K^2}{C_p'} \tag{3}$$

Here the parameter that relaxes back to equilibrium is the Kth Fourier component S_K of the entropy fluctuation. Its relaxation rate (Γ_{S_K}) is measured by measuring the spectral width of the light scattered an angle θ away from the forward direction i.e.

$$K = \frac{4\pi}{\lambda_o} n \operatorname{Sin}(\theta/2) \tag{4}$$

where $\frac{\lambda_o}{n}$ is the light wavelength inside the medium.

In Eq. (3) $\frac{1}{C_p'} = \left(\frac{\partial^2 S}{\partial h^2}\right)$ where C_p' is the specific heat at constant pressure per unit volume. Λ is the thermal conductivity of the fluid.

In a two component mixture the order parameter is the Kth Fourier component of the concentration fluctuation and its relaxation rate is given by:

$$\Gamma_{c_K} = \alpha K^2 \left(\frac{\partial \mu}{\partial c}\right) \tag{5}$$

where $\left(\frac{\partial \mu}{\partial c}\right)$ is the concentration derivative of the difference of chemical potential between the two constituents. The kinetic coefficient is α the socalled "concentration conductivity."

Eq. (2), (3), and (5) have the property that as the critical point is approached the quantities $\partial^2 S/\partial h^2$, $1/C_p'$, and $\partial\mu/\partial c$ go to zero. Thus the order parameter relaxes back to its undisturbed value ever more slowly as the critical point is approached. Near T_c the amplitude of the fluctuations grow very large and the time for these to return to equilibrium also grows very long.

This association of opalescence with slowing down produces along with the extraordinary divergence in the intensity of scattered light a concomitant narrowing of its spectral width. In fact the width of the spectrum of light scattered from pure fluids and two component systems near their critical points ranges from a few cycles per second or less to a few kilocycles. The detection of such narrow lines requires spectroscopic methods having a resolving power of ~ 10^{14} which is 6 orders of magnitude higher than what is available in the best optical devices such as the Fabry-Perot spectrometers.

To provide sufficient resolution it is necessary to use the techniques of optical mixing spectroscopy. These techniques are the extension to the optical frequency region of the techniques of superheterodyne and monodyne or low level detection commonly used in radio frequency or microwave receivers. In the heterodyne method, the spectral composition of the signal is found by mixing the signal with a monochromatic local oscillator in a square law device. In the optical region the square law mixer is the photocathode of a photomultiplier. In the monodyne, or "self beating" method no local oscillator is used. The spectral composition of the signal appears as beat notes in the photomultiplier output between each of the spectral components in the incident signal. The spectrum of the beat notes, either heterodyne or self beat is found from a spectral analysis of the fluctuations in the output photocurrent from the photomultiplier.

A.T. Forrester, R.A. Gudmundson and P.O. Johnson [3] conducted in 1955 a photoelectric mixing experiment on the Zeeman components of light produced by a mercury source. The beat notes were detected in a 3 cm cavity built into the phototube. With the invention of lasers Forrester [4] presented a calculation of the signal to shot noise ratios to be expected in photoelectric mixing experiments and suggested that these methods be used to measure the spectral components in the output of lasers.

Professor C. H. Towns [5] suggested that heterodyne mixing spectroscopy be used to study the spectrum of light produced in Brillouin, Rayleigh and Raman scattering. The present author independently suggested its use in Brillouin scattering [6, 7].

Both the heterodyne beat and the self beat methods have been used to study the spectrum of light scattered quasielastically from pure fluids and from mixtures. S. Alpert, Y. Yeh and E. Lipworth [8, 9] studied the quasielastic scattering from a critical mixture using a heterodyne beat spectrometer [8]. N.C. Ford, Jr. and G.B. Benedek introduced a self beating spectrometer to study the quasielastic spectrum of light scattered from sulphur hexafluoride near its critical point [10, 11]. Also Alpert and his collaborators used a heterodyne beat spectrometer to study the quasielastic scattering from CO_2. Since those early experiment's data from a number of laboratories has provided increasingly more detailed information on the temperature and density dependence of Γ_{S_K} and Γ_{c_K}. I should like to describe briefly the principle results that are now emerging.

First we deal with data on pure liquids. These data show that along the critical isochore and along the coexistence curve that Γ_{S_K} can be expressed as a function of temperature in the form Γ_{S_K} = const $(T - T_c)^{\sigma \pm}$. σ_+ being the exponent appropriate for temperatures above the critical temperature T_c, and σ_- that for below T_c along the coexistence curve. H.Z. Cummins and H.L. Swinney [13] have made a very careful measurement of σ_+ and σ_- in the case of CO_2 and find $\sigma_+ = 0.73 \pm 0.02$ while $\sigma_- = 0.66 \pm 0.05$ along the liquid side of the coexistence curve and $\sigma_- = 0.72 \pm 0.05$ along the gas side. Their results have the beautiful property of symmetry i.e. σ_+ and σ_- are essentially the same. This is consistent with widely held views as to the symmetry of the critical properties along the critical isochore and coexistence curve. These results on CO_2 also have the satisfying feature that it is in agreement with the theoretical prediction of Kadanoff and Swift that Γ_{S_K} is proportional to the correlation range, which according to scaling law ideas vary with temperature like $(T - T_c)^{0.66}$. By combining Cummins and Swinney's data on Γ_{S_K} with static measurements of C_p it is possible to deduce that the thermal conductivity, a transport coefficient, diverges at $T \rightarrow T_c$ like $(T - T_c)^{-0.6}$.

In contrast to these CO_2 results the data of A. Saxman and G.B. Benedek [15-17] on sulphur hexafluoride indicates very clearly that $\sigma_+ = 1.27 \pm 0.02$ and $\sigma_- = 0.633 \pm 0.003$. These values are quite astonishing as they imply a great asymmetry in the divergence of the thermal conductivity as $T \rightarrow T_c$ from above or from below, provided of course that $C_p{}'$ diverges symmetrically around T_c as is believed. These results also suggest that at least insofar as the transport coefficients are concerned all fluids do not behave similarly. Both Cummins and Swinney at Johns Hopkins and M. Giglio at M.I.T. are now investigating the monatomic fluid xenon near its critical point. In this fluid there should be no internal molecular degrees of freedom to contribute to the thermal conductivity. Professor Cummins has informed me that their early results indicate that σ_+ for xenon is definitely less than unity, in line with the behavior they find in CO_2.

Work on two component mixtures is being carried out by B. Chu [18], by P. Berge and B. Volochine [19] and by S.H. Chen and N. Polonski [20]. These workers have used self beating spectroscopy in a study of isobutyric acid-water system, cyclohexane-aniline system and nitrobenzene - N hexane system respectively. The results in all these mixtures are quite similar. They studied Γ_{c_K} at the critical concentration as a function of temperature for $T \geq T_c$ and find $\Gamma_{c_K} = \text{const}\,(T - T_c)^{\mu_+}$ where $\mu_+ \sim 0.6$. Since $(\partial\mu/\partial c)$ goes to zero like $(T - T_c)^{1.24}$, as B. Chu has shown from intensity measurements of the scattered light, it follows that the "concentration conductivity" α diverges like $(T - T_c)^{-0.6}$. This is very much like the divergence in thermal conductivity for a pure fluid. Let me conclude my discussion of the critical scattering from fluids at this point. I hope this has given some sense of the kind of information this sort of experiment is now yielding.

How is this behavior in the fluid related to the spectrum of light scattered by a solid near its critical point? To deal with this let us return to Ginzburg's treatment. Departing from the fluid analogy, Ginzburg pointed out that the onset of the order in a ferroelectric was intimately connected with the collapse of an optical mode in the vibration spectrum of the solid as the lattice approaches its polarized state. He discusses the Landsberg - Mandlestam experiment on quartz and concludes that the 207 cm^{-1} line is produced by the softening mode and that this line is the one that approaches zero frequency as $T \to T_c$. What Ginzburg does not do however is to examine that feature of the spectrum which seems to me to most directly bear on the establishment of order. That is the spectral width of the central or Rayleigh component. If we have a critical opalescence in quartz, or any solid for that matter we must expect that the feature in the spectrum whose intensity grows large will also become extremely narrow. Indeed in the case of NH_4Cl near its order disorder phase transition at - 30°C the central component intensity grows large. In fact P. Lazay, N. Clark and J.H. Lunacek have reported on this in a previous paper [21]. If this "opalescence" is really due to fluctuations in the order parameter, the spectral width of the central component should grow narrow as $T \to T_c$. How narrow can we expect it to be, and can this width be detected? I think it can be detected and would like to give some estimates. Using Eq. (3) for the width of the quasielastic component of the scattered spectrum and the room temperature values $C_p = 1.59$ joules/gm°K, $\lambda \sim 6 \times 10^{-3}$ watts/cm °K [22], furthermore, using as the index of refraction 1.635 and the light wavelength 6328 Å we find that for 90° scattering the half width at half height $\Delta\nu_{1/2}$ is equal to ~ 60 MHz. In the vicinity of the order disorder transition in NH_4Cl the specific heat C_p' grows very large. Unpublished data by C. Stephenson of M.I.T. indicates that near T_c, C_p' can be as large as 100 times that at room temperature. This indicates that if Λ does not diverge near T_c that the line width could be as small as ~ 600 kHz. This should be measurable using optical mixing techniques provided that an argon ion laser is used. If such a laser were operated with an output of 500 m.w., the increase in quantum efficiency and the $1/\lambda^4$ factor in the scattering power indicates that the total detectible scattered power would be 60 times greater than that scattered from toluene at room temperature. In toluene the post detection signal to noise ratio for a 3kHz wide line is 10:1 [23]. Thus one could expect in the NH_4Cl case to detect with the same signal to noise ratio a line as wide as 180 kHz. The experiment therefore appears feasible even at 90°. Of course the power per unit bandwidth can be increased by scattering closer to the forward direction. It would appear then worthwhile to make such a study to determine whether or not the "opalescence" observed in NH_4Cl is in fact connected with a critical slowing down. I also submit that this critical narrowing of the spectrum of the scattered light is the essential feature of critical scattering.

May I conclude then by simply reading to you the conclusion of Ginzburg's paper. It seems particularly prophetic: remember he writes in 1961.

"But why are there no experimental researches on the spectrum of the scattered light near second-order phase transition points? It is hard to give a completely definite answer to this question. In our opinion there are three facts to be mentioned in explanation. First, there is still too little recognition of the potentialities of measurements made along the lines we have suggested. Second, these measurements are by no means simple. Third, the study of the scattering of light is, so to speak, an old classical problem and has "gone out of style."

There are, however, many examples of "old" and "unstylish" fields of research which have again come to the center of attention as the result of the appearance of more modern experimental resources, new ideas, or new objects of study. It may be that this will also be the fate of the study of the scattering of light, especially in solids. Here there are many new materials which have second-order phase transitions (ferroelectrics, so-called antiferroelectrics, and others). We hope that it is clear that the study of the scattering of light in these substances is important. Finally, remarkable new light sources have been developed (quantum generators-lasers), which are as it were especially predestined for the spectral analysis of scattered light.

Will all of these favorable conditions be enough to make work on the scattering of light again lead to great advances in the study of crystals and molecules? We shall of course know the answer to this question only in the future."

REFERENCES

1. V.L. Ginzburg, Sov. Phys. Uspekhi 5, 649 (1963).
2. L.D. Landau and E.M. Lifschitz, Statistical Physics, Addison Wesley, Reading, Mass., 1958.
3. A.T. Forrester, R.A. Gudmundson, and P.O. Johnson, Phys. Rev. 99, 1691 (1955).
4. A.T. Forrester, J. Opt. Soc. Am. 51, 253 (1961).
5. C.H. Townes, "Advances in Quantum Electronics" pp. 3-11, J.R. Singer (ed.), Columbia, New York, 1961.
6. G.B. Benedek, "Research in Materials Sciences and Engineering," M.I.T. Annual Report, 1962 - 1963.
7. J.B. Lastovka and G.B. Benedek, "Physics of Quantum Electronics," p. 231, P. Kelley, B. Lax, and P.E. Tannenwald (eds.), McGraw Hill Co., New York, 1966.

8. S.S. Alpert, Y. Yeh, and E. Lipworth, Phys. Rev. Letters 16, 639 (1966).
9. S.S. Alpert, Proc. Conf. on Phenomena in the Neighborhood of Critical Points NBS, Misc. Publ. 273, Washington D.C., NBS 1965.
10. N.C. Ford, Jr. and G.B. Benedek, Phys. Rev. Letters 15, 649 (1965).
11. N.C. Ford, Jr. and G.B. Benedek, Proc. Conf. on Phenomena in the Neighborhood of Critical Points NBS, Misc. Publ. 273, Washington D.C., NBS 1965.
12. S.S. Alpert, D. Balzarini, R. Novick, L. Siegel, and Y. Yeh, "Phys. of Quantum Electronics," p. 253, P.L. Kelly, B. Lax, and P. Tannenwald (eds.) McGraw Hill Book Co., New York, 1966.
13. H.L. Swinney and H.Z. Cummins, Phys. Rev. 171, 152 (1968).
14. L. Kadanoff and J. Swift, Phys. Rev. 165, 310 (1968) and 166, 89 (1968).
15. G.B. Benedek, "Thermal Fluctuations and the Scattering of Light", Brandeis Summer Institute for Theoretical Physics 1966, Gordon and Breach (in press).
16. P. Heller, Repts. on Prog. in Phys. XXX, 731 (1967).
17. G.B. Benedek, "Optical Mixing Spectroscopy with Applications to Physics, Chemistry, Biology and Engineering." A Kastler Jubilee Volume, Paris (in press).
18. B. Chu and F.J. Schoenes, Phys. Rev. Letters 21, 6 (1968).
19. P. Berge and B. Volochine, Phys. Rev. Letters 26A, 267 (1968).
20. S.H. Chen and N. Polonsky, Bull. Am. Phys. Soc. 13, 183 (1968).
21. P. Lazay, J.H. Lunacek, N.A. Clark, and G.B. Benedek, Paper G-5 this conference.
22. K. Apel and C. von Simson, Zeitschrift für Phys. Chem. N.F. 25, 393 (1960).
23. J.B. Lastovka and G.B. Benedek, Phys. Rev. Letters 17, 1039 (1966).

H-2: CRITICAL HARMONIC SCATTERING IN POWDERED NH_4CL

Isaac Freund
Bell Telephone Laboratories, Incorporated
Murray Hill, New Jersey

Optical frequency doubling in powdered NH_4Cl near the order-disorder transition temperature of $T_c \approx 242.4°K$ has recently been reported[1,2]. These experiments have been interpreted in terms of the pair correlation function, G(r), for angular reorientation of NH_4^+ ions, and have been found to lead to results that are in apparent conflict with present critical point theory. As has been emphasized[1,2], such intensity measurements cannot yield G(r) directly, but if the form given by theory is assumed, i.e., $G(r) = \text{const} \times r^{-(1+\eta)} \exp(-\varkappa r)$, then the temperature dependence of the inverse correlation length, $\varkappa$, may be determined. Theory predicts that for an Ising lattice $\varkappa = A(T/T_c - 1)^{\nu}$, with $\nu = 0.64$[3], while analysis of the experimental data yielded $\nu = 2.2$. Such an analysis is dependent upon a sufficiently correct description of the influence of multiple scattering of both the laser and harmonic radiation by the powdered sample. Multiple scattering of the laser radiation has been implicitly neglected previously [1,2], since it was already known that its inclusion could not greatly affect the value of ν derived from the data. We consider this problem explicitly here and verify this.

The measured harmonic intensity, σ, may be written[2],

$$\sigma = \text{const} \times \int_{\underset{\sim}{k}_{min}}^{\underset{\sim}{k}_{max}} d\underset{\sim}{k} \int_{0}^{\infty} 4\pi r^2 \frac{\sin kr}{kr} G(r)dr \tag{1}$$

where $\underset{\sim}{k} = \underset{\sim}{k}_{2\omega} - \underset{\sim}{k}_{NLP}$ is the difference between the wave vector of the harmonic radiation, $\underset{\sim}{k}_{2\omega}$, and that of the induced nonlinear polarization, $\underset{\sim}{k}_{NLP}$. The integral over $\underset{\sim}{k}$ derives from the effects of multiple scattering by the powdered sample. Because of this scattering, there exists in the medium many different waves at the fundamental frequency, ω, traveling with different wave vectors, $\underset{\sim}{k}_\omega$ [4]. For N such waves there are N frequency doubling, but N(N-1)/2 pairwise wave vector mixing, processes, so that the latter occurrence dominates. The relevant wave vectors are shown in Fig. 1, while $|k|$ is given by

$$k^2 = \frac{16\pi^2 n_1 n_2 \cos\varphi}{\lambda_2^2} \sin^2\left(\frac{\theta}{2}\right) + \frac{4\pi^2}{\lambda_2^2}(n_2 - n_1\cos\varphi)^2 \tag{2}$$

where n_1, n_2 are the refractive indices of the medium at the fundamental and harmonic (λ_2) wavelengths, respectively, and $0 \leq \varphi \leq \pi/2$. The (presently uninteresting) constant in Eq. 1 contains the results of the necessary averaging of the k vector directions with respect to crystallite axes, the averaging of the orientation of these with respect to the laboratory frame, the form of the nonlinear tensor, integrals over crystallite sizes and shapes, a description of the scattering properties of the powder and instrumental effects. None of these factors depend on $\underset{\sim}{k}$ (except to the extent retained in our treatment) or on $\varkappa$ and are temperature independent.

σ is then given by ($v = \cos\varphi$)

$$\sigma(\varkappa) = \sigma_{max} \int_O^1 \mathrm{Arg}(v,\varkappa)dv \Big/ \int_O^1 \mathrm{Arg}(v,0)dv \tag{3}$$

$$\mathrm{Arg}(v,\varkappa) = \frac{1}{v}\ln\left[\frac{\lambda_2^2\varkappa^2 + 4\pi^2(n_1 v + n_2)^2}{\lambda_2^2\varkappa^2 + 4\pi^2(n_1 v - n_2)^2}\right]$$

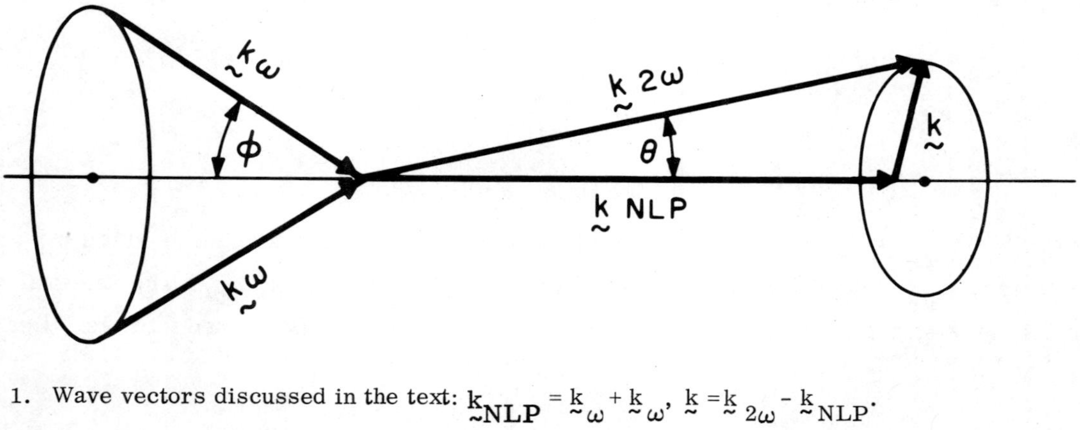

Fig. 1. Wave vectors discussed in the text: $\underset{\sim}{k}_{NLP} = \underset{\sim}{k}_\omega + \underset{\sim}{k}_{\omega'}$, $\underset{\sim}{k} = \underset{\sim}{k}_{2\omega} - \underset{\sim}{k}_{NLP}$.

Here σ_{max} is the value of σ at $\varkappa = 0$ ($T = T_c$) and must be determined experimentally. In computing (3) the presently nearly isotropic molecular form factor, F(k), for harmonic scattering with depolarized laser radiation has been neglected[5];

$$F(k) = \frac{1 + 3\rho + (1-\rho)\cos^2\theta}{2+2\rho} \tag{4}$$

where ρ (here equal to $\frac{2}{3}$)[6] is the depolarization of the harmonic radiation when excited with linearly polarized laser light. The small dependence of F(k) on φ is similarly neglected. We also neglect η, which theory[3] indicates to be small (~ 1/18) and which has been shown[2] to increase the apparent value of ν by ~ $2/(2-\eta)$.

A least squares fit[7] of the data to Eq. (3) (the dashed line in Fig. 2), in which

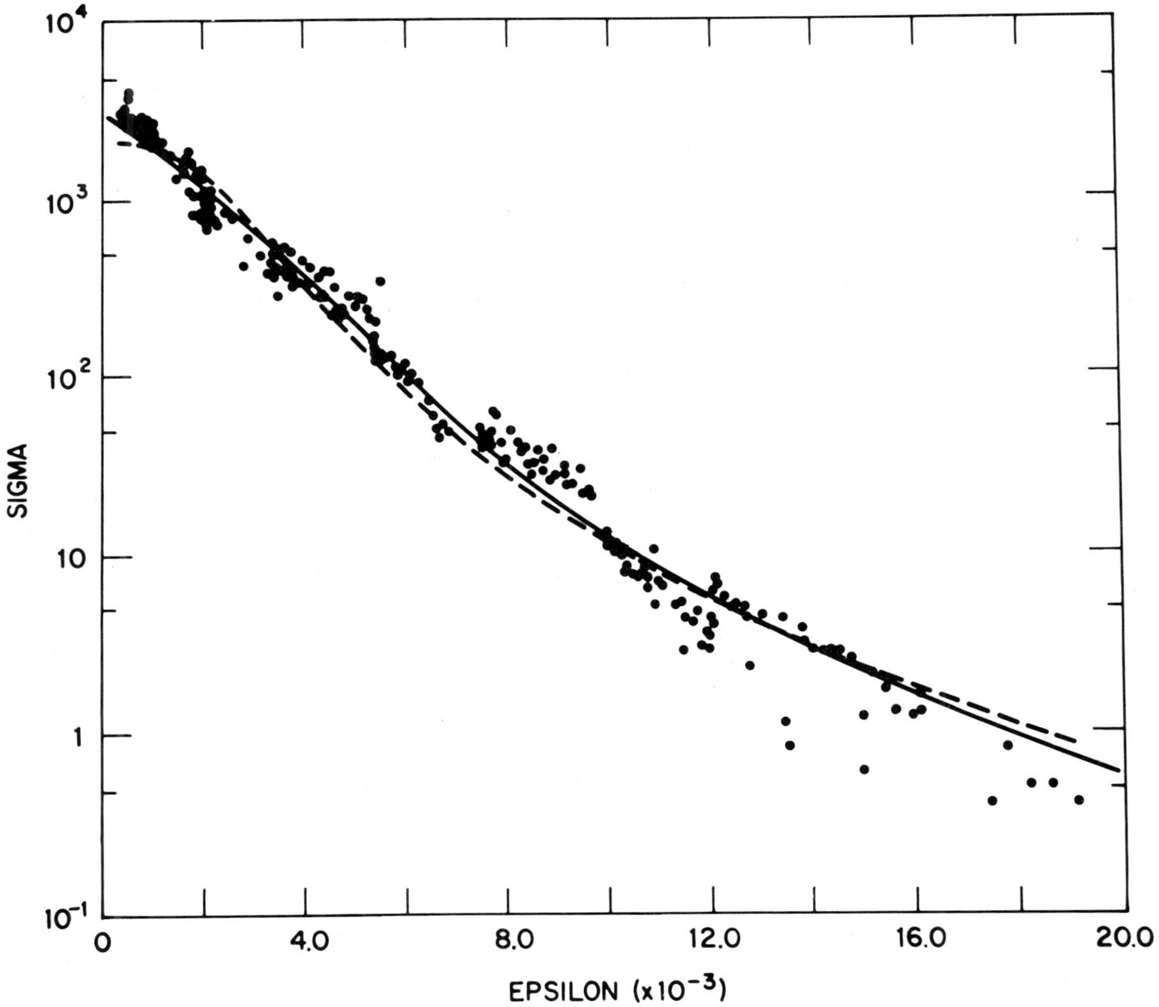

Fig. 2. Relative harmonic intensities, σ, vs $\epsilon = T/T_c - 1$. Circles - experimental data, ---- Eq. (3) of this paper, ——— Eq. (7) of Ref. [2].

A, σ_{max}, and ν were simultaneously adjusted leads to $A = 2.6\times10^6\mu^{-1}$, and $\nu = 2.01$ for $T_c = 242.35$, as compared with $A = 4.7\times10^6\mu^{-1}$ and $\nu = 2.20$ found previously (the solid line in Fig. 2) for this choice of T_c. The relatively small variation of ν with T_c is also rather similar to that found previously.

The introduction of wave vector mixing leads mainly to an effective reduction in the coherence length, ℓ_c ($\approx 30\mu$), for harmonic generation and can only affect the calculations for $1/\varkappa \sim \ell_c$. For $1/\varkappa \ll \ell_c$, the nonlinear susceptibility always will look like $1/\varkappa^{2-\eta}$, the form that the linear susceptibility takes. The difference between the linear and nonlinear susceptibilities, which necessitates the kind of calculation given here, is that the nonlinear susceptibility does not diverge uniformly as $T \to T_c$, since the region of $\underset{\sim}{k} \equiv 0$ is not generally accessible.

ACKNOWLEDGMENT

I am pleased to acknowledge the assistance of L. Lopf in performing the experiments and the programming.

REFERENCES

1. (a) I. Freund, Phys. Rev. Letters 19, 1288 (1967); (b) Critical harmonic scattering has also been considered by J. Lajzerowicz, Solid State Comm. 3, 369 (1965).
2. I. Freund, Chem. Phys. Letters 1, 551 (1968).
3. (a) M.E. Fisher and R.J. Burford, Phys. Rev. 156, 583 (1967); (b) M.E. Fisher, J. Math. Phys. 5, 944 (1964).
4. The isotropic polarizability of the NH_4^+ ion eliminates, in first order, the problem of critical opalescence arizing from fluctuations in angular orientation. Scattering by density fluctuations is a possibility, but we find experimentally that the transmission of our powdered samples does not change significantly (for light at 6328 Å) over the temperature region studied. This is in accord with the results of O.A. Shustin, J.E.T.P. Letters 3, 320 (1966), who observed a marked, but still relatively small (less than 10x) increase in right-angle scattering at T_c.
   In contrast, M. Barbaron, Annales de Physique 6, 934 (1951), reported that a single crystal of NH_4Cl became "opaque" between (-29.5)-(-30.0)$^\circ$C due to critical opalescence. This corresponds to a range in ϵ of $\pm1\times10^{-3}$, and even if this had occurred in our samples it would (±1) not greatly alter our findings. Our treatment, which tacitly assumes that the distribution of radiation inside the powdered sample arising from multiple scattering is independent of temperature, is thus justified. We neglect, also, spectral broadening of the harmonic radiation due to the finite lifetime of the fluctuations considered here, because of the large (~ 50 Å) optical bandwidth of our detection system. We neglect, in addition, the large (~ 30 Å) spectral width of the laser light. Our procedure is equivalent to

$$\sigma = \text{const} \times \int d(\omega) \int d(2\omega)\cdot\sigma(\omega, 2\omega).$$

5. I. Freund, J. Chem. Phys. 45, 3882 (1966). Eq. (4) follows fairly directly from the content of this paper.

6. S. J. Cyvin, J. E. Rouch, and J. C. Decius, J. Chem. Phys. 43, 4083 (1965).

7. The (machine) calculation proceeded as follows: a 200 line table of $\sigma(\varkappa)/\sigma_{max}$ vs $\varkappa$ was constructed using an available Simpson's rule numerical integration routine and Eq. (3). 4-point Lagrangian interpolation in this table was used to obtain $\sigma(\varkappa)$ and $\varkappa(\sigma)$, as required. A 10 point coarse grid of σ_{max} was established, $\varkappa(\sigma)$ computed for the first point on this grid, and the corresponding values of $A(\sigma_{max})$ and $\nu(\sigma_{max})$ determined using standard least squares procedures by linearizing $\varkappa = A(T/T_c - 1)^{\nu}$. The quantity $\delta(\sigma_{max}) = \Sigma[(\sigma_{calc} - \sigma_{exp})/\sigma_{calc}]^2$ was computed and the process repeated for all values of σ_{max} on the grid. The value of σ_{max} leading to a minimum in δ was chosen as the best estimate of this quantity, a new, finer, grid established about this point, and the whole procedure repeated. Convergence was smooth, and the processing of 310 data points consumed slightly less than 1 minute on a GE 645 computer.

H-3: RAMAN STUDY AND THE EVOLUTION OF ORDER IN NH_4Br AT THE λ TRANSITION

C. H. Wang and P. A. Fleury
Bell Telephone Laboratories, Incorporated
Holmdel, New Jersey

INTRODUCTION

NH_4Br is known to experience a λ-type phase transition from a disordered cubic (CsCl structure) to an ordered tetragonal structure at $T_\lambda = 235°K$[1]. Above T_λ the NH_4^+ ions are randomly distributed between two energetically equivalent orientations. (This is often called phase II.) Below T_λ the crystal is tetragonally distorted along one of the cube axes and at the same time the NH_4^+ ions assume ordered orientations. (This is often called phase III.) The identically oriented NH_4^+ ions form chains along the tetragonal direction with neighboring chains oppositely oriented. In this sense the NH_4Br ordered phase is analogous to an antiferromagnet and is therefore more complicated than the ordered phase of NH_4Cl, which is analogous to ferromagnetic order. As a consequence, the NH_4Br unit cell contains two formula units below T_λ.

In the ordered phase (III) the Raman spectrum of NH_4Br is expected to be significantly different from that in phase II for two reasons. First, in the disordered phase there is, strictly speaking, no translational invariance so that first-order Raman scattering should receive contributions from phonons of all k vectors - similar to the F-center induced Raman spectrum of the alkali halides[2]. That is, no sharp lattice mode lines are expected. (The one exception is the F_2 limiting lattice mode.) Second, in the ordered phase, new zone-center phonons appear as a consequence of the unit cell doubling so that a more conventional lattice mode spectrum is expected. Further, group theory analysis indicates that the librational mode of the NH_4^+ ion is Raman active below T_λ. Effects of the onset of order on the modes of vibration internal to the NH_4^+ ion should also be evident in the Raman spectrum.

In this paper we shall describe results of Raman scattering experiments at various temperatures in phases II and III with particular emphasis on those aspects affected by the onset of order. Our basic conclusions are that the order evolves rather slowly and that in the disordered phase there is some residual short range order. To support these conclusions we concentrate on the anomalous low frequency lines observed in the disordered phase, the unusual temperature dependence of the linewidth of the librational mode (ν_6) and the temperature dependence of the linewidth of an internal vibrational mode (ν_4'). Some data on the low frequency spectrum of NH_4Cl are presented for contrast to the NH_4Br results.

EXPERIMENTAL TECHNIQUES

The Raman spectra were obtained using an argon ion laser source (200mW at 4880A or 5145A), a double monochromator for dispersion, and photoelectric detection. The samples were mounted in a glass dewar and their temperature varied by flowing He or N_2 gas at a controlled rate. Sample temperature was monitored by a calibrated Pt resistence thermometer and could be maintained to ±0.2°K. Since the resistor was not mounted directly on the sample absolute temperature measurements are estimated accurate to within ±2°K.

The crystals used here were grown by the method of slow evaporation from aqueous solutions of NH_4Br or NH_4Cl thermostated at 38±0.1°C. In order to promote the growth of (100) phases, urea was added as a habit modifier.

RAMAN SPECTRA AND SELECTION RULES

Although the Raman spectra in both phases II and III are complex, they can be conveniently divided into two classes: (1) scattering from internal vibrations of the NH_4^+ ion, and (2) scattering from the "lattice" modes in which the NH_4^+ ion is considered a rigid tetrahedron. The fundamental frequencies of the two classes are separated by more than 1000 cm^{-1} due mainly to the fact that the N-H bond is stronger than the NH_4^+-Br^- bond.

In the ordered phase (III) the Raman selection rules are easily worked out for the two classes of modes. With two formula units/unit cell (space group D_{4h}^7)[1] there are 36 degrees of freedom. Eighteen are associated with the internal modes and eighteen with the lattice modes - including the librational motion of the NH_4^+ tetrahedron. The lattice modes are

$$1A_{1g} + 1A_{2g} + 1B_{2g} + 3E_g + 2A_{2u} + 1B_{2u} + 3E_u$$

for a total of eighteen. The Raman active modes are the A_{1g}, B_{2g} and E_g phonons. The Raman tensor elements are $\alpha_{xx} + \alpha_{yy}$, α_{zz} for A_{1g}; α_{xy} for B_{2g}; and α_{xz}, α_{yz} for E_g. Because of the formation of domains, the "Z" direction in the crystal has no unique direction in the laboratory and so the only distinction that can be made experimentally is between diagonal and off-diagonal elements. The assignments for the lattice modes in phase III given in Table I arise from the observation (see Fig. 1) that the F_2 mode

TABLE I

Observed Frequencies and Symmetries of NH_4Br in Phases II and III

	Phase II (300°K)			Phase III		
		Freq (cm^{-1})			Freq (cm^{-1})	
					(150°K)	(173°K)
A - Lattice Modes		Present	Krishnan		Present	Krishnan
		56	56	A_{1g}	67	63
		76	75	E_g	76	75
	F_2	130	128	E_g	134	133
			136			156
			155	B_{2g}	181	178
			175			258
				$E_g(\nu_6)$	332	328
	$2\nu_6$	641		$2\nu_6$	660	
B - Internal Modes of NH_4^+	$F_2(\nu_4)$	1408	1398	$E_g(\nu_4)$	1400	1399
			1429	$B_{2g}(\nu_4')$	1420	1416
			1462			1432
						1480
	$E(\nu_2)$	1689	1685	$A_{1g}(\nu_2)$	1693	1690
				$B_{1g}(\nu_2')$	1697	
	$\nu_2+\nu_6$	1959	1960	ν_2(or ν_2')$+\nu_6$	1980	1970
	$2\nu_4$	2810	2806	$2\nu_4$	2822	2814
						2885
						2987
	$A_1(\nu_1)$	3040	3037	$A_{1g}(\nu_1)$	3038	3037
						3076
	$F_2(\nu_3)$	3135	3121	$B_{2g}(\nu_3')$	3117	
				$E_g(\nu_3)$	3126	3122

Note: Additional broad weak peaks appear at 443 cm^{-1}, 1129 cm^{-1} and 1364 cm^{-1} at 300°K. These all disappear as temperature is lowered. While their assignment is difficult, we ascribe them to combination bands.

(at 130 cm^{-1} in phase II) splits into and E_g and a B_{2g} as T is lowered into phase III. This fixes the assignments of the 76 cm^{-1} and 332 cm^{-1} (librational) modes as E_g. Incidentally one can show group theoretically that the librational motion has E_g character. Hence this assignment is consistent.

The internal modes which are Raman active are listed in Table I as well. All of those predicted by group theory in phase III have been identified, in addition to some combination lines. Comparison throughout the table is made with the results of Krishnan[3] at nearly the same temperatures. Agreement is quite good except that Krishnan finds more lines in both types of spectra. Without exception our experimental signal-to-noise ratio should have permitted observation of these extra lines if the relative strengths assigned them by Krishnan were correct.

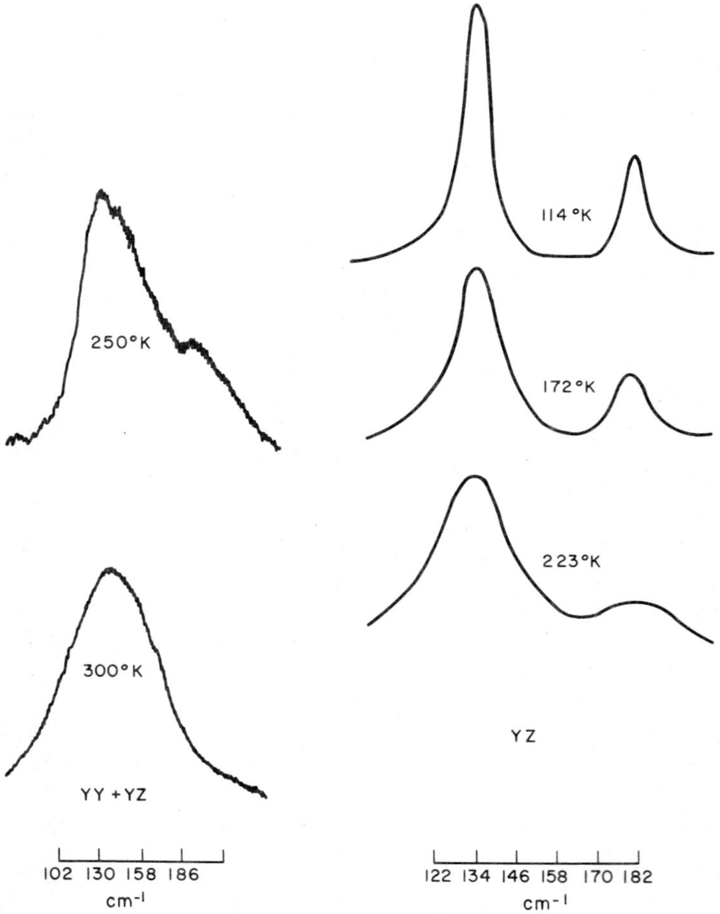

Fig. 1. Temperature variation of lattice modes [F_2 in phase II; E_g (134 cm^{-1}) and B_{2g} (180 cm^{-1}) in phase III]. The stronger line is assigned as E_g although the present experiment cannot distinguish between E_g and B_{2g}. The letters in parentheses indicate the elements of the Raman tensor observed.

The operation of selection rules in the cubic, disordered phase (space group T_d^1) is somewhat less clear than in the ordered phase. With regard to the internal modes, one merely has some additional degeneracies due to the cubic symmetry. According to group theory there is one Raman active lattice mode of symmetry F_2 in phase II. In Table I this is identified as the 130 cm^{-1} line, correctly exhibiting off-diagonal scattering elements.

SHORT RANGE ORDER ABOVE T_λ; THE LOW FREQUENCY SPECTRUM

The puzzling features of the low frequency spectrum in phase II are the lines at 56 cm^{-1} and 76 cm^{-1}. The 56 cm^{-1} line appears clearly in Fig. 2a) and is polarized. The depolarized spectrum at 235°K and above exhibits the 76 cm^{-1} line clearly. The temperature dependence of the spectra in this region is shown in Fig. 3. The spectrum at 235°K is essentially the same as at 300°K in the 0-80 cm^{-1} region. As T is lowered below 235°K the 56 cm^{-1} peak decreases and a sharp peak at 67 cm^{-1} emerges. Notice the 211°K spectrum of Fig. 3. The 76 cm^{-1} peak merely sharpens. These are Raman allowed zone center phonons (along with the others in Table I at 134, 181, and 332 cm^{-1}).

The question to be answered is why are the symmetric and rather sharp lines at 56 and 76 cm^{-1} present above T_λ? As indicated above, the lack of translational invariance accompanying the random orientation of NH_4^+ ions should have the effect of relaxing the wave vector conservation requirement. Hence one should observe scattering from even parity phonons without restriction to k = 0. The resulting Raman spectrum should mirror the one phonon density of states - which is in general <u>not</u> a symmetrically shapted nor very sharp function. If as we contend, the 56 and 76 cm^{-1} features are due scattering from large wave vector phonons then there must be some mechanism which weights the density of states about a fairly narrow region of $\underline{k}$ values. Such a mechanism is clearly operative at the transition temperature where the zone edge k vector in the [100] direction becomes a reciprocal lattice point. The first order lattice spectrum below T_λ could be viewed as scattering from a phonon at the old zone boundary (with its appropriate frequency) in combination with an oppositely directed zero frequency phonon (the frozen in distortion). Such a picture may be extended above T_λ where one imagines that short range order persists corresponding to a fairly narrow range of k-vector. Phonons near the zone boundary are then more heavily weighted in contributing to the Raman spectrum. Phonon dispersion curves for CsI by Karo and Hardy[4] on the basis of several models all share the common features that (a) all modes are degenerate at the R point $\left(\frac{1}{2}, \frac{1}{2}, \frac{1}{2}\right)\frac{\pi}{a}$ and (b) there are between two and four distinct frequencies at the X point $\left(\frac{1}{2}, 0, 0\right)\frac{\pi}{a}$ (depending on values of parameters in the calculation). Since the CsCl structure is the same as the CsI structure, these conclusions are of qualitative significance for NH_4Br in phase II.

From Fig. 2a) it is evident that a similar effect may be present in NH_4Cl, with the lines at 94 cm^{-1} and 138 cm^{-1}. In the chloride the ordered phase does not have the

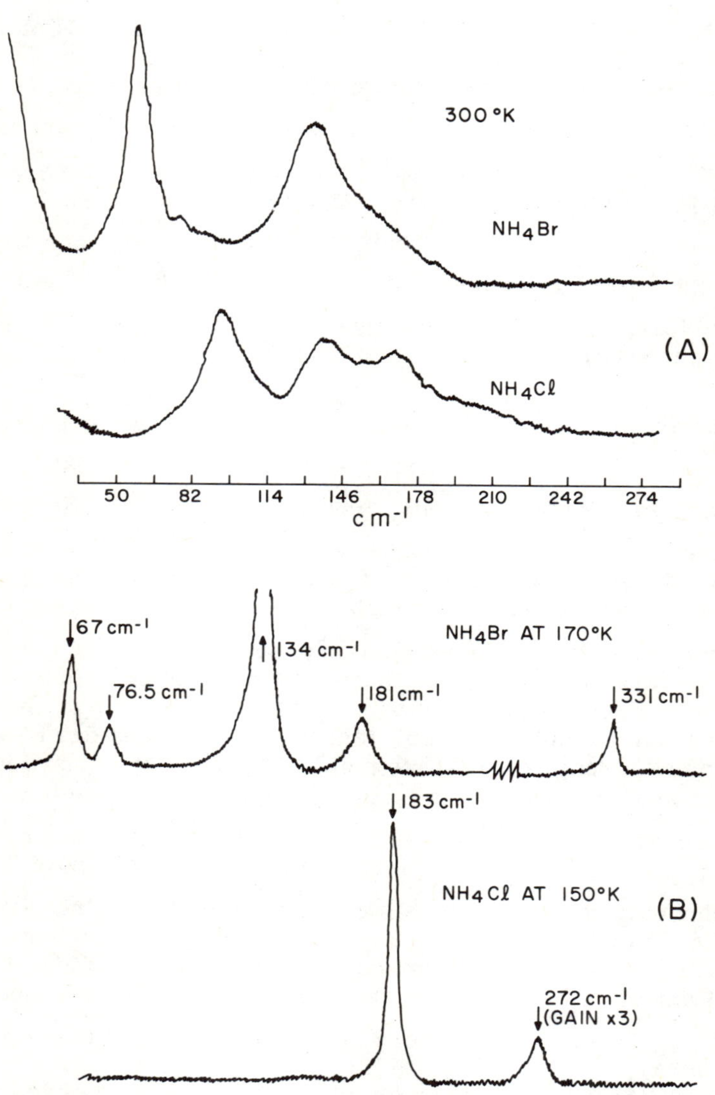

Fig. 2. a) The low frequency Raman spectra of NH_4Br and NH_4Cl at $300^{\circ}K$. b) The low frequency spectra of NH_4Br at $170^{\circ}K$ and NH_4Cl at $150^{\circ}K$. The numbers at the bottom of spectra correspond to the position of the peaks in cm^{-1}. Both polarized and depolarized components are shown superimposed.

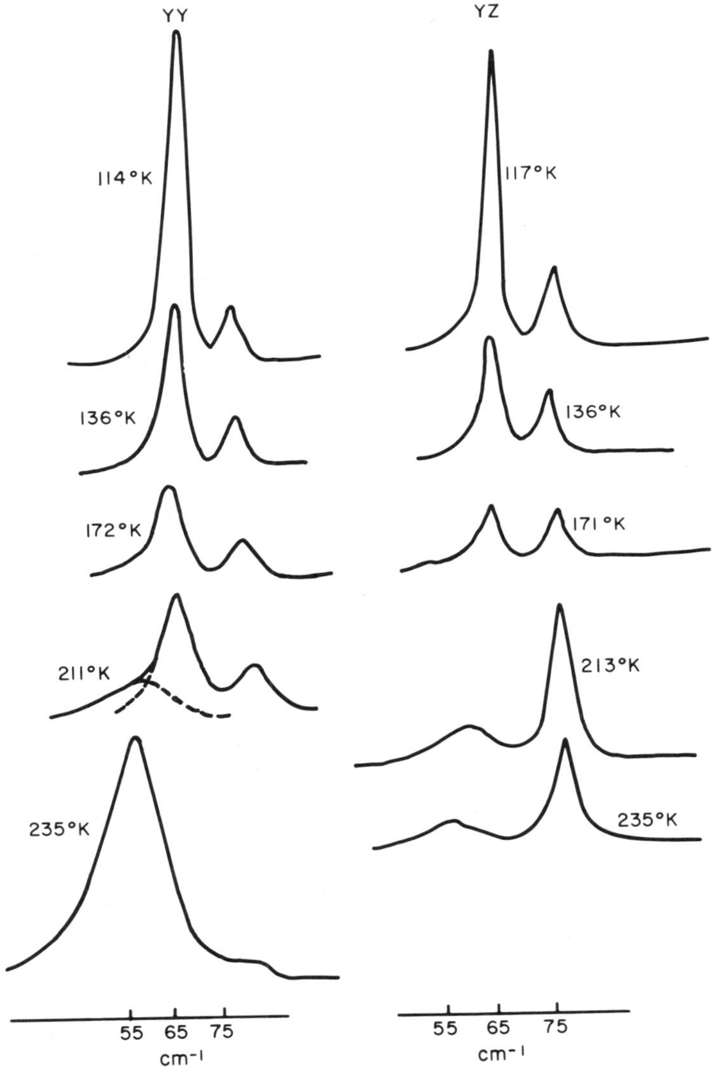

Fig. 3. The spectra of the lowest frequency lattice lines. The dashed lines for 211°K indicate the emergence of the 56 cm^{-1} line as the 67 cm^{-1} line disappears as T increases.

doubled unit cell and hence only a single lattice mode F_2 (observed at 183 cm^{-1}) is Raman active. A comparison between the chloride and bromide in their ordered phases appears in Fig. 2b). The line at 272 cm^{-1} was also observed by Krishnan[3]. We have not made sufficiently extensive measurements in NH_4Cl to identify this line.

One independent bit of evidence for order in the disordered phase, the slow evolution of order in NH_4Br, is provided by the specific heat measurements of Sorai et al. [5]. The entropy change associated with the λ transistion is 0. 74 eu, substantially less than the expected $R\ell n2$ = 1. 2 eu for such a transition. This result suggests that the change from complete order to complete disorder is spread out over a much wider temperature range than for the chloride and is consistent with the above explanation of our low frequency spectra.

Unfortunately the picture of short range order given above is only speculation at this point. One possible verification would be a measurement of the phonon dispersion curves in NH_4Br to see if the X point phonon frequencies agree with the 56 and 76 cm^{-1} values we observed.

EFFECTS OF ORDER ON THE INTERNAL MODES

Physically one expects the ordering process to exert only small effects on the internal modes. The tetragonal distortion below T_λ may remove degeneracies and cause some small frequency shifts. Space limitations forbid our discussing all the internal modes; we shall discuss only the behavior of ν_4' for illustration.

Earlier IR work has shown the existence of an anomalous component at 1444 cm^{-1} (ν_4') near the bending mode (ν_4)of NH_4^+ in NH_4Cl[1]. The IR intensity of this line has been associated by Garland and Schumaker[6] with the gradual appearance of order as T_λ is approached from above. They observed a similar component at 1435 cm^{-1} in NH_4Br only with a somewhat more complicated temperature behavior[7]. We have also studied the 1400 cm^{-1} region via Raman scattering. The results are summarized in Fig. 4a). The line at 1400 cm^{-1} is the fundamental bending mode ν_4. Above T_λ, the line shape is asymmetrical, and in the vicinity of T_λ there appears a hint of a line at 1430 cm^{-1} (Fig. 4B) which disappears below T_λ.

As the temperature is lowered below T_λ, ν_4 sharpens and a definite split off component (ν_4^*) emerges at 1420 cm^{-1}. We concentrate our attention on the 1420 cm^{-1} feature. Notice the marked increase in peak intensity as T is lowered. Since the integrated intensity of the ν_4^* remains essentially constant with T we have plotted the inverse peak intensity of ν_4^* in Fig. 5. This is proportional to the width of ν_4^* and illustrates qualitatively the increase in linewidth with decreasing order. That is, it suggests the presence of disorder in the ordered phase, just as Schumaker and Garland results[7] suggest the presence of disorder in the ordered phase in NH_4Cl.

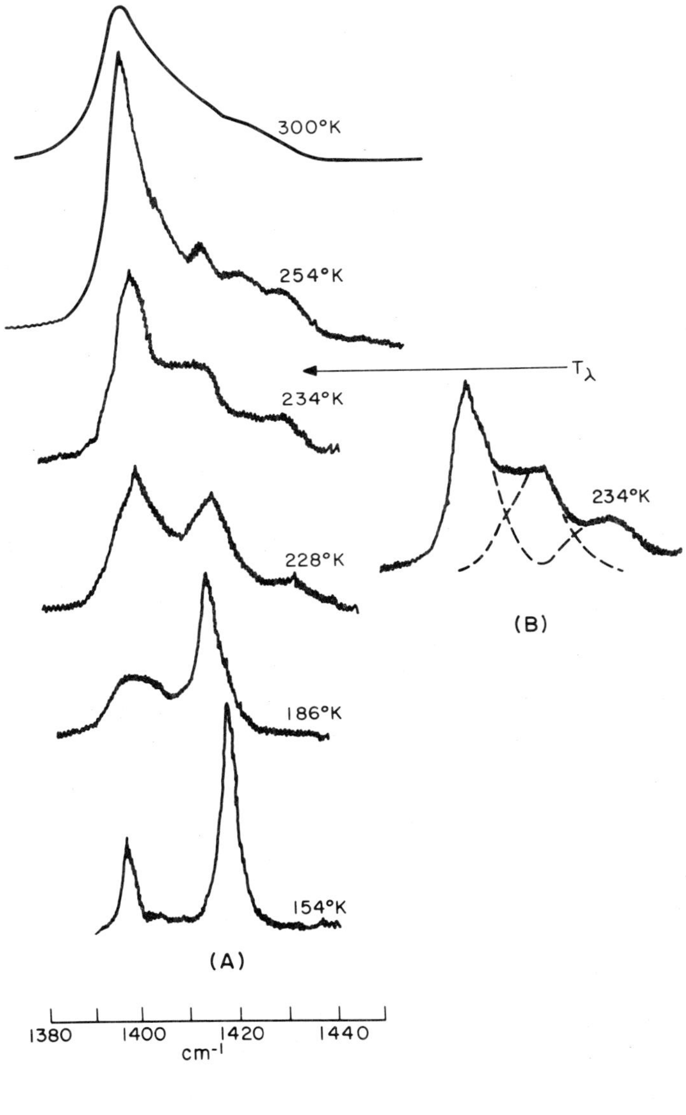

Fig. 4. a) The bending modes (ν_4 and ν_4') of the internal NH_4^+ vibration at different temperatures.
b) The dashed lines in the spectrum at 234°K indicate the possible presence of another line at 1430 cm^{-1}.

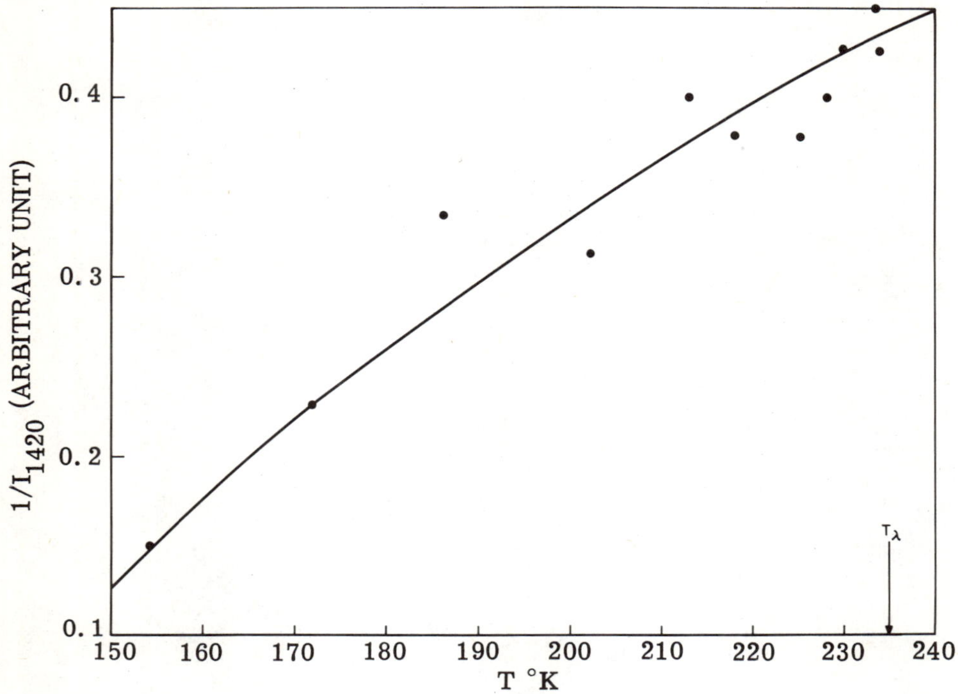

Fig. 5. The inverse of peak intensity as function of temperature for ν_4 at 1420 cm^{-1}.

TEMPERATURE EFFECTS ON THE LIBRATIONAL MODE, ν_6

As mentioned in Sec. 3 the librational mode (torsional oscillation of the NH_4^+) in NH_4Br is Raman active (see Table I) in the ordered phase. This mode is of special importance because its motion is closely associated with the motion accompanying the λ transition. It is not however exactly the same and the librational mode is not the "soft mode" appropriate to the order-disorder transitions. Fig. 6 shows this mode at temperatures between 115°K and 200°K. As T increases the frequency decreases slightly (ν_6 decreases from 333 cm^{-1} at 115°K to 326 cm^{-1} at 200°K) and the width increases dramatically. The remarkable exponential behavior of the librational linewidth is shown in Fig. 7. Over a wide temperature range the linewidth Γ follows the form

$$\Gamma = \Gamma_0 e^{-\epsilon/RT}$$

where Γ_0 = 73.7 cm^{-1} and ϵ = 0.84 Kcal/mole (ϵ/R = 421°K).

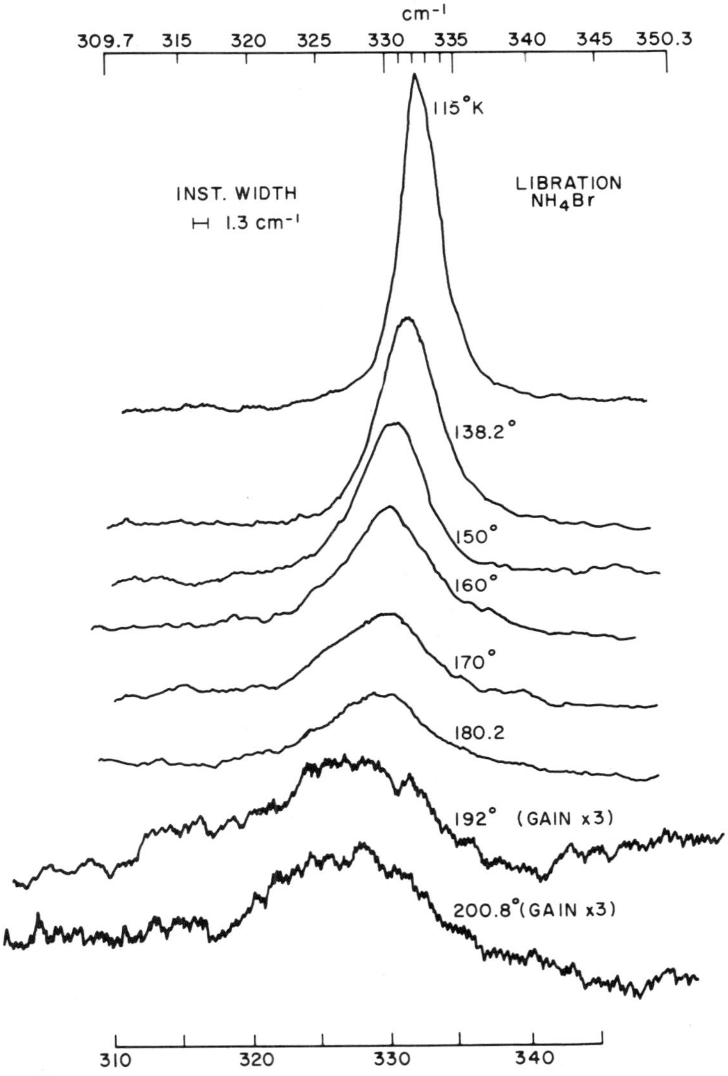

Fig. 6. The E_g librational mode (ν_6) at different temperatures in the ordered phase of NH_4Br.

This temperature dependence is quite different from that due to damping via phonon interactions and suggests some sort of rotational diffusion thermal activation process as the mechanism for broadening the line. At first glance the obvious activation energy to associate with ϵ might be the barrior height between two wells of the NH_4^+ orientation potential. The latter has been deduced from NMR studies[8], neutron diffration[9], and IR absorption[6] as approximately 3.3 Kcal/mole. This is about a factor of 4 larger than ϵ of Eq. (1). It is likely that the ϵ which determines the ν_6 linewidth is the activation energy associated with a change in axis of rotation of NH_4^+ ion. Group theory indicates that the libration about the z axis (C_4 axis of the tetragon) is Raman active (E_g). Our suggested mechanism for broadening is the reorientation of the librational axis from x to y (or vice versa). It is likely that the barrier height associated with such a reorientation is considerably smaller than that required for the order-disorder reorientation. However, more quantitative statements must await detailed calculations.

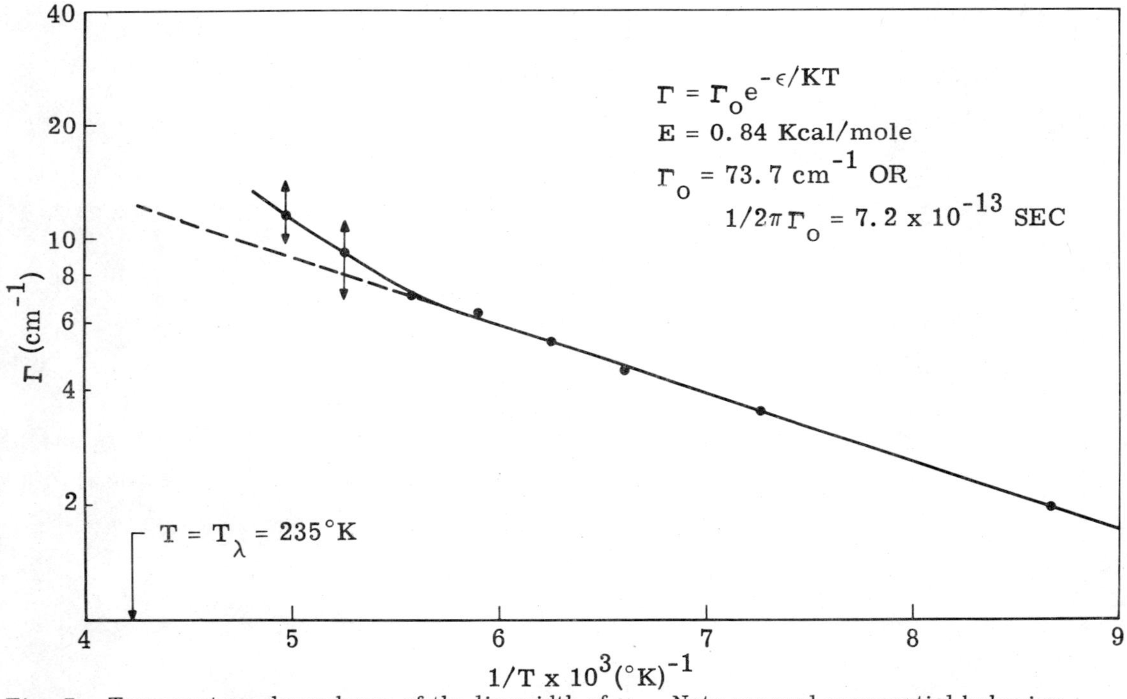

Fig. 7. Temperature dependence of the linewidth of ν_6. Note unusual exponential behavior.

ACKNOWLEDGMENTS

We thank J. M. Worlock for helpful discussions and H. L. Carter for expert technical assistance, and N. E. Schumaker for sending us NH_4Br crystals for preliminary experiments.

REFERENCES

1. E.L. Wagner and D.F. Hornig, J. Chem. Phys. 18, 305 (1950).
2. J.M. Worlock and S.P.S. Porto, Phys. Rev. Letters 15, 697 (1965).
3. R.S. Krishnan, Proc. Ind. Acad. Sci. A27, 321 (1948).
4. A.M. Karo and J.R. Hardy, Phys. Rev. 160, 702 (1967).
5. M. Sorai, H. Suga, and S. Seki, Bull. Chem. Soc. (Japan) 38, 1125 (1965).
6. C.W. Garland and N.E. Schumaker, J. Phys. Chem. Solids 28, 799 (1967).
7. C.W. Garland and N.E. Schumaker (to be published).
8. H.S. Gutowsky, G.E. Pake, and R. Bersohn, J. Chem. Phys. 22, 643 (1954).
9. G. Venkataraman, et al., J. Phys. Chem. Solids 27, 1107 (1966).

H-4: RAMAN SPECTRA OF LATTICE VIBRATIONS AND THE ORDER-DISORDER PHASE TRANSITION IN AMMONIUM HALIDES

L. Rimai, T. Cole and J. Parsons
Scientific Laboratory, Ford Motor Company
Dearborn, Michigan

INTRODUCTION

Ammonium chloride and bromide undergo an order-disorder phase transition at T_c = 243°K and 235°K[1, 2] respectively. For the corresponding deuterated crystals the critical temperatures are only slightly shifted. Above this transition (Phase II) the structure of both is tetrahedral (point group T_d) but disordered: the NH_4 tetrahedra are randomly distributed between the two inverted configurations. At temperatures sufficiently below T_c such that complete order is attained, (Phase III) NH_4Cl becomes rigorously tetrahedral with all NH_4 ions similarly oriented forming domains[3]. Phase III of NH_4Br is tetragonal (point group D_{4h}), with a pair of inverted molecules per unit cell[2]. A macroscopic crystal consists of tetragonal domains. The bromide undergoes yet another transition at about 108°K, but with little information available about its nature. We do have evidence in this work that such a transition indeed occurs.

In earlier work on these materials Raman spectra were measured at a few discrete temperatures[4-6], from which assignments were made for some vibrational frequencies of zero wave vector ($\underline{q}$ = 0) in phases II and III. The main purpose of this paper is to report on the detailed temperature dependence of certain features in the spectra in hope of learning about the microscopic aspects of the order-disorder transformation. The results indicate that there is a large temperature range around T_c in which microscopic changes are continuously occurring[7]. There are three categories of information that we present. a) Spectroscopic information in the ordered phases (III) from which we have obtained more complete assignments of normal modes than previously published. b) Static effects of the disordering process, i. e., study of new spectral lines absent in the completely ordered phase. These effects yield information about the spatial correlations in the distribution of different molecules, and their interactions. c) Dynamic effects due to the onset of disorder in the lattice. NMR studies have already shown the presence of large amplitude molecular motion associated with the disordering process[1, 2]. Depending on the frequency spectrum of such a random process, and on the strength of its interaction with the vibrational modes one can expect temperature effects on the widths of the Raman lines.

Concerning a), in Section 2 we confirm the earlier assignments in NH_4Cl and extend these in NH_4Br, although the latter are still incomplete. We also give corresponding results for deuterated crystals. In Section 3 we discuss spectra in the presence of disorder focusing mainly on the low frequency region in NH_4Cl. The dynamical effects c) are presented in Section 4.

The measurements were performed with an experimental setup described previously[8], with laser excitation either at 6328 Å (He-Ne) or 5145 Å (A ion), at a power level of about 35 mw. The crystals were grown by slow evaporation from urea containing saturated water solutions.

SPECTRA IN THE ORDERED PHASES

To classify the $\underline{q} = 0$ vibrations we use the approximate separation into internal ammonium vibrations and lattice modes. The latter include the acoustic waves (ν_7) threefold degenerate with $\nu = 0$ at $\underline{q} = 0$, the NH_4 libration mode (ν_6) which is roughly described as a small angle rigid rotation of the H_4 tetrahedron, and the optic mode (ν_5) which corresponds to the optic mode in the CsCl structure with Cs^+ replaced by NH_4^+. In Phase III of the chlorides ν_6 is threefold degenerate, transforming as the T_1 representation of the T_d ($\bar{4}3$ m) point group and being both Raman and absorption inactive. ν_5 transforms as T_2 but the threefold degeneracy is removed by the electrostatic interaction between ions giving longitudinal and transverse waves. Only the latter have been observed in the Raman spectrum. The internal modes correspond to the following representations: $\nu_1(A_1)$ nondegenerate symmetric N-H stretch; $\nu_2(E)$ twofold bending motion; $\nu_3(T_2)$ asymmetric N-H stretch and $\nu_4(T_2)$ threefold degenerate bending motion. All but ν_6 are Raman active. ν_7, at $\underline{q}$ near zero, gives rise to the very low frequency Brillouin spectra[9]. ν_1 and ν_2 have diagonal Raman tensors. It is purely off-diagonal for the others. Neglecting small sublattice displacements in the [100] or $\hat{z}$ direction, the tetragonal (D_{4h}) structure for Phase III of NH_4Br is as follows. The N and Br atoms form a b. c. c. CsCl structure with simple cubic sublattices while the proton tetrahedra alternate orientation in the (100) plane. The unit cell consists of two molecules with oppositely oriented NH_4, paired either along the [010] or [001] direction. Assuming that the shifts in mode frequencies due to the small tetragonal displacements are small one can relate the $\underline{q} = 0$ modes of the D_{4h} structure to the modes of a crystal with T_d point group at $\underline{q} = 0$ and at a point on the surface of the first Brillouin zone, with $\underline{q}\,||\,[010]$ or [001]. For each mode in the T_d crystal, NH_4Br will have two $\underline{q} = 0$ modes of distinct frequency, corresponding to a symmetric and an antisymmetric linear combination of displacements of the paired molecules.

The splitting of these two frequencies comes from the intermolecular interactions. In addition, for modes degenerate in the T_d structure there will be further splitting due to the small tetragonal displacements. For each of the ν_i (i = 1 .... 6) we expect now to see a group of lines in the spectrum. In addition new lines will appear due to the zone surface mode of ν_7 (acoustic branch). The electromagnetic splitting of the IR active T_2 modes is only expected to be significant for ν_5, where we have only been able to observe the transverse mode. The representations for the various modes in tetragonal NH_4Br are summarized below: $\nu_1 \rightarrow B_{1u} + A_{1g}$; $\nu_2 \rightarrow A_{1u} + B_{1u} + A_{1g} + B_{1g}$; $(\nu_3, \nu_4, \nu_5) \rightarrow A_{2u} + E_u + B_{2g} + E_g$; $\nu_6 \rightarrow B_{1u} + E_u + A_{2g} + E_g$; $\nu_7 \rightarrow A_{1g} + E_g$ (zone surface point of acoustic branch). The Raman active modes have the following nonzero tensor components ($\hat{z}$ = [100], $\hat{x}$ = [010], $\hat{y}$ = [0,0,1]).

$$A_{1g} \rightarrow (\alpha_{xx} = \alpha_{yy}, \alpha_{zz});\ B_{1g} \rightarrow (\alpha_{xx} = -\alpha_{yy});\ B_{2g} \rightarrow \alpha_{xy};\ E_g \rightarrow (\alpha_{xz} = \alpha_{yz})$$

We should point out that the domain structure makes it impossible to distinguish between the $\hat{z}$ and $\hat{x}$ or $\hat{y}$ axes. Our data for the more important transitions in Phase III and the corresponding assignments are included in Table I. Two new lines observed in the lattice region for the lowest temperature phase of NH_4Br are also included. Some strong transitions presumably due to second order processes have been omitted for compactness' sake. The libration and internal mode frequencies are, to reasonable approximation, consistent with the H to D mass ratio between normal and deuterated crystals.

STATIC EFFECTS OF DISORDER: SPECTRA IN PHASE II

As the temperature rises, even well below T_c, one observes changes in the structure of the spectrum near the internal modes. More importantly, in the chlorides new lines appear in the low frequency region, which become quite intense around room temperature. Fig. 1 illustrates the appearance of the line at 345 cm^{-1} which corresponds to the

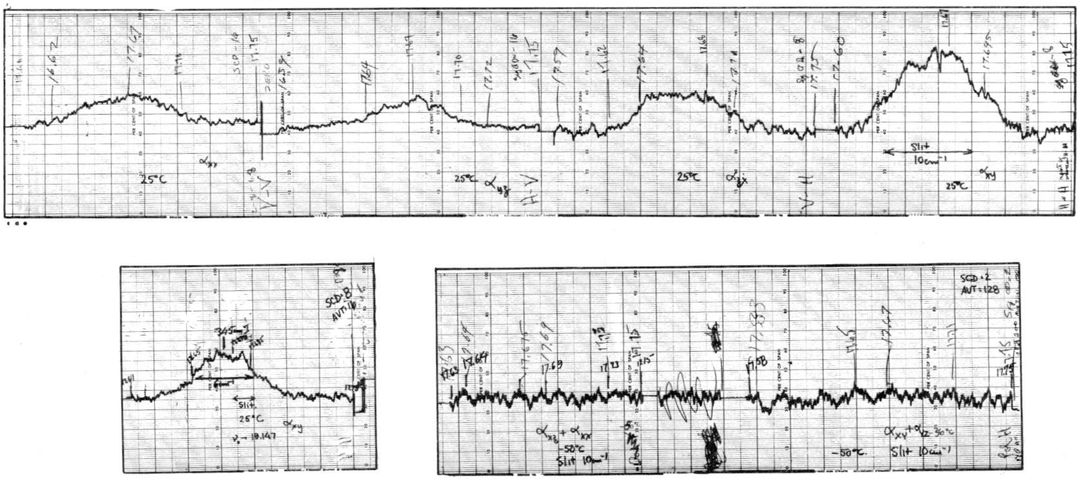

Fig. 1. Top-NH_4Cl Libration line (ν_6) at room temperature-four different polarization combinations. Bottom left-Room temperature high resolution trace of same line. Bottom right-Same region of spectrum, at -50°C, and with considerably higher gain with no visible indication of a line.

TABLE I

Main Raman Transitions in Ammonium Halides

	NH_4Cl		NH_4Br		ND_4Cl		ND_4Br	
	300°K	170°K	300°K	170°K	300°K	170°K	300°K	170°K
ν_7(acoust)	90(A_{1g})		52.5(A_{1g})	65(A_{1g})	90.5(A_{1g})		57	69(A_{1g})
	73(E_g)		76		80(E_g)		73	
	171(T_2)	182(T_2)			176(T_2)	182(T_2)		
ν_5	145, 192 (B_{2g} or E_g)		136, 179	138, 180 (B_{2g} or E_g)	129, 157.5, 205 (B_{2g} or E_g)		131, 165	138, 180 (B_{2g} or E_g)
ν_6	345			330 E_g	220			250 E_g
ν_4	1400	1420(T_2)	1392	1430 (B_{2g} or E_g)	1065	1075(T_2)	980	1020 (B_{2g} or E_g)
ν_2	1705	1720(E)	1662	1600 A_{1g}	1220	1225(E)	1198	1200(A_{1g})
				1700 B_{1g}				
ν_1	3040	3038(A_1)	3040	3040 A_{1g}	2215	2215(A_1)	2228	2220(A_{1g})
ν_3	3165	3165(T_2)		3110, 3135 (B_{2g} or E_g)	2350	2350(T_2)	2312	2280
	3185		3151	3290?				

libration frequency in NH_4Cl. It is Raman inactive in the ordered phase, and it is activated in Phase II by the lowering of the local symmetry, due to the onset of molecular disorder. This assignment agrees with results on hyper-Raman spectrum of NH_4Cl[10] where it is active in the ordered phase and by the frequencies obtained for it in ND_4Cl and NH_4Br. In the latter it has a Raman active component (E_g) even in the ordered phase. Fig. 2 shows the NH_4Cl lattice spectrum of two Raman tensor components. The intensity of the TO mode (around 180 cm^{-1}) characteristic of Phase III decreases with increasing T as one progressively finds smaller and smaller homogeneous regions of order in the crystal. To explain the origin of the new high T lattice lines we propose the following argument. Table I includes the main transitions found in the lattice region at room temperature for which it is possible to draw a correspondence between the chloride and the bromides (the TO lattice vibration characteristic of the ordered Phase III of NH_4Cl has to be of course excluded). In both there is a strong lowest frequency line in the parallel (α_{xx}) spectrum (90 cm^{-1} in NH_4Cl and 52.5 cm^{-1} in NH_4Br), and a pair of strong lines in the perpendicular (α_{xz}) spectrum (145 and 192 for NH_4Cl; 136 and 179 for NH_4Br). In contrast to NH_4Cl, the ordered phase of NH_4Br also has corresponding lines; they are slightly shifted, which is expected just from the change in lattice parameters. This leads us to associate the high temperature lattice lines characteristic of disorder in both crystals with pairs of inverted molecules, as those forming the unit cell of ordered tetragonal NH_4Br. From another point of view one can say that in disordered NH_4Cl, lack of periodicity eliminates wave vector conservation requirements and first order scattering becomes possible throughout the Brillouin zone. The active regions will depend on the spatial Fourier spectrum of the random distribution and of the Raman polarizability in the disordered phase. In view of the discussion in 2.1 we may associate these sharp lines with critical points on the surface of the zone for the ordered T_d structure. The following assignments seem most plausible. The lowest frequency (α_{xx}) is an A_{1g} vibration coming from the [010] point of the acoustic branch. The two α_{xz} lines (B_{2g} or E_g) involve vibrations at the same point in the Brillouin zone but belonging to the lattice optic branches.

DYNAMIC EFFECTS

At the lower temperatures the internal vibrations exhibit a linewidth that increases slowly, linearly with temperature. At about 200°K one observes additional broadening, which increases rapidly with T in a nonlinear manner. This effect is strongest for the stretching vibrations ν_1 and ν_3. Fig. 3 shows pertinent experimental data on the $\nu_1(A_{1g})$ line in NH_4Br. These internal vibrations can be approximated as excitations localized on individual molecules. Assuming that this extra broadening is due to relaxation by the random jumping of the NH_4^+ between inverted configurations one can predict its temperature dependence by an argument borrowed from the theory of magnetic resonance relaxation[11]. The inverse relaxation time, and therefore the additional linewidth will be proportional to the Fourier component of the time correlation

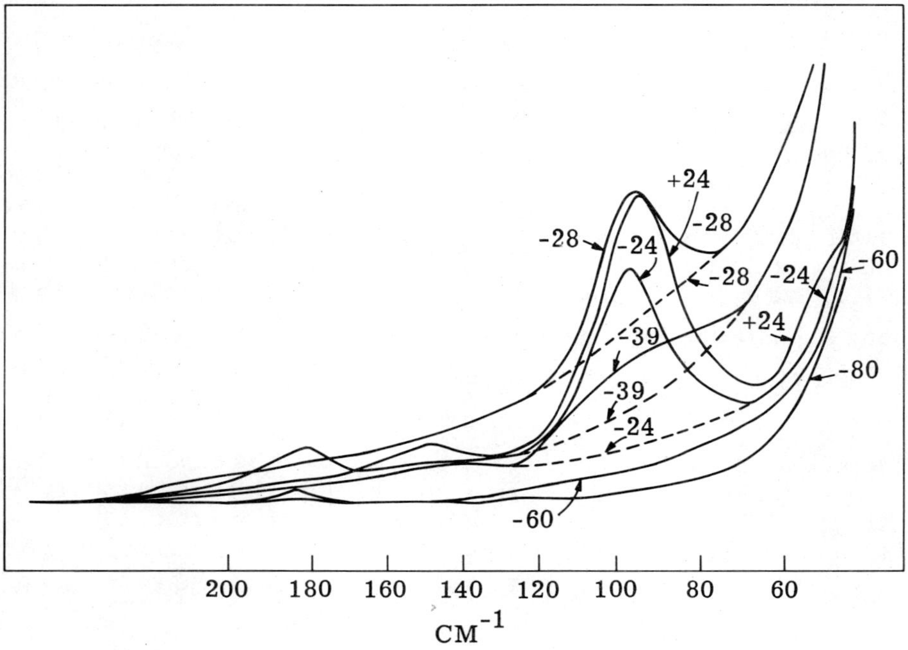

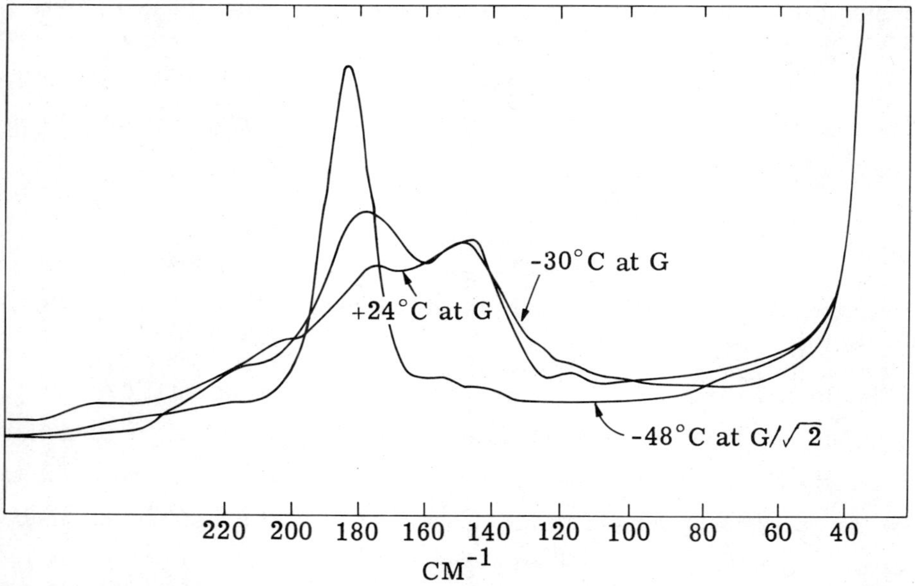

Fig. 2. Top-Lattice spectrum of α_{xx} in NH_4Cl at various temperature-indicated in ^{o}C next to traces. Bottom-Lattice spectrum of α_{xz} in NH_4Cl at 3 temperatures-notice change in gain from G to $G/\sqrt{2}$.

function associated with the random process, evaluated at the frequency $\omega_1 = 2\pi\nu_1$ of the vibration. For a Debye process with exponential correlation function K exp $(-t/\tau)$, the broadening will have the form

$$\Delta\nu_{rand} = CK\tau\,[1 + (2\pi\,\nu_1\,\tau)^2]^{-1} \tag{1}$$

C is a coupling constant measuring the interaction between the vibration ν_1 and the displacements for the jumping motion of the NH_4 ion. From the NMR work[1, 2] we know that the latter motion is performed against a potential barrier ΔE. Thus

$$\tau = \tau_0 \exp(\Delta E/kT) \tag{2}$$

with $\tau_0 \simeq 10^{-15}$, $\Delta E/k \sim 10^3$ °K for order of magnitude, yielding $\tau \sim 10^{13}$ and $(2\pi\,\nu_1\,\tau)^2 \sim 10^3$ in the temperature range of interest. We therefore have to good accuracy:

$$\Delta\nu_{rand} = KC(2\pi\,\nu_1)^{-2}\tau^{-1} = A\exp(-\,\Delta E/kT) \tag{3}$$

Fig. 4 shows the temperature dependence of the linewidth of $\nu_1(A_{1g})$ in NH_4Cl and NH_4Br; by linear extrapolation from the lower temperatures one can subtract off the

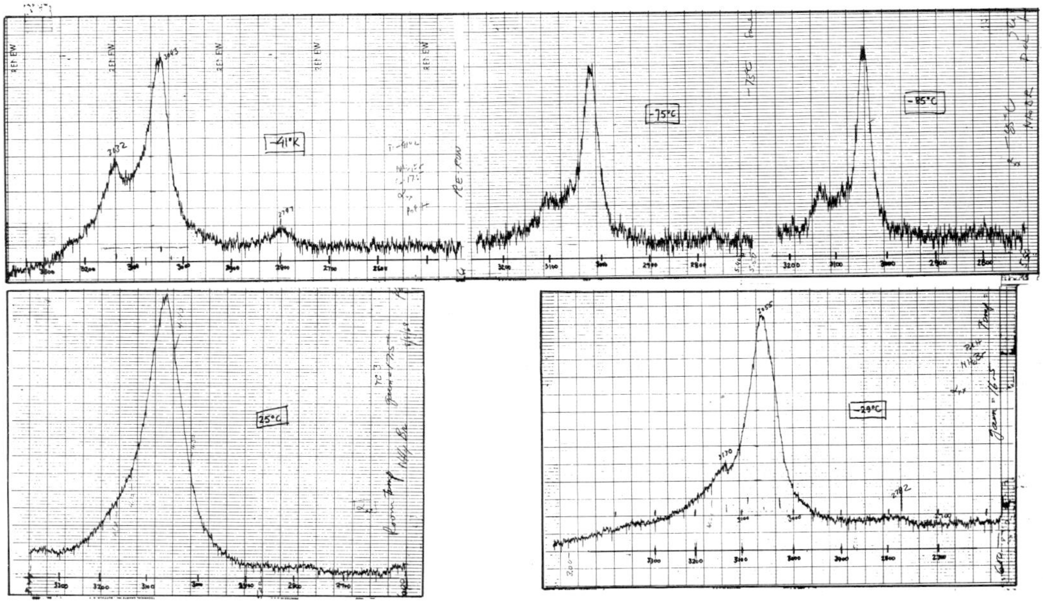

Fig. 3. The $\nu_1(A_{1g})$ symmetric stretch line in NH_4Br at various temperatures (Top left to right: -41°C, -75°C, -85°C; Bottom left to right: + 25°C - 29°C).

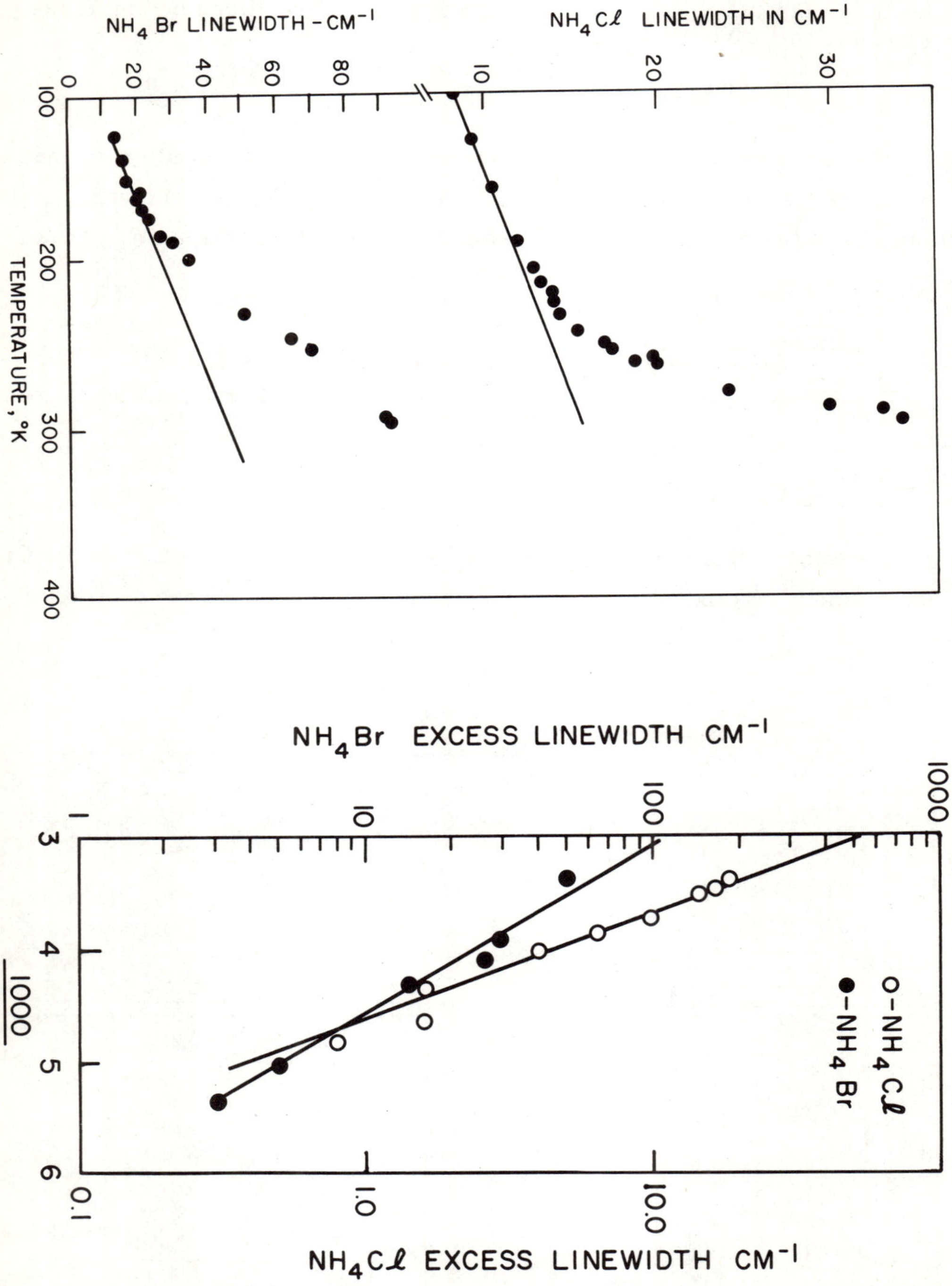

Fig. 4. Left-Linewidths of the ν_1 vibration in NH_4Cl and NH_4Br as a function of temperature. Right-Logarithmic plots of the contribution from the NH_4 jumping motion to the widths of the same lines.

slowly varying linewidth component. This figure also shows a semilog plot (against 1/T) of the remaining component, which gives approximately a straight line, as predicted by (3). The activation energies $\Delta E/k$ thus obtained compare favourably with those obtained from NMR[2].

	This Paper	NMR
NH_4Cl	2500 ± 300	2380
NH_4Br	1500 ± 200	1710

For lattice modes, in contrast to the case of $\nu_1(A_{1g})$ we have approximately $(2\pi\ \nu\ \tau) \sim 1$, in the same T range, and according to (1) we may expect, in this range, a maximum of any relaxation contribution to the linewidth. Fig. 5 shows linewidth data on two lines in NH_4Cl which indeed exhibit maxima. They occur at slightly

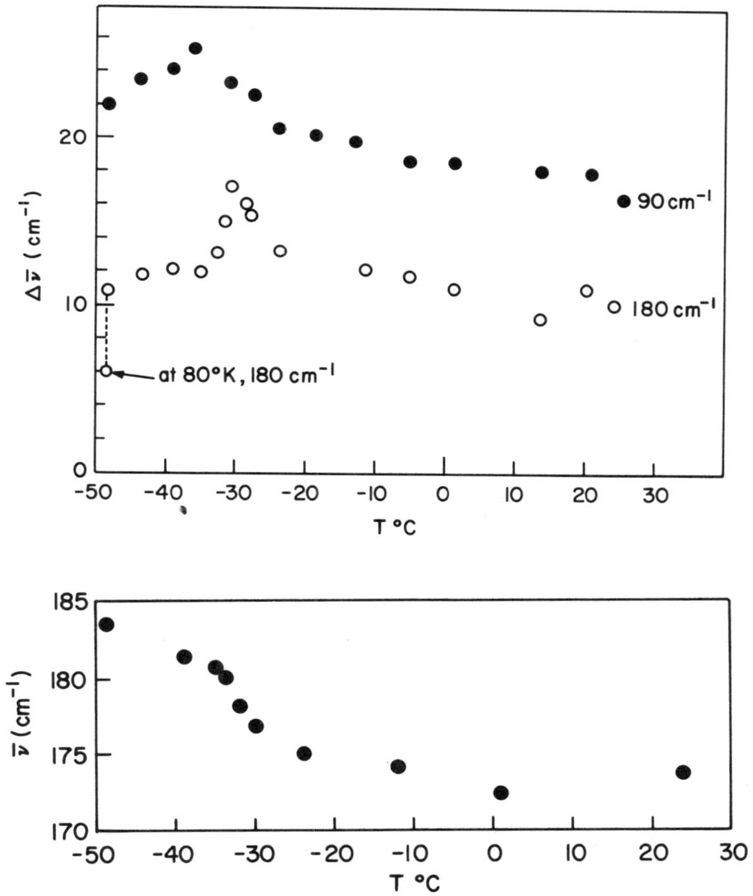

Fig. 5. Top-Linewidths of two lattice lines in the NH_4Cl spectrum Bottom-Frequency of the TO lattice line in NH_4Cl.

different temperatures, as expected from the difference in frequencies. Applying the relation $2\pi\nu\tau = 1$ at the maxima for both lines and using (2) we come up with the following: $\tau \simeq 10^{-15}$ sec and $\Delta E/k \simeq 3000°K$. These are rough estimates (mainly due to the closeness in temperature of the two maxima). However, they further support the assumption that there is a strong interaction between lattice vibrations and the random jumping of the NH_4 ions. Fig. 5 also illustrates the temperature dependence of ν_5 in the region of linewidth maximum which exhibits qualitatively the expected dispersive character.

SUMMARY

In addition to presenting new information concerning assignments of Raman active vibrations in Phase III of normal and deuterated ammonium halides, we have studied the low frequency lines characteristic of the disordered Phase II. These lines start appearing at temperatures far below T_c, and increase continuously in intensity with the gradual loss of order. They are ascribed to vibrations of neighbouring pairs of inverted molecules. We also show that the random jumping motion of the NH_4 tetrahedra involved in the disordering process is responsible for a significant contribution to the temperature dependent linewidths of many transitions.

ACKNOWLEDGMENTS

We acknowledge enlightening discussions with Dr. P. Maker, an informal conversation with Drs. Fleury and Wang and the expert help of Mr. R. Kilponen in taking the spectra.

REFERENCES

1. J. Itoh, Kusaka, and Saito, J. Phys. Soc. Japan, 17, 463 (1962).
2. J. Itoh and Y. Yamagata, J. Phys. Soc. Japan 17, 481 (1962).
3. I. Freund, paper H-2 this conference.
4. R.S. Krishnan, Proc. Ind. Acad. Sci. 27A, 321 (1948).
5. L. Couture and J.P. Mathieu, Proc. Ind. Acad. Sci. 28A, 401 (1948).
6. J. de Chimie Physique 49, 226 (1952).
7. C.W. Garland and N.E. Schumaker, J. Phys. Chem. Solids 28, 799 (1967).
8. L. Rimai, J. Parsons, J. Hickmott, and T. Nakamura, Phys. Rev. 168, 623 (1968).
9. P. Lazay, J.H. Lunacek and G.B. Benedek, paper G-5 this conference.
10. P. Maker (to be published).
11. Bloembergen, Purcell, and Pound, Phys. Rev. 73, 679 (1948).

H-5: LIGHT SCATTERING BY POLARIZATION FLUCTUATIONS IN KH_2PO_4

I. P. Kaminow
Bell Telephone Laboratories, Incorporated
Holmdel, New Jersey

ABSTRACT

Experimental results on the temperature dependence of the ferroelectric mode in KH_2PO_4 as measured by Raman scattering are discussed in the light of current dynamical theories. The coupled proton-phonon model of Kobyashi seems most reasonable. Other experiments on the longitudinal mode in KH_2PO_4 and the ferroelectric mode in KD_2PO_4 and $KD_2H_{2(1-x)}PO_4$ with x = .35 are mentioned briefly.

INTRODUCTION

Dynamical theories of the ferroelectric transition in KH_2PO_4 (KDP) lead to the notion of a polar vibrational mode whose characteristic frequency tends to vanish as the temperature T approaches the Curie point T_c from the paraelectric region. Such a mode was observed recently by Kaminow and Damen[1] by means of a Raman scattering technique. This experiment along with some previously unreported work is described below and the implications of the ferroelectric mode observations are discussed.

Spontaneous Raman scattering in a crystal is produced by thermally excited fluctuations in the optical polarizability tensor α_{ij}. In the Placzek approximation, the polarizability fluctuation is proportional to lattice polarization fluctuation δP. Thus we can write the first half of (1)

$$\begin{aligned} J(\omega) &= K|\partial \alpha_{xy}/\partial P_z|^2 \left\langle \delta P_z^2(\omega) \right\rangle \\ &= K' \bar{n}(\omega)\, \chi''_z(\omega) \end{aligned} \qquad (1)$$

where $J(\omega)$ is the scattered power at frequency shift ω (positive for anti-Stokes and negative for Stokes scattering) due to a fluctuation in the α_{xy} component induced by δP_z,

K is a constant of the experiment and $\langle \delta P_z^{\,2}(\omega) \rangle$ is the power spectrum of polarization fluctuations. There are 6 optic and 1 acoustic modes that transform like α_{xy} and P_z; the differential $(\partial \alpha_{xy}/\partial P_z)$ may be different for each mode although for a given well defined mode it should be independent of T and ω. Now $\langle \delta P_z^2(\omega) \rangle$ is directly related to the imaginary part of the dielectric susceptibility $\chi_z''(\omega)$ by the fluctuation-dissipation theorem[2]

$$\left\langle \delta P_z^2(\omega) \right\rangle = \frac{2\hbar\epsilon_o}{\pi V} \bar{n}(\omega) \chi_z''(\omega) \tag{2}$$

where ϵ_o is the permittivity of vacuum, V the sample volume and the population factor $\bar{n}(\omega) = (\exp(\hbar\omega/kT)-1)^{-1}$. Substitution of (2) yields the second half of (1) with K' a constant over a particular vibrational mode. A formal derivation of (1), which also applies to inelastic neutron scattering, is given by Kwok[3] and others. This relation was applied in the classical limit $\hbar\omega/kT \ll 1$ by DiDomenico, Porto and Wemple[4] in their Raman scattering observation of the soft mode in $BaTiO_3$.

The fluctuation-dissipation theorem reflects the fact that a system in thermal equilibrium with its surroundings radiates as much noise power as it absorbs. Only a lossy system, one capable of absorbing energy when driven by a non-equilibrium force, can deliver noise power. For thermally excited transverse vibrational modes the macroscopic electric field $E(\omega)$ vanishes and the appropriate transfer function for $E(\omega) \neq 0$ as driving force is the dielectric function $\varkappa_z(\omega)$, with power dissipation proportional to $\varkappa_z'' \equiv \chi_z''$. For longitudinal modes it is the dielectric displacement $D(\omega)$ that is absent when no external field is applied, and $1/\varkappa_z(\omega)$ is the appropriate transfer function for $D(\omega) \neq 0$ as driving force, with power dissipation proportional to $\mathrm{Im}(1/\varkappa_z)$. Thus, for longitudinal modes one merely replaces χ_z'' by $\mathrm{Im}(1/\varkappa_z)$ in (1) to find the Raman spectrum.

EXPERIMENTS

A typical room temperature Raman spectrum $J(\omega)$ for KDP using an Argon laser source at 4880Å is shown in Fig. 1. The incident and scattered beams travel along x and y axes, respectively, while the incident and scattered polarizations are along y and x, respectively; only transverse modes transforming like α_{xy} and P_z produce the scattering. The variation in $\bar{n}(\omega)$ over the well defined lines is small and, neglecting instrumental corrections, the spectrum is proportional to $\chi''(\omega)$. However, for the low frequency scattering, which appears as a skirt around the unshifted laser scattering, the frequency variation in $\bar{n}(\omega)$ must be taken into account in deriving $\chi''(\omega)$ from $J(\omega)$. It is found that except for dramatic changes in the low frequency region the remaining spectrum is temperature independent. The temperature dependent $\chi_z''(\omega)$ obtained from $J(\omega)$ for the low frequency region is shown in Fig. 2. A similar band has also been observed in infrared reflection[5] and transmission[6] and in inelastic neutron scattering[7]. However, these techniques do not permit an accurate characterization of the mode.

The curves in Fig. 2 can be fitted within experimental uncertainty by the damped harmonic oscillator function

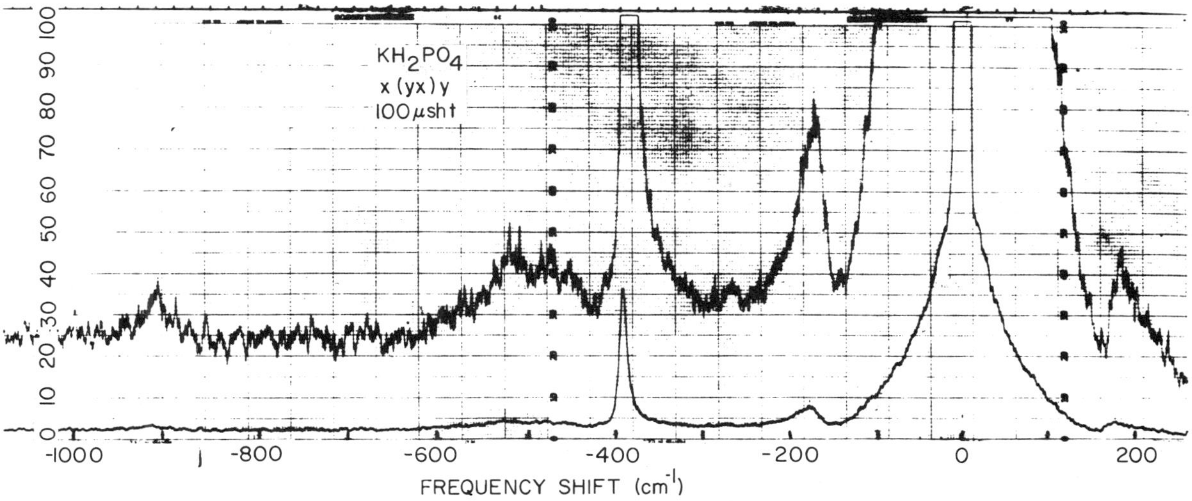

Fig. 1. Raman spectrum $J(\omega)$ for KH_2PO_4 at room temperature for transverse α_{xy} modes. Upper trace has 10x gain.

$$\chi_z(\omega) = \frac{\omega_o^2 \chi(0)}{\omega_o^2 - \omega^2 + i\omega 2\Gamma} \tag{3}$$

Note for example that in Fig. 2 as in (3) $\chi_z'' \sim \omega$ for small ω and $\chi_z'' \sim \omega^{-3}$ for large ω and high T; for low T where damping is stronger, $\chi_z'' \sim \omega^{-1}$ for large ω. The three parameters obtained by fitting (3) to the measurements in Fig. 2 are plotted in Fig. 3. It is seen that $\chi(0) = C/(T-T_c)$ to good accuracy. Here C is an undetermined Curie constant and $T_c = 117$ °K, which is 5°K lower than the actual phase transition temperature. The phase transition is caused by the acoustic mode which becomes unstable (i.e., its velocity vanishes) before the optic mode has an opportunity to do so[8]. Although there is considerable scatter in the measurements, the damping factor Γ appears to be temperature independent with a value of 85 cm^{-1}.

The characteristic frequency ω_o tends to zero as $T \to T_c$ and approaches a constant value $(2\Omega) = 99$ cm^{-1} at high temperature. The measurements are well fitted by $\omega_o^2 = (2\Omega)^2 (T-T_c)/T$, a form which arises naturally from several ferroelectric models. The single mode Lydanne-Sachs-Teller relation $(\omega_o/\omega_L)^2 = \varkappa(\infty)/\varkappa(0)$ using microwave measurements of $\varkappa(0)$ vs. T does not give as good a fit when it is assumed that both $\varkappa(\infty)$ and ω_L are temperature independent.

The longitudinal mode spectrum was computed using the parameters ω_o, Γ, T_c reported here and $\varkappa(\infty) = 4.7$, C=2820°K from microwave measurements[9]. The spectrum has a broad but well defined peak that shifts from 100 to 200 cm^{-1} as T varies from 417 to 127°K. The Stokes peak height is not very sensitive to temperature and is about 1/300 the peak, which occurs at $\omega=0$, in the transverse mode at room temperature. The macroscopic field $E(\omega)$ can also contribute to the scattering intensity through the electrooptic effect from higher frequency modes and electronic processes[10]. Using the spectrum of Fig. 1, the infrared strengths of the modes[5], the second harmonic coefficient d_{36}[11], and the electrooptic coefficient r_{63}[12] one can calculate[13] that the low frequency mode itself is responsible for 90% of r_{63} so that the remaining contribution is small.

With P. A. Fleury, several attempts were made to observe the longitudinal mode, both with 90° and with forward scattering. This turned out to be difficult because of the weakness of the mode and the presence of nearby modes of other symmetry in 90° scattering and instrumental interference in forward scattering. Although a weak mode of correct nature might have been present, it could not be identified with confidence.

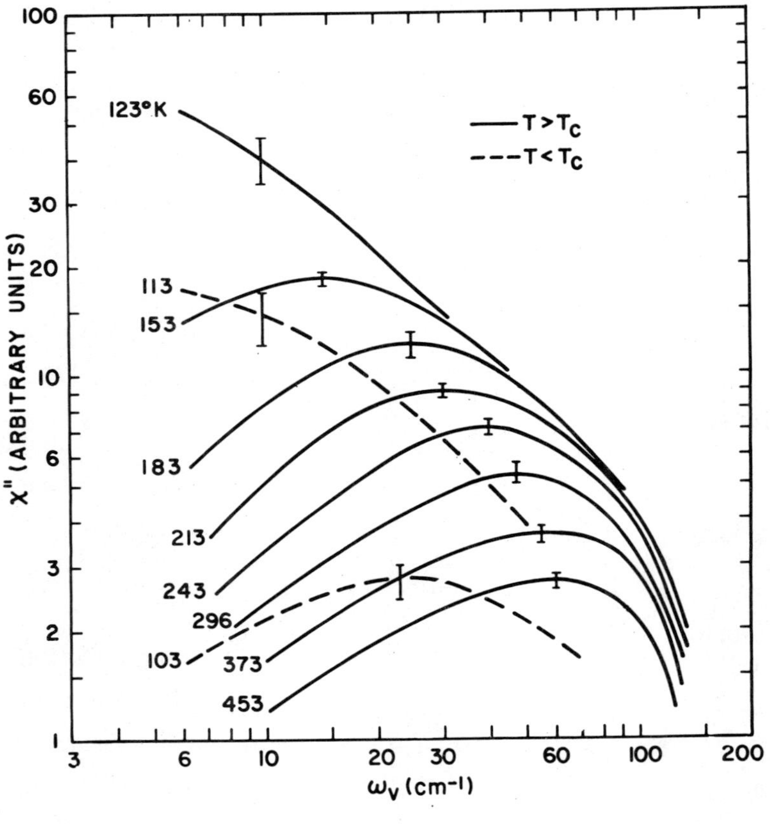

Fig. 2. Imaginary part of susceptibility $\chi_z''(\omega)$ for KDP obtained from $J(\omega)$. Typical error bars are shown. (from Ref. 1).

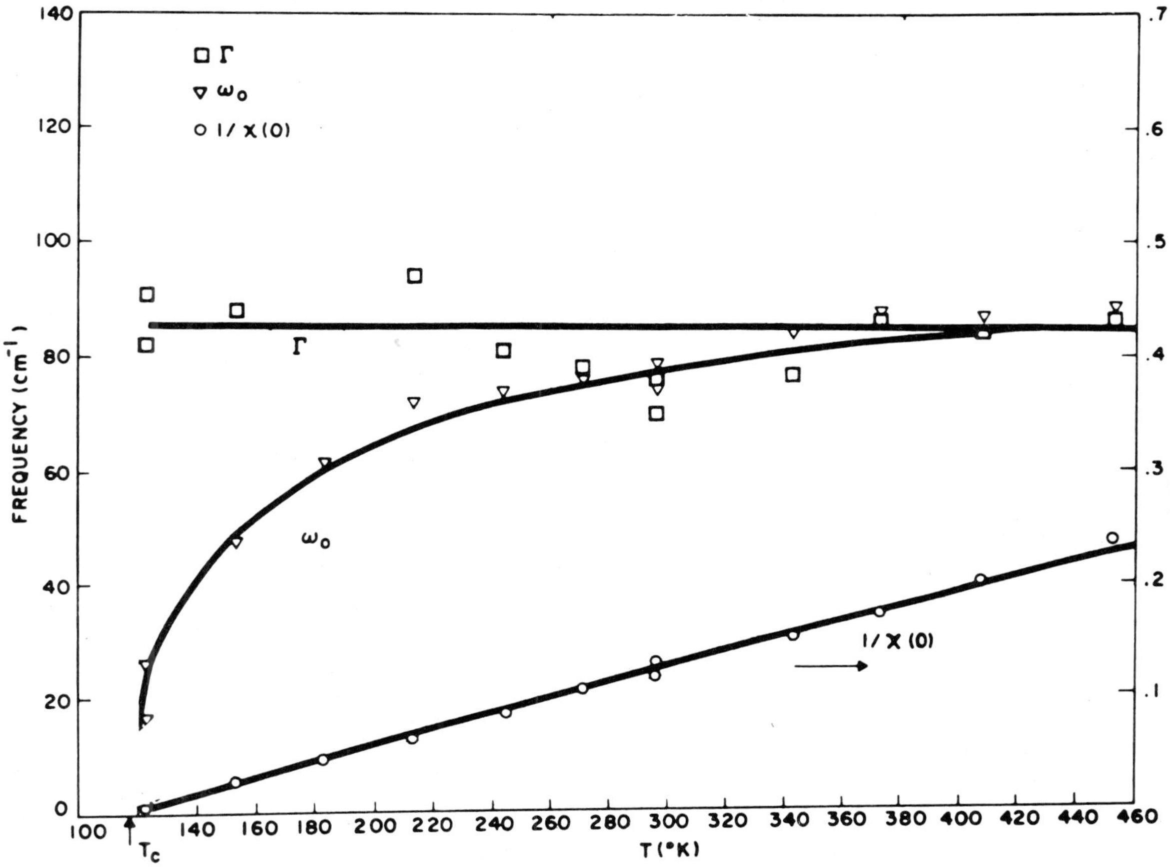

Fig. 3. Parameters of susceptibility function obtained by fitting measured $\chi_z''(\omega)$. (from Ref. 1).

Preliminary observations of the transverse mode in KD_2PO_4 (DKDP) and $KD_{2x}H_{2(1-x)}PO_4$ with x=0.35 (.35 DKDP)[14] were attempted with T.C. Damen. Microwave measurements of Hill and Ichiki[15] on DKDP show that χ_z'' has a Debye form with a 300°K peak at ~1 cm^{-1}, which is beyond the range of the current Raman technique. However, the high frequency tail falling off like ω^{-1} as for a Debye function is observed, although not enough of the curve is visible for fitting purposes (see Fig. 4). The χ_z'' for .35 DKDP has a broad peak at 25 cm^{-1} at room temperature that moves toward lower frequency as T approaches T_c(~ 159°K) and to slightly higher frequency at T = 427°K (see Fig. 5). However, χ_z'' cannot be fitted by the simple harmonic oscillator function of (3); a more complex function, perhaps a mixture of KDP and DPDP functions, is required.

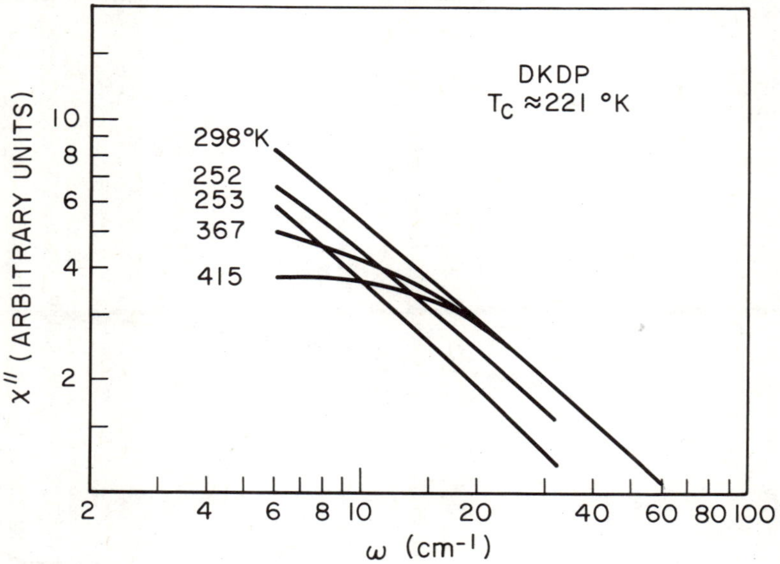

Fig. 4. Imaginary part of susceptibility $\chi_z''(\omega)$ for DKDP. (Scale not related to Fig. 2)

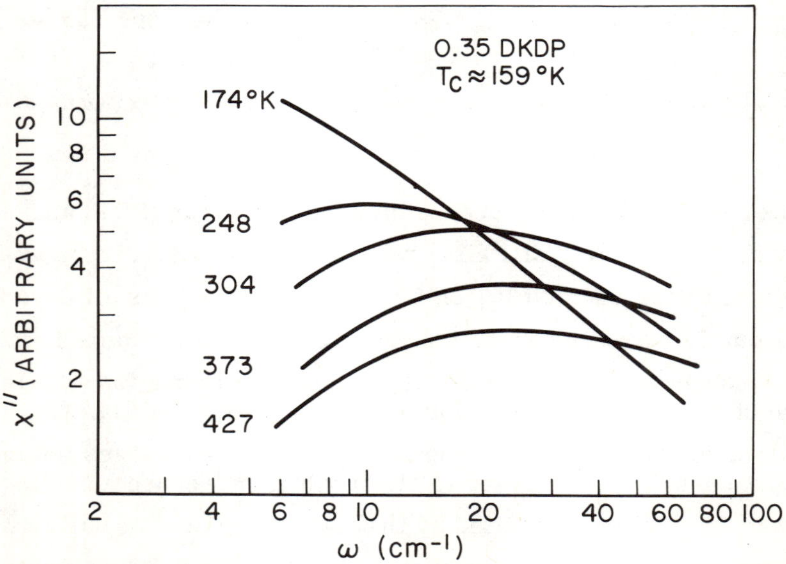

Fig. 5. Imaginary part of susceptibility $\chi_z''(\omega)$ for 0.35 DKDP. (Scale not related to Figs. 2 and 4)

DISCUSSION

Slater[16] introduced the idea of two equivalent minima for the proton configuration in paraelectric KH_2PO_4. Protons make random jumps between these positions by hopping the barrier separating them when they receive sufficient thermal activation energy. Such a model leads naturally to a Debye susceptibility[17]. Our experiment, however, shows the existence of a temperature dependent characteristic frequency for KDP (if not for DKDP). Such a result is consistent with a soft phonon[18] and/or a collective tunneling mode[19-21] model. The latter model can be described formally in terms of a spin Hamiltonian, and following Tokunaga[21]

$$\begin{aligned} H &= -(2\Omega)\sum_i X_i - \frac{1}{2}\sum_{ij} J_{ij}Z_iZ_j \\ &= \sum_i -[(2\Omega)X_i + J\langle Z\rangle Z_i] - \frac{1}{2}\sum_{ij} J_{ij}\zeta_i\zeta_i \end{aligned} \tag{4}$$

where $Z_i = \pm\frac{1}{2}$ represents a dipole along the ferroelectric c-axis, (2Ω) is the tunneling frequency, J_{ij} represents the dipole-dipole and short-range Coulomb interaction tending to align neighboring dipoles, $J = \Sigma J_{ij}$ [not to be confused with $J(\omega)$], $\langle Z\rangle$ is a thermal average, and $\zeta_i = Z_i - \langle Z\rangle$ is the fluctuation about $\langle Z\rangle$

The ground state of the tunneling term alone is the symmetric linear combination of $+\frac{1}{2}$ and $-\frac{1}{2}$ states, which has no mean dipole moment along z. The ground state of the interaction term is the ferroelectric state with all dipoles aligned along z. Entropy considerations require disorder in the interaction term at high temperature. At T_c the reduction in free energy obtained by ordering the dipoles just equals the increase in energy incurred by stopping the tunneling. The substance becomes ferroelectric at $T_c > 0$ provided $4\Omega/J<1$. When Ω is small, as for DKDP, the temperature at which the interaction term dominates the tunneling term is high. Comparing our measurements with the solution of (4), we find $(2\Omega) = 99\ cm^{-1}$ and $J/2 = 183\ cm^{-1}$, where $4\Omega/J = \tanh(\Omega/kT_c)$.

Tokunaga shows that the fluctuation term in (4) is responsible for a line broadening of order $\sim J$ with no anomaly at T_c. However, his $\chi_z(\omega)$ has a form that is a satisfactory approximation to the observed harmonic oscillator function only when $(\omega_0/2\Gamma) \gg 1$. Inspection of Fig. 2 shows that the linewidth (measured at half-power points) decreases as T approaches T_c, roughly as $[(T-T_c)/T]^{1/2}$. However 2Γ, which is the linewidth when $(\omega_0/2\Gamma) \gg 1$, does remain constant at 170 cm^{-1}.

The theory of paraelectric line broadening based on the spin Hamiltonian (4) is analogous to theories of paramagnetic broadening[22], which also lead to widths of order J at high temperature. (A field in the x direction will give the tunneling term.) Physically, the effective field at a particular site fluctuates about its mean due to the variety of up and down dipole configurations that can be assumed by near neighbors.

At high temperature, $T \gg T_c$, all configurations are equally likely. However, the probability of a given frequency depends on the number of equivalent configurations that can produce it, the peak occurring at $\sim \langle 2\Omega \rangle$. (For DKDP, $\langle 2\Omega \rangle \sim 0$.) As T is reduced the low energy configurations are favored and clusters of aligned dipoles tend to form, thereby reducing the distribution of frequencies as in Fig. 2. Experimentally, we have seen that this distribution is well represented by an overdamped harmonic oscillator with constant frictional damping.

The harmonic oscillator form supports the coupled proton-phonon model of Kobyashi[23], in which the tunneling mode is coupled to the harmonic K-PO_4 displacements that produce P_z. Further support is provided by the agreement between our observed room temperature $\omega_o \approx 80\ cm^{-1}$ and Cochran's estimate[18] of 85 cm^{-1} based on the harmonic lattice model (with $\omega_o^2 \sim T - T_c$). Hence, the ferroelectric mode in KH_2PO_4 appears to be a cross between the soft phonon and the collective tunneling modes[23].

REFERENCES

1. I.P. Kaminow and T.C. Damen, Phys. Rev. Letters 20, 1105 (1968).
2. "Fluctuation, Relaxation and Resonance in Magnetic Systems," D. Ter Haar (ed.), Oliver and Boyd, London, England, 1962.
3. P.C.K. Kwok, "Solid State Physics," F. Seitz, D. Turnbull, and H. Ehrenreich (ed.), vol. 20, Academic Press, New York, 1967.
4. M. DiDomenico, Jr., S.P.S. Porto, and S.H. Wemple, Phys. Rev. Letters 19, 855 (1967).
5. A.S. Barker, Jr. and M. Tinkham, Phys. Rev. 38, 2257 (1963).
6. I.M. Aref'ev, P.A. Bazhulin, and T.V. Mikhal'tseva, Fiz. Tverd. Tela 1, 2413 (1965): (Translation: Soviet Phys. - Solid State 7, 1948 (1966)).
7. C.H. Schenk, E. Wiener, B. Weckermann, and W. Kley Phys. Rev. 172, 576 (1968).
8. E.M. Brody and H.Z. Cummins, Phys. Rev. Letters 21, 1263 (1968); C.W. Garland and D.B. Novotny, Phys. Rev. (to be published).
9. I.P. Kaminow and G.O. Harding, Phys. Rev. 129, 1562 (1963).
10. H. Poulet, Ann. Phys. (Paris) 12, 908 (1955); R. Loudon, Proc. Roy. Soc. (London) A275, 223 (1963).
11. J.E. Bjorkholm, IEEE J. Quantum Electronics, Nov. 1968.
12. R.D. Rosner, E.H. Turner, and I.P. Kaminow, Appl. Optics 6, 778 (1967).
13. I.P. Kaminow and W.D. Johnston, Phys. Rev. 160, 5191 (1967).
14. I.P. Kaminow, Phys. Rev. 138, A1539 (1965).
15. R.M. Hill and S.K. Ichiki, Phys. Rev. 130, 150 (1961).
16. J.C. Slater, J. Chem. Phys. 9, 16 (1941).
17. V.V. Daniel, "Dielectric Relaxation," Academic Press, New York, 1967.
18. W. Cochran, Phil. Mag. Suppl. 10, 401 (1961).
19. R. Blinc, J. Phys. Chem. Sol. 13, 204 (1960).
20. P.G. deGennes, Solid State Commun. 1, 132 (1963).
21. M. Tokunaga, Prog. Theor. Phys. (Japan) 36, 857 (1966).
22. M.F. Collins and C.G. Windsor, Proc. Phys. Soc. 90, 1015 (1967); M.F. Collins, J. Appl. Phys. 39, 533 (1968).
23. K.K. Kobyashi, J. Phys. Soc. (Japan) 24, 497 (1968).

H-6: BRILLOUIN SCATTERING STUDY OF THE FERROELECTRIC PHASE TRANSITION OF KH_2PO_4

Edward M. Brody and Herman Z. Cummins
The Johns Hopkins University
Baltimore, Maryland

The ferroelectric phase transition of potassium dihydrogen phosphate, KDP, which occurs at 122°K, is generally considered to result from the ordering of protons which occupy double minima potential wells[1]. Blinc and Kobayashi[2] have included the effects of interaction between the tunnelling protons and lattice distortion to predict a collective excitation having the Cochran behavior $\omega^2 \propto T-T_0$; we shall henceforth refer to it as the ferroelectric soft mode.

In this paper we concern ourselves with a presentation of our Brillouin scattering study and a discussion of its relationship to the soft mode. The apparatus which we have used is a Spectra Physics Model 125 helium neon laser, and a pressure scanned plane Fabry Perot interferometer with "photon counting" electronics. The optical arrangement used limits the observed scattering volume to a cylinder having a length of 1 mm and a diameter of .1 mm. Since the experiment measures the properties of a region much smaller than the actual volume of the crystal, it is rather insensitive to temperature gradients throughout the crystal volume.

The scattered light wavevector, $\vec{k}_s$, is at 90° to the incident light wavevector, $\vec{k}_0$. The crystallographic axes of the KDP are then oriented so that the wavevector of the acoustic modes which scatter the incident light is along x. If the incident light is polarized vertically with respect to the scattering plane (V), then only the Xy shear mode scatters light polarized in the scattering plane (H)[3]. By using the VH polarization selection, only light scattered from the Xy transverse shear modes is observed.

This mode is chosen for study since it is the only acoustic mode which is coupled piezoelectrically to the ferroelectric soft mode. Recently, Kaminow and Damen[4] have observed a heavily damped feature in the Raman spectrum of KDP which they have identified as the ferroelectric soft mode. The soft mode is observed to have a characteristic frequency, $\omega_0/2\pi$, which goes to zero at $T_0 = \sim 117°K$. However, the coupling to the Xy acoustic mode produces a "level repulsion," or anticrossing interaction, which drives the acoustic mode frequency to zero, causing the transition to occur at $T_c = 122°K$, a temperature higher than T_0. Our observed Brillouin spectra clearly show the frequency of the acoustic mode tending to 0 at T_c.

In addition, the intensity of the Brillouin components increases as T_c is approached, since the intensity of the scattered light is inversely proportional to the square of the

Brillouin shift. The effect is quite dramatic; the Brillouin components actually become visible a few degrees above T_c. Because of this large increase in scattering intensity, the incident laser power can be reduced without having to increase integration time. Hence, the slight laser induced heating of the sample can be reduced, thereby increasing temperature resolution and accuracy near T_c, where the Brillouin shifts are most strongly temperature dependent.

The temperature dependence of the Brillouin shifts from 118°K to 150°K is shown in Fig. 1. We have computed the elastic constant C_{66}^E for our shorted crystal (E = 0) from the measured acoustic frequencies, $\omega_2/2\pi$. Our measurements, the solid dots, agree with ultrasonic measurements of Garland and Novotny[5], the solid triangles. The non-anomalous elastic constant C_{66}^P (P = 0) is added for comparison. (According to measurements of Baumgartner[6], it has no discontinuity at T_c.) The very rapid increase in C_{66}^E with decreasing temperature below T_c, compared with the much slower increase as the temperature is raised above T_c, is evident. As an example, $C_{66}^E = 6\cdot 10^{10}$ dynes/cm^2 at 150°K ($T-T_c$ = 28°K), and at 120°K ($T-T_c \approx -2$°K). Looking at the insert, we observe that C_{66}^E approaches 0 at the same temperature from above and below. Within experimental error, domains are seen to appear at this same temperature. We identify this temperature as T_c. In a second order transition C_{66}^E extrapolates to 0 at the same temperature at which the transition occurs, therefore, we conclude that the transition is second order, with a maximum uncertainty of .03°K, in which first order effects cannot be ruled out by this experiment. In the ferroelectric phase, C_{66}^E rises quite rapidly, but continuously, over the first .01°K. Then the rate of rise decreases slowly with temperature.

To analyze the region above the phase transition, we can use a formalism of coupled modes similar to that employed by Dvorak in treating barium titanite[7]. First a Lagrangian density function is constructed to include the kinetic, T, and potential energy densities, U, of the Xy shear strains, x_6, and the polarization, P, which is along the ferroelectric z axis.

$$\mathcal{L} = T - U$$
$$T = 1/2\, \bar{m}\, \dot{P}^2 + 1/2\, (\rho/q^2)\, \dot{x}_6^2$$
$$U = 1/2\, \chi\, P^2 + 1/2\, C_{66}^P\, x_6^2 + a_{36}\, P\, x_6$$
$$\bar{m} = m/Ne^2$$
$$\chi = K/Ne^2$$

To explain the terminology, $\bar{m}$ is an "effective mass density," N is the number of dipoles per unit volume, m is the effective mass per dipole unit, e is the effective charge per dipole unit, χ is the inverse susceptibility of the clamped crystal, K is an effective force constant of the ferroelectric mode, q is the wavevector of the strain and polarization modes, ρ is the KDP density and a_{36} is the piezoelectric stress constant. The Lagrangian equations of motion are shown in Eq. (1), with a phenomenological damping term, $2\Gamma\dot{P}$, included.

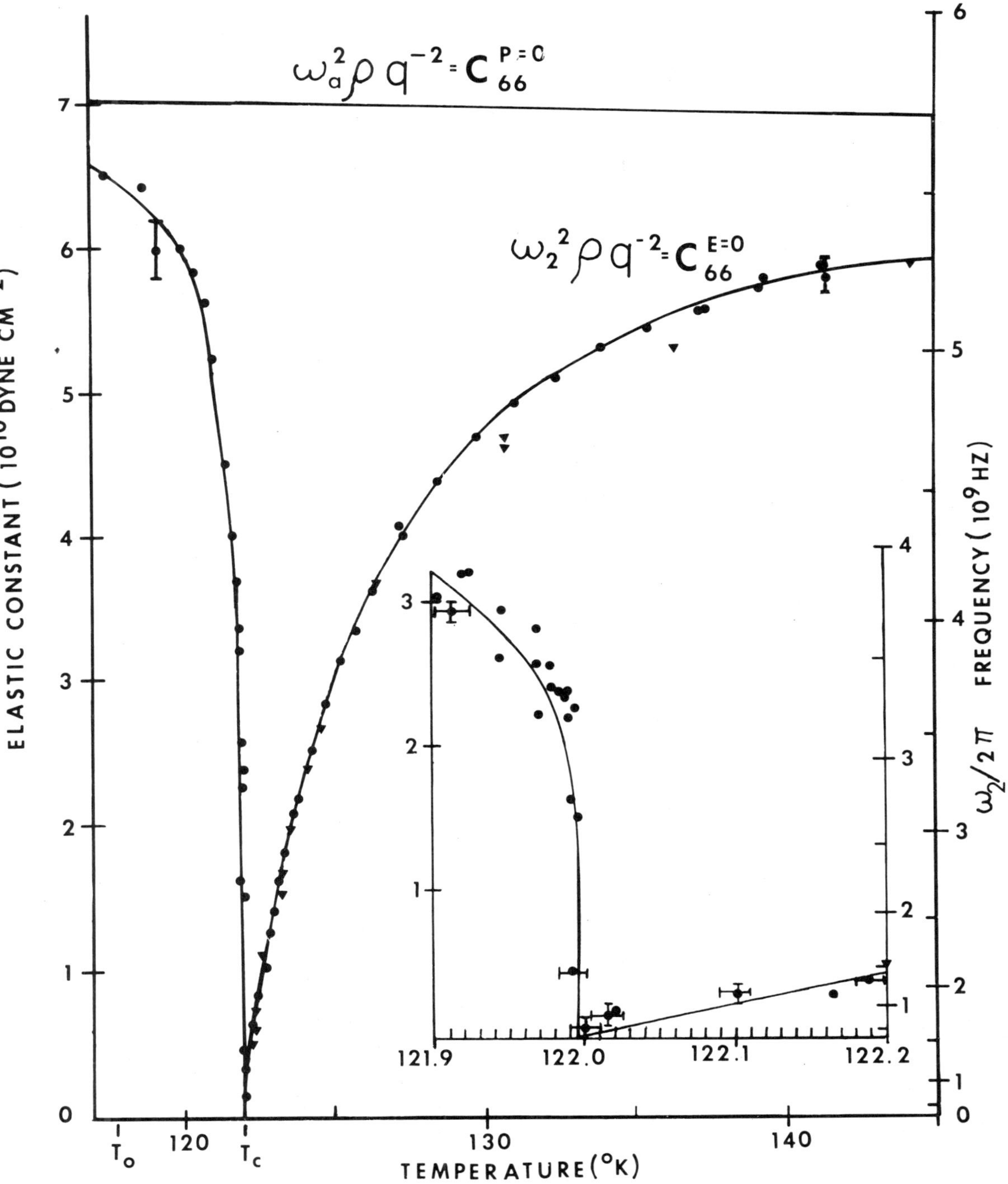

Fig. 1. Elastic constants C_{66}^E and C_{66}^P. Solid circles: Brillouin shifts ($\omega_2/2\pi$). Solid triangles: Ultrasonic measurements of C_{66}^E (see Ref. 5). C_{66}^P is from piezoelectric resonance data (see Ref. 6 and 8).

$$\ddot{P} = -(a_{36}/\bar{m})\, x_6 - (\chi/\bar{m})\, P - 2\Gamma \dot{P}$$
$$x_6 = -(q^2/\rho)\, C^P_{66}\, x_6 - (a_{36}\, q^2/\rho)\, P \qquad (1)$$

The notation is changed by introducing the angular frequencies, ω_o and ω_a, which have the following physical significance. ω_o would be the angular frequency of the ferroelectric mode in the absence of damping and coupling; ω_a would be the angular frequency of the acoustic mode in the absence of coupling.

$$\omega_o^2 = \frac{\chi}{\bar{m}} \qquad \omega_a^2 = q^2\, C^P_{66}/\rho$$

Harmonic solutions to the coupled Eq. (1) are given by the roots of the secular Eq. (2). In our experiment, q is small enough, about $2 \cdot 10^5$ cm^{-1}, that $\omega_o \gg \omega_a$ and the solutions of Eq. (2) separate into a high-frequency pair which are "mostly polarization" and are essentially identical to the uncoupled polarization modes (Eq. (3)) and a low frequency "mostly acoustic" pair, (Eq. (4)).

$$(\omega^2 - \omega_a^2)\,(\omega^2 - \omega_o^2 - 2i\Gamma\omega) - (q^2 a_{36}/\rho\bar{m}) = 0 \qquad (2)$$

$$\omega^2 - \omega_o^2 - 2i\Gamma\,\omega = 0 \qquad (3)$$

$$\omega_2^2 = \omega_a^2 - q^2 a_{36}^{\;2}/\rho\bar{m}\,\omega_o^2 \qquad (4)$$

Eq. (4) is recast into a form, Eq. (5), explicityly showing the relationship between the Brillouin measurements of ω_2 and the characteristic angular frequency, ω_o, of the ferroelectric mode.

$$\bar{m}\,\omega_o^2 = a_{36}^2\, q^2\, \rho^{-1}\, (\omega_a^2 - \omega_2^2)^{-1} \qquad (5)$$

Using our Brillouin data $\bar{m}\,\omega_o^2$ can be computed and the result is shown in Fig. 2. We have fit our data to a function proportional to T-T_o, the solid line, giving us $\bar{m}\,\omega_o^2 = 4.14 \cdot 10^{-3}$ (T - 117.7). The dashed line represents ω_o^2 proportional to (T-T_o)/T, the functional dependence observed in the Raman experiment.

$$(\omega_o/2\pi)^2 = (99\ \text{cm}^{-1})^2\ \frac{(T-117)}{T}$$

The two experimental observations are normalized at 140°K. The linear fit appears superior in describing our data.

The form of the adiabatic dielectric constant at constant strain, ϵ_x, and at constant stress, ϵ_σ, can be derived from our potential energy density U and then evaluated using the Brillouin data with the linear fit. Remember that $\bar{m}\,\omega_o^2 = \chi$.

$$\epsilon_X = 4\pi/\chi = 3030/(T - 117.7)$$

$$\epsilon_\sigma = 4\pi/(\chi - a_{36}^2\, C_{66}^P) = 3030/(T - 122.0)$$

The result is in good agreement with a dielectric measurement of the Curie constant of 3122 by Mason[8], Baumgartner's determination of the difference in the Curie temperatures of the clamped and free crystals of 4.3°K[9], and our independent measurement of T_c as 122.0°K.

Setting the results of the Brillouin and Raman experiments equal at 140°K gives us $\bar{m}$, and assuming that there is one dipole unit per phosphate group gives us $m/e^2 = 2.34$ (proton mass/(electron charge)^2.

For $T < T_c$ we must add a nonlinear term F(P) to the potential energy, U, to stabilize the ferroelectric phase. If the transition were basically of the lattice displacement type considered by Cochran[10], F(P) would be determined by the anharmonic forces and should be of the Devonshire form:

$$F(P) = 1/4\,\xi\,P^4 + 1/6\,\zeta\,P^6$$

Alternatively, the transition could be controlled by the statistical mechanics of the order-disorder transition of the protons, as proposed by Slater. Slater's original theory was modified by Takagi and Senko, and the combined theories discussed by Silsbee, Uehling

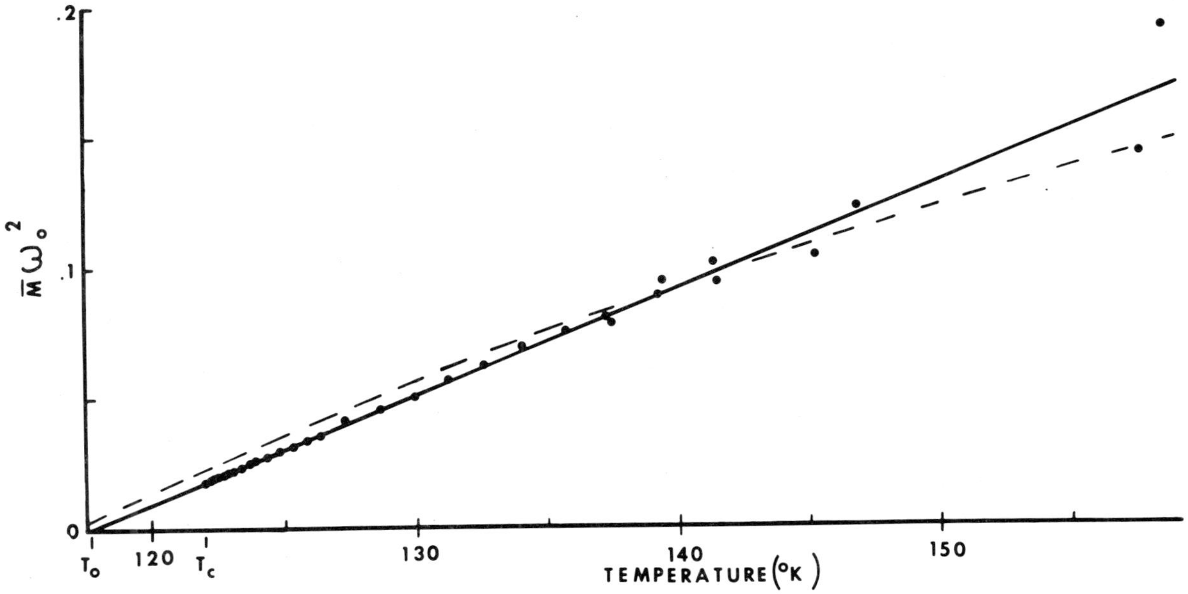

Fig. 2. $\bar{m}\omega_o^2$ from Eq. (5) and the data of Fig. 1. Solid line: Fit to linear function of T. Dashed line: Fit to $\bar{m}\omega_o^2 \propto (T - T_o)/T$ matched at 140°K.

and Schmidt (SUS)[11], who derived equations from which all thermodynamic properties could be deduced in terms of 3 energy parameters, ϵ_1, ϵ_0, β. Although the SUS formulation gives a superior fit to our Brillouin results than does the Devonshire free energy, it does not appear to completely describe the immediate vicinity of the transition. Our analysis is continuing in order to deduce a phenomenological F(P) which will describe our experimental data in the ferroelectric phase.

REFERENCES

1. J.C. Slater, J. Chem. Phys. 9, 16 (1941).
2. R. Blinc, Phys. Rev. Letters 26A, 8 (1967); K. Kobayashi, Ibid, 26A, 55 (1967).
3. H.Z. Cummins, "Ferroelectricity," E.F. Weller (ed.), p. 197, Elsevier Publishing Co., Amsterdam, 1967.
4. I.P. Kaminow and T.C. Damen, Phys. Rev. Letters 20, 1105 (1968), and paper H-5 this conference.
5. C.W. Garland and D.B. Novotny (to be published).
6. H. Baumgartner, Helv. Phys. Acta 23, 651 (1950).
7. V. Dvorak, Phys. Rev. 167, 525 (1968).
8. W.P. Mason, Phys. Rev. 69, 173 (1946).
9. H. Baumgartner, Helv. Phys. Acta 24, 326 (1951).
10. W. Cochran, Advanc. Phys. 10, 401 (1961).
11. H.B. Silsbee, et al., Phys. Rev. 133A, 165 (1964).

H-7: SOFT PHONON MODES AND THE 110°K PHASE TRANSITION IN $SrTiO_3$

J. M. Worlock, J. F. Scott, and P. A. Fleury
Bell Telephone Laboratories, Incorporated
Holmdel, New Jersey

The structure of $SrTiO_3$ has been the subject of X-ray, [1] infrared, [2] ESR, [3-5] neutron, [6] and Raman [7-10] spectroscopic investigations since 1962. While the high-temperature phase is acknowledged to be simple cubic perovskite O_h^1 with one formula group per unit cell, there have been, until recently, apparent inconsistencies in the diverse data and a lack of agreement concerning the symmetry of the crystal at low temperatures. Following discovery of anomalies in the sound velocity of $SrTiO_3$ near 110°K [11], ESR studies [3] showed the presence of a cubic to tetragonal phase transition. This was verified by means of X-ray analysis [1]; in the latter study the c/a ratio was determined to be 1.000 56, and two other phase transitions were inferred to be at 65 and 35°K. Infrared and neutron studies [2, 6] supported the X-ray symmetry assignment of structure at 77°K: the crystal was viewed as having domain structure with one of the equivalent (100) axes in each domain slightly elongated in the tetragonal phase; below 110°K the crystal was thought to retain a single formula group per unit cell and all atoms at inversion centers. The distortion from cubic symmetry was apparently so slight, in fact, that even the predicted $F_{1u} \rightarrow E_u + A_{2u}$ doubling below 110°K of each of the O_h-phase IR-active vibrations was unobservable in infrared studies. [2]

All of this was a puzzlement to the Raman spectroscopists [7-10], who found features at low temperatures with sharpness and temperature dependences indicative of first-order processes. Finally, in our last paper [12], we reconciled the existing data by verifying the ESR-based structure given recently by Unoki and Sakudo: D_{4h}^{18} with two formula groups per unit cell. Since our publication, related experiments have been performed. Shirane and Yamada [13] have directly verified our Raman conclusions with neutron scattering studies. And Muller et al. [5] have shown that a dynamically analogous phase transition occurs in $LaAlO_3$ [14], and that the transitions can be described by very similar order parameters. Since the analog of the 110°K $SrTiO_3$ transition seems now to be typical of perovskites, we wish in this paper to elaborate upon our earlier letter and to review the experimental data which led to our conclusions.

The basic features of our model are (a) that as $T \rightarrow T_o$ from above, the triply degenerate phonon F_{2u} (Γ_{25}) at the R point of the Brillouin zone softens (approaches zero frequency); (b) for $T \leq T_o$ the point R is then a reciprocal lattice point so that the unit cell is twice as large as in the cubic phase and the number of zone-center excitations is doubled; (c) as T is lowered from T_o, the two new zone-center phonons, whose progenitor was the F_{2u} zone-corner phonon, increase in frequency or "harden."

Our experimental evidence for these conclusions consists of (1) the appearance of several sharp lines in the Raman spectrum below 110°K, (2) a definite softening of two of these phonon frequencies as $T \rightarrow T_o$ from below (see Fig. 1), and (3) the direct interaction of these two phonons with components of the "ferroelectric" mode, observed using the technique of electric-field-induced Raman scattering [15] (EFIRS) and by tuning the "ferroelectric" mode frequencies with an electric field [16].

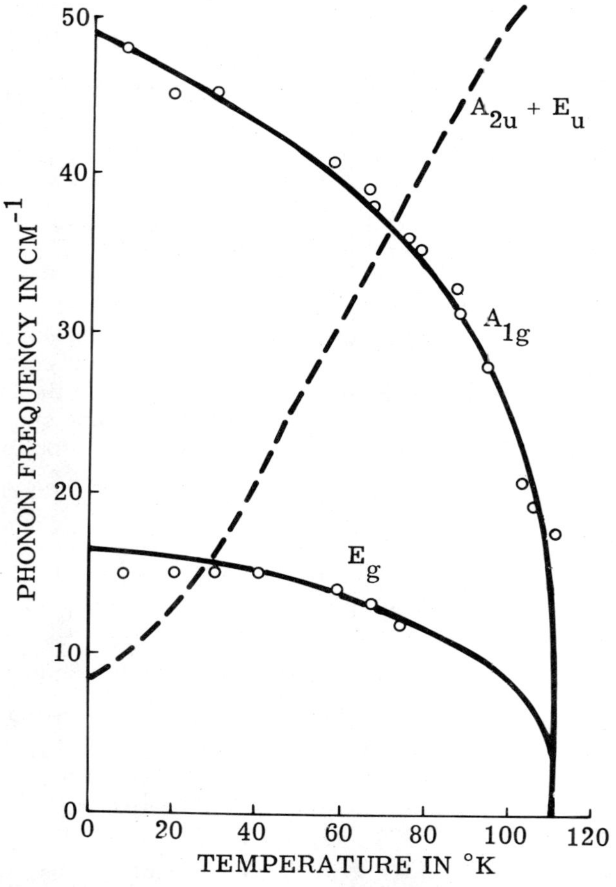

Fig. 1. Temperature dependence of the soft-phonon modes in the tetragonal phase with zero applied field.

The experimental apparatus is as described elsewhere [16-17]. The intrinsic (no external electric field) Raman spectra for various temperatures have been presented elsewhere and need not be reproduced here [6, 11-14]. Above 110°K the spectrum consists of several broad bands which have been interpreted, we believe correctly, by Nilsen and Skinner as entirely due to second-order scattering. Below 110°K five additional sharp lines appear at 15, 48, 146, 235, and 460 cm^{-1}. (These are the frequencies at 15°K.) While most of the features have been noticed by other workers [7-10], they have always been, in our view, misinterpreted. Previous interpretations range from strain-induced and shifted versions of the infrared (IR) modes [12-13] ("ferroelectric" and other TO's at 170 and 560cm^{-1}) to local modes induced by impurities [6]. Our earlier work with EFIRS has shown clearly the presence of the IR modes at their accepted frequencies [16-17], none of which exhibits noticeable change at 110°K.

We interpret these sharp lines in the Raman spectrum as scattering from the additional zone-center phonons, resulting from the doubled unit cell. They are made Raman active by the B_{2u} distortion accompanying the slight rotation of the oxygen octahedra in the (001) plane [7]. The Raman tensor components predicted by our model are listed in Table I and compared with our experimental observations. Column I lists the irreducible

TABLE I

CUBIC O_h		TETRAGONAL D_{4h}				TRIGONAL D_{3d}	
I	II	III	IV	V	VI	VII	VIII
$A_{2u}(\Gamma_2')$	~ 800 cm^{-1}	A_{2g}	silent	—	—	A_{2g}	silent
$F_{2g}(\Gamma_{25}')$	~ 500 cm^{-1}	E_u	IR	—	—	E_u	IR
		A_{1u}	silent	—	—	A_{1u}	silent
$F_{1u}(\Gamma_{15})$	~ 400 cm^{-1}	E_g	xz, yz	—	—	E_g	xx,-yy,xz,yz,xy
		B_{1g}	xx,-yy	460 cm^{-1}	D	A_{1g}	xx, yy, zz
$E_u(\Gamma_{12}')$	200-300 cm^{-1}	B_{2g}	xy	235 cm^{-1}	N,D(?)	E_g	xx,-yy,xz,yz,xy
		A_{2g}	silent	—	—		
$F_{1u}(\Gamma_{15})$	~110 cm^{-1}	E_g	xx,yz	143 cm^{-1}	N,D	E_g	xx,-yy, xz,yz,xy
		B_{1g}	xx,-yy	—	—	A_{1g}	xx, yy, zz
$F_{2u}(\Gamma_{25})$	SOFT	E_g	xz,yz	15 cm^{-1} "A"	N	E_g	xx,-yy, xz, yz, xy
		A_{1g}	xx, yy, zz	48 cm^{-1} "D"	D	A_{1g}	xx, yy, zz

I. Comparison of phonons in perovskites undergoing $O_h \rightarrow D_{4h}$ distortion. Column I, zone-corner phonon characters $\left(\frac{1}{2}, \frac{1}{2}, \frac{1}{2}\right)\pi/a$. Column II, zone-corner phonon frequencies in cm^{-1} estimated by Cowley (Reference 6). Column III, zone center symmetries below 110°. Column IV, Raman tensor elements predicted. Column V, Raman frequencies observed in cm^{-1}. Column VI, Raman tensor symmetries observed: D for diagonal elements, N for nondiagonal. Columns VII and VIII show the symmetry and selection rules of modes when the $O_h \rightarrow D_{3d}$ transition occurs, as in $LaAlO_3$.

representations of the R-point phonons in the cubic phase. Their approximate frequencies based on Cowley's calculations [6] from neutron-scattering data in the [100] direction appear in Column II. The irreducible representations of the zone-center phonons below 110°K are listed in Column III. These are the symmetries at zone center resulting from the distortion at 110°K acting on the corresponding phonons listed in Column I. Column IV lists the IR or Raman activity and the Raman tensors predicted by the model. In Columns V and VI we list the observed frequencies and Raman tensor elements. Columns V and VI are to be compared with Columns II and IV, respectively. The agreement between observations and the predictions of the model is excellent. Because of domain structure [1] the tetragonal axis of the crystal has no unique direction in the laboratory. Therefore, we can distinguish only between diagonal and off-diagonal elements in the Raman tensor. Appreciable off-diagonal elements were observed unambiguously only for the 143-cm^{-1} line (E_g) and the 15-cm^{-1} line (E_g). The E_g component at 460 cm^{-1} was not observed. Also the 235-cm^{-1} feature can only tentatively be identified as first order because this peak sits just at the top of a large second-order peak and the sharp component exhibits diagonal scattering in contrast to the model predictions.

Table I also presents a similar mode symmetry correlation for perovskites such as $LaAlO_3$, in which the soft R point phonon produces a trigonal rather than tetragonal distortion. The same R point soft phonon may produce in the simplest cases either an $O_h \rightarrow D_{4h}$ or D_{3d} transition according to the relative admixtures of the three components of the soft triply degenerate vibration, as discussed by Shirane and Yamada. The relative stability of D_{4h} and D_{3d} phases in $SrTiO_3$ and $LaAlO_3$ is determined by the details of the lattice anharmonicities.

Since the two lowest-frequency Raman-active modes (designated A and D in our previous work) originate from the soft, triply degenerate F_{2u} (Γ_{25}) mode in the cubic phase, one might expect the frequencies of both A and D to soften as $T \rightarrow T_o$ in the tetragonal phase [18]. This behavior is clearly shown in Fig. 1. The solid lines describe a temperature dependence of the form $\omega = \text{const} \times (T_o - T)^n$, where T_o = 110 K° and n = 0.31. This is only a convenient form. There is no a priori requirement that both ω_A and ω_D exhibit the same form of temperature dependence; nor is it necessary that their frequencies precisely vanish at T_o.

Of special interest is the interaction of modes in the D_{4h} phase. As shown in Fig. 1, two odd-parity modes in $SrTiO_3$ - the components of the so-called "ferroelectric" mode - soften as the temperature is lowered toward 0°K. At the same time, two even-parity modes - components of the soft R point mode in the cubic phase - harden as $T \rightarrow 0$°K. Consequently, there exist several temperatures at which crossings occur - two modes have exactly the same frequency. Of these two modes one will have odd parity and the other, even. When a macroscopic electric field $\mathcal{E}$ is applied to the crystal, odd and even parity states are mixed (e.g. E_g and E_u, A_{1g} and A_{2u}), as shown in the matrix below for the case of $\mathcal{E}$ along the four-fold axis.

$$\langle i | H_o + V(\varepsilon) | j \rangle =$$

	$\Psi(E_g)$	$\Psi(E_u)$	$\Psi(A_{1g})$	$\Psi(A_{2u})$
$\Psi(E_g)$	$E_1(T)$	$\beta(\varepsilon)$	0	0
$\Psi(E_u)$	$\beta(\varepsilon)$	$E_2(T,\varepsilon)$	0	0
$\Psi(A_{1g})$	0	0	$E_3(T)$	$\alpha(\varepsilon)$
$\Psi(A_{2u})$	0	0	$\alpha(\varepsilon)$	$E_4(T,\varepsilon)$

Another way of saying this is that the crystal symmetry is in the presence of the field reduced from D_{4h} to C_{4v}. Experimentally we have found that for ε of a few KV/cm the off-diagonal terms linking phonon states in the Hamiltonian matrix above are of the order of 1cm^{-1}. Ordinarily, such terms would produce only a small perturbation. However, at the mode crossings mentioned above these small off-diagonal terms will produce substantial mixing of wave functions, hence of intensities. (A similar coupling of soft phonons in $LiNbO_3$ and $LiTaO_3$, attributable to strain rather than electric fields, has been reported by Johnston and Kaminow [19].)

Since the frequencies of the odd-parity vibrations have been discovered to be intrinsically field dependent [16-17], such mode crossings can be produced at any temperature by tuning the E_u and A_{2u} phonons with the field.

The matrix above may be used to describe the four phonon modes in the 8-50 cm^{-1} region of $SrTiO_3$. The empirical forms of E_2 (T,ε) and E_4 (T, ε) have been detailed by Worlock and Fleury [17]. The forms of $E_1(T)$ and $E_3(T)$ have been observed by O'Shea and by us [12, 18], and some information concerning β and α was presented in our letter [12]. Attempts to obtain the field dependence of $\beta(\varepsilon)$, $\alpha(\varepsilon)$ have not yet been successful. The matrix describes coupled oscillators, of which one of each pair is field independent. However, our data cannot be fitted by this model, and it is necessary to include some explicit ε-dependence in E_1, and probably E_3, to explain the high field data. While E_2 and E_4 contain terms in ε, E_1 and E_3 can only contain field dependence with leading ε^2 terms, for symmetry reasons. Progress in this phase of the coupled mode analysis can be made by extending our measurements to higher fields; while we succeeded in applying ~20KV/cm without breakdown, much higher fields across $SrTiO_3$ have been reported.

Intensity transfers near the 8°K anticrossings of A_u/A_g pairs and E_u/E_g pairs are shown in Fig. 2. The relative intensities of the modes at ~50 cm^{-1} as a function of frequency are nearly of the form expected for classical coupled oscillators, one of which operates at constant frequency, the other of which is of variable driven frequency.

Fig. 3 presents ω versus E at 8 °K. Fig. 1 and 3 represent perpendicular planes in the three-dimensional graph of $\omega(T, E)$.

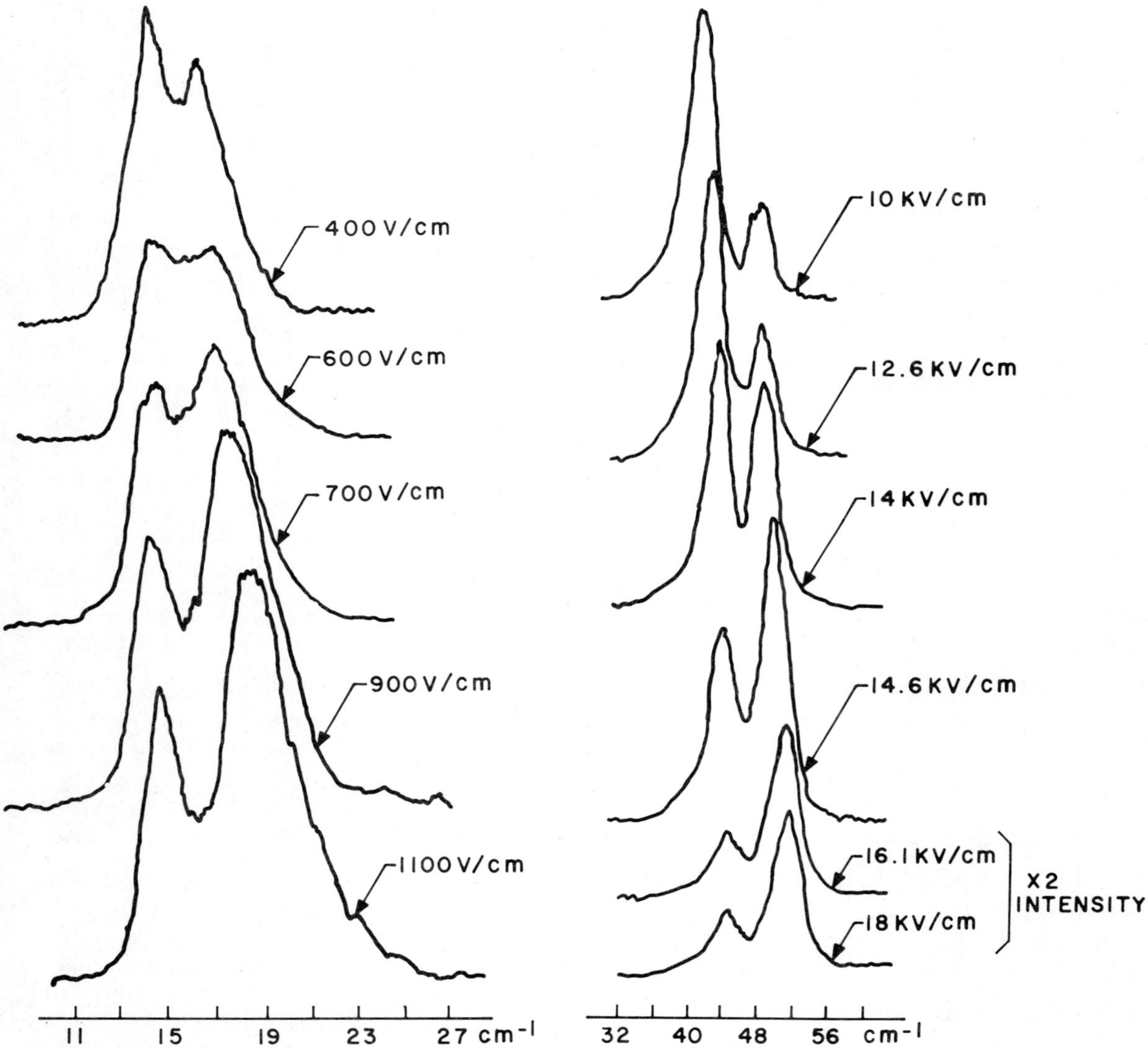

Fig. 2. Raman scattering intensities for A symmetry coupled modes near 15 cm^{-1} and E symmetry modes near 48 cm^{-1}

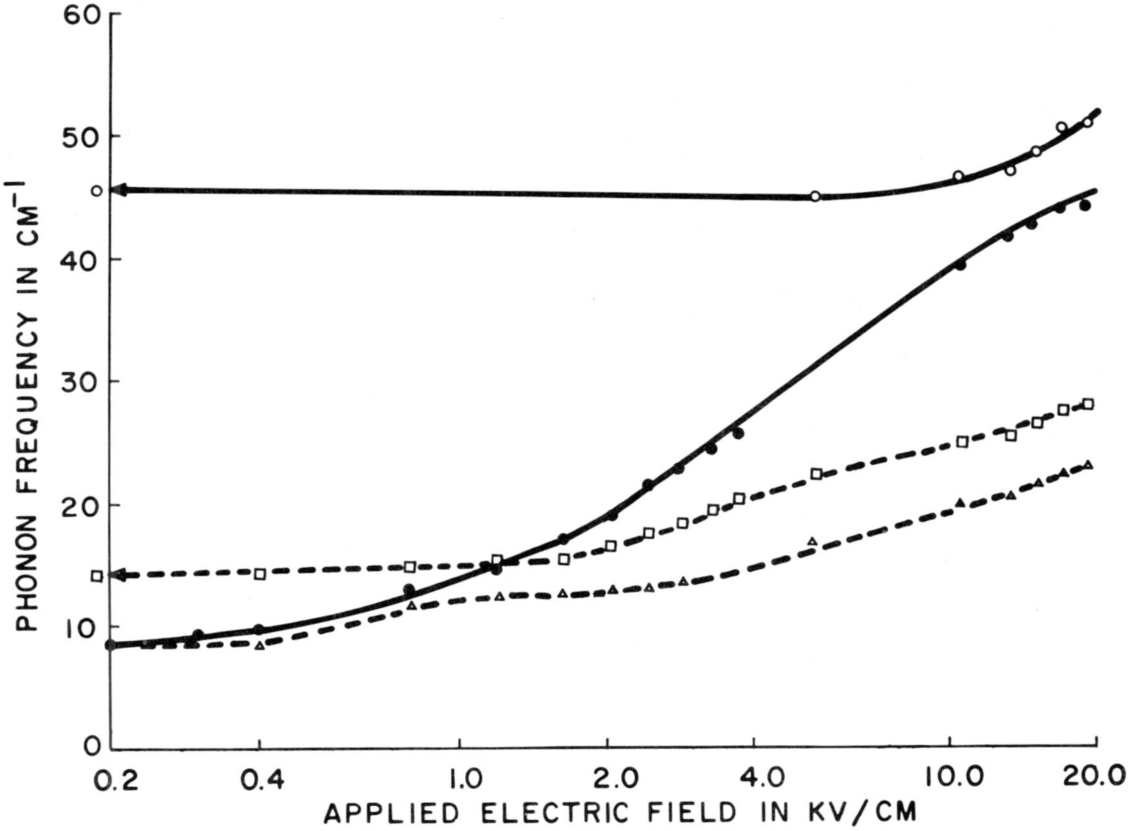

Fig. 3. Electric field dependence of four low-frequency phonons in $SrTiO_3$ at $10^{o}K$; illustrating the interactions between modes of E symmetry (dashed lines) and between modes of A symmetry (solid lines). The modes labeled B and C are components of the IR-active "ferroelectric" mode discussed in Reference 16. The left-directed arrows on the curves for A and D indicate frequencies measured at zero field.

The implications of our model for experiments other than Raman scattering will now be briefly discussed. Of these one of the most interesting is the sound-velocity behavior near 110°K. Experimentally [3, 11], both the longitudinal and transverse sound velocities remain fairly constant as T is decreased toward 110°K. Within about a one-degree interval of 110°K all of the velocities decrease suddenly by several percent. Below 110°K the temperature dependence differs for differing acoustic branches, but is generally less severe than immediately above 110°K. Cowley [6] has associated the temperature dependence of the sound velocity with that of the "ferroelectric" phonon by means of three-phonon interactions. Our suggestion is essentially different: Above 110°K the relevant soft mode is at the zone corner and no interaction with the zone-center acoustic phonons is operative. As 110°K is reached the soft optic phonon mode appears at zone center, where it interacts with the acoustic phonon – depressing its frequency. As T is lowered from 110°K the optic-phonon frequency increases, relieving the pressure on the acoustic branch and allowing the sound velocity to increase gradually. However, it is not possible for us to say definitely whether the interaction which lowers the sound velocity is harmonic, taking place in the tetragonal phase, as described above, or anharmonic in the cubic phase. The distinction is somewhat academic since the allowed harmonic interactions in the tetragonal phase can be viewed as anharmonic interactions involving the frozen zero-frequency distortion of the B_{2u} phonon at the zone corner. More detailed work is necessary to make quantitative the relation between the optic and acoustic phonon and their temperature dependences, but this model should provide a good basis for such work.

Our model predicts new experimental effects in several other areas. First, a new IR-active mode (E_u) is predicted below 110°K. Because the angle of rotation of the oxygen octahedron increases as T is lowered [4], we expect the IR oscillator strength to increase as T is lowered. We suggest that the low-temperature IR spectrum of $SrTiO_3$ be examined with particular attention paid to the 500 cm^{-1} region (see Table I). Second, inelastic neutron scattering experiments in the vicinity of the R point of the Brillouin zone have been undertaken [13].

Third, the effects of phonons on the semiconducting and superconducting properties [20] of doped $SrTiO_3$ should be re-examined, especially in view of the presence of additional soft phonon modes in the tetragonal phase. Fourth, any crystal properties associated with band structure, such as optical absorption in the uv, should be affected by the halving of the Brillouin zone at 110°K. For example, we expect that the indirect band gap, as calculated by Kahn and Leyendecker [21] in the cubic phase, should become direct below 110°K.

REFERENCES

1. F.W. Lytle, J. Appl. Phys. 35, 2212 (1964).
2. A.S. Barker, Jr., Phys. Rev. 145, 391 (1966).
3. L. Rimai and G. deMars, Phys. Rev. 127, 702 (1962).
4. H. Unoki and T. Sakudo, J. Phys. Soc. Japan 23, 546 (1967).
5. K.A. Muller, W. Berlinger, and F. Waldner, Phys. Rev. Letters 21, 814 (1968).
6. R.A. Cowley, Phys. Rev. 134, A981 (1964).
7. W.G. Nilsen and J.G. Skinner, J. Chem. Phys. 48, 2240 (1968).
8. D.C. O'Shea, R.V. Kolluri, and H.Z. Cummins, Solid State Commun. 5, 387 (1967).
9. R.F. Schaufele and M.J. Weber, J. Chem. Phys. 46, 2859 (1967).
10. L. Rimai and J.L. Parsons, Solid State Commun. 5, 381 (1967).
11. R.O. Bell and G. Rupprecht, Phys. Rev. 125, 1915 (1962); see also W. Kaiser and R. Zurek, Phys. Rev. Letters 23, 668 (1966); D.C. O'Shea, "Thesis," John Hopkins University, 1968 (unpublished).
12. P.A. Fleury, J.F. Scott, and J.M. Worlock, Phys. Rev. Letters 21, 16 (1968).
13. G. Shirane and Y. Yamada, Phys. Rev. (to be published).
14. W. Cochran and Z. Zia, Phys. Stat. Sol. 25, 273 (1968).
15. P.A. Fleury and J.M. Worlock, Phys. Rev. Letters 18, 665 (1967).
16. J.M. Worlock and P.A. Fleury, Phys. Rev. Letters 19, 1176 (1967).
17. P.A. Fleury and J.M. Worlock, Phys. Rev. 174, 613 (1968).
18. O'Shea, see Ref. [11].
19. W.D. Johnston and I.P. Kaminow, Phys. Rev. 168, 1045 (1968).
20. J.F. Schooley, W.R. Hosler, E. Ambler, J.H. Becker, M.L. Cohen, and C.S. Koonce, Phys. Rev. Letters 14, 305 (1965).
21. A.H. Kahn and A.J. Leyendecker, Phys. Rev. 135, A1321 (1964).

H-8: SPATIAL VARIATION IN THE RAMAN SPECTRUM OF $SrTiO_3$

D. C. O'Shea* and H. Z. Cummins
The Johns Hopkins University
Baltimore, Maryland

INTRODUCTION

With traditional Raman sources the observed spectra were spatial averages of the Raman scattering over a large volume of the sample. Since the advent of the laser with its high degree of spatial coherence, the beam can be easily focused to a 10μ cross-section, allowing one to explore possible spatial variation in the Raman scattering[1]. This technique is particularly useful for the study of crystals which exhibit domain structure since it permits distinction between the true bulk spectrum and possible effects associated with inhomogeneities at domain boundaries.

Crystals which undergo phase transitions from higher to lower symmetry frequently become multidomain in the lower symmetry phase. An example of this behavior can be seen in $SrTiO_3$ which has a cubic-tetragonal phase transition at 110°K[2]. The photographs in Fig. 1 show a $SrTiO_3$ crystal between crossed polarizers (a) at room temperature, (b) at 77°K, and (c) the same crystal viewed by unpolarized light at 18°K. The crystal is cut with the edges parallel to cube or tetragonal axes ((100) cut). The fine pattern of parallel lines are tetragonal domains on the order of 20μ to 200μ thick separated by walls parallel to (110) planes. A number of inclusions are seen as a grainy patch at the right center of the crystal. At 18°K the difference in the refractive indices for the two tetragonal directions is apparently large enough that the domains can be observed by unpolarized light.

When $SrTiO_3$ is cooled below 110°K several sharp features not present in the Raman spectrum above the phase transition appear in the tetragonal phase. During recent experiments on these sharp low temperature features[3,4] it was noticed that the line intensities varied when the beam was moved about in the crystal. Further observation appeared to show that there was some relation between the presence or absence of the low temperature features and the position of the beam with respect to the domain walls. It seemed possible that the features were first order lines due to a breaking of crystal

* Present address: Gordon McKay Laboratory, Harvard University, Cambridge, Massachusetts.

symmetry at the domain boundaries. In order to gain some insight into this effect a system was constructed to allow us to observe the position of the beam in the crystal while recording a Raman spectrum.

INSTRUMENTATION

The system consists of three sections: (1) a beam focusing and steering section which locates the narrow scattering column in the crystal, (2) the optical train for Raman scattering which images the scattering column onto the spectrometer slit and (3) the optical train for photographing the beam position in the crystal. The entire system is shown in Fig. 2. The beam must be focused down to a diameter somewhere near or below domain thicknesses (20μ - 200μ) to get any result other than an average of the domains. A 5 cm lens (item 7, Fig. 2) is used to focus the beam (item 4) from a Spectra-Physics Model 140 A^+ laser (nominal beam diameter is ~ 1.6 mm): the diffraction limited beam diameter is calculated to be 20μ [1]. Measurements from photographs show that it is less than 50μ. By mounting the lens on an x-y microscope stage, the beam can be directed to various parts of the crystal with comparative ease. When a synchronous motor is connected to one of the microscope stage screws, the beam can be continuously scanned across the crystal.

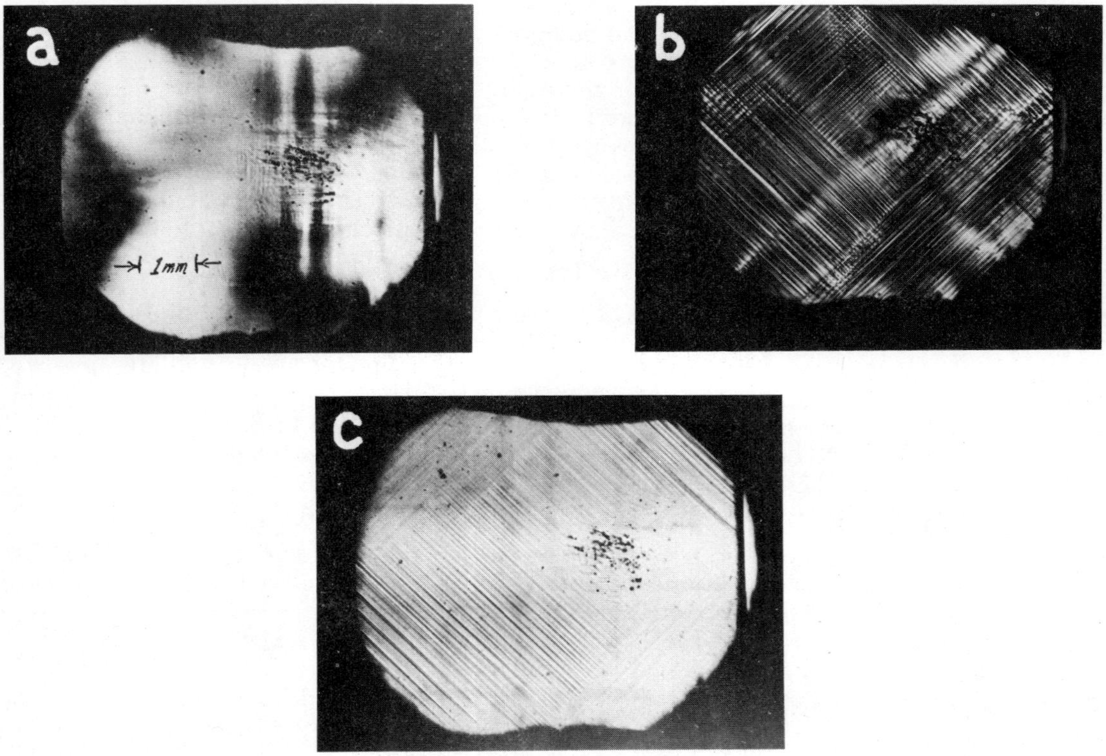

Fig. 1. Strontium titanate single crystal seen through crossed polarizers at (a) $300^{\circ}K$, (b) $77^{\circ}K$, and (c) without crossed polarizers at $18^{\circ}K$. Grainy patch at right center of crystal is a number of inclusions. (Horizontal and vertical axes and axis out of page are all principal axes [100] .

The optical train for the Raman scattering is also shown in Fig. 2. In most cases the laser beam traversed the crystal horizontally. To align the horizontal scattering column onto the vertical spectrometer slit, the image of the column is rotated through 90° by two high reflectivity mirrors mounted on a common bracket (item 16). To increase the flexibility of the system a double lens optical system is used (items 15 and 17). By positioning collecting and focusing lenses at distances equal to their focal lengths from the sample (item 8) and slit (item 19), respectively, we obtain parallel light between the lenses which allows one to locate the sample at any convenient point, to obtain a reasonable collection speed and to match the speed to the spectrometer (f/8). The polarization of the scattered light is analyzed with a Polaroid HN-38 filter (item 18). The image of the scattering column on the slit is masked to cut down parasitic scattering from the edges of the crystal and, by using only 2mm slit heights, to select portions of the scattering column for study.

Since we want to photograph the beam in relation to the crystal and its domains, the directions of polarized illumination and the laser beam must be colinear. This is accomplished by drilling a hole in a first surface mirror (item 5) and allowing the beam to pass through the hole, while the mirror surface at 45° to the beam reflected the output of a microscope illuminator (item 1) along the beam path. To render the white

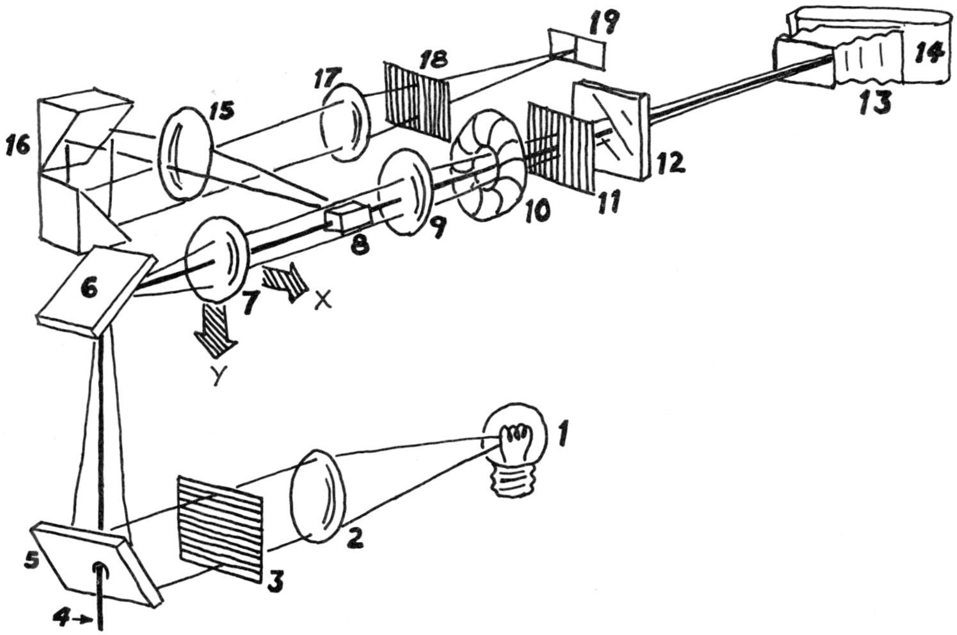

Fig. 2. System for exploration of spatial variation of Raman intensities. 1. Microscope illuminator, 2. White light focusing lens, 3. Polarizer, 4. Laser beam, 5. First surface mirror, tilted at 45° to beam (hole in mirror to permit passage of beam), 6. First surface mirror, 7. Beam focusing lens mounted on movable x-y stage, 8. Crystal, 9. Imaging lens, 10. Iris diaphraghm, 11. Polarizer, 12. Colored glass filter (beam attenuator), 13. Graflex camera with focal plane shutter, 14. Polaroid-Land adapter, 15. Raman scattering collecting lens, 16. 90° image rotation mirrors, 17. Raman scattering focusing lens, 18. Polarizer, 19. Spectrometer slit.

light parallel through the crystal, while focusing the laser beam into the crystal, the polarized white light must be focused at the back focal point of the beam focusing lens. The image of the beam and the crystal is focused onto the focal plane of a camera about 1 m distant, giving a 10:1 magnification. A polarizer (item 11) is inserted after the lens to analyze the light which passes through the crystal. During the photographic exposure the laser beam was attenuated by an appropriate Corning filter (item 12) (No. 3384 for λ = 488 nm).

The camera is a Graflex with a focal plane shutter and a Polaroid-Land Film adapter (items 13 and 14). Focusing is checked with a ground glass screen which can be mounted at the film plane. Photographs are recorded on Type 47 Polaroid film (ASA speed: 3000) with shutter speeds in the range of 1/100 to 1/1000 second.

RESULTS AND DISCUSSION

Rotating the incident polarization with respect to the domain walls failed to show any anisotropy in the Raman intensity of the low temperature features for polarization of the incident beam parallel and perpendicular to the walls. All attempts to obtain a set of selection rules with respect to domain wall orientations failed. The probable explanation of these features has been given in the preceding paper by Worlock, Scott and Fluery[5].

Some effects we have observed with this system illustrate the use of the technique. When the beam focusing lens is moved along the x-direction (see Fig. 2), the image of the scattering column remains centered on the slit. Therefore, it is possible using the synchronous motor attached to the micrometer screw to scan the scattering column across domain walls, while the spectrometer is set at some wavelength. (We call this procedure a crystal scan.) The results are shown in Fig. 3 and photographs taken at certain points in the scan are shown in Fig. 4. The distances between the dark features in the photographs were measured on a comparator and divided by the magnification. These were found to be equal to the separations between peaks on the traces converted to the distance traveled by the beam.

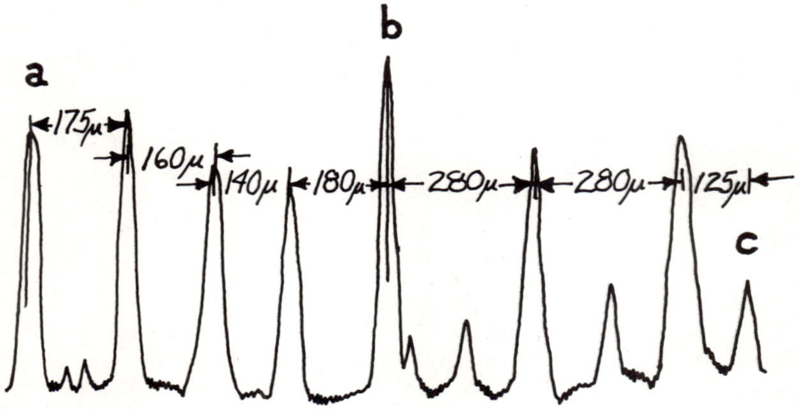

Fig. 3. Raman intensity ($\Delta\nu = 44\ cm^{-1}$) variation with position of beam in the crystal. (Letters above peaks correspond to beam location shown in photos in Fig. 4; same crystal orientation as in Fig. 1.)

A crystal scan across a discontinuity produced by applying an 8 KV/cm electric field to a crystal in the [110] direction was made with the spectrometer set at a point ($\Delta\nu = -100\ cm^{-1}$) on the second order spectrum. During this scan the beam and chart motors were stopped at maxima and minima in the trace and photographs of the beam position were taken. The trace and photographs are shown in Fig. 5. The entire scan is about 400μ; the distance between minima: 100μ. When the incident polarization vector was rotated 90°, the maxima and minima were interchanged.

Another attack used to investigate the Raman scattering from selected portions of the crystal is what we have termed a column scan. The variation of Raman intensities along the scattering column is observed by reducing the slit height to 2 mm (which corresponds to 1 mm of the scattering column at the crystal) and moving the image of the scattering column parallel to the slit stepwise while recording the output on a particular feature of the spectrum.

Fig. 6 shows column scans for incident light polarized parallel (H) and perpendicular (V) to the scattering plane incident upon a (110) cut crystal along [001] and an unanalyzed (T) scatterd light parallel to [110]. The column scans were taken at a Raman shift of $\Delta\nu = -100\ cm^{-1}$. Note the quasiperiodic and complementary variation of the VT and HT traces. Application of an electric field did not change the sinusoidal character of the traces, but it did change the period. It would appear that the polar-

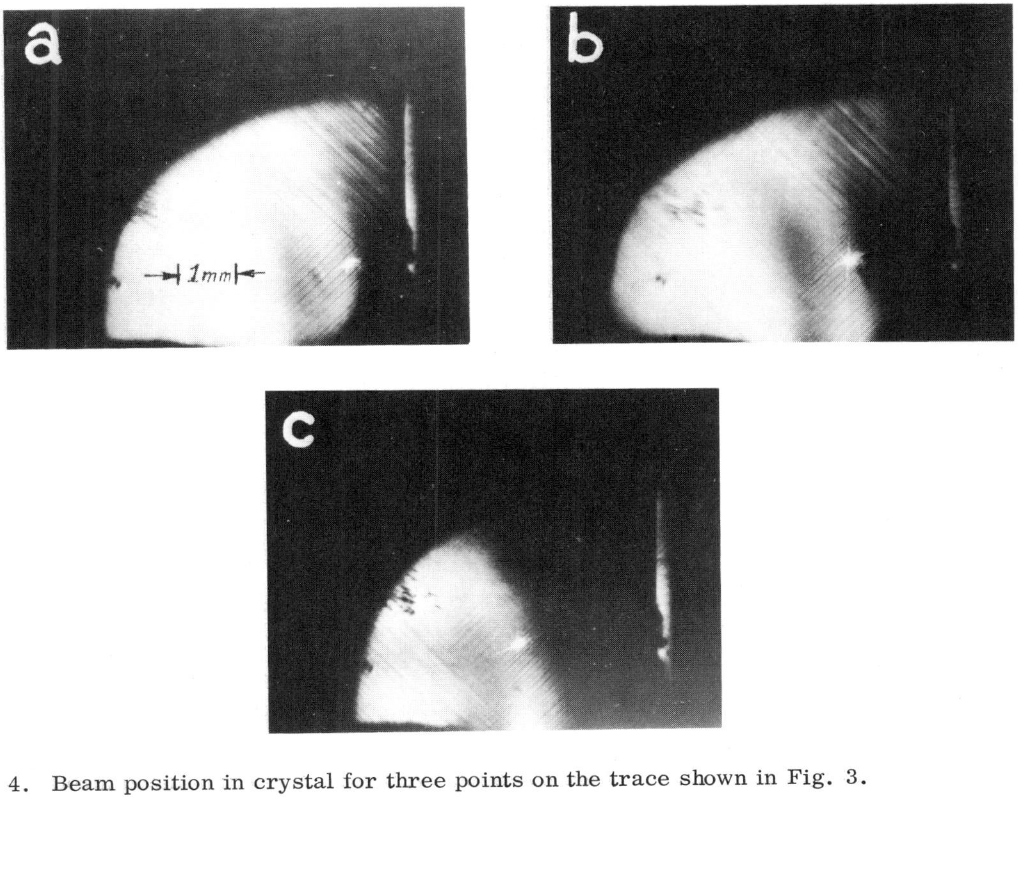

Fig. 4. Beam position in crystal for three points on the trace shown in Fig. 3.

ization of the incident beam is being rotated in the crystal as it passes through the crystal possibly due to the net birefringence from the uniaxial crystal domains.

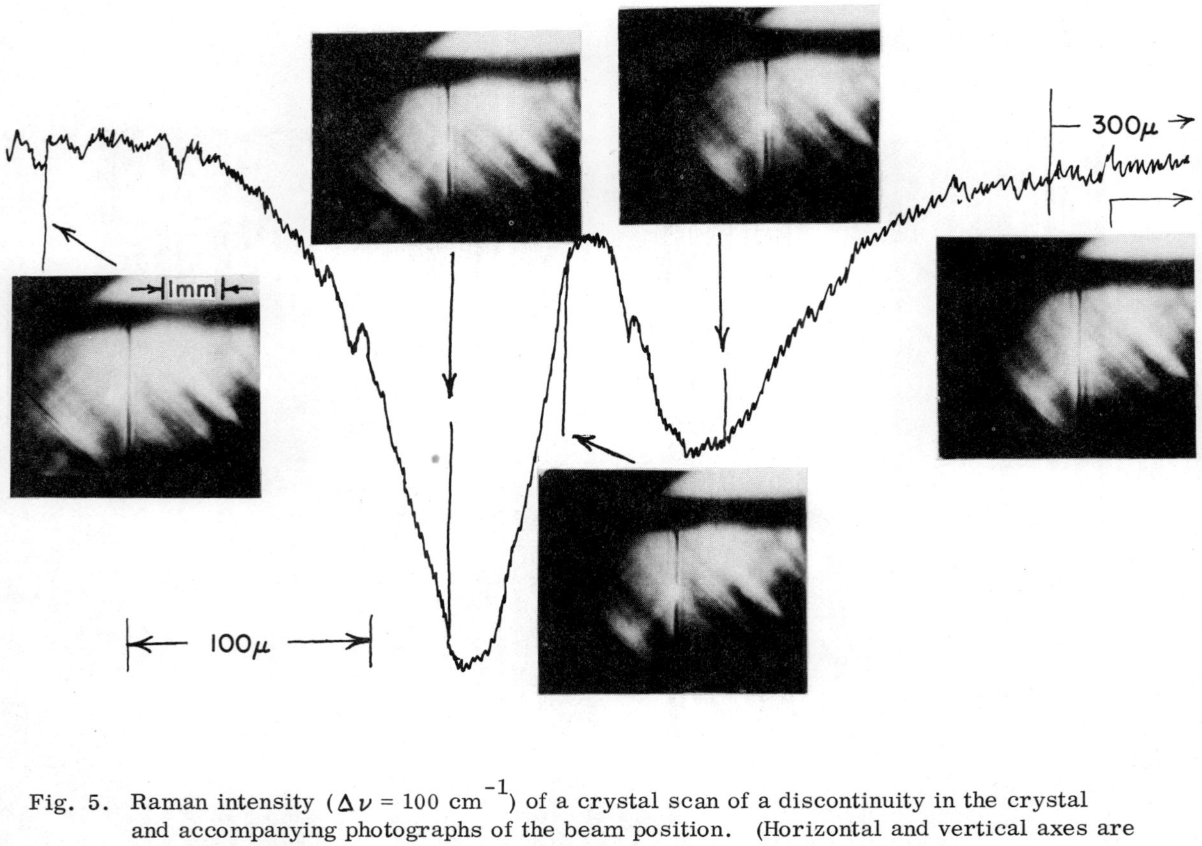

Fig. 5. Raman intensity ($\Delta\nu = 100\ \mathrm{cm}^{-1}$) of a crystal scan of a discontinuity in the crystal and accompanying photographs of the beam position. (Horizontal and vertical axes are [110] and [110], axis out of page, [001].)

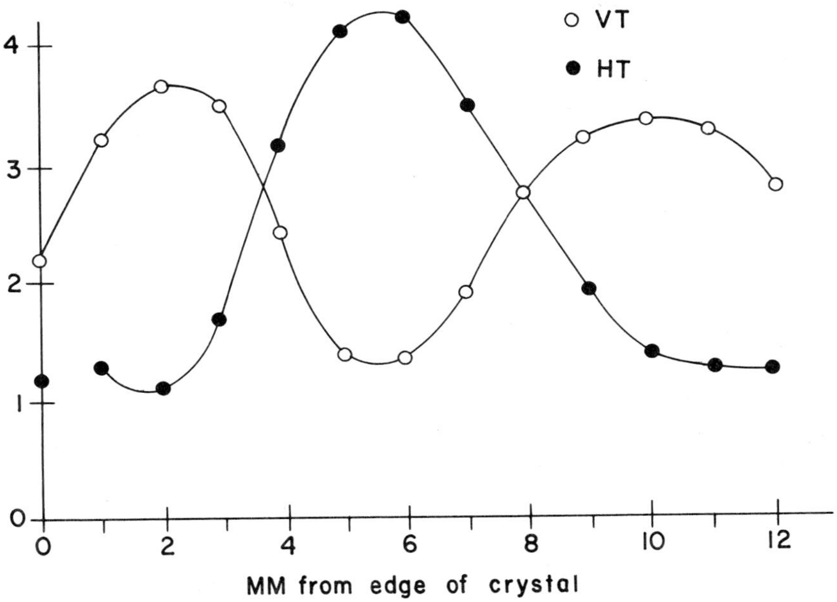

Fig. 6. Raman intensity ($\Delta \nu = 100\ cm^{-1}$) of a beam scan of a crystal for two incident beam polarizations. (V: incident polarization perpendicular to the scattering plane, H: parallel incident polarization, T: unanalyzed polarization.)

CONCLUSION

While the laser beam probe technique seems to indicate a fairly simple solution to the Raman scattering intensity variations in $SrTiO_3$, these experiments illustrate the usefulness of Raman scattering coupled with the observation of the beam position in the crystal. Beside detecting inhomogeneities in crystals, surface effects could be investigated by this technique. Obviously, variations in dopant concentrations and other such effects utilizing laser-excited flourescence could be investigated with this instrumentation.

REFERENCES

1. M. Born and E. Wolf, "Principles of Optics," 3rd edition, section 8.8, Pergamon, London, 1965.
2. F.W. Lytle, Jour. Appl. Phys. 35, 2212 (1964).
3. D.C. O'Shea, unpublished Ph.D. dissertation, 1968, The Johns Hopkins University, Baltimore, Maryland.
4. P.A. Fleury, J.F. Scott, and J.M. Worlock, Phys. Rev. Letters 21, 16 (1968).
5. J.M. Worlock, J.F. Scott, and P.A. Fleury, paper H-7 this conference.

H-9: TEMPERATURE DEPENDENCE OF THE RAMAN, BRILLOUIN AND RAYLEIGH SCATTERING BY CRYSTALLINE QUARTZ.

Stephen M. Shapiro and Herman Z. Cummins
The Johns Hopkins University
Baltimore, Maryland

INTRODUCTION

Crystalline quartz undergoes a phase transition at 573°C from its low temperature (α) phase with point symmetry D_3 to its high temperature (β) phase with point symmetry D_6. At the transition temperature there are anomalous changes in many of its properties [1]. In 1956, Yakovlev et al. [2] reported that as quartz underwent its transition it exhibited opalescence very similar to the critical opalescence observed in a liquid-vapor transition. Almost concurrently, Ginzburg[3] applied Landau's theory of second order phase transitions to quartz and was able to calculate the increase in the intensity of light scattered at the transition temperature, which was in agreement with Yakovlev's experiment. Ginzburg also predicted that the order parameter used to describe the quartz transition would be connected with one branch of the optic vibrations. The frequency of this mode should go to zero as the transition temperature is approached, while the fluctuations of the order parameter become very large; thus the intensity of the scattered light should increase. Early experiments[4] indicated that the mode exhibiting this behavior was the Raman active vibration with a frequency of 207 cm^{-1} at room temperature.

In the experiments reported here, we attempted to clarify the mechanism of the transition by studying the Raman and Brillouin spectra, and to determine the origin of the observed opalescence.

RAMAN SCATTERING

In each unit cell of quartz there are three SiO_2 units. There are 27 normal modes of vibration, three of which are acoustic. The remaining 24 are optic modes which can be divided, by group theory, into three species: four totally symmetric A_1 modes which are Raman active; four A_2 modes which are infrared active; and the eight doubly degenerate E modes, which are both Raman and infrared active. Of the four A_1 modes in α quartz, only one is Raman active in the beta phase. The eight E modes go into $4E_1 + 4E_2$ modes in beta quartz. Fig. 1 shows the Raman spectrum of α quartz. The degeneracies

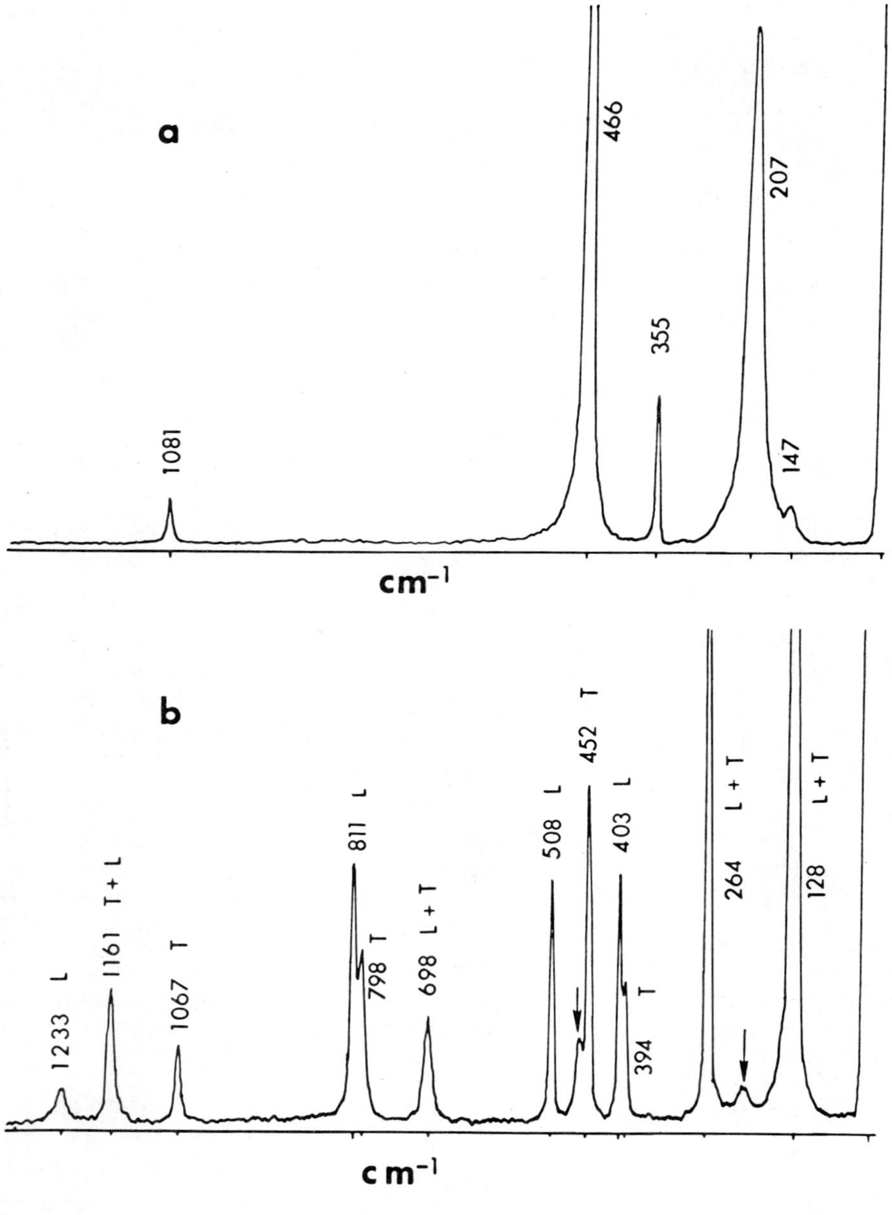

Fig. 1. Room temperature Raman spectra of quartz with measured values of frequencies in cm^{-1}, and Scott and Porto's (Ref. 5) polarizations assignments. (a) the A_1 modes; (b) the E modes. The arrows indicate intense A_1 modes being transmitted due to imperfect alignment and non-ideal polarizers.

of some of the E modes are lifted by the long-range electrostatic interaction which was first recognized by Scott and Porto[5]. The interesting point in the A_1 spectrum is the appearance of five A_1 lines instead of four, the additional line having a frequency at room temperature of 147 cm^{-1} [5, 6]. As the temperature is raised, it is the 147 cm^{-1} line which exhibits the strongest temperature dependence (Fig. 2); its intensity increases while its frequency decreases from 147 cm^{-1} at room temperature to 30 cm^{-1} in the transition region, where it suddenly disappears from our spectra. On cooling from the β phase, this line suddenly reappears at a temperature approximately 1°C lower than that at which it disappeared on heating. In the β phase, two A_1 lines are seen instead of one predicted by group theory. Fig. 3 shows a plot of ν^2 vs. $T-T_c$ for the 147 cm^{-1} and 207 cm^{-1} A_1 lines. The temperature dependence of the 147 cm^{-1} line is given by

$$\nu^2 \approx |T - T_c|^{\gamma}$$

where $.4 \le \gamma \le .5$.

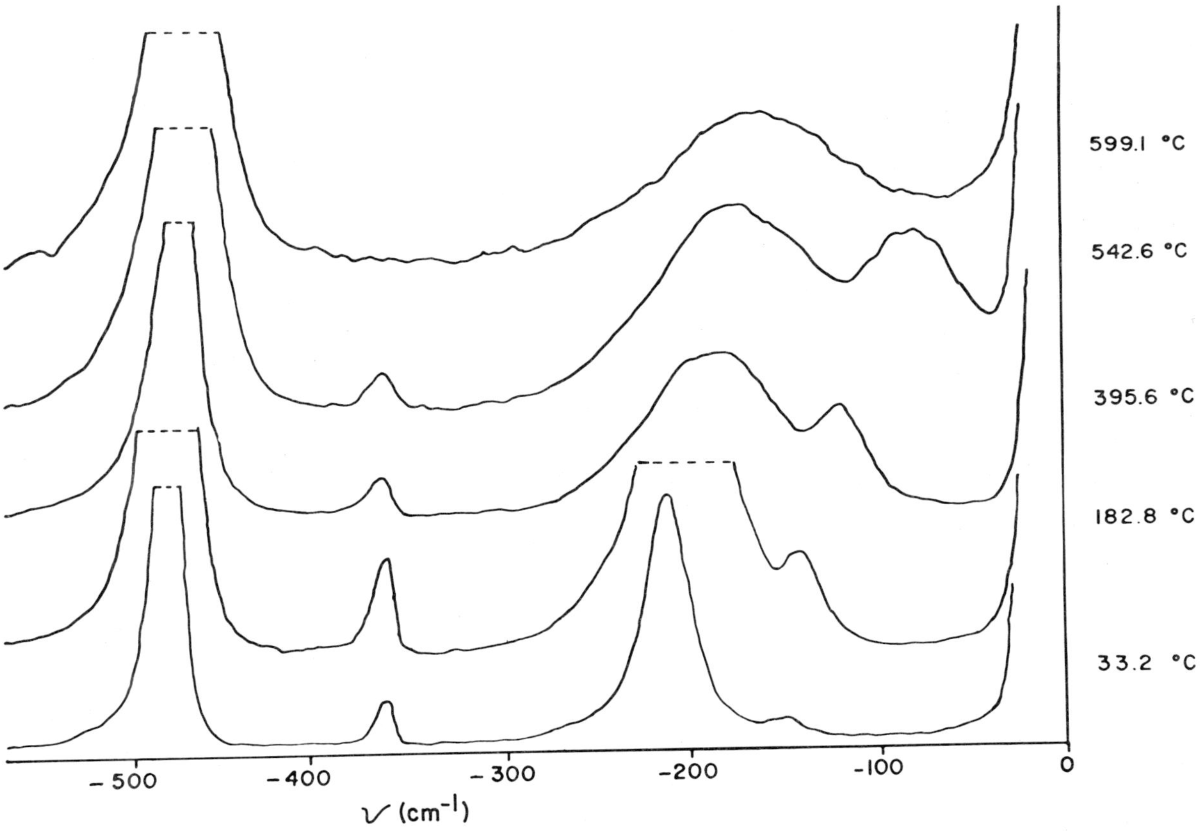

Fig. 2. 0 to 550 cm^{-1} portion of the A_1 Raman spectrum at different temperatures. The upper curve is for β quartz. The gain of the lowest temperature curve is one-half that of the others.

Two theories have been proposed to explain the temperature dependent Raman spectrum:

1. An ad hoc theory has been proposed[6] which interprets the experimental observations but is lacking firm theoretical support. The Si atoms are assumed to be moving in asymmetric double potential wells. There would be a strong normal vibration (207 cm^{-1}) associated with the lower well, and a satellite at a different frequency (147 cm^{-1}). As the temperature is increased, the energy difference between the two configurations decreases and the intensity of the satellite increases, approaching that of the parent line. If, in addition, the height of the barrier decreases faster than the energy difference between the two minima, the frequency associated with the higher well (147 cm^{-1}) would decrease toward zero as the transition is approached.

2. The second explanation, recently proposed by Scott[7], considers the 147 cm^{-1} line as a second order Raman line, produced by the excitation of two oppositely

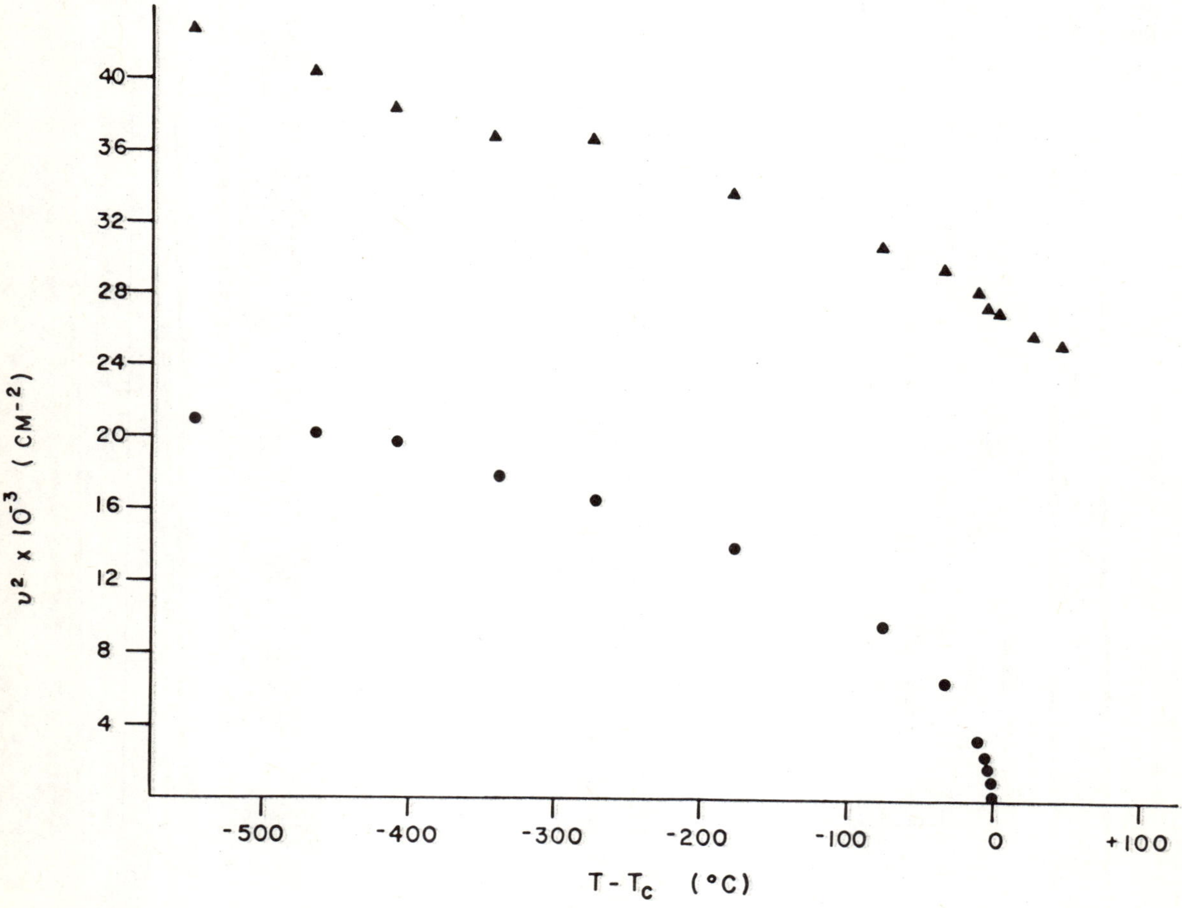

Fig. 3. ν^2 vs. $T-T_c$. Triangles, 207 cm^{-1} (room temperature frequency) mode; Circles, 147 cm^{-1} (room temperature frequency) mode.

directed zone edge acoustic phonons. The anomalous frequency and intensity dependence on temperature are a result of anharmonic coupling of the two phonon line with the "soft" zone center phonon with frequency of 207 cm^{-1}. As the temperature approaches the transition temperature, this coupling produces a "level repulsion," which decreases the frequency of the 147 cm^{-1} line and prevents the "soft" mode from going to zero. In the β phase, the extra A_1 mode (162 cm^{-1}) is totally second order.

BRILLOUIN SCATTERING

In the Brillouin scattering experiments, acoustic phonons (momentum $\hbar\vec{q}$) propagating in the [100], [010], [001], and [110] directions were studied. Fig. 4 shows a room-temperature spectrum with $\vec{q}$ along [110]. (Note the small Rayleigh component, which is indicative of the purity of our crystals.) When the sample was placed in the oven, the intensity of the Rayleigh line increased due to extraneous scattering, but in most cases it was less than five times the intensity of the Brillouin peaks. Thus, any strong opalescence could be observed in the Brillouin experiments. Fig. 5 is a plot of the Brillouin shift for the [100] longitudinally polarized mode. 5a shows the behavior from

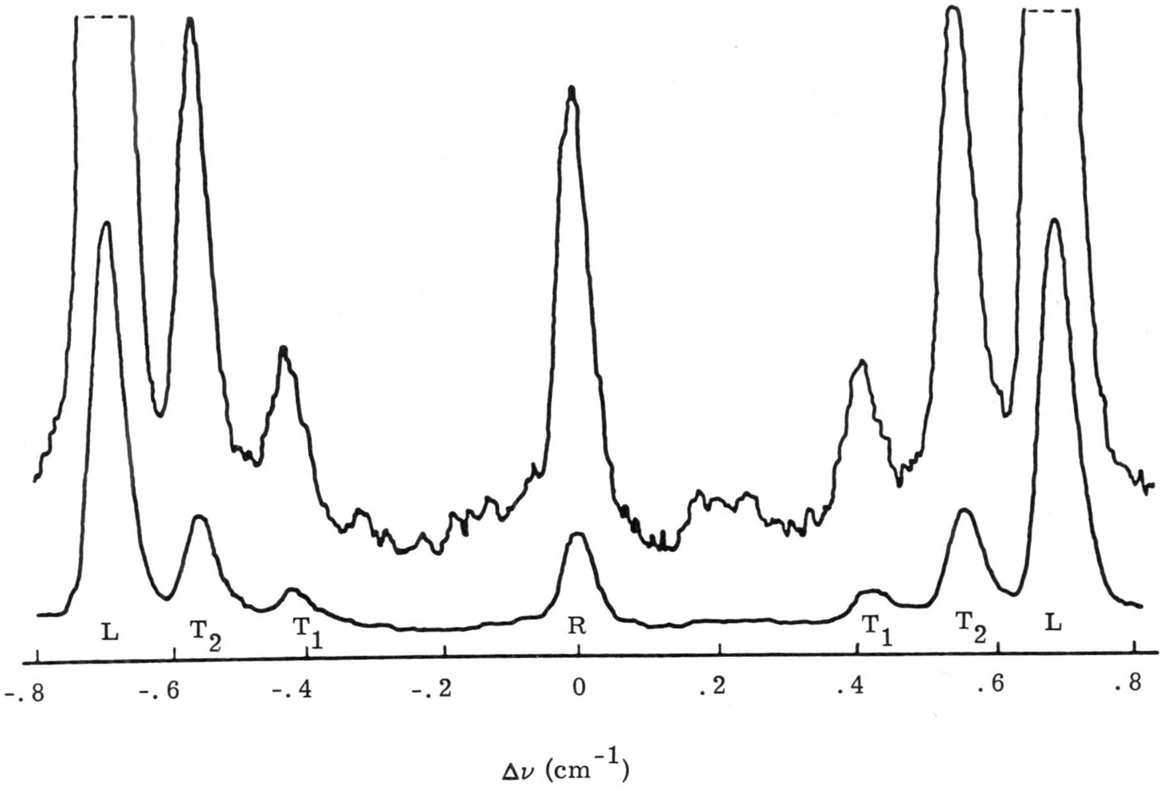

Fig. 4. The room temperature Brillouin spectra of the [110] phonon of natural crystalline quartz showing the Rayleigh peak (R), the longitudinal (L), and the two transverse (T_1, T_2) at two different gains.

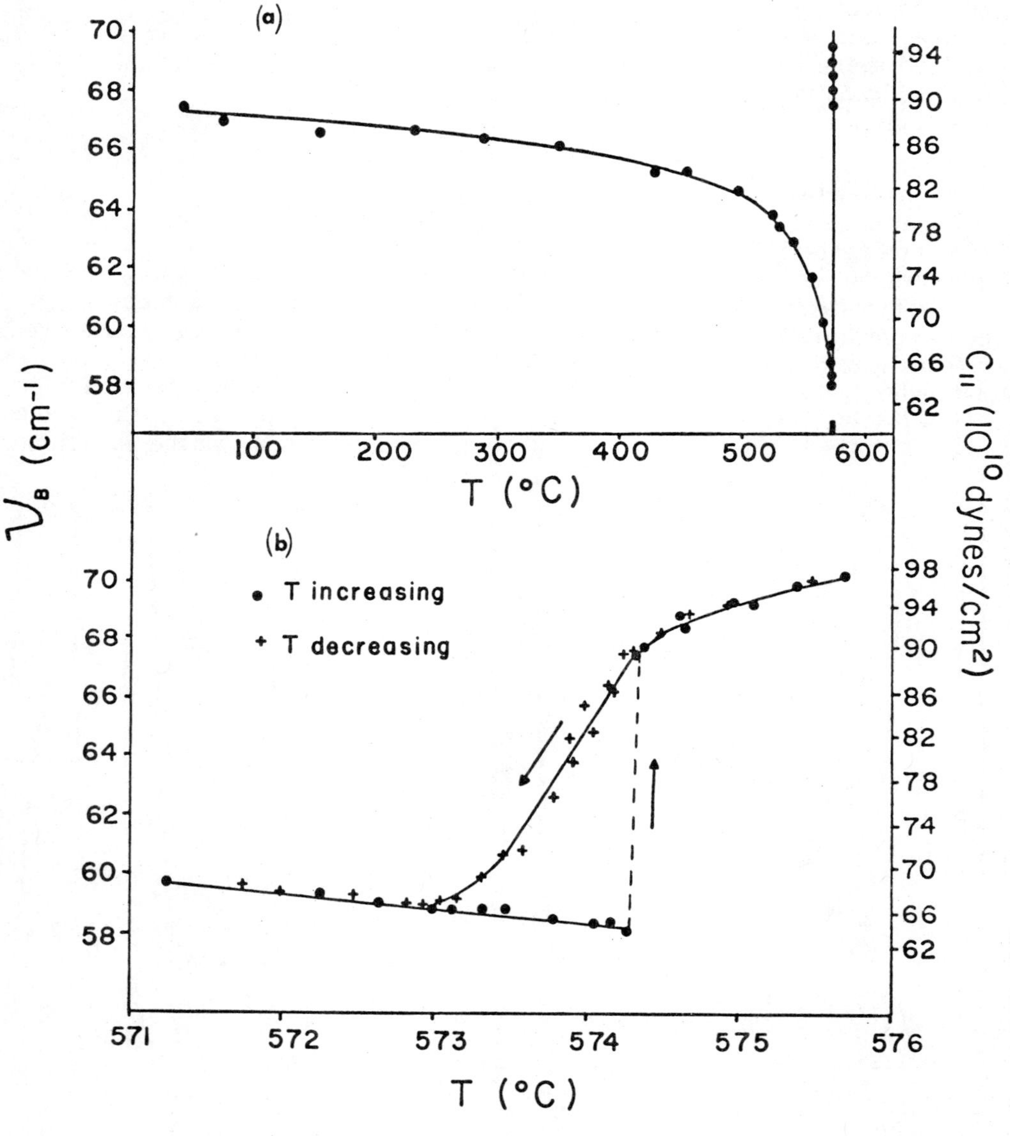

Fig. 5. Frequency shift and elastic constant, C_{11}, as a function of temperature for natural crystalline quartz. (a) The temperature scale is from $0^{\circ}C$ to $600^{\circ}C$; (b) the data is plotted on an expanded temperature scale in the transition region. The solid circles represent T increasing and the crosses are for T decreasing. The arrows also show which way the temperature is changing.

room temperature through the transition temperature and into the β phase. 5b is a plot of the Brillouin shift on an expanded temperature scale. The longitudinally polarized component for each direction studied exhibited this temperature hysteresis with a sharp discontinuity on heating and a gradual change on cooling. Fig. 6 shows the linewidth of the longitudinal components for the [100] mode, measured by subtracting the instrumental width from the measured Brillouin width at half maximum. Hysteresis is again observed, in addition to a large increase of linewidth on heating.

At 574.3°C on heating, and at 573°C on cooling, the Rayleigh scattered light increased suddenly by 10^4 and the crystal appeared to opalesce under white light illumination. In the opalescing region, the Brillouin peaks were still visible. Thus the sudden increase in scattered light was totally within the Rayleigh line.

The important facts deduced from Brillouin scattering experiments were:

1. For all directions studied, no velocity approached zero. This implies that the adiabatic compressibility remained finite, in agreement with Yakovlev's assertion[1].

2. The temperature hysteresis observed in the Brillouin shifts near the transition temperature indicates that quartz undergoes a first order phase transition, as opposed to a second order transition upon which Ginzburg's theory is based.

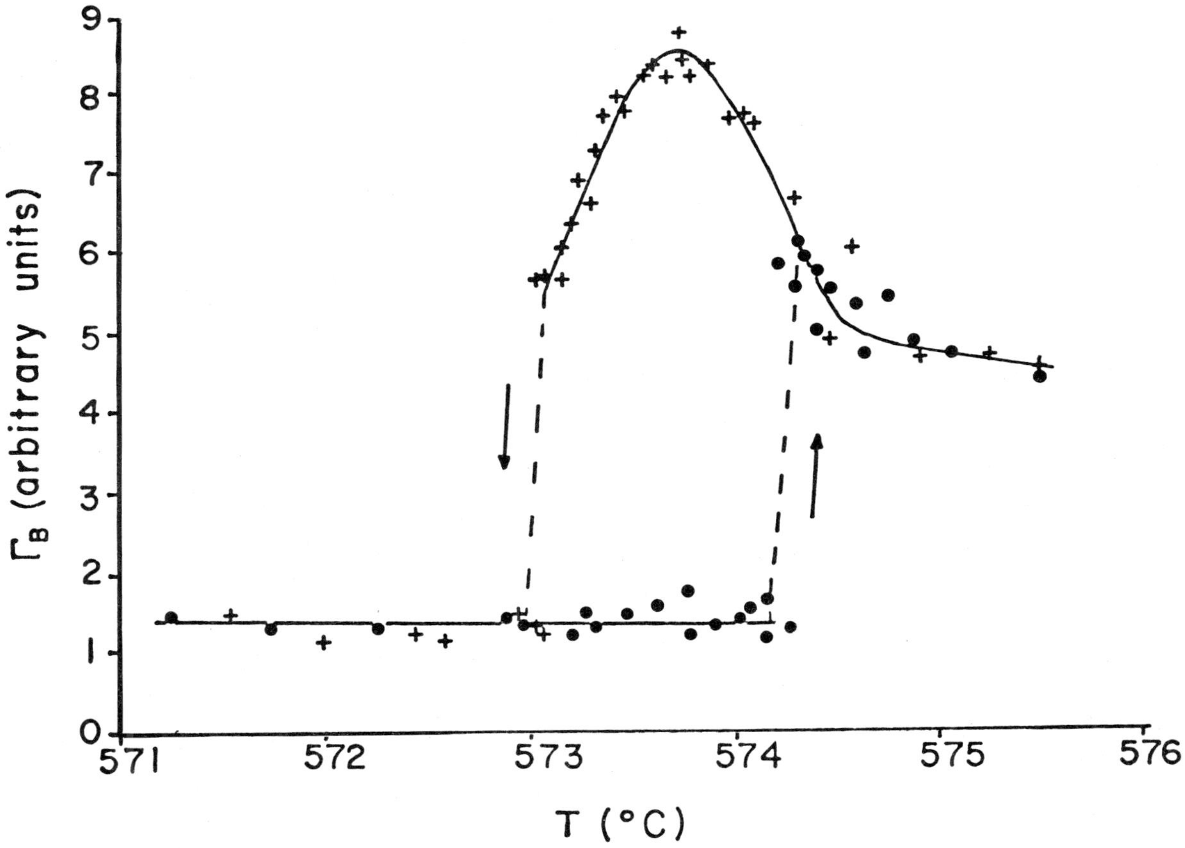

Fig. 6. The linewidth (in arbitrary units) of the longitudinal components for the [100] phonon of natural crystalline quartz in the transition region. The symbols have the same meaning as those in Fig. 5(b).

3. The elastic constants calculated from the Brillouin shifts generally agree with the ultrasonic experiments [8]; thus, there is no frequency dispersion in the elastic constants.

RAYLEIGH SCATTERING

Observations were made of the elastically scattered light as the crystal underwent its transition. At the transition temperatures, the intensity of the Rayleigh line increased abruptly by a factor of 10^4. (This is the effect which Yakovlev and Ginzburg interpreted as critical opalescence.)

Fig. 7 shows photographs taken of the laser beam traversing the crystal. In 7a, the crystal is in the low-temperature phase, just below the transition temperature. The bright specks are due to dirt or imperfections in the crystal. The exposure time was 15 minutes. The scattering is true molecular scattering, as evidenced by the homogeneous appearance of the scattered light. In 7b the crystal is undergoing its transition to the β phase. The exposure time is 2 sec. Notice that the intensity of the scattered light has increased greatly, and that the beam no longer appears homogeneous but has a granular structure. This suggests a stationary rather than a dynamic scattering process.

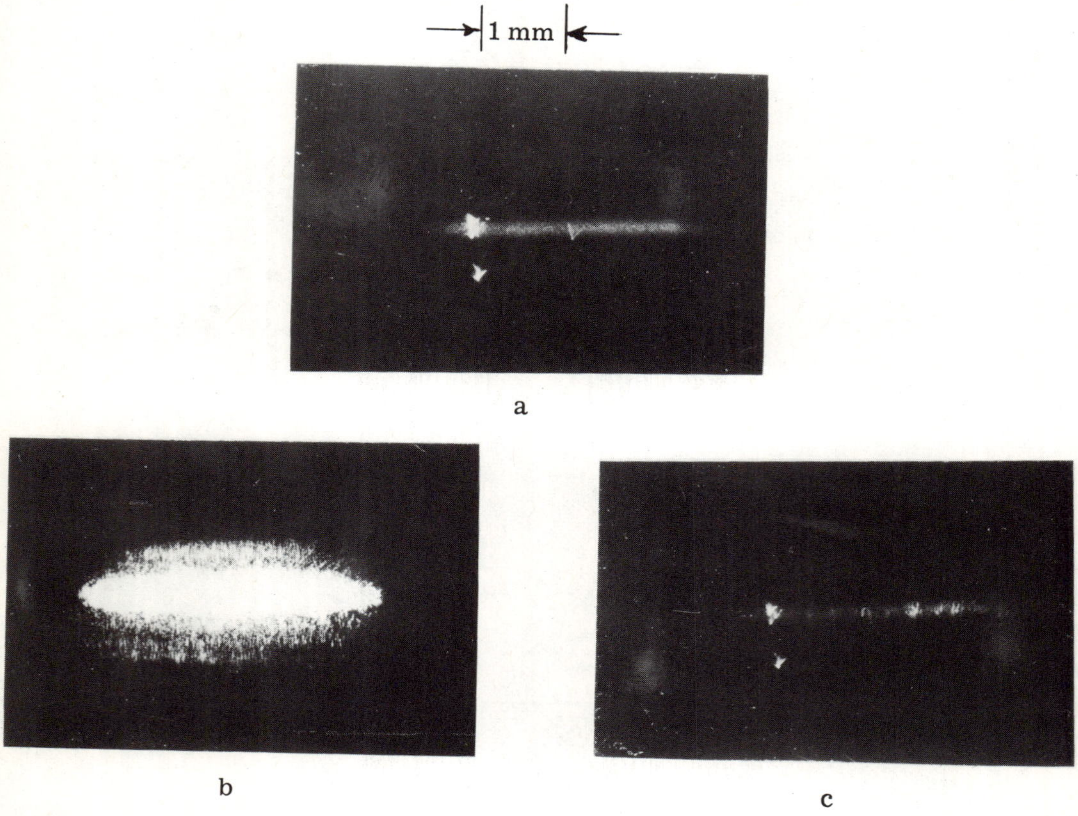

Fig. 7. The scattered laser light traveling through quartz at different temperatures. (a) $T < T_c$, exposure time: 15 min.; (b) $T = T_c$, exposure time: 2 sec.; (c) $T > T_c$, exposure time: 5 min.

If there were true opalescence, the frequency of the fluctuations would be equal to that of one of the "soft" optic modes. Since the minimum value of this frequency ($30\ cm^{-1}$) is much higher than the frequency resolution of the eye, the beam would appear homogeneous. Since we see a nonhomogeneous beam whose graininess can be followed by the eye, the scattering process is essentially static. We can deduce that because of this granular appearance, the anomalous scattering is not critical opalescence but is due to some other cause.

In 7c, which is a 5 minute exposure, the crystal is now in the beta phase. In addition to the specks caused by some imperfection within the crystal, there is some structure to the beam. On heating the crystal further into the beta phase, this structure disappears and the beam looks the same as in Fig. 7a. On cooling, the above sequence of photos is reversed and the brightly scattered beam appears at a lower temperature than on heating.

The probable origin of the anomalous scattering was suggested by Young[9]. In his X-ray analysis of the quartz inversion, he observed extensive microtwinning of the Dauphiné, or electrical, type in the transition region. We believe that the increased scattering of light is the result of stresses set up in the twin boundaries which perturb the index of refraction and cause light to be scattered[10]. As the critical temperature is reached, the amount of twinning is seen to increase, along with a reduction of twin size. The light scattered would also increase, since there are then more boundaries which can scatter. In Fig. 7c the structure seen is probably due to light being scattered by domains persisting into the beta phase.

SUMMARY

In the Raman scattering experiments, five A_1 components are observed instead of four predicted by group theory for α quartz. The newly observed mode exhibits the greatest temperature dependence and its frequency goes to zero as the critical temperature is approached. The behavior of this mode can be explained either by the atoms moving in an asymmetric double potential well or by anharmonic coupling of a "soft" zone center phonon and two zone edge acoustic phonons.

The Brillouin shifted lines were studied with high temperature resolution. A temperature hysteresis was observed in the frequency and the linewidth, which suggests a first order transition.

Observations of the anomalous scattering region of the crystal indicate that the large increase in Rayleigh scattered light is not opalescence but a result of extensive microtwinning of the Dauphiné type [11].

REFERENCES

1. R.B. Sosman, "The Properties of Silica," The Chemical Catalog Co., New York, 1927.
2. I.A. Yakovlev, L.F. Mikheeva and T.S. Velichkina, Kristallografiya 1, 123 (1956). (Translation: Soviet Phys., Crystallography 1, 91 (1956)).
3. V.L. Ginzburg and A.P. Levanyuk, J.Phys. Chem. Solids 6, 51 (1958); V.L. Ginzburg, Usp. Fiz. Nauk. 77, 621 (1962). (Translation: Soviet Phys. Usp. 5, 649 (1963).
4. P.K. Narayanaswamy, Proc. Indian Acad. Sci. A26, 521 (1947); A28, 417 (1948).
5. J.F. Scott and S.P.S. Porto, Phys. Rev. 161, 903 (1967).
6. S.M. Shapiro, D.C. O'Shea and H.Z. Cummins, Phys. Rev. Letters 19, 361 (1967).
7. J.F. Scott, Phys. Rev. Letters 21, 907 (1968).
8. V.G. Zubov and M.M. Firsova, Kristallografiya 7, 374 (1962). (Translation: Soviet Phys. Crystallography 1, 374 (1962)).
9. R.A. Young, U.S. Air Force Office of Scientific Research Final Report No. AFOSR-2569, Defense Documentation Center Report No. AD276235.
10. S.M. Shapiro, R.W. Gammon and H.Z. Cummins, Appl. Phys. Letters 10, 113 (1967).
11. S.M. Shapiro and H.Z. Cummins, Phys. Rev. Letters 21, 1578 (1968).

H-10: STUDIES OF VISCOELASTIC RELAXATION PROCESSES IN LINEAR POLYMERS BY MEANS OF BRILLOUIN SCATTERING*

J. N. Gayles and W. L. Peticolas†
IBM Research Division
San Jose, California

ABSTRACT

Unlike crystalline solids, polymeric solids do not undergo first order phase changes upon being heated but they undergo relaxational transitions which exhibit rather sudden changes in both the distribution of mechanical relaxation times as well as changes in certain thermodynamic properties. These transitions appear to involve either changes in the motion of the backbone or the side chains and their determination is strongly dependent upon the frequency of the detection method and other extrinsic factors. This paper deals with the detection of these transitions in amorphous solids in the gigahertz range using acoustical phonons as a probe. The relationship of observed phonon velocities and scattering intensities to bulk properties is investigated and in certain cases it has also been possible to associate relaxational transitions with internal degrees of freedom in the sub-skeletal or side chain regions. A theory is employed which accounts for changes observed in the Landau-Placzek ratio, absorption coefficients, and phonon velocities as temperature and phonon frequencies are varied.

INTRODUCTION

Amorphous polymeric solids are composed of extremely long molecular chains which possess sufficient disorder to prevent crystallization. Such materials cannot undergo ordinary first-order phase transitions involving changes in the structure of the unit cell such as are characteristically observed in crystalline solids. However, as the temperature of polymers is raised from near 0°K, they have been observed to undergo rather sudden changes in their physical properties including certain thermodynamic and viscoelastic properties[1,2]. Because these changes involve distinct changes in the relaxation spectrum of the sample they are generally referred to as relaxational transitions[1,2].

*Partially supported by U.S.A.F. Rocket Propulsion Laboratory, Edwards A.F. base contract No. FO4611-68-C-0019
†Present address: University of Oregon, Dept. of Chemistry, Eugene, Oregon

These transitions are believed to be due to the onset of specific molecular motions involving the backbone, side-chains or other specific groups within the polymer molecule.

Just as an idealized model of a crystalline solid is a set of coupled harmonic oscillators with regular repeating masses and potential constants (the Born-Von Karman model), an idealized model of an amorphous polymer may be considered to be a coupled but highly damped set of harmonic oscillators with randomly varying masses and force constants. In view of the marked contrast in the behavior of amorphous and crystalline solids, it seemed worthwhile to explore the possibility of studying these relaxational transitions by means of Brillouin scattering in order to obtain additional insight into the changes in the thermodynamic properties and relaxation spectrum of these important and interesting class of solids. In view of the close chemical and physical relation of amorphous synthetic polymers to biological polymers it is hoped that the techniques developed here will have future application to biological polymers.

Relaxation transitions in synthetic polymers have been intensively studied. They have been detected by changes in the dynamic mechanical properties of polymers[3, 4], the broadening of the NMR lines on specific groups[5], dilatometry and other methods[1]. The measured changes usually occur over an interval of a few degrees. The exact temperature of the transition will depend on a number of variables including the rate of heating[6] the thermal history of the sample[7] and, perhaps most important, the frequency of the test method[3]. Of all the transitions in solid polymers samples, the one which is most widely known and studied is the glass transition which appears to involve the onset of motion of 30 to 40 atoms in the backbone of the chain[1]. As the sample temperature is increased above the glass transition temperature, T_g, the material undergoes a change from a brittle glass to a viscoelastic liquid. A controversy has existed for some time regarding whether or not the glass transition is actually a second order thermodynamic transition[8, 9], or whether it is a purely relaxation transition[2-7].

In general, rather abrupt changes in the isothermal compressibility β_T, the thermal expansivity, α, and the specific heat, C_p, have been noticed at the glass transition[8]. The changes in the second order thermodynamic quantities have been used to support the theory that below or near the glass transition, T_g, there exists another transition temperature, T_2, which is a true second order transition but whose identity is masked by the related transition T_g lying just above T_2[10].

Recently a study was made of the Brillouin spectrum of the light scattered from polyethylmethacrylate in the range of 20 - 90°C. At a temperature of 61.4°C, which is close to the reported 63°C glass transition temperature, a remarkably sharp increase in the intensity of the Brillouin lines was observed. This produced a sharp discontinuity in the ratio of the intensity of the central to that of the side Brillouin peaks which was interpreted as additional evidence of a second-order thermodynamic transition underlying a relation transition, in terms of the well-known Landau-Placzek formula which, uncorrected for dispersion becomes[12, 13],

$$\frac{I_c}{2 I_B} = (\gamma - 1) \tag{1}$$

where γ is the ratio of heat capacities, C_p/C_v.

Thus the discontinuity of $I_c/2I_B$ at or near T_g was interpreted[11] as being due to changes in β_T and C_p with no latent heat. This is characteristic of an Ehrenfest-second order transition[8-10]. However, when relaxation phenomena are involved, one must be cautious in making assignments of Ehrenfest-second order thermodynamic transitions. If we treat the amorphous solid as a collection of normal mode oscillators the specific heat will be given by[14],

$$C_v = k \int_0^\infty (x^2/4) \operatorname{cosech}^2 (x/2)\, g(\nu)\, d\nu \tag{2}$$

where

$$x = h\nu / kT$$

$g(\nu)$ is the density of vibrational states, and the other symbols have their usual significance. If the temperature is lowered, the relaxation times of the viscoelastic liquid suddenly become much longer (as they do at the glass transition temperature). Then there must be a corresponding sudden drop in C_v. This happens because many of the lower frequency motions are damped out below the glass transition. If we assume only a single relaxation time, τ, then all those frequencies which are higher than $1/\tau$ will tend to be damped out, and thus will not contribute to the heat capacity. Eq. (2) must then be replaced by

$$C_v = k \int_0^\infty \theta(\nu)\, (x^2/4) \operatorname{cosech}^2 (x/2)\, g(\nu)\, d\nu \tag{3}$$

where $\theta(\nu)$ will be of the form,

$$\theta = 0 \qquad \nu\tau \gg 1 \tag{4a}$$

$$\theta = 1 \qquad \nu\tau \ll 1 \tag{4b}$$

Thus, any abrupt change in the relaxation times of a polymer must necessarily be accompanied by a change in the second order thermodynamic properties. This follows since some of the modes of oscillation or rotation are frozen out and the specific heat must be correspondingly reduced. This phenomenon is well known for relaxation processes in liquids which involve the coupling of energy between external and internal degrees of freedom.

For example, if the total specific heat of a liquid is broken up into that due to the external (acoustical) modes $\tilde{C}_v$ and that due to the internal modes C', Herzfeld and Litovitz[15] have shown that the effective specific heat is given by,

$$(C_v)_{\text{eff.}} = \tilde{C}_v + C' / (1 + i\omega\tau) = C_v - \frac{C' i\omega\tau}{1 + i\omega\tau} \tag{5}$$

so that for $\omega \gg 1$,

$$(C_v)_{\text{eff.}} = \tilde{C}_v = C_v - C' \tag{6}$$

where C_v is the total specific heat. Hence, cooperative phenomena leading to an increase in τ would necessarily lead to a decrease in $(C_v)_{eff.}$ and there is both experimental and theoretical evidence that the glass transition is involved by such a cooperative phenomena in the statistical mechanical sense[9, 10].

BRILLOUIN SCATTERING AND RELAXATION TRANSITIONS IN POLYMERS

The theory and experimental techniques for studying the light scattered from thermally-driven density fluctuations in solid or liquid media are now well developed[16-19]. Longitudinal compression waves are present in every solid or liquid, occurring with wavelengths of from a few angstroms to the size of the sample. With each wavelength Λ, there may be associated a wave vector $k = 2\pi/\Lambda$ and a frequency $\Omega = v/\Lambda$ where v is the velocity of the wave.

The thermally driven oscillations are responsible for contributing to the internal energy and heat capacity of the modes of vibration. In addition to these correlated pressure fluctuations, there are uncorrelated entropy fluctuations. Thus Brillouin scattering arises from Doppler shifted light scattered from the acoustic waves while Rayleigh (elastic) scattering arises from random entropy fluctuations. Brillouin's equation states that:

$$|\underset{\sim}{k}| = \pm\, 2\, |\underset{\sim}{k}_o|\, n \sin \theta/2 \tag{7}$$

where $\underset{\sim}{k}_o$ is the wave vector of the incident laser light, n is the refractive index of the medium, and θ is the scattering angle. Neglecting the sign the sound wavelength is given by,

$$\Lambda = \lambda/2 \sin(\theta/2) \tag{8}$$

where λ is the wavelength of the light. Thus for the light of wavelength λ the wavelengths of the acoustic waves which can be studied are

$$\lambda/2 \le \Lambda \le 10\,\lambda$$

if the scattering angle is in the range

$$180^\circ \le \theta \le 5^\circ$$

Since hypersonic velocities are characteristically of the order of 10^5 to 10^6 cm sec^{-1}, this means that one can obtain measurements of relaxation processes somewhere in the range 10^9 - 10^{11} sec^{-1}. Thus Brillouin scattering extends the frequency range for the study of relaxation phenomena to a rather high frequency range. In addition, from a measurement of the total light scattered including the Rayleigh component, one can obtain information about static mechanical properties.

Table I gives a list of the methods used for studying relaxation phenomena in polymers with their characteristic time.

TABLE I

Method	Frequency Range
Dilatometry	0.001 Hz
Resilience	1 Hz
NMR linewidth	10 Hz
d.c. Transient	10^{-4} - 10^{-1} Hz
Ultra-low frequency bridge	10^{-2} - 10^{+2} Hz
Schering bridge	10 - 10^{-7} Hz
Resonance circuits	10^{5} - 10^{8} Hz
Coaxial line	10^{8} - 10^{9} Hz

From Eq. (7) one can obtain Brillouin's formula for Doppler frequency shift,

$$\Omega = \Delta\omega = \pm (2 n \omega_o v/c) \sin \theta/2 \tag{9}$$

Thus by measuring the frequency shift of the light of the Brillouin components from the central components as a function of the scattering angle, θ, one can determine the velocity, v, as a function of the circular frequency. This temporal measurement of velocities should be distinguished from the standard spatial measurement using acoustic transducers. Furthermore, as we shall see from the intensity of the Brillouin and central bands, one can determine the amount of correlated acoustical oscillatory motion of the polymer sample for the given frequency. Since the smallest oscillatory wavelength is given by,

$$\Lambda = \lambda/2 \sim 2500 - 3000 \text{ Å}$$

one can determine those dynamical properties depending on oscillations whose length is the order of magnitude of the polymer molecule itself.

Recently the theory of Mountain[17] has been extended[18, 19] to show how the real and imaginary part of the complex longitudinal modulus, $M^*(\omega)$, may be obtained from Brillouin scattering measurements. However, Brillouin scattering is not without limitations as a means of investigating polymer transitions. Since the shortest wavelength oscillation is ~ 2500 - 2000 Å one cannot directly study the correlated motion of shorter intra-chain oscillation. If the conversion between inter-chain and intra-chain modes is exceptionally fast, however, it is possible to ascertain relaxation times due to the intra-chain motions. Also, the technical difficulties of obtaining Brillouin spectra are such that one must use polymers of the very highest optical quality. This tends to limit one to amorphous polymer samples such as the polymethacrylates which are prepared from highly purified dust-free monomers.

THEORY OF LIGHT SCATTERING IN POLYMERS

The scattering spectrum for a relaxing liquid is given by:[18, 19]

$$I(k, \omega) = I_o < \rho(k)\rho(-k) >_T \sigma(k, \omega) (\partial n/\partial \rho)_T^2 \tag{10}$$

where k is the wave vector of the density fluctuation, i. e., the acoustic waves; ω, is the frequency shift from the pseudomonochromatic source frequency, I_o is the intensity of the incident light and $< \rho(k)\rho(-k) >_T$ is the isothermal ensemble average of density fluctuations of wave vector k, $(\partial \eta/\partial \rho)_T$ is the isothermal change in refractive index with density, and $\sigma(k, \omega)$, the line shape function, is composed of two terms:

$$\sigma(k, \omega) = \sigma_c(k, \omega) + 2\sigma_B(k, \omega) \tag{11}$$

where

$$\sigma_c = \frac{[1 - {}^1/\gamma]\, 2ak^2\gamma}{(ak^2/\gamma)^2 + \omega^2} \tag{12}$$

and

$$\sigma_B = \frac{(M_o/\gamma)\left(\eta(\omega) + \eta_o\right)}{[\omega\left(\eta(\omega) + \eta_o\right)]^2 + [\rho_o\omega^2/k^2 - M'(\omega)]^2} \tag{13}$$

σ_c and σ_B are the line shapes of the central and the Brillouin peaks respectively; $a = \lambda/\rho_o C_v$ and λ is the thermal conductivity; γ is the ratio of specific beats; C_ρ/C_v; $\eta(\omega)$ is the frequency dependent longitudinal viscosity, M_o is the static modulus, $M'(\omega)$ is the real part of the longitudinal modulus of elasticity at frequency, ω, and ρ_o is the static or average density. Montrose, et al. [18], have shown that

$$< \rho(k)\rho(-k) >_T = \rho_o^2 k_B T \beta_{T,o}/v_k \tag{14}$$

where k_B is the Boltzman constant and $\beta_{T,o}$ is static isothermal compressibility, and v_k is the volume associated with a fluctuation of wave vector k.

The total intensity of the light scattered from liquids, I_T, including both Rayleigh and Brillouin components is obtained by integrating $I_T(\omega, k)$ over all ω to obtain[19],

$$I_T = I_c + 2I_B = A\beta_{T,o} \tag{15a}$$

$$A = 2\pi I_o (\partial n/\partial \rho)_T^2 \rho_o^2 k_B T/v_k \tag{15b}$$

Eq. (15a) shows that the total intensity of the scattered light is directly proportional to the isothermal compressibility at static frequencies. Since the isothermal compressibility is greater in a liquid polymer rather than in a glassy polymer, this means that the total intensity, I_T, including both the Brillouin and central components must increase if a polymer sample is heated through the glass transition. This has now been observed in samples of polyethymethacrylate, polymethylmethacrylate and polycyclohexylmethacrylate. This paper deals primarily with the latter material.

For a non-relaxing material, or for a fluid whose relaxation times τ are very small compared to the frequency of the hypersonic wave ($\omega\tau \ll 1$) the intensities of the central and both Brillouin components are given by:

$$I_c = A\left(1 - \frac{1}{\gamma}\right)\beta_{T,o}; \text{ or} \tag{16}$$

$$I_c = A[\beta_{T,o} - \beta_{S,o}], \text{ and} \tag{17}$$

$$2I_B = A\beta_{T,o}/\gamma = A\beta_{S,o} \tag{18}$$

where $\beta_{S,o}$ is the adiabatic compressibility at static frequency. The Landau-Placzek ratio is given by,

$$\frac{I_c}{2I_B} = \frac{I_T}{2I_B} - 1 = \gamma - 1 = \frac{\beta_{T,o} - \beta_{S,o}}{\beta_{S,o}} \tag{19}$$

For either structural or thermal relaxation these equations are not adequate since in the case of high frequency sound waves the relaxation times of the polymer molecules may be so large that many of the modes of motion cannot follow the very rapid fluctuations of the hypersonic wave. In this case $\beta_{S,o}$ must be broken into that part which is damped out due to relaxation effects and therefore is frequency dependent, and that part which is due to modes of motion which can respond to any compressional wave, no matter how high its frequency. This may be done if we realize that the adiabatic compressibility measured at low frequencies, $\beta_{S,o}$, contains contributions from all modes of motion,

$$\beta_{S,o} = \int_0^\infty \beta_S'(\tau)\, d\tau \tag{20}$$

where $\beta_S'(\tau)\, d\tau$ is the contribution to the adiabatic compressibility from all of those modes motion which possess relaxation times in the range between τ and $\tau + d\tau$.

All those modes of motion which are oscillatory at the frequency of the hypersonic wave, ω, will contribute to the intensity of the Brillouin lines scattered from these waves. The non-oscillatory modes will not contribute to the intensity of the Brillouin lines and thus this contribution will show up in the central peak. Thus $\beta_{S,o}$ may be broken up into that part which is oscillatory at frequency ω, $\beta_S(\omega)$, and that which non-oscillatory $\beta_{S,o} - \beta_S(\omega)$.

Where

$$\beta_{S,o} - \beta_S(\omega) = \int_0^\infty \frac{(\omega\tau)^2 \beta'(\tau)\, d\tau}{1 + \omega^2 \tau^2} \tag{21}$$

$$\beta_S(\omega) = \int_0^\infty \frac{\beta'(\tau)\, d\tau}{1 + \omega^2 \tau^2} \tag{22}$$

These equations are easily obtained by assuming that the polymer consists of a distribution of Maxwell elements[3].

If the distribution of relaxation time is $G(\tau)$ then we have the relation

$$\beta'_S(\tau) = \beta_{S,o}\, G(\tau) \tag{23}$$

We note that $G(\tau)$ is a normalized distribution function so that $\int_0^\infty G(\tau)\, d\tau = 1$. There will always be some contribution from G(o) which is due to those modes which can respond to quasi-infinitely high frequencies.

For a structurally relaxing liquid such as a polymer, the intensity of the central peak will be high compared to ordinary liquids because of the nonoscillatory contribution of adiabatic fluctuations. That is, the density of acoustic phonon states is relatively low in both the rubber and glassy polymer. This is what we have observed. In view of Eq. (17) the Rayleigh component of the scattered light will be given by

$$I_c = A \left[\beta_{T,o} - \beta_{S,o} \int_0^\infty \frac{(\omega\tau)^2\, G(\tau)\, d\tau}{1 + (\omega\tau)^2} \right] \tag{24}$$

Correspondingly the Brillouin component will be given by

$$2I_B = A \left[\beta_{S,o} \int_0^\infty \frac{G(\tau)\, d\tau}{1 + (\omega\tau)^2} \right] \tag{25}$$

and the total intensity will be the sum of the two,

$$I_T = A \beta_{T,o} \tag{26}$$

As we discussed in Section I, if a system of oscillators undergoes a sudden change in its relaxation spectrum such that certain frozen-in modes of motion are freed, then one must expect an increase in the heat capacity, C_v, the static isothermal compressibility, $\beta_{T,o}$, and other second-order thermodynamic functions. Thus from Eq. (26) we would expect an increase in the total amount of scattered light. Furthermore, we would expect that because $G(\tau)$ is different above the glass transition there will be a difference in the ratio $I_c/2I_B$.

EXPERIMENTS

The requirements of high optical quality mentioned previously are compounded in Brillouin scattering from amorphous materials for two reasons. Unlike pure liquids at room temperature the density of states for isentropic pressure fluctuations is low, hence the intensity of the Brillouin doublets is correspondingly low both above and below the glass transition temperature. Also, in common with other Brillouin spectral measurements, the resolution requirements are such that only devices with resolution on the order of 10^5 to 10^6 suffice to separate the central (Rayleigh) component from the doppler-shifted Brillouin doublets. To deal with the first of these problems we have employed photon-counting detection with an Amperex photomultiplier cooled to dry-ice temperatures. The operating voltage for the PMT was supplied from a filtered, stable power supply and the output of the PMT was coupled to a preamplifier through which the signal was passed to a single channel analyzer. This configuration produced a noise-limited response of one photoelectron per second. The device employed allowed us to discriminate against extremely low and extremely high dark pulse heights. We were thus able to optimize our signal to noise ratio for the sensitivity required in these measurements. The effectiveness of this procedure is demonstrated in Fig. 1, which shows a 5×10^5 counts/min scan of the entire spectrum below and a x 50 amplification of the Brillouin component in the upper portion.

The high resolution requirement was satisfied by using fairly low laser powers (~ 30 mW) so as to reduce the problem of multimoding, in conjunction with a pressure-scanned Fabry-Perot interferometer, with the scan speed linearized to within 1% with a Whitey fine metering valve. The mirrors on our Fabry-Perot were precisely positioned 1.0 cm apart and had dielectric coatings which were 95% reflecting. The effective finesse of the interferometer was about 30. Frequent adjustment of the interferometer-pinhole combination was necessary in order to maintain symmetric band shapes for the instrument scanning function.

The observation angle was set to within ± 0.5 degrees using a micrometer screw which rotated a lever arm about a 2 cm diameter cell. The Fabry-Perot, the PMT, preamplifier, and a collimating lens system were mounted on the lever arm. This arrangement produced frequency shifts measurable to ± 0.5%.

The samples employed in this study were all doubly distilled monomers with 0.1% azo-bis-butyronitrile added as initiator. The liquids were filtered through a 0.01 μ Millipore filters until their Brillouin spectra showed no evidence of Tyndall scattering, whereupon they were outgassed and sealed in evacuated scattering cells. Polymerization was then initiated just below the glass transition temperature for 24 hours and completed at about 20°C above the glass transition temperature for another 24 hours. The samples were then annealed until they showed reproducible phonon velocity and scattering intensity behavior. The bulk polymer was housed in a thermostatted jacket and the temperature was continuously monitored to within a degree by two thermocouples imbedded in the polymer.

The reduction of data was carried out on an IBM 360/50 computer. This consisted of averaging over six of the eight free spectral range orders, and calculating phonon velocities and Rayleigh and Brillouin line widths. Deconvolution of the spectra were carried out using computer programs which are similar to procedures previously described[20].

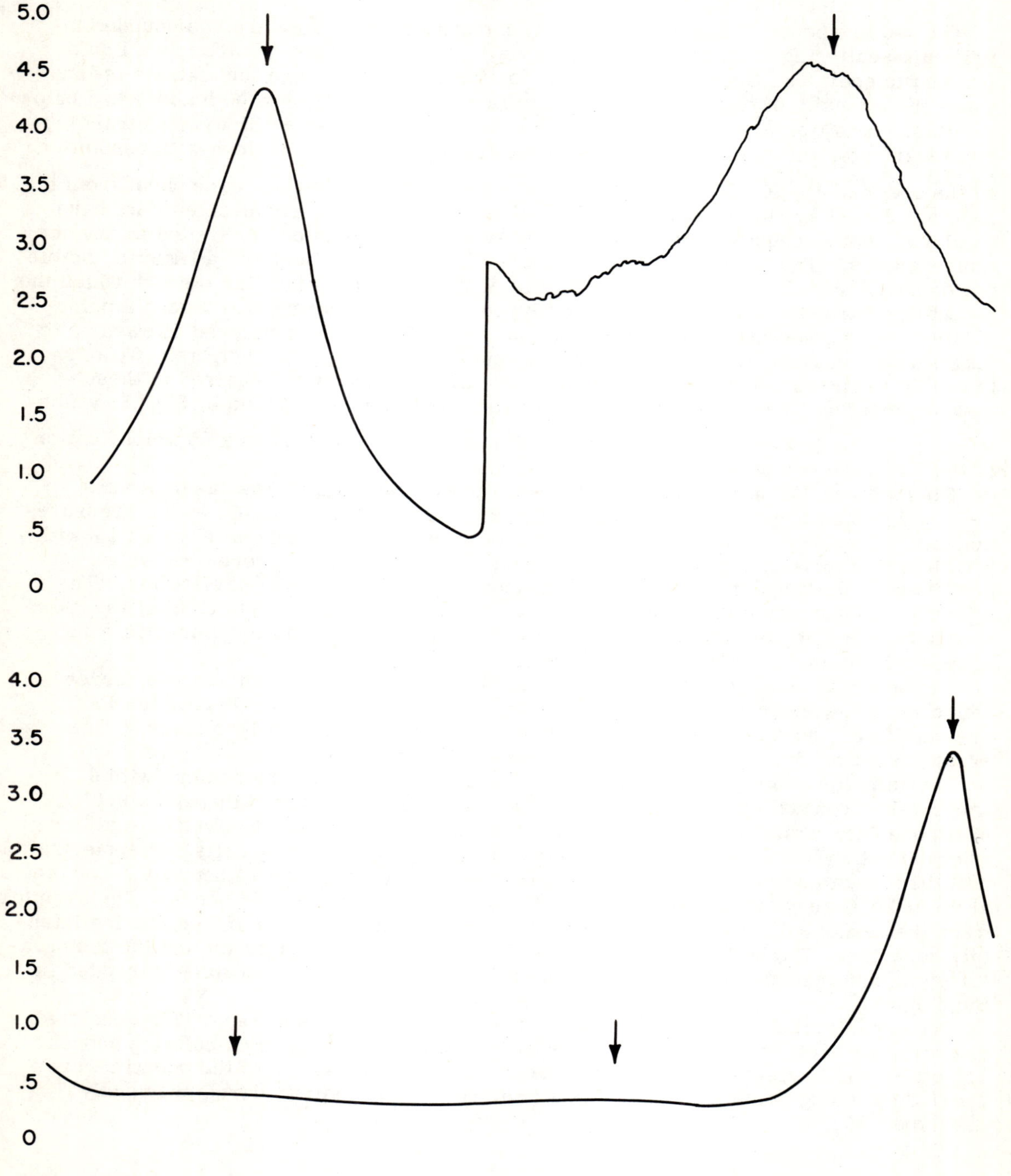

Fig. 1. Brillouin spectra of polycyclohexylmethacrylate. The upper part indicates a Rayleigh line on the left and a x 50 amplified Brillouin component on the right. The lower trace indicates sequentially unamplified Brillouin components and a Rayleigh line.

RESULTS

The Landau-Plazcek ratio (R_{L-P}) for PCHM has been observed to undergo two types of transitions in this frequency range over the temperature interval 20° C to 110° C. A spectrum of the polymer above and below its glass transition is shown in Fig. 2. The results for R_{L-P} are summarized in Fig. 3. There is an abrupt decrease in the ratio over the 50 - 60°C internal followed by a transition at about 90° C. The dashed line in Fig. 3 corresponds to the result observed previously for polyethylmethacrylate[11]. The phonon velocity and frequency shift behavior over the same range are shown in Figs. 4 and 5, respectively. The data shown in both figures refer to fixed scattering angle of $2\pi/3$ radians.

The dependence of phonon velocity may be best understood by referring to the following relation for the longitudinal modulus, G' (19),

$$G'(\omega) = \rho_0 (\omega/k)^2 (1 + 2\delta) \tag{27}$$

where ρ_0 is the average static density and δ is approximately given by $(\Delta/\omega)^2$, where Δ is the band width at half height. The value of δ on the order of 10^{-2} at most and may be neglected here. The attenuation in velocity at 52°C and 90°C represents then an abrupt change in the elastic modulus at wave vectors of about 2.4×10^5 cm^{-1}. It is also possible with this technique to measure the increase in the loss modulus, G'', in the dispersive range by examining the following relation[19],

$$G''(\omega) \cong G'(\omega) [\, 2\, \Delta/\omega\, (1 + \Delta/2\omega)\,] \tag{28}$$

$$\cong G'(\omega)\, (2\, \Delta/\omega)$$

The deconvolution the Brillouin line shape produces values of Δ. The decrease in ω and increase in Δ (absorption) allows for a direct determination of relative loss modulus or loss angle over the accessible temperature range. It has been difficult to apply this analysis to the methacrylate polymers because of the low Brillouin intensities and associated uncertainties in the deconvoluted line-shape, however, the line widths are in the range 244 to 54 MHz and the absorption coefficients corresponding to this range are 32.9 to 6.9×10^4 cm^{-1}.

Aside from the Eq. (19) presented earlier, it is possible to show that the Landau-Placzek ratio is given by

$$R_{LP} = \gamma_0 [\, G'(\omega) / G_0\,] - 1 \tag{29}$$

in general, where γ_0 is the ratio of heat capacities, and G_0 is the adiabatic bulk modulus of the material. Our results then indicate an abrupt decrease in $G'(\omega)$ at T_g. This result applies to the methyl, ethyl and cyclohexyl derivatives of the methacrylate polymers.

It is apparent from Fig. 3 that the glass transition at about 60°C is followed by a second transition at about 100°C. The dashed line in Fig. 3 indicates the behavior

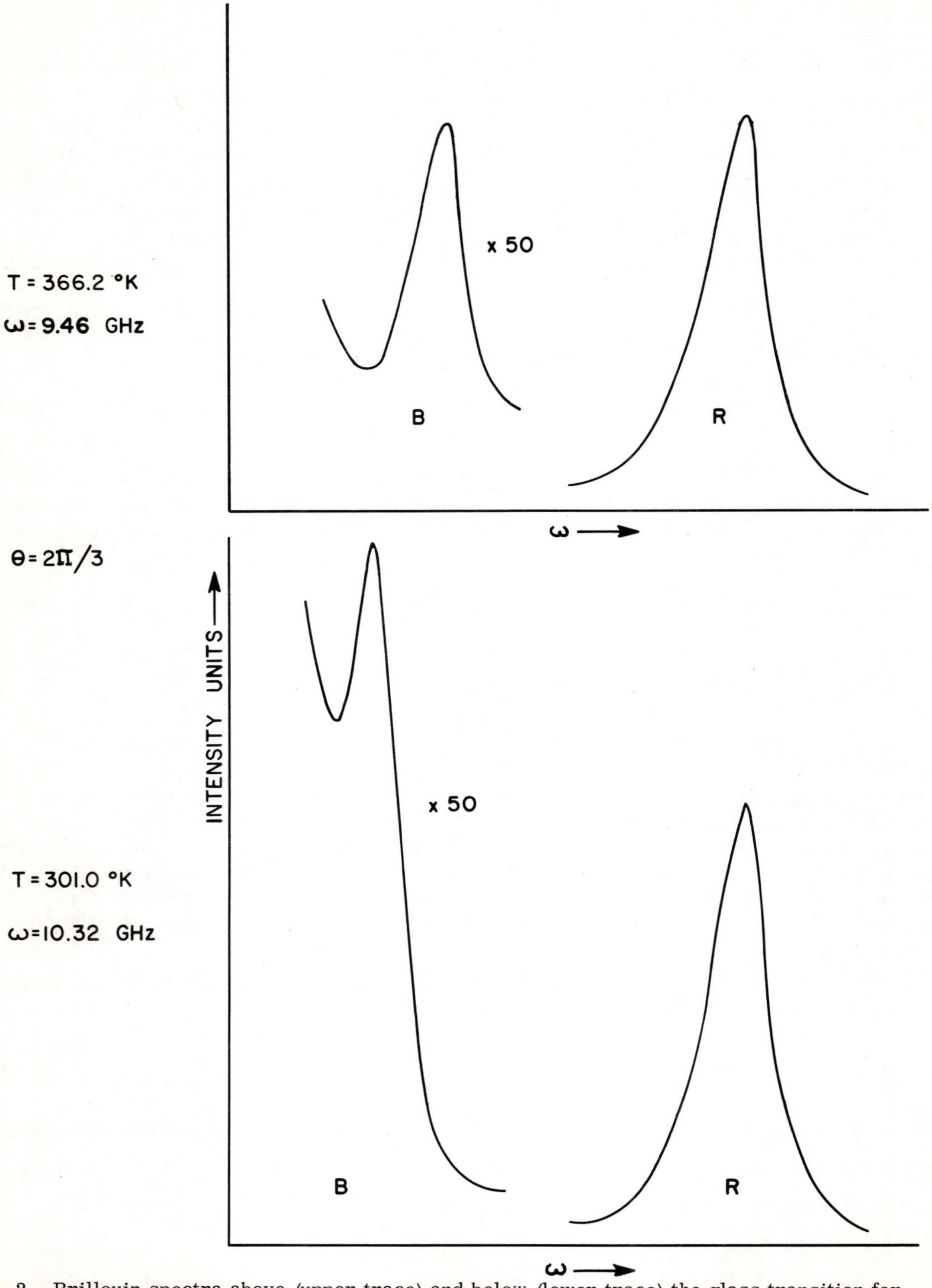

Fig. 2. Brillouin spectra above (upper trace) and below (lower trace) the glass transition for polycyclohexylmethacrylate. The doppler shift (ω) and scattering angle (θ) are indicated.

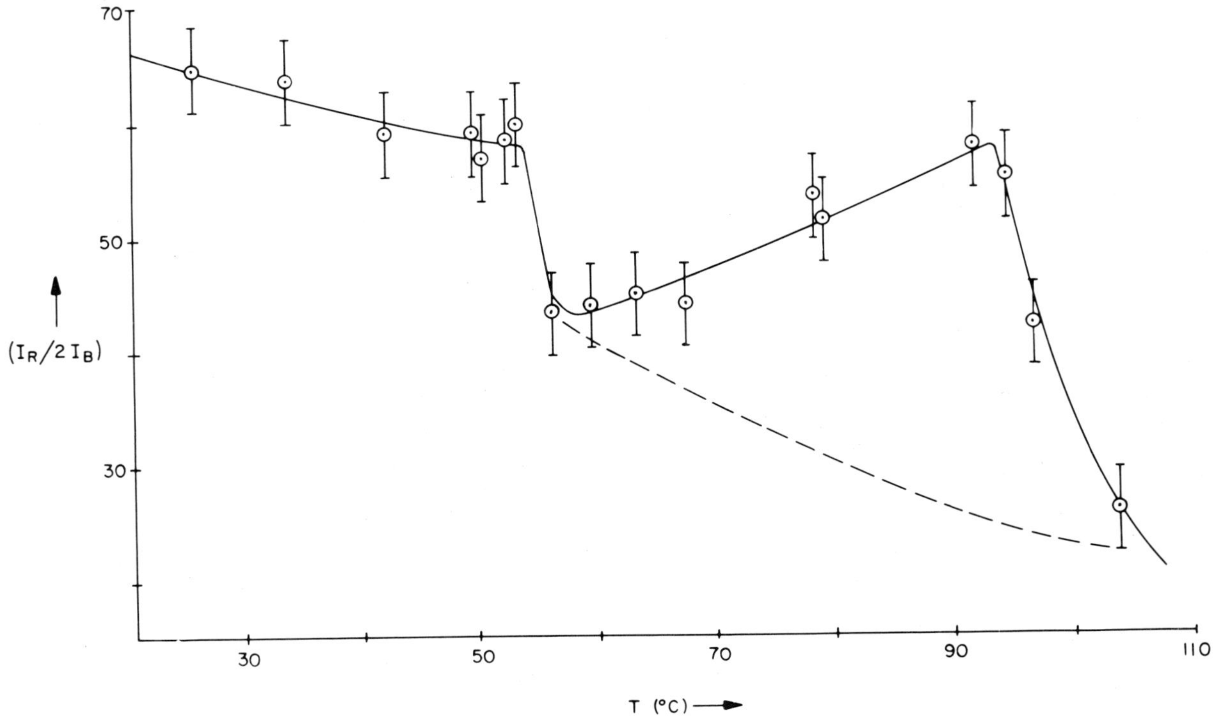

Fig. 3. Variation in the Landau-Placzek ratio for polycyclohexylmethacrylate with temperature. The dashed line indicates the variation observed for the methyl and ethyl methacrylate polymers.

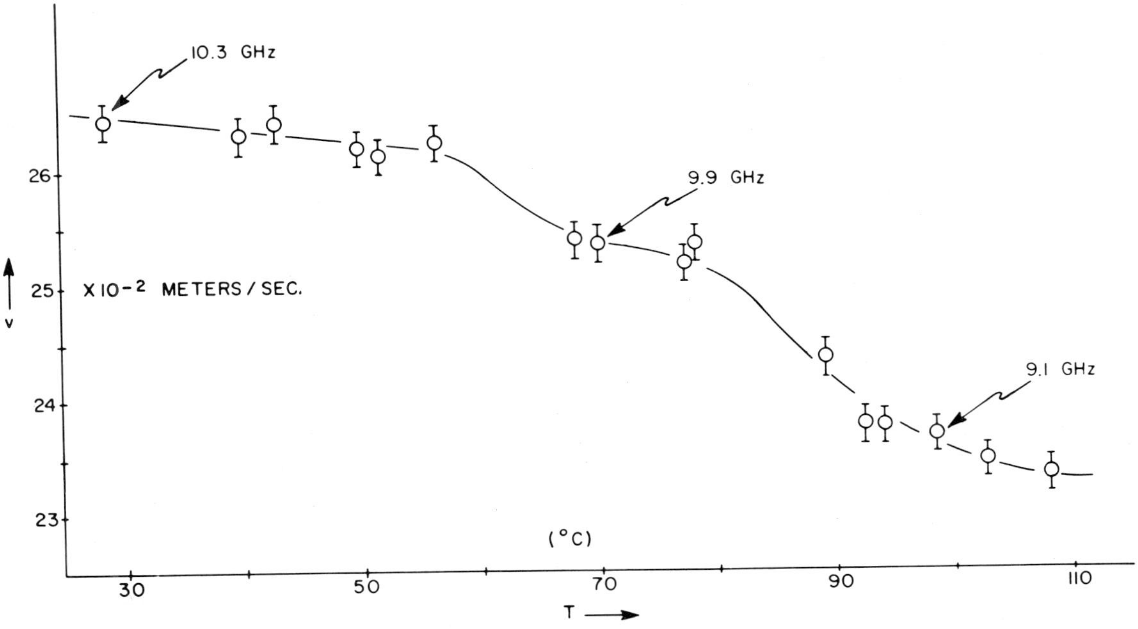

Fig. 4. Variation of polymer phonon velocity with temperature for polycyclohexylmethacrylate.

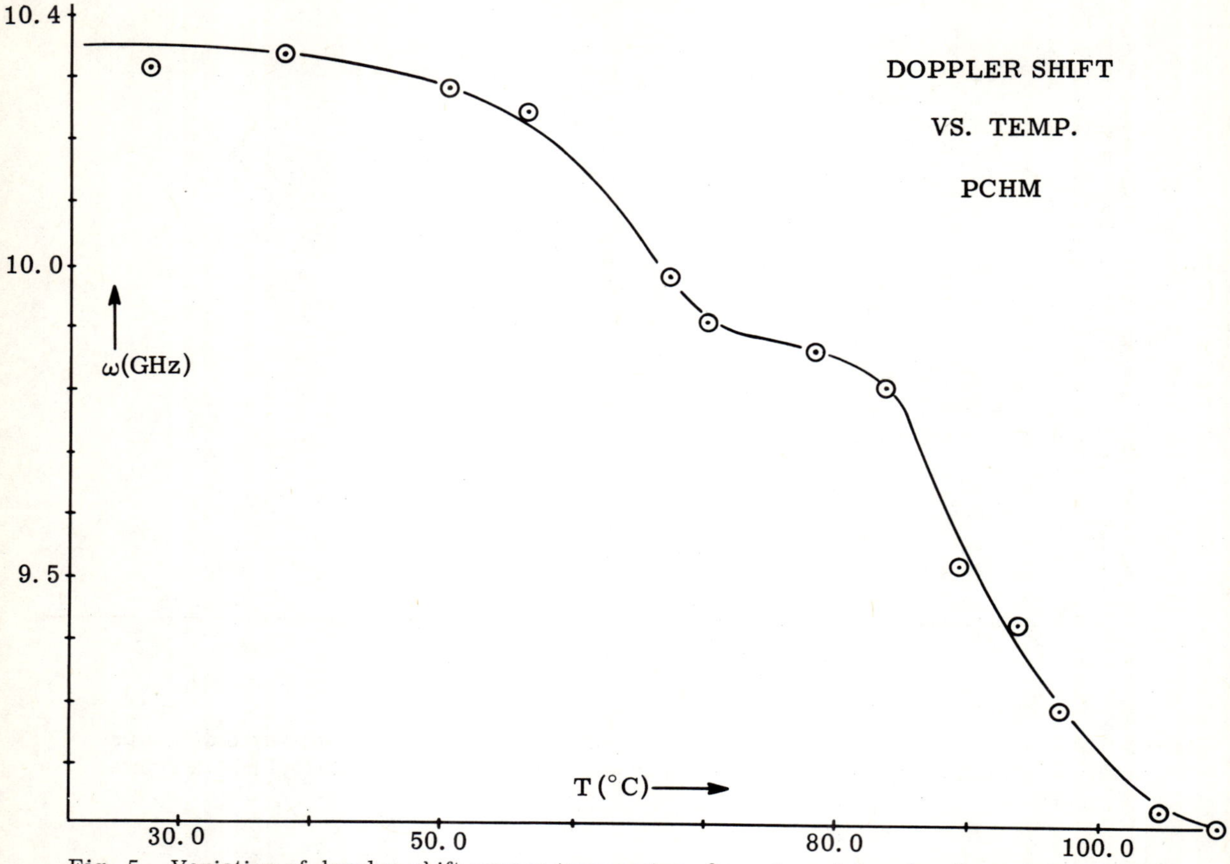

Fig. 5. Variation of doppler shift versus temperature for polycyclohexylmethacrylate (PCHM).

observed in the case of the methyl and ethyl polymethacrylates. This second transition also appears to show a pronounced frequency dependence, shifting to lower temperatures with decreasing frequency. The frequency dependence of this relaxation process is accounted for by the following form of the storage modulus,

$$G'(\omega) = \rho_0 \frac{v_o^2 + v_\infty^2 (\omega\tau)^2}{1 + (\omega\tau)^2} \tag{30}$$

where τ refers to a particular internal relaxation process, and v_o and v_∞ refer to the limiting high and low frequency phonon velocity, respectively.

The source of the secondary transition in polycyclohexylmethacrylate is likely the β transition alluded to by previous workers[22]. This refers to the side chain chair-to-chair inversion mode of the six-membered cyclohexyl ring. Heijboer has found that this relaxation mode shifts from -90°C to -20°C as the dielectric frequency is varied over four decades from 10^{-1} to 10^3 Hz[22]. An extention of his results to the range

10^9 Hz indicates that one should expect this transition to occur at approximately $91 \pm 10°C$ in reasonable agreement with our data. It should be pointed out, however, that this transition temperature is compatible with the low frequency (and low temperature) dielectric data only if there is a sharp increase in the activation energy for the side chain chair-to-chair transition as one goes to hypersonic frequencies and higher temperatures.

REFERENCES

1. R.F. Boyer, Rubber Reviews 36, 1303 (1963).
2. N. Saito, K. Okamo, S. Iwayangi, and T. Hideshima, "Solid State Physics," F. Sietz and D. Turnbull (eds.), 14, p.344, Academic Press, Inc., New York, 1963.
3. J.D. Ferry, "Viscoelastic Properties of Polymers," John Wiley and Sons, New York, 1921.
4. E.A. Hoff, D.W. Robinson, and A.H. Willbourn, J. Polymer Science 13, 161 (1955).
5. W.P. Slichter, E.R. Mandele, J. Appl. Phys. 29, 1438 (1958).
6. A.J. Kovacs, Fortschr. Hochpolymer. Forsch. 3, 394 (1963).
7. Y.A. Sharonov and M.V. Vol'kenshtein, Soviet Physics - Solid State 6, 992 (1964).
8. N. Hirai and H. Eyring, J. Poly. Sci. 37, 51 (1955).
9. J.H. Gibbs and E. Dimarzio, J. Chem. Phys. 28, 373 (1958).
10. J.H. Gibbs, "Modern Aspects of the Vitreous State," Butterworths, London, 1960.
11. W.L. Peticolas, G.I.A. Stegeman, and B.P. Stoicheff, Phys. Rev. Lett. 18, 1130 (1967).
12. H.Z. Cummins and R.W. Gammon, J. Chem. Phys. 44, 2785 (1966).
13. I.L. Fabelinskii, Dokl. Akad. Nauk. SSSR 106, 822 (1956) (Transl: Soviet Physics - Doklady 1, 115 (1956)).
14. G.H. Wannier, "Statistical Physics," p.260, John Wiley and Sons, Inc., 1966.
15. K.F. Herzfeld and T.A. Litovitz, "Absorption and Dispersion of Ultrasonic Waves," Chapter 2, Academic Press, 1959.
16. R.Y. Chiao and B.P. Stoicheff, J. Opt. Soc. Amer. 54, 1286 (1964).
17. R.D. Mountain, J. Res. Natl. Bur. Std. 70A, 207 (1966).
18. C.J. Montrose, V.A. Solovyev, and T.A. Litovitz, J. Acoust. Soc. Amer. 43, 117 (1967).
19. D.A. Pinnow, S.J. Candau, K.T. LaMacchi, and T.A. Litovitz, ibid., p. 131.
20. E.H. Eberhardt, App. Opt. 6, 251 (1967).
21. H.W. Leidecker, Jr. and J.T. LaMacchia, J. Acoust. Soc. Am. 43, 143 (1967).
22. J. Heijboer, "Physics of Non-Crystalline Solids," North Holland, Amsterdam, 1965.

H-11: LATTICE DYNAMICS IN KNO_3, PHASES I, II AND III

M. Balkanski, M. K. Teng and M. Nusimovici
Laboratoire de Physique des Solides,
Faculte des Sciences
Paris, France

INTRODUCTION

Since potassium nitrate is known to possess ferroelectric properties in one of its three phases, a renewal of interest of this material has been recently stimulated. The appearance of the ferroelectric phase transition in some materials has been related to the lattice dynamical properties of the crystal[1] and one may therefore expect that a detailed study of the lattice dynamics of the different crystallographic varieties of KNO_3 would lead to a better understanding of its ferroelectric phase transistion.

Under atmospheric pressure, KNO_3 can be obtained in three different crystallographic phases by varying the temperature. The temperature range extending from room temperature to about 130°C, corresponds the stable phase II in which the crystalline structure is associated with the Aragonite structure having D_{2h}^{16} space group and 4 molecules per unit cell. About 130°C by heating and down to 125°C by cooling, the crystal exhibits a high temperature phase commonly called phase I. The question of the structure of this phase has not been definitively clarified: Tahvonen[2] suggested that in this phase, the nitrate ions can rotate freely about the trigonal axis, this allows the assignment of a one molecule unit cell associated with D_{3d}^{5} space group. Furthermore, by interpreting the X-ray diagrams, Shinnaka[3] proposed that the rotation of nitrate ions would occur only by an angle of 60° about this axis, then one can emphasize that the structure of this phase may be closely related to the Calcite structure having D_{3d}^{6} space group, and accordingly, the unit cell contains two molecules. In the temperature range extending from 125°C to 105°C by cooling, is obtained the ferroelectric phase commonly called phase III. The crystal structure of this phase is supposed to have C_{3v}^{5} space group with one molecule per unit cell.

From the experimental point of view, infrared absorption studies have already been reported by several authors[4-6]. These experimental results concerned mainly the high frequency region. Some Raman scattering data have also been reported in either low and high frequency ranges[7-10].

A theoretical investigation of the lattice vibrations of Aragonite and Calcite structures based on group theory considerations has been done by Bhagavantam[11] and recently, Nusimovici[12] has also determined the normal modes of vibrations of KNO_3 in phases I, II and III.

In this paper, we review in detail all the normal modes of vibrations of KNO_3 in its three phases by comparison with available experimental data. An attempt is made to follow each of the normal modes from one phase to another leading to a correspondance of the different modes when the crystal undergoes transitions from one crystallographic structure to another. For the phase I, calculations are made for both D_{3d}^5 and D_{3d}^6 space groups, the comparison with experimental results favoured the assignment of the D_{3d}^6 space group for this paraelectric phase.

In the next section we give some general considerations concerning the classification of the normal modes in KNO_3. Following, we present the study of the symmetry properties and the selection rules for both infrared absorption and Raman scattering of KNO_3 in its three phases. Section 3 and 4 are devoted to the results obtained in this work, and finally a comparison between theoretical predictions and available infrared and Raman data is discussed in the last section.

GENERAL CONSIDERATIONS

For an ionic crystal structure composed of complex ions like NO_3^- in potassium nitrate, and in the case when the interacting forces are much weaker between ions than the bonding forces between atoms in the complex ion, one can classify the lattice vibrations into three categories as illustrated in Fig. 1:

Covalent modes are those describing the relative displacement of the atoms in the nitrate ion. The atoms which constitute the ion are bound by covalent forces and then, the normal modes describing the relative displacements of these atoms are called covalent modes. Such a mode, since involving a deformation of the nitrate ion, would occur at high frequencies.

Ionic modes for which the potassium and nitrate ions are considered as rigid spheres and the normal modes describing the relative motions of the ions with respect to each other. Such modes would generally occur at low frequencies.

Rotational modes which describe the rotation of the nitrate ion around its center of mass. The nitrate ion is considered as a rigid rotator. Since these modes induce neither displacements of the center of mass of the ion, nor its deformation, the rotational modes would generally occur at very low frequency.

We would like to mention that this classification of normal modes is suggested by a physical analogy, suitable to interpret the experimental results in some particular cases, rather than by a rigorous mathematical formulation. Some of the true normal modes may be a linear combination of the first two categories of modes of same symmetry.

Bhagavantam has classified the normal modes into two categories: the external modes corresponding to ionic and rotational modes and the internal modes corresponding to covalent modes, whereas Nusimovici has included the rotational modes into the covalent category.

SYMMETRY PROPERTIES AND SELECTION RULES

In the present paper, we shall confine our study to the center of the Brillouin zone, that is, only phonons with nearly zero wave vector are considered. Furthermore, we shall consider that for a given space group, all translations (primitive or non primitive) implied in the group of symmetry operations are equivalent to the identity, the factor group thus obtained is isomorphic with the corresponding point group and in our calculations, this assumption allows us to use the character tables of the point groups under consideration.

The knowledge of the character table, together with the knowledge of the transformed position of all atoms or ions in the unit cell after being acted by the group operations, permits us to obtain the decomposition of the representation of the different categories of modes into irreducible representations of the crystal group.

Group theory considerations led us to deduce the following sets of optically and Raman active modes in the three phases of KNO_3.

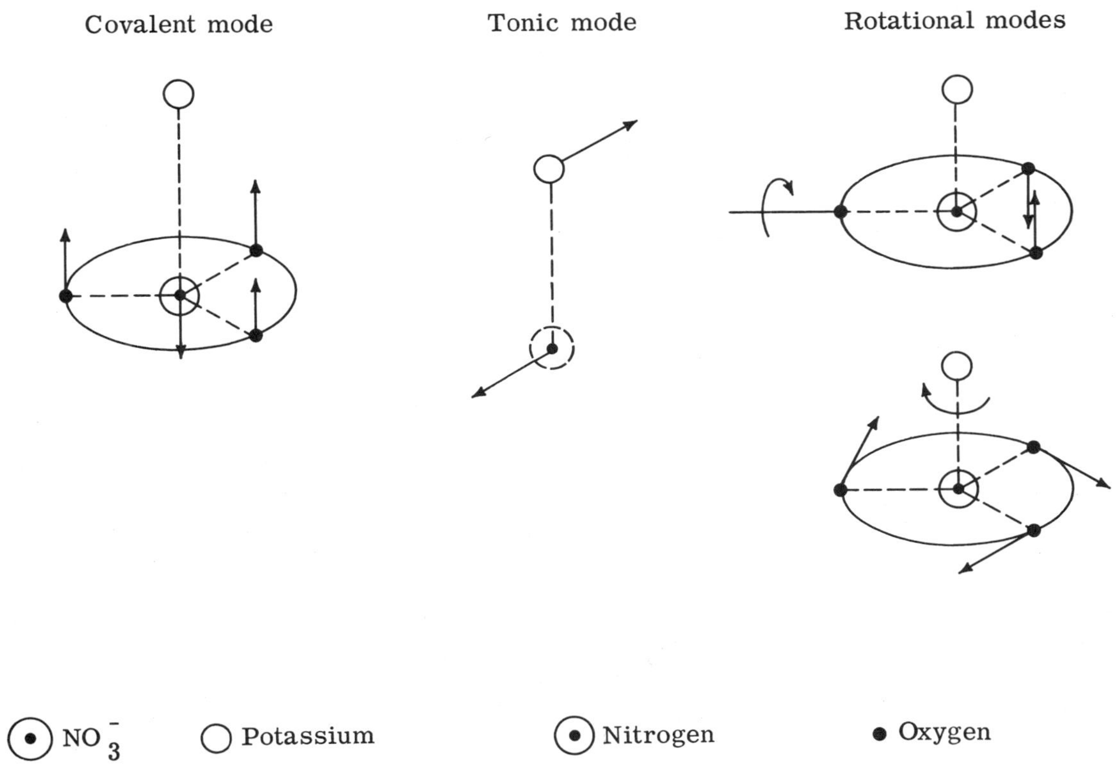

Fig. 1. The three categories of normal modes of vibration in KNO_3.

a) PHASE III: This ferroelectric phase of KNO_3 belongs to the ditrigonal pyramidal class and its space group is C^5_{3v} (R 3m). The description of the different categories of modes is then given by the equations:

$$\Gamma_{ionic} = 2\Gamma_I + 2\,{}^2\Gamma_3 \qquad (\text{with } \Gamma_{acoustic} = \Gamma_I + {}^2\Gamma_3)$$

$$\Gamma_{covalent} = 2\Gamma_I + 2\,{}^2\Gamma_3$$

$$\Gamma_{rotational} = \Gamma_2 + {}^2\Gamma_3.$$

The subscripts indicate the degeneracy of the mode. The modes Γ_I and ${}^2\Gamma_3$ are both infrared and Raman active, whereas Γ_2 is forbidden (Table I).

TABLE I

Decomposition in Terms of Irreducible Representations of the Electric Dipole Moment and the Polarizability Tensor [α] of KNO_3 In its Three Phases

	PHASE III (C^5_{3v})	PHASE I (D^6_{3d})	PHASE II (D^{16}_{2h})
Components of the electric dipole moment			
X	Γ_3	Γ_6	Γ_8
Y	Γ_3	Γ_6	Γ_4
Z	Γ_1	Γ_4	Γ_6
Polarizability tensor [α]	$2\Gamma_I+2\Gamma_3$	$2\Gamma_I+2\Gamma_3$	$3\Gamma_I+\Gamma_3+\Gamma_5+\Gamma_7$

b) PHASE I: As discussed previously, two crystal structures have been proposed for this paraelectric phase. The Calcite structure with space group D^6_{3d} ($R\bar{3}c$) and a structure closely related to it with space group D^5_{3d} ($R\bar{3}m$). In the former case, the unit cell is trigonal and contains two molecules.

For the space group D_{3d}^6, the decomposition of the three categories of modes is given by the equations:

$$\Gamma_{ionic} = \Gamma_2 + {}^2\Gamma_3 + 2\Gamma_4 + \Gamma_5 + 3\,{}^2\Gamma_6. \qquad (\text{with } \Gamma_{acoustic} = \Gamma_4 + {}^2\Gamma_6)$$

$$\Gamma_{covalent} = \Gamma_1 + \Gamma_2 + 2\,{}^2\Gamma_3 + \Gamma_4 + \Gamma_5 + 2\,{}^2\Gamma_6.$$

$$\Gamma_{rotational} = \Gamma_2 + {}^2\Gamma_3 + \Gamma_4 + {}^2\Gamma_6.$$

For the space group D_{3d}^5, the decomposition of the different categories of modes is given by:†

$$\Gamma_{ionic} = 2\Gamma_4 + 2\,{}^2\Gamma_6.$$

$$\Gamma_{covalent} = 3\Gamma_4 + 3\,{}^2\Gamma_6.$$

For both D_{3d}^6 and D_{3d}^5 space groups, the modes Γ_4 and ${}^2\Gamma_6$ are infrared active, whereas Γ_1 and ${}^2\Gamma_3$ are Raman active (Table I).

It is evident from the above decomposition that if the D_{3d}^5 space group is the accurate one, no Raman lines would be found in this phase, this fact is inconsistent with the experimental results[10] and has decouraged us to pursue our study in this way. Hence, we will suppose that the phase I would have D_{3d}^6 space group with two molecules per unit cell, this assumption seems to be supported satisfactorily by Raman data[10].

c) <u>PHASE II</u>: This phase is known to have the Aragonite structure, its space group is D_{2h}^{16} (Pnma). The unit cell is orthorhombic and contains four molecules. The following equations represent the decomposition in terms of irreducible representations of the different categories of modes:

$$\Gamma_{ionic} = 4\Gamma_1 + 2\Gamma_2 + 4\Gamma_3 + 2\Gamma_4 + 2\Gamma_5 + 4\Gamma_6 + 2\Gamma_7 + 4\Gamma_8.$$

$$(\text{with } \Gamma_{acoustic} = \Gamma_4 + \Gamma_6 + \Gamma_8).$$

$$\Gamma_{covalent} = 4\Gamma_1 + 2\Gamma_2 + 4\Gamma_3 + 2\Gamma_4 + 2\Gamma_5 + 4\Gamma_6 + 2\Gamma_7 + 4\Gamma_8.$$

$$\Gamma_{rotational} = \Gamma_1 + 2\Gamma_2 + \Gamma_3 + 2\Gamma_4 + 2\Gamma_5 + \Gamma_6 + 2\Gamma_7 + \Gamma_8.$$

†These results are deduced on the assumption that the position of oxygen atoms have an equal probability of distribution on a circle. It is possible that in this case, a formal application of group theory considerations may not be appropriate.

The modes Γ_4, Γ_6 and Γ_8 are allowed in infrared absorption whereas the modes Γ_1, Γ_3, Γ_5 and Γ_7 are Raman active (Table I).

In the attempts to follow the different modes through the phase transformation one may point out that to a given mode of phase III, there may now correspond two modes in phase I in such a way that for the first one, the two molecules in the unit cell vibrate in the same manner and with the same sign that the original mode, we will call this mode R type; for the second one, although the two molecules in the unit cell move in the same manner that the corresponding mode in phase III, the vibrations are of opposite sign from one molecule to another, this kind of mode will be called AR type. Nevertheless, an exception may be granted for the ionic modes Γ_2 and Γ_5 which must be considered as derived from a combination of the mode E and the acoustic mode K of phase III; similarly, the ionic modes Γ_3 and Γ_6 must be considered as derived from a combination of the modes F and the acoustic mode J of phase III.

EXPERIMENTA PROCEDURE

The experimental apparatus consists of a 50 mW He-Ne gas laser source (λ = 6, 328 Å), double grating monochromator CODERG and a system of focusing lens and mirrors allowing the collection of at least three passages of the incident beam through the crystal. The scattered light is collected at right angle from the incident direction and is focused into the spectrometer entrance slit and detected by an S20 photomultiplier. The signal of the photomultiplier is finally amplified and recorded. Measurements are performed with both incident and scattered lights polarized. The slitwidth used in our investigation corresponds to a spectral separation of about 3, 6, or 20 cm^{-1} according to the intensity of the observed lines.

The temperature of the sample whose dimension is about 7x5x5 mm is controlled during both heating and cooling. The sample is sandwiched between two metal plates heated by an electric resistor. The experimental error of the temperature of the heating plates is estimated to about 0. 5°C and the temperature of the sample is thus determined with an error of about 1°C.

In order to determine the actual temperature range corresponding to the ferroelectric phase III in our crystal, we measured the D. E. hysteresis loops at various temperatures on cooling. Fig. 2 shows the results obtained in this experiment. The crystal is first heated gradually from room temperature to 115°C, then slowly at the rate of 1°C per hour from 115°C to 140°C, and finally with a faster speed from 140°C to 180°C. The cooling process is quite similar to the heating process: the crystal is cooled gradually from 180°C to 132°C, then very slowly between 132°C to 100°C, and finally with a faster speed down to room temperature.

RESULTS

In order to establish the correlation between vibrational modes in the three phases under consideration, we start from the simplest of them, say phase III.

PHASE III: Since the symmetry of an isolated nitrate ion corresponds to D_{3h} point group and the interacting forces are much stronger in the nitrate ion than between the nitrate and potassium ions, the identification of the covalent modes A, B, C, and D, (Fig. 3) is easy; they are closely related to the well-known ν_2, ν_1, ν_3, and ν_4 modes of the

D_{3h} point group[13] with: $\omega_A = 836\ cm^{-1}$ (Γ_1), $\omega_B = 1054\ cm^{-1}$ (Γ_1), $\omega_C = 1352\ cm^{-1}$ (Γ_3), and $\omega_D = 716\ cm^{-1}$ (Γ_3).

At 120 cm^{-1} we observe a broad band having a halfwidth of about 40 cm^{-1} and a Γ_3 symmetry character (Fig. 4). Two degenerate Γ_3 modes are Raman active in the low frequency region: the ionic mode F and the rotational mode G. The relatively strong intensity of the observed band and the very close variation of its halfwidth with temperature seems to encourage the assignment of the rotational mode G. From another point of view, the mode G which corresponds to the vibration of the nitrate plane around an axis perpendicular to the z direction would occur at lower frequency in phase II (about 80 cm^{-1}); now it happens that no shifting toward low frequency region of the observed band has been noted, even in the temperature range covering the III → II phase transition: the frequency of the band remains always constant and centered at 120 cm^{-1}. Moreover, some partial

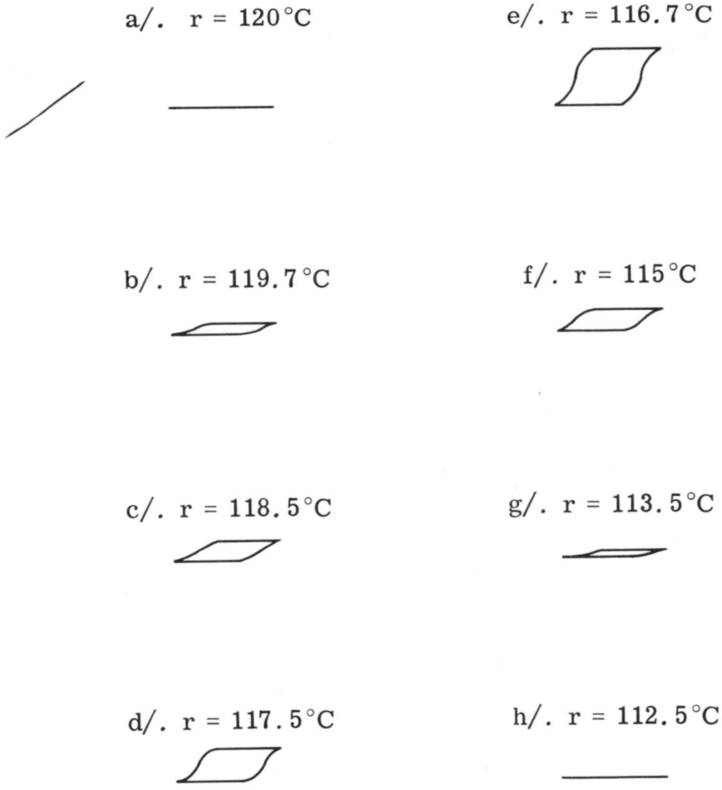

Fig. 2. D.E. hysteresis loops of KNO_3 measured along the c axis when the crystal is cooled (E_{Max} = 3.7 kV/cm).

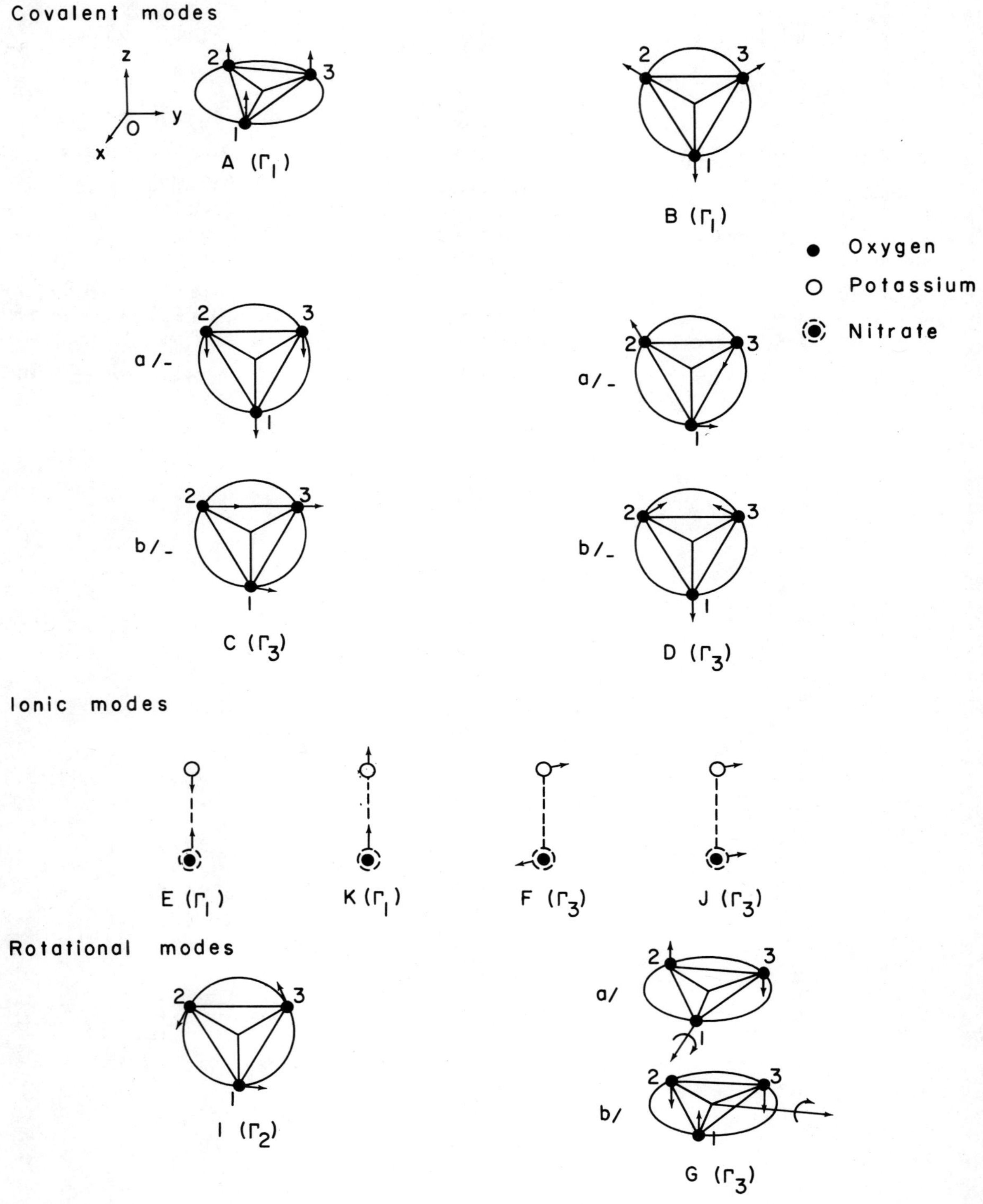

Fig. 3. Scheme of normal modes in the ferroelectric phase III.

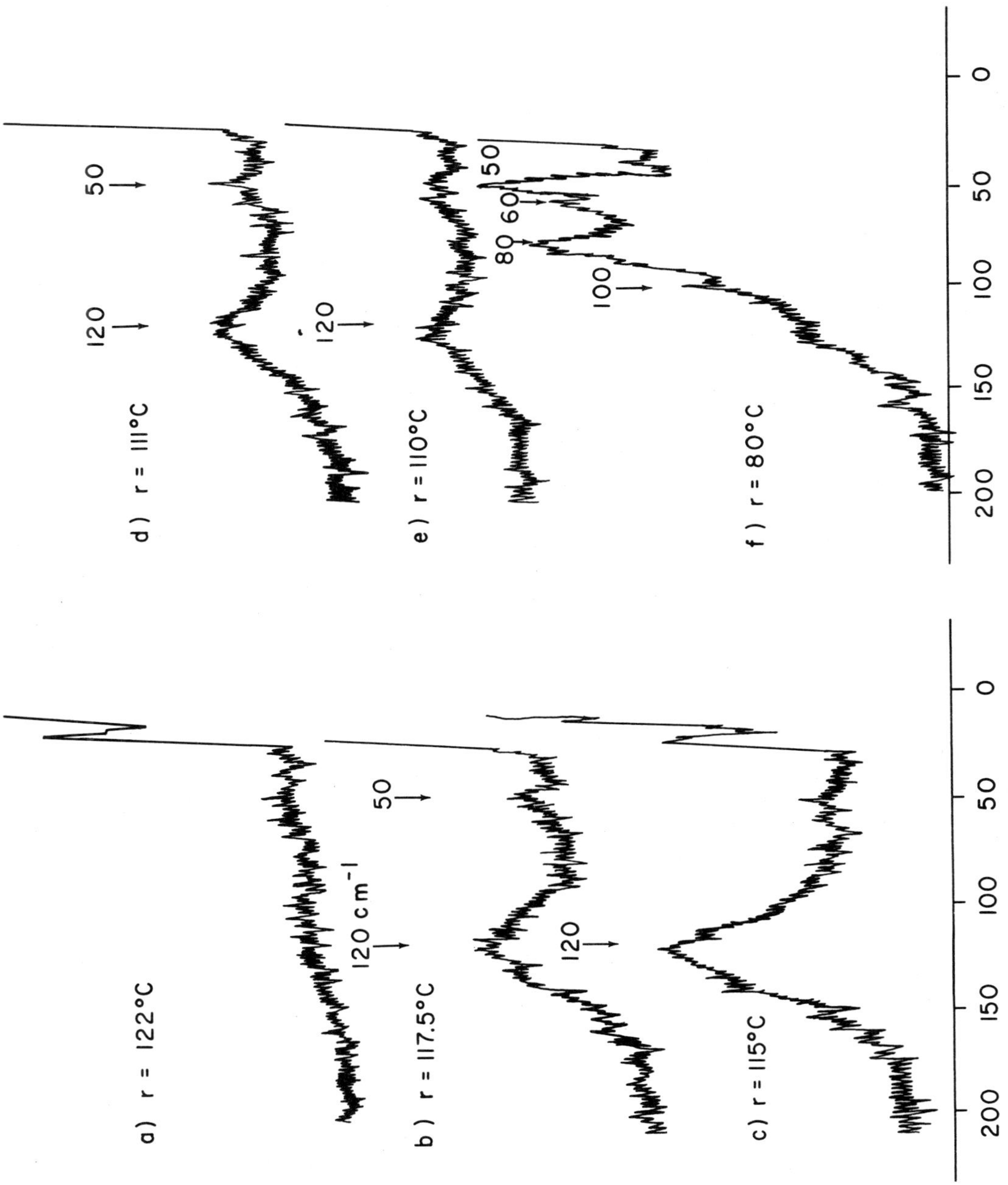

Fig. 4. Temperature dependence of the low frequency mode at 120 cm^{-1}.

infrared data indicate that an ionic mode is observed near 120 cm^{-1} in the high tempera-range[6]. Therefore, the assignment of the ionic mode F seems favoured by these last arguments. Just as for the mode E which will be discussed in the next paragraph, further infrared results in low frequency range will be useful to settle definitively this question.

PHASES I AND II: Starting from the modes A, B, C, D, E, F, G, and I, in phase III, we try to deduce the representations of normal modes in the two other phases which correspond to each of the above modes. Let us consider that the crystal passes from phase III to phase II via phase I, the number of molecules per unit cell passes from one to four and its structure changes from lower to higher symmetry. Then, a given mode in phase III can give rise to several modes in phase I and II. Among them, one can distinguish two types of modes; the first is R type and the second AR type. An example of these two types of modes is illustrated in Fig. 3 for the mode B.

Based on the idea that the modes of type AR would give rise to relatively weak intensity even though they are allowed by selection rules, we try to interpret the results in phases I and II by using this assumption. In Table II is shown the correspondence between the normal modes of the ferroelectric phase and the normal modes of R type for the two other phases of KNO_3. The predicted normal modes by group theory considerations are compared with the experimental results.

Table III gives the comparison with the experimental results and the correspondence between covalent modes of the ferroelectric phase and the covalent modes of AR type in the two other phases of KNO_3.

The correspondence between ionic modes of the ferroelectric phase and the ionic modes of AR type in the two other phases is displayed in Table IV where a comparison with the experimental results is also given.

The rotational modes are listed on Table V where the correspondence between the ferroelectric phase and the rotational modes of AR type in the two other phases of KNO_3 is given as well as the comparison with the experimental results.

The inspection of the Table II shows that for covalent modes all the predicted Raman active modes of R type are observed, and that the correspondence between all these modes is firmly evidenced. Furthermore, Table IV shows that three ionic modes of AR type are also observed in the room temperature phase II.

DISCUSSION

In order to develop a more general discussion, some available infrared data are also listed in Tables II and III for comparison. It is evidenced from this comparison that the agreement between theoretical previsions and experimental results concerning the covalent modes of R type is perfect. The absence of infrared data for the mode C in phase III is due to the fact that no specific studies are available until now for the spectral region where this mode should be observed. Besides, some covalent modes of AR type are also observed in infrared experiment with

TABLE II

Correspondence Between Normal Modes of the Ferroelectric Phase and Normal Modes of R Type in the Two Other Phases of KNO_3. Comparison With Experimental Results

	Phase III (C_{3v}^5)		Phase I (D_{3d}^6)		Phase II (D_{2h}^{16})	
Covalent modes	Symmetry	Observed frequency in cm^{-1}	Symmetry	Observed frequency in cm^{-1}	Symmetry	Observed frequency in cm^{-1}
A	Γ_1(R, I)	$836^+(\Gamma_1)$ $836^o(\Gamma_1)$	Γ_4(I)	$836^+(\Gamma_4)$	Γ_6(I)	$829^+(\Gamma_6)$ $830^*(\Gamma_6)$
B	Γ_1(R, I)	$1053^+(\Gamma_1)$ $1054^o(\Gamma_1)$	Γ_1(R)	$1056^o(\Gamma_1)$	Γ_1(R)	$1054^o(\Gamma_1)$
C	Γ_3(R, I)	$1352^o(\Gamma_3)$	Γ_3(R)	$1428^o(\Gamma_3)$	Γ_1(R) Γ_5(R)	$1362^o(\Gamma_1)$ $1348^o(\Gamma_5)$
D	Γ_3(R, I)	$717^+(\Gamma_3)$ $716^o(\Gamma_3)$	Γ_3(R)	$714^o(\Gamma_3)$	Γ_1(R) Γ_5(R)	$714^o(\Gamma_1)$ $714^o(\Gamma_5)$
Ionic modes						
E	Γ_1(R, I)		Γ_4(I)		Γ_1(R)	$138^o(\Gamma_1)$
F	Γ_3(R, I)	$120^o(\Gamma_3)$?	Γ_6(I)		Γ_4(I) Γ_8(I)	
J	Γ_3(Ac)		Γ_6(Ac)		Γ_4(Ac) Γ_8(Ac)	
K	Γ_1(Ac)		Γ_4(Ac)		Γ_6(Ac)	
Rotational modes						
G	Γ_3(R, I)	$120^o(\Gamma_3)$?	Γ_3(R)	$115^o(\Gamma_3)$	Γ_3(R) Γ_7(R)	$83^o(\Gamma_3)$ $50^o(\Gamma_7)$
I	Γ_2(f)		Γ_2(f)		Γ_5(R)	$65^o(\Gamma_5)$

R : Allowed in Raman scattering.
I : Allowed in Infrared absorption.
f : Forbidden in both Raman and Infrared processes.
Ac : Acoustic mode.

+ : Observed in Infrared experiment. Ref. 5.
* : Observed in Infrared experiment. Ref. 4.
o : Observed in Raman experiment. Ref. 10.

TABLE III

Correspondence Between Covalent Modes of the Ferroelectric Phase and Covalent Modes of AR Type in the Two Other Phases of KNO_3. Comparison With Experimental Results.

Phase III C_{3v}^{5}	Phase I D_{3d}^{6}		Phase II D_{2h}^{16}	
Covalent Modes	Symmetry	Observed frequency in cm^{-1}	Symmetry	Observed frequency in cm^{-1}
A	Γ_2(f)		Γ_1(R) Γ_3(R) Γ_8(I)	 $846^{+}(\Gamma_8)$
B	Γ_5(f)		Γ_3(R) Γ_6(I) Γ_8(I)	 $1051^{+}(\Gamma_6)$, $1050^{*}(\Gamma_6)$ $1049.9^{+}(\Gamma_8)$, $1050^{*}(\Gamma_8)$
C	Γ_6(I)		Γ_2(f) Γ_3(R) Γ_7(R) Γ_4(I) Γ_6(I) Γ_8(I)	 $1390^{*}(\Gamma_4)$ $1390^{*}(\Gamma_6)$ $1390^{*}(\Gamma_8)$
D	Γ_6(I)	$716^{+}(\Gamma_6)$	Γ_2(f) Γ_3(R) Γ_7(R) Γ_4(I) Γ_6(I) Γ_8(I)	 $714.5^{+}(\Gamma_4)$, $715^{*}(\Gamma_4)$ $715^{*}(\Gamma_6)$ $715.2^{+}(\Gamma_8)$, $715^{*}(\Gamma_8)$

R : Allowed in Raman scattering.
I : Allowed in Infrared absorption.
f : Forbidden in both Raman and Infrared processes.
\+ : Observed in Infrared experiment. Ref. [5].
* : Observed in Infrared experiment. Ref. [4].

TABLE IV

Correspondence Between Ionic Modes of the Ferroelectric Phase and Ionic Modes of AR Type in the Two Other Phases of KNO_3. Comparison With Experimental Results

Phase III C_{3v}^5	Phase I D_{3d}^6		Phase II D_{2h}^{16}	
Ionic Modes	Symmetry	Observed frequency in cm^{-1}	Symmetry	Observed frequency in cm^{-4}
E			Γ_3(R) Γ_6(I) Γ_8(I)	$103^o(\Gamma_3)$
F			Γ_2(f) Γ_1(R) Γ_3(R) Γ_5(R) Γ_7(R) Γ_6(I)	$122^o(\Gamma_3)$ $133^o(\Gamma_5)$
J (Ac)			Γ_2(f) Γ_1(R) Γ_3(R) Γ_5(R) Γ_7(R) Γ_6(I)	
K (Ac)			Γ_1(R) Γ_3(R) Γ_8(I)	
E + K	Γ_2(f) Γ_5(f)			
F + J	Γ_3(R) Γ_6(I)			

R : Allowed in Raman scattering
I : Allowed in Infrared absorption
f : Forbidden in both Raman and Infrared processes.
Ac : Acoustic mode.
o : Observed in Raman experiment. Ref. [10].

TABLE V

Correspondence Between Rotational Modes of the Ferroelectric Phase and Rotational Modes of AR Type in the Two Other Phase of KNO_3. Comparison With Experimental Results

Phase III C_{3v}^5	Phase I D_{3d}^6		Phase II D_{2h}^{16}	
Rotational modes	Symmetry	Observed frequency in cm^{-1}	Symmetry	Observed frequency in cm^{-1}
G	Γ_6(I)		Γ_2(f) Γ_1(R) Γ_5(R) Γ_4(I) Γ_6(I) Γ_8(I)	
I	Γ_4(I)		Γ_2(f) Γ_7(R) Γ_4(I)	

R : Allowed in Raman scattering.
I : Allowed in Infrared absorption.
f : Forbidden in both Raman and Infrared processes.

weaker intensity. As it is expected, the covalent modes, which are essentially related to internal vibrations of the nitrate ion, would produce no significant change when the crystal goes from one phase to another and one can easily follow such modes through different phase transitions as shown in Fig. 5.

Finally, several remarks concerning the different phase transitions studied in this work must be mentioned. On heating, all the ionic and rotational modes in phase II vanish abruptly at 130°C suggesting that the transition II → I is of a rather sudden nature. A notable and continuous broadening of the rotational mode at 83 cm^{-1} is also observed when the temperature approaches the transition point.

Between 110°C and 100°C on cooling, the general aspect of the spectra suggests that in this region the limits between phase III and phase II may be less precise and the III → II phase transition probably occurs more gradually than for the other phase transitions.

When cooling from 110°C to room temperature, the spectra show the reappearance of all the ionic and rotational modes which are previously observed in phase II before the phase transition cycle. The intensity measurements in this temperature range cannot be compared quantitatively with those given by phase II before heating since the crystal begins to exhibit some crackings and fractures at about 90°C; nevertheless, they are sufficiently strong to confirm the existence of these modes.

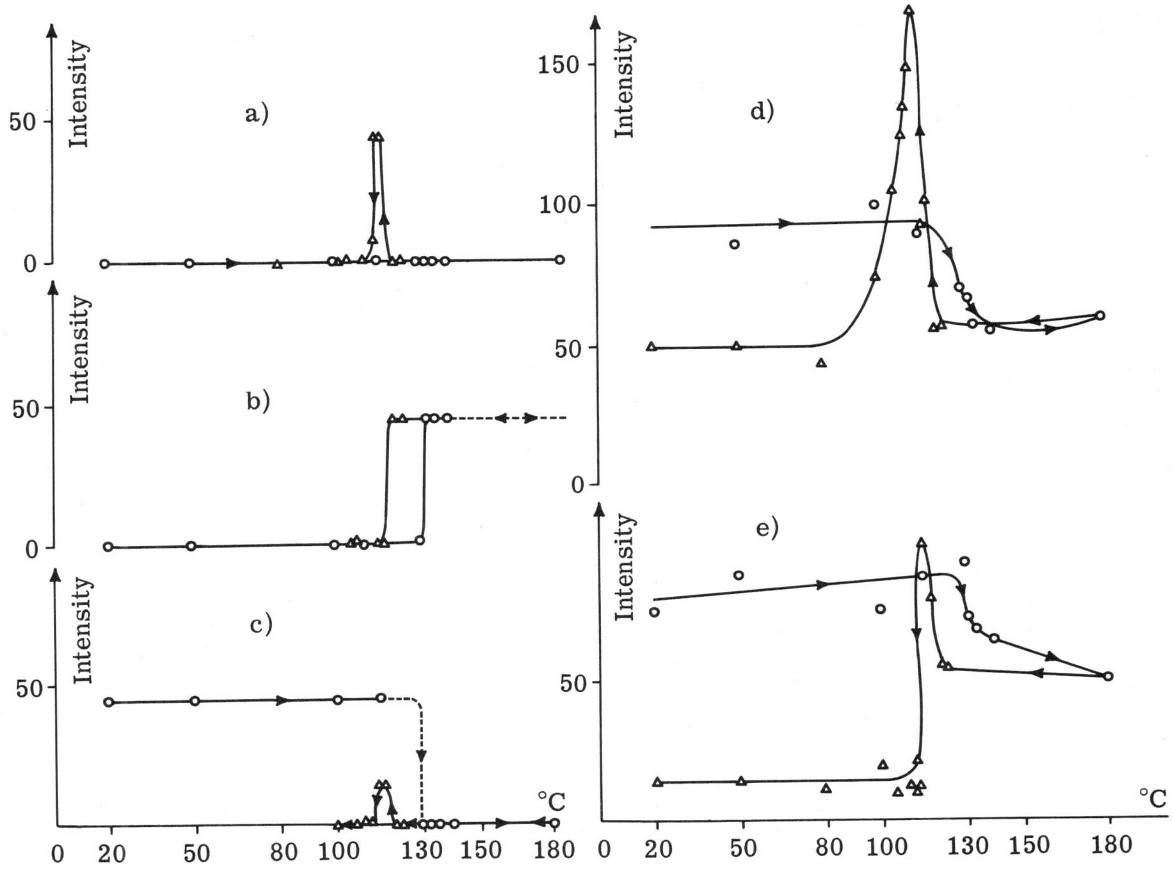

Fig. 5. Intensity diagram of the covalent modes of KNO_3 as a function of the temperature. The open circles represent data when the crystal is heated, and the triangles when the crystal is cooled. The arrow indicates the sense of increasing or decreasing temperature: a) Line at 1352 cm^{-1}, b) Line at 1428 cm^{-1}, c) Line at 1348 cm^{-1}, d) Line at 1054 cm^{-1}, and e) Line at 714 cm^{-1}.

CONCLUSION

The comparison with the experiments can be briefly summarized as follows: For the normal modes of R type, all the predicted covalent and rotational modes are found and the agreement between group theory predictions and experimental results is almost perfect, wheras for the ionic modes, the lack of experimental results, principally in infrared absorption, does not allow us to pursue the comparison.

For normal modes of AR type, one can note that they occur generally with weaker intensity than that of the corresponding modes of R type. Except for the ionic modes whose behavior would be rather different from the covalent and rotational modes when the crystal changes from one structure to another.

This review of the lattice dynamics of the three phases of KNO_3 suggest close correspondence between normal modes of the different phases. The results obtained by using this deduction are verified satisfactorily by available infrared and Raman data. Nevertheless, more complete experimental results are expected to pursue this comparison, specially for the ionic modes which constitute the most interesting part in the study of the phase transition problems.

REFERENCES

1. W. Cochran, Advances in Physics, 9, 387, (1960).
2. P.E. Tahvonen, Ann. Acad. Sci. Fennicae, Serie Math-Phys. 44, (1947).
3. Y. Shinnaka, J. Phys. Soc. Japan 17, 820 (1962).
4. T. Yanagi, J. Phys. Soc. Japan 20, 1351 (1965).
5. S.V. Karpov and A.A. Shultin, J. Phys. Chem. Solids 29, 475, (1968).
6. A. Grynwald (private communication).
7. T.M.K. Nedungadi, Proc. Ind. Acad. Sci. A14, 242, (1942).
8. L. Couture, C.R. Acad. Sci. Paris 656, (April 1945).
9. E.V. Chisler, Sov. Phys. Solid State 8, 1534, (1966).
10. M. Balkanski, M.K. Teng and M. Nusimovici (to be published).
11. S. Bhagavantam, Proc. Ind. Acad. Sci. 13, 543, (1941).
12. M. Nusimovici, Proc. Conf. on Ferroelectricity, Gen. Motors (1966).
13. G. Herzberg, "Infrared and Raman spectra," D. Van Nostrand Co. Inc., New York, 1945.

DISTINGUISHED GUEST

Professor Leon Brillouin, France and U. S. A.

PARTICIPANTS

Alfano, R., General Telephone and Electronics Laboratories, U. S. A.
Anastassakis, E., University of Pennsylvania, U. S. A.
Anderson, C. H., R. C. A. Laboratories, U. S. A.
Ashkin, M., Westinghouse Research Laboratories, U. S. A.
Balkanski, M., Faculté des Sciences de Paris, France
Bendow, B., New York University, U. S. A.
Benedek, George B., Massachusetts Institute of Technology, U. S. A.
Benedek, Giorgio, Instituto di Fisica dell'Università, Italy
Berge, P., Commissariat a l'Énergie Atomique, France
Birnboim, N. H., Rensselaer Polytechnic Institute, U. S. A.
Birman, Joseph L., New York University, U. S. A.
Bobb, L. C., Frankford Arsenal, U. S. A.
Boghosian, C., United States Army Research Office (Durham), U. S. A.
Boon, J. P., Free University of Brussels, Belgium
Brafman, O., University of Rhode Island, U. S. A.
Bray, R., Purdue University, U. S. A.
Brody, E., The Johns Hopkins University, U. S. A.
Brueck, S., Massachusetts Institute of Technology, Lincoln Laboratory, U. S. A.
Bruun, M., Technical University of Denmark, Denmark
Brya, W. J., Bell Telephone Laboratories, U. S. A.
Bucci, C., University of Parma, Italy
Buchenauer, C. J., Cornell University, U. S. A.
Burns, G., IBM Research Laboratory (Yorktown Heights), U. S. A.
Burstein, E., University of Pennsylvania, U. S. A.
Cahill, K. E., National Bureau of Standards, U. S. A.
Callender, R., Harvard University, U. S. A.
Carleton, H. R., State University of New York (Stony Brook), U. S. A.
Chang, R. K., Yale University, U. S. A.
Chen, S. H., Harvard University, U. S. A.
Chinn, S., Massachusetts Institute of Technology, Lincoln Laboratory, U. S. A.
Chirico, S., University of Palermo, Italy
Cohen, E., Bell Telephone Laboratories, U. S. A.
Cole, T., Ford Motor Company, U. S. A.
Coleau, H., Commissariat a l'Énergie Atomique, France
Comly, J., General Electric Research Laboratory, U. S. A.
Courtens, E., IBM Research Laboratory, Switzerland
Crooker, P. P., Massachusetts Institute of Technology, U. S. A.
Cummins, H. Z., The Johns Hopkins University, U. S. A.
Damen, T. C., Bell Telephone Laboratories, U. S. A.
Da Silva, E., Société de Conversion des Énergies, France
De Wames, R. E., North American Rockwell Corp., U. S. A.
Di Domenico, M., Jr., Bell Telephone Laboratories, U. S. A.
Dietz, R., Bell Telephone Laboratories, U. S. A.
Dolino, G., Domaine Universitaire, France
Dransfeld, K., Technischen Hochschule (Munich), Germany
Durand, G., Université de Paris, France

Enz, C. P., Université de Geneve, Switzerland
Feldman, D. W., Westinghouse Research Laboratories, U. S. A.
Fenner, W. R., University of Illinois, U. S. A.
Ferraro, J. R., Argonne National Laboratories, U. S. A.
Fleury, P. J., Bell Telephone Laboratories, U. S. A.
Ford, N. C., University of Massachusetts, U. S. A.
Freedman, S. J., Polytechnic Institute of Brooklyn, U. S. A.
Freund, I., Bell Telephone Laboratories, U. S. A.
Friedman, E. A., Stevens Institute of Technology, U. S. A.
Fritz, B., University of Stuttgart, Germany
Gammon, R. W., Catholic University, U. S. A.
Ganesan, S., City College (New York), U. S. A.
Ganguly, A. K., General Telephone and Electronics Laboratories, U. S. A.
Gayles, J. N., IBM Research Laboratory (San Jose), U. S. A.
Gerlach, J., University of Stuttgart, Germany
Geschwind, S., Bell Telephone Laboratories, U. S. A.
Giordmaine, J. A., Bell Telephone Laboratories, U. S. A.
Goldstein, B., Fairleigh Dickinson University, U. S. A.
Green, M. S., National Bureau of Standards, U. S. A.
Griffin, A., University of Toronto, Canada
Grob, K., Pädagogische Hochschule (Ludwigsburg), Germany
Gunthard, H. H., Eidgenossische Technische Hochschule (Zurich), Switzerland
Haas, M., Naval Research Laboratory, U. S. A.
Hacker, H., Universitat Munchen, Germany
Haidemenakis, E., Paris, France
Halley, J. W., University of California (Berkeley), U. S. A.
Hamilton, D. C., Massachusetts Institute of Technology, U. S. A.
Harbeke, G., R. C. A. Laboratories, Switzerland
Hardy, J. R., University of Nebraska, U. S. A.
Hart, T., Massachusetts Institute of Technology, U. S. A.
Hartmann, H., New York University, U. S. A.
Henry, C. H., Bell Telephone Laboratories, U. S. A.
Henry, D. L., The Johns Hopkins University, U. S. A.
Henry, D. L., Bell Telephone Laboratories, U. S. A.
Hexter, R. M., Carnegie-Mellon University, U. S. A.
Hope, L. L., General Telephone and Electronics Laboratory, U. S. A.
Hopfield, J., Princeton University, U. S. A.
Huang, J. S., Cornell University, U. S. A.
Huber, D. L., University of Wisconsin, U. S. A.
Imry, Y., Cornell University, U. S. A.
Jacucci, G., University of Rome, Italy
Jha, S. S., Harvard University, U. S. A.
Johnson, F. A., Radar Research Establishment Laboratory, England
Kahane, A., Faculté des Sciences de Grenoble, France
Kaminow, I., Bell Telephone Laboratories, U. S. A.
Karo, A. M., Lawrence Radiation Laboratory, U. S. A.
Kashnow, R., Natick Laboratory, U. S. A.
Katyl, R. H., Massachusetts Institute of Technology, U. S. A.
Kiel, A., Bell Telephone Laboratories, U. S. A.
Klein, M. V., University of Illinois, U. S. A.
Klein, R., R. C. A. Laboratories, U. S. A.
Koenig, J. L., Case Western Reserve University, U. S. A.
Koningstein, J. A., Carleton University, Canada
Korenman, V., University of Maryland, U. S. A.

Krauzman, M., Faculté des Sciences de Paris, France
Kravitz, L. C., General Electric Company, U. S. A.
Krishnan, R. S., Indian Institute of Science, India
Kruer, M., Catholic University of America, U. S. A.
Kurtz, S. K., Bell Telephone Laboratories, U. S. A.
Lacina, W., Harvard University, U. S. A.
Lalor, E., University of Rochester, U. S. A.
Laubereau, A., Technische Hochschule (Munich), Germany
Lajzerowicz, J., Faculté des Sciences de Grenoble, France
Langer, D. W., Wright Patterson Air Force Base, U. S. A.
Langley, K. H., University of Massachusetts, U. S. A.
Lastovka, J. B., Massachusetts Institute of Technology, U. S. A.
Lax, B., Massachusetts Institute of Technology, U. S. A.
Lazay, F. D., Massachusetts Institute of Technology, U. S. A.
Leigh, R. S., University of Reading, England
Leite, R. C. C., Bell Telephone Laboratories, U. S. A.
Leonard, D. A., Avco Everett Research Laboratory, U. S. A.
Lin-Chung, P. J., Naval Research Laboratory, U. S. A.
Lippmann, B., New York University, U. S. A.
Litster, J. D., Massachusetts Institute of Technology, U. S. A.
Lockwood, D. J., University of Canterbury, New Zealand
Loebner, E. E., Hewlett-Packard Company, U. S. A.
Logothetis, E. M., Ford Motor Company, U. S. A.
Loudon, R., University of Essex, England
Low, W., Hebrew University, Israel
Lowndes, R. P., Massachusetts Institute of Technology, U. S. A.
Macrakis, M. S., NASA Electronics Research Center, U. S. A.
Magiros, D. G., General Electric Company, U. S. A.
Marcuvitz, N., New York University, U. S. A.
Martinke, T. L., Hercules Research Center, U. S. A.
Mavroyannis, C. A., National Research Council of Canada, Canada
May, A. D., University of Toronto, Canada
McTague, J. P., North American Rockwell Corp., U. S. A.
McWhorter, A. L., Massachusetts Institute of Technology, Lincoln Laboratory, U. S. A.
Miller, S. A., Corona Laboratories, U. S. A.
Mills, D. L., University of California, U. S. A.
Mireille, A., Commissariat a l'Énergie Atomique, France
Moch, P., Faculté des Sciences de Paris, France
Mollow, B. R., Brandeis University, U. S. A.
Mon, J. P., Faculté des Sciences de Paris, France
Montgomery, P., Jr., University of Illinois, U. S. A.
Mooradian, A., Massachusetts Institute of Technology, Lincoln Laboratory, U. S. A.
Moriya, T., University of Tokyo, Japan
Mortensen, O. S., University of Copenhagen, Denmark
Mountain, D., National Bureau of Standards, U. S. A.
Mulazzi, E., Università di Fisica Milano, Italy
Muldawer, L., Temple University, U. S. A.
Nilsen, W. G., Bell Telephone Laboratories, U. S. A.
Nordland, W. A., Jr., Bell Telephone Laboratories, U. S. A.
Nusimovici, M. Faculté des Sciences de Paris, France
O'Brien, R. J., Fordham University, U. S. A.
Ockman, N., General Telephone and Electronic Laboratory, U. S. A.
Oseroff, A., Harvard University, U. S. A.

O'Shea, D. C., The Johns Hopkins University, U. S. A.
Page, J. B., Jr., Cornell University, U. S. A.
Papon, P., Rutgers—The State University (New Jersey), U. S. A.
Parker, J. H., Jr., Westinghouse Research Laboratories, U. S. A.
Parsons, J. L., Ford Motor Company, U. S. A.
Perry, C. H., Massachusetts Institute of Technology, U. S. A.
Pershan, P., Harvard University, U. S. A.
Pfeuty, P., Faculté des Sciences de Paris, France
Pickart, S. J., National Bureau of Standards, U. S. A.
Pike, E. R., Royal Radar Establishment, England
Pine, A. S., Massachusetts Institute of Technology, U. S. A.
Pinczuk, A., University of Pennsylvania, U. S. A.
Platzman, P. S., Bell Telephone Laboratories, U. S. A.
Polucci, A. J., Duke University, U. S. A.
Porto, S. P. S., University of Southern California, U. S. A.
Ralston, J., Yale University, U. S. A.
Ramdas, A. K., Purdue University, U. S. A.
Rast, H. E., Corona Laboratories, U. S. A.
Rebane, K. K., Tartu Academy of Sciences of the E. S. S. R., U. S. S. R.
Reese, R. L., The Johns Hopkins University, U. S. A.
Rimai, L., Ford Motor Company, U. S. A.
Riseberg, L. A., Bell Telephone Laboratories, U. S. A.
Rosasco, G. J., Fordham University, U. S. A.
Rowe, I., Office of Naval Research, U. S. A.
Ruppin, R., Israel Atomic Energy Laboratory, Israel
Russell, J. P., R. C. A. Laboratories, Switzerland
Scott, J. F., Bell Telephone Laboratories, U. S. A.
Schröder, U., New York University, U. S. A.
Shapiro, S., The Johns Hopkins University, U. S. A.
Shepherd, I. W., Cornell University, U. S. A.
Shiren, N., IBM Research Laboratory (Yorktown Heights), U. S. A.
Silverman, B. D., NASA Electronics Research Center, U. S. A.
Singh, S., Bell Telephone Laboratories, U. S. A.
Sizemore, T., University of California, U. S. A.
Sjolander, A., Chalmers Tekniska Hogskola, Sweden
Smith, A. W., Research Laboratory (Yorktown Heights), U. S. A.
Smith, R. W., R. C. A. Laboratories, U. S. A.
Solin, S. A., Purdue University, U. S. A.
Spears, D. L., Purdue University, U. S. A.
Stoicheff, B. P., University of Toronto, Canada
Strahm, N. D., Massachusetts Institute of Technology, U. S. A.
Swinney, H. L., The Johns Hopkins University, U. S. A.
Szigeti, B., University of Reading, England
Tait, W. C., 3M, Central Research Laboratories, U. S. A.
Taylor, W., University of Edinburgh, Scotland
Tell, B., Bell Telephone Laboratories, U. S. A.
Teng, M. K., Faculté des Sciences de Paris, France
Thorpe, M., University of Oxford, England
Timusk, T., McMaster University, Canada
Toa-Ning NG, Carleton University, Canada
Tornberg, N. T., Massachusetts Institute of Technology, U. S. A.
Tzoar, N., City College (New York), U. S. A.
Ushioda, S., University of Pennsylvania, U. S. A.

Vassell, M. O., General Telephone and Electronics Laboratories, U. S. A.
Vella-Coleio, G., Bell Telephone Laboratories, U. S. A.
Volochine, B., Commissariat a l'Énergie Atomique, France
Wadsack, R., Yale University, U. S. A.
Wagner, R. J., Naval Research Laboratory, U. S. A.
Walker, L., Bell Telephone Laboratories, U. S. A.
Wallace, P. R., McGill University, Canada
Wallis, R. F., Naval Research Laboratory, U. S. A.
Walls, D. F., Harvard University, U. S. A.
Wang, C. H., Bell Telephone Laboratories, U. S. A.
Wang, F. F. Y., State University of New York (Stony Brook), U. S. A.
Ward, A. T., Xerox Research Laboratory, U. S. A.
Watson, G., University of Kent (Canterbury), England
Webb, W. W., Cornell University, U. S. A.
Weber, A., Fordham University, U. S. A.
Weissmann, A. N., Universitatea Babes-Bolyai, Rumania
Welsh, H. L., University of Toronto, Canada
Wemple, S. H., Bell Telephone Laboratories, U. S. A.
Wieder, S., Fairleigh Dickinson University, U. S. A.
Wilkinson, G. R., King's College, England
Wilson, C. M., The Johns Hopkins University, U. S. A.
Wolff, P. A., Bell Telephone Laboratories, U. S. A.
Woolston, J. E., International Atomic Energy Agency, Austria
Worlock, J. M., Bell Telephone Laboratories, U. S. A.
Wright, G. B., Massachusetts Institute of Technology, Lincoln Laboratory, U. S. A.
Xinh, N. X., University of Colorado, U. S. A.
Zeiger, H., Massachusetts Institute of Technology, Lincoln Laboratory, U. S. A.
Zucker, J., General Telephone and Electronics Laboratories, U. S. A.

AUTHOR INDEX

TOPICAL REFERENCES*

*Owing to the nature of these Proceedings and the fact that particular topics appear many times in a given paper, it has been decided to give Topical References, rather than a conventional Subject Index. Consequently, reference to each topic is given by citing the paper-number and the first page of the paper, or papers, in which the topic appears.